PARTICLE PHYSICS AND COSMOLOGY

HIGH ENERGY PHYSICS

PARTICLE PHYSICS AND COSMOLOGY

First Tropical Workshop

HIGH ENERGY PHYSICS

Second Latin American Symposium

San Juan, Puerto Rico   April 1998

EDITOR
José F. Nieves
Department of Physics, University of Puerto Rico

AIP CONFERENCE PROCEEDINGS 444

American Institute of Physics   Woodbury, New York

Editor:

José F. Nieves
Department of Physics
University of Puerto Rico
P.O. Box 23343
Río Piedras, Puerto Rico 00931-3343

Email: nieves@ltp.upr.clu.edu

Authorization to photocopy items for internal or personal use, beyond the free copying permitted under the 1978 U.S. Copyright Law (see statement below), is granted by the American Institute of Physics for users registered with the Copyright Clearance Center (CCC) Transactional Reporting Service, provided that the base fee of $15.00 per copy is paid directly to CCC, 222 Rosewood Drive, Danvers, MA 01923. For those organizations that have been granted a photocopy license by CCC, a separate system of payment has been arranged. The fee code for users of the Transactional Reporting Service is: 1-56396-775-8/ 98 /$15.00.

© 1998 American Institute of Physics

Individual readers of this volume and nonprofit libraries, acting for them, are permitted to make fair use of the material in it, such as copying an article for use in teaching or research. Permission is granted to quote from this volume in scientific work with the customary acknowledgment of the source. To reprint a figure, table, or other excerpt requires the consent of one of the original authors and notification to AIP. Republication or systematic or multiple reproduction of any material in this volume is permitted only under license from AIP. Address inquiries to Office of Rights and Permissions, 500 Sunnyside Boulevard, Woodbury, NY 11797-2999; phone: 516-576-2268; fax: 516-576-2499; e-mail: rights@aip.org.

L.C. Catalog Card No. 98-87300
ISBN 1-56396-775-8
ISSN 0094-243X
DOE CONF- 980455

Printed in the United States of America

CONTENTS

Preface ix
Dean's Message xii

NEUTRINO PHYSICS, DARK MATTER, AND COSMOLOGY

Results from Super-Kamiokande on Atmospheric Neutrinos 3
    J. Stone
Neutrino Oscillations 17
    S. Pakvasa
Electroweak Measurements and Neutrino Oscillations: The NuTeV and
BooNE Experiments 28
    M. H. Shaevitz
Alternative Mechanisms for Neutrino Oscillations 40
    A. Halprin
Invariant Box-Parameterization of Neutrino Oscillations 46
    T. J. Weiler and D. J. Wagner
A New Cosmological Paradigm: The Cosmological Constant and
Dark Matter 59
    L. M. Krauss
The Search for Axion Dark Matter 70
    P. Sikivie
Neutrino Mass and Dark Matter 82
    D. O. Caldwell
The Pierre Auger Project: An Observatory for Measuring
Extremely High-Energy Cosmic Rays 95
    D. Zavrtanik
Connection Between Relic Neutrinos and Cosmic Rays at $\gtrsim 10^{20}$ eV? 105
    T. J. Weiler

FLAVOR PHYSICS AND *CP* VIOLATION

Breaking Lorentz Invariance 119
    S. L. Glashow
Gravity Couplings in the Standard Model: CPT Nonconservation 130
    L. N. Chang and C. Soo
Experimental Tests of *CP*, *T* and *CPT* Symmetries using K^0 and $\bar{K}^0$ 138
    D. Zavrtanik
Lepton Flavour Violation Experiments—Some Recent Developments 148
    K. P. Jungmann
Spontaneous *CP* Violation 160
    P. H. Frampton
New Physics Models and *CP* Violation Experiments 165
    D. Silverman

SUPERSYMMETRY AND NEW PHYSICS

Pre-LHC SUSY Searches: An Overview 179
    A. Masiero and L. Silvestrini
R-Symmetry in MSSM and Beyond 196
    Q. Shafi
Searches for SUSY Particles at LEP2 211
    S. Braibant
Searches for Leptoquarks at Fermilab 223
    D. Hedin
Searches for SUSY at the Tevatron 228
    A. Nomerotski
Higgs Boson Searches at LEP2 237
    U. Schwickerath
Standard Model Tests at Very High Q^2 at HERA 253
    G. Eckerlin

W PHYSICS AND STANDARD MODEL TESTS

Standard Model Tests and New Physics at LEP 271
    R. Faccini
The Measurement of the Mass of the W Boson from the Tevatron 282
    R. Thurman-Keup
W Couplings Measurements at LEP 294
    M. Verzocchi
Measurement of $|V_{cs}|$ with the DELPHI Experiment 306
    B. Golob

QCD AND τ PHYSICS

QCD at the Tevatron: W, Z, and Photon Production 317
    D. P. Casey
High Precision Tests of QCD at LEP 329
    J. Fuster Verdú
Jet Production at the Tevatron 345
    F. Nang
QCD at HERA 357
    N. H. Brook
Tau Physics from LEP 368
    D. Y. Kim
Recent Tau Results from CLEO 380
    R. G. Baker

b AND t PHYSICS

b Physics 395
    P. Perret
Tevatron Results on the Top Quark 416
    N. Sotnikova
B Physics at the Tevatron Collider 429
    J. F. de Troconiz

FIELD THEORY

Chiral Symmetry Breaking in an External Field 443
    C. N. Leung
Yukawa Interactions and Dynamical Generation of Mass in an External
Magnetic Field 452
    E. J. Ferrer and V. de la Incera
Magnetic Response in Anyon Fluid at High Temperature 458
    E. J. Ferrer and V. de la Incera

FUTURE ACCELERATORS

CMS: Concept and Physics Potential 467
    C.-E. Wulz
The BTeV Program at Fermilab 479
    S. Stone
Muon Colliders: New Prospects for Precision Physics and
the High Energy Frontier 503
    B. J. King
B Physics Expected Performances with the Compact Muon Solenoid
Detector 520
    F. Charles

SEMINARS

Upper Bound on the Neutrino Magnetic Moment from Collisions
Induced by Landau Damping in Supernovae 533
    A. Ayala
Asymmetry Studies in $\Lambda^0/\bar{\Lambda}^0$, Ξ^-/Ξ^+ and Ω^-/Ω^+ Production 540
    J. C. Anjos, J. Magnin, F. R. A. Simão, and J. Solano
Λ^0 Polarization in Exclusive pp Reactions at 27.5 GeV/c 547
    J. Félix, C. Avilez, D. C. Christian, M. D. Church, M. Forbush,
    E. E. Gottschalk, G. Gutierrez, E. P. Hartouni, S. D. Holmes, F. R. Huson,
    D. A. Jensen, B. C. Knapp, M. N. Kreisler, G. Moreno, J. Uribe, B. J. Stern,
    M. H. L. S. Wang, A. Wehmann, L. R. Wiencke, and J. T. White

A Model for Baryon Structure and its Application to
Magnetic Moments and Semileptonic Decays............................ 553
   V. Gupta, R. Huerta, and G. Sánchez-Colón

SPECIAL TALKS

The National Astronomy and Ionosphere Center's (NAIC)
Arecibo Observatory in Puerto Rico..................................... 563
   D. R. Altschuler
Radio Astronomy Highlights at Arecibo................................. 571
   C. J. Salter

List of Participants... 577
Workshop and Symposium Schedule...................................... 583
Author Index... 589

PREFACE

The First Latinamerican Symposium on High Energy Physics (I-SILAFAE) was held in Mérida, Mexico in October 1996. On that occasion, Puerto Rico was chosen as the organizing country for the next edition of the Symposium, which would be held in 1998. Simultaneously, but independently, the idea of holding a workshop on Particle Physics and Cosmology in Puerto Rico was born, to a large extent as an offspring of the Winter School that had being held on two previous occasions.

In a field in which the concepts of naturalness and simplicity play guiding roles, it is not surprising that the decision was made to hold the two activities back to back. In this way, the First Tropical Workshop on Particle Physics and Cosmology (Tropical98) and the Second Latinamerican Symposium on High Physics (II-SILAFAE) were organized to be held from April 1-7 and from April 8-11, 1998, respectively. Taking these guiding principles one step further, it was decided also to combine the proceedings of both activities, which this publication represents.

This volume thus contains the written version of the talks that were presented in the Workshop and the Symposium. In the preparation of these proceedings, we have divided the articles thematically into various sets, independently of whether they were presented in one or the other of the two activities. We feel that this arrangement will facilitate its reading. For those interested in knowing which talk was presented in which activity, we have included in the book the original schedules. Unfortunately, some of the talks that were presented have no corresponding article in this book, because we had not received them by the time that the manuscripts had to be sent to the publisher. However, we believe that these omissions will be compensated by the outstanding quality of the articles that are included here, which should set an example for the next editions of these meetings.

It is gratifying to have a place to acknowledge the help and support from various people and institutions. For the Symposium, Tom Ferbel and his colleagues from the D0 and CDF experiments at Fermilab, as well the leaders of the various LEP collaborations, were instrumental in the design of the scientific program. For the Tropical Workshop, it simply would not have taken place without the intervention of Arthur Halprin, Terry Leung and Qaisar Shafi, who served as the Organizing Committee. The scientific session of Saturday April 4 took place in the imposing premises of the Arecibo Observatory. That afternoon, which was highlighted by Glashow's talk, we were the subjects of a wonderful reception, for which we thank the Observatory's Director Daniel Altschuler and also the director of its Visitors Center, Jose Luis Alonso. Joe Taylor (of the Arecibo Hall of Fame) was to deliver his talk on pulsars that same afternoon. However, unforeseen commitments forced him to change his travel plans at the last minute, and we thank him for inserting the stop in San Juan during his trip in order to deliver his lecture in the Workshop.

By far, the major financial support for these meetings came from the Office of the Chancellor of the University of Puerto Rico and from the office of the Dean of the Faculty of Natural Sciences, Gladys Escalona. Needless to say, neither one of the meetings would have taken place without her agreement and support. Supplemental

The (non-local) Organizing Committee

funds were received from the Arecibo Observatory for the Workshop, and from the US Department of Energy for the Symposium through a grant to Angel López.

For handling the inevitable administrative chores that the organization of activities like these brings, we are grateful to Ileana Desiderio and Carmelo Figueroa from the Department of Physics. The Conference Secretary was Maria Eugenia Rodriguez but, because at the time of the meetings she had already become my wife, I will refrain here from making any further comments.

I hope that you will enjoy reading these articles as much as we enjoyed hearing the talks.

José F. Nieves
July 16, 1998

PREFACE TO TROPICAL WORKSHOP ON PARTICLE PHYSICS AND COSMOLOGY

The first Tropical Workshop on Particle Physics and Cosmology took place April 1-7, 1998 in San Juan, Puerto Rico and was hosted by the University of Puerto Rico. The meeting was intended to provide a forum for discussion of the rapidly changing experimental situation in this most fundamental field of physics. The organizers are grateful to all the speakers for their substantial talks and to all participants for lively discussions. A special thanks goes to Jim Stone, who provided us with nearly two hours of information on the exciting evidence concerning neutrino oscillations obtained at the Japanese underground facilities, Kamiokande and Superkamiokande, which became front page international news in June of this year. Progress in this field is rapidly accelerating, and we can expect to know very much more at the next Tropical Workshop.

The environment provided by the University of Puerto Rico played a large role in the success of the meeting. We are grateful for the financial support provided by its Chancellor and Dean of Natural Sciences, and we are especially grateful for the warm welcome we received from Dean Gladys Escalona.

Wonderful local arrangements (and a few others) were made by Jose Nieves, and executed as a team effort with the Conference Secretary, Maria Eugenia Rodriguez and members of the Department of Physics, Ileana Desiderio and Carmelo Figuroa. We are grateful for all their efforts.

Arthur Halprin C.N. Leung Qaisar Shafi
July 1998

Message from the Dean Gladys Escalona de Motta for the First Tropical Workshop on Particle Physics and Cosmology

April 1, 1998

Members of the Organizing Committee, Ladies and Gentlemen:

I am honored to welcome you to San Juan for the celebration of the First Tropical Workshop on Particle Physics and Cosmology. It is an exciting opportunity for the University of Puerto Rico and our Faculty of Natural Sciences to have as visitors some of the most prestigious scientists in these fields of Particle Physics and Cosmology. On two previous occasions the Physics Department and the University of Puerto Rico had a similar opportunity sponsoring the First and the Second Winter School of Physics on these topics.

I hope that this time we will provide the arena for lively interaction among participants and for the announcement of new discoveries and results, to make this workshop an even more successful event.

Although the main aim of physics research is to satisfy human curiosity about the universe, history shows that advances made in these basic studies have had, sooner or later, beneficial impact on our everyday lives. Your discipline is not an exception as it has already contributed to the development of numerous industrial and technological applications.

I sincerely wish that this workshop will contribute significantly to increase the knowledge in your field. May all of you have a very pleasant stay in San Juan.

Thank you.

Message from the Dean Gladys Escalona de Motta for the Second Latin American Symposium on High Energy Physics

April 8, 1998

Members of the Organizing Committee, Ladies and Gentlemen:

I am honored to welcome you to San Juan for the celebration of the Second Latin American Symposium on High Energy Physics. Hosting this Symposium is an important event for the University of Puerto Rico and our Faculty of Natural Sciences as it gives us the occasion to have as visitors some of the most prestigious scientists in this field. It is also a way to give new strength to an important tradition of our Department of Physics which, on two previous occasions, acted as the sponsor of the Winter Schools of Physics on the same topics. Though this time the format is somewhat different, I am sure there will be ample opportunities for lively interactions among the participants and also for the announcement of new discoveries and results.

Investigation in Physics, above all, aims to satisfy our human curiosity about the mysteries of our universe but history shows that advances made in these basic studies have, sooner or later, affected our everyday lives. Many findings in your discipline of high energy physics, have already contributed to communications and medical technologies, to mention only two important and recent areas of its applications.

It is, thus, my wish that the presentations and discussions of this symposium will make a significant contribution to the basic and applied knowledge being generated in this field. May all of you have a very pleasant stay in San Juan.

Thank you.

NEUTRINO PHYSICS, DARK MATTER, AND COSMOLOGY

Results from Super-Kamiokande on Atmospheric Neutrinos

James Stone

Department of Physics, Boston University
590 Commonwealth Ave.
Boston, MA 02215

Abstract. This paper reports the latest indications of an anomaly in the measurements of atmospheric neutrinos. New results from Soudan-2 and Super-Kamiokande provide evidence that the ratio of ν_μ to ν_e interactions is not as expected. High energy Super-Kamiokande data indicates the cause is a deficit of upward-going ν_μ, and the zenith angle dependence of the effect is consistent with neutrino oscillations. Upward-going muon measurements by several detectors are discussed, but in total they provide inconclusive evidence for the anomaly.

I INTRODUCTION

Large underground detectors originally built to search for proton decay are also exposed to a flux of neutrinos created by cosmic ray showers in the upper atmosphere. Neutrino interactions in the detector can mimic proton decay, and therefore a great effort has gone into predicting the rate and topology of the neutrino background. A byproduct of this effort has been the recognition of an anomaly: the relative rate of ν_μ and ν_e interactions disagrees with expectation, and the baseline and energy dependence of the disagreement suggests that neutrino oscillation may be the culprit.

The nature of these experiments is to measure the rate of neutrino interactions in the detector and to compare that rate with theoretical prediction. The theoretical task is to predict the neutrino flux as a function of energy, direction, and flavor, taking into account measurements of cosmic ray flux, geomagnetic cutoff, and production and decay of secondary mesons. Detailed Monte Carlo programs are used to estimate the cross section for neutrino interaction and simulate the response of the detector. The experimental task is to measure as much as possible about the neutrino interaction, in particular, the energy, direction, and flavor of the final state lepton, from which the neutrino properties are inferred.

II FLUX PREDICTION

One of the tenets of the atmospheric neutrino anomaly is than the flux ratio[1] (ν_μ/ν_e) is more accurately predicted that the ν_μ or ν_e flux alone. The principal effect is that cosmic ray showers consist mostly of pions, which decay to $\mu+\nu_\mu$, and the μ decays to to $e + \nu_e + \nu_\mu$, resulting in a flux ratio $(\nu_\mu/\nu_e) \sim 2$. The authors of detailed calculations of the flux models have collaborated to compare results [1] and reached an understanding of many of the differences between their earlier publications. Two updated calculations [2,3] cover a wide energy range (10 MeV–10 TeV) and are used by current experiments. There are now also an assortment of new high altitude μ measurements available for comparison [4–6]. Further details of flux calculations were presented recently by T. Stanev [7]. However, there is currently no indication that poor knowledge of the predicted flux could be responsible for the experimental anomalies described below.

III SUMMARY OF RESULTS

To study the (ν_μ/ν_e) flux ratio, most experiments calculate the double-ratio or ratio-of-ratios:

$$R = \frac{(N_\mu/N_e)_{DATA}}{(N_\mu/N_e)_{MC}}, \qquad (1)$$

where N refers to the number of events where the final state lepton is classified as μ-like or e-like by some identification algorithm. As mentioned above, the ratio of ν_μ to ν_e flux is accurately predicted; in addition, other theoretical and experimental uncertainties largely cancel. Table 1 lists previous measurements[2] of R for $E_\nu \sim$ 1 GeV, including the new results from Soudan and Super-Kamiokande discussed in this paper. The kinematic limits differ somewhat from experiment to experiment, with minimum lepton momenta requirements from 100 to 200 MeV/c. Kamiokande restricted their sample to[3] $E_{vis} < 1.33$ GeV, IMB to $p < 1.5$ GeV; the other experiments did not specify an upper limit, but all results are dominated by ~ 1 GeV neutrinos.

Kamiokande also studied events with $E_{vis} > 1.33$ GeV (multi-GeV) and included partially contained (PC) events where a track was detected exiting the inner detector [15]. They measured a low value for R of $0.57^{+0.08}_{-0.07} \pm 0.07$, but more interesting was the dependence of R on zenith angle. Neutrinos that travelled $\sim 10^4$ km from below showed a small value of R, but those that travelled ~ 10 km from above agreed with expectation, suggesting an oscillation length somewhere in between.

---

[1] technically $(\nu_\mu + \bar{\nu}_\mu)/(\nu_e + \bar{\nu}_e)$
[2] The first uncertainty quoted is statistical, the second is systematic; this convention will be used throughout the paper.
[3] Visible energy (E_{vis}) is defined as the energy of an electromagnetic shower that produces a given amount of Cherenkov light.

TABLE 1. Summary of R measurements, $E_\nu \sim 1$ GeV.

Experiment	kt-yr	events	R (data/MC)
Super-K [8]	25.5	1853	$0.61 \pm .03 \pm .05$
IMB [9]	7.7	610	$0.54 \pm .05 \pm .11$
Kam. [10]	7.7	482	$0.60^{+.06}_{-.05} \pm .05$
Soudan 2 [12]	3.2	~200	$0.61 \pm .15 \pm .05$
Fréjus [13]	2.0	200	$1.00 \pm .15 \pm .08$
NUSEX [14]	0.7	50	$0.96^{+.32}_{-.28}$

Most of the IMB multi-GeV exposure had a restriction on the maximum number of PMT hits (to concentrate on proton decay); the restriction was eventually removed, so they made a separate analysis [16] of their last 2.1 kt·years for $E_{vis} > .95$ GeV and measured $R = 1.40^{+0.45}_{-0.34} \pm 0.14$ with no zenith dependence. The caveats are: limited statistics of 72 events (some overlapping with Ref. [9]); no outer detector to help identify PC muons; coarser sampling (4% photon coverage), resulting in only 90% correct e/μ identification.

IV SOUDAN-2

Until recently, atmospheric neutrino results seemed to be divided between water Cherenkov detectors [9,10] (anomalous) and iron calorimeters [13,14] (as expected). The Soudan-2 collaboration, which operates a fine-grained iron tracking calorimeter in Minnesota, U.S.A., has recently published results [11] which support the anomaly seen in the water Cherenkov experiments.

Recently, T. Kafka has updated the results from Soudan-2 to 3.2 kt·years [12]. They measure 91 single-prong track events (mostly charged current (CC) ν_μ) with $p > 100$ MeV/c, and 137 shower events (mostly CC-ν_e) with $p > 150$ MeV/c. The interaction vertex is allowed as close as 20 cm from the edge of the detector; with only 32 gm/cm^2 of shielding they observe a significant (25-30%) background from gamma rays and neutrons associated with nearby cosmic rays. However, they use an active shield of proportional tube planes lining the detector hall to veto most nearby cosmic rays, as well as separately estimate the remaining background rate as a function of flavor, depth into the detector, and energy.

After background subtraction, the number of shower events matches their Monte Carlo prediction; however they observe 37% fewer tracks than they predict. Since the total flux is uncertain, it is better to consider the double-ratio, which they measure to be $R = 0.61 \pm 0.15 \pm 0.05$. Regarding the atmospheric neutrino anomaly, this is a considerable new piece of information, as the systematics are very different from the water Cherenkov detectors. Although there are no demonstrated nuclear effects that would change the ratio of ν_μ to ν_e cross sections [17], it interesting that the anomaly has also been seen in Fe as well as in H_2O. What the difference is between these results and those of Fréjus and NUSEX (beyond what may be encompassed by large uncertainties) remains to be explained.

V SUPER-KAMIOKANDE

Super-Kamiokande is the next generation water Cherenkov experiment after IMB and Kamiokande. The detector resides nearby the old Kamiokande detector in a mine near Kamioka, Japan. However, it is much larger (22.5 kton fiducial mass, versus 1 kton for Kamiokande and 3.3 kton for IMB). It is instrumented with 11,146 PMTs, each 50 cm across, such that 40% of the inner surface area is active photocathode. The PMTs and electronics are of advanced design, with 2.5 ns RMS timing for single photoelectrons. The inner detector is surrounded by an outer volume of water ~2.7 m thick that shields against incoming radioactivy and is instrumented with PMTs to tag penetrating muons. Further description of the detector, as well as new measurements of the ^8B solar neutrino flux are presented by K. Inoue in ref. [18].

The measurement from Super-Kamiokande is the result of a 414.2 live-day exposure (25.5 kton·yrs) during the period from May 1996 to October 1997. The data is reduced from approximately 800K events per day to about 30 events per day by a series of software cuts. The most powerful requirement is the absence of hits in the outer detector, which indicates a fully contained (FC) interaction. The remaining events are then filtered by a visual scan, where the principal backgrounds are: (1) cosmic ray muons that evade the outer detector veto, typically by entering along cable bundles and then stopping in the detector, and (2) "flashing" PMTs that emit light due to internal corona discharge. The partially contained sample is formed by a different reduction program from the same exposure[4], since outer detector hits are now expected, and the background rejection of entering cosmic rays is different. A 10.0 live-year Monte Carlo sample of ν interactions is passed through the same reduction chains, except for the visual scan.

The remaining events, both data and Monte Carlo, are passed through the same reconstruction code to: (a) fit the vertex of the interaction, by residual PMT timing, (b) count the number of Cherenkov rings, (c) estimate the direction of each ring, (d) estimate the energy of each ring, (e) determine the particle type (μ-like,e-like) for each ring, and (f) count the number of μ-decay electrons that follow each event. Most of the analysis is then done with the sample of events in which the number of rings found in (b) is exactly one. In most cases, this is the final state lepton from a charged current neutrino interaction; the principal contamination is single pion production associated with neutral current (NC) interactions.

---

[4] The results quoted here are updated from previous conference presentations, where PC livetime was somewhat less than FC livetime, and PC data was scaled (by ~ 1.1) when FC+PC results were plotted.

The absolute energy scale was determined to ∼ 2.5% accuracy using several calibration signatures: LINAC electrons, radioactive sources, π^0s and cosmic ray muons. About 9 photoelectrons are recorded for 1 MeV of visible energy. Conversion to lepton momentum takes into account the Cherenkov cutoff for muons.

Data samples are defined using the same kinematic criteria as in the Kamiokande experiment: $p_e > 100$ MeV/c, $p_\mu > 200$ MeV/c, and $E_{vis} \leq 1.33$ GeV for sub-GeV; $E_{vis} > 1.33$ GeV for multi-GeV FC. The partially contained sample is specified by a vertex in the inner detector and correlated hits in the outer detector; the minimum visible energy required is ∼ 350 MeV. Because $\langle E_\nu \rangle$ is >> 1 GeV for the PC sample, these data are added to to the FC multi-GeV sample; the CC lepton is assumed to be a muon, and no single-ring is required. The fiducial sample is restricted to events with vertex 2 m from the PMT wall (22.5 kton).

TABLE 2. Event summary for 25.5 kt·year sample of fully-contained atmospheric neutrinos in Super-Kamiokande. Monte Carlo breakdown uses Honda flux.

	Data	Monte Carlo prediction		Monte Carlo breakdown		
		Bartol [3]	Honda [2]	CC-ν_μ	CC-ν_e	NC
sub-GeV e-like	983	788.9	812.2	2%	88%	10%
sub-GeV μ-like	900	1185.4	1218.3	96%	0.5%	4%
sub-GeV multi-ring	696	753.7	759.2	43%	24%	33%
multi-GeV e-like	218	190.9	182.7	7%	84%	9%
multi-GeV F.C. μ-like	176	229.7	229.0	99%	0.5%	0.4%
multi-GeV P.C. (μ-like)	230	305.0	287.7	98%	1.5%	0.6%
multi-GeV multi-ring	398	450.1	433.6	55%	30%	15%

The event totals are listed in Table 2 along with the totals for the Monte Carlo samples, scaled to 25.5 kton·yrs. These yield the following values of $R = (N_\mu/N_e)_{DATA}/(N_\mu/N_e)_{MC}$:

$$\text{sub-GeV} \begin{cases} 0.610^{+0.029}_{-0.028} \pm 0.049 & \text{(Honda flux)} \\ 0.609^{+0.029}_{-0.028} \pm 0.049 & \text{(Bartol flux)} \end{cases}$$

$$\text{multi-GeV} \begin{cases} 0.659^{+0.058}_{-0.053} \pm 0.081 & \text{(Honda flux)} \\ 0.665^{+0.059}_{-0.053} \pm 0.082 & \text{(Bartol flux)}. \end{cases}$$

For both the high and low energy samples there is a significant deviation of the μ/e ratio from the expected value of 1. The leading contributions to the systematic uncertainty in R are: (ν_μ/ν_e) flux (5%), neutrino cross section (4.6% for sub-GeV and 5.8% for multi-GeV), and single-ring selection (3% for sub-GeV and 6% for multi-GeV).

It is informative to check the relative rate of μ-decay associated with the event sample. The decay electrons are detected as time separated hits from the neutrino interaction; most come either from associated $\pi^+ \to \mu^+ \to e^+$, or directly from $\mu^\pm \to e^\pm$ in CC-ν_μ interactions. The performance of e/μ identification algorithms is no longer in question [19], but the measured μ-decay fractions check that the associated pion production is reasonably modeled in the Monte Carlo. Table 3 shows that the expected fraction of μ-decay agrees well with the prediction; the fraction of μ-decay found in stopping cosmic ray muons verifies the efficiency of the reconstruction.

TABLE 3. μ-decay fractions.

	Percentage of events with ≥ 1 μ-decay	
	Data	Monte Carlo
stopping CR μ	$74.0 \pm 0.3\%$	$72.9 \pm 0.4\%$
μ-like	$67.6 \pm 1.6\%$	$68.1 \pm 0.1\%$
e-like	$9.3 \pm 0.9\%$	$8.7 \pm 0.3\%$
	Percentage of events with ≥ 2 μ-decay	
	Data	M.C.
stopping CR μ	$0.0 \pm 0.0\%$	$0.0 \pm 0.0\%$
μ-like	$2.9 \pm 0.6\%$	$4.1 \pm 0.1\%$
e-like	$0.2 \pm 0.1\%$	$0.1 \pm 0.0\%$

The Super-Kamiokande group had two independent analysis efforts that were used to check each other and minimize the possibility that some mistake would be made. The data were separated after electronics calibration of the PMT data to photoelectrons and nanoseconds. Otherwise, everything was coded independently, including event reduction and reconstruction, Monte Carlo generation, and estimation of the energy scale. Beyond the independent code, the major differences in the second analysis were: (a) data reduction involved no scanning, (b) single-ring selection was based on an algorithm that classified events as single-ring or multiple-ring without attempting to count the number of rings, (c) e/μ identification was somewhat simpler and less efficient (97% vs >99%), (d) some details of vertex and direction reconstruction were different.

Upon comparison, the independent analyses were in exceptional agreement. Of the sub-GeV events found in the fiducial volume by the second analysis, 99.9% were found in the data sample of the first analysis, with 89% in the fiducial volume, consistent with the vertex fit resolution. Single-ring classification was in agreement 90% of the time. Comparing common events in the fiducial single-ring sub-GeV sample, vertices agreed to 84 cm RMS, direction agreed to $2.5°$, momentum agreed to 0.5%, 97% of the events agreed in particle identification. The value of R for the sub-GeV sample of the second analysis is: $0.65 \pm 0.03 \pm 0.05$; the difference in value from the first analysis is understood to be due to differences in analysis methods and within their systematic uncertainties. In sum, the independent analysis provides reassurance that the deviation of R from unity is not due to experimental mistakes.

To consider that the anomalous R is due to neutrino oscillation, one looks for a path length or energy dependence of the effect. The probability of two-flavor neutrino oscillation from ν to ν' is given by:

$$P_{\nu\nu'} = \sin^2 2\theta \sin^2 1.27 \frac{\Delta m^2 (\text{eV}^2/c^2) L(\text{km})}{E(\text{GeV})}, \qquad (2)$$

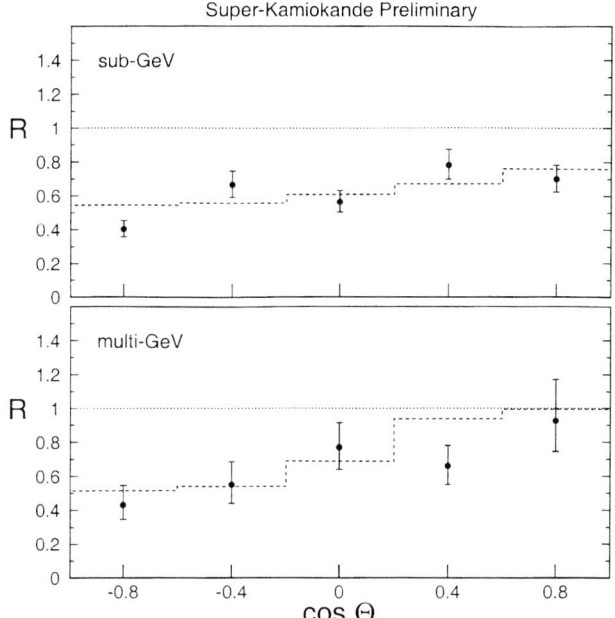

FIGURE 1. The zenith angle dependence of R for sub-GeV and multi-GeV atmospheric neutrino samples from Super-Kamiokande. The dashed line shows the expected shape for $\nu_\mu \to \nu_\tau$ oscillation with $\sin^2 2\theta = 1$ and $\Delta m^2 = .005$ eV$^2/c^2$.

where θ and $\Delta m^2 \equiv |m_\nu^2 - m_{\nu'}^2|$ are fundamental parameters that govern the neutrino mixing, and L and E are the path length and energy of the neutrino. The final state lepton direction and energy are correlated with the incoming neutrino; for the sub-GeV sample, the mean opening angle for ν_μ–μ is 54°, for ν_e–e it is 62°; for the multi-GeV sample it improves to $< 15°$.

The samples are divided into 5 $\cos\Theta$ bins where Θ is the angle between the outgoing lepton direction and the nadir[5]; so down-going neutrinos that are produced directly overhead, with short travel distance, populate the bin near $\cos\Theta = 1$. Calculating R for each zenith bin results in Fig. 1. A slight asymmetry is evident in the sub-GeV sample, and a strong asymmetry is evident in the multi-GeV sample. If there were no anomaly, the R values would be around 1; for hypothetical oscillation parameters of $\sin^2 2\theta = 1$ and $\Delta m^2 = .005$ eV$^2/c^2$, the dashed line is expected, and is a better match to the data.

It is interesting to check the result as a function of position in the detector because (a) some reconstruction algorithms are less certain for vertices close to the PMT wall, and (b) possible backgrounds, such as neutrons from nearby cosmic rays

---

[5] Caveat: the IMB collaboration used the opposite definition, so down-going neutrinos are near $\cos\Theta = -1$ in their publications.

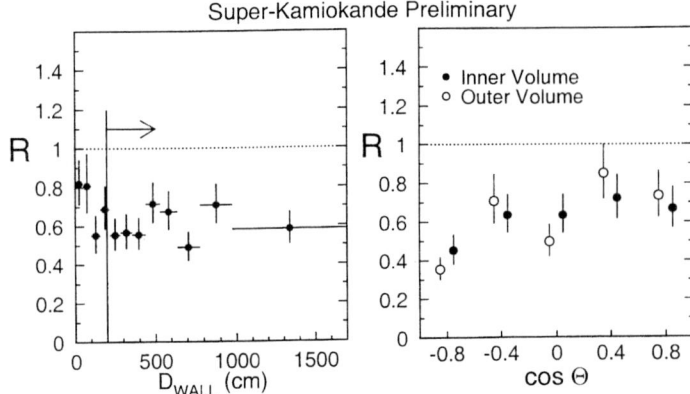

FIGURE 2. (A) R for the sub-GeV samples as a function of distance from the PMT wall (D_{WALL}). (B) R versus zenith angle for two concentric fiducial volumes: $5 > D_{WALL} > 2$ meters (outer) and $D_{WALL} > 5$ meters (inner).

[21,22], or entering events that evade the outer detector veto, would accumulate near the fiducial boundary. Figure 2a shows R versus distance from the PMT wall, where the fiducial volume is found at 2 meters. Figure 2b shows the zenith angle dependence of R, dividing the data into two approximately equal fiducial volumes: an outer volume between 2–5 m from the PMT wall and an inner volume greater than 5 m from the wall. There is no variation of the result due to the fiducial boundary evident in either figure.

The double-ratio of $(\mu/e)_{DATA}$ to $(\mu/e)_{MC}$ is useful to illustrate the effect, but it does not indicate whether μ or e rates (or both) are affected. Furthermore, R is not so practical for statistical tests. Figure 3 shows the μ and e rates separately for sub-GeV and multi-GeV (FC+PC) compared to Monte Carlo prediction. The solid bands are the absolute prediction, where the height of the band is equal to the Monte Carlo statistical uncertainty. Not shown is the $\pm 20\%$ normalization uncertainty, which is highly correlated bin-to-bin, between μ and e, and between sub-GeV and multi-GeV. Even accounting for this uncertainty, it is apparent that the anomalous R is dominated by a deficit of μ-like events coming from below ($\cos \Theta < 0$).

TABLE 4. Up-down asymmetry for multi-GeV data.

Sample	N_{up}	N_{down}	A
μ-like data	102	195	0.313±0.055
e-like data	76	90	0.084±0.077
μ-like MC	1669	1707	0.013±0.017
e-like MC	596	589	-0.006±0.029

The significance of this result can be easily evaluated by calculating the up-down

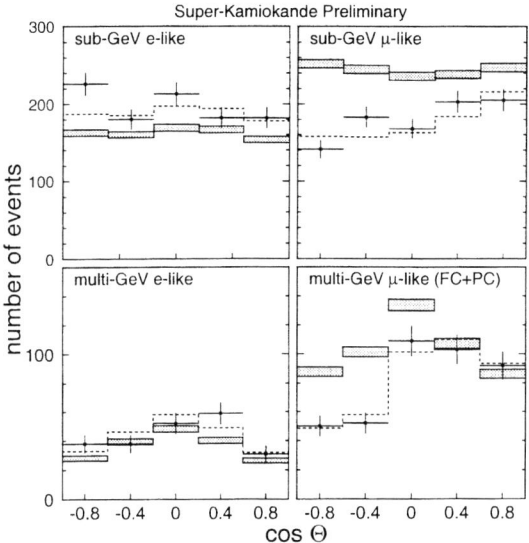

FIGURE 3. The number of μ-like and e-like events as a function of zenith angle. The solid histograms are the Monte Carlo expectation with no neutrino oscillation; the thickness represents the statistical uncertainty in the Monte Carlo sample. The dashed line shows the expected shape for $\nu_\mu \to \nu_\tau$ oscillation with $\sin^2 2\theta = 1$ and $\Delta m^2 = .005 \text{eV}^2/c^2$.

asymmetry $A = (N_{down} - N_{up})/(N_{down} + N_{up})$ where up and down are defined by $\cos \Theta < -.2$ and $> .2$ respectively (Tab.4). Besides other interesting possibilities [20], the distribution of A is nicely described by a gaussian variance. For multi-GeV events, N_{up} and N_{down} should be nearly symmetrical, whereas A for μ-like data (FC+PC) differs from expectation by greater than 5σ.

The dashed line in Fig.3 represents an oscillation hypothesis[6] of ($\sin^2 2\theta = 1, \Delta m^2 = .005$) for ν_μ disappearance. The overall normalization is adjusted upward (thus the e-like rate increases to better match the data, even when ν_e mixing is not considered) while (2) is used to calculate the probability of ν_μ disappearance and reweight the Monte Carlo. The ν travel distance L is calculated as a function of energy and flavor based on a production height model [23]. The oscillation hypothesis provides a reasonable fit to the data, certainly better than the null hypothesis.

The exact details of fitting the data to estimate possible mixing parameters are still being evaluated. Using a method similar to that used by Kamiokande [15], χ^2 terms are formed between the data and Monte Carlo prediction [modified by $P_{\nu\nu'}(\sin^2 2\theta, \Delta m^2)$], binned in zenith angle, energy, and flavor (values of R are not used directly in the fit). The normalization, N_μ/N_e ratio and systematic terms are

---

[6] The values $(1, .005)$ represent a test point; the exact best fit location can change when the technique or data sample changes since the minimum is fairly flat.

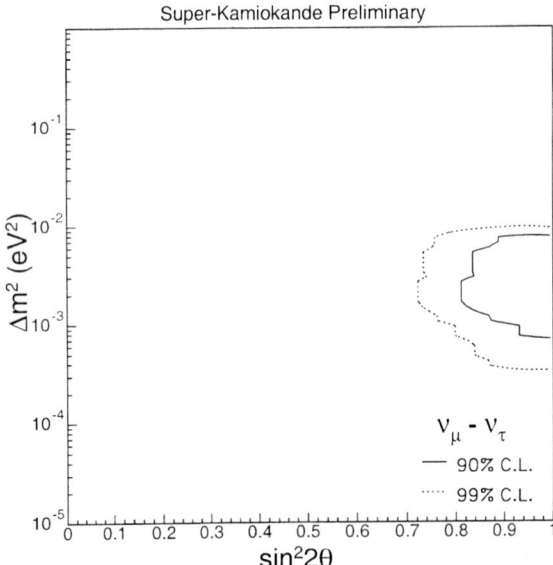

FIGURE 4. Confidence intervals for $\sin^2 2\theta$ and Δm^2 based on a χ^2 fit to Super-Kamiokande atmospheric neutrino data binned by lepton identification, lepton momentum and $\cos\Theta$. The solid line is 90% CL, the dashed line is 99% CL.

allowed to adjust and contribute to the χ^2. For the Super-Kamiokande data, the minimum χ^2 is found to be rather likely, $\sim 30\%$ depending on the details. ¿From the fit, a confidence interval is drawn based on $\chi^2_{min} + 4.6$ (90% CL) such as shown in Fig. 4.

The preferred interval for $\sin^2 2\theta$ is found near 1 because the upward zenith bin with $\langle L \rangle \sim 10,000$ km, presumably has averaged over several oscillation lengths and $P_{\nu\nu'}$ is one half. In a scenario where more than two ν flavors are mixing, the average value can be less than one half. The Δm^2 range is determined by the shape of the zenith angle, also considering dependence on E_ν. This 90% confidence interval from Super-Kamiokande prefers a lower range in Δm^2 than that found by the Kamiokande collaboration [15], which had a minimum Δm^2 of .005.

VI UPWARD-GOING MUONS

The above discussion covered neutrino interactions in the fiducial volume of the detector. The other class of atmospheric neutrino event studied is that of ν_μ interactions in the rock around the detector, where the final state muon enters the sensitive region of the detector. To separate these from ordinary cosmic ray muons, the muon must be upward-going, or come from the direction of a known thick overburden. The parent neutrino energy is 10–1000 GeV, significantly higher than for contained events.

There are several current measurements of the total rate of upward-going muons. In addition to the water Cherenkov detectors described elsewhere in this paper, MACRO and Baksan are large area scintillator detectors that distinguish upward-going muons by time-of-flight. The measured and expected event rates for various experiments are compared in Table 5; the experiments are listed in order of increasing absorber depth for directly vertical muons. The uncertainty in the measured number of events is statistical only; some experiments have estimated that the uncertainty due to experimental systematics could be as large as 8%. The uncertainty in the expected number of events is a common 15–20%, due mostly to the uncertainty in the absolute flux. Considering this, in no case is a significant deficit of muons measured, but the measurements are generally low compared to expectation.

TABLE 5. Summary of through-going upward-μ totals.

Experiment	Number of Events	
	Measured	Expected
MACRO [24]	350 ± 19	472
Baksan [25]	558 ± 24	557
Kamiokande [26]	373 ± 19	414
IMB [27]	539 ± 23	550 or 625
Super-K [28]	410 ± 20	445

There are two other approaches that probe neutrino oscillations [29] using upward-going muons. IMB measured the ratio of stopping upward muons ($\langle E_\nu \rangle \sim 10$ GeV) to through-going upward muons ($\langle E_\nu \rangle \sim 100$ GeV) to be $0.16 \pm .02$. Although much of the flux uncertainty cancels out, the usual analytic integration using deep inelastic scattering and parton distribution functions must be handled with care [27,30], especially at low energy [31]. After these considerations, the predicted rate was .14 or .18 depending on the flux model [27]. Based on the agreement of data with prediction, a small excluded region in $\sin^2 2\theta$, Δm^2 was drawn around $\Delta m^2 = 10^{-3} \text{eV}^2/c^2$, where the strongest deviation would have been found; this happens to be in conflict with the region favored by the Super-K contained vertex data. Super-Kamiokande will also measure the stopping ratio, with the advantage of a very thick detector that stops a large number of upward-going muons; the analysis is in progress.

The second approach uses the shape of the zenith distribution, which may be distorted as the baseline varies from 500 km at the horizon to 12,000 km at the nadir. Because the energy spectrum of the parent neutrinos is broad, $\sim 10 - 1000$ GeV, the change in shape is gradual, with some steepening at the horizon as the probability decreases for high energy ν_μ to oscillate. Figure 5 shows the Super-K measurement of the flux versus zenith angle compared to expectation; recall that the normalization of the prediction is uncertain to $\pm 20\%$. When the normalization of the Monte Carlo is decreased by a factor of $\alpha = 0.83$ to match the data, the χ^2 is 12.7; alternatively, when the normalization is increased by a factor of $\alpha = 1.2$,

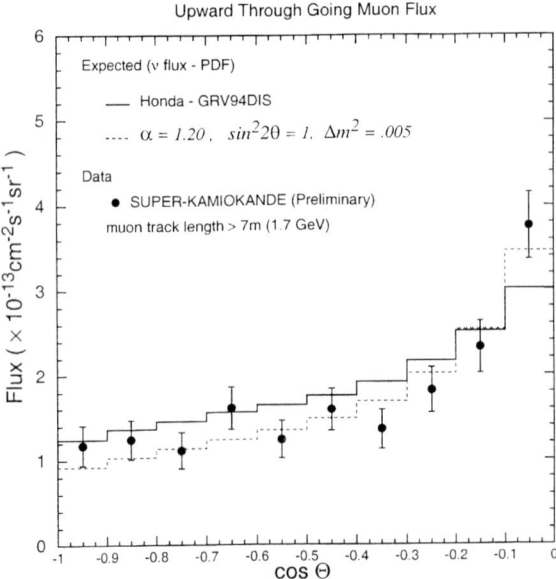

FIGURE 5. The through-going upward-going μ flux as a function of zenith angle as measured by Super-Kamiokande. The data points are compared against expectation (solid histogram) and the same expectation with normalization $\alpha = 1.2$ and ν-mixing $\sin^2 2\theta = 1$, $\Delta m^2 = .005$.

and ν_μ disappearance oscillations are applied with $\sin^2 2\theta = 1$, $\Delta m^2 = .005$, the fit is somewhat better, with a χ^2 of 8.3.

The other experiments listed in Table 5 can make this comparison as well. The scintillator detectors MACRO and Baksan unfortunately have reduced acceptance at the horizontal, so those bins require significant geometric correction. Curiously, even the well-measured upward bins of those two experiments are not smooth and suffer from poor agreement in shape with: (a) no oscillation, (b) any 2-flavor oscillation parameters[7], and (c) each other. Both experimental groups have made extensive checks for a systematic error, but have found no cause [25,32].

VII CONCLUSIONS

Currently, the evidence for an anomalous ratio of atmospheric neutrino flavors is inconsistent across experiments. Of course, prior results remain intact, and either support or disagree with the anomaly. But the latest results support that a significant discrepancy exists between experiment and prediction. The measurement by Soudan-2 shows that the anomaly is not specific to water Cherenkov detectors. A zenith angle measurement of the Soudan-2 data could be very interesting. Sig-

---

[7] A recent preprint considers that matter oscillation with a sterile neutrino crossing the Earth's core may modulate the prediction with features similar to the MACRO data [33].

nificant new information is taken from the high statistics Super-Kamiokande data: the anomaly is strongly confirmed in R and the zenith angle dependence of R. The shape of the zenith angle dependence is very suggestive of neutrino oscillations. The multi-GeV μ-like rate as a function of zenith angle indicates that ν_μ disappearance is favored over $\nu_\mu \to \nu_e$ oscillation. Even though 1.1 years of Super-Kamiokande running has surpassed the prior world statistics, more livetime will be welcome and allow finer subdivision of the data for cross-checks and estimation of possible mixing parameters.

VIII ACKNOWLEDGEMENTS

The author thanks the conference organizers and the participants listed in this paper. He is especially appreciative of his collaborators on Super-Kamiokande for preparing the latest results, which are presented here on their behalf. He gratefully acknowledges financial support by the U.S. Department of Energy.

REFERENCES

1. T.K. Gaisser et al., Phys. Rev. **D54** (1996) 5578.
2. M. Honda et al., Phys. Lett. **B248** (1990) 193; M. Honda et al., ibid.**D52** (1995) 4985.
3. G. Barr et al., Phys. Rev. **D39**(1989) 3532; V. Agrawal et al., ibid.**D53**(1996) 1314; T.K. Gaisser and T. Stanev, Proc. 24th Int. Cosmic Ray Conf.(Rome) Vol.1 (1995) 694.
4. M. Circella et al., Proc. 25th Int. Cosmic Ray Conf.(Durban) Vol.7 (1997) 117.
5. G. Barbiellini et al., Proc. 25th Int. Cosmic Ray Conf.(Durban) Vol.6 (1997) 317.
6. G. Tarlé et al., Proc. 25th Int. Cosmic Ray Conf.(Durban) Vol.6 (1997) 321.
7. T. Stanev, TAUP97, 5th Int'l Workshop on Topics in Astroparticle and Underground Physics, Sept.7-11, 1997, Laboratori Nazionali del Gran Sasso, Asergi, Italy.
8. Y. Fukuda et al., submitted to Phys. Lett., ICRR-Report-411-98-7, hep-ex/9803006.
9. D. Casper et al., Phys. Rev. Lett. **66** (1991) 2561; R. Becker-Szendy, et al., Phys. Rev. **D46** (1992) 3720.
10. K.S. Hirata et al., Phys. Lett. **B205** (1988) 416; K.S. Hirata et al., ibid.**B280** (1992) 146.
11. W.W.M. Allison et al., Phys. Lett. **B391** (1997) 491
12. T. Kafka, TAUP97, 5th Int'l Workshop on Topics in Astroparticle and Underground Physics, Sept.7-11, 1997, Laboratori Nazionali del Gran Sasso, Asergi, Italy.
13. K. Daum et al., Z. Phys. **C66** (1995) 417.
14. M. Aglietta et al., Europhys. Lett. **8** (1989) 611.
15. Y. Fukuda et al., Phys. Lett. **B335** (1994) 237.
16. R. Clark et al., Phys. Rev. Lett. **79** (1997) 345.
17. J. Engel et al., Phys. Rev. **D48** (1993) 3048.
18. K. Inoue, TAUP97, 5th Int'l Workshop on Topics in Astroparticle and Underground Physics, Sept.7-11, 1997, Laboratori Nazionali del Gran Sasso, Asergi, Italy.

19. S. Kasuga et al., Phys. Lett. **B374** (1996) 238.
20. J. Flanagan et al., UH-511-880-97, hep-ph/9709438
21. O.G. Ryaznskaya, JETP Lett. **61** (1995) 237.
22. Y. Fukuda et al., Phys. Lett. **B388** (1996) 397.
23. T.K. Gaiser and T. Stanev, BRI-97-28, astro-ph/9708146.
24. T. Montaruli, TAUP97, 5th Int'l Workshop on Topics in Astroparticle and Underground Physics, Sept.7-11, 1997, Laboratori Nazionali del Gran Sasso, Asergi, Italy.
25. S. Mikheyev, TAUP97, 5th Int'l Workshop on Topics in Astroparticle and Underground Physics, Sept.7-11, 1997, Laboratori Nazionali del Gran Sasso, Asergi, Italy.
26. Y. Totsuka, 28th Int. Symposium on Lepton Photon Interactions, Hamburg, 1997; Kamiokande Collaboration.
27. R. Becker-Szendy et al., Nucl. Phys. **B38** (Proc. Supp.) (1995) 331.
28. Super-Kamiokande collaboration, To be published.
29. P. Lipari and M. Lusignoli, ROME-1190-97, hep-ph/9712278.
30. W. Frati et al., Phys. Rev. **D48** (1993) 1140.
31. P. Lipari et al., Phys. Rev. Lett. **74** (1995) 4384.
32. MACRO collaboration, In preparation.
33. Q.Y. Liu and A.Yu. Smirnov, IC-97-211, hep-ph/9712493.

Neutrino Oscillations

Sandip Pakvasa

*Department of Physics and Astronomy, University of Hawaii
Honolulu, HI 96822 USA*

Abstract. The current status of neutrino oscillations is reviewed.

The current high level of interest in neutrino properties is well justified. Neutrino properties (such as masses, mixings, magnetic moments etc.) are of interest for a variety of reasons: (i) in their own right as fundamental parameters and (ii) as harbingers of new physics beyond the standard model (if e.g. $m_i \neq 0, \theta_i \neq 0, \mu_i \neq 0$ etc.).

I will not review here the kinematic limits on masses but concentrate on the current evidence for mixing and oscillations. First we summarize some salient features of neutrino oscillations. For two flavor mixing (say ν_e and ν_μ), the standard forms for survival probability and conversion probability are given by

$$P_{ee}(L) = 1 - \sin^2 2\theta \sin^2\left(\frac{\delta m^2 L}{4E}\right) \tag{1}$$

$$P_{e\mu}(L) = \sin^2 2\theta \sin^2\left(\frac{\delta m^2 L}{4E}\right)$$

for a neutrino starting out as ν_e. Here θ is the mixing angle, $\delta m^2 = m_2^2 - m_1^2$, L=ct and the ultra-relativistic limit $E_i \approx p + \frac{m_i^2}{2p}$ has been taken. Although these formulae are usually derived in plane wave approximation with $p_1 = p_2$, it has been shown that a careful wave-packet treatment yields the same formulae [1]. When the argument of the oscillating term ($\frac{\delta m^2 L}{4E}$) is too small, no oscillations can be observed. When it is much larger than one, then due to the spread of E at the source or finite energy resolution of the detector, the oscillating term effectively averages out to 1/2.

There are some obvious conditions to be met for oscillations to take place. As the beam travels, the wave packet spreads and the mass eigenstates separate. If the width Δx remains greater than the separation, then oscillations will occur; but if the separation is greater then two separate pulses of ν_1 (mass m_1) and ν_2 (mass m_2) register in the detector with intensities $\cos^2\theta$ and $\sin^2\theta$ separated by

$\Delta t = (\delta m^2/2E^2)(L/c)$. In principle, the intensities as well as oscillation expressions should reflect the slightly different decay widths for different mass eigenstates but this is of no practical importance [1]. The same expressions remain valid if the mixing is with a sterile neutrino with no weak interactions. With 3 flavors mixing, the mixing matrix can have a phase (á la Kobayashi and Maskawa) [2] and the oscillations have a CP non-conserving term leading to

$$P_{\alpha\beta}(L) \neq P_{\beta\alpha}(L), \quad P_{\alpha\beta}(L) \neq P_{\bar{\alpha}\bar{\beta}}(L) \qquad (2)$$

etc [3].

An old observation which has become relevant recently is the following: it is possible for neutrinos to be massless but not be orthogonal [4]. For example, with three neutrino mixing we have

$$\nu_e = U_{e1}\nu_1 + U_{e2}\nu_2 + U_{e3}\nu_3 \qquad (3)$$
$$\nu_\mu = U_{\mu 1}\nu_1 + U_{\mu 2}\nu_2 + U_{\mu 3}\nu_3$$

Suppose $m_1 = m_2 = 0$ but m_3 is non-zero and $m_3 > Q$ where Q is the energy released in β-decay or π-decay producing ν_e and ν_μ beams. Then ν_e and ν_μ will have zero masses but will not be orthogonal:

$$< \nu_e \mid \nu_\mu > = -U_{e3}^* U_{\mu 3} \neq 0 \qquad (4)$$

(Scenarios similar to this are realized in combined fits [5], to solar and LSND neutrino anomalies). Incidentally, the "ν_e" and "ν_μ" produced in Z decay will not be massless and will be nearly orthogonal! This example illustrates the fact that neutrino flavor is not a precise concept and is process dependent.

There are two other ways in which massless neutrinos can mix and even oscillate. One is when there exist Flavor Changing couplings (FCNC) and Non-Universal couplings(NUNC) of neutrinos to light quarks [6]:

FCNC: $\quad \epsilon_q \bar{\nu}_e \nu_\tau \bar{q} q$

NUNC: $\quad \epsilon'_q (\bar{\nu}_e \nu_e - \bar{\nu}_\tau \nu_\tau) \bar{q} q$

In presence of matter this can lead to an analog of MSW [7] effect and can, in principle, account for solar neutrino results without masses or conventional mixings of neutrinos [6]. Oscillations of massless neutrinos can also occur in models of flavor violating gravity (violation of equivalence principle) [8] and of violation of Lorentz invariance [9]. Here the mixing is due to mis-alignment of the gravity eigenstate or of the velocity eigenstate with the flavor eigenstate. In both cases the dependence of the oscillating term on energy is very different from the usual $\delta m^2 L/4E$, it goes as $\frac{1}{2}f\delta\phi EL$ or $\frac{1}{2}\delta v \, EL$. Here $\delta f = 2\delta\gamma = 2(\gamma_2 - \gamma_1)$ is the small number parameterizing the violation of equivalence principle, ϕ is the gravitational potential and $\delta v = v_2 - v_1$ is the difference between the two maximum speeds of velocity eigenstates in case of violation of Lorentz invariance.

ATMOSPHERIC NEUTRINOS

The cosmic ray primaries produce pions which on decays produce $\nu'_\mu s$ and $\nu'_e s$ by the chain $\pi \to \mu\nu_\mu$, $\mu \to e\nu_e\nu_\mu$. Hence, one expects a ν_μ/ν_e ratio of 2:1. As energies increase the $\mu's$ do not have enough time (decay length becomes greater than 15-20 km) and the ν_μ/ν_e ratio increases. Also at low energies the ν flux is almost independent of zenith angle; at high energies due to competition between π-decay and π-interaction the famous "sec θ" effect takes over. Since the absolute flux predictions are beset with uncertainties of about 20%, it is better to compare predictions of the ratio (which may have only a 5% uncertainty) ν_μ/ν_e to data in the form of the famous double ratio $R = (\nu_\mu/\nu_e)_{data}/(\nu_\mu/\nu_e)_{mc}$.

For the so-called "contained" events which for Kamiokande and IMB correspond to visible energies below about 1.5 GeV, the weighted world average (before SuperKamiokande) is $R = 0.64 \pm 0.06$ [10]. This includes all the data from IMB, Kamiokande, Frejus, Nusex and Soudan. As we heard, the new SuperK results are completely consistent with this [10]. It may be worthwhile to recall all the doubts and concerns which have been raised about this anomaly (i.e. deviation of R from 1) in the past and their resolution. (i) Since initially the anomaly was only seen in Water Cerenkov detectors, the question was raised whether the anomaly was specific to water Cerenkov detectors. Since then, it has been seen in a tracking detector i.e. SOUDAN II. (ii) Related to the above was the concern whether e/μ identification and separation was really as good as claimed by Kamiokande and IMB. The beam tests at KEK established that this was not a problem [11]. (iii) The ν_e and ν_μ cross-sections at low energies are not well known; however $e - \mu$ universality should hold apart from known kinematic effects. (iv) If more $\pi^{+'}s$ than $\pi^{-'}s$ are produced, then even though the ratio of 2/1 is preserved there is an asymmetry in $\bar{\nu}_e/\nu_e$ versus $\bar{\nu}_\mu/\nu_\mu$. Since ν cross-sections are larger than $\bar{\nu}$ cross-sections, the double ratio R would become smaller than 1 [12]. However, to explain the observed R, $\pi^{+'}s$ would have to dominate over $\pi^{-'}s$ by 10 to 1, which is extremely unlikely and there is no evidence for such an effect. (v) Cosmic ray muons passing thru near (but outside) the detector could create neutrals (especially neutrons) which enter the tank unobserved and then create $\pi^{0'}s$ faking "e" like events [13]. Again this effect reduces R. However, Kamiokande plotted their events versus distance from wall and did not find any evidence for more "e" events near the walls [14]. It seems that the anomaly is real and does not have any mundane explanation. The new data from SuperK that we just heard about extends the anomaly to higher energies than before and shows a clear zenith angle dependence as well. This rules out most explanations offered except for the ones based on neutrino oscillations.

If the atmospheric neutrino anomaly is indeed due to neutrino oscillations as seems more and more likely; one would like to establish just what the nature of oscillations is. There have been several proposals recently. One is to define an up-down asymmetry for $\mu's$ as well as $e's$ as follows:

$$A_\alpha = (N_\alpha^d - N_\alpha^u)/(N_\alpha^d + N_\alpha^u) \tag{5}$$

where $\alpha = e$ or μ, d and u stand for downcoming ($\theta_Z = 0$ to $\pi/2$) and upcoming ($\theta_Z = \pi/2$ to π) respectively. A_α is a function of E_ν. The comparison of $A_\alpha(E_\nu)$ to data can distinguish various scenarios for ν-oscillations rather easily [15]. This asymmetry has the advantage that absolute flux cancels out and that statistics can be large. It can be calculated numerically or analytically with some simple assumptions. One can plot A_e versus A_μ for a variety of scenarios: (i) $\nu_\mu - \nu_\tau$ (or $\nu_\mu - \nu$ sterile) mixing, (ii) $\nu_\mu - \nu_e$ mixing, (iii) three neutrino mixing (iv) massless ν mixing etc.

The general features of the asymmetry plot are easy to understand. For $\nu_\mu - \nu_\tau$ (or $\nu_\mu - \nu_{st}$) case, A_μ increases with energy, and A_e remains 0; for $\nu_\mu - \nu_e$ mixing, A_e and A_μ have opposite signs; the three neutrino cases interpolate between the above two; for the massless case the energy dependence is opposite and the asymmetries decrease as E_ν is increased; when both ν_μ and ν_e mix with sterile ν's, both A_μ and A_e are positive etc. With enough statistics, it should be relatively straightforward to determine which is the correct one. As we heard [10], data from SuperK clearly point to $\nu_\mu - \nu_\tau$ (or $\nu_\mu - \nu_{st}$) as the culprit [16]. There are suggestions which can in principle distinguish $\nu_\mu - \nu_\tau$ from $\nu_\mu - \nu_{st}$ mixing. If one considers the total neutral current event rate divided by the total charged current event rate; the ratio is essentially the n.c. cross section divided by the c.c. cross section. With $\nu_\mu - \nu_{st}$ oscillations the ratio remains unchanged since ν_{st} has neither n.c. nor c.c. interactions and the numerator and denominator change equally ($\nu_\mu - \nu_e$ case is even simpler: nothing changes); however, in $\nu_\mu - \nu_\tau$ case the denominator decreases and the ratio is expected to increase by $\left(\frac{1+r}{P+r}\right) \approx 1.5$, (here $r = N^0_{\nu e}/N^0_{\nu \mu} \approx 1/2$ and $P = 1/2 = \nu_\mu$ survival probability). Of course, it is difficult to isolate neutral current events; but it is proposed to select $\nu N \to \nu \pi^0 N$ and $\nu N \to \ell \pi^\pm N$ events and the old Kamiokande data seem to marginally favor $\nu_\mu - \nu_\tau$ over $\nu_\mu - \nu_{st}$ [17]. Another possibility is to consider the up-down asymmetry (A_N) in neutral current events e.g. $\nu_N \to \nu \pi^0 N$. For high enough energies the π^0 direction tends to track the neutrino direction [18]. It is easy to show that for $\nu_\mu - \nu_\tau$ case this asymmetry A_N is essentially zero, while for $\nu_\mu - \nu_{st}$ case $A_N \approx \frac{2}{3} A_\mu \approx 0.2$ [19]. Finally one can also use the asymmetry of all multi-ring events where the asymmetry is rather diluted and the difference in the two cases rather small but has the advantage of higher statistics [20].

If we scale L and E each by the same amount, say ~ 100, we should again see large effects. Hence, upcoming thrugoing μ's which correspond to $E \sim 100$ GeV on the average, with path lengths of $L \overset{\sim}{>} 10,000$ km should be depleted. There are data from Kolar Gold Fields, Baksan, Kamiokande, IMB, MACRO, SOUDAN and now SuperK. It is difficult to test the event rate for ν_μ depletion since there are no ν'_es to take flux ratios and the absolute flux predictions have 20% uncertainties. However, there should be distortions of the zenith angle distribution and there seems to be some evidence for this [21]. Very recently, both Super-Kamiokande [16] and MACRO [22] (with lower statistics) have presented analysis of these muons and found the zenith angle distributions to agree well with oscillations of ν'_μs with

δm^2 and $sin^2 2\theta$ consistent with the contained events. Furthermore, SuperK has also presented an analysis of ratio of stopping $\mu's (<E_\nu> \sim 10$ GeV$)$ to thrugoing $\mu's$ and again found consistency with oscillations [16]. The overlap of δm^2 from the SuperK and the Kamiokande analysis is at about $5.10^{-3} eV^2$ and mixing angle is rather large, $\sin^2 2\theta \sim 0.8$ to 1.

SOLAR NEUTRINOS

The data from four solar neutrino detectors (Homestake, Kamiokande, SAGE and Gallex) have been discussed extensively [23]. The SuperK data are consistent with those from Kamiokande but increase the statistics by an order of magnitude in one year [10]. To analyze these data one makes the following assumptions: (i) the sun is powered mainly by the pp cycle, (ii) the sun is in a steady state, (iii) neutrino masses are zero and (iv) the β-decay spectra have the standard Fermi shapes. Then it is relatively straightforward to show using these data with the solar luminosity that the neutrinos from ^7Be are absent or at least two experiments are wrong [24]. ^7Be is necessary to produce ^8B and the decay of ^8B has been observed; and the rate for $^7Be + e^- \to \nu + Li$ is orders of magnitude greater than $^7Be + \gamma \to ^8B + p$ and hence it is almost impossible to find a "conventional" explanation for this lack of ^7Be neutrinos. The simplest explanation is neutrino oscillations.

Assuming that neutrino oscillations are responsible for the solar neutrino anomaly; there are several distinct possibilities. There are several different regions in $\delta m^2 - sin^2 2\theta$ plane that are viable: (i) "Just-so" with $\delta m^2 \sim 10^{-10} eV^2$ and a large $\sin^2 2\theta$ [24], (ii) MSW small angle with $\delta m^2 \sim 10^{-5} eV^2$ and $\sin^2 2\theta \sim 10^{-2}$ and (iii) MSW large angle with $\delta m^2 \sim 10^{-7} eV^2$ (or $\delta m^2 \sim 10^{-5} eV^2$) and $\sin^2 2\theta$ large [25]. The "just-so" is characterized by strong distortion of ^8B spectrum and large real-time variation of flux, especially for the ^7Be line; MSW small angle also predicts distortion of the ^8B spectrum and a very small $^7Be\nu$ flux and MSW large angle predicts day-night variations. These predictions (especially spectrum distortion) will be tested in the SuperK as well as SNO detectors. In particular SNO, in addition to the spectrum, will be able to measure NC/CC ratio thus acting as a flux monitor and reducing the dependence on solar models [26]. The oscillations could be either ν_e to another flavor ν_α or to ν_{st}. The parameter space for the MSW regions is slightly different for the two cases. The absence of any day-night effect in the most recent SuperK data seems to disfavor the MSW large angle region with $\delta m^2 \sim 10^{-5} eV^2$ [27].

The only way to directly confirm the absence or depletion of ^7Be neutrinos is by trying to detect them with a detector with a threshold low enough in energy. One such detector under construction is Borexino, which I describe below [28].

Borexino is a liquid scintillator detector with a fiducial volume of 300T; with energy threshold for 0.25MeV, energy resolution of 45 KeV and spatial resolution of $\sim 20cm$ at 0.5 MeV. The PMT pulse shape can distinguish between $\alpha's$ and $\beta's$. Time correlation between adjacent events of upto 0.3 nsec is possible. With these features, it is possible to reduce backgrounds to a low enough level to be able to extract a signal from ^7Be $\nu'_e s$ via $\nu - e$ scattering. Radioactive impurities such as ^{238}U, ^{232}Th and ^{14}C have to be lower than $10^{-15}, 10^{-16} g/g$ and $10^{-18}(^{14}C/^{12}C)$ respectively. In the test tank CTF (Counting Test Facility) containing 6T of LS, data were taken in 1995-96 and these reductions of background were achieved. As of last summer, funds for the construction of full Borexino have been approved in Italy (INFN), Germany (DFG) and the U.S. (NSF); and construction should begin soon. The Borexino collaboration includes institutions from Italy, Germany, Hungary, Russia and the U.S..

With a FV of 300T, the events rate from ^7Be $\nu's$ is about 50 per day with SSM, and if $\nu'_e s$ convert completely to $\nu_\alpha (\alpha = \mu/\tau)$ then the rate is reduced by a factor $\sigma_{\nu\mu e}/\sigma_{\nu ee} \sim 0.2$ to about 10 per day, which is still detectable. Since the events in a liquid scintillator have no directionality, one has to rely on the time variation due to the $1/r^2$ effect to verify the solar origin of the events. If the solution of the solar neutrinos is due to "just so" oscillations with $\delta m^2 \sim 10^{-10} eV^2$, then the event rate from ^7Be $\nu's$ shows dramatic variations with periods of weeks to months.

Borexino has excellent capability to detect low energy $\bar{\nu}'_e s$ by the Reines-Cowan technique: $\bar{\nu}_e + p \rightarrow e^+ + n, n + p \rightarrow d + \gamma$ with 0.2 msec separating the e^+ and γ. This leads to possible detection of terrestial and solar $\bar{\nu}'_e s$. The terrestial $\bar{\nu}'_e s$ can come from nearby reactors and from ^{238}U and ^{232}Th underground. The Geo-thermal $\bar{\nu}_e s'$ have a different spectrum and are relatively easy to distinguish above reactor backgrounds. Thus one can begin to distinguish amongst various geophysical models for the U/Th distribution in the crust and mantle [29]. Solar $\bar{\nu}'_e s$ can arise via conversion of ν_e to $\bar{\nu}_\mu$ inside the sun when ν_e passes thru a magnetic field region in the sun (for a Majorana magnetic moment) and then $\bar{\nu}_\mu \rightarrow \bar{\nu}_e$ by the large mixing enroute to the earth [30]. The Kamland detector will also detect Geo-thermal $\bar{\nu}'_e s$ at a higher rate due to the larger volume [31].

THREE NEUTRINO MIXING.

In addition to the atmospheric and solar neutrino anomalies, there are also the LSND observations which require $\nu_e - \nu_\mu$ mixing with δm^2 in the range $0.3 - 8eV^2$ and $\sin^2 2\theta$ in the range 10^{-2} to 10^{-3} [33]. With the atmospheric anomaly requiring ν_μ mixing with a $\delta m^2 \sim 5.10^{-3} eV^2$ and solar neutrinos a δm^2 in the range $10^{-4} - 10^{-6} eV^2$ (or $10^{-10} eV^2$) for ν_e mixing; it is clear that one needs 4 neutrino states to mix in order to account for the three separate δm^2's. There have been two proposals to account for the three effects with just three flavors. One was by Acker and Pakvasa [33] which uses the same $\delta m^2 \sim 5.10^{-3}$ with large $\nu_e - \nu_\mu$ mixing to account for both solar and atmospheric neutrinos; and a small mixing with $\nu_\tau (\delta m^2 \sim 1 eV^2)$ to account for the LSND. The other, by Cardall and Fuller [34] employs a δm^2 of $\sim 0.3 eV^2$ to account for both atmospheric and LSND with solar neutrinos driven by either MSW ($\delta m^2 \sim 10^{-5} eV^2$) or "just so" ($\delta m^2 \sim 10^{-10} eV^2$). Both of these are now ruled out: by the CHOOZ results [35] which saw no $\nu_e - \nu_\mu$ oscillations at a δm^2 of $5.10^{-3} eV^2$ with large mixing and by the SuperK data which require a δm^2 of $5.10^{-3} eV^2$. It thus seems inescapable that the three anomalies together require four light neutrino states; and thus at least one sterile neutrino. Some advantages of a light sterile neutrino mixing with the flavor neutrinos were described here [36].

There are three ways to account for all the anomalies with just three neutrino flavors by appealing to new physics. One is to blame the LSND observation on a lepton number violating decay mode:

$$\mu^+ \to e^+ \bar{\nu}_e \nu_\mu \tag{6}$$

with a branching ratio of order of 10^{-3} [37]. Then only the atmospheric neutrino and solar-neutrino observations are due to neutrino oscillations. Another is to have FCNC and NUNC of neutrinos with light quarks with a resulting MSW like phenomena in the sun: e.g. values of $\epsilon_u \sim 10^{-2}$ and $\epsilon'_u \sim 0.43$ can easily account for the solar neutrino data [6,38] thus leaving only the atmospheric and LSND effects to be explained by neutrino oscillations. The third proposal is very imaginative: Ma and Roy [39] envision a mass-mixing pattern á la Cardall-Fuller with $\delta m^2 \sim 0.3 eV^2$ for LSND and atmospheric neutrinos and a $\delta m^2 \sim 10^{-5} eV^2$ for small angle MSW for solar neutrinos. As mentioned above, this fails to account for the zenith angle dependence observed in the Superk data. To this end, they invoke new diagonal ν_τ neutral current interactions:

$$\frac{\delta \, G_F}{\sqrt{2}} \bar{\nu}_{\tau_L} \gamma_\mu \nu_{\tau_L} \{\bar{q}_L \gamma_\mu q_L + \alpha \bar{q}_R \gamma_\mu q_R\} \tag{7}$$

where q is a light quark. In their picture, $\nu_2 \approx \nu_e$ mixes with $\nu_1 (\approx (\nu_\mu + \nu_\tau)/\sqrt{2})$ and $m_2 > m_1$. The MSW in the sun is restored (in spite of the reversal of mass hierarchy) by the extra ν_τ interactions, but the parameter δ has to be quite large, of order 5 to 10. Now for atmospheric neutrinos, at low energies, an $R \sim 0.6$ is guaranteed by the large mixing of $\nu_\mu - \nu_\tau$ and a $\delta m^2 \sim 0.3 eV^2$. The zenith angle dependence is primarily a matter effect which arises as follows. The matter effects due to the strong new n.c. interactions of ν_τ are different for ν and $\bar{\nu}$; and since $\sigma_{\bar{\nu}}$ and $\phi_{\bar{\nu}}$ are also different from σ_ν and ϕ_ν there is a cumulative effect in $(\mu + \bar{\mu})$ event rate which gives a zenith angle dependence. It is not clear whether this would give precisely the observed zenith angle(or L/E) dependence and furthermore, the predicted $\nu_\tau N$ cross-section may be so large as to be in trouble already. In any case the enhanced $\nu_\tau N$ n.c. cross-section should be easily detectable in SNO (via the $\nu'_\tau s$ converted from $\nu'_e s$ in the model).

NEUTRINO OSCILLATIONS AND PULSAR VELOCITIES

A novel application of neutrino oscillations is to account for pulsar velocities. It has been known for a long time that pulsars have "kick" velocities which are rather large $\sim 450 km/s$ (upto $1000\ km/s$) and difficult to account for conventionally (non-spherical collapse etc). A possible particle physics origin is suggested on dimensional grounds. The pulsar kick momentum is of order of $m_\odot v \sim 10^{41} gm - cm/sec$; the energy carried off by neutrinos is about $3.10^{53} ergs$; this corresponds to an equivalent momentum magnitude kick (i.e. $E_\nu/c = | \underline{k}_\nu | = \Sigma_i | \underline{k}_{\nu i} |$) of 10^{43} gm-cm/sec. Hence, a small 1% anisotropy in neutrino emission would be sufficient to account for the observed velocities. The basic idea of the proposal, due to Segre and Kusenko [40], goes as follows. The neutrino-sphere of ν_τ (or ν_μ) is expected to be inside the ν_e because of the longer mean free path of ν_τ; and as a result $\nu'_\tau s$ come out at higher energies than $\nu'_e s$. Now suppose there is $\nu_e - \nu_\tau$ mixing with (i) $m_{\nu_\tau} > m_{\nu_e}$ and (ii) small θ_{vac} such that an adiabatic MSW resonance occurs at r_0 where r_0 is between the ν_e and ν_τ spheres ($R_e < r_0 < R_\tau$). Now r_0 is defined by the MSW condition:

$$\delta m^2 (cos2\theta) = 2\sqrt{2}\ G_F\ N_e(r_0) k \qquad (8)$$

where $k = | \underline{k}_\nu |$.

For typical values of N_e, one needs $\delta m^2 \sim (100 eV)^2$ to satisfy the above condition. Effectively r_0 becomes the new radius of ν_τ sphere since this is where $\nu'_e s$ will convert to $\nu'_\tau s$ and escape. Now consider the effect of a large magnetic field B inside the star. The MSW resonance condition is modified to [41]:

$$\frac{\delta m^2}{2k} \cos 2\theta = \sqrt{2}\ G_F N_e + \frac{e\ G_F}{\sqrt{2}} \left(\frac{3N_e}{\pi^4} \right)^{1/3} \underline{k}.\underline{B}. \qquad (9)$$

Because of the presence of the $\underline{k}.\underline{B}$ term the ν_τ sphere is distorted and r_0 is a function of the polar angle ϕ. An estimate of the resultant asymmetry in neutrino momentum yields values at a 1% level as desired if $| \underline{B} |$ is of order 10^{15} to 10^{16} Gauss [41].

This is an elegant, imaginative proposal with two problems: one needs (i) a ν_τ mass of about 100eV and (ii) a magnetic field close to 10^{16} Gauss.

There have been related proposals to modify this basic idea: (i) replace $\nu_e - \nu_\tau$ mixing by $\nu_e - \nu_{st}$ mixing [42]; in that case $m_{\nu_{st}}$ has to be a few KeV but B can be lowered to $10^{15}G$; (ii) generalize the discussion to $3 - \nu$ mixing [43], (iii) invoke resonant spin flavor precession [44] (in this case one needs a transition magnetic moment of order $10^{-14} \mu_B$, a m_{ν_τ} of 16 KeV and a B of 4.10^{15}); (iv) invoke massless but non-orthogonal neutrinos with $< \nu_e | \nu_\tau > \sim 10^{-2}$ and $B \sim 10^{15}$ [45].

Can this general idea be tested? One would expect a correlation between v_p (the pulsar velocity) and $\underline{B}$ the magnetic field in both magnitude and direction. Empirically there is some correlation for slow pulsars [46]. The question seems to be open at the moment.

CONCLUSION

It seems clear that the evidence for neutrino oscillations (of ν_μ) in the SuperK atmospheric data is quite convincing. In the solar neutrino data from all the detectors there is impressive indirect evidence for neutrino oscillations (of ν_e). In the next few(5?) years we will have data available from SuperK, SNO and Borexino on solar neutrinos; from K2K, MINOS, CERN-GSL, and JHF-SuperK Long Baseline experiments; from Short baseline experiments such as KARMEN, Mini-Boone, and the future incarnations of CHORUS, NOMAD and COSMOS. With these in hand it should be possible to settle the following questions: (i) ν_τ versus ν_{st} in the atmosperic neutrinos; (ii) ν_{flavor} versus ν_{st} in the solar neutrinos; (iii) whether the relevant δm^2 for solar neutrinos is $\sim 10^{-5} eV^2$ or $\sim 10^{-10} eV^2$; (iv) whether the mixing for solar neutrinos is large or small; (v) whether the LSND effect is confirmed or (vi) whether there is some other indication of oscillations at yet another L/E? If, in addition, there are any positive results from either the Tritium end-point measurements [47] or the new round of neutrinoless double-beta decay experiments [48], we may well have neutrino masses(rather than just mass differences) in addition to having all the mixings in hand. This is a challenge to theorists: to predict this mass matrix before all the data is in hand.

ACKNOWLEDGMENTS

I thank Jose Nieves for the outstanding hospitality and the organisation, and Vernon Barger, David Caldwell, Plamen Krastev, John Learned, Sergei Petcov, Jim Stone and Tom Weiler for useful and enjoyable discussions. This research was supported in part by the US Department of Energy Grant No. DE-FG-03-94ER40833.

REFERENCES

1. C.W. Kim and A. Pevsner, *Neutrinos in Physics and Astrophysics*, Harwood, 1994 and references therein.
2. M. Kobayashi and T. Maskawa, Progress Theoret. Phys. (Kyoto), 49, 652 (1973).
3. S. Pakvasa, in *Proc. of the XX International Conference of High Energy Physics*, Madison, Wisconsin, USA, 1980, ed. L. Durand and L. G. Pondrom (AIP, New York 1980) Part 2, p. 1165; V. Barger, K. Whisnant and R.J.N. Phillips, *Phys. Rev. Lett.* 45 (1980) 2084; *Phys. Rev.* D23 (1981) 2773; S. M. Bilenky and F. Nidermayer, *Sov. J. Nucl. Phys.* 34 (1981) 606.
4. B. W. Lee, S. Pakvasa and H. Sugawara, *Phys. Rev. Lett.* 38, 937 (1977); S. B. Treiman, F. Wilczek and A. Zee, *Phys. Rev.* D16, 152 (1977).
5. K. S. Babu, J. Pati, and F. Wilczek, *Phys. Lett.* B359, 351 (1997).
6. L. Wolfenstein, it Phys. Rev. D17, 2369(1978); M. M. Guzzo, A. Masiero and S. Petcov, *Phys. Lett.* B260, 154 (1991); E. Roulet, *Phys. Rev.*D38, 2635 (1991); V. Barger, R. J. N. Philips and K. Whisnant, *Phys. Rev.* D44, 1569 (1991); P. Krastev and J.N. Bahcall, hep-ph/9703267.
7. L.Wolfenstein, Ref.6; S. P. Mikheyev and A. Yu. Smirnov, *Zh. Eksp. Teor. Fiz.* 91, 7 (1986).

8. M. Gasperini, *Phys. Rev.* D38, 2635 (1988); A. Halprin and C.N. Leung, *Phys. Rev. Lett.* 67, 1833 (1991); A. Halprin, C.N. Leung and J. Pantaleone, *Phys. Rev.* D53, 5365 (1996).
9. S. Coleman and S.L. Glashow, *Phys. Lett.* B405, 249 (1997); S. L. Glashow; A. Halprin, P. Krastev, C.N. Leung and J. Pantaleone, *Phys. Rev.* D56, 2433 (1997).
10. J. L. Stone, *These proceedings*.
11. S. Kasuga et al. *Phys. Lett.* B374, 238 (1994).
12. L. Volkova, *Phys. Lett.* B316, 178 (1993).
13. O. G. Ryashkaya, JETP Lett. 61, 237 (1995).
14. Kamiokande Collaboration, Y. Fukuda et al., *Phys. Lett.* B388, 397 (1996).
15. J. Flanagan, J. G. Learned and S. Pakvasa, *Phys. Rev.* D57, R2649 (1998); J. Bunn, R. Foot and R. Volkas, *Phys. Lett.* B413, 109 (1997); R. Foot, R. Volkas and O. Yasuda, *Phys. Lett.* B421, 245(1998).
16. T. Kajita, *Talk at Neutrino 98*; Y. Fukuda, hep-ex/9805006.
17. F. Vissani and A. Smirnov, hep-ph/9710565.
18. M. Diwan and M. Goldhaber, Unpublished.
19. J.G. Learned, S. Pakvasa and J. L. Stone, hep-ph/9805343.
20. L. Hall and H. Murayama, hep-ph/9806218.
21. J.G. Learned (private communication).
22. F. Ronga, *Talk at Neutrino 98*, Takayama, June 4-9, 1998.
23. Proceedings of Neutrino 96, June 1996, Helsinki. ed. J. Malaampi and M. Roos (World Scientific) to be published; K. Lande, T. Kirsten, V.N. Gavrin, *Talks at Neutrino 98*, Takayama, June 4-9 (1998).
24. N. Hata and P. Langacker, *Phys. Rev.* D56, 6107 (1997), and references therein.
25. J. N. Bahcall, Lectures at SLAC Summer Institute on Particle Physics, Aug. 1997 (to be published); hep-ph/9711358 and references therein.
26. A. McDonald, *Talk at Neutrino 98*, Takayama, June 4-9, 1998.
27. Y. Suzuki,*Talk at Neutrino 98*, Takayama, June 4-9, 1998.
28. C. Arpesella et al., *The Borexino Proposal* Vol. 1 and 2., ed. G. Bellini and R.S. Raghavan (Univ. of Milan) 1991; L. Oberauer, it Talk at Neutrino 98, Takayama, June 4-9, 1998.
29. R. S. Raghavan et al., *Phys. Rev. Lett.* 80, 635 (1998); C. S. Rothschild, M. Chen and F. Calaprice, Nucl.-ex/9710001.
30. R.S. Raghavan, A. B. Balantekin, F. Loreti, A.J. Baltz, S. Pakvasa and J. Pantaleone, *Phys. Rev.* D44, 3786 (1991).
31. A. Suzuki, *Talk at Neutrino 98*, Takayama, June 4-9, 1998.
32. The LSND Collaboration: C. Athanassopoulos et al, *Phys. Rev. Lett.* 7512650 (1995); *Phys. Rev.* C54, 2685 (1996); H. White,*Talk at Neutrino 98*, Takayama, June 4-9, 1998.
33. A. Acker and S. Pakvasa, *Phys. Lett.* B397, 209 (1997).
34. C. Cardall and G. Fuller, *Phys. Rev.* D53, 4421 (1996).
35. The CHOOZ Collaboration, M. Apollonio et al., *Phys. Lett.* B420, 597 (1998).
36. D. Caldwell,*These proceedings*.
37. L. M. Johnson and D, McKay, hep-ph/9805311.
38. P. Krastev (private communication).

39. E. Ma and P. Roy, *Phys. Rev. Lett.* 80, 4637 (1998).
40. A. Kusenko and G. Segre, *Phys. Rev. Lett.* 77, 4872 (1996); ibid 79, 2751 (1997); Y.Z. Qian,*ibid.* 79, 2750 (1997).
41. J. Nieves and P. Pal, *Phys. Rev.*, D40, 1693 (1989); J. C. D'Olivo, J. Nieves and P. Pal,*ibid.*, D40, 3679 (1989).
42. A. Kusenko and G. Segre, *Phys. Lett.* B396, 197 (1997).
43. C. W. Kim, J.D. Kim and J. Song, *Phys. Lett.* B419, 279 (1998).
44. E. Kh. Akhmedov, A. Lanza and D. W. Sciama, *Phys. Rev.* D56, 6117 (1997).
45. D. Grasso, H. Nunokawa and J. Valle, hep-ph/9803002.
46. G. Segre and A. Kusenko, hep-ph/9608103; M. Birkel and R. Toldra, Astron. Astrophys. 326, 995 (1997).
47. C. Weinheimer, V. M. Lobashev, *Talks at Neutrino 98*, Takayama, June 4-9,1998.
48. A. Morales, H. Ejiri, O. Cremonesi, H. V. Klapdor-Kleingrothaus, *Talks at the Neutrino 98*, Takayama, June 4-9,1998.

Electroweak Measurements and Neutrino Oscillations: The NuTeV and BooNE Experiments

Michael H. Shaevitz [1]

Columbia University
New York, New York 10027

Abstract. A preliminary measurement of $\sin^2\theta_W$ in $\nu - N$ deep inelastic scattering from the NuTeV experiment is presented. Using separate neutrino and antineutrino beams, NuTeV is able to determine $\sin^2\theta_W$ with low systematic errors. NuTeV measures $\sin^2\theta_W^{(\text{on-shell})} = 0.2253 \pm 0.0019(\text{stat}) \pm 0.0010(\text{syst})$, which implies $M_W = 80.26 \pm 0.11$ GeV. Expectations are also presented for the NuTeV search for $\nu_\mu \to \nu_\tau$ and $\nu_\mu \to \nu_e$ oscillations using the measured neutral to charged-current ratio. The future BooNE experiment is described which searches for $\nu_\mu \to \nu_e$ appearance and ν_μ disappearance oscillations using a new neutrino beam created from the Fermilab 8 GeV Booster accelerator. The experiment is designed to cover definitively the $\delta m^2/\sin^2 2\theta$ region associated with the possible oscillation signal from the LSND experiment and to measure accurately the oscillation parameters if a signal is observed.

INTRODUCTION

In the past, neutrino scattering experiments have played a key role in establishing the validity of the electroweak Standard Model. Today, even with the large samples of on-shell W and Z bosons at e^+e^- and $p\bar{p}$ colliders, precision measurements of neutrino-nucleon scattering still play an important role. Not only are these measurements competitive in precision with direct probes of weak boson parameters, but they also test the validity of the electroweak theory in different processes and over many orders of magnitude in Q^2. In this respect, if neutrino scattering observed deviations from expectations based on direct measurements from W and Z bosons, this would be an exciting hint of new physics entering in tree-level processes or in radiative corrections. In this way, neutrino scattering would be sensitive to a different menu of non-Standard Model effects [1] and, in particular, to neutrino oscillations [2].

---

[1] Representing the NuTeV and BooNE Collaborations

Neutrino oscillations provide one of the best "laboratories" for probing the effects associated with neutrino mass. Sensitivity to neutrino masses in the range below direct measurements can be accomplished using the relatively pure neutrino beams from accelerators. The Fermilab NuTeV and BooNE experiments are examples of such measurements. The recently completed NuTeV experiment searches for indications of $\nu_\mu \to \nu_\tau$ and $\nu_\mu \to \nu_e$ oscillations through the precision measurement of the weak mixing angle, $\sin^2 \theta_W$. The future BooNE experiment will search for $\nu_\mu \to \nu_e$ appearance and ν_μ disappearance oscillations using a very pure, low energy ν_μ beam from the Fermilab 8 GeV booster combined with a highly sensitive, totally active detector.

THE NUTEV $\sin^2 \theta_W$ MEASUREMENT

Experimental quantities sensitive to electroweak physics that are most precisely measured in neutrino scattering are the ratios of charged-current (W exchange) to neutral-current (Z exchange) scattering cross-sections from quarks in heavy nuclei. The ratio of these cross-sections for either neutrino or anti-neutrino scattering from isoscalar targets of u and d quarks can be written as [3]

$$R^{\nu(\bar{\nu})} \equiv \frac{\sigma(\overset{(-)}{\nu}_\mu N \to \overset{(-)}{\nu}_\mu X)}{\sigma(\overset{(-)}{\nu}_\mu N \to \mu^{-(+)} X)} = (g_L^2 + r^{(-1)} g_R^2), \quad (1)$$

where

$$r \equiv \frac{\sigma(\bar{\nu}_\mu N \to \mu^+ X)}{\sigma(\nu_\mu N \to \mu^- X)} \sim \frac{1}{2}, \quad (2)$$

and $g_{L,R}^2 = u_{L,R}^2 + d_{L,R}^2$, the isoscalar sums of the squared left or right-handed quark couplings to the Z. At tree level in the Standard Model, $q_L = I_{\text{weak}}^{(3)} - Q_{\text{EM}} \sin^2 \theta_W$ and $q_R = -Q_{\text{EM}} \sin^2 \theta_W$; therefore, R^ν is particularly sensitive to $\sin^2 \theta_W$.

In a real target, there are corrections to Eqn. 1 resulting from the presence of heavy quarks in the sea, the production of heavy quarks in the target, non leading-order quark-parton model terms in the cross-section, electromagnetic radiative corrections and any isovector component of the light quarks in the target. In particular, in the case where a charm-quark is produced from scattering off of low-x sea quarks, the uncertainties resulting from the effective mass suppression of the heavy final-state charm quark are large. The uncertainty in this suppression ultimately limited the precision of previous νN scattering experiments which measured electroweak parameters [4–6].

To eliminate the effect of uncertainties resulting from scattering from sea quarks, one can instead form a quantity suggested by Paschos and Wolfenstein [7],

$$R^- \equiv \frac{\sigma(\nu_\mu N \to \nu_\mu X) - \sigma(\bar{\nu}_\mu N \to \bar{\nu}_\mu X)}{\sigma(\nu_\mu N \to \mu^- X) - \sigma(\bar{\nu}_\mu N \to \mu^+ X)} = \frac{R^\nu - r R^{\bar{\nu}}}{1 - r} = (g_L^2 - g_R^2). \quad (3)$$

Since $\sigma^{\nu q} = \sigma^{\bar{\nu}\bar{q}}$ and $\sigma^{\bar{\nu}q} = \sigma^{\nu\bar{q}}$, the effect of scattering from sea quarks, which is symmetric under $q \leftrightarrow \bar{q}$, cancels in the difference of neutrino and anti-neutrino cross-sections. While allowing substantially reduced uncertainties, R^- is a more difficult quantity to measure than R^ν, primarily because neutral current neutrino and anti-neutrino scattering have identical observed final states and can only be separated by *a priori* knowledge of the initial state neutrino.

The NuTeV Experiment and Neutrino Beam

The NuTeV detector consists of an 18 m long, 690 ton target calorimeter with a mean density of 4.2 g/cm^3, followed by an iron toroid spectrometer. The target calorimeter consists of 168 iron plates, 3m × 3m × 5.1cm each. The active elements are liquid scintillation counters spaced every two plates and drift chambers spaced every four plates.

In the detector $\nu_\mu/\bar{\nu}_\mu$ charged-current events are identified by the presence of an energetic muon in the final state which travels a long distance in the target calorimeter. Quantitatively, a length is measured for each event based on the number of neighboring scintillation counters above a low threshold. Charged-current candidates are those events with a length of greater than 20 counters (2.1 m of steel-equivalent), and all other events are neutral-current candidates.

NuTeV's target calorimeter sits in the Sign-Selected Quadrupole Train (SSQT) neutrino beam at the FNAL TeVatron. The observed neutrinos result from decays of sign-selected pions and kaons produced from the interactions of 800 GeV protons in a production target. Focusing and sign-selection magnets direct the mesons into a 0.5 km decay region which ends 0.9 km upstream of the NuTeV detector. The resulting beam is either almost purely neutrino or anti-neutrino, depending of the selected sign of mesons. Anti-particle backgrounds are observed at a level of less than 1–2 parts in 10^3. The beam is almost entirely muon neutrinos, with electron neutrinos creating 1.3% and 1.1% of the observed interactions from the neutrino and anti-neutrino beams, respectively.

Extraction of $\sin^2\theta_W$

Events are selected for this analysis using a 20 GeV visible energy cut along with fiducial volume cuts. Small backgrounds from cosmic-ray and muon induced events are subtracted from the sample. After all cuts, 1.3 million and 0.30 million events are observed in the neutrino and anti-neutrino beam, respectively. The ratios of neutral-current candidates (short events) to charged-current candidates (long events), R_{meas}, are 0.4198 ± 0.0008 in the neutrino beam and 0.4215 ± 0.0017 in the anti-neutrino beam.

R_{meas} is related to the ratios of cross-sections and $\sin^2\theta_W$ using a detailed detector, cross-section, and flux Monte Carlo simulation tuned to the observed distributions. This Monte Carlo must predict the substantial cross-talk between the samples. In the neutral-current sample, the backgrounds in the neutrino and antineutrino beam from $\nu_\mu/\overline{\nu}_\mu$ charged-current events are 19.3% and 7.4%, and the backgrounds from $\nu_e/\overline{\nu}_e$ charged-currents are 5.3% and 5.8%. The charged-current sample has only a 0.3% background from neutral-current events for each beam.

The cross-section model is of paramount importance to this analysis. Neutrino-quark deep-inelastic scattering processes are simulated using a leading-order cross-section model. Neutrino-electron scattering and quasi-elastic scattering are also included. Leading-order parton momentum distributions come from a modified Buras-Gaemers parameterization [8] of structure function data from the CCFR experiment [9] which used the same target-calorimeter and cross-section model as NuTeV. The parton distributions are modified to produce u and d valence and sea quark asymmetries consistent with muon scattering [10] and Drell-Yan [11] data. The shape and magnitude of the strange sea come from an analysis of events in CCFR with two oppositely charged muons (e.g., $\nu q \to \mu^- c$, $c \to \mu^+ X$) [12]. Mass suppression from heavy quark production is generated in a slow-rescaling model whose parameters are measured from the same dimuon data. The charm sea is taken from the CTEQ4L parton distribution functions [13]. The magnitude of the charm sea is assigned a 100% uncertainty and the slow-rescaling mass for $(\nu/\overline{\nu})c \to (\nu/\overline{\nu})c$ is varied from m_c to $2m_c$. Our parameterization of $R_{\text{long}} = \sigma_L/\sigma_T$ is based on QCD predictions and data [14] and is varied by 15% of itself in order to estimate uncertainties. Electroweak and pure QED radiative corrections to the scattering cross-sections are applied using computer code supplied by Bardin [15], and uncertainties are estimated by varying parameters of these corrections. Possible higher-twist corrections are considered with a 100% uncertainty using a VMD-based model which is constrained by lepto-production data [16].

The key test of the Monte Carlo is its ability to predict the length distribution of events in the detector. Fig. 1 shows good agreement between the data and Monte Carlo within the systematic uncertainties.

To compute $\sin^2\theta_W$, a linear combination of R^ν_{meas} and $R^{\overline{\nu}}_{\text{meas}}$ was formed,

$$R^-_{\text{meas}} \equiv R^\nu_{\text{meas}} - \alpha R^{\overline{\nu}}_{\text{meas}}, \qquad (4)$$

where α is calculated using the Monte Carlo such that R^-_{meas} is insensitive to small changes in the slow-rescaling parameters for charm production. For this measurement, an $\alpha = 0.5136$ was used. This technique is similar to an explicit calculation of R^-, but here the background subtractions, the cross-section corrections to Eqn. 3, and the dependence on $\sin^2\theta_W$ are calculated by Monte Carlo. This approach explicitly minimizes uncertainties related to the suppression of charm production, largely eliminates uncertainties related to scattering from sea quarks, and reduces many of the detector uncertainties common to both the ν and $\overline{\nu}$ samples. Uncertainties in this measurement of $\sin^2\theta_W$ are shown in Table 1.

The preliminary result[2] from the NuTeV data is

---

[2] The weak radiative correction applied to extract $\sin^2\theta_W^{\text{on-shell}}$ from the measured quantities

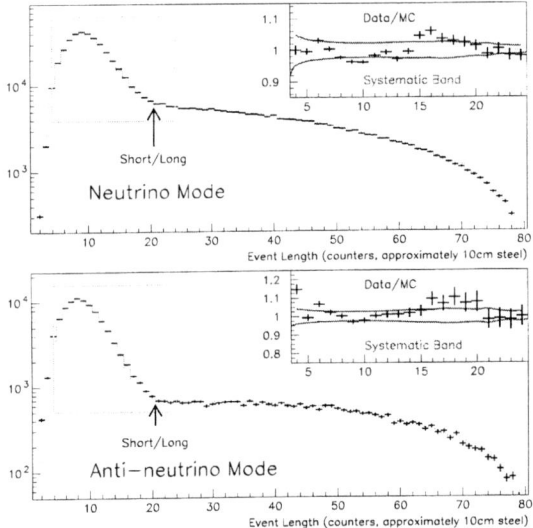

FIGURE 1. Length distributions in the data from the neutrino and anti-neutrino beams. Neutral-current/charged-current separation is made at a length of 20 counters, approximately 2.1 m of steel.

$$\sin^2 \theta_W^{(\text{on-shell})} = 0.2253 \pm 0.0019(\text{stat}) \pm 0.0010(\text{syst})$$
$$-0.00142 \times \left(\frac{M_{\text{top}}^2 - (175 \text{ GeV})^2}{(100 \text{ GeV})^2}\right) + 0.00048 \times \log_e \left(\frac{M_{\text{Higgs}}}{150 \text{ GeV}}\right).$$

The small residual dependence of our result on M_{top} and M_{Higgs} comes from the leading terms in the electroweak radiative corrections [15]. Since $\sin^2 \theta_W^{(\text{on-shell})} \equiv 1 - M_W^2/M_Z^2$, this result is equivalent to

$$M_W = 80.26 \pm 0.10(\text{stat}) \pm 0.05(\text{syst})$$
$$+0.073 \times \left(\frac{M_{\text{top}}^2 - (175 \text{ GeV})^2}{(100 \text{ GeV})^2}\right) - 0.025 \times \log_e \left(\frac{M_{\text{Higgs}}}{150 \text{ GeV}}\right).$$

A comparison of this result with direct measurements of M_W is shown in Figure 2.

NuTeV Neutrino Oscillation Measurements

By comparing the ratio of neutral to charged current scattering events to expectations, NuTeV can perform an indirect search for neutrino oscillations. The

---

has changed since the presentation at the conference due to an error in the implementation of the Bardin code for radiative corrections. Two other small experimental corrections, for muon energy deposition and for charm semi-leptonic decays, were improved as well. The net shift in the result, 0.0054, is dominated by the fix in the implementation of the radiative corrections.

TABLE 1. Uncertainties in $\sin^2\theta_W$

SOURCE OF UNCERTAINTY	$\delta\sin^2\theta_W$
Statistics: Data	**0.00188**
Monte Carlo	0.00028
TOTAL STATISTICS	0.00190
$\nu_e/\bar{\nu}_e$	0.00045
Energy Measurement	0.00051
Event Length	0.00036
TOTAL EXP. SYST.	0.00078
Radiative Corrections	0.00051
Strange/Charm Sea	0.00036
Charm Mass	0.00009
u/d, $\bar{u}/\bar{d}$	0.00027
Longitudinal Structure Function	0.00004
Higher Twist	0.00011
TOTAL PHYSICS MODEL	0.00070
TOTAL UNCERTAINTY	0.0022

absence of a muon in most charged current ν_τ and all ν_e scattering events makes them appear experimentally as "neutral current" event. In addition, the final state electron or τ adds additional energy in the calorimeter that would not be present in an muon neutral or charged current event. The observed effect of ν_μ oscillations is, then, an increase in R_{meas} that depends on the visible energy in the calorimeter. A comparison of the observed R_{meas} distribution to the Monte Carlo expectation determined using a $\sin^2\theta_W$ value from the LEP/SLC experiments can then be used to search for ν_μ oscillations

Using this technique, CCFR has published [2] one of the most restrictive oscillation limits at certain δm^2 for $\nu_\mu \to \nu_\tau$ (Fig. 3). The NuTeV analysis is underway, and $\nu_\mu \to \nu_\tau$ results should be available in Summer 1998. The NuTeV SSQT beam improves the sensitivity because of the reduced ν_e beam contamination and uncertainty compared to CCFR. NuTeV expects a sensitivity to $\nu_\mu \to \nu_\tau$ at high Δm^2 of $\sin^2 2\theta \approx 1.3 \times 10^{-3}$ as shown in Fig. 3. This technique is similar and, therefore, an important proof-of-principle for the analysis proposed by the Minos experiment at Fermilab.

THE BOOSTER NEUTRINO OSCILLATION EXPERIMENT (BOONE) AT FERMILAB

The BooNE neutrino oscillation experiment is a new program to run at Fermilab using the 8 GeV Booster to produce a low energy neutrino beam. The experiment will search for $\nu_\mu \to \nu_e$ appearance and ν_μ disappearance with sensitivity to measure the oscillation parameters, Δm^2 and $\sin^2 2\theta$, over a broad range of parameters. Both neutrino and antineutrino beams will be available making the experiment

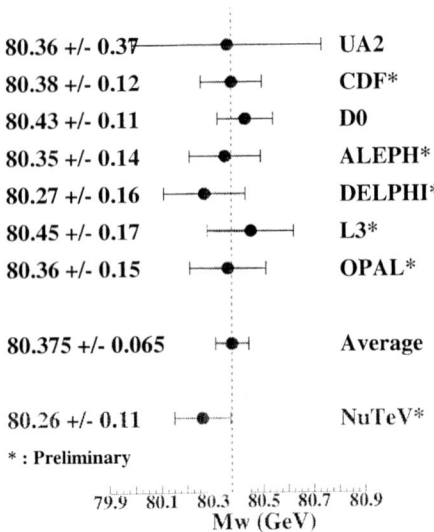

FIGURE 2. Current direct M_W measurements compared with this result

also sensitive to CP violation in the lepton sector.

The proposed BooNE program would start with a single detector (MiniBooNE) with the goal of covering the entire LSND mass region and establish definitively whether there are neutrino oscillations. If a positive signal is observed, a two detector experiment would be proposed where a second detector at a different distance is added to the MiniBooNE setup.

Physics Motivation of the BooNE Program

In 1996, the LSND experiment at Los Alamos first reported evidence for $\bar{\nu}_\mu \to \bar{\nu}_e$ oscillations with an oscillation probability of $\sim 0.3\%$ [17] . Previous oscillation searches have not seen oscillations in the LSND allowed region for $\Delta m^2 > 4$ eV^2, as shown in Fig. 4, restricting Δm^2 to below that value.

Since the original LSND publication, the KARMEN experiment at the ISIS facility and the LSND experiment have obtained further results that strengthen the case for a $\bar{\nu}_\mu \to \bar{\nu}_e$ oscillation signal. KARMEN observes an excess of events in their data taken through 1996 which is consistent with the LSND $\bar{\nu}_\mu \to \bar{\nu}_e$ oscillation signal, but is only at the 1σ level due to low luminosity and high rates of

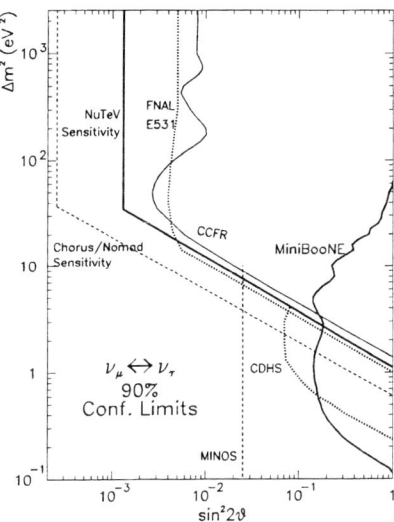

FIGURE 3. Predicted sensitivity for the NuTeV $\nu_\mu \to \nu_\tau$ oscillation measurement along with the results from several other experiments. The ultimate sensitivity for the Chorus/Nomad and Minos experiments is also shown. The curve on the right corresponds to the disappearance sensitivity for the MiniBooNE experiment.

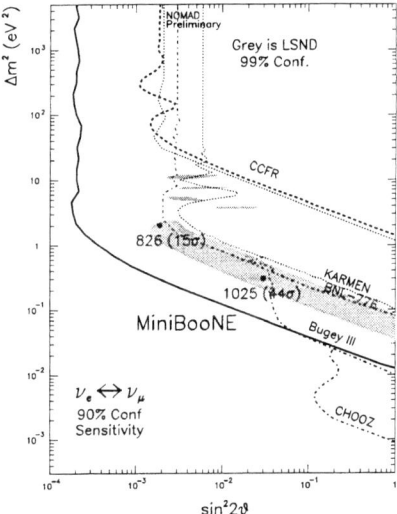

FIGURE 4. Sensitivity for the MiniBooNE experiment. The two points indicate the number of oscillation events that MiniBooNE will detect in one year along with a estimate of the significance of the signal. Thin and dashed lines indicate regions excluded by past experiments. The LSND allowed region is shown in grey.

background. [18] Therefore, KARMEN prefers to quote a limit which appears on Fig. 4. LSND has released preliminary results from the 1996-1997 run which are consistent with the previous LSND results. [19] The preliminary $\bar{\nu}_\mu \to \bar{\nu}_e$ oscillation probability for the entire data sample is $(0.31 \pm 0.09 \pm 0.05)\%$ corresponding to the grey band region in Fig. 4. LSND has published a complementary $\nu_\mu \to \nu_e$ decay-in-flight oscillation search, which has completely different systematics and backgrounds than the $\bar{\nu}_\mu \to \bar{\nu}_e$ oscillation search and which indicates a signal, although of lesser significance, in the same favored regions of Δm^2 and $\sin^2 2\theta$ as the decay at rest analysis. [20]

If the LSND signal is due to neutrino oscillations, MiniBooNE expects between 800 and 1000 events per year, depending on the Δm^2 and $\sin^2 2\theta$ of the oscillation. The MiniBooNE systematics are significantly different than those of the LSND experiment. Thus MiniBooNE will be able to verify or disprove the LSND result. If a signal is observed, the full BooNE two-detector system (phase II of this program) will then accurately measure the oscillation parameters. If the LSND signal is not observed by MiniBooNE, then the expected sensitivity is shown in Fig. 4. The experiment will extend significantly the region probed for oscillations beyond previous limits in Δm^2 and $\sin^2 2\theta$.

BooNE Capabilities and Design Issues

As shown above, there is a need for experiments to probe $\nu_\mu \to \nu_e$ oscillations in the $0.01 - 10.0$ eV^2 mass region with mixing down to $\sin^2 2\theta \approx 10^{-4} - 10^{-2}$. For $\Delta m^2 = 1$ eV^2, an experiment needs an L/E value of about 1.0. Since the rate from a neutrino source falls as $1/L^2$, the most cost effective way to probe this region is with the smallest L for the available E_ν value. A neutrino beam from the 8 GeV Fermilab Booster is almost optimal for this region using an L value of ~ 500 m combined with $0.1 < E_\nu < 1.0$ GeV.

The BooNE program will use a magnetic horn to focus the copiously produced 3 GeV pions into a parallel secondary beam. The secondary beam then traverses a relatively short decay length which can be varied from 25 to 50 m. The short decay length keeps the fraction of ν_e in the beam from the $\pi \to \mu \to e$ decay chain at a low level. Varying the decay length provides a test of signal versus background if an excess of ν_e events is observed. The focusing system will be capable of operation in either positive or negative polarity, which opens further opportunities for verifying a signal and studying possible CP violation effects.

For the initial single detector experiment, MiniBooNE, accurate determination of the ν_e beam background is needed. The observed ν_μ event spectrum is highly correlated with the decaying pion spectrum due to the small solid angle subtended by the detector. The observed ν_μ events constrain the pion decay spectrum and subsequent muon decay spectrum to the 5% level.

The MiniBooNE experiment needs a detector with a large fiducial mass and good particle identification for neutrino events in the $0.10 < E_\nu < 2.0$ GeV energy region. At these low energies, a totally active detector is necessary. A detector based on a large volume of mineral oil is both cost effective and very powerful for particle identification. Many of the critical detector components are available from the LSND experiment including the 1220 eight-inch photomultiplier tubes (PMTs) with readout and data acquisition system. The mineral oil will be contained in a spherical tank 12 m in diameter leading to a fiducial volume corresponding to ~ 445 tons. The phototubes will be located at a radius of 5.5 m, and the 50 cm veto region between the outer wall and phototubes will be optically isolated from the main volume and viewed by an additional 292 phototubes facing outward.

Particle mis-identification can be reduced to the $\approx 10^{-3}$ level while keeping the ν_e and ν_μ efficiency at $\sim 60\%$. The identification techniques use the spatial and time correlation of the detected Čerenkov and scintillation light by the PMTs. The level of these backgrounds can be measured directly from the data using the preponderance of events which are identified unambiguously.

BooNE Oscillation Measurement Sensitivity

The 8 GeV Booster ν beam using the horn focusing system and running for 2×10^7 seconds first with a 25 m and then with a 50 m decay pipe will provide $\sim 1,000,000\ (200,000)$ identified $\nu_\mu(\bar{\nu}_\mu)$ events/yr in the 445 ton MiniBooNE detector. For the $\nu_\mu \to \nu_e$ oscillation measurement, the beam ν_e background is expected to be at the 0.3% level with a systematic uncertainty of 5(10)% for the $\mu(K)$-decay source. The mis-identification backgrounds can be held to less than $\sim 0.2\%$ and be known with an uncertainty of 5%. These statistics and systematic uncertainty estimates were used to obtain the sensitivity curves and the significance shown in Fig. 4. (For the ν_μ disappearance sensitivity calculation shown in Fig. 3, a 25% uncertainty in the overall normalization and a 10% bin-to-bin shape uncertainty in the energy distribution was assumed.)

The ability of the MiniBooNE experiment to isolate a ν_e appearance oscillation signal and measure the parameters for the signal can be estimated from energy dependent fits to fake data generated with given input parameters. For one year of running with 5×10^{20} protons on target, Table 2 shows the results for two sets of parameters that span the allowed edge of the LSND 90% CL region. In both cases, MiniBooNE will establish an oscillation signal at greater than 15 standard deviations.

TABLE 2. Energy dependent fits for two example oscillation signals. The statistical sample corresponds to one year of MiniBooNE running with 5×10^{20} protons on target.

Δm_0^2	$\sin^2 2\theta_0$	$\delta\left(\Delta m^2\right)$	$\delta\left(\sin^2 2\theta\right)$	Signal Significance
0.3 eV^2	0.03	0.10 eV^2	0.02	44 σ
2.0 eV^2	0.002	0.10 eV^2	0.0002	15 σ

From the above analysis, it is clear that the MiniBooNE experiment can establish a signal in the LSND region. If a signal is observed, Δm^2 will be determined with sufficient accuracy (≈ 0.1 eV^2) to indicate where a second detector should be placed for the follow-up two detector BooNE experiment. For example, if $\Delta m^2 \approx 0.3$ eV^2 (2.0 eV^2) then the second detector should be placed at 2 km (0.25 km) to best determine the oscillation parameters as indicated in Fig. 5. As an example of the expectations for BooNE (the full 2 detector experiment), consider the case of low Δm^2 where the second detector is placed at 2 km. For this example, BooNE will measure Δm^2 to ± 0.014 eV^2 and $\sin^2 2\theta$ to ± 0.002 for one year of running with 5×10^{20} protons on target.

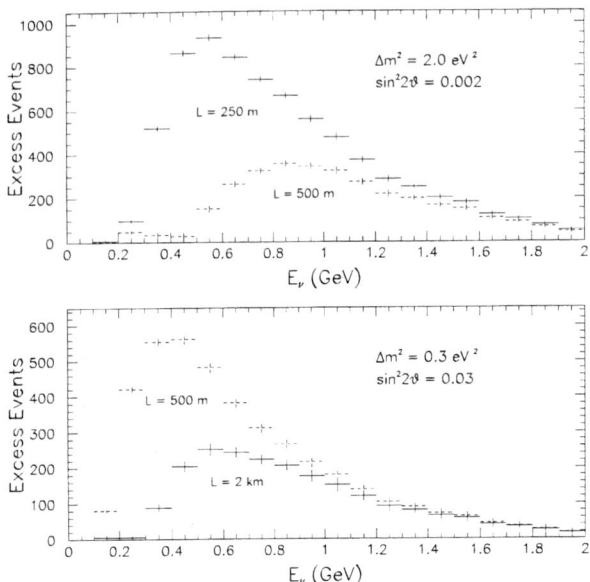

FIGURE 5. The energy distribution of excess events over background for two example oscillation signals and several distances for the detector. The solid (dashed) points in the top plot are for $\Delta m^2 = 2$ eV^2, $\sin^2 2\theta = 0.002$ at a distance of 250 m (500 m). The solid (dashed) points in the bottom plot are for $\Delta m^2 = 0.3$ eV^2, $\sin^2 2\theta = 0.03$ at a distance of 2 km (500 m). The statistical sample corresponds to an increased BooNE sample corresponding to 2×10^{21} protons on target.

As shown above, the BooNE program has the capabilities to cover the entire LSND range with a greater than 10σ signal expectation. In this way, BooNE will either observe a signal consistent with the LSND excess or definitively rule out the excess as being related to neutrino oscillations. Further, through the energy dependence of the signal, BooNE will very accurately determine the oscillation parameters if a signal is observed. Other future experiments also plan to address oscillations in this mass region but without the sensitivity and capability to isolate and study an observed signal; the BooNE program is unique in its "discovery potential" for $\nu_\mu \to \nu_e$ oscillation in this important region.

REFERENCES

1. P. Langacker, et al., Rev. Mod. Phys. **64**, 87. (1991)
2. K.S. McFarland, D. Naples, et al., Phys. Rev. Lett. **75**, 3993 (1995).
3. C.H. Llewellyn Smith, Nucl. Phys. **B228**, 205 (1983).
4. K.S. McFarland, et al., Eur. Phys. Jour. **C1**, 509 (198).
5. A, Blondel, et al., Zeit. Phys. **C45**, 361 (1990).
6. J. Allaby, et al., Zeit. Phys. **C36**, 611 (1985).
7. E.A. Paschos and L. Wolfenstein, Phys. Rev. **D7**, 91 (1973).
8. A.J. Buras and K.J.F. Gaemers, Nucl. Phys. **B132**, 249 (1978).
9. W.G. Seligman, et al, Phys. Rev. Lett. **79**, 1213 (1997).
10. M. Arneodo, et al., Nucl. Phys. **B487**, 3 (1997).
11. E.A. Hawker, et al., Phys. Rev. Lett. **80**, 3715 (1998).
12. S.A. Rabinowitz, et al., Phys. Rev. Lett. **70**, 134 (1993).
13. CTEQ Collaboration, Phys. Rev. **D55**, 1280 (1997).
14. L.W. Whitlow, SLAC-Report-357, 109 (1990).
15. D.Yu. Bardin, V.A. Dokuchaeva, JINR-E2-86-260 (1986); and private communication.
16. J. Pumplin, Phys. Rev. Lett. **64**, 2751 (1990). $S_0 \leq 2$ GeV^2 is allowed by data summarized in M. Virchaux and A. Milsztajn, Phys. Lett. **B274**, 221 (1992).
17. C. Athanassopoulos et. al., Phys. Rev. Lett. **75**, 2650 (1995); C. Athanassopoulos et. al., Phys. Rev. Lett. **77**, 3082 (1996); C. Athanassopoulos et. al., Phys. Rev. C **54**, 2685 (1996).
18. B. Bodmann et. al., Phys. Lett. B **267**, 321 (1991); B. Bodmann et. al., Phys. Lett. B **280**, 198 (1992); B. Zeitnitz et. al., Prog. Part. Nucl. Phys., **32** 351 (1994).
19. W. C. Louis, to appear in the Proceedings of the Erice School on Nucl. Physics, 19th course, "Neutrinos in Astro, Particle and Nuclear Physics", 16-24 September 1997.
20. C. Athanassopoulos et. al., LA-UR-97-1998, nucl-ex/9709006, submitted to Phys. Rev. C.

Alternative Mechanisms for Neutrino Oscillations

A. Halprin

Department of Physics and Astronomy, University of Delaware
Newark, DE 19716
USA

Abstract.
The energy dependence of the neutrino oscillation length in vacuum is examined in a theoretically unbiased way by expanding it in a Laurent series. The three simplest power laws are compared with experimental data and potential physical mechanisms beyond the usual mass mixing mechanism are discussed for the two survivors.

The purpose of this talk is to confront you with even more exciting paradigms for neutrino oscillations than are allowed by conventional wisdom. We begin with an unbiased parameterization of experimental data as follows.

Two neutrino vacuum oscillations can always be described by an effective hamiltonian which is diagonal in some particle basis, *e.g.* the usual mass basis, and can be put into the democratic form

$$H_o^{eff} = A \begin{pmatrix} -1 & 0 \\ 0 & 1 \end{pmatrix} \qquad (1)$$

When described in terms of the particle basis of the weak interactions, obtained by rotating through an angle θ, there will be off diagonal elements which give rise to oscillations between the states of this basis. The dynamics of such an effective hamiltonian always leads to a survival probability of an initially pure weak eigenstate, say ν_e, given by

$$P(\nu_e \to \nu_e) = 1 - \sin^2 2\theta \sin^2 \frac{\pi L}{\lambda(E)} \qquad (2)$$

where $\pi/\lambda = A$ and L is the distance traveled since birth.
To extend the analysis to oscillations in matter, one simply adds in the matter hamiltonian, H_{matter}. The task for experiment is to determine $\lambda(E)$. A theoretically unbiased parameterization of this quantity is

$$\lambda = \sum_{n=-\infty}^{+\infty} A_n E^n \qquad (3)$$

Simplicity compells us to look at the three pure power laws $n = -1, 0, +1$. First consider the case $n = 0$. This power law is produced by the anzatz of a neutral flavor changing current. But a recent analysis of solar neutrino data by Petcov and Krastev [1] rules out $A_n \sim \delta_{n,0}$. They have calculated χ^2 values for a two parameter fit of data from the chlorine,Kamiokande, GALLEX and SAGE experiments while varying fractions of boron and berylium within sensible limits. This was done for t hree different solar models and in no case were they able to produce a χ^2 value less than 6.

The power law $n = +1$ has been studies by many people, since it is produced by the conventional mass mixing mechanism. In a recent study by Krastev [2], the χ^2 value is calculated for a two parameter fit with the Standard Solar Model [3]. For the small angle solution he finds $\chi^2 = 2.9$ and for the large angle solution $\chi^2 = 4.8$. Therefore, the case $A_n \sim \delta_{n,1}$ is alive and well.

A similar analysis has been done for the remaining power law, $n = -1$ with the result that there is only a large angle solution for which χ^2 has a similar value to that above. Thus, it too is alive and well.

In order to choose between the two power laws $n = \pm 1$, we fold in all of the accelerator data save LSND. But these data are also compatible with both $n = \pm 1$. Next we fold in the published data on the anomolus ratio of ν_e to ν_μ [6]. For the case $n = +1$, one needs an additional neutrino to which ν_μ can oscillate (ν_τ or a sterile neutrino). In contrast, the power law $n = -1$ still works with just two neutrinos with mixing parameters compatible with those required fro m the solar neutrino data [4]. A further distinction may be possible from the up/down asymmetry data being accumulated at Superkamiokande, which was discussed in the Tropical Workshop companion meeting to this symposium by Pakvasa [5] and by Stone [6].

If the LSND data is include, which may be premature, then an additional neutrino is needed in both cases. In any event, experiment has not yet been able to distinguish between these two simplest power laws for the energy dependence of the neutrino oscil lation length in vacuum and it behooves us to look for theoretical interpretations of both these possibilities.

A simple theoretical interpretation of the $n = +1$ power law is found in the hypothesis of a neutrino flavor changing interaction with a nearly constant scalar field. The conventional field of choice is the vacuum expectation value of the Higgs field resp onsible for hypothetical neutrino masses,i.e. mass mixing. A more exciting albeit provocative interpretation of this same power law, and one which ,unlike mass mixing, allows oscillations between degenerate neutrinos, comes from a string theory scenario put forth by Damour and Polyakov [7]. They argue that it is quite possible for the string dilaton field to remain massless, producing a scalar contribution to gravitation as in the well know Brans-Dicke [8] proposal. That is, the stat ic gravitational potential between particles A and B is given by

$$V_{AB}(R) = \frac{-G_N m_A m_B (1 + \alpha_A \alpha_B)}{R} \quad (4)$$

Moreover, they point out that while string theory demands that the coupling strength of the spin-2 exchange contribution be universal, the scalar exchange contribution coming from the $\alpha_A \alpha_B$ term may be species dependent.

One can write down a local interaction that gives rise to the species dependent interaction. If neutrinos participate in such an interaction we define an α-basis distinct from the weak basis but related by a mixing angle completely analogous to mass mixing. For two neutrinos, the two bases are then related by

$$\nu_e = \nu_1 \cos\theta + \nu_2 \sin\theta$$
$$\nu_\mu = -\nu_1 \sin\theta + \nu_2 \cos\theta \tag{5}$$

Assuming for simplicity that the mass and α-bases are equivalent, it one finds [9] for degenerate neutrinos of mass m_ν that the analog of the conventional Δm^2 is an effective mass squared difference given by

$$\Delta m^{*2} = -2 m_\nu \alpha_{ext} \Phi_N \Delta\alpha \tag{6}$$

where α_{ext} refers to the dilaton coupling to the matter which produces the Newtonian potential Φ_N and $\Delta\alpha^2 = \alpha_2 - \alpha_1$.

There is the experimental constraint $\alpha_{ext}^2 \leq 10^{-2}$ coming from solar system gravitational experiments [10]. If we assume that $\Delta\alpha \sim \alpha_{ext}$, then we find

$$\Delta m^{*2} \leq 3 \times 10^{-9} eV \frac{\Phi_N}{3 \times 10^{-5}} \tag{7}$$

We take the proper Newtonian potential to be the sum of $1/r$ contributions from all known masses in the universe. The major contribution in the vicinity of the earth or the sun comes from the Great Attractor and has the value $\Phi_N = 3 \times 10^{-5}$. The contributions from other sources at various locations is given in Table(1). Using

TABLE 1. Contributions to the gravitational potential at various locations.

POSITION	SOURCE	$\Phi_N/(3 \times 10^{-5})$
Earth	Earth	2×10^{-5}
Earth	Sun	3×10^{-4}
Solar System	Galaxy	2×10^{-2}
Solar System	Virgo Cluster	3×10^{-2}
Sun	Sun	2×10^{-1}
Solar System	Great Attractor	1

the above numerical value for the Newtonian potential, yields $\Delta m^{*2} \sim 10^{-9} eV^2$. Such a value is satisfactory for a vacuum oscillation solution to the solar neutrino deficit. However, in that case one cannot distinguish this string mechanism from conventional mass mixing. If, on the other hand, $\Delta\alpha \gg \alpha_{ext}$, an MSW solution to the solar neutrino problem would be possible. In that case, the two mechanisms might be distinguishable through the day/night effect, since the sun's contribution to Φ_N is about 20% of the great attractor's contribution near the sun but negligible at the earth.

We now look for theoretical mechanisms to generate the alternate power law, $n = -1$. A simple but extremely heretical way to generate this power law is to postulate that the limiting velocity for neutrinos is not the velocity of light, but differs from it s lightly in a species dependent manner, thereby violating Lorentz invariance [11]. More specifically, the energy-momentum relationship for a neutrino of species j is postulated to be

$$(p/E)_j = 1 + v_j \tag{8}$$

The velocity eigenstates and the weak eigenstates are two basis that can be related by a mixing angle, θ in the usual fashion. The momentum difference for states of the same energy is then

$$\Delta p = E\delta v, \quad \delta v = v_2 - v_1 \tag{9}$$

and therefore the oscillation length is

$$\lambda = \frac{\pi}{E\delta v} \tag{10}$$

Thus, Lorentz invariance violaton (LIV) produces the E^{-1} power law. Limits on the LIV parameters from solar neutrino data are [12]

i) Small angle solution (90% CL): $0.002 \leq \sin^2 2\theta \leq 0.003, \delta v \sim 6 \times 10^{-19}$

ii) Large angle solution (90% CL): $0.38 \leq sin^2 2\theta \leq 0.81, 4 \times 10^{-22} \leq \delta v \leq 4 \times 10^{-21}$

Lorentz invariance violation can also lead to anomalous decays such as $\gamma \to \nu\nu$. Normally such a decay is forbidden by lack of phase space, i.e. 4-momentum conservation implies

$$p_\nu = (1 - \cos\theta_{\nu\gamma}) = 0 \tag{11}$$

But if $v_\nu \neq 0$, this relaxes to the condition

$$p_\nu = (1 - \cos\theta_{\nu\gamma}) = -v_\nu p_\gamma \tag{12}$$

and the phase space no longer vanishes. In his talk at Arecibo, Glashow discussed several other normally forbidden decays [13].

A slightly less violent way to achieve essentially the same result is, instead, to postulate a violation of the equivalence principle in the tensor sector of gravity. The way such a violation is introduced [14] is to modify the conventional gravitational interaction lagrangian (linearized theory)

$$L(x) = f \sum_{j=1}^{2} T^j_{\alpha\beta} G^{\alpha\beta} \tag{13}$$

by letting $f \to f\gamma_j$ in the gravitational neutrino basis. Here, $f = \sqrt{8\pi G_N}$, $T^j_{\alpha\beta}$ is the neutrino energy momentum tensor for species j, and $G_{\alpha\beta}$ is the gravitational tensor field. This leads to the energy momentum relation

$$E = (1 - 2\Phi_N\gamma_j)p \tag{14}$$

The speed of ν_j is therefore species dependent and its departure from the speed of light is given by

$$v_j = 2\Phi_N\gamma_j \tag{15}$$

To the extent that the Newtonian potential is a constant, as discussed earlier, the phenomenological results of this mechanism are identical to that of a possible violation of Lorentz invariance violation. In fact, one might even say that an equivalence principle violation can provide an explanation of any observed Lorentz invavariance violation.

Conclusions:

1) Neutrino oscillations may be entirely unrelated to neutrino mass differences.

2) If the vacuum oscillation length follows the E^{+1} power law, as in the conventional mass mixing mechanism, an attempt should be made to detect any possible spatial dependence of the effective mass difference.

3) Since the E^{-1} power law still seems viable, very high energy neutrino oscillation experiments should be vigorously pursued.

Acknowledgements

I would like to thank Jose Nieves for providing the primary organization for this meeting and for the companion Tropical Workshop. This work was supported in part by the U.DS. Department of Energy under contract DE-FG02-84ER40163.

REFERENCES

1. P.J. Krastev, S.T. Petcov, Phys. Lett. **B395**, 69 (1997).
2. P.J. Krastev, private communication.
3. J.N. Bahcall and M.H. Pinsonneault, Rev. Mod. Phys. **67** (1995).
4. A. Halprin, C. N. Leung and J. Pantaleone, Phys. Rev **D53**, 5365 (1996).
5. S. Pakvasa, Proceedings of Tropical Workshop on Particle Physics and Cosmology, San Juan Puerto Rico, April 1-7, 1998.
6. J. Stone, ibid.
7. T. Damour and A.M. Polyakov,, Gen. Rel. Grav. **26**, 1171 (1994); Nuc. Phys. **B423**, 532 (1994).
8. C. Brans and R. Dicke, Phys. Rev. **124**, 9221 (1961); R. Dicke, Phys. Rev. **125**, 2163 (1962).
9. A. Halprin and C.N. Leung, Phys. Lett. **B416**, 361 (1998).
10. R.D. Reasenberg, et al, Astrophys. J. **234**, L219 (1979).
11. S. Coleman and S.L. Glashow, Phys. Lett. **B405**, 249 (1997)

12. S.L. Glashow, A. Halprin, P.I. Krastev, C.N. Leung, J.Pantaleone, Phys. Rev **D56**,2433 (1997).
13. S.L. Glashow, proceedings of Tropical Workshop on Particle Physics and Cosmology, San Juan, Puerto Rico, April 1-7, 1998.
14. A. Halprin and C.N. Leung, Phys. Rev. Lett. **67**,1833 (1991).

Invariant Box–Parameterization of Neutrino Oscillations

Thomas J. Weiler[1] and DJ Wagner[2]

[1] *Department of Physics and Astronomy, Vanderbilt University, Nashville, TN 37235*
[2] *Department of Physics and Astronomy, Angelo State University, San Angelo, TX 76909*

Abstract. The model-independent "box" parameterization of neutrino oscillations is examined. The invariant boxes are the classical amplitudes of the individual oscillating terms. Being observables, the boxes are independent of the choice of parameterization of the mixing matrix. Emphasis is placed on the relations among the box parameters due to mixing–matrix unitarity, and on the reduction of the number of boxes to the minimum basis set. Using the box algebra, we show that CP-violation may be inferred from measurements of neutrino flavor mixing even when the oscillatory factors have averaged. General analyses of neutrino oscillations among $n \geq 3$ flavors can readily determine the boxes, which can then be manipulated to yield magnitudes of mixing matrix elements.

I INTRODUCTION

If neutrinos have mass and are non-degenerate, then their flavors may oscillate as they propagate. Resonant oscillations for the sun [1], oscillations for the atmosphere [2], and the LSND data [3] each require a different neutrino mass-squared difference if neutrino oscillations are to account for all features of the data [4]. Since three-neutrino models can have at most two independent mass-squared differences, a sterile neutrino is apparently needed to reconcile all the data while retaining consistency with LEP measurements of $Z \to \nu\bar{\nu}$ [5]. Several four-neutrino analyses appear in the literature [4,6]. It is also possible that some data will turn out to have an explanation other than neutrino oscillations, in which case three-neutrino oscillations may be sufficient. So our task is to examine the physics of neutrino oscillations with three or more mixed flavors.

Oscillation probabilities depend on products of four mixing-matrix elements. Several parameterizations of the mixing matrix in terms of rotation angles have been introduced, beginning with the pioneering work of Kobayashi and Maskawa [7]. With three or more neutrino generations, the oscillation probabilities are complicated functions of the neutrino mixing angles. But oscillations are observable and therefore parameterization-invariant. One must ask if there is not a better description of oscillations which avoids the arbitrariness of angular-parameterization schemes. Recently, we introduced a "box" parameterization of neutrino mixing valid for any number of neutrino generations [8]. Oscillation probabilities are linear in the boxes, enabling a straighforward description of oscillation data. Here we present the algebra of the boxes and the unitarity constraints on that algebra. Then we illustrate the boxes' reduction to a basis in the case of three generations, thereby setting the framework for a future phenomenological analysis.

The probability for a neutrino to oscillate from ν_α to ν_β is given by the square of the transition amplitude:

$$P_{\nu_\alpha \to \nu_\beta}(x) = \left| \sum_{i=1}^{n} V_{\alpha i} V_{\beta i}^* e^{-i\phi_i} \right|^2 = \sum_{i=1}^{n} \sum_{j=1}^{n} (V_{\alpha i} V_{\beta i}^* V_{\alpha j}^* V_{\beta j}) e^{-i(2\Phi_{ij})}, \quad (1)$$

where n is the number of neutrino generations,

$$\Phi_{ij} \equiv \frac{1}{2}(\phi_i - \phi_j) = \frac{1}{2}(E_i t_i - p_i x_i - E_j t_j + p_j x_j), \quad (2)$$

and $V_{\alpha i}$ is the mixing–matrix element which connects the αth charged lepton mass eigenstate and the ith neutrino mass eigenstate. For relativistic neutrinos, Φ_{ij} is given by

$$\Phi_{ij} \approx \frac{\Delta m_{ij}^2}{4p} x, \quad \text{where} \quad \Delta m_{ij}^2 \equiv m_i^2 - m_j^2. \quad (3)$$

With a little bit of algebra, the oscillation probability may be brought into the form

$$P_{\nu_\alpha \to \nu_\beta}(x) = -2 \sum_i \sum_{j \neq i} \operatorname{Re}(V_{\alpha i} V_{\beta i}^* V_{\alpha j}^* V_{\beta j}) \sin^2(\Phi_{ij}) \quad (4)$$
$$+ \sum_i \sum_{j \neq i} \operatorname{Im}(V_{\alpha i} V_{\beta i}^* V_{\alpha j}^* V_{\beta j}) \sin(2\Phi_{ij}) + \delta_{\alpha\beta}.$$

The probability for an antineutrino to oscillate from $\bar\nu_\alpha$ to $\bar\nu_\beta$ is obtained by replacing V with V^*. This is equivalent to changing the sign of Φ_{ij}, or the second term in equation (4).

With the familiar case of two neutrino flavors, the mixing matrix V has the simple form of a rotation matrix (phases cancel in oscillation probabilities for Majorana neutrinos, and may be absorbed into the definitions of Dirac fermion fields):

$$V = \begin{pmatrix} \cos\theta & -\sin\theta \\ \sin\theta & \cos\theta \end{pmatrix}. \quad (5)$$

The oscillation probability in the two-flavor case is simply

$$P_{\nu_\alpha \to \nu_\beta}(x) = \delta_{\alpha\beta} + \sin^2 2\theta \, \sin^2\left(\frac{\Delta m_{12}^2}{4p} x\right), \quad n = 2. \quad (6)$$

The mixing-angle parameterization is a natural choice in the two-flavor situation.

The formalism becomes more complicated with three flavors. An arbitrary 3×3 unitary matrix has three real degrees of freedom and six phases, but $2n - 1 = 5$ phases may be absorbed into field redefinitions. The original choice of the four remaining parameters, due Kobayashi and Maskawa to describe quark mixing, is [7]

$$\begin{pmatrix} c_1 & s_1 c_3 & s_1 s_3 \\ -s_1 c_2 & c_1 c_2 c_3 - s_2 s_3 e^{i\delta} & c_1 c_2 s_3 + s_2 c_3 e^{i\delta} \\ -s_1 s_2 & c_1 s_2 c_3 + c_2 s_3 e^{i\delta} & c_1 s_2 s_3 - c_2 c_3 e^{i\delta} \end{pmatrix}, \tag{7}$$

where $c_a \equiv \cos\theta_a$, and $s_a \equiv \sin\theta_a$. There is arbitrariness associated with the placement of the phase, since we absorb five relative phases into the field definitions. Because of this arbitrariness, the phases of individual matrix elements are not observable.

The observable oscillation probabilities are quite complicated functions of the angle-based parameterizations. As an example, consider the product $V_{22} V_{23}^* V_{32}^* V_{33}$ appearing in the $\nu_\mu \to \nu_\tau$ oscillation probability:

$$V_{22} V_{23}^* V_{32}^* V_{33} = c_3^2 s_3^2 \left[s_2^2 c_2^2 (s_1^4 + 6c_1^2 + 2c_1^2 \cos 2\delta) - c_1^2 \right]$$
$$+ \frac{\mathcal{J}}{s_1^2} (1 + c_1^2)(c_2^2 - s_2^2)(s_3^2 - c_3^2) \cot\delta + i\mathcal{J}, \quad n = 3. \tag{8}$$

where the Jarlskog invariant $\mathcal{J}$ [9] has the form $\mathcal{J} = c_1 s_1^2 c_2 s_2 c_3 s_3 \sin\delta$ in this parameterization.

The expression (but not its value) on the right-hand side of equation (8) is convention-dependent, as well as being unwieldy. Our development of a model-independent parameterization is motivated by the arbitrariness and complexity of this traditional approach.

II THE BOX PARAMETERIZATION

The immeasurability of the individual complex mixing–matrix elements in the quark sector has been addressed by numerous authors [9–13]. Measurable quantities include only the magnitudes of mixing matrix elements, the products of four mixing-matrix elements appearing in the oscillation probabilities, and particular higher-order functions of mixing-matrix elements [11,14]. As evidenced in equations (1) and (4), neutrino oscillation probabilities depend linearly on the fourth-order objects,

$$^{\alpha i}\square_{\beta j} \equiv V_{\alpha i} V_{i\beta}^\dagger V_{\beta j} V_{j\alpha}^\dagger = V_{\alpha i} V_{\alpha j}^* V_{\beta i}^* V_{\beta j}, \tag{9}$$

which we call "boxes" since each contains as factors the corners of a submatrix, or "box," of the mixing matrix. For example, the upper left 2×2 submatrix elements produce the box

$$^{11}\square_{22} = V_{11}V_{12}^*V_{21}^*V_{22}. \tag{10}$$

The name "box" also seems appropriate in light of the Feynman box–diagram which describes the oscillation process. Examination of equation (9) reveals a few symmetries in the indexing:

$$^{\alpha i}\square_{\beta j} = {}^{\beta j}\square_{\alpha i} = {}^{\beta i}\square_{\alpha j}^* = {}^{\alpha j}\square_{\beta i}^*. \tag{11}$$

If the order of either set of indices is reversed (*id est*, $j \leftrightarrow i$ or $\beta \leftrightarrow \alpha$), the box turns into its complex conjugate; if both sets of indices are reversed, the box returns to its original value [10]. And if V is replaced by $V^\dagger$, then $^{\alpha i}\square_{\beta j} \to {}^{i\alpha}\square_{j\beta}^*$.

Boxes with $\alpha = \beta$ or $i = j$, are real, given from equation (9) as

$$^{\alpha i}\square_{\alpha j} = |V_{\alpha i}|^2 |V_{\alpha j}|^2, \quad {}^{\alpha i}\square_{\beta i} = |V_{\alpha i}|^2 |V_{\beta i}|^2, \quad \text{and} \quad {}^{\alpha i}\square_{\alpha i} = |V_{\alpha i}|^4. \tag{12}$$

We call boxes with one and two repeated indices "singly-degenerate" and "doubly-degenerate," respectively. Boxes with $\alpha \neq \beta$ and $i \neq j$ are called "nondegenerate". As can be seen from equation (4), singly-degenerate boxes with repeated flavor indices enter into the formulae for flavor-conserving survival probabilities, but not for flavor-changing transition probabilities. Degenerate boxes with repeated mass indices (including the doubly-degenerate boxes) do not appear in any oscillation formula. Degenerate boxes may be expressed in terms of the nondegenerate boxes, as will be shown shortly. This possibility and the symmetries expressed in equation (11) allow us to express combinations of boxes in terms of only the nondegenerate "ordered" boxes for which $\alpha < \beta$ and $i < j$.

Using the symmetries expressed in equation (11), the oscillation probabilities (4) in terms of boxes become

$$P_{\nu_\alpha \to \nu_\beta}(x) = \delta_{\alpha\beta} - 2\sum_{i=1}^n \sum_{j>i} \left[2\,{}^{\alpha i}R_{\beta j} \sin^2\Phi_{ij} - {}^{\alpha i}J_{\beta j} \sin 2\Phi_{ij}\right], \tag{13}$$

where we have defined the shorthand $^{\alpha i}R_{\beta j} \equiv \text{Re}({}^{\alpha i}\square_{\beta j})$ and $^{\alpha i}J_{\beta j} \equiv \text{Im}({}^{\alpha i}\square_{\beta j})$. ¿From equation (11) we deduce that the Js are antisymmetric in both flavor indices and mass indices; Rs are symmetric in both. Survival probabilities $P_{\nu_\alpha \to \nu_\alpha}(x) = 1 - \sum_{\beta \neq \alpha} P_{\nu_\alpha \to \nu_\beta}(x)$ are more simply expressed in terms of degenerate boxes, or $|V|$s, rather than nondegenerate boxes. From equations (13) and (12), they are

$$P_{\nu_\alpha \to \nu_\alpha}(x) = 1 - 4\sum_{i=1}^n \sum_{j>i} {}^{\alpha i}\square_{\alpha j} \sin^2\Phi_{ij} = 1 - 4\sum_{i=1}^n \sum_{j>i} |V_{\alpha i}|^2 |V_{\alpha j}|^2 \sin^2\Phi_{ij}. \tag{14}$$

Interchanging $\alpha \leftrightarrow \beta$ in equation (13) gives the time-reversed reactions $P_{\nu_\beta \to \nu_\alpha}(x)$:

$$P_{\nu_\beta \to \nu_\alpha}(x) = \delta_{\alpha\beta} - 2\sum_{i=1}^n \sum_{j>i} \left[2\,{}^{\alpha i}R_{\beta j} \sin^2\Phi_{ij} + {}^{\alpha i}J_{\beta j} \sin 2\Phi_{ij}\right]. \tag{15}$$

Ignoring possible CP-violating phases in the mixing matrix, the number of real parameters determining V is the number of rotational planes available in n–dimensions, $N \equiv \frac{1}{2}n(n-1)$. Determining these N parameters determines the complete mixing matrix. Conveniently, there are N transition probabilities $P_{\nu_\alpha \to \nu_\beta}(x) = P_{\nu_\beta \to \nu_\alpha}(x)$. Thus, all of the information in the mixing matrix is contained in the N transition probabilities. In this sense, they form a convenient basis for determining all oscillation parameters. Of course, if the same transition probability is measured at two or more different distances, then all N transition probabilities may not be needed to determine V.

Allowing CP-violation in the mixing matrix, there are N real parameters and $\frac{1}{2}(n-1)(n-2)$ phases, for a total of $(n-1)^2$ parameters. With CP-violation, however, there are $2N = n(n-1)$ independent transition probabilities $P_{\nu_\alpha \to \nu_\beta}(x)$. The number of transition probabilities exceeds the number of independent parameters, so they again form a convenient basis for determining the mixing matrix. In reality, only the three flavor indices e, μ, τ are easily accessible. Moreover, some of the N parameters in the mixing matrix, namely those which rotate sterile states for $n \geq 5$, are not accessible at all, which complicates the counting.

The transition probabilities for which $\alpha \neq \beta$ in equation (13) may be conveniently expressed in matrix form. The matrix of boxes is an $N \times N$ matrix. For three flavors, we have

$$\mathcal{P}(n=3) \equiv \begin{pmatrix} P_{\nu_e \to \nu_\mu}(x) \\ P_{\nu_\mu \to \nu_\tau}(x) \\ P_{\nu_e \to \nu_\tau}(x) \end{pmatrix} = -4 \, \mathrm{Re}(\mathcal{B}) \, S^2(\Phi) + 2 \, \mathrm{Im}(\mathcal{B}) \, S(2\Phi), \tag{16}$$

where

$$\mathcal{B} \equiv \begin{pmatrix} {}^{e1}\square_{\mu 2} & {}^{e2}\square_{\mu 3} & {}^{e1}\square_{\mu 3} \\ {}^{\mu 1}\square_{\tau 2} & {}^{\mu 2}\square_{\tau 3} & {}^{\mu 1}\square_{\tau 3} \\ {}^{e1}\square_{\tau 2} & {}^{e2}\square_{\tau 3} & {}^{e1}\square_{\tau 3} \end{pmatrix}, \quad \text{and} \quad S^k(\Phi) \equiv \begin{pmatrix} \sin^k \Phi_{12} \\ \sin^k \Phi_{23} \\ \sin^k \Phi_{13} \end{pmatrix}, \quad n=3. \tag{17}$$

For the time–reversed channels, or for the antineutrino channels, the sign of the $\mathrm{Im}(\mathcal{B})$ term is reversed. The box parameterization is especially well-suited for considering higher numbers of generations. The matrix $\mathcal{B}$ merely acquires extra columns when new flavors are introduced; extra rows are not accessible at energies below new charged-lepton thresholds. Furthermore, oscillation probabilities are linear in boxes, no matter how many generations.

Neutrino oscillation experiments will directly measure the boxes in equation (13), not the individual mixing matrix elements, $V_{\alpha i}$. But one would like to obtain the fundamental $V_{\alpha i}$ from the measured boxes. We develop here an algebra relating boxes and mixing matrix elements.

Some tautologous relationships between the degenerate and nondegenerate boxes are easily confirmed using equation (9); they hold for any number of generations:

$$|V_{\alpha i}|^2 |V_{\alpha j}|^2 = {}^{\alpha i}\square_{\alpha j} = \frac{{}^{\alpha i}\square_{\eta j}^* \; {}^{\alpha i}\square_{\lambda j}}{{}^{\eta}\square_{\lambda j}}, \qquad (\eta \neq \lambda \neq \alpha), \qquad (18)$$

$$|V_{\alpha i}|^2 |V_{\beta i}|^2 = {}^{\alpha i}\square_{\beta i} = \frac{{}^{\alpha i}\square_{\beta x}^* \; {}^{\alpha i}\square_{\beta y}}{{}^{\alpha x}\square_{\beta y}}, \qquad (x \neq y \neq i), \quad \text{and} \qquad (19)$$

$$\frac{|V_{\alpha i}|^2}{|V_{\beta j}|^2} = \frac{{}^{\alpha i}\square_{\eta j}^* \; {}^{\alpha i}\square_{\beta x}}{{}^{\alpha j}\square_{\beta x} \; {}^{\beta i}\square_{\eta j}^*}, \qquad (\eta \neq \alpha \neq \beta, \text{ and } x \neq i \neq j). \qquad (20)$$

Equations (18) and (19) are themselves special cases of the more general

$$\begin{aligned}
{}^{\alpha i}\square_{\beta j} \; {}^{\gamma i}\square_{\delta j} &= \left[V_{\alpha i}V_{\alpha j}^*V_{\beta j}V_{\beta i}^*\right]\left[V_{\gamma i}V_{\gamma j}^*V_{\delta j}V_{\delta i}^*\right] \\
&= \left[V_{\alpha i}V_{\alpha j}^*V_{\delta j}V_{\delta i}^*\right]\left[V_{\gamma i}V_{\gamma j}^*V_{\beta j}V_{\beta i}^*\right] = {}^{\alpha i}\square_{\delta j} \; {}^{\gamma i}\square_{\beta j},
\end{aligned} \qquad (21)$$

and the analogous relation ${}^{\alpha i}\square_{\beta j} \; {}^{\alpha k}\square_{\beta l} = {}^{\alpha i}\square_{\beta l} \; {}^{\alpha k}\square_{\beta j}$. The relations above hold for both degenerate boxes and nondegenerate boxes.

Due to the symmetry ${}^{\alpha i}\square_{\beta j} \to {}^{i\alpha}\square_{j\beta}^*$ when $V \to V^\dagger$, there will generally be analogous but distinct pairing of our box equations, differing only in whether the degeneracy or sum is over a flavor index or a mass index. In the following we will mainly show only one equation per analogous pair, for reasons of space limitations in this proceeding.

We may express $|V_{\alpha i}| = \left({}^{\alpha i}\square_{\alpha i}\right)^{\frac{1}{4}}$ in terms of three singly-degenerate boxes by setting $\alpha = \beta$ in equation (19). Then, using equation (18) to substitute for the singly-degenerate boxes yields an expression for the doubly-degenerate box in terms of nine nondegenerate boxes:

$$|V_{\alpha i}|^4 = {}^{\alpha i}\square_{\alpha i} = \frac{{}^{\alpha i}\square_{\alpha x} \; {}^{\alpha i}\square_{\alpha y}}{{}^{\alpha x}\square_{\alpha y}} = \frac{{}^{\alpha x}\square_{\tau i} \; {}^{\alpha i}\square_{\sigma x} \; {}^{\alpha y}\square_{\rho i} \; {}^{\alpha i}\square_{\zeta y} \; {}^{\omega x}\square_{\mu y}}{{}^{\tau i}\square_{\sigma x} \; {}^{\rho i}\square_{\zeta y} \; {}^{\alpha y}\square_{\omega x} \; {}^{\alpha x}\square_{\mu y}}, \qquad (22)$$

where the index constraints are $\tau \neq \sigma \neq \alpha$, $\zeta \neq \rho \neq \alpha$, $\mu \neq \omega \neq \alpha$, and $x \neq y \neq i$. In the three-generation case, equation (22) is uniquely specified by the index constraints. For example,

$$|V_{11}|^4 = \frac{{}^{11}\square_{22} \; {}^{11}\square_{23}^* \; {}^{11}\square_{33} \; {}^{11}\square_{32}^* \; {}^{22}\square_{33}}{{}^{12}\square_{23}^* \; {}^{12}\square_{33} \; {}^{21}\square_{33} \; {}^{21}\square_{32}^*} \qquad (23)$$

holds with any number of generations, but it is the unique 5 on 4 box representation of $|V_{11}|^4$ in three generations.

We note that all of the relationships in this section follow from the definitions of the boxes in equation (9) and so are valid for any matrix, unitary or otherwise. The constraints of unitarity will provide us with expressions for $|V_{\alpha i}|^4$ which are easier to manage than the expression in (22) above.

III UNITARITY RELATIONS AMONG THE BOXES

Unitarity requires that

$$\sum_{\eta=1}^{n} V_{\eta i} V_{\eta j}^* = \delta_{ij}, \quad \text{and} \quad \sum_{y=1}^{n} V_{\alpha y} V_{\beta y}^* = \delta_{\alpha\beta}. \tag{24}$$

Multiplying the first equation in (24) by $V_{\lambda i}^* V_{\lambda j}$ and the second by $V_{\alpha x}^* V_{\beta x}$ gives the unitarity constraints for the boxes:

$$\sum_{\eta=1}^{n} {}^{\eta i}\square_{\lambda j} = \sqrt{{}^{\lambda i}\square_{\lambda i}}\, \delta_{ij}, \tag{25}$$

and its analogue. Isolating the manifestly degenerate boxes from the nondegenerate boxes, equation (25) becomes

$$\sum_{\eta \neq \lambda} {}^{\eta i}\square_{\lambda j} = \sqrt{{}^{\lambda i}\square_{\lambda i}}\, \delta_{ij} - {}^{\lambda i}\square_{\lambda j}. \tag{26}$$

Summing equation (25) over λ in the $i \neq j$ case, we find

$$0 = \sum_{\lambda=1}^{n}\sum_{\eta=1}^{n} {}^{\eta i}\square_{\lambda j} = \sum_{\lambda=1}^{n} {}^{\lambda i}\square_{\lambda j} + 2\sum_{\lambda=1}^{n}\sum_{\eta<\lambda} {}^{\eta i}\mathrm{R}_{\lambda j}. \tag{27}$$

The double sum is over Rs only, since the first term is manifestly real. The resulting conditions on the Js are found in equation (29) below. Comparison of equation (27) with equations (17) and (18) reveals an interesting property of the matrix $\mathcal{B}$:

$$\sum_{\text{column of } \mathcal{B}} \mathrm{Re}\,(\mathcal{B}) = -\frac{1}{2}\sum_{\lambda=1}^{n} |V_{\lambda i}|^2 |V_{\lambda j}|^2, \tag{28}$$

where the sum is over a column of $\mathcal{B}$ specified by fixed i and j. There is an analogue relation for the sum over a row of $\mathcal{B}$.

The unitarity constraint (25) holds independently for the real and imaginary parts of the sum. We will first explore the implications of these constraints for the imaginary parts of boxes, before turning to the more complicated constraints for the real parts. The right-hand side of (25) is manifestly real, so the imaginary constraints are simply

$$\sum_{\eta \neq \lambda} {}^{\eta i}\mathrm{J}_{\lambda j} = 0, \quad \text{and} \quad \sum_{y \neq x} {}^{\alpha y}\mathrm{J}_{\beta x} = 0. \tag{29}$$

Equation (11) indicates that ${}^{\eta i}\mathrm{J}_{\lambda j}$ is an antisymmetric matrix in the indices η and λ for fixed i and j, and vice versa. Equation (29) shows that the sum of elements along any row or column of that antisymmetric matrix equals zero, whether the sum is over mass indices or flavor indices.

Summing the first equation in (29) over λ gives zero trivially since a sum of all elements of an antisymmetric matrix vanishes by definition. Hence, for fixed (i,j), the first equation in (29) expresses $n-1$ constraints. Thus, the number of independent flavor pairs on Js after implementing the constraints of equation (29) is $N - (n-1) = \frac{1}{2}(n-1)(n-2)$. Ditto for independent mass pairs, so the number of independent Js after implementing both sets of constraints is the product $\frac{1}{4}(n-1)^2(n-2)^2$.

In three generations, this number of independent Js is one. Each sum in equation (29) contains only two terms, leading to

$$\text{Im}(\mathcal{B}) = \begin{pmatrix} \mathcal{J} & \mathcal{J} & -\mathcal{J} \\ \mathcal{J} & \mathcal{J} & -\mathcal{J} \\ -\mathcal{J} & -\mathcal{J} & \mathcal{J} \end{pmatrix}, \quad n=3, \tag{30}$$

with $\mathcal{J} \equiv {}^{11}J_{22}$ [9]. One consequence of the equality of all $|J|$s in three generations is that if any one $V_{\alpha i}$ is zero, then all ${}^{\alpha i}J_{\beta j}$ vanish and there can be no CP-violation.

We now consider the real parts of the constraint (25), focusing first on the homogeneous constraint for which the Kronecker delta is zero. This constraint gives the singly-degenerate boxes as sums of ordered boxes:

$$|V_{\lambda i}|^2 |V_{\lambda j}|^2 = {}^{\lambda i}\square_{\lambda j} = -\sum_{\eta \neq \lambda} {}^{\eta i}\square_{\lambda j} = -\sum_{\eta \neq \lambda} {}^{\eta i}R_{\lambda j}, \quad i \neq j. \tag{31}$$

This linear relation complements the relation expressed in equation (18). For three generations, each of the sums contains two terms, allowing us to express the singly-degenerate boxes in terms of two nondegenerate boxes which are measurable in neutrino appearance oscillation experiments.

The real unitarity constraint (31) greatly simplifies our expressions for a doubly-degenerate box ${}^{\alpha i}\square_{\alpha i} = |V_{\alpha i}|^4$:

$${}^{\alpha i}\square_{\alpha i} = \frac{{}^{\alpha i}\square_{\alpha x} \, {}^{\alpha i}\square_{\alpha y}}{{}^{\alpha x}\square_{\alpha y}} = \frac{\left(-\sum_{\eta \neq \alpha} {}^{\alpha i}R_{\eta x}\right)\left(-\sum_{\lambda \neq \alpha} {}^{\alpha i}R_{\lambda y}\right)}{\left(-\sum_{\tau \neq \alpha} {}^{\alpha x}R_{\tau y}\right)}, \quad x \neq y \neq i, \tag{32}$$

where the first equality is due to equation (18) with $j = i$. Applying equation (32) to three generations, one finds that doubly-degenerate boxes are expressible in terms of the real parts of six ordered boxes, rather than the nine complex boxes used in equation (22). For example,

$$|V_{11}|^4 = {}^{11}\square_{11} = \frac{-({}^{11}R_{22} + {}^{11}R_{32})({}^{11}R_{23} + {}^{11}R_{33})}{{}^{12}R_{23} + {}^{12}R_{33}}, \quad n = 3. \tag{33}$$

In cyclic coordinates α, β, γ, and with $x \neq y \neq i$,

$${}^{\alpha i}\square_{\alpha i} = |V_{\alpha i}|^4 = \frac{-({}^{\alpha i}R_{\beta x} + {}^{\alpha i}R_{\gamma x})({}^{\alpha i}R_{\beta y} + {}^{\alpha i}R_{\gamma y})}{{}^{\alpha x}R_{\beta y} + {}^{\alpha x}R_{\gamma y}}, \quad n = 3. \tag{34}$$

When considering $n > 3$, each sum has more terms, but all terms in the numerator in equation (32) always contain Rs to the second order, while the denominator terms contain only the first order of Rs. Thus these expressions will be much more manageable than equation (22) which exhibits the fifth order of complex boxes in the numerator and the fourth order in the denominator.

For fixed (i,j) in equation (31), λ can take n possible values, implying n constraint equations. N ordered nondegenerate boxes appear in these n equations. Thus, for $N \leq n$, which is true for $n \leq 3$, the unitarity constraint (31) may be inverted to find a nondegenerate box in terms of singly-degenerate boxes. Manipulation of equation (31) gives an expression in term of the flavor triad (α, β, γ):

$$^{\alpha i}R_{\beta j} = -\frac{1}{2}\left(|V_{\alpha i}|^2|V_{\alpha j}|^2 + |V_{\beta i}|^2|V_{\beta j}|^2 - |V_{\gamma i}|^2|V_{\gamma j}|^2\right), \quad n = 3. \tag{35}$$

It is known that knowledge of four $|V|$s completely specifies the three-generation mixing matrix, provided no more than two $|V|$s are taken from the same row or same column [15]. Here, we can use three-generation unitarity and equation (35) to re-write $^{\alpha i}R_{\beta j}$ in terms of just four $|V|$s. The result is

$$^{\alpha i}R_{\beta j} = \frac{1}{2}\left[1 - |V_{\alpha i}|^2 - |V_{\alpha j}|^2 - |V_{\beta i}|^2 - |V_{\beta j}|^2 + |V_{\alpha i}|^2|V_{\beta j}|^2 + |V_{\alpha j}|^2|V_{\beta i}|^2\right], \tag{36}$$

which for $n = 3$ expresses the real part of the box $\text{Re}\left[V_{\alpha i}V_{\alpha j}^*V_{\beta j}V_{\beta i}^*\right]$ in terms of the magnitudes of the four complex Vs which define the box. Three-generation unitarity may be used again to replace the first five terms on the right-hand side of equation (36) with $-|V_{\gamma k}|^2$.

Summing equation (31) over $j \neq i$ yields another expression for $|V_{\lambda i}|^2$ in terms of nondegenerate boxes, which further complements equations (32) and (22):

$$|V_{\lambda i}|^2 \sum_{j \neq i} |V_{\lambda j}|^2 = |V_{\lambda i}|^2 \left(1 - |V_{\lambda i}|^2\right) = -\sum_{j \neq i}\sum_{\eta \neq \lambda} {}^{\eta i}R_{\lambda j}. \tag{37}$$

The explicit solution of this equation, valid for any number of generations, is,

$$|V_{\lambda i}|^2 = \frac{1}{2}\left[1 \pm \sqrt{1 + 4\sum_{j \neq i}\sum_{\eta \neq \lambda} {}^{\eta i}R_{\lambda j}}\right], \tag{38}$$

which yields $|V_{\lambda i}|^2$ in terms of $(n-1)^2$ Rs, but subject to a two-fold ambiguity.

We may use the real homogeneous unitarity condition (31) along with the tautology (18) to obtain constraints between nondegenerate boxes, thereby reducing the number of real degrees of freedom. Substituting the tautology (18) into the unitarity constraint (31) gives

$$^{\eta i}\square_{\alpha j} \; ^{\alpha i}\square_{\lambda j} + {}^{\eta i}\square_{\lambda j}\sum_{\tau \neq \alpha} {}^{\tau i}R_{\alpha j} = 0. \tag{39}$$

This unitarity constraint interrelates imaginary and real parts of n different boxes for any number of generations. For example, taking the imaginary part of equation (39) leads to

$$^{\eta i}J_{\alpha j}\,^{\alpha i}R_{\lambda j} + {}^{\alpha i}J_{\lambda j}\,^{\eta i}R_{\alpha j} + {}^{\eta i}J_{\lambda j}\sum_{\tau \neq \alpha}{}^{\tau i}R_{\alpha j} = 0, \quad \eta \neq \lambda \neq \alpha, \quad i \neq j. \tag{40}$$

Taking the real part of equation (39) leads to

$$^{\eta i}R_{\alpha j}\,^{\alpha i}R_{\lambda j} + {}^{\eta i}R_{\lambda j}\sum_{\tau \neq \alpha}{}^{\tau i}R_{\alpha j} = {}^{\eta i}J_{\alpha j}\,^{\alpha i}J_{\lambda j}, \quad \eta \neq \lambda \neq \alpha, \quad i \neq j. \tag{41}$$

One may also use the pairs of equations (40) and (41) to eliminate the sums and isolate a single R:

$$^{\alpha i}R_{\beta j} = \frac{{}^{\alpha i}J_{\beta j}\,^{\alpha i}R_{\lambda j}\,^{\beta i}R_{\lambda j} + {}^{\alpha i}J_{\beta j}\,^{\alpha i}J_{\lambda j}\,^{\beta i}J_{\lambda j}}{{}^{\alpha i}J_{\lambda j}\,^{\beta i}R_{\lambda j} - {}^{\alpha i}R_{\lambda j}\,^{\beta i}J_{\lambda j}}, \tag{42}$$

with $\beta \neq \lambda \neq \alpha$, $i \neq j$. Input from these unitarity relations among Rs and Js is necessary to establish the minimum set of independent box parameters.

IV INDIRECT MEASUREMENT OF CP-VIOLATION

Suppose CP is conserved. Then ${}^{\alpha i}J_{\beta j} = 0$ for all index choices. The inference from equation (41) is that

$$^{\alpha i}R_{\eta j}\,^{\alpha i}R_{\lambda j} + {}^{\eta i}R_{\lambda j}\sum_{\tau \neq \alpha}{}^{\alpha i}R_{\tau j} = 0, \qquad (\eta \neq \lambda \neq \alpha, \text{ and } i \neq j). \tag{43}$$

If this relation is violated, then so is CP.

For three generations, ${}^{\eta i}J_{\alpha j} = {}^{\alpha i}J_{\lambda j}$ by equation (29), and equation (41) may be solved for $\mathcal{J}^2$ directly:

$$\mathcal{J}^2 = {}^{\alpha i}R_{\beta j}\,^{\beta i}R_{\lambda j} + {}^{\alpha i}R_{\beta j}\,^{\alpha i}R_{\lambda j} + {}^{\alpha i}R_{\lambda j}\,^{\beta i}R_{\lambda j}, \quad n = 3. \tag{44}$$

Equation (44) says that the three real elements in any row (or any column in the analogue equation) of the matrix $\mathcal{B}$ may be summed in their three pairwise products to yield the CP-violating invariant $\mathcal{J}^2$. These real elements on the right-hand side of this equation are measurable with CP-conserving averaged neutrino oscillations. Thus, even if CP violating asymmetries are not directly observable in an experiment, the effects of CP violation may be seen through the relationships among the real parts of different boxes, which are determinable from averaged flavor–mixing measurements! Note that if CP is conserved and $\mathcal{J}$ is zero, then equation (44) also tells us that all three Rs in any row (or column) cannot have the same sign.

V INHOMOGENEOUS UNITARITY CONSTRAINTS AND A BOX–BASIS

The inhomogeneous unitarity constraints with the Kronecker delta nonzero in equation (25) are necessary to provide the desired normalization of the $V_{\alpha i}$ or the boxes. The inhomogeneous constraints are functions of degenerate boxes and therefore purely real:

$$^{\alpha i}\square_{\alpha i} + \sum_{\eta \neq \alpha} {}^{\alpha i}\square_{\eta i} = \sqrt{^{\alpha i}\square_{\alpha i}}, \qquad (45)$$

This equation can be rewritten strictly in terms of nondegenerate boxes by using the homogeneous unitarity constraints (31) to replace the singly-degenerate boxes, and equation (32) to replace the doubly-degenerate box:

$$\Sigma_\lambda \Sigma_\sigma + \sqrt{-\Sigma_\lambda \Sigma_\sigma \Sigma_\tau} + \Sigma_\tau \Sigma_{\eta z} = 0, \qquad (46)$$

with $\Sigma_\lambda \equiv \sum_{\lambda \neq \alpha} {}^{\alpha i}R_{\lambda x}$, $\Sigma_\sigma \equiv \sum_{\sigma \neq \alpha} {}^{\alpha i}R_{\sigma y}$, $\Sigma_\tau \equiv \sum_{\tau \neq \alpha} {}^{\alpha x}R_{\tau y}$, $\Sigma_{\eta z} \equiv \sum_{\eta \neq \alpha} \sum_{z \neq i} {}^{\alpha z}R_{\eta i}$, and $x \neq y \neq i$. These inhomogeneous unitarity constraints do not involve the Js. Isolating the square root and squaring the equation, we get polynomial equations of degrees three and four in the Rs, each relating $n(n-1)$ Rs.

We provide here an example of a basis construction, obtained by substituting in the unitarity equations derived above. The unitarity constraints among the Js, given in equation (29), are linear and therefore the simplest to implement. These constraints may be used first to reduce the number of independent Js to $\frac{1}{4}(n-1)^2(n-2)^2$. Further reduction to independent Js and Rs requires the nonlinear constraints. The homogeneous constraints (40) and (41) are much simpler than the inhomogeneous constraints (46), but the inhomogeneous constraints must be invoked at least once (Otherwise, the the boxes and the matrix element magnitudes $|V|$ could not be normalized.).

For three generations, one begins with nine Rs and one J, and seeks a basis of just four elements. Rearranging the three-generation equation (44) yields expressions for one R in terms of two other Rs and $\mathcal{J}$. This equation may be used three times to eliminate $^{12}R_{23}$, $^{11}R_{32}$ and $^{21}R_{33}$. The utility of the analogue equation is exhausted to eliminate $^{12}R_{33}$ and $^{11}R_{33}$. As expected, one must next turn to the inhomogeneous constraints (46) to eliminate the last degree of freedom. We are left with a constraint which is quartic in all five of its parameters $A \equiv {}^{11}R_{22}$, $B \equiv {}^{11}R_{23}$, $C \equiv {}^{21}R_{32}$, $D \equiv {}^{22}R_{33}$, and $\mathcal{J}^2$:

$$\begin{aligned}
0 = &(A+B)^2(A+C)^2\left(BC+BD-CD+\mathcal{J}^2\right)^2 \\
&+\left(AD+BD-AB+\mathcal{J}^2\right)^2\left[(A+B)(C+D)+C^2+\mathcal{J}^2\right]^2 \\
&+(A+B)(A+C)\left(BC+BD-CD+\mathcal{J}^2\right)\left(AD+BD-AB+\mathcal{J}^2\right) \\
&\times\left[C+D+2\left((A+B)(C+D)+C^2+\mathcal{J}^2\right)\right].
\end{aligned} \qquad (47)$$

We may eliminate any one parameter by either algebraic or numerical means, leaving us with the desired four parameters as the basis.

VI SUMMARY

Neutrino physics has entered a golden age of research. New experiments all over the globe promise an unequaled amount of data from the sun, the atmosphere, accelerators, supernovae, and other cosmic sources. The latest data suggests that more than three neutrino flavors may participate in neutrino oscillations [4]. Analyzing such refined data requires a consistent, model-independent approach which may be easily applied, and easily extended to higher generations. Here we have discussed such an approach, wherein one works directly with the observable coefficients of the oscillating terms. ¿From unitarity of the mixing matrix, we derived relations among these CP–conserving and CP–violating coefficients for the various oscillation channels. One result which we view as particularly noteworthy is that high-statistics data on *averaged* oscillations are sufficient to determine the conservation or non-conservation of CP in the lepton mixing matrix. This indirect test of CP can be traced back to unitarity of the mixing matrix, but in the present formulation there is no need to even mention the mixing matrix.

Acknowledgements: This work was supported in part by the U.S. Department of Energy, Division of High Energy Physics, under Grant No. DE-F605-85ER40226, and the Vanderbilt University Research Council.

REFERENCES

1. J. Bahcall and M. H. Pinsonneault, Rev. Mod. Phys. **67**, 781 (1995); J.N. Bahcall, S. Basu, and M.H. Pinsonneault, astro-ph/9805135; B. T. Cleveland, *et al.*, Nucl. Phys. B (Proc. Suppl.) **38**, 47 (1995); Kamiokande collaboration, Y. Fukuda *et al.*, Phys. Rev. Lett. **77**, 1683 (1996); GALLEX collaboration, W. Hampel *et al.*, Phys. Lett. **B388**, 384 (1996); SAGE collaboration, J. N. Abdurashitov *et al.*, Phys. Rev. Lett. **77**, 4708 (1996).

2. Super–Kamiokande Collaboration, Y. Fukuda et al., hep–ex/9803006 and Phys. Lett. **B**, to appear; hep–ex/9805006; talks by Y. Suzuki and Y. Totsuka at *Neutrino–98*, Takayama, Japan, June 1998; Kamiokande collaboration, K. S. Hirata *et al.*, Phys. Lett. **B280**, 146 (1992); Y. Fukuda *et al.*, Phys. Lett. **B335**, 237(1994); IMB collaboration, R. Becker-Szendy *et al.*, Nucl. Phys. Proc. Suppl. **38B**, 331 (1995); Soudan-2 collaboration, W. W. M. Allison *et al.*, Phys. Lett. **B391**, 491 (1997); J. G. Learned, S. Pakvasa, and T. J. Weiler, Phys. Lett. **B207**, 79 (1988); V. Barger and K. Whisnant, Phys. Lett. **B209**, 365 (1988); K. Hidaka, M. Honda, and S. Midorikawa, Phys. Rev. Lett. **61**, 1537 (1988); M. C. Gonzalez-Garcia, H. Nunokawa, O. Peres, T. Stanev, and J. W. F. Valle, hep-ph/9801368.

3. Liquid Scintillator Neutrino Detector (LSND) collaboration, C. Athanassopoulos *et al.*, Phys. Rev. Lett. **75**, 2650 (1995); *ibid.* **77**, 3082 (1996); nucl-ex/9706006.

4. V. Barger, T.J. Weiler, and K. Whisnant, hep-ph/9712495; V. Barger, S, Pakvasa, T.J. Weiler, and K. Whisnant, hep-ph/9806328.
5. LEP Electroweak Working Group and SLD Heavy Flavor Group, D. Abbaneo *et al.*, CERN-PPE-96-183, December 1996.
6. D.O. Caldwell and R.N. Mohapatra, Phys. Rev. **D 48**, 3259 (1993); R. Foot and R.R. Volkas, Phys. Rev. **D 52**, 6595 (1995); J.J. Gomez-Cadenas and M.C. Gonzalez-Garcia, Z. Phys. **C 71**, 443 (1996); S.M. Bilenky, C. Giunti, and W. Grimus, hep-ph/9711416; R.N. Mohapatra, hep-ph/9711444.
7. M. Kobayashi and T. Maskawa, Prog. Theor. Phys. **49**, 652 (1972).
8. D.J. Wagner and T.J. Weiler, hep-ph/9801327, and Phys. Rev. **D**, to appear.
9. C. Jarlskog, Z. Phys. C **29**, 491 (1985); C. Jarlskog, Phys. Rev. D **35**, 1685 (1987); here the maximum value for $J = c_1 s_1^2 c_2 s_2 c_3 s_3$ in three generations is easily found to be $\pm \frac{1}{6\sqrt{3}}$.
10. J. F. Nieves and P. B. Pal, Phys. Rev. D **36**, 315 (1987).
11. Dan-di Wu, Phys. Rev. D **33**, 860 (1986).
12. J. Bjorken and I. Dunietz, Phys. Rev. D **37**, 2109 (1987).
13. D. Du, I. Dunietz, and Dan-di Wu, Phys. Rev. D **34**, 3414 (1986); O. W. Greenberg, Phys. Rev. D **32**, 1841 (1985); I. Dunietz, O. W. Greenberg, and D. Wu, Phys. Rev. Lett. **55**, 2935 (1985); C. Hamzaoui and A. Barroso, Phys. Lett. **154B**, 202 (1985).
14. A. Kusenko and R. Shrock, Phys. Rev. D **50**, R30 (1994); Phys. Lett. B **323**, 18 (1994).
15. C. Hamzaoui, Phys. Rev. Lett. **61**, 35 (1988); C. Jarlskog, in *CP Violation*, edited by C. Jarlskog (World Scientific, Singapore, 1989).

A New Cosmological Paradigm: the Cosmological Constant and Dark Matter

Lawrence M. Krauss[1]

Departments of Physics and Astronomy
Case Western Reserve University
10900 Euclid Ave.
Cleveland, OH 44106-7079

Abstract. The Standard Cosmological Model of the 1980's is no more. I describe the definitive evidence that the density of matter is insufficient to result in a flat universe, as well as the mounting evidence that the cosmological constant is not zero. I finally discuss the implications of these results for particle physics and direct searches for non-baryonic dark matter.

I INTRODUCTION

One of the great developments of the 1980's was the creation of a Standard Model of Cosmology based on ideas arising from Particle Theory. This model involved the following trilogy of ideas:

$$(1) \quad \Omega \equiv 1$$

$$(2) \quad \Lambda \equiv 0$$

$$(3) \quad \Omega_{\text{matter}} \approx \Omega_{\text{CDM}^{\text{WIMP}}_{\text{axion}}} \geq 0.9$$

A decade later observational cosmology has made tremendous strides, and we now know that at least two of these fundamental notions must be incorrect. Either

$$(1) \quad \Omega \neq 1$$

$$(2) \quad \Lambda \equiv 0$$

$$(3) \quad \Omega_{\text{matter}} \approx \Omega_{\text{CDM}^{\text{WIMP}}_{\text{axion}}} \approx 0.1 - 0.3$$

---

[1] Research supported in part by the DOE. email:krauss@theory1.phys.cwru.edu

or

(1) $\quad \Omega \equiv 1$

(2) $\quad \Lambda \neq 0$

(3) $\quad \Omega_{\text{matter}} \approx \Omega_{\text{CDM}_{\text{axion}}^{\text{WIMP}}} \approx 0.1 - 0.3$

In either case the implications for both cosmology and particle physics are profound. In the first place,

Either: $\Omega \neq 1$ or $\Lambda \neq 0$

Whichever is true, this implies we don't understand something very fundamental about the microphysics of the Universe—a very exciting prospect! If $\Omega \neq 1$ then the canonical prediction of inflation is incorrect, and we have to understand how inflation, or another theory, might address the fine tuning required to solve the flatness problem without actually resulting in a flat universe today. If $\Lambda \neq 0$ then the situation is in a sense even more exciting, as there is no theory of the cosmological constant at the present time, and the supposition that this quantity is indeed zero rests primarily on *a priori* theoretical prejudice at this point. (I here include in the term "cosmological constant" those models which involve a very slowly varying scalar field, which in effect mimics a cosmological constant over long time intervals.)

At the same time, we have:

$$\Omega_{\text{matter}} \approx \Omega_{\text{CDM}_{\text{axion}}^{\text{WIMP}}} \approx 0.1 - 0.3$$

This also has dramatic implications, not only for our understanding of the role dark matter plays in the formation of large scale structure, but also for our propsects for direct detection of non baryonic dark matter. Contrary to one's naive expectations however, the implications are quite positive. Dark matter may not contribute 90% of the mass of the Universe, as previously envisaged, but it still appears to outweigh baryonic matter. Moreover, as I will demonstrate, in all cases these results suggest that the interaction strength of dark matter with normal matter will be *INCREASED*, and thus in principle direct detection should be *easier* than it would otherwise be. As long as the dark matter contribution to the fraction of the closure density is larger than 0.1, so that it can account for essentially the entire inferred dark matter content of galactic halos in general, and our galactic halo in particular, the increase in interaction cross section is not counterbalanced by a decrease in the dark matter flux on earth, so that the net event rate in detectors should be larger than would be the case if $\Omega_{\text{CDM}} \approx 0.9$.

II THE CASE FOR A COSMOLOGICAL CONSTANT

Over the past 5 years a variety of indirect observables, involving the three fundamental independent observables in cosmology, the expansion rate, the matter content, and large scale structure, have all suggested that either the universe is open, or the cosmological constant is not zero [1–3]. In the past year, the indirect evidence has been strengthened by new large scale structure measurements, and for the first time, striking new direct measurements suggest that the Hubble expansion is accelerating. I first review the older, indirect evidence, and then describe the most recent results.

A The Age Problem

The Hubble constant, by a very simple argument, gives an upper limit on the age of a matter dominated universe. Matter causes a deceleration of the universal expansion over time. Thus, at earlier times the universe would have been expanding faster than it is at the present time. One can, in turn, therefore derive an upper limit on the age of the universe by considering the fact that all galaxies were once located together, and using the relation for a constant velocity to determine the length of time a galaxy at a given distance, moving away at a constant velocity took to get there, i.e. $d = vt \to t = d/v = H_0^{-1}$, where the definition of the Hubble constant, H_0 has been used. In fact, of course, this upper limit on the age of the Universe is an overestimate of its age, and with a given cosmological model one can derive a specific relation between the Hubble constant today and the age of the universe. One has:

Flat matter dominated $\quad t = (2/3)H_0^{-1} = 9.7 \text{Gyr}(65/H_0)$

Open ($\Omega > 0.2$) $\quad\quad\quad t < .85 H_0^{-1} = 12.5 \text{Gyr}(65/H_0)$

Flat ($\Omega_\Lambda < 0.8$) $\quad\quad\quad t < 1.08 H_0^{-1} = 16 \text{Gyr}(65/H_0)$

Thus, if one could definitively demonstrate that the Universe were older than either of the first two relations allowed, given the allowed range of H_0, one would have strong evidence that $\Lambda \neq 0$, since a non-zero cosmological constant would allow a universal *acceleration*, and hence allows an older universe for a fixed Hubble Constant.

While precisely such a situation seemed to prevail as recently as 1996 [4], more recent estimates for the age of globular clusters have suggested that the age of our galaxy is younger than previously estimated. At the same time, estimates of the Hubble constant are now somewhat lower than previously claimed, so that a range of 65- 75 kms^{-1}Mpc^{-1} is now preferred [5]. Nevertheless, the new quoted 95 % lower limit, of approximately 9.8 Gyr with a best estimate of the age of 11.5 Gyr [6], strongly disfavors a flat universe, even if it remains compatible with an open universe.

B   The Baryon Problem

Big Bang Nucleosynthesis (BBN) has for some time provided an upper limit on the total density of baryonic matter in the univere [7,8]:

$$\Omega_B h^2 \leq .026 \tag{1}$$

Most recently, claimed observations of the deuterium fraction in primordial hydrogen clouds illuminated by the light of distant quasars [9] suggest that the actual baryon abundance is near the upper limit of this range. While this puts pressure on BBN analyses, more germaine for this discussion is the fact that when combined with direct observations of the baryon fraction on large scales today, it effectively rules out the possibility of a flat universe.

X-Ray Observations of Clusters of Galaxies, the largest bound structures known in the Universe suggest that the dominant baryonic material in these systems exists in the form of hot X-Ray emitting gas. Assuming this material is in hydrostatic equilibrium with the gravitational potential of these systems one can, by observing both the X-Ray Luminosity and temperature as a function of radius, perform an inversion which gives an estimate of this potential, and hence the total mass, M_T, of these . At the same time, direct observations of the luminosity yield an estimate for the total baryonic mass in hot gas, and hence the total baryonic mass M_B. Thus, one derives the ratio $R = M_B/M_T$ for these sytems. Now, as these systems are the largest bound objects known, it is reasonable to assume that they are good probes of the distribution of all gravitating matter on large scales. Thus, the ratio R is expected to be not just the ratio of baryon to total mass of clusters, but the ratio of baryonic to total mass in the Universe. Thus, *if* the Universe is flat, so that the density corresponding to M_T yields $\Omega = 1$ then one has precisely the relation $R = \Omega_B$. Therein lies the problem. Observations, combined with theoretical models of clusters yield the constraint [10]

$$R > .043 h^{-3/2} \tag{2}$$

If $R = \Omega_B$ then clearly this equation is inconsistent with the BBN bound.

This problem can be simply resolved however, if $\Omega_{M_T} < 1$ so that the ratio R is in fact larger than Ω_B. There are then two possibilities. Either $\Omega_{M_T} = \Omega < 1$, or $\Omega_{M_T} + \Omega_\Lambda = 1$.

C   Large Scale Structure

The growth of structure in the Universe, if gravity is responsible for such growth, provides an excellent probe of the universal mass density, based largely on issues associated with causality alone. The basic idea is the following: If primordial density fluctuations have no preferred scale, then one can express their Fourier transform as a simple power of the wavenumber k. At the same time, if this power is much greater than unity, density fluctuations will blow up for large wavenumber, or small wavelength, and too many primordial black holes will be created. If the power is much less than unity, then fluctuations on large scales (small wavenumbers) will be inconsistent with the observed isotropy of the Cosmic Microwave Background radiation. Thus, we expect the exponent, n to be near one, and inflationary models happen to predict precisely this behavior.

The primordial power spectrum, however, is not what we observe today, as density fluctuations can be affected by causal microphysical processes once the scale of these fluctuations is inside the horizon scale—the distance over which light can have travelled between t=0 and the time in question. One can show that in an expanding universe, as long as the dominant form of energy resides in radiation, gravity is ineffective at causing the growth of density fluctuations. In fact, such primordial fluctuations in baryons will be damped out due to their coupling to the radiation gas. Once the universe becomes matter dominated, however, primordial fluctuations on scales smaller than the horizon size can begin to grow.

Fluctuation Spectrum

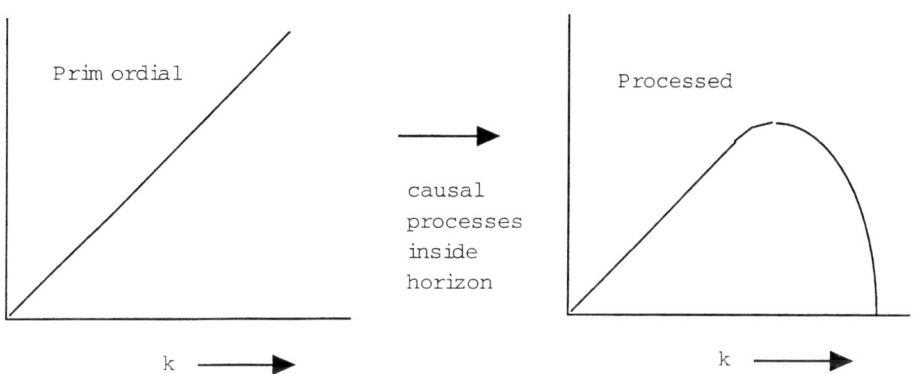

These arguments suggest that an initial power law spectrum of fluctuations will "turn over" as shown in Figure 1 above for large wavenumbers which entered inside the causal horizon during the early period of radiation domination in the Universe. By exploring the nature of the clustering of galaxies today over different scales, including measurements of the two point correlation function of galaxies, the angular correlation of galaxies across the sky on different scales, etc, one can hope to probe the location of this turn-around, and from that probe the time, and thus the scale which first entered the horizon when the universe became matter dominated. Clearly this time will depend upon the ratio of matter to radiation in the Universe today (if this ratio is increased, then matter, whose density decreases at a slower rate than radiation as the universe expands, will begin to dominate the expansion at an earlier time, and vice versa. In turn, knowing this ratio today gives us a handle on Ω_{matter}. A recent compilation of large scale galaxy clustering data [11,12]

restricts this quantity to be in the range:

$$0.25 \leq \Omega_{\text{matter}} h \leq 0.35 \quad (3)$$

Since h appears to lie in the range 0.65-0.75, this suggests $\Omega_{\text{matter}} < 1$. Note, however that this argument does not restrict the component of Ω which might reside in a cosmological constant today, since this energy density is fixed, so that even if it dominates today, it was irrelevant compared to the energy density of matter and radiation at early times.

By combining these three independent sets of constraints, one can, for either an open universe, or a flat universe with a cosmological constant, constrain the parameter space of h versus Ω_{matter} [1,3]. These constraints are shown in the figures below [3], which clearly indicate that a flat matter dominated universe appears to be inconsistent with observations.

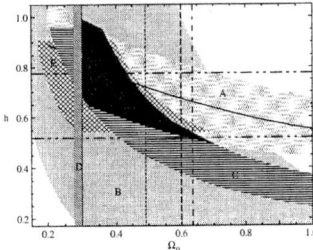 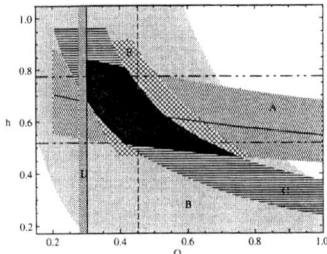

FIGURE 1. Constraints on h vs Ω_{mat} for a flat universe. The constraints discussed in the text (A, B, and C). The other constraints are discussed in (Krauss 98).

FIGURE 2. Constraints on h vs Ω_{matter} for an open universe. Constraints are as described in previous figure.

D Recent Observations: Galaxy Clustering, the CMB and Supernova Standard Candles

The situation described above has been recognized for over two years. Within the last year or so, however, new observations have made the case for a cosmological constant even stronger. I briefly describe these observations below.

(1) Cluster Evolution: In a Universe in which there is not a precisely critical density of matter, small density fluctuations on large scales cease to grow due to gravity once either the total density begins to deviate significantly from that associated with a flat universe, or when the density in a cosmological term begins to dominate over the density of matter. Thus, if the universe is not flat today, or if the cosmological constant dominates, then structure has ceased to continue to grow on large scales. If, however, the universe is flat and matter dominated, structures on

every larger scales are continuing to form by gravitational collapse. This suggests that if one examines out to high redshifts one should see significantly fewer rich clusters of galaxies then one sees today. The difference is significant. As Neta Bahcall and colleagues have recently pointed out [13], the probability of finding a rich cluster at a redshift of 0.7 is perhaps 100 times smaller for a flat matter dominated universe than for a universe in which the growth of large structures stopped some time ago. A single large cluster observed at such high redshift can then provide, largely independent of detailed modelling, damning evidence against a flat matter dominated universe. By cataloguing such clusters out to redshifts of this order, these authors have recently claimed to present definitive evidence ruling out a flat, cold dark matter dominated universe. Both open, or flat cosmological constant dominated universes are consistent with this data.

(2) Type 1a Supernova at High Redshift: As anyone who has glanced at a paper in the past six months knows, two groups have recently and hopefully independently claimed to measure the relation between redshift and distance out to redshifts in excess of 0.5, and in so doing have been able to probe for cosmic deceleration or acceleration. The probes used have been Type 1a Supernovae. These have been claimed to be superb Standard Candles for two reasons: (i) Type 1a supernovae occur when a white dwarf, through accretion, passes the Chandrsekhar limit, and undergoes a detonation explosion. The physics of this process should not depend significantly on the evolutionary status of the galaxy in which the star is housed. (ii) Detailed studies of the luminosity profile of such supernovae suggest a strong relation between the width of the light curve, and the absolute luminosity of the supernova. This allows one, in principle, to accurately determine this absolute luminosity. Based on these features both groups have now claimed to report definitive evidence for a non-zero cosmological constant. Moreover, they claim, at the 99 % confidence level, to be able to rule out both a flat, and an open universe with zero cosmological constant [14,15]. Even more remarkably, the favored region, for a flat universe, is precisely in the range favored by the other constraints in the figures shown above (which I remind you were drawn before these results appeared), namely $\Omega_{mat} \approx 0.3 - 0.4$, $\Omega_\Lambda \approx 0.6 - 0.7$. It remains to be seen if further data taken at high redshift confirm these results, and more importantly confirm the assumption that evolution is negligible for such supernovae.

(3) CMB preliminary studies: If one decomposes the observed CMB anistropies on the sky into multipoles on the sky, it is well known that CDM cosmologies predict a rise in the power spectrum as a function of multipole approaching a large peak, followed by smaller peaks. The position, in multipole space, of this first peak is a probe of the geometry of the universe, as it is related to the angular size of the horizon at the last scattering surface, as seen today. *Very* preliminary results from terrestrial observations of high multipole anisotropies in the CMB tend to confirm the existence of such a peak, and moreover the position of the peak appears to favor a flat, versus an open universe. If this is the case, then the existence of a non-zero cosmological constant, in light of all the other direct and indirect evidence, seems assured.

It is almost unnerving that all existing cosmological data appears to point in the same direction—towards a non-zero cosmological constant. As someone who has been promoting the idea that the cosmological constant might be non-zero for some time, I frankly found myself more comfortable when some of the data argued against this possibility. In any case, the wealth of cosmological data now available appears to unambiguously point to the fact that $\Omega_{\text{mat}} < 1$, whether or not the cosmological constant is non-zero. This fact may have profound implications for dark matter detection.

III IMPLICATIONS FOR PARTICLE PHYSICS AND THE SEARCH FOR DARK MATTER

The magnitude of the cosmological constant which would be required by the present data is remarkably small. Before proceeding to examine the consequences of the above results for dark matter detection, it is worth pausing for a moment to reflect on this feature. As we do, we can turn to the pocket calendar distributed to attendees of this meeting, and we find in the month of May, the quotation:

"To see what is in front of one's nose requires a constant struggle"
George Orwell

What, you may ask, does this have to do with the topic at hand. Plenty, I claim. For it reminds us that we can put remarkably stringent limits on certain quantities by using macroscopic amounts of material. In particular, it harkens back to another famous quotation, this time from Maurice Goldhaber, who put one of the first limits on proton decay by declaring that if the proton had a lifetime less than about 10^{17} years, *"You could feel it in your bones!"*. By this he meant that proton decays in our body would be so frequent that we would die from the radiation exposure.

In this spirit we can perform a similar experiment. Look at the end of your nose. Now, in a universe dominated by a cosmological constant, space begins to expand exponentially. One can calculate than for distances separated by larger than an amount $R > M_{Pl}/3\Lambda^{1/2}$, points will have a relative velocity exceeding that of light, and thus will remain out of causal contact. Thus, the fact that you can see the end of your nose implies a bound $\Lambda < 10^{-68} M_{Pl}^4$!

Of course, the fact that we can see distant galaxies gives us an even stronger bound. And, the fact that the cosmological constant affects dynamics on larger scales no more than it is claimed to by the present observations gives a bound $\Lambda < 10^{-123} M_{Pl}^4$. What makes this small number so hard to understand, in a cosmological context is not merely the "naturalness" problem of which particle physicists are aware, but rather, if this has been constant over cosmological time, this is the first time in the history of the universe when the energy density in a cosmological constant is comparable to the energy density of matter and radiation! It is for this reason that some cosmologists are driven to the idea that what is being observed is not really a cosmological constant, but something perhaps more exotic [16].

Be that as it may. Particle physicists will never measure a quantity of this magnitude directly in the laboratory. However, they may one day directly measure non-baryonic particles which presumably make up our galactic halo. And the new data brings good tidings in this regard. For the only well motivated candidates for Cold Dark Matter, axions and WIMPS one can write down a general relation:

$$\sigma_{detection} \approx \frac{1}{\Omega_{DM}} \qquad (4)$$

The reasons for this are different for each candidate. For axions, one can understand the origin of this relation as follows: Axions are dark matter because at early times thier potential (considered as a function of an angular variable which can be taken to go from $-\pi$ to π) changes form:

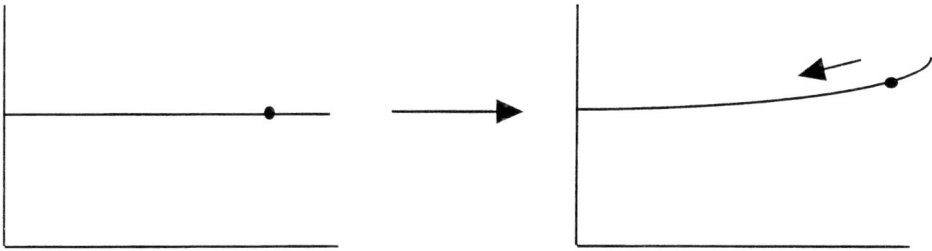

In the former case, no energy is stored in the axion field. However, once the axion gets a mass, energy is stored in the axion field, which then dynamically rolls to the bottom of its potential. However, the time it takes to begin rolling is inversely proportional to the curvature of its potential, and is thus inversely proportional to the axion mass. Thus, the smaller the axion mass, the longer the energy gets stored before it begins to redshift and the greater the remnant axion density. Since the axion couplings are inversely proportional to the axion mass, one therefore obtains the relation above.

For WIMPs, the situation is more direct. Remnant WIMPs results from incomplete annihilations in an initial thermal poulation, so that

$$\Omega_X h^2 \approx \frac{10^{-37} \text{cm}^2}{<\sigma_{ann} v>} \qquad (5)$$

By crossing symmetry, the WIMP annihilation cross section is roughly proportional to the WIMP scattering cross section. Thus, as the WIMP abundance decreases, its scattering cross section generally increases.

Astute experimentalists may argue that this is a scam, because as the WIMP (axion) density decreases, the flux on Earth also decreases, so even if there are larger cross sections, the event rate will not change! However, this is wrong. Until the density decreases to the point (below about $\Omega_x < 0.1$) when WIMPs (axions) do not have sufficient densities to account for all galactic halo dark matter, it is natural to assume that their galactic density is given by the halo density. Just because their overall cosmic density is insufficient to close the universe, this need not imply that their flux on earth is reduced!

IV CONCLUSIONS

It is time to throw in the towel and accept the paradigm shift in Cosmology. All evidence suggests that $\Omega_{DM} < 1$. The dominant energy density in the universe may be far darker than that stored in dark matter —it may be stored in empty space itself! Nevertheless the news is good for direct detection of non-baryonic dark matter. Cross sections may be higher than previously invisaged when it was felt that Cold Dark Matter must result in a closure density all by itself.

Of course, as a theorist one tries to think beyond the next set of experiments. What if the next generation of WIMP detectors detects a signal, for example? What then? How will we be sure that it is from the galactic halo, and how can we learn about the halo properties, and/or the properties of the dark matter particles? I will close this lecture by advertising some new work we have been involved in which may shed some light on new WIMP signatures. First, by exploring the angular variation of the predicted WIMP signals which might arise from a variety of different models for our galactic halo, we have recently demonstrated [17] that as few as 15-20 events would be needed in a detector having directional sensitivity before a halo induced signal could be distinguished from a flat background of noise. Next, with T. Damour, I have recently demonstrated that there is likely to be a new solar system population of WIMPs existing in trapped Solar orbits intersecting the earth if the WIMP cross sections on matter are large enough to be detected at the next generation of detectors [18]. This population will produce a dramatically different signal in cryogenic detectors, and could be used as a discriminant to verify any previoulsy detected WIMP signal, or could be searched for independently.

I thank my collaborators involved in various aspects of the work described here, Michael Turner, Brian Chaboyer, Pierre Demarque, Peter Kernan, Craig Copi, Junseong Heo, and Thibault Damour. I also thank the organizers for a very stimulating and enjoyable meeting.

REFERENCES

1. L.M. Krauss and M. S. Turner, J. Gen. Rel. Grav., **27**,1137 (1995)
2. J.P. Ostriker and P. Steinhardt, Nature, **377**, 600 (1995)
3. L.M. Krauss, Ap. J., **501**, in press, (1998)
4. B. Chaboyer, P. Demarque, P. Kernan, and L. M. Krauss, Science, **271**, 957 (1996)
5. W. L. Freedman, Proc. Nat. Acad. Sci., **95**, 2 (1998)
6. B. Chaboyer, P. Demarque, P. Kernan, and L. M. Krauss, Ap. J. , **494**, 96 (1998)
7. L. M. Krauss and P. J. Kernan, Phys. Lett. **B347**, 347 (1995)

8. C. Copi, D.N. Schramm, and M.S. Turner, Science, **267**, 192 (1995)
9. D. Tytler, X. M. Fan, and S. Burles, Nature, **381**, 207 (1996)
10. A. E. Evrard, MNRAS, in press (1998)
11. J.A. Peacock and S.J. Dodds, MNRAS, **267**, 1020 (1994)
12. A. R. Liddle, D.H. Lyth, P.T.P. Viana, M. White, MNRAS, **282**, 281 (1996)
13. N.A. Bahcall, X. Fan, Ap. J. **504**, in press, (1998); Proc. Nat. Acad. Sci., **95**, 5956 (1998)
14. S. Perlmutter *et al*, LBNL preprint 41801 (1998)
15. A. Reiss *et al*, preprint, submitted to Ap. J.
16. R. Caldwell *et al*, Phys. Rev. Lett., **80**, 1582 (1998)
17. L.M. Krauss, Phys. Rep. (Proc. Workshop on Dark Matter Detection), in press; see also C. Copi, J. Heo and L.M.Krauss, in preparation.
18. T. Damour and L.M. Krauss, astro-ph/9806165 (1998)

The search for axion dark matter

Pierre Sikivie

Department of Physics
University of Florida
Gainesville, FL 32611

Abstract. This talk reviews the original motivation for the axion as a solution to the strong CP problem and the constraints that have been placed on the axion by experimental searches and by astrophysical and cosmological considerations. As a result of the bounds, the axion mass is presently restricted to a window extending from about 10^{-2} eV to about 10^{-6} eV. In this window, axions are a form of cold dark matter. It is possible to detect galactic halo axions by stimulating their conversion to photons in a laboratory magnetic field. I'll report on two experiments of this type, one at Lawrence Livermore National Laboratory and the other at Kyoto University. I'll also discuss what can be learned about the structure of our galactic halo if a signal is found.

THE STRONG CP PROBLEM

The axion was postulated nearly two decades ago [1] to explain why the strong interactions conserve P and CP in spite of the fact that the weak interactions violate those symmetries. Consider the Lagrangian of QCD:

$$\mathcal{L}_{QCD} = -\frac{1}{4}G^a_{\mu\nu}G^{a\mu\nu} + \sum_{j=1}^{n}\left[\bar{q}_j\gamma^\mu i D_\mu q_j - (m_j q^+_{Lj} q_{Rj} + \text{h.c.})\right] + \frac{\theta g^2}{32\pi^2}G^a_{\mu\nu}\tilde{G}^{a\mu\nu} \ . \quad (1)$$

The last term is a 4-divergence and hence does not contribute in perturbation theory. That term does however contribute through non-perturbative effects [2] associated with QCD instantons [3]. Such effects can make the physics of QCD depend upon the value of θ. Using the Adler-Bell-Jackiw anomaly [4], one can show that θ dependence must be there if none of the current quark masses vanishes. If θ dependence were absent QCD would have a $U_A(1)$ symmetry and would predict the mass of the η' pseudo-scalar meson to be less than $\sqrt{3}m_\pi \approx 240$ MeV [5], contrary to observation. One can further show that QCD depends upon θ only through the combination of parameters:

$$\bar{\theta} = \theta - \arg(m_1, m_2, \ldots m_n) \quad (2)$$

If $\bar{\theta} \neq 0$, QCD violates P and CP. The absence of P and CP violations in the strong interactions therefore places an upper limit upon $\bar{\theta}$. The best constraint follows from the experimental bound [6] on the neutron electric dipole moment which yields: $\bar{\theta} < 10^{-9}$.

The question then is: why is $\bar{\theta}$ so small? In the Standard Model of particle interactions, the quark masses originate in the electroweak sector of the theory which violates P and CP. There is no reason why the overall phase of the quark mass matrix should exactly match the value of θ from the QCD sector to yield $\bar{\theta} < 10^{-9}$. In particular, if CP violation is introduced in the manner of Kobayashi and Maskawa [7], the Yukawa couplings that give masses to the quarks are arbitrary complex numbers and hence arg det m_q and $\bar{\theta}$ are expected to be of order one.

The problem why $\bar{\theta} < 10^{-9}$ is usually referred to as the "strong CP problem". The existence of an axion solves this problem in a simple elegant manner which is rich in implications for experiment, for astrophysics and for cosmology. There are other solutions however. Setting $m_u = 0$ removes the θ-dependence of QCD and hence solves the strong CP problem. The well-known calculation of the pseudo-scalar meson masses in lowest order of chiral perturbation theory yields $m_u \simeq 4$ MeV. This calculation also predicts the successful Gell-Mann - Okubo relation among the pseudo-scalar masses squared. It is possible to have $m_u = 0$ by invoking second order effects [8]. This a reasonable proposition because m_s happens to be of order the QCD scale. However, when second order effects are included [9], the Gell-Mann - Okubo relation is in general violated. Thus the price for having $m_u = 0$ through higher order effects is that the Gell-Mann - Okubo relation becomes an accident. Another way to solve the strong CP problem is to assume that CP and/or P is spontaneously broken but is otherwise a good symmetry. In this case, $\bar{\theta}$ is calculable and may be arranged to be small [10]. Finally, the strong CP problem need not be solved in the low energy theory. Indeed, as Ellis and Gaillard [11] pointed out, if in the standard model $\bar{\theta} = 0$ near the Planck scale then $\bar{\theta} \ll 10^{-9}$ at the QCD scale.

Peccei and Quinn [12] proposed to solve the strong CP problem by postulating the existence of a global $U_{PQ}(1)$ quasi-symmetry. $U_{PQ}(1)$ must be a symmetry of the theory at the classical (i.e., at the Lagrangian) level, it must be broken explicitly by those non-perturbative effects that make the physics of QCD depend upon θ, and finally it must be spontaneously broken. The axion [13] is the quasi-Nambu-Goldstone boson associated with the spontaneous breakdown of $U_{PQ}(1)$. One can show that, if a $U_{PQ}(1)$ quasi-symmetry is present, then

$$\bar{\theta} = \theta - arg(m_1 \ldots m_n) - \frac{a(x)}{f_a}, \qquad (3)$$

where $a(x)$ is the axion field and f_a, called the axion decay constant, is of order the vacuum expectation value (VEV) which spontaneously breaks $U_{PQ}(1)$. It can further be shown [14] that the non-perturbative effects that make QCD depend upon $\bar{\theta}$ produce an effective potential $V(\bar{\theta})$ whose minimum is at $\bar{\theta} = 0$. Thus, by postulating an axion, $\bar{\theta}$ is allowed to relax to zero dynamically and the strong CP problem is solved.

The properties of the axion can be derived using the methods of current algebra [15]. The axion mass is given in terms of f_a by

$$m_a \simeq 0.6 \ eV \ \frac{10^7 GeV}{f_a}. \tag{4}$$

All the axion couplings are inversely proportional to f_a. Of particular interest here is the axion coupling to two photons:

$$\mathcal{L}_{arr} = -g_\gamma \frac{\alpha}{\pi} \frac{a(x)}{f_a} \vec{E} \cdot \vec{B} \tag{5}$$

where $\vec{E}$ and $\vec{B}$ are the electric and magnetic fields, α is the fine structure constant, and g_γ is a model-dependent coefficient of order one. $g_\gamma = 0.36$ in the DFSZ model [16] whereas $g_\gamma = -0.97$ in the KSVZ model [17]. A priori the value of f_a, and hence that of m_a, is arbitrary. However, searches for the axion in high energy and nuclear physics experiments combined with astrophysical constraints, the latter derived by considering the effect of the axion upon the lifetimes of red giants and SN1987a, rule out $m_a \gtrsim 10^{-2}$ eV [1,18]. In addition, as is discussed below, cosmology places a lower limit on m_a of order 10^{-6} eV from the requirement that axions do not overclose the universe.

DARK MATTER AXIONS

For small masses, axion production in the early universe is dominated by a novel mechanism [19]. The non-perturbative QCD effects that produce the effective potential $V(\bar{\theta})$ are suppressed at temperatures high compared to Λ_{QCD} [20]. At these high temperatures, the axion is massless and all values of $\langle a(x) \rangle$ are equally likely. At $T \simeq 1$ GeV, the potential V turns on and the axion field starts to oscillate about a CP conserving minimum of V. These oscillations do not dissipate into other forms of energy because, in the relevant mass range, the axion is too weakly coupled for that to happen. The oscillations of the axion field may be described as a fluid of axions. The typical momentum of the axions in the fluid is the inverse of the correlation length of the axion field. Because that correlation length is of order the horizon when the axion mass effectively switches on, we have $p_a \sim (10^{-6}$ sec$)^{-1} \sim 10^{-9}$ eV at $T \simeq 1$ GeV, and $p_a \sim R^{-1}$ afterwards. R is the cosmological scale factor here. Thus the axion fluid is very cold compared to the ambient temperature.

Let me briefly indicate how the present cosmological energy density of this axion fluid is estimated. Let $\varphi(x)$ be the complex scalar field whose VEV v spontaneously breaks $U_{PQ}(1)$. At extremely high temperatures, the $U_{PQ}(1)$ symmetry is restored. It becomes spontaneously broken when the temperature drops below a critical value

T_{PQ} of order v. Below T_{PQ}, the axion field $a(x)$ appears as the phase of the VEV of φ: $\langle\varphi(x)\rangle = ve^{ia(x)/v}$. We must now distinguish two cases. Either inflation occurs with reheat temperature above T_{PQ} (equivalently, for our purposes, it does not occur at all) or it occurs with reheat temperature below T_{PQ}. In the second case, inflation homogenizes the axion field and there is only one contribution to the axion cosmological energy density, the contribution from 'vacuum realignment'. In the first case, there are additional contributions from axion string and axion domain wall decay. Only the contribution from vacuum realignment is discussed in any detail here.

When the axion mass turns on near the QCD phase transition, the axion field starts to oscillate about one of the CP conserving minima of the effective potential. The oscillation begins approximately at cosmological time t_1 such that $t_1 m_a(T(t_1)) = 0(1)$, where $m_a(T)$ is the temperature dependent axion mass. Soon after t_1, the axion mass changes sufficiently slowly that the total number of axions in the oscillations of the axion field is an adiabatic invariant. $T_1 \equiv T(t_1)$ has been estimated to be of order 1 GeV. The number density of axions at time t_1 is

$$n_a(t_1) \simeq \frac{1}{2} m_a(t_1) \langle a^2(t_1) \rangle \simeq \pi f_a^2 \frac{1}{t_1} \qquad (6)$$

where $f_a = \frac{v}{N}$ is the axion decay constant introduced earlier. N is an integer which expresses the color anomaly of $U_{PQ}(1)$. N also equals the number of degenerate vacua [21] at the bottom of the 'Mexican hat' potential, i.e. in the interval $0 \leq \frac{a}{v} < 2\pi$. In Eq. (6), we used the fact that the axion field $a(x)$ is approximately homogeneous on the horizon scale t_1. Wiggles in $a(x)$ which entered the horizon long before t_1 have been red-shifted away [22]. We also used the fact that the initial departure of $a(x)$ from the nearest minimum is of order $\frac{v}{N} = f_a$. The axions of Eq. (6) are decoupled and non-relativistic. Assuming that the ratio of the axion number density to the entropy density is constant from time t_1 till today, one finds [19]

$$\Omega_a \simeq \left(\frac{0.6 \; 10^{-5} \text{ eV}}{m_a}\right)^{\frac{7}{6}} \left(\frac{200 \text{ MeV}}{\Lambda_{QCD}}\right)^{\frac{3}{4}} \left(\frac{75 \text{ km/s} \cdot \text{Mpc}}{H_0}\right)^2 \qquad (7)$$

for the ratio of the axion energy density to the critical density for closing the universe. H_0 is the present Hubble rate. Eq. (7) implies the bound $m_a \gtrsim 10^{-6}$ eV.

If $N > 1$, there is a domain wall problem [21]. The walls, which appear when the axion acquires mass at the QCD phase transition, end up dominating the cosmological energy density. After this happens the universe expands as $R \sim t^2$ and is incompatible with what we observe today. The problem is avoided if inflation occurs with reheat temperature below the PQ phase transition temperature. Another way to avoid the problem is to have a small violation of the $U_{PQ}(1)$ symmetry which lowers one of the N vacua with respect to all the others [21]. There is little room in parameter space for this to happen but it is a logical possibility. Finally, if $N = 1$ the domain wall problem is avoided because in this case the domain walls appear at the QCD phase transition in finite size pieces bounded by string [23]. These wall pieces decay away before they dominate the energy density.

It should be emphasized that there are many sources of uncertainty in the estimate of Eq. (7). The axion energy density may be diluted by the entropy release from heavy particles which decouple before the QCD epoch but decay afterwards [24], or by the entropy release associated with a first order QCD phase transition. On the other hand, if the QCD phase transition is first order [25], an abrupt change of the axion mass at the transition may increase Ω_a. If inflation occurs with reheat temperature less than T_{PQ}, there may be an accidental suppression of Ω_a because the homogenized axion field happens to lie close to a CP conserving minimum. Because the RHS of Eq. (7) is multiplied in this case by a factor of order the square of the initial vacuum misalignment angle $\frac{a(t_1)}{v}N$ which is randomly chosen between $-\pi$ and $+\pi$, the probability that Ω_a is suppressed by a factor x is of order $\sqrt{x}$. This rule cannot be extended to arbitrarily small x however because quantum mechanical fluctuations in the axion field during the epoch of inflation do not allow the suppression to be perfect [26]. If inflation occurs with reheat temperature larger than T_{PQ} or if there is no inflation, there are contributions to Ω_a from axion string [27] and axion domain wall decay [28,29] in addition to the contribution, Eq. (7), from vacuum realignment. The author and his collaborators [30,28,31] have estimated each of these additional contributions to be of the same order of magnitude as that from vacuum realignment. Others [27,32] have estimated that the contribution from axion string decay dominates over that from vacuum realignment by a factor of order 100.

The axions produced when the axion mass turns on during the QCD phase transition are cold dark matter (CDM) because the axions are non-relativistic from the moment of their first appearance at 1 GeV temperature. Studies of large scale structure formation support the view that the dominant fraction of dark matter is CDM. Moreover, any form of CDM necessarily contributes to galactic halos by falling into the gravitational wells of galaxies. Hence, there is excellent motivation to look for CDM candidates as constituent particles of our galactic halo, even after some fraction of our halo has been demonstrated to be in MACHOs [33] or some other form.

Finally, let's mention that there is a particular kind of clumpiness [34,31] which affects axion dark matter if there is no inflation after the Peccei-Quinn phase transition. This is due to the fact that the dark matter axions are inhomogeneous with $\delta\rho/\rho \sim 1$ over the horizon scale at temperature $T_1 \simeq 1$ GeV, when they are produced at the start of the QCD phase-transition, combined with the fact that their velocities are so small that they do not erase these inhomogeneities by free-streaming before the time t_{eq} of equality between the matter and radiation energy densities when matter perturbations can start to grow. These particular inhomogeneities in the axion dark matter are immediately in the non-linear regime after time t_{eq} and thus form clumps, called 'axion mini-clusters' [34]. These have mass $M_{mc} \simeq 10^{-13} M_\odot$ and size $l_{mc} \simeq 10^{13}$ cm.

THE CAVITY DETECTOR OF GALACTIC HALO AXIONS

Axions can be detected by stimulating their conversion to photons in a strong magnetic field [35]. The relevant coupling is given in Eq. (5). In particular, an electromagnetic cavity permeated by a strong static magnetic field can be used to detect galactic halo axions. The latter have velocities β of order 10^{-3} and hence their energies $E_a = m_a + \frac{1}{2} m_a \beta^2$ have a spread of order 10^{-6} above the axion mass. When the frequency $\omega = 2\pi f$ of a cavity mode equals m_a, galactic halo axions convert resonantly into quanta of excitation (photons) of that cavity mode. The power from axion $\to$ photon conversion on resonance is found to be [35,36]:

$$P = \left(\frac{\alpha}{\pi} \frac{g_\gamma}{f_a}\right)^2 V B_0^2 \rho_a C \frac{1}{m_a} \text{Min}(Q_L, Q_a)$$

$$= 0.5\ 10^{-26} \text{Watt} \left(\frac{V}{500\ \text{liter}}\right) \left(\frac{B_0}{7\ \text{Tesla}}\right)^2 C \left(\frac{g_\gamma}{0.36}\right)^2$$

$$\cdot \left(\frac{\rho_a}{\frac{1}{2} \cdot 10^{-24} \frac{gr}{cm^3}}\right) \left(\frac{m_a}{2\pi (\text{GHz})}\right) \text{Min}(Q_L, Q_a) \qquad (8)$$

where V is the volume of the cavity, B_0 is the magnetic field strength, Q_L is its loaded quality factor, $Q_a = 10^6$ is the 'quality factor' of the galactic halo axion signal (i.e. the ratio of their energy to their energy spread), ρ_a is the density of galactic halo axions on Earth, and C is a mode dependent form factor given by

$$C = \frac{\left|\int_V d^3 x \vec{E}_\omega \cdot \vec{B}_0\right|^2}{B_0^2 V \int_V d^3 x \epsilon |\vec{E}_\omega|^2} \qquad (9)$$

where $\vec{B}_0(\vec{x})$ is the static magnetic field, $\vec{E}_\omega(\vec{x}) e^{i\omega t}$ is the oscillating electric field and ϵ is the dielectric constant.

Because the axion mass is only known in order of magnitude at best, the cavity must be tunable and a large range of frequencies must be explored seeking a signal. The cavity can be tuned by moving a dielectric rod or metal post inside it. Using Eq. (8), one finds that to perform a search with signal to noise ratio s/n, the scanning rate is:

$$\frac{df}{dt} = \frac{12 \text{GHz}}{\text{year}} \left(\frac{4n}{s}\right)^2 \left(\frac{V}{500\ \text{liter}}\right)^2 \left(\frac{B_0}{7\ \text{Tesla}}\right)^4 C^2 \left(\frac{g_\gamma}{0.36}\right)^4$$

$$\cdot \left(\frac{\rho_a}{\frac{1}{2} \cdot 10^{-24} \frac{gr}{cm^3}}\right)^2 \left(\frac{3K}{T_n}\right)^2 \left(\frac{f}{\text{GHz}}\right)^2 \frac{Q_L}{Q_a}, \qquad (10)$$

where T_n is the sum of the physical temperature of the cavity plus the noise temperature of the microwave receiver that detects the photons from $a \to \gamma$ conversion. Eq. (10) assumes that $Q_L < Q_a$ and that some strategies have been followed which optimize the search rate. The best quality factors attainable at present, using oxygen free copper, are of order 10^5 in the GHz range.

Eq. (10) shows that a galactic halo search with the required sensitivity is feasible with presently available technology, provided the form factor C can be kept at values of order one for a wide range of frequencies. For a cylindrical cavity and a longitudinal magnetic field, $C = 0.69$ for the lowest TM mode. The form factors of the other modes are much smaller. The resonant frequency of the lowest TM mode of a cylindrical cavity is $f = 115$ MHz $\left(\frac{1m}{R}\right)$ where R is the radius of the cavity. Since 10^{-6} eV $= 2\pi$ (242 MHz), a large cylindrical cavity is convenient for searching the low frequency end of the range of interest. To extend the search to high frequencies without sacrifice in volume, one may power-combine many identical cavities which fill up the available volume inside a magnet's bore [37,38]. This method allows one to maintain $C = 0(1)$ at high frequencies, albeit at the cost of increasing engineering complexity as the frequency, and hence the number of cavities, is increased.

Pilot experiments were carried out at Brookhaven National Laboratory [39] and at the University of Florida [40]. These experiments used relatively small magnets and hence the limits they placed on the local axion dark matter density are not severe (see Fig. 1). However they developed the various aspects of the cavity

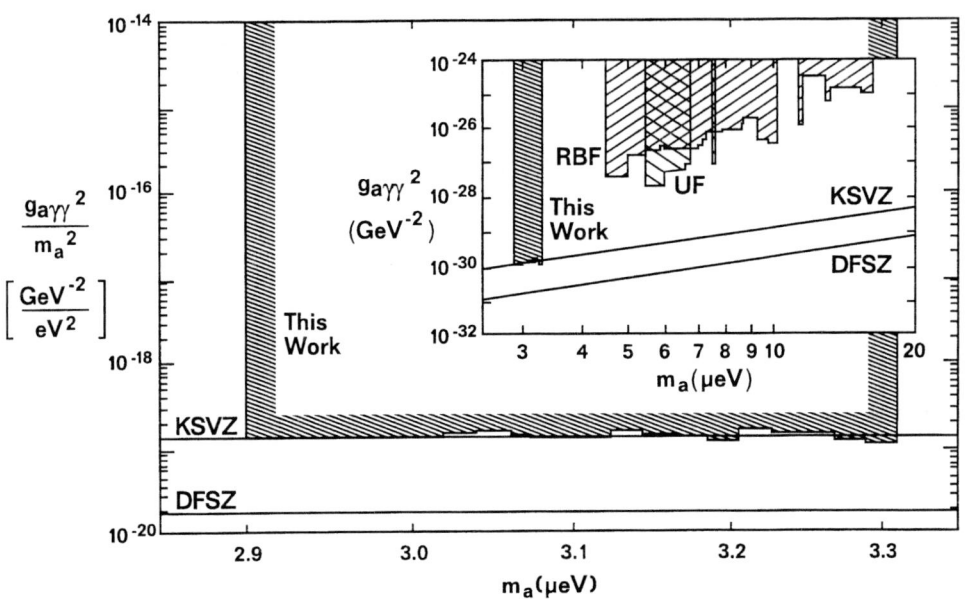

FIGURE 1. Axion couplings and masses excluded by the Large Scale US Dark Matter Axion Search at LLNL. Also shown are the KSVZ and DFSZ model predictions. Indicated on the insert are the regions excluded by the pilot experiments at Brookhaven National Lab. (RBF) and the University of Florida (UF). All results are scaled to $\rho_a = 7.5 \ 10^{-25}$ g/cm^3.

detection technique and demonstrated its feasibility.

Second generation experiments are presently under way at Lawrence Livermore National Laboratory (LLNL) [41] and at Kyoto University [45]. The present LLNL experiment is similar in concept to the UF pilot experiment but uses a much larger magnet ($B_0^2 V = 12 T^2 m^3$). It is very well designed and runs with a near 100% duty cycle. It recently reported the results from its first year of running [42]. It searched over a relatively narrow mass range, $2.9\ 10^{-6}$ eV to $3.3\ 10^{-6}$ eV, but with sufficient sensitivity to actually detect dark matter axions under realistic assumptions. The exclusion plot is shown in Fig. 1. By definition, $g_{a\gamma\gamma} = \frac{\alpha}{\pi} \frac{g_\gamma}{f_a}$. The limits shown assume that the local halo density, estimated to be $7.5\ 10^{-5}$ g/cm^3 [43], is entirely in axions. The experiment ruled out the hypothesis that 100% of the local halo density is in KSVZ axions with mass in the covered range.

The LLNL experiment will search the mass range $1.3\ 10^{-6}$ eV to $13\ 10^{-6}$ eV at KSVZ sensitivity during the next few years. It will use a 4-cavity array to cover the upper half of this range. That will be the first time this technique is used in an actual search.

A development project is under way to equip the LLNL detector with SQUID microwave receivers. These would replace the HEMT receivers presently in use. The HEMT receivers have noise temperature $T_n \sim 3\ K$. It appears that $T_n \sim 0.3\ K$, and possibly better, will be reached with the SQUIDs [44]. If this development project is successful, the LLNL detector will acquire sufficient sensitivity to detect DFSZ axions at the local halo density and below.

The Kyoto experiment has a magnet of size similar to that of the pilot experiments but uses a beam of Rydberg atoms to count the photons from $a \to \gamma$ conversion. The $a \to \gamma$ conversion part is the same as in the other experiments. Single photon counting constitutes a dramatic improvement in microwave detection sensitivity. With HEMT amplifiers one needs to have thousands of $a \to \gamma$ conversions per second and integrate for about 100 sec to find a signal in the noise. With single photon counting, a few $a \to \gamma$ conversions suffice in principle. To build a beam of Rydberg atoms capable of single photon counting is a considerable achievement in itself. In addition, the cavity will be cooled by a dilution refrigerator down to a temperature (~ 10 mK) where the thermal photon background is negligible. The Kyoto experiment will first search near $m_a = 10^{-5}$ eV. Its projected sensitivity is sufficient to discover DFSZ axions even if their local density is only $\frac{1}{5}$ of the local halo density.

THE PHASE SPACE STRUCTURE OF COLD DARK MATTER HALOS

If a signal is found in the cavity detector of galactic halo axions, it will be possible to measure the energy spectrum of cold dark matter particles on Earth with great precision and resolution. As mentioned above, the spread ($\Delta f/f = \Delta E_a/m_a \simeq 10^{-6}$) of the axion signal is due to the kinetic energy of motion of the axions through the galactic halo. The LLNL experiment is equiped with a high resolution spectrometer that resolves $\Delta f/f \sim 10^{-11}$ and hence divides the axion signal width into 10^5 bins. If a signal is found, data on the energy spectrum will accumulate very quickly because all the time previously spent searching for the signal can now be used to gather data. Let us ask what can be learned about our galaxy from analyzing the signal.

In many past discussions of dark matter detection on Earth, it has been assumed that the dark matter particles have an isothermal distribution. Thermalization has been argued to be the result of a period of "violent relaxation" following the collapse of the protogalaxy. If it is strictly true that the velocity distribution of dark matter particles is isothermal, which seems to be a strong assumption, then the only information that can be gained from its observation is the corresponding virial velocity and our own velocity relative to its standard of rest. If, on the other hand, thermalization is incomplete, a signal in a dark matter detector may yield additional information.

J.R. Ipser and I discussed [46] the extent to which the phase-space distribution of cold dark matter particles is thermalized in a galactic halo and concluded that there are substantial deviations from a thermal distribution in that the highest energy particles have discrete values of velocity. There is one velocity peak on Earth due

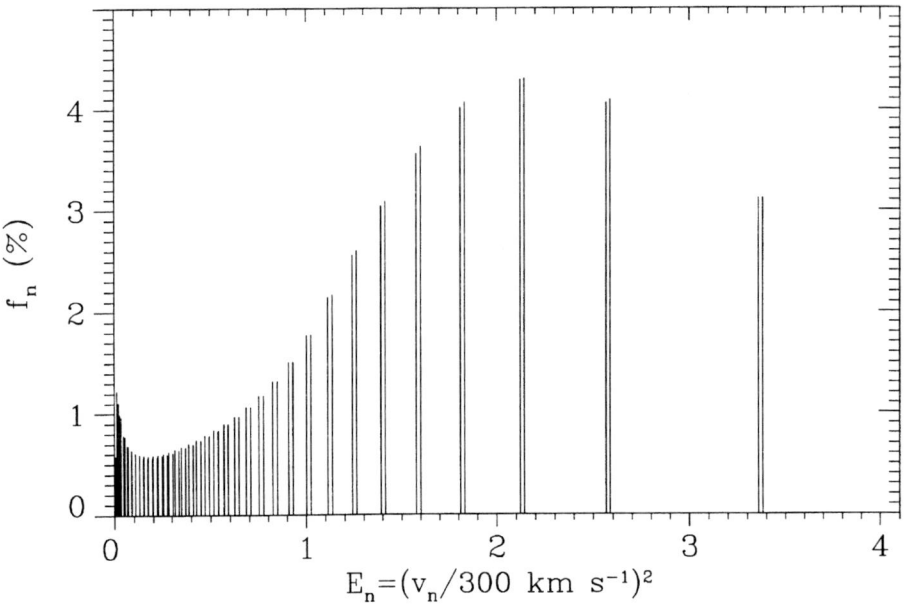

FIGURE 2. The spectrum of velocity peaks in a typical case ($\epsilon = 0.2, h = 0.7$ and $\bar{j} = 0.2$) studied in ref. [40]. f_n and E_n are defined in the text.

to dark matter particles falling onto the galaxy for the first time, one peak due to particles falling out of the galaxy for the first time, one peak due to particles falling into the galaxy for the second time, etc. The peaks due to particles that have fallen in and out of the galaxy a large number of times in the past are washed out because of scattering in the gravitational wells of stars, globular clusters and large molecular clouds. But the peaks due to particles which have fallen in and out of the galaxy only a small number of times in the past are not washed out.

I. Tkachev, Y. Wang and I have used the secondary infall model of galactic halo formation to estimate the local densities and the velocity magnitudes of the dark matter particles in the velocity peaks [47]. We generalized the existing version of that model to take account of the angular momentum of the dark matter particles. In the absence of angular momentum, the model produces flat rotation curves for a large range of values of a parameter ϵ whose average value may be inferred from the spectrum of primordial density perturbations. We find that the presence of angular momentum produces an effective core radius, i.e., it makes the contribution of the halo to the rotation curve go to zero at zero radius. The model provides a detailed description of the large scale properties of galactic halos including their density profiles, their extent and their total mass. Fig. II shows the predictions of the model for the average density fractions $f_n = \rho_n/\rho$ and the kinetic energies per unit mass E_n of the particles in the highest energy peaks for the case where $\epsilon = 0.2$, $H_0 = 70$ km/sec·Mpc and the average amount of angular momentum, in the dimensionless units defined in ref. [47], is $\bar{j} = 0.2$. The density fractions f_n are averages over all locations at the same distance (8.5 kpc) from the galactic center as we are. The E_n are measured in a frame of reference which is not co-rotating with the disk.

More recently, it was realized that the flows of dark matter in and out of a galaxy produce caustic rings in the halo distribution. [48]. Caustics are places in physical space where the density is very large. There is one caustic ring for every pair of velocity peaks. Bumps in the rotation curve of our own galaxy may be interpreted as due to caustic rings [48]. If this interpretation holds up, it will yield values for the local densities and the velocity vectors of the first dozen or so pairs of flows.

ACKNOWLEDGEMENTS:

This work was supported in part by a Fellowship grant from the J.S. Guggenheim Memorial Foundation and by the US Department of Energy under grant No. DEFG05-86ER40272.

REFERENCES

1. Axion reviews include: J.E. Kim, Phys. Rep. **150** (1987) 1; H.-Y. Cheng, Phys. Rep. **158** (1988) 1; R.D. Peccei, in 'CP Violation', ed. by C. Jarlskog, World Scientific

Publ., 1989, pp 503-551; M.S. Turner, Phys. Rep. **197** (1990) 67; G.G. Raffelt, Phys. Rep. **198** (1990) 1.

2. G. 't Hooft, Phys. Rev. Lett. **37** (1976) 8 and Phys. Rev. **D14** (1976) 3432; R. Jackiw and C. Rebbi, Phys. Rev. Lett. **37** (1976) 172; C. G. Callan, R. F. Dashen and D. J. Gross, Phys. Lett. **B63** (1976) 334.
3. A.A Belavin, A.M. Polyakov, A.S. Shvarts and Yu.S. Tyupkin, Phys. Lett. **59B** (1975) 85.
4. S. Adler, Phys. Rev. **177** (1969) 2426; J.S. Bell and R. Jackiw, Nuov. Cim. **60A** (1969) 47.
5. S. Weinberg, Phys. Rev. **D11** (1975) 3583.
6. I.S. Altarev et al., Phys. Lett. **B276** (1992) 242; K.F. Smith et al., Phys. Lett. **B234** (1990) 191.
7. M. Kobayashi and K. Maskawa, Progr. Theor. Phys. **49** (1973) 652.
8. H. Georgi and I. McArthur, preprint HUTP-81/A011; D.B. Kaplan and A.V. Manohar, Phys. Rev. Lett. **56** (1986) 2004; K. Choi, Nucl. Phys. **B383** (1992) 58.
9. J. Gasser and H. Leutwyler, Nucl. Phys. **B250** (1985) 465.
10. M.A.B. Beg and H.S. Tsao, Phys. Rev. Lett. **41** (1978) 278; R.N. Mohapatra and G. Senjanovic, Phys. Lett. **B79** (1978) 283; H. Georgi, Hadronic J. **1** (1978) 155; A.E. Nelson, Phys. Lett. **B136** (1984) 387; S.M. Barr, Phys. Rev. **D30** (1984) 1805; K.S. Babu and R.N. Mohapatra, Phys. Rev. **D41** (1990) 1286; P.H. Frampton and T.W. Kephart, Phys. Rev. Lett. **66** (1991) 1666; P.H. Frampton and D. Ng, Phys. Rev. **D43** (1991) 3043; S.M. Barr, D. Chang and G. Senjanovic, Phys. Rev. Lett. **67** (1991) 2765; R. Kuchimanchi, Phys. Rev. Lett. **76** (1996) 3486; R.N. Mohapatra and A. Rasin, Phys. Rev. Lett. **76** (1996) 3490.
11. J. Ellis and M.K. Gaillard, Nucl. Phys. **B150** (1979) 141.
12. R. D. Peccei and H. Quinn, Phys. Rev. Lett. **38** (1977) 1440 and Phys. Rev. **D16** (1977) 1791.
13. S. Weinberg, Phys. Rev. Lett. **40** (1978) 223; F. Wilczek, Phys. Rev. Lett. **40** (1978) 279.
14. C. Vafa and E. Witten, Phys. Rev. Lett. **53** (1984) 535.
15. S. Weinberg in ref. [13]; W.A. Bardeen and S.-H.H. Tye, Phys. Lett. **B74** (1978) 229; J. Ellis and M.K. Gaillard, Nucl. Phys. **B150** (1979) 141; T.W. Donnelly et al., Phys. Rev. **D18** (1978) 1607; M. Srednicki, Nucl. Phys. **260** (1985) 689; P. Sikivie, in 'Cosmology and Particle Physics', ed. E. Alvarez et al., World Scientific, 1987, pp 143-169.
16. M. Dine, W. Fischler and M. Srednicki, Phys. Lett. **B104** (1981) 199; A. P. Zhitnitskii, Sov. J. Nucl. **31** (1980) 260.
17. J. Kim, Phys. Rev. Lett. **43** (1979) 103; M. A. Shifman, A. I. Vainshtein and V. I. Zakharov, Nucl. Phys. **B166** (1980) 493.
18. G.G. Raffelt, astro-ph/9707268 and references therein.
19. L. Abbott and P. Sikivie, Phys. Lett. **B120** (1983) 133; J. Preskill, M. Wise and F. Wilczek, Phys. Lett. **B120** (1983) 127; M. Dine and W. Fischler, Phys. Lett. **B120** (1983) 137.
20. D. J. Gross, R. D. Pisarski and L. G. Yaffe, Rev. Mod. Phys. **53** (1981) 43.
21. P. Sikivie, Phys. Rev. Lett. **48** (1982) 1156.

22. A. Vilenkin, Phys. Rev. Lett. **48** (1982) 59.
23. Vilenkin and Everett, Phys. Rev. Lett. **48** (1982) 1867.
24. P. J. Steinhardt and M. S. Turner, Phys. Lett. **B129** (1983) 51; G. Lazarides, R. Schaefer, D. Seckel and Q. Shafi, Nucl. Phys. **B346** (1990) 193.
25. W. G. Unruh and R. M. Wald, Phys. Rev. **D32** (1985) 831; M. S. Turner, Phys. Rev. **D32** (1985) 843; T. DeGrand, T. W. Kephart and T. J. Weiler, Phys. Rev. **D33** (1986) 910; M. Hindmarsh, Phys. Rev. **D45** (1992) 1130.
26. A. D. Linde, JETP Lett. **40** (1984) 1333 and Phys. Lett. **B158** (1985) 375; D. Seckel and M. Turner, Phys. Rev. **D32** (1985) 3178; D. H. Lyth, Phys. Lett. **B236** (1990) 408; A. D. Linde and D. H. Lyth, Phys. Lett. **B246** (1990) 353; M. Turner and F. Wiczek, Phys. Rev. Lett. **66** (1991) 5; A. Linde, Phys. Lett. **B259** (1991) 38; D.H. Lyth, Phys. Rev. **D45** (1992) 3394; D.H. Lyth and E.D. Stewart, Phys. Lett. **B283** (1992) 189 and Phys. Rev. **D46** (1992) 532.
27. R. Davis, Phys. Rev. **D32** (1985) 3172 and Phys. Lett. **B180** (1986) 225.
28. C. Hagmann and P. Sikivie, Nucl. Phys. **B363** (1991) 247.
29. D. Lyth, Phys. Lett. **B275** (1992) 279; M. Nagaswa and M. Kawasaki, Phys. Rev. **D50** (1994) 4821.
30. D. Harari and P. Sikivie, Phys. Lett. **B195** (1987) 361.
31. S. Chang, C. Hagmann and P. Sikivie, to be published.
32. A. Vilenkin and T. Vachaspati, Phys. Rev. **D35** (1987) 1138; R.L. Davis and E.P.S. Shellard, Nucl. Phys. **B324** (1989) 167; A. Dabholkar and J. Quashnock, Nucl. Phys. **B333** (1990) 815; R.A. Battye and E.P.S. Shellard, Nucl.Phys.**B423** (1994) 260, Phys. Rev. Lett. **73** (1994) 2954 and erratum-ibid. **76** (1996) 2203.
33. C. Alcock et al., Phys. Rev. Lett. **74** (1995) 2867, and astro-ph/9606165.
34. C. J. Hogan and M. J. Rees, Phys. Lett. **B205** (1988) 228; E. Kolb and I. I. Tkachev, Phys. Rev. Lett. **71** (1993) 3051, Phys. Rev. **D49** (1994) 5040, and Ap. J. **460** (1996) L25.
35. P. Sikivie, Phys. Rev. Lett. **51** (1983) 1415 and Phys. Rev. **D32** (1985) 2988.
36. L. Krauss, J. Moody, F. Wilczek and D. Morris, Phys. Rev. Lett. **55** (1985) 1797.
37. C. Hagmann et al., Rev. Sci. Inst. **61** (1990) 1076.
38. C. Hagmann, Ph. D. thesis, unpublished.
39. S. DePanfilis et al., Phys. Rev. Lett. **59** (1987) 839 and Phys. Rev. **40** (1989) 3153.
40. C. Hagmann et al., Phys. Rev. **D42** (1990) 1297.
41. K. Van Bibber et al., Int. J. Mod. Phys. **D3** Suppl. (1994) 33.
42. C. Hagmann et al., Phys. Rev. Lett. **80** (1998) 2043.
43. E.I. Gates, G. Gyuk and M.S. Turner, Astroph. J. **449** (1995) 123.
44. M. Andre, J. Clarke and M. Mueck, to be published.
45. S. Matsuki and K. Yamamoto, Phys. Lett. **B263** (1991) 523; S. Matsuki, I. Ogawa, K. Yamamoto, Phys. Lett. **B336** (1994) 573.
46. J.R. Ipser and P. Sikivie, Phys. Lett. **B291** (1992) 288.
47. P. Sikivie, I.I. Tkachev and Y. Wang, Phys. Rev. Lett. **75** (1995) 2911 and Phys. Rev. **D56** (1997) 1863.
48. P. Sikivie, astro-ph/9705038, to be published in Phys. Lett. B.

Neutrino Mass and Dark Matter

David O. Caldwell

Institute for Nuclear and Particle Astrophysics and Cosmology and[1]
Physics Department, University of California, Santa Barbara, CA 93106-9530, USA

Abstract. Having a neutrino component of dark matter provides the best existing model for structure formation in the universe. If at least one light sterile neutrino exists, this hot dark matter can be accommodated, along with the solar ν_e deficit, the anomalous μ/e ratio produced by atmospheric neutrinos, and the candidate events for $\nu_\mu \to \nu_e$ (or $\bar{\nu}_\mu \to \bar{\nu}_e$) from the LSND experiment. This neutrino mass pattern also provides a robust solution to problems presently making heavy-element synthesis by suppernovae seem impossible, and it could help resolve a possible discrepancy between big bang nucleosynthesis theory and observations.

INTRODUCTION

A neutrino component of dark matter has long been an attractive means of reducing the overproduction of small-scale structure by the majority cold component, and very recently an extensive analysis [1] has shown that such a cold + hot mixture provides the only known model statistically consistent with observations. It is necessary, then, to see whether such neutrinos can be compatible with particle-physics constraints. Some time ago there appeared to be only two ways [2] hot dark matter could be accommodated along with the neutrino oscillation explanations of the paucity of electron neutrinos (ν_e) from the sun and the anomalous ratio of ν_μ/ν_e neutrinos produced in the atmosphere: either all three neutrinos had to be almost degenerate in mass, with all contributing about equally to the dark matter, or the ν_μ and ν_τ shared the dark matter role, with $\nu_\mu \to \nu_\tau$ explaining the atmospheric problem, and $\nu_e \to \nu_s$ (a sterile neutrino not having the normal weak interaction) accounting for the observed reduction in solar ν_e. A fourth neutrino is required because of the need for three distinct mass differences, but it must be sterile since the width of the Z^0 boson allows only three light active neutrinos.

The first scheme became more difficult with the evidence for $\nu_\mu \to \nu_e$ oscillations from the LSND experiment, although it could be kept viable [3] by invoking indirect oscillations; other difficulties appeared later. The second scheme was reinforced by

---

[1] This work was partially supported by the U.S. Department of Energy.

the LSND results and by the success of simulations [4] utilizing this two-neutrino hot dark matter.

After a brief introduction to mixed dark matter, recent developments in the three other areas indicating neutrino mass will be presented, followed by supplementary information for the second scheme of neutrino masses, that which requires an inactive, or sterile, neutrino. There are severe problems with the production of heavy elements by supernovae which the sterile neutrino can solve, and no other robust solution is known. The sterile neutrino may also aid in the apparent discrepancy between primordial ^{4}He abundance and the D/H ratio.

DARK MATTER

There is now a tremendous amount of information on structure in the universe over three orders of magnitude in spatial scale. Using all this information with appropriate errors, a χ^2 fit was made [1] of these data to 10 popular cosmological models. Nine of these fits had probabilities of $\sim 10^{-5}$ or less, whereas a model with 20% neutrinos had a probability of 9%. If one piece of information, which none of the models fit, was excluded, the cold + hot model had χ^2 per degree of freedom of 66/62.

This analysis [1] used only one cold + hot model, but for the purposes of this discussion, it is worth noting the genesis of these models. The cold dark matter model (CDM), which was a fair approximation to the structure of the universe, when normalized to the COBE data produced too much structure on small scales, since baryons readily clump around the cold dark matter. The first cold + hot dark matter models (CHDM) had $\sim 30\%$ neutrinos and fit structure on all scales very well [5] because the free streaming of the neutrinos reduced density fluctuations on small scales. Unfortunately, this damping of density perturbations also caused structure to form too late. Reducing the neutrino content to $\sim 20\%$ allowed early enough structure formation [6]. With all the mass in one neutrino species, this otherwise successful model (CνDM) overproduced clusters of galaxies. In other words, the CνDM model worked well at all distance scales except $\sim 10h^{-1}$Mpc, where h is the Hubble constant in units of 100 km·s^{-1}·Mpc^{-1}. If $h = 0.5$ and $\Omega = 1$ (i.e., a critical density universe), the mass of the neutrino required in the CνDM model is $94h^2\Omega F_\nu = 4.7$ eV, for a neutrino fraction of Ω, $F_\nu = 0.20$. If instead this mass is divided between two nearly degenerate neutrino species, the motivation for which is one of the two neutrino mass patterns mentioned in the Introduction, this Cν^2DM model turned out to have a remarkable property [4]. While 4.7 eV in one neutrino species or two makes essentially no difference at very large or very small scales, at $\sim 10h^{-1}$Mpc the larger free-streaming length of the 2.4 eV neutrinos tends to wash out fluctuations and lowers the abundance of clusters. Thus the model (Cν^2DM) with two, 2.4 eV neutrinos fits structure information on all scales. In every aspect of simulations done subsequently to Ref. 4 the two-neutrino dark matter gives the best results. For example, a single neutrino species (as well as low-

Ω models) overproduce void regions between galaxies, whereas the $C\nu^2$DM model agrees with observations.

So far these simulations have been done with a critical density universe with no cosmological constant. While a universe with matter density much less than critical has become very popular at this time, note that the analysis of Ref. 1 disagrees with this view, which is based on a number of dynamical measurements included in that analysis. Nevertheless, the cold + hot model could be in some trouble with the early formation of sufficient bright galaxies, and this would be less of a problem if the matter density were reduced.

INDICATIONS FOR NONZERO NEUTRINO MASS

A neutrino component of dark matter must be present if neutrinos have mass, since $\sim 100/\text{cm}^3$ of each flavor of neutrino remains from the early universe. A brief summary follows of the evidence for such mass, with emphasis on recent developments.

Solar Neutrino Deficit

All solar neutrino experiments observe fewer electron neutrinos (ν_e) than solar models predict. In addition, because the three types of experiments cover different ν_e energy ranges and hence sample differently the contributions from the various nuclear processes producing neutrinos, there is an energy-dependent discrepancy well illustrated in Fig. 1(a). This figure, from a very complete review [7], shows the relationship between neutrino fluxes from ^{7}Be and ^{8}B neutrinos as measured in the three types of experiments. The SAGE [8] and GALLEX [9] radiochemical experiments go to the lowest energy and hence measure all of both fluxes (designated "Ga"), while the Homestake [10] radiochemical experiment measures all of the ^{8}B spectrum but only part of the ^{7}Be flux (labeled as "Cl"), and the Kamiokande [11] and Super-Kamiokande [12] scattering experiments measure only ^{8}B flux (designated as "Kam"). Results from all three actually intersect at a negative value of the ^{7}Be flux, yet ^{8}B is produced from ^{7}Be + p $\rightarrow$ ^{8}B + γ. This problem cannot be avoided by one of the experiments being wrong. The discrepancy between a standard solar model [13] and all three types of experiments is shown by the point with error bars in the upper right-hand corner indicating predicted fluxes. Solar models which drastically change solar properties (some of which are shown) do not solve the problem, and these models are severely constrained by very accurate helioseismology measurements.

A good solution to the solar ν_e deficit is provided by oscillation into ν_μ, ν_τ, or ν_s, a sterile neutrino. While this can be a vacuum oscillation, requiring a mass-squared difference $\Delta m^2 \sim 10^{-10}$ eV^2 and large mixing between ν_e and the other neutrino, more favored is a matter-enhanced MSW [14] type of oscillation. For a ν_μ or ν_τ final state, $\Delta m_{ei}^2 \sim 10^{-5}$ eV^2 and mixings either $\sin^2 2\theta_{ei} \sim 6 \times 10^{-3}$ or

~ 0.6 are possible, while only the former is allowed for ν_s. The main change as a result of the new Super-Kamiokande data is that the lack of a day-night effect has reduced the parameter space for the large-angle solution for the ν_μ or ν_τ final state [15]. The Super-Kamiokande result which will become of prime importance as the error bars are reduced is shown in Fig. 1(b). This energy spectrum could not only choose among the oscillation solutions—and it is well fit at this stage by the MSW small-angle solution—but also it may be the one of the few means of proving that oscillations are occurring.

Atmospheric Neutrino Deficit

Pions produced in the atmosphere would decay via $\pi \to \mu + \nu_\mu, \mu \to e + \nu_\mu + \nu_e$, so that one would expect $N(\nu_\mu + \bar{\nu}_\mu) = 2N(\nu_e + \bar{\nu}_e)$, with a small correction for K decays. The $(\nu_\mu + \bar{\nu}_\mu)/(\nu_e + \bar{\nu}_e)$ ratio would be observed in underground experiments as $\mu^\pm/e^\pm$, and the result is far from the expected value. Because the calculated $\mu^\pm$ and $e^\pm$ individual fluxes are known to $\sim 15\%$, whereas much of the uncertainty drops out in the ratio, the experiments utilize $R = (\mu/e)_{\text{Data}}/(\mu/e)_{\text{Calc}}$. Values of R for many experiments are shown in Fig. 2(a). While it once appeared that there was a discrepancy between water Cherenkov detectors and tracking calorimeters [16], the Soudan II results [17] agree with those from IMB [18], Kamiokande [19], and Super-Kamiokande [20].

While the statistical evidence for R being less than unity is now quite compelling, it is the angular distributions of the μ and e events which provide the primary evidence that this deviation of R from unity is explained by neutrino oscillations. This non-flat distribution with angle of R was first observed in the high-energy (> 1.3 GeV) event sample from Kamiokande, but has now been confirmed with better statistics in the similar data sample from Super-Kamiokande, as shown in Fig. 2(b). The data fit an oscillation hypothesis, using $\Delta m^2 = 5 \times 10^{-3}$ eV^2, $\sin^2 2\theta = 1$ (as a sample, but not a best, fit) and is far from a non-oscillation, flat distribution. The low-energy (< 1.3 GeV) sample also agrees with the same oscillation parameters, but this should be a much shallower angle dependence, and hence it is statistically less compelling, as is also shown in the figure.

The disappearance of the muon neutrinos could be due to $\nu_\mu \to \nu_\tau$ or $\nu_\mu \to \nu_e$, with $\nu_\mu \to \nu_s$ being unlikely because the large mixing angle would bring the ν_s into equilibrium in the early universe, possibly providing too many neutrinos to get agreement between predictions of nucleosynthesis and observed light element abundances. The Super-Kamiokande observations of e and μ compared to calculated fluxes, as well as the individual e and μ angular distributions, as shown in Fig. 3(a), makes $\nu_\mu \to \nu_e$ very unlikely. Note the e distributions are like the non-oscillation Monte Carlo, whereas those for μ agree with the oscillation prediction. The recent results of the CHOOZ nuclear reactor experiment [17] shown in Fig. 3(b), which does not see evidence of ν_e disappearing in the appropriate region of Δm^2 and $\sin^2 2\theta$, confirms that the atmospheric effect is very unlikely to

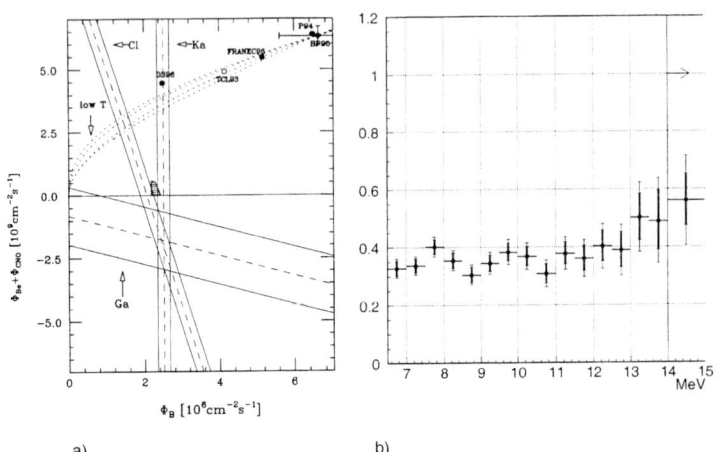

FIGURE 1. a). The ^{8}B and ^{7}Be (+ CNO) neutrino fluxes for standard neutrinos. The dashed (solid) lines correspond to central ($\pm 1\sigma$) experimental values for the chlorine (Cl), gallium (Ga) and Kamiokande (Ka) experiments. The hatched area corresponds to a region within 2σ from each experimental result. The predictions of solar models are shown, with the one [13] with error bars being most often referenced. The dotted lines indicate the behavior of non-standard solar models with low central temperature. b).The Super-Kamiokande solar neutrino energy spectrum with dark error bars for statistics and light error bars including systematics, mainly determined by the energy calibration.

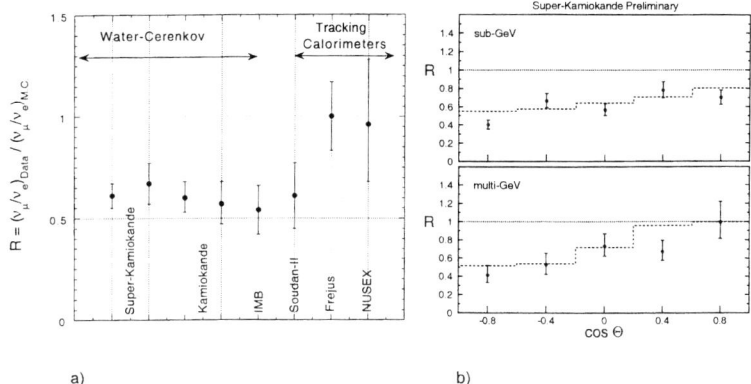

FIGURE 2. a) The double ratio of ν_μ/ν_e from data to that from calculations for atmospheric neutrinos for various detectors. Note that Super-Kamiokande and Kamiokande each have two independent data sets, above and below 1.3 GeV; b) The ratio $(\mu/e)_{\text{DATA}}/(\mu/e)_{\text{MC}}$ for sub-GeV and multi-GeV atmospheric neutrino samples from Super-Kamiokande, as a function of zenith angle. Neutrinos coming from below are at $\cos\Theta = -1$. The dashed line shows the expected shape for $\nu_\mu \to \nu_\tau$ oscillation with $\sin^2 2\theta = 1$ and $\Delta m^2 = .005$ eV^2.

be $\nu_\mu \to \nu_e$. On the basis that the Super-Kamiokande observed values of R and angular distributions of R are due to $\nu_\mu \to \nu_\tau$, the likely value of Δm^2 is definitely much larger than that required for an explanation of the solar neutrino deficit, and the flavors of neutrinos cannot be the same in the two cases. Turning now to the third possible manifestation of neutrino mass, we shall see that the atmospheric Δm^2 is much smaller than that required for the LSND experiment, and hence that three distinctly different values of neutrino mass differences are required.

Evidence for Neutrino Oscillations from the LSND Experiment

The LSND accelerator experiment uses a decay-in-flight ν_μ beam of up to ~ 180 MeV from $\pi^+ \to \mu^+ \nu_\mu$ and a decay-at-rest $\bar\nu_\mu$ beam of less than 53 MeV from the subsequent $\mu^+ \to e^+ \nu_e \bar\nu_\mu$. The 1993+1994+1995 data sets included 22 events of the type $\bar\nu_e p \to e^+ n$, based on identifying an electron between 36 and 60 MeV using Cherenkov and scintillation light and tightly correlated with a γ ($< 0.6\%$ accidental rate) from $np \to d\gamma$ (2.2 MeV), whereas only 4.6 ± 0.6 such events were expected from backgrounds [22]. The chance that these data, using a water target, result from a fluctuation is $< 10^{-7}$. Subsequent data sets from 1996+1997 taken with an iron target gave a similar oscillation probability with much worse statistical accuracy. More importantly, the first data sets (1993-5) yielded events from π decay in flight consistent with being from $\nu_\mu \to \nu_e$. These were similar in number to those from $\bar\nu_\mu \to \bar\nu_e$, but with about twice the background, since the observed process ($\nu_e C \to e^- X$) gave only one signal instead of two. While the fluctuation probability in this case is only $\sim 10^{-2}$, the two ways of detecting oscillations are essentially independent [23].

While the $\nu_\mu \to \nu_e$ results are consistent with those from $\bar{\nu}_\mu \to \bar{\nu}_e$, only the latter have sufficient statistics to provide restrictions on the value of Δm^2. These $\bar{\nu}_\mu$ results interpreted as a two-generation oscillation have been presented [22] in a plot like Fig. 4(a), except that comparisons were made to limits from other experiments. Figure 4(a) is the correct way to determine favored regions of Δm^2 as a function of the mixing angle, θ. The plot utilizes all the information about the events, in particular the neutrino energy, E, and the distance of the event from the source, L. In order to increase the range of L/E, values of E down to 20 MeV were used. Figure 4(a) shows contours at 2.3 and 4.5 log-likelihood units from the maximum. If this were a gaussian distribution, which it is not (its integral being infinite), the contours would correspond to 90% and 99% likelihood levels, but in addition they have been smeared to account for some systematic errors. Comparison to the KARMEN experiment [24], which presents results in a similar way, shows no conflict, but if limits are plotted (as they are in Ref. 22) on this graph from E776 at BNL [25] and the Bugey reactor experiment [26], then one might conclude that the only allowed Δm^2 region is 0.2–3 eV^2. If instead an 80% confidence level band is calculated to compare with the 90% confidence level limits of those experiments using, as they do, just numbers of events (i.e., not using the L/E information) and using only the 36–60 MeV region with its much lower background, then there is no conflict with other experiments above 0.2 eV^2, up to the recent limit of about 10 eV^2 from the NOMAD experiment [27], as shown in Fig. 4(b). Note that the prominent 6–8 eV^2 region is that required for a critical matter density universe.

PATTERN OF NEUTRINO MASSES REQUIRED BY EXPERIMENTS

Because measurements of the width of the Z^0 boson require that there be only three light neutrinos coupled to the Z^0, it would be desirable to explain the phenomena described in the previous section in terms of oscillations among those three neutrinos. Since the flavors are constrained, one has to invoke indirect neutrino oscillations, so LSND could be observing $\nu_\mu \to \nu_\tau \to \nu_e$, for example, with the largest (dominant) Δm^2 being between ν_μ and ν_τ or ν_e and ν_τ, with a small $\Delta m^2_{e\mu}$. There is still the problem that three neutrinos provide only two mass-squared differences, so one might assume $\Delta m^2_{\text{solar}} \approx \Delta m^2_{\text{atmos.}}$, as did Acker and Pakvasa [28], requiring both processes to be dominantly $\nu_e \rightleftharpoons \nu_\mu$. This leads to requiring the solar ν_e deficit to be energy independent, in conflict with the data (see, e.g., Fig 1). This is a problem besetting most three-neutrino schemes. The Acker-Pakvasa model is essentially ruled out by the CHOOZ result of Fig. 3(b), as well as the angular distributions of Fig. 3(a).

The only other alternative for three neutrinos is making $\Delta m^2_{\text{atmos.}} = \Delta m^2_{\text{LSND}}$.

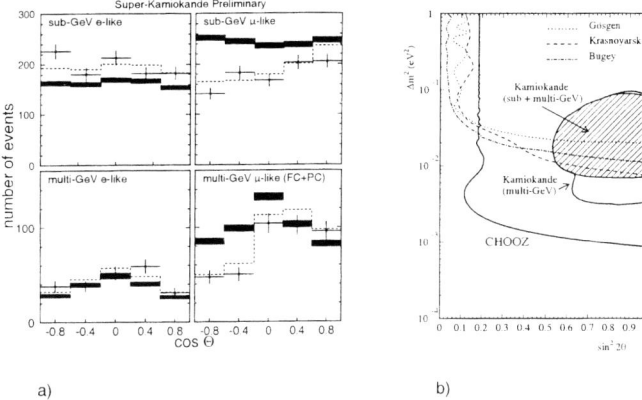

FIGURE 3. a) The rate of μ-like and e-like events in the sub-GeV and multi-GeV atmospheric neutrino samples from Super-Kamiokande, as a function of zenith angle. The solid histograms are the Monte Carlo expectation with no neutrino oscillation; the thickness represents the statistical uncertainty in the Monte Carlo sample. The dashed line shows the expected shape for $\nu_\mu \to \nu_\tau$ oscillation with $\sin^2 2\theta = 1$ and $\Delta m^2 = .005$ eV^2. b)The 90% C.L. exclusion plot for CHOOZ, compared with previous experimental limits and with the Kamiokande allowed region for ν_e disappearance.

This was suggested by Cardall and Fuller [3], who used the indirect oscillation for LSND and made $\Delta m^2_{e\tau} \approx \Delta m^2_{\mu\tau} \approx 0.3$ eV^2, since the solar $\Delta m^2_{e\mu} \approx 10^{-5}$ eV^2 is so small. This scheme has no difficulty with the solar data, but could be in some conflict with limits from neutrinoless double beta decay, if one wants to provide a hot dark matter component, since $m_{\nu_e} \approx m_{\nu_\mu} \approx m_{\nu_\tau} \approx 1.6$ eV, in the first scheme in the Introduction. This scheme, the three-neutrino pattern having the least conflict with data [29], is definitely ruled out by the Super-Kamiokande data shown in Figs. 2(b) and 3(a). There is no way the Δm^2 required by LSND can be the same as that needed for the atmospheric anomaly.

Either the experiments are wrong or one is forced to invoke another light neutrino, a sterile one which does not have the normal weak interactions and hence does not couple to the Z^0. Then the solar ν_e deficit is a result of $\nu_e \to \nu_s$, with $\Delta m^2_{es} \lesssim 10^{-5}$ eV^2, the atmospheric ν_μ/ν_e ratio is explained by $\nu_\mu \to \nu_\tau$, with $\Delta m^2_{\mu\tau} \sim 10^{-2}-10^{-3}$ eV^2, and the LSND observation is caused by $\nu_\mu \to \nu_e$, with 0.2 eV$^2 \lesssim \Delta m^2_{e\mu} \lesssim 10$ eV^2. This is just the second mass scheme in the Introduction, which gave such good results with the ν_μ and ν_τ sharing the role of hot dark matter.

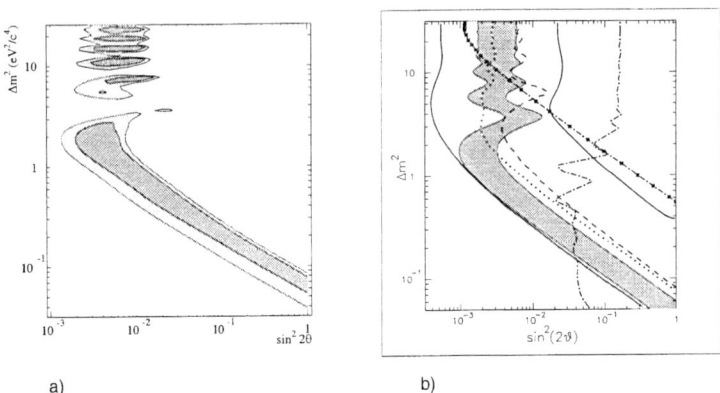

FIGURE 4. a) Mass-squared difference (Δm^2) vs. degree of mixing ($\sin^2 2\theta$) assuming a two-neutrino oscillation explanation of the LSND beam-excess data. Shown are regions of Δm^2 favored using the energy (from 20 to 60 MeV) and distance from the source of each event.b) As in (a) but the LSND $\bar{\nu}_\mu$ data here provide an 80% C.L. band, which is to be compared with the LSND ν_μ result (solid lines), KARMEN (dashes), E776 (dots), Bugey (dash-dot), and NOMAD.

HEAVY-ELEMENT NUCLEOSYNTHESIS IN SUPERNOVAE

Further evidence for two-neutrino dark matter is provided indirectly if a sterile neutrino exists, and the likelihood of this is enhanced by such a neutrino's providing the only robust solution to the problems posed by the production of heavy elements by the r process of rapid neutron capture in the outer neutrino-heated ejecta of Type II supernovae. There are three problems discussed below in inverse order of their importance.

First, the LSND results could be in conflict with the existence of the r process, which would seem to place a limit on the mixing of ν_μ and ν_e. Energetic ν_μ ($\langle E \rangle \approx 25$ MeV) coming from deep in the supernova core could convert by an MSW transition [14] to ν_e inside the region of the r process, producing ν_e of much higher energy than the thermal ν_e ($\langle E \rangle \approx 11$ MeV). The latter, because of their charge-current interactions, emerge from farther out in the supernova where it is cooler. Since the cross section for $\nu_e n \to e^- p$ increases as the square of the energy, these converted energetic ν_e would deplete neutrons, stopping the r process. Calculations [30] of this effect limit $\sin^2 2\theta$ for $\nu_\mu \to \nu_e$ to $\lesssim 10^{-4}$ for $\Delta m^2 \gtrsim 2$ eV^2, in possible conflict with compatibility between the LSND result and a neutrino component of dark matter.

The other two problems are with the existence of the r process itself. The less serious of these is that recent calculations have shown that the region in which the process needs to occur is not sufficiently neutron rich. To this difficulty of too low an entropy (i.e., too few neutrons per seed nucleus, like iron) has very recently been added knowledge of an apparently fatal problem [31]: all available protons swallow up neutrons to form α particles, following which $\nu_e n \to e^- p$ reactions both destroy more neutrons and create protons, which again create more α particles, and so on. Essentially no nuclei above $A = 95$ can be produced.

The sterile neutrino would produce two effects [32]. First, there is a zone, outside the neutrinosphere (where neutrinos can readily escape) but inside the $\nu_\mu \to \nu_e$ MSW ("LSND") region, where the ν_μ interaction potential goes to zero, so a $\nu_\mu \to \nu_s$ transition can occur nearby, depleting the dangerous high-energy ν_μ population. Second, because of this ν_μ reduction, the dominant process in the MSW region reverses, becoming $\nu_e \to \nu_\mu$, dropping the ν_e flux going into the r-process region, hence reducing $\nu_e n \to e^- p$ reactions and allowing the region to be sufficiently neutron rich. This rescuing scenario—the only robust one which has been found after many attempts—works even better if the MSW region is inside the radius at which the weak interactions freeze out, which is certainly the case if Δm^2 is as large as 6 eV^2.

LIGHT-ELEMENT NUCLEOSYNTHESIS IN THE EARLY UNIVERSE

The sterile neutrino might also ameliorate the apparent incompatibility of determinations of the baryon-to-photon ratio from ^{4}He abundance and D/H ratio [33]. If the potential is appropriate so that $\bar{\nu} \to \bar{\nu}_s$ transitions occur instead of $\nu \to \nu_s$, such an MSW transition could lead to a significant excess of ν_e over $\bar{\nu}_e$, so that the n/p ratio (and hence ^{4}He) would be depleted prior to the decoupling of the $\nu_e n \to e^- p$ reaction. The scarcity of initial sterile neutrinos, which are produced only via mixing with active ones, makes the dominant MSW transition active→sterile and not the other way around. The small mass difference of the solar case makes $\bar{\nu}_e \to \bar{\nu}_s$ have a negligible effect, but $\bar{\nu}_\mu \to \bar{\nu}_s$ and $\bar{\nu}_\tau \to \bar{\nu}_s$ with $\Delta m^2 \sim 6$ eV^2 could create a large lepton asymmetry which would be transferred to ν_e via $\nu_\mu \to \nu_e$ and $\nu_\tau \to \nu_e$. Calculations [34] show that this effect helps, but a bigger effect would be produced if there were a sterile neutrino for each generation of active neutrinos.

CONCLUSIONS

If all present indications for neutrino mass are correct—solar and atmospheric neutrino deficits, the LSND results, and the need for some hot dark matter—the most likely scenario requires the ν_μ and ν_τ to be that component of dark matter and for at least one light sterile neutrino to exist. Further indications for the sterile neutrino, and hence for this hot dark matter, come from its role in rescuing the r process of heavy element production by supernovae and possibly in helping the present lack of concordance in the primordial abundance of ^{4}He and the D/H ratio.

ACKNOWLEDGEMENTS

This work was supported in part by the U.S. Department of Energy. Special thanks go to collaborators on various portions of this material, George Fuller, Rabindra Mohapatra, Joel Primack, and Yong-Zhang Qian.

REFERENCES

1. E. Gawiser and J. Silk (to appear May 1998 in *Science*).
2. D.O. Caldwell, *Perspectives in Neutrinos*, Atomic Physics and Gravitation, Editions Frontières, Gif-sur-Yvette, France p. 187; D.O. Caldwell and R.N. Mohapatra, *Phys. Rev.* D **48**, 3259 (1993); *Phys. Rev.* D **50**, 3477 (1994); J.T. Peltoniemi and J.W.F. Valle, *Nucl. Phys.* B **406**, 409 (1993).
3. C. Cardall and G. Fuller, *Phys. Rev.* D **53**, 4421 (1996).
4. J.R. Primack, J. Holtzman, A. Klypin, and D.O. Caldwell, *Phys. Rev. Lett.* **74**, 2160 (1995).

5. A. Klypin, J. Holtzman, J.R. Primack, and E. Regös, *Astrophys. J.* **416**, 1 (1993).
6. A. Klypin, S. Borgani, J. Holtzman, and J.R. Primack, *Astrophys. J.* **444**, 1 (1995).
7. V. Castellani *et al.*, *Phys. Rep.* **281**, 309 (1997).
8. J.N. Abdurashitov *et al.*, *Phys. Lett.* B **328**, 234 (1994); *Nucl. Phys.* B **38**, 60 (1995); *Phys. Rev. Lett.* **77**, 4708 (1996).
9. P. Anselmann *et al.*, *Phys. Lett.* B **327**, 377 (1994); *Phys. Lett.* B **342**, 440 (1995); *Phys. Lett.* B **357**, 237 (1995); W. Hampel *et al.*, *Phys. Lett.* B **388**, 384 (1996).
10. B.T. Cleveland *et al.*, *Astrophys. J.* **496**, 505 (1998).
11. Y. Fukuda *et al.*, *Phys. Rev. Lett.* **77**, 1683 (1996).
12. Y. Totsuka, *Proc. Lepton-Photon Symposium*, Hamburg 1997 (in press).
13. J.N. Bahcall and M.H. Pinsonneault, *Rev. Mod. Phys.* **67**, 781 (1995).
14. L. Wolfenstein, *Phys. Rev.* D **17**, 2369 (1978); *Phys. Rev.* D **20**, 2634 (1979); S.P. Mikheyev and A. Yu. Smirnov, *Sov. J. Nucl. Phys.* **42**, 913 (1985); *Nuovo Cimento* **9C**, 17 (1986).
15. N. Hata and P. Langacker, *Phys. Rev.* D **56**, 6107 (1997); N. Hata has done an update for $\nu_e \to \nu_s$, showing little change.
16. K. Daum *et al.*, *Z. Phys.* C **66**, 417 (1995); M. Aglietta *et al.*, *Europhys. Lett.* **8**, 611 (1989).
17. W.W.M. Allison *et al.*, *Phys. Lett.* B **391**, 491 (1997); H. Gallagher, *WIN '97* (Proc. of Weak Interactions and Neutrino Workshop, Capri, Italy, 1997, in press).
18. R. Becker-Szendy *et al.*, *Phys. Rev.* D **46**, 3720 (1992).
19. Y. Fukuda *et al.*, *Phys. Lett.* B **335**, 237 (1994).
20. Y. Fukuda *et al.*, *Phys. Lett. B* (1998, accepted for publication); *ibid.* hep-ex/9805006. The figures used here are from E. Kearns and will appear in *Proc. of TAUP '97*, LNGS, Assergi, Italy (1997, in press).
21. M. Apollonio *et al.*, *Phys. Lett.* B **420**, 397 (1998).
22. C. Athanassopoulos *et al.*, *Phys. Rev.* C **54**, 2685 (1996); *Phys. Rev. Lett.* **77**, 3082 (1996).
23. C. Athanassopoulos *et al.*, nucl-ex/9706006 (submitted to *Phys. Rev. C*), nucl-ex/9709006 (submitted to *Phys. Rev. Lett.*).
24. G. Drexlin *et al.*, *Prog. in Part. and Nucl. Phys.* **32**, 375 (1994); B. Armbruster *et al.*, *Nucl. Phys.* B (Proc. Suppl.) **38**, 235 (1995).
25. L. Borodovsky *et al.*, *Phys. Rev. Lett.* **68**, 274 (1992).
26. B. Achkar *et al.*, *Nucl. Phys.* B **434**, 503 (1995).
27. L. DiLella, *Proc of TAUP '97*, LNGS, Assergi, Italy (1997, in press).
28. A. Acker and S. Pakvasa, *Phys. Lett.* B **397**, 209 (1997).
29. P. Krastev and S. Petcov, *Phys. Lett.* B **397**, 69 (1997); G.L. Fogli, E. Lisi, D. Montanino, and G. Scioscia, *Phys. Rev.* D **56**, 4365 (1997).
30. Y-Z. Qian *et al.*, *Phys. Rev. Lett.* **71**, 1965 (1993); Y.-Z. Qian and G.M. Fuller, *Phys. Rev.* D **51**, 1479 (1995); G. Sigl, *Phys. Rev.* D **51**, 4035 (1995).
31. G.M. Fuller and B.S. Myer, *Astrophys. J.* **453**, 202 (1995); B.S. Myer, G.C. McLaughlin, and G.M. Fuller (in preparation, 1998).
32. D.O. Caldwell, G.M. Fuller, and Y.-Z. Qian (to be submitted to *Phys. Rev. Lett.* 1998).
33. For a recent review, see G. Steigman, astro-ph/9803055 (1998, unpublished).

34. N.F. Bell, R. Foot, and R.R. Volkas, preprint UM-P-98/17 (1998); R. Foot and R.R. Volkas, *Phys. Rev.* D **56**, 6653 (1997).

The Pierre Auger Project: An Observatory for Measuring Extremely High-Energy Cosmic Rays

D. Zavrtanik

School of Environmental Sciences, Nova Gorica and
J. Stefan Institute, Ljubljana, Slovenia
(danilo.zavrtanik@ses-ng.si)

for the P. AUGER Collaboration

CONAE - IAFE, Buenos Aires; CRICyT, Mendoza; Universidad Nacional de La Plata; IAR, Villa Elisa; TANDAR-CNEA, Buenos Aires; Centro Atómico Bariloche; Yerevan Physics Institute; University of Adelaide; University of La Paz; University of Campinas; University of Sao Paulo; CBPF-Lafex, Rio de Janeiro; Federal University Rio de Janeiro; IHEP, Beijing; ENST, Paris; LTF, Observatoire de Besancon; Collège de France Paris; LPNHE, Université Paris 6; IK1 Forschungszentrum Karlsruhe; National Technical University of Athens; University of Milano; University of Roma 2; University of Torino; ICRR Tokyo; BUAP Puebla; CINVESTAV-IPN, Mexico; UNAM, Mexico; UMSNH, Morelia; Universidad de Guanajuato, León; Jagiellonian University, INP, Krakow; University of Lodz; MEPhI Moscow; J. Stefan Institute, Ljubljana; School of Environmental Sciences, Nova Gorica; University of Leeds, University of Colorado, Boulder; E. Fermi Institute, University of Chicago; Fermilab, Batavia; Louisiana State University, Baton Rouge; University of Michigan, Ann Arbor; University of New Mexico, Albuquerque; Notheastern University, Boston; Pennsylvania State University, University Park; University of Utah, Salt lake city; DNRI-VAEC, Dalat; HINCST-VAEC, Hanoi.

Abstract. We present the scientific motivation and conceptual design of the P. Auger Observatory. Two giant ground arrays of water Čerenkov tanks overlooked by fluorescence detectors will cover an area of 3000 km^2 each. They will be build in the Southern and Northern hemisphere to provide full sky coverage. The total aperture of 14000 $km^2 sr$ will allow to study all observable aspects of cosmic rays from below 10 EeV up to arbitrarily high energies with an unprecedented accuracy.

Introduction

Cosmic rays were discovered more than 80 years ago by V. Hess. Since that time experiments have shown that cosmic rays are predominantly atomic nuclei ranging in energy from less than 10^6 eV to more than 10^{20} eV. The fluxes of cosmic rays range from more than 1 $cm^{-2}s^{-1}$ at 100 MeV to approximately 1 $km^{-2}century^{-1}$

above 100 EeV. With the help of Earth's magnetic field it was established that cosmic rays are positively charged particles.

In 1938 P. Auger demonstrated that cosmic rays spectrum extends beyond 10^{15} eV [1]. At that time the highest energy particles produced by radioactivity or artificial acceleration were less than 10^7 eV. Today there is a plausible understanding that cosmic rays up to 10^{15} eV are produced by shock waves of supernova explosions and the sources lie in our galaxy.

Following the Auger's work the energy spectrum of the cosmic rays was measured to higher energies by Rossi, Clark and co-workers [2]. The experimental efforts culminated in the large cosmic rays detector built by J. Linsley and collaborators at Volcano Ranch in New Mexico, USA. This detector was an array of scintillation counters spread over an area of 10 km^2. The cosmic rays were detected by the air showers produced by the primary particle in the atmosphere. In 1963 they reported a shower produced by a cosmic ray of energy 10^{20} eV [3].

Recently a dozen of cosmic ray events have been reported well in excess of 10^{20} eV [4], [5]. Since there have been no published criticism of the determination of the energy of these events it is almost impossible that the detected events could result from experimental biases, artifacts or calibration errors. The fact that there is no easy explanation of these events calls for new high statistics experiments.

At present the largest operating detector is AGASA array in Japan [5] with an acceptance of 125 $km^2 sr$ for cosmic rays above 10^{19} eV. Now under construction in Utah is the High Resolution Fly's Eye, which when complete in 1999 will have a time averaged acceptance of 1000 $km^2 sr$ above 10^{20} eV. The P. Auger collaboration is aiming at building two observatories (one in the northern and one in the southern hemisphere) for a high statistics measurement of extremely high-energy cosmic rays with an overall acceptance of 14000 $km^2 sr$ in order to resolve the cosmic ray puzzle in the energy region above the 10^{20} eV.

Scientific Motivation

The energy spectrum of cosmic rays expands over 13 orders of magnitude in energy and 34 orders of magnitude in flux (Fig. 1) and exhibits a simple power law almost over the whole energy range. However, at energies above $10^{18.5}$ eV an unexpected structure in the spectrum can be observed (see Fig. 2) despite the fact that statistics is rather poor due to extremely low fluxes. Matching the observations to production and propagation models has proven to be very difficult, however. In the known astrophysical context, it is a challenge to explain the acceleration of particles to the energies above 10^{20} eV. The suggested models for acceleration fall into the following classes:

- Diffusive shock acceleration (first order Fermi acceleration) in extended objects such as in the lobes of radio galaxies or in clusters of galaxies;

- Acceleration in strong fields associated with accretion disks and compact rotating objects;
- Acceleration in catastrophic events.

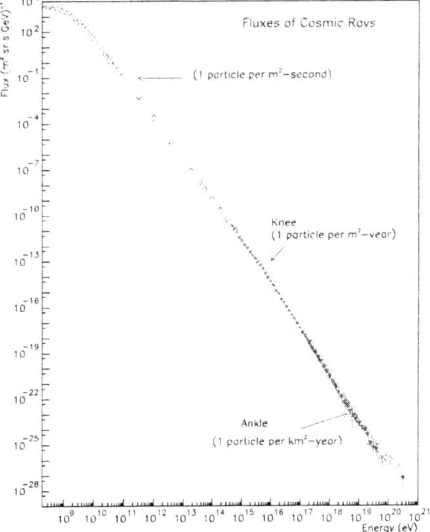

FIGURE 1. Differential energy spectrum of the cosmic rays with some indications of integrated fluxes per steradian [6].

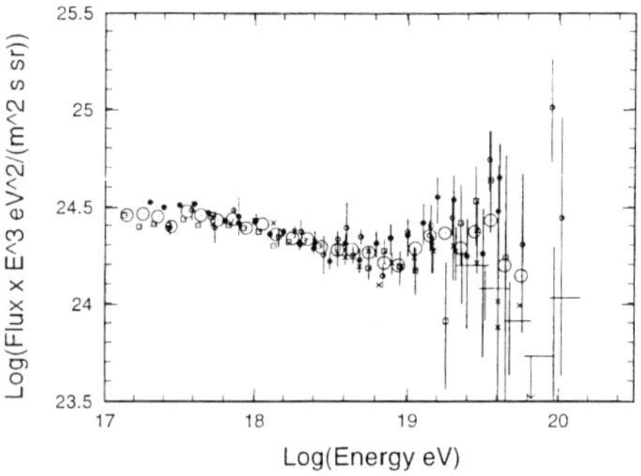

FIGURE 2. Detailed cosmic ray flux multiplied by E^3 at extremely high energies [7].

Perhaps, the favored mechanism is the Fermi acceleration operating in extended sources [8]. For a variety of acceleration mechanisms, including Fermi acceleration, the maximum possible energy obtained by a particle of charge Ze is:

$$E_{max} \approx \beta \times Ze \times B \times L \qquad (1)$$

where β is a characteristic relative velocity within the system, and B and L field strength and the size of the accelerating region, respectively (see Ref. [9]). Fig. 3 shows B and L for a number of candidate astrophysical sources of high-energy cosmic rays. If one looks for the sites where a proton can reach the energy of 100 EeV, one can see that very few possibilities remain above the corresponding diagonal line on Fig. 3. These are: radio galaxies in the large L - low B region, neutron stars in low L - large B region and active galactic nuclei (AGN) in between. Although the acceleration mechanism is different, these are the most powerful candidate accelerators in our Universe. However, there is no general agreement in the community that any one of these accelerators is able to reach energies in the ZeV region. The most extreme predictions usually limit the accelerating power of such systems to a maximum of 10 to 100 EeV. However, some authors do not exclude the possibility that diffusive shock acceleration in quasars and radio galaxies may give a proton an energy around 1 ZeV [10].

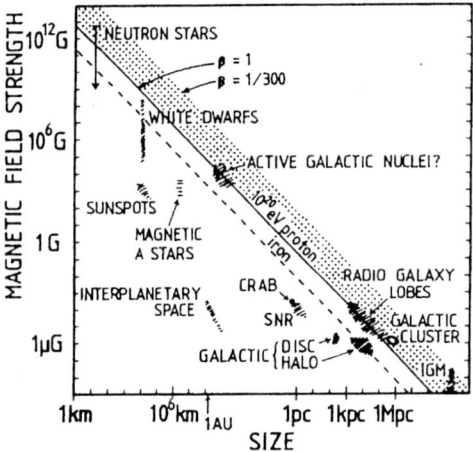

FIGURE 3. Various candidate astrophysical particle accelerators presented as a function of their magnetic fields and spatial extensions [9].

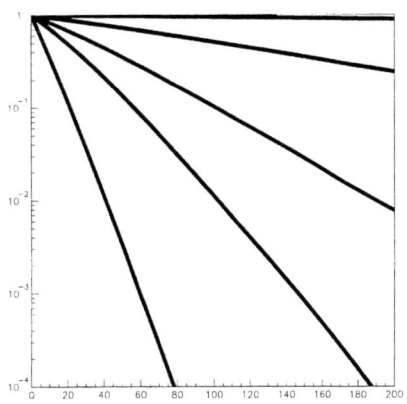

FIGURE 4. Probability (y axis) that a detected particle has traveled farther than a distance D in Mpc (x axis). The curves in descending order pertain to primary energies of E = 40, 60, 80, 100, 300 Eev.

The maximum possible energy of a particle exiting the source may be lower than suggested above due to energy loss processes. For example, synchrotron radiation is important for protons of extremely high energies in the regions where the magnetic field strength is high. Losses due to photo-pion production are important in compact regions with intense radiation fields.

Soon after the discovery of the cosmic microwave background (CMB), it was pointed out by Greisen [11], Zatsepin and Kuz'min [12] that CMB may render the Universe opaque to cosmic rays of sufficiently high-energies. The threshold for pion photo-production (via Δ resonance) through the reaction of protons or neutrons on CMB photons occurs at the energy of approximately $10^{19.7}$ eV. Successive interactions reduce the energy of a nucleon until it falls below threshold. This effect is known as a GZK cutoff. Characteristic attenuation lengths for the protons at the highest energies are of the order of 10 Mpc or approximately 30 million light-years which is a small distance on the cosmological scale. Energetic nuclei and photons also loose energy by photo-disintegration and pair production, respectively. Only neutrinos or unknown neutral weakly interacting particles would be immune but their interaction probability with atoms in the atmosphere would also be negligible unless the cross-section becomes strong [13]. Thus, there exist limits to the distance to the sources of the most energetic particles. For any observed cosmic ray energy, the distance limit is nearly independent of the initial particle energy, if it is above the GZK cutoff. Fig. 4 shows the probability that a cosmic ray detected at energy E would have traveled a distance D (in Mpc). In particular, for particles detected with the energy equal to 300 EeV, only 4 % of them should have traveled farther than 30 Mpc, and fewer than 0.4 % of them farther than 50 Mpc. GZK cutoff should be manifested in the observed spectrum as a strong suppression near $10^{19.7}$ eV, with a possible pile-up in the differential spectrum just below that energy threshold. The detailed shape of the spectrum will depend on the source distribution in space and time as well as the initial production spectra. On the other hand, the evidence of beyond GZK cosmic rays raises questions on how the cutoff could be violated.

A way to account for super-GZK cosmic rays without invoking dominance by nearby sources is to suppose that particles are produced with an extremely hard source spectrum. Such scenarios have not so far been proposed in the context of bottom-up acceleration mechanisms. But, they do occur in top-down acceleration processes like topological defect annihilations [10]. In that picture, the spectrum results from the decay of super-massive X particles that are radiated by the topological effects formed in the early universe as a product of spontaneous symmetry breaking implicit in some Grand Unified Theories [14].

Auger Observatory

The Auger Observatory is designed to make a definitive study of the mysterious cosmic ray events in the energy region above 10^{20} eV. The most important requirement of a dedicated experiment is to accumulate statistics. The detector must be able to measure the energy of the incoming cosmic rays with a resolution of approximately 10 %. It must also have a good resolution on measuring the direction of the cosmic ray since we would like to open the window to cosmic ray astronomy. And finally, the identity of the incident particles must be determined at least on the statistical, if not individual, basis.

Method

An incoming cosmic ray interacts in the upper atmosphere producing a hadronic cascade which can build up to some 10^{11} particles when the primary particle has an energy of 10^{20} eV. At each generation about 1/3 of the available energy is transferred to an electromagnetic cascade through the prompt decay of neutral pions. The hadronic cascade proceeds until the charged pion energy is degraded to the point ($\approx$ 20 GeV) that decay to muons become more probable than further interactions. This means that the predominant particles in the shower are electrons, photons and muons. At approximately 1500 m above the sea level more than 50 % of the shower particles are within the Molière radius of 80 m from the shower core. Far from the core ($\geq$ 500 m) the energies of the EM particles are around 10

MeV while the muons carry approximately 1 GeV of energy. At large distances from the shower axis the particles arrival is spread out in time by more than 1 μs. An array of particle detectors measures a 2 dimensional spatial distribution of particles in the shower and a time distribution of the arrival of the particles in the shower as it intersects the ground plane. The surface detectors are placed far apart from one another, sampling the shower far from its core. The ground array is in fact a sampling calorimeter but with a single detector layer located near the shower maximum X_{max}. In this case the atmosphere is analogous to the absorber of the calorimeter. The determination of the shower energy and the direction of the shower axis, i.e. the energy and direction of the primary cosmic ray, is obtained by simultaneously fitting the measured lateral distribution of EM particles and muons and their arrival times. In addition, the proportion of muons with regard to the EM compound provides information about the identity of the primary cosmic ray.

Charged particles in the shower produce fluorescence light in the range between 300 and 400 nm as they pass through the atmosphere. The amount of light produced at each atmospheric depth is proportional to the shower size, i.e. the number of charged particles. A fluorescence detector which images the sky onto an array of photomultipliers can directly measure the longitudinal development of the shower. The integral of the longitudinal profile is a calorimetric measure of the total shower energy. Fluorescence detector can also measure the position of shower axis and provide information on the identity of the primary particle through the observation of the depth (X_{max}) at which the shower maximum occurs.

Experimental setup

The Auger Observatories are designed to work in a hybrid mode, employing fluorescence detectors overlooking ground arrays. During clear moon-less nights ($\approx$ 10 % of the time) all registered events will be observed by the fluorescence light given off by nitrogen in the atmosphere and also by particle detectors at ground level. The fluorescence detectors will be similar in design to the Hi-Res Fly's Eye detector [15]. The surface array will be made up of water Čerenkov tanks placed approximately 1.5 km apart, and resembles the array successfully operated by the Haverah Park group for more then 20 years [16]. Fig. 5 shows the reference layout of one of the P. Auger Observatory sites. The Auger Observatory will be installed in the southern (province of Mendoza, Argentina) and northern (Millard County, Utah, USA) hemispheres in order to provide full sky coverage. The southern hemisphere detector is particularly interesting since very few detectors took data in the past in this part of the world from where the direction of the center of our galaxy is visible. Each site will be equipped with an array of ground detectors covering an area of about 3000 km^2, each of them being overlooked by three fluorescence detectors in order to provide full hybrid detection of incoming high energy cosmic rays.

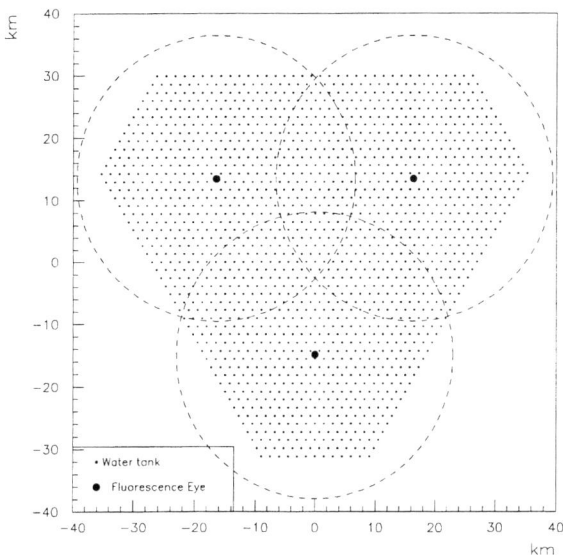

FIGURE 5. Reference layout for one of the Auger sites.

Water Čerenkov Ground Array. The individual stations of the ground array will be water Čerenkov tanks (see Fig. 6) with a round surface of 10 m^2 and a height of 1.2 meters.

Čerenkov radiation emitted by charged particles penetrating the detector will be read out by three photomultipliers. The output signal will be digitized by flash ADCs, with the aim to separate the EM signal (low energy electrons and photons) from the muons crossing the tank. The relative synchronization of the detectors will be done using the GPS (Global Positioning System) satellites with a precision of a few ns. The communication between the stations and the data transfer to a central computer will be achieved by using radio signals by methods similar to cellular telephone techniques. The stations will be powered by solar panels and batteries which will allow the operation in a stand-alone mode.

Fluorescence detector. The reference design for the fluorescence detector is based on tested Fly's Eye detector [15]. However, there are some minor differences. The pixels are expected to have a 1.5^0 diameter instead of 1.0^0. Each telescope will use 121 pixels instead of 256. With the use of FADC electronics, the pixel size will not impact the longitudinal profile segmentation. The conceptual design of one of the many fluorescence telescopes needed for a 60^0 coverage is seen on Fig. 7. Full description of the P. Auger fluorescence detectors and their simulated performances can be obtained from the Pierre Auger Project Design Report [17].

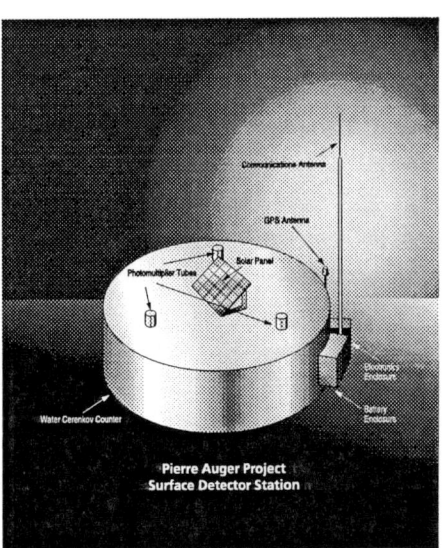 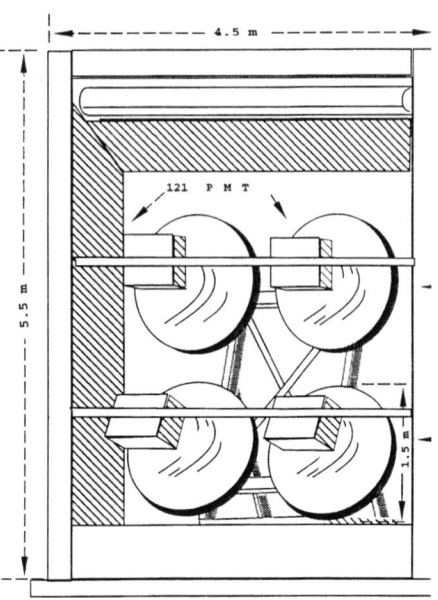

FIGURE 6. Artist's view of a Čerenkov tank. The drawing shows the phototubes, the solar panel, the GPS and telecom antennae.

FIGURE 7. Schematic view of one segment of the fluorescence detector (with its mirrors and PM cameras).

Variations of the above reference design are under study. A promising option is to use Schmidt optics for each telescope. A diaphragm at the mirror's centre of curvature can eliminate the coma aberration that otherwise distorts the spot on the focal surface for off-axis parallel light rays. The elimination of coma enables the field of view to be extended from $15^0 \times 15^0$ to $30^0 \times 30^0$. This would reduce the number of telescopes by a factor of 4.

Performances

We have carried out extensive simulations to predict the performances of the proposed detector. The simulated air showers data were passed through detailed simulations of ground array and fluorescence detectors to determine the experimental resolutions and acceptance of the detector. Results are summarized in Table 1.

The hybrid data set will yield a subset of events which are particularly well measured, having independent data on energy and direction. Resolutions on this events are significantly improved over that of the surface array alone. When coupled with the determination of X_{max}, these will greatly enhance the analysis of cosmic ray composition.

	10^{19} eV		10^{20} eV	
	Surface	Hybrid	Surface	Hybrid
$\Delta\Theta$	2.0^0	0.35^0	1.0^0	0.35^0
$\Delta core$	80 m	29 m	40 m	29 m
$\Delta E/E$	18 %	4.3 %	7 %	2.5 %
ΔX_{max}		17 gcm^{-2}		15gcm^{-2}

TABLE 1. Summary and comparison of reconstruction resolution for the surface array alone and for hybrid data.

Aperture of the Auger detector depends on trigger conditions. It has a nice feature that it is constant above an energy level which is determined by the alerted station multiplicity. In case the trigger requires at least five alerted stations a constant aperture of 14000 $km^2 sr$ above $10^{19.1}$ eV is obtained and it falls to only 10 % at $10^{18.3}$ eV. The spectrum measurement can be easily extended to lower energies by simply changing the trigger conditions. Details can be found in P. Auger Project Design Report [17].

Summary

The reference design of P. Auger Observatory is now fixed. A detailed description can be found elsewhere [17]. All the technical issues are now under the process of optimization and testing. The funding of the project is under way in most of the 20 participating countries. The two sites have been chosen and preparatory work as well as monitoring have already started. The beginning of full construction of the Observatories is planned for 1999 and we expect first hybrid data in the year 2001.

P. Auger Observatory will have the capability to distinguish between currently favored scenarios in astrophysics and cosmology. It is likely that this issues will otherwise remain unsolved.

REFERENCES

1. P. Auger et al., Comptes Rendus **207** (1938) 228;
   P. Auger, Rev. Mod. Phys. **11** (1939) 288.
2. G.W. Clark et al., Phys. Rev. **122** (1961) 637.
3. J. Linsley, Phys. Rev. Lett. **10** (1963) 146.
4. D.J. Bird et al., Astrophys. J. **441** (1995) 144;
   M.A. Lawrence et al., J. Phys. **G17** (1991) 733;
   N. Efimov et al.; Proc. ICRR Int. Symp. on Astroph. of the Most Energetic Cosmic Rays (eds. M. Nagano and F. Takahara) (1991) p20.

5. N. Hayashida et al., Phys. Rev. Lett. **73** (1994) 3491.
6. S. Swordy, private communication. The points are from LEAP, Proton, Akeno, AGASA, Fly's EYE, Haverah Park, and Yakutsk experiments.
7. M. Nagano, private communication. The figure is an update of a similar plot from: S. Yoshida et al., Astropart. Phys. **3** (1995) 105.
8. E. Fermi, Phys. Rev. **75** (1949) 1169.
9. A.M. Hillas, Ann. Rev. Astron. Astrophys. **22** (1984) 425.
10. G. Sigl, D.N. Schramm and P. Bhattacharjee, Astropart. Phys. **2** (1994) 401.
11. K. Greisen, Phys. Rev. Lett. **16** (1966) 748.
12. G.T. Zatsepin and V.A. Kuz'min, JETP Lett. **4** (1966) 78.
13. J. Bordes et al., preprints RAL-TR-97-067 and hep-ph/9711438.
14. C.T. Hill, Nucl. Phys. **B224** (1983) 469.
15. R. Baltrusaitis et al., Nucl. Instr. Meth. **A240** (1985) 410.
16. M.A. Lawrence, R.J.O. Reid and A.A. Watson, J. Phys. **G17** (1991) 733.
17. The P. Auger Project Design Report is available in the World Wide Web at (http://www.ses-ng.si/public/pao/design.html).

Connection between Relic Neutrinos and Cosmic Rays at $\gtrsim 10^{20}$ eV?

Thomas J. Weiler

Department of Physics & Astronomy, Vanderbilt University
Nashville TN 37235
email: weilertj@ctral1.vanderbilt.edu

Abstract. The observation of cosmic–ray events above the Greisen–Kuzmin–Zatsepin (GZK) cutoff of 5×10^{19} eV challenges orthodox modeling. We discuss a possible solution which uses Standard Model (SM) physics augmented only by $\lesssim$ eV neutrino masses as suggested by solar, atmospheric, and terrestrial neutrino detection and the comological preference for a hot dark matter component. In this scheme, cosmic ray neutrinos from distant, highest–energy sources annihilate relatively nearby on the relic–neutrino background to produce "Z-bursts", highly-collimated, highly-boosted ($\gamma_Z \sim 10^{11}$) hadronic jets. ¿From the SM and hot Big Bang cosmology, the probability for each neutrino flavor at its resonant energy to annihilate within the halo of our galactic supercluster is likely within an order of magnitude of 1%. The kinematics of the annihilation are completely determined by the neutrino masses and the properties of the Z boson. The burst energy is $E_R = 4 \,(\text{eV}/m_\nu) \times 10^{21}$ eV, and the burst content includes, on average, thirty photons and 2.7 nucleons with super–GZK energies. Several tests of the neutrino annihilation hypothesis are indicated.

THE COSMIC RAY PUZZLE ABOVE 100 EEV

It has long been anticipated that the highest–energy cosmic primaries would be protons from outside the galaxy, perhaps produced in active galactic nuclei (AGNs) [1]. Since the mid–sixties it was also anticipated that the highest energies for protons arriving at earth would be $\sim 5 \times 10^{19}$ eV. The origin of this Greisen–Kuzmin–Zatsepin (GZK) cutoff [2] is degradation of the proton energy by the resonant scattering process $p + \gamma_{2.7K} \to \Delta^* \to N + \pi$ when the proton is above the resonant threshold for Δ^* production; $\gamma_{2.7K}$ denotes a photon in the $2.7K$ cosmic background radiation. A proton produced at its cosmic source with an initial energy E_p will on average arrive at earth with only a fraction $\sim (0.8)^{D/6\,\text{Mpc}}$ of its original energy. Proton energy is not lost significantly only if the highest–energy protons come from rather nearby sources, $\lesssim 50$ to 100 Mpc [3]. However, no AGN sources are known to exist within 100 Mpc of earth.[1] Any observation of air–shower events above 5×10^{19} eV would challenge the standard theory. The recent discoveries by the AGASA [5], Fly's Eye [6], Haverah Park [7], and Yakutsk [8] collaborations of air–shower events with energies near and above 10^{20} eV challenge the standard theory.

A primary nucleus mitigates the cutoff problem (energy per nucleon is reduced by $1/A$), but it has additional problems: above $\sim 10^{19}$ eV nuclei may be photo-dissociated by the 2.7K background [9], and possibly disintegrated by the particle density ambient at the astrophysical source. Gamma–rays and neutrinos are other possible primary candidates for the highest–energy events. The gamma–ray possibility appears inconsistent with the time–development of the Fly's Eye event, but is not ruled out for this event [10]. However, the mean free path for a $\sim 10^{20}$ eV photon to annihilate to e^+e^- on the radio background is believed to be only 10 to 20 Mpc based on recent estimates of the IR background [11], and the density profile of the Yakutsk event [8] showed a large number of muons which may argue against gamma–ray initiation. Concerning the neutrino possibility, the Fly's Eye event occured high in the atmosphere, whereas the expected event rate[2] for early development of a neutrino–induced air–shower is down from that of an electromagnetic or hadronic interaction by six orders of magnitude [10].

New ideas are welcomed. Here we develop one [12] which uses standard model (SM) particle physics, the relic neutrino background predicted by Big Bang cosmology, and neutrino mass values suggested by oscillation data.

NEUTRINO–RELIC NEUTRINO ANNIHILATION AND Z–BURSTS

It was noted some time ago [14] that the mean free path $\lambda_j = [n_{\nu_j} \sigma_{ann}(\nu_j + \bar{\nu}_j \to Z)]^{-1}$ for a cosmic ray neutrino to annihilate at the Z resonance on a background of nonrelativistic relic antineutrinos (and vice versa) having mass m_j and density n_{ν_j} is only slightly larger than the Hubble size of the Universe, $D_H \equiv c\, H_0^{-1} = 0.9\, h_{100}^{-1} \times 10^{28}$ cm, where h_{100}^{-1} is the Hubble parameter in units of 100 km/s/Mpc. This means that the annihilation probability per cosmic distance of travel may be significant.

The invariant energy–averaged annihilation cross section for the process $\nu_j + \bar{\nu}_j \to Z$ is given by the integral over the Z pole: $\langle \sigma_{ann} \rangle \equiv \int \frac{ds}{M_Z^2} \sigma_{ann}(s)$, with s the square of the energy in the center of momentum frame. The standard model value for this cross section is $\langle \sigma_{ann} \rangle = 4\pi G_F/\sqrt{2} = 4.2 \times 10^{-32} \mathrm{cm}^2$ for each neutrino type j (flavor or mass basis), independent of any neutrino mixing–angles since the annihilation mechanism is a neutral current process. The energy of the neutrino

---

[1]) The suggestion has been made that hot spots of radio galaxies in the supergalactic plane at distances of tens of megaparsecs may be the sources of the super–GZK primaries [4]. Statistical support for this hypothesis is weak at present.

[2]) A dissident neutrino cross–section, tailored to solve the origin of super–GZK air–showers, is given in [13].

annihilating at the peak of the Z-pole is $E^R_{\nu_j} = M^2_Z/2m_{\nu_j} = 4\,(\mathrm{eV}/m_{\nu_j}) \times 10^{21}$ eV. The energy–averaged annihilation cross section $\langle \sigma_{\mathrm{ann}} \rangle$ is the effective cross section for all neutrinos within $\frac{1}{2}\delta E_R/E_R = \Gamma_Z/M_Z = 3\%$ of their peak annihilation energy. We will refer to neutrinos with energy in the range 0.97 E_R to 1.03 E_R as being at the resonant energy. (We will sometimes use E_R generically for resonant energy, as we do here, with the understanding that there are really three different resonant energies, one for each neutrino mass.) At a given resonant energy $E^R_{\nu_j}$, only relic neutrinos with the jth mass m_j may annihilate.

Each resonant neutrino annihilation produces a Z boson which immediately decays (its lifetime is 3×10^{-25} s in its rest frame). 70% of these decays are hadronic, consisting of a particle burst known to include on average fifteen neutral pions and 1.35 baryon–antibaryon pairs [15]. The fifteen π^0's decay to produce thirty high–energy photons. We will refer to the end product of this Z production and hadronic decay as a "Z-burst." The nucleons and photons in the Z-burst are candidates for the primary particles with energy above the GZK–cutoff [12].

Let us call the distance over which a stable particle can propagate without losing more than an order of magnitude of its energy the GZK distance. For a photon it is 10 to 20 Mpc, with the exact number depending on the strengths of the diffuse radio and infrared background. For a proton it is $D_{\mathrm{GZK}} \sim -[\ln(1/10)/\ln(0.8)] \times 6$ Mpc $\sim$ 50 to 100 Mpc. If the Z-burst points in the direction of earth and occurs within the GZK distance, then one or more of the photons and nucleons in the burst may initiate a super–GZK air–shower at earth. The mean multiplicity in Z decay is about 40 [15]. This dilutes the energy per hadron somewhat compared to $E^R_{\nu_j}$, but it also provides a larger flux of photons (from π^0–decays) and a few baryons per burst. Shown in Figure 1 is a schematic of the Z-burst mechanism within the GZK zone.

The annihilation/Z-burst rate depends on the relic–neutrino density. The mean neutrino density of the universe is predicted by hot Big Bang cosmology. The density of neutrinos with mass below an MeV (the neutrino decoupling temperature) is given by a relativistic Fermi-Dirac distribution characterized by a single temperature parameter. The distribution is that of relativisitic neutrinos, even though the neutrinos are nonrelativistic today, because the distribution is determined at the decoupling epoch. As a result of photon reheating from the era of $e^+e^- \to \gamma\gamma$ annihilation, the neutrino temperature $T_\nu \sim 1.95$K turns out to be a factor of $(4/11)^{1/3}$ less than that of the photon temperature, $T_\gamma = 2.73$K. The resulting mean neutrino number density is $\langle n_{\nu_j} \rangle = (3\zeta(3)/4\pi^2)T^3_\nu = 54\,\mathrm{cm}^{-3}$ for each light flavor j with an equal number density for each antineutrino flavor.[3] Note that the predicted relic–neutrino density is normalized via the temperature relation to the relic photon density which is measured. Consequently, the predicted mean density of $\langle n_{\nu_j} \rangle = 54\,\mathrm{cm}^{-3}$ must be considered firm.

---

[3] We do not consider here the possibility of unequal numbers of relic neutrinos and antineutrinos. For unpolarized Majorana neutrinos, the two numbers must be equal.

Probability for Annihilating to a Z-burst

In a universe where the neutrinos are nonrelativistic and uniformly distributed, the mean annihilation length for neutrinos at their resonant energy would be $\lambda = (\langle\sigma_{\rm ann}\rangle\langle n_{\nu_j}\rangle)^{-1} = 4.4 \times 10^{29}$ cm, which is $50\,h_{100}$ times the Hubble distance. A cosmic ray neutrino arriving at earth from a cosmically distant source will have traversed approximately a Hubble distance of space, so its annihilation probability on the relic–neutrino sea is roughly $D_H/\lambda_{\rm ann} = 2.0\,h_{100}^{-1}\%$ (neglecting cosmic expansion).

For a more careful derivation, we let $F_{\nu_j}(E_\nu, x)$ denote the flux of the $j^{\rm th}$ neutrino flavor, as would be measured at a distance x from the source, with energy within dE of E_ν. The units of $F_{\nu_j}(E_\nu, x)$ are neutrinos/energy/area/time/solid angle. This flux may be quasi–isotropic ("diffuse"), as might arise from a sum over cosmically–distant sources such as AGNs; or it may be highly directional, perhaps pointing back to sources within our supergalactic plane. The production rate of Z's with energy within dE of E_ν, per unit length, per area, time, and solid angle, is therefore $dF_{\nu_j}(E_\nu, x)/dx = \sigma_{\rm ann}(E_\nu)n_{\nu_j}(x)F_{\nu_j}(E_\nu, x)$. Integrating this equation over the distance D from the emission site to earth, and integrating over neutrino energy then gives the total rate of resonant annihilation, *i.e.* Z–burst production (in the narrow resonance approximation), within the distance D:

$$\delta F_{\nu_j}(D) = E_R F_{\nu_j}(E_R, 0) \int \frac{ds}{M_Z^2}\left[1 - \exp(-\sigma_{\rm ann}(s)\,S_j(D))\right],\ \ S(D) \equiv \int_0^D dx\, n_{\nu_j}(x).$$

$S(D)$ is the neutrino column density from earth to the distance D. If $\sigma_{\rm ann}(s)S(D)$ is small compared to one, then

$$\delta F_{\nu_j}(D) \approx E_R \langle\sigma_{\rm ann}\rangle S_j(D) F_{\nu_j}(E_R, 0).$$

For $S(D) \ll 1/\sigma_{ann}(s)$, the rate for Z-burst generation depends linearly on the relic–neutrino column density. Writing $S(D)$ as $\langle n_{\nu_j}\rangle \times D$, we make contact with our previous simple estimate.

For each 50 Mpc of travel through the mean neutrino density, the probability for a neutrino with resonant energy to annihilate to a Z–boson is 3.6×10^{-4}. Since the branching fraction for a Z to decay to hadrons is 70%, one part in 4000 of the resonant neutrino flux will be converted into a Z-burst containing ultrahigh–energy photons and nucleons within the 50 Mpc GZK–distance of earth in this unclustered universe. We live in a matter–rich portion of the universe. Consequently, it is expected that our local relic–neutrino density is somewhat enhanced compared to the universal average, due to the local potential well. Therefore, the value of 0.025% for the probability of a resonant neutrino creating a Z–burst within the GZK zone, obtained without neutrino clustering, is the <u>absolute minimum</u> annihilation probability in a Big Bang universe.

The annihilation probability within the GZK volume is enhanced due to neutrino trapping by the factor $\xi \equiv \langle n_\nu(< D_{\rm GZK})\rangle/\langle n_\nu\rangle$, where $\langle n_\nu\rangle$ is the average value for the neutrino density throughout the entire Universe, and $\langle n_\nu(< D_{\text{GZK}})\rangle$ is the average value within the GZK zone. How large is the local relic–neutrino enhancement expected to be? The mean baryon density of our Galaxy (within a sphere of radius ~ 15 kpc) compared to the baryon density averaged over the visible universe is enhanced by about 10^6. However, one expects the relicneutrino density to be much less enhanced than the baryon density for several reasons. First of all, neutrinos do not dissipate energy as easily as electrically charged baryons and elctrons. Secondly, since gravity couples universally to energy and is blind to particle quantum numbers, it is unlikely that the mass fraction of neutrinos in any halo is larger than the universal mass fraction, $f_\nu \sim \Omega_\nu = \sum_j m_{\nu_j}/(92 h_{100}^2\,\text{eV})$, which is of the order of a per cent for $m_\nu \sim 1$ eV. Thirdly, Pauli blocking presents a significant barrier to clustering of light–mass fermions on the scale of our Galactic halo. As a crude estimate of Pauli blocking, one may use the zero temperature Fermi gas as a model of the gravitationally–bound halo neutrinos. Requiring that the Fermi momentum of the neutrinos not exceed the virial velocity σ of the Galaxy, one gets $\xi = n_{\nu_j}/54\,\text{cm}^{-3} \lesssim 10^3 (m_{\nu_j}/\text{eV})^3 (\sigma/220\,\text{kms}^{-1})^3$. Finally, neutrinos have a much larger Jeans ("free–streaming") length than do baryons at the crucial time when galaxies start to grow nonlinearly. This free–streaming length of $\lambda_{FS} \sim 100\,(10\text{eV}/m_\nu)$ Mpc is the minimum size of an overdensity which can gravitationally contain the collisionless neutrinos, and should roughly correspond to the expected size of a neutrino halo today. Within the uncertainties of the estimate, this length is the size of our local supergalactic cluster. Taking 20 Mpc as the supercluster size, and 100 Mpc as the characteristic distance between superclusters, one gets a density enhancement of $\sim 5^3 \sim 100$. If neutrinos manage to cluster on the smaller galactic or galactic cluster scales, then their density may be larger by another factor of 10 or so, but probably not more. We multiply the probability above for converting a resonant neutrino into a hadronic Z–burst within 50 Mpc of earth, obtained for a smooth relic background, by the density enhancement factor ~ 100 to get our best conversion estimate of about 2%.

This is our main result, which we repeat for emphasis: *the probability for neutrinos at their resonant energy to annihilate within the halo of our Super Galactic Cluster is likely within an order of magnitude of 1%, with the exact value depending on unknown aspects of g clustering.*

Note that two crucial elements are required for this mechanism to produce super–GZK air–showers: the existence of a neutrino flux at $\gtrsim 10^{21}$ eV, and the existence of a neutrino mass in the 0.1 to 10 eV range. Concerning the flux, it is not unlikely that whatever mechanism produces the most energetic hadrons also produces charged pions of comparable energy. Thus, one may expect neutrino production at ultrahigh energy, coming from pion decay and subsequent muon decay [16]. The opacity difference between neutrino and proton in dense sources such as AGNs makes credible the possibility of a neutrino flux considerably above the proton flux at highest energies.[4] Quantitatively, the requirement

---

[4] There is also the possibility that the highest–energy neutrinos originate in quark jets produced

on the neutrino flux at the resonant energy is that the product of this flux per flavor, times the annihilation probability within the GZK zone which we have estimated to be $10^{-2\pm1}$, times the photon and nucleon multiplicity per burst which we will show to be ~ 30, is comparable to the flux of events at 10^{20} eV; this is, $F_{\nu_j}(\sim 4 \times 10^{21} \text{eV}) \sim 10^{0.5\pm1} F_p(10^{20}\text{eV})$. Such a possibility does not seem unnatural, although a straightforward extrapolation using a standard, falling power–law spectrum is not promising [18]. Concerning possible neutrino masses, the simplest explanation for the anomalous atmospheric–neutrino flavor–ratio [19] is neutrino oscillations driven by a mass–squared difference of $\sim 10^{-3} \text{eV}^2$ [20], which implies a mass of at least 0.03 eV. Also, the recent LSND measurement appears [21] to indicate a neutrino mass of 0.5 to 2 eV. Furthermore, according to Big Bang cosmology, the fraction of closure density provided by possible neutrino masses is $\Omega_\nu = \sum_j m_{\nu_j}/(92 h_{100}^2 \text{ eV})$, with $0.5 \leq h_{100} \equiv H_0/(100\text{km/s/Mpc}) \leq 0.75$. One sees that $\sim$ eV neutrino masses are consistent, and even required if neutrino hot dark matter is to contribute in any significant way to the evolution of large–scale structures.

Before considering specific signatures of the annihilation mechanism for generating 100 EeV particles, we briefly mention two subtleties:

As their momenta red–shifted in the expanding universe, the relic neutrinos evolved to the unpolarized nonrelativistic state which they occupy today. As a result, if the neutrino is a Dirac particle, then the sterile right–handed neutrino and the sterile left–handed antineutrino fields are populated equally with the two active fields. Therefore, for Dirac neutrinos the active densities available for annihilation with the incident high–energy neutrino are half of the total densities, and the Z–burst production probability is half of what we quote in this article. For Majorana neutrinos, there are no sterile fields and the total densities are active in annihilation. Majorana neutrinos are favored over Dirac neutrinos in currently popular theoretical models with nonzero neutrino mass [22].

It is possible, even probable, that massive neutrinos exhibit mixing in analogy to the quark sector of the standard model of particle physics. In such a case, the flavor states are unitary mixtures of the mass states. Letting $\alpha = e, \mu, \tau$ label flavor and $j = 1, 2, 3$ label mass, one writes $|\nu_\alpha> = \sum_j U_{\alpha j} |\nu_j>$. Then each neutrino flavor at the resonant energy of a given mass state has a nonzero probability to annihilate, but with an extra probability factor of $|U_{\alpha j}|^2$. For example, the ν_μ's and ν_e's from pion and mu decay will annihilate at the resonant energy of m_2 with the probability factors $|U_{\mu 2}|^2$ and $|U_{e2}|^2$, respectively, times what we calculate below. If there is mixing, the factors $|U_{\alpha j}|^2$ can be easily multiplied in. The phenomenon of neutrino oscillations is irrelevant in our context of annihilation, because the $\nu - \bar{\nu}$ annhilation process requires just a single transformation from flavor to mass basis, so the phase differences induced between mass states are not observed.

---

when some supermassive relic particles decay, in which case the neutrino flux greatly exceeds the proton flux [17].

The p, n, γ Flux Above 100 EeV from Z–bursts

The decay products of the Z are well-known from the millions of Z's produced at LEP and at the SLC. [15]. The respective branching fractions for Z–decay into hadrons, neutrino-antineutrino pairs, and charged lepton pairs are 70%, 20%, and 10%. The mean multiplicity $\langle N \rangle$ in hadronic Z decay is about 40 particles, of which, on average, 17 are charged pions, 9 are neutral pions, and importantly, 1.35 are baryon-antibaryon pairs which become 2.7 nucleons and antinucleons. Unstable hadrons decay to produce more pions. We estimate that the total pion and nucleon count for the Z-burst is then 15 π^0's, 28 $\pi^\pm$'s, and 2.7 nucleons (we now mean "nucleons" to include the antinucleons as well). What is of major interest are the 15 π^0's and the 2.7 nucleons. The 2.7 nucleons may be protons or neutrons. (In fact, a nucleon above the GZK cutoff energy will on average spend half of its transit time as a neutron, and half as a proton.)

Among the 15 π^0's, 9 are produced directly in Z-decay, while the other 6 arise from decays of various hadronic resonances. A comparison of the data [15] for the momentum spectra for direct pions and for protons produced in Z–decay reveals that the boosted mean energy per proton is larger than that of a direct pion by a factor ~ 3.5. We expect the energy of the six π^0's produced through resonance decays to be softer yet, by another factor of ~ 3. Weighting the direct and secondary pions appropriately, we arrive at a factor of about six for the softness of the mean pion energy compared to the mean nucleon energy. Since the photon on average carries half of the parent pion energy, the mean energy of a photon in a Z–burst is expected to be less than that of a nucleon by an order of magnitude.

The 30 photons and the 2.7 nucleons are the candidate primary particles for inducing super–GZK air–showers in the earth's atmosphere. The *a priori* photon-to-nucleon ratio is about 30. However, the hardness of the nucleon spectrum compared to the photon spectrum mitigates this ratio if a selection is made for the very highest-energy particles. Of course, these average values for the multiplicities and energies must be used with some caution, since fluctuations are large in multiplicity and particle-types per event, and in energy per individual particle.

Further Signatures from Z–bursts

The particle spectrum in Z–decay and the Lorentz factor of the Z, $\gamma_Z = E_R/M_Z = M_Z/2m_\nu = 0.9\,(m_\nu/\text{eV})^{-1} \times 10^{11}$, determine the possible signatures of Z-bursts. We comment on some of the possible Z-burst observables, beyond the photon– and nucleon–initiated super–GZK air–showers:

(i) The Z–decay products which in the Z rest frame lie within the forward hemisphere are boosted into a highly-collimated lab-frame cone of half-angle $1/\gamma_Z = 2(m_{\nu_j}/\text{eV}) \times 10^{-11}$ radians. Z-bursts originating within $20(\text{eV}/m_{\nu_j})$ parsecs of earth, if directed toward the earth, arrive with a transverse spatial spread of less than one earth diameter. It is possible for the decay products of a single

Z–burst to initiate multiple air–showers. A large area surface array (*e.g.* the Auger project) or an orbiting all–earth observing satellite (*e.g.* the OWL or AIRWATCH proposals) [23] could search for these nearly coincident showers.

(ii) The mean number of baryon–antibaryon pairs per hadronic Z–decay is 1.35. Baryon number conservation requires each hadronic Z decay to contain an integer number (possibly zero) of baryon–antibaryon pairs. If the number of baryon pairs per hadronic shower is governed by Poisson statistics, then the probabilities for 0, 1, 2, 3, 4, and 5 pairs are 26%, 35%, 24%, 10%, 4%, and 1%, respectively.

(iii) The energy of the Z–bursts are fixed at $4\,(\mathrm{eV}/m_{\nu_j}) \times 10^{21}$ eV by the neutrino mass(es). The energy of individual particles produced in the burst can approach this value but cannot exceed it. This may serve to distinguish the Z–burst hypothesis from some recent speculations for super–GZK events based on SUSY or GUT–scale physics [17], in which cutoff energies are expected to be much higher.

(iv) From the highest super–GZK event energy E^{max}, one can deduce an upper bound on the neutrino mass of $m_\nu < M_Z^2/2E^{\mathrm{max}} = 4\,(10^{21}\mathrm{eV}/E^{\mathrm{max}})$ eV. Similarly, from the mean energy $\langle E \rangle$ of super–GZK events one can estimate the mass of the participating neutrino flavor via $m_\nu = M_Z^2/2E_R \sim M_Z^2/(2\langle N \rangle \langle E \rangle) \sim 0.5\,(10^{20}\mathrm{eV}/\langle E \rangle)$ eV; if there is a selection bias toward events at higher energy, then this formula gives a lower bound on the neutrino mass.

(v) The most significant time–scale for the Z–burst is the hadronization time in the lab frame, which is $\gamma_Z \times \mathrm{fm}/c \sim 10^{-12}$ s. However, some hard photons will arrive "late." These are the photons arising from pions produced in kaon decays (with rest frame lifetimes of 0.9×10^{-10} s, 1.2×10^{-8} s, and 5×10^{-8} s, respectively, for the K_S, $K^\pm$, and K_L.) These particles contribute 3 π^0's per Z–burst. There are also 0.25 π^0's from Λ decay and 0.08 π^0's from $\Sigma^\pm$ decay. Λ and $\Sigma^\pm$ lifetimes in the rest frame are $\mathcal{O}(10^{-10})$ s. In the boosted lab frame, the kaon and strange baryon lifetimes will be $\mathcal{O}(20$ to $10^4)$ s. We therefore expect ~ 7 hard photons per event to straggle by this amount of time. Long time–scales are also available from the decays in–flight of the charged pions and muons, and from the inverse Compton and synchrotron processes of the $e^\pm$'s. (Recall that there are about 28 $\pi^\pm$'s per Z–burst, which cascade through 28 $\mu^\pm$'s to 28 $e^\pm$'s.) It is not clear to us whether these processes can generate an observable photon yield.

(vi) There could be a "neutrino pile–up" at two to three decades of energy below E_R. These pile–up neutrinos are the result of the hadronic decay chain which includes $Z \to \sim 28\,(\pi^\pm \to \nu_\mu + \mu^\pm \to \nu_\mu + \bar{\nu}_\mu + \nu_e/\bar{\nu}_e + e^\pm)$; *i.e.* each of the 70% of the resonant neutrino interactions which yield hadrons produces about 85 neutrinos with mean energy $\sim E_\pi/4 \sim E_R/160$. These neutrinos are in addition to the neutrinos piling up from the decay of pions photo–produced by any super–GZK nucleons scattering on the 2.7K background. [24]

(vii) A very interesting issue is to what extent the boosted Z–decay products will contain a copious amount of observable gamma–rays produced by internal brehmsstrahlung during the decay process. It seems possible that Z–bursts may generate short duration gamma–ray bursts observable at earth. Even extreme infrared radiation becomes observable after boosting by $\gamma_Z \sim 10^{11}$. A 10^{-5} radio photon becomes an MeV gamma–ray, and a 10^{-2} eV infrared photon becomes a hard GeV gamma–ray. The glib statement that "the 1/E brehmsstrahlung singularity produces photons with such low energy that they cannot be observed" may not hold for Z–bursts. As we have stated, the hadronization time when boosted to the lab frame, $\sim 10^{-12}$ s, sets the basic time–scale for particle emission and brehmsstrahlung.

There are a few more observations that should be made concerning the hypothesis under discussion:

(i) If the highest–energy neutrino cosmic ray flux points back to discrete sources of origin, then the super–GZK event arrival directions should point back to these same sources. In fact, neutrinos with differing energies emitted from the same source will all point back to the common source. This is in contrast to nucleon primaries, which are bent in space by intervening cosmic magnetic fields (and therefore dispersed in arrival time), by an amount varying inversely with their energy. On the other hand, if the highest–energy neutrino flux is diffuse, then the super–GZK event directions should correlate with the spatial distribution of the relic neutrino density. Perhaps the angular distribution of super–GZK events can be used to perform halo tomography of our supergalactic cluster.

(ii) If the super–GZK events are due to neutrino annihilation on relics as hypothesized here, and if the high–energy neutrino flux is eventually measured, then an estimate of the relic–neutrino column density out to $D_{\rm GZK} \sim 50$ Mpc may be made. Let $\mathcal{L}(> E_{\rm GZK})$ be the luminosity of super–GZK events (in units of events/area/time/solid angle). Then the column density of the annihilating neutrino flavor out to $D_{\rm GZK}$ is $S_{\rm GZK}^{\nu_j} \sim \mathcal{L}(> E_{\rm GZK})/\langle \sigma_{\rm ann}\rangle F_{\nu_j}(E_R)\,\delta E_R = 4.5\times 10^{32}[\mathcal{L}(> E_{\rm GZK})/E_R F_{\nu_j}(E_R)]\,{\rm cm}^{-2}$. If $F_{\nu_j}(E)$ is measured below the resonant energy, an estimate of the neutrino column density can still be made by extrapolating the flux to E_R. For example, if a power law is assumed with a spectral index α, then $F_{\nu_j}(E_R) = (E/E_R)^\alpha F_{\nu_j}(E)$.

SUMMARY AND PROSPECTUS

In summary, if one or more neutrino mass is within about an order of magnitude of an eV, and if there is a sufficient flux of cosmic ray neutrinos at $\gtrsim 10^{21}$ eV, then $\nu_{\rm cr} + \bar\nu_{\rm relic}$ (or vice versa) $\to Z \to hadrons \to nucleons\ and\ photons$ within the GZK volume $\sim (50{\rm Mpc})^3$ of earth may be the origin of air–shower events observed above the GZK cutoff. Possible signatures to validate or invalidate this hypothesis are abundant. If the hypothesis is correct, then air–shower observations may show the existence of the relic–neutrino gas liberated from the primordial early–universe plasma when the universe was only one second old!

There are good prospects for more cosmic ray data at the highest energies. Present cosmic–ray detection efforts are ongoing, and the "Hi-Res", "Telescope Array", "Auger", "OWL", and "Airwatch" projects has been formed to coordinate international efforts to collect air–shower data from ever–larger parts of our sky.

Acknowledgements: This work was supported in part by the U.S. Department of Energy, Division of High Energy Physics, under Grant No. DE-F605-85ER40226, and the Vanderbilt University Research Council.

REFERENCES

1. P. L. Biermann and P. A. Strittmatter, *Astrophys. J.* **322**, 643 (1987); J. P. Rachen and P. L. Biermann, *Astron. & Astrophys.* **272**, 161 (1993); G. Sigl, D. N. Schramm, and P. Bhattacharjee, *Astropart. Phys.* **2**, 401 (1994); an excellent overview of models proposed to generate ultrahigh energy primaries is given by P. L. Biermann, *J. Phys. G:* **23**, 1 (1997).
2. K. Greisen, *Phys. Rev. Lett.* **16**, 748 (1966); G. T. Zatsepin and V. A. Kuzmin, *Pisma Zh. Eksp. Teor. Fiz.* **4**, 114 (1966); F. W. Stecker, *Phys. Rev. Lett.* **21**, 1016 (1968); J. L. Puget, F. W. Stecker and J. H. Bredekamp, *Astrophys. J.*, **205**, 638 (1976); V. S. Berezinsky and S. I. Grigoreva, *Astron. & Astrophys.*, **199**, 1 (1988).
3. S. Yoshida and M. Teshima, *Prog. Theor. Phys.* **89**, 833 (1993); F. A. Aharonian and J. W. Cronin, *Phys. Rev.* **D50**, 1892 (1994); G. Sigl, D. N. Schramm, and P. Bhattacharjee, *Astropart. Phys.* **2**, 401 (1994); J. W. Elbert and P. Sommers, *Astrophys. J.* **441**, 151 (1995);
4. T. Stanev, P. Biermann, J. Lloyd-Evans, J. Rachen and A. Watson, *Phys. Rev. Lett.* **75**, 3056 (1995).
5. S. Yoshida, et al., (AGASA Collab.) *Astropart. Phys.* **3**, 105 (1995); N. Hayashida et al., *Phys. Rev. Lett.* **73**, 3491 (1994).
6. D. J. Bird et al., (Fly's Eye Collab.) *Phys. Rev. Lett.* **71**, 3401 (1993); *Astrophys. J.* **424**, 491 (1994); *ibid.* **441**, 144 (1995).
7. G. Brooke et al. (Haverah Park Collab.), Proc. 19th Intl. Cosmic Ray Conf. (La Jolla) **2**, 150 (1985); reported in M. A. Lawrence, R. J. O. Reid, and A. A. Watson (Haverah Park Collab.), *J. Phys. G* **17**, 733 (1991).
8. N. N. Efimov et al., (Yakutsk Collab.) ICRR Symposium on Astrophysical Aspects of the Most Energetic Cosmic Rays, ed. N. Nagano and F. Takahara, World Scientific pub. (1991); and Proc. 22nd ICRC, Dublin (1991).
9. F. W. Stecker, *Phys. Rev.* **180**, 1264 (1969); *Phys. Rev. Lett.* **80**, 1816 (1998).
10. F. Halzen, R. A. Vazquez, T. Stanev, and V. P. Vankov, *Astropart. Phys.*, **3**, 151 (1995).
11. M.A. Malkan and F.W. Stecker, astro-ph/9710072 and *Astrophys. J.*, in press; F.W. Stecker and O.C. De Jager, astro-ph/9804196.
12. T. J. Weiler, hep–ph/9710431 and *Astropart. Phys.*, to appear; D. Fargion, B. Mele and A. Salis, astro-ph/9710029.
13. J. Bordes et al., *Astropart. Phys.* **8**, 135 (1998).
14. T. J. Weiler, *Phys. Rev. Lett.* **49**, 234 (1982); *Astrophys. J.* **285**, 495 (1984); and in *High Energy Neutrino Astrophysics*, ed. V. J. Stenger, J. G. Learned, S. Pakvasa, and X. Tata, Honolulu HI, 1992, pub. World Scientific; E. Roulet, *Phys. Rev.* **D47**, 5247 (1993); P. Gondolo, G. Gelmini, and S. Sarkar, *Nucl. Phys.* **B393**, 111 (1993); S. Yoshida, H. Dai, C. Jui, and P. Sommers, *Astrophys. J.* **479**, 547 (1997).

15. Particle Data Group, *Phys. Rev.* **D54**, pp. 187–8 (1996).
16. K. Mannheim and P. L. Biermann, *Astron. & Astrophys.* **221**, 211 (1989); R. J. Protheroe and A. P. Szabo, *Phys. Rev. Lett.* **69**, 2885 (1992); *Astropart. Phys.* **2**, 375 (1994); K. Mannheim, *Astropart. Phys.* **3**, 295 (1995); F. W. Stecker and M. H. Salamon, *Space Sci. Rev.* **75**, 341 (1996); R. J. Protheroe, in *Towards the Millennium in Astrophysics: Problems and Prospects*, Erice 1996, eds. M.M. Shapiro and J.P. Wefel, pub. World Scientific (astro-ph/9612213), and in *Accretion Phenomena and Related Outflows*, ed. D. Wickramashinghe et al., 1996 (astro-ph/9607165); F. Halzen and E. Zas, astro–ph/9702193.
17. P. Bhattacharjee, C. T. Hill, and D. N. Schramm, *Phys. Rev. Lett.* **69**, 567 (1992); P. Bhattacharjee and G. Sigl, *Phys. Rev.* **D51**, 407 (1995); G. Sigl, S. Lee, D. N. Schramm, and P. Coppi, *Phys. Lett.* **B392**, 129 (1997); V. Berezinsky, X. Martin and A. Vilenkin, *Phys. Rev.* **D56**, 2024 (1997); V.Berezinsky and A. Vilenkin, *Phys. Rev. Lett.* **79**, 5202 (1997); V. Berezinsky, M. Kachelriess, and A.Vilenkin, *Phys. Rev. Lett.* **79**, 4302 (1997); V. Berezinsky and M. Kachelriess, *Phys. Lett.* **B422**, 163 (1998); P. H. Frampton, B. Keszthelyi, Y. J. Ng, astro-ph/9709080; V. A. Kuzmin and V. A. Rubakov, astro-ph/9709187; P. Bhattachatjee, Q. Shafi, and F. Stecker, *Phys. Rev. Lett.* **80**, 3698 (1998); M. Birkel and S. Sarkar, hep-ph/9804285.
18. E. Waxman, astro-ph/9804023.
19. Super–Kamiokande Collaboration, Y. Fukuda et al., hep–ex/9803006 and Phys. Lett. **B**, to appear; hep–ex/9805006; talks by Y. Suzuki and Y. Totsuka at *Neutrino–98*, Takayama, Japan, June 1998; Kamiokande collaboration, K. S. Hirata *et al.*, *Phys. Lett.* **B280**, 146 (1992); Y. Fukuda *et al.*, *Phys. Lett.* **B335**, 237(1994); IMB collaboration, R. Becker-Szendy *et al.*, *Nucl. Phys. Proc. Suppl.* **38B**, 331 (1995); Soudan-2 collaboration, W. W. M. Allison *et al.*, *Phys. Lett.* **B391**, 491 (1997).
20. J. G. Learned, S. Pakvasa, and T. J. Weiler, *Phys. Lett.* **B207**, 79 (1988); V. Barger and K. Whisnant, *Phys. Lett.* **B209**, 365 (1988); K. Hidaka, M. Honda, and S. Midorikawa, *Phys. Rev. Lett.* **61**, 1537 (1988).
21. C. Athanassopoulos et al. (LSND Collaboration), *Phys. Rev. Lett.* **77**, 3082 (1996); *Phys. Rev.* **C54**, 2685 (1997); nucl–ex/9709006 (1997);
22. Particle physics possibilities for neutrino mass, and the phenomenological implications of nonzero mass for terrestrial experiments and for cosmology are reviewed in G. Gelmini and E. Roulet, *Rept. Prog. Phys.* **58**, 1207 (1995).
23. Presentations by the "Telescope Array", "Hi–Res", "Auger", "OWL" and "AIR-WATCH" are available in these proceedings. In addition, the Auger Project has a home page at
http://www–td–auger.fanl.gov:82/; and the "Orbiting Wide–angle Light–collectors" collaboration has a homepage at
http://lheawww.gsfc.nasa.gov/docs/gamcosray/hecr/OWL/.
24. C. T. Hill and D. N. Schramm, *Phys. Rev.* **D31**, 564 (1985); see also F. A. Aharonian and J. W. Cronin in ref. [3].

FLAVOR PHYSICS AND *CP* VIOLATION

Breaking Lorentz Invariance

Sheldon L. Glashow [1]

Department of Physics
Harvard University
Cambridge, MA

I INTRODUCTION

This is a preliminary, incomplete, and unauthorized version of work in progress with Sidney Coleman. We seek to characterize possible departures from special relativity and explore their consequences [1]. We posit a preferred frame (the rest frame of the CBR?) in which rotations and translations are exact symmetries. Nature is nearly Lorentz invariant, so violations are treated as perturbations to the standard-model Lagrangian. Tiny Lorentz-violating terms (rotationally and translationally invariant in the preferred frame) must respect the gauge symmetries of the standard model. Imagining these effects to arise from unknown physics at the Planck or unification scale, we limit ourselves to renormalizable terms (of dimension ≤ 4). Terms of higher dimension should be suppressed by reciprocal powers of the large energy scale.

We begin by examining QED, whose unperturbed (Lorentz invariant) Lagrangian is

$$\mathcal{L}_0 = \tfrac{1}{2}(E^2 - B^2) + \bar\psi(\gamma\Pi + m_0)\psi,$$

where Π is the 4-vector $p - eA$. In the preferred frame, we enumerate all Lorentz-violating perturbations of $\mathcal{L}_0$ meeting the above criteria.

Aside from bosonic perturbations proportional to E^2, B^2 and $\vec{A}\cdot\vec{B}$, there are fermionic perturbations (bracketed between $\bar\psi$ and ψ) consisting of the twelve possible products between $\{\vec\gamma\cdot\vec\Pi,\ \gamma_0\Pi_0,\ m_0\}$ and $\{1,\ \gamma_5,\ \gamma_0,\ \gamma_0\gamma_5\}$:

$$\begin{aligned}
&(r_1 + r_2\gamma_5 + ir_3\gamma_0 + r_4\gamma_0\gamma_5)\,\vec\gamma\cdot\vec\Pi \\
&(s_1 + s_2\gamma_5 + s_3\gamma_0 + is_4\gamma_0\gamma_5)\,\gamma_0\Pi_0 \\
&(t_1 + it_2\gamma_5 + t_3\gamma_0 + t_4\gamma_0\gamma_5)\,m_0,
\end{aligned} \quad (1)$$

where the r_i, s_i and t_i are small parameters and hermitivity of the perturbation requires that they be real. We may redefine ψ so as to simplify this expression:

$$\psi \longrightarrow (u_1 + u_2\gamma_5 + u_3\gamma_0 + u_4\gamma_0\gamma_5)\,\psi.$$

---

[1] Research supported partly by the National Science Foundation, grant NSF-PHYS-92-18167.

Judicious choice of the small complex parameters u_i (and a redefinition of the scale of A) allows us to eliminate many terms. A canonical form for the perturbed Lagrangian is:

$$\mathcal{L} = \tfrac{1}{2}(E^2 - c_\gamma^2 B^2 + A \cdot B/l)$$
$$+ \bar{\psi}(\gamma_0 \Pi_0 - \vec{\gamma} \cdot \vec{\Pi}\,[c_l \tfrac{1}{2}(1-\gamma_5) + c_r \tfrac{1}{2}(1+\gamma_5)] + m(1 + a\gamma_0 + b\gamma_0\gamma_5))\psi\,. \quad (2)$$

Six independent Lorentz-violating terms remain: c_γ is the photon velocity and $c_{l,r}$ are the maximal attainable velocities (MAV) of electrons of each helicity. (The MAV of a left-handed electron coincides with that of a right-handed positron, and conversely.) Because Lorentz symmetry is well tested, the three velocities are nearly equal. The remaining terms, with dimensional couplings a, b, and $1/l$, are TCP odd.

We may generalize this analysis to the entire standard model. Each fermion family consists of five $SU(3) \times SU(2) \times U(1)$ multiplets: a doublet of quarks, one of leptons, and three singlets, e.g., u_r, d_r and e_r. Each multiplet is assigned a different 3×3 Hermitean matrix whose eigenvalues are the MAV's of their eigenvectors. [The TCP-violating a and b terms are also replaced by matrix generalizations, but we shall not discuss TCP violation in this report.] In addition, each gauge boson multiplet (and the Higgs boson) is assigned its own MAV. In Sections 2 and 3 we neglect possiible flavor violation and explore purely kinematic consequences of different MAV for different particles. A final Section discusses some dynamical sequelæ of Lorentz violations that change flavor.

II NEARLY LORENTZ INVARIANT KINEMATICS

Tiny Lorentz violating terms can dramatically affect reaction kinematics at high energy. Ordinarily unstable particles (such as neutrons or neutral pions) can become stable as they approach their maximal velocities in the preferred frame. Conversely, ordinarily stable particles (such as protons or photons) can become unstable. Furthermore, the kinematics of particle reactions at high energy may be surprisingly altered.

The net effect of Lorentz non-invariant terms conserving TCP and preserving quark and lepton flavor is to provide each particle species its own MAV c_a in the preferred frame. To be sure, a fermion's MAV may depend on its helicity—a complication ignored for simplicity of presentation but easily taken into account. Particle energies, momenta, and velocities are expressed in the preferred frame. The dispersion relation of a particle may be written $E^2 = c_a^2 p^2 + m_a^2 c_a^4$, where its velocity is $v = dE/dp = c_a^2 p/E$. [Note also that $E = m_a c_a^2 / \sqrt{1 - v^2/c_a^2}$ and $p = m_a v / \sqrt{1 - v^2/c_a^2}$.] Because Lorentz symmetry is nearly exact, $|c_a - c_b| \ll 1$ for any particle pair. Indeed, prior tests of special relativity indicate velocity differences less than 10^{-20} among first-family fermions and photons. For any pair of particle species a and b, we define the small parameter $\delta_{ab} \equiv c_a^2 - c_b^2$.

Consider the elementary-particle decay, $A_0 \to \sum A_i$. Let the mass of particle i be m_i. The decay is allowed at rest iff $m_0 c_0^2 \geq \sum m_i c_i^2$. (Summations, here and hereafter, are over all daughters, i.e., for $i = 1, \ldots n$ with $n \geq 2$.) Because departures from Lorentz symmetry are tiny, we say that decay at rest is allowed or forbidden if m_0 is greater or less than $\sum m_i$. However, ordinarily negligible departures from Lorentz symmetry become important at high energy, where a kinematically permitted decay mode may become forbidden, and conversely, a forbidden mode may become allowed.

Let $\vec{p}_0$ and E_0 be the momentum and energy of A_0. The decay is allowed iff there are $\vec{p}_i$ such that $\vec{p}_0 = \sum \vec{p}_i$ and $E_0 = \sum E_i$. A more useful criterion is obtained by restricting ourselves to *collinear* configurations of momenta for which $p_0 = \sum p_i$. Let $E_{\min}(p_0)$ be the smallest value of $\sum E_i$ for all such configurations with initial momentum p_0.

- *Theorem I:* The decay is allowed iff $E_0 \geq E_{\min}$. *Proof:* If $E_0 \geq E_{\min}$, we may secure energy balance by adding appropriate transverse components to the daughters' collinear momenta. Conversely, if the decay is allowed for any (not necessarily collinear) momentum assignment, we obtain the inequality by deleting all transverse momentum components. QED. Let c_s be the least c_i of all the daughters. For $p_0 \gg \sum m_i c_i$, we may put all the initial momentum on particle s to obtain $E_{\min} \leq c_s p_0 +$ constant. Several conclusions follow immediately:

Ii. Iff $\delta_{i0} < 0$ for at least one daughter, there is an E_0 above which the decay is always allowed

Iii. Iff $\delta_{i0} > 0$ for all daughters, there is an E_0 above which the decay is always forbidden

Iiii. We define $Y(p_0) \equiv E_{\min}^2(p_0) - c_0^2 p_0^2$. The decay $A_0 \to \sum A_1$ is allowed iff $m_0 c_0^2 \geq Y$. [For A_0 at rest, this condition is $m_0 c_0^2 > Y(0) = \sum m_i c_i^2$.]

There are four possibilities: a decay mode may be either allowed or forbidden at low energy and (quite independently) either allowed or forbidden in the limit of high energy. Can there be more than one transition energy separating allowed and forbidden domains? The following theorems address and answer this question.

- *Theorem II:* If $\{p_i\}$ satisfy $p_0 = \sum p_i$ and minimize $\sum E_i$, the daughters have a common velocity given by $v = dE_{\min}/dp_0$. *Proof:* To characterize $E_{\min}$, we introduce a Lagrange multiplier v to enforce $p_0 = \sum p_i$. Differentiating $[\sum E_i + v(p_0 - \sum p_i)]$ by p_j, we identify v as the common velocity: $v_j \equiv dE_j/dp_j = v > 0$. The relation $dE_{\min} = \sum(dE_i/dp_i)dp_i = v\sum dp_i = v dp_0$ proves the second part of the theorem.

- *Theorem III:* $Y(p_0)$ is constant or monotone except under the following circumstance. If $\sum \delta_{i0} m_i > 0$, and furthermore $\delta_{j0} < 0$ for at least one j, then $Y(p_0)$ has one maximum and no other extrema. *Proof:* Using $v = dE_i/dp_i = c_i^2 p_i/E_i$, we express $E_{\min}$ in terms of momenta, $E_{\min} = \sum c_i^2 p_i/v$, to find: $\frac{1}{2}dY/dp_0 =$

$vE_{\min} - c_0^2 p_0 = \sum \delta_{i0} p_i$. If $Y(p_0)$ is extremal at $p_0 = \hat{p}$, then $\sum \delta_{i0} p_i = 0$. This requires $\delta_{j0} < 0$ for at least one j. dY/dp_0 is a continuous and differentiable function of p_0, so that the extrema of Y are characterized by d^2Y/dp_0^2 at $\hat{p}$:

$$\frac{1}{2}\frac{d^2Y}{dp_0}\bigg|_{\hat{p}} = \frac{dv}{dp_0^2}\sum \delta_{i0}\frac{dp_i}{dv}\bigg|_{\hat{p}} .$$

The factor of dv/dp_0 is positive because $p_0 = \sum p_i$ is a sum of increasing functions of v. We use $dp_i/dv = (1/v) p_i/(1 - v^2/c_i^2)$, to obtain for the sum:

$$\frac{1}{v}\sum \delta_{i0} p_i \left\{ \frac{c_i^2}{c_i^2 - v^2} - \frac{c_0^2}{c_0^2 - v^2} \right\} = -v \sum \frac{\delta_{i0} p_i}{(c_i^2 - v^2)(c_0^2 - v^2)} .$$

The second term in curly brackets contributes nothing to the sum at $p_0 = \hat{p}$. The denominator on the rhs is positive because $v < c_i$, at least one of which is smaller than c_0. Thus, $d^2Y/dp_0^2|_{\hat{p}} < 0$. Because every extremum must be a maximum, $Y(p_0)$ has at most one, which occurs iff $\delta_{j0} < 0$ for at least one j and $dY/dp_0|_0 > 0$. The latter condition is $\sum \delta_{i0} m_i > 0$. QED. Our results are summarized as follows:

- $Y(p_0)$ is constant if $\delta_{i0} = 0$ for all i. Decay is either allowed at all energies or forbidden at all energies.

- $Y(p_0)$ is strictly increasing and unbounded above if $\delta_{i0} \geq 0$ for all i (and at least one difference is positive). If forbidden at rest, decay is forbidden at all energies. Otherwise, there is a transitional energy below which decay is allowed, above which it is forbidden.

- $Y(p_0)$ is strictly decreasing and unbounded below if $\sum \delta_{i0} m_i < 0$. If allowed at rest, decay is allowed at all energies. Otherwise, there is a transitional energy below which the decay is forbidden, above which it is allowed.

- For the remaining possibility, $\sum \delta_{i0} m_i > 0$ and $\delta_{j0} < 0$ for some j, $Y(p_0)$ rises to $Y_{\max}$ and is thereafter strictly decreasing and unbounded below. For $m_0^2 c_0^2 > Y_{\max}$, decay is allowed at all energies. For $m_0^2 c_0^2 < \sum m_i^2 c_i^2$, there is a transitional energy below which the decay is forbidden, above which it is allowed. For $Y_{\max} > m_0^2 c_0^2 > \sum m_i^2 c_i^2$, the decay is allowed both at low energy and in the high-energy limit. However, the decay is forbidden within a single interval of intermediate energies.

Example I: Consider the two-body decay $A_0 \to A_1 + A_2$, with $c_0 = 1$ (by convention), $\delta_{20} > 0$ and $\delta_{10} < 0$. For $Y(p) = E_{\min}^2 - c_0^2 p_0^2$ to have a maximum, we must have $\delta_{10} p_1 + \delta_{20} p_2 = 0$. Putting $p_i = m_i v/\sqrt{1 - v^2/c_i^2}$ and solving for the square of the common velocity v, we find:

$$v^2 = c_1^2 c_2^2 \left(\frac{m_2^2 \delta_{20}^2 - m_1^2 \delta_{10}^2}{m_2^2 c_2^2 \delta_{20}^2 - m_1^2 c_1^2 \delta_{10}^2} \right) .$$

This is a physically realizable result (rather than a spurious root) iff $v < c_1$, which is equivalent to the condition encountered in Theorem *III*, $\delta_{20}m_2 + \delta_{10}m_1$. If this condition is satisfied, Y has a maximum whose value and argument are:

$$Y_{\max}(\hat{p}) = (\delta_{20} - \delta_{10})\left(\frac{m_1^2 c_1^2}{\delta_{20}} - \frac{m_2^2 c_2^2}{\delta_{10}}\right), \qquad \hat{p}^2 = c_1^2 c_2^2 (\delta_{20} - \delta_{10})\left(\frac{m_2^2}{\delta_{10}^2} - \frac{m_1^2}{\delta_{20}^2}\right).$$

Massless Particles: The preceding analysis applies if some daughters are massless. Nonetheless, further discussion of this special case seems in order. Consider the decay $A_0 \to \sum A_i$ where $m_i > 0$ for $i \leq n$ and $m_i = 0$ for $q \geq i > n$. Let $\hat{c}$ be the least velocity of the $q-n$ massless daughters. The minimum of $\sum E_i$ for collinear momenta is obtained with $E_i = 0$ for those massless particles with $c_i > \hat{c}$. (If several massless paticles have velocity $\hat{c}$, we may set $E_i = 0$ for all but one of them.) We denote by p_l the momentum carried by that particle. The problem reduces to one involving $n+1$ daughters, one of them massless: to the determination of $E_{\min}^{n+1}(p_0)$. We also define $E_{\min}^n(p_0)$ as the solution to the problem without the massless particle. There are two possibilities: (1) $\delta_{il} > 0$ for all $n \geq i \geq 1$, and (2) $\delta_{jl} < 0$ for at least one daughter.

At $E_{\min}^{n+1}$, all daughters have common velocity v. If $p_l \neq 0$, then $v = c_l < c_i$ for all i. For case (2), this condition cannot be met, so that $p_l = 0$ and $E_{\min}^{n+1} = E_{\min}^n$. The massless particles play no role in determining whether the decay is allowed. Case (1) admits two choices: $p_l = 0$, or $p_l > 0$ and $v = c_l$. We define $p_t \equiv \sum m_i c_l / \sqrt{1 - c_l^2/c_i^2}$, summed from $i = 1$ to n. If $p_0 \leq p_t$, the second choice is unphysical and $E_{\min}^{n+1}(p_0) = E_{\min}^n(p_0)$. Otherwise, we have:

$$E_{\min}^{n+1}(p_0) = \min\left\{E_{\min}^n(p_0);\; E_t + c_l(p_0 - p_t)\right\},$$

where $E_t \equiv E_{\min}(p_t) = \sum m_i c_i^3 / \sqrt{\delta_{il}}$.

III KINEMATIC CONSEQUENCES OF BROKEN LORENTZ SYMMETRY

The preceding analysis is applied to various potentially observable phenomena induced by departures from exact Lorentz invariance. Because these departures are tiny at best, kinematic effects may be relevant for ultra high energy cosmic rays, but cannot be relevant elsewhere. However, the dynamical effects of Lorentz symmetry violation (discussed later on and resulting from terms violating flavor) may lead to observable effects in precision laboratory experiments, in neutrino physics, or at accelerators.

A Stable Neutral Pion? The dominant mode of neutral pion decay, $\pi^0 \to 2\gamma$, is allowed at rest and known to occur at TEV energies. However, if $c_\gamma > c_\pi$ ($\delta_{\gamma\pi^0} > 0$), the decay is forbidden for pions with energies exceeding $E_{\gamma\pi} \equiv m/\sqrt{\delta_{\gamma\pi^0}}$, where m is the pion mass. Rare decay modes (such as $e^+ + e^-$ or $\nu + \bar{\nu}$) may be forbidden at high energy if c_ν or c_e exceeds c_π. If $\delta_{\gamma\pi_0} \geq 10^{-20}$, stable neutral pions may be present in the cosmic-ray flux beyond 10^{18} eV.

An Unstable Photon? The 'decay' process $\gamma \to \gamma + \pi^0$ is forbidden at all energies at which photons have been observed, up to $E \sim 20$ TeV in cosmic rays. However, for $E > E_{\gamma\pi}$ (as defined above), an energetic photon will lose energy rapidly via pion emission. Thus, the stabilization of energetic π^0's is accompanied by the *destabilization* of energetic photons! Indeed, there are other mechanisms by which photons may decay, such as $\gamma \to l + \bar{l}$, with l one of the six known leptons. The threshold energy for this process is $E_{\gamma l} = 2m_l/\sqrt{\delta_{\gamma l}}$. The decay rate of a photon with energy E well above threshold is $\Gamma \simeq \frac{1}{2}\alpha\, \delta_{\gamma l} E$. Its range is $R \simeq [10^{-20}/\delta_{\gamma l}][TeV/E] \times 3$ km — terrestrially large, but astrophysically tiny. From the observation of primary cosmic-ray photons with energies up to 20 TeV, we deduce the following relatively weak constraints:

$$\delta_{\gamma e} < 3 \times 10^{-15}, \quad \delta_{\gamma\mu} < 3 \times 10^{-11}, \quad \delta_{\gamma\pi^0} < 5 \times 10^{-11}, \quad \delta_{\gamma\tau} < 8 \times 10^{-9}.$$

It might be imagined that more stringent tests of Lorentz invariance could result from the stability of photons under decay into neutrinos, for which the threshold is merely $E_{\gamma\nu} = 2m_\nu/\sqrt{\delta_{\gamma\nu}}$. Although this decay mode is forbidden in the standard model (which attributes neither masses nor magnetic moments to neutrinos), there is a considerable body of empirical evidence for neutrino oscillations, and hence for neutrino masses. Thus it is not beyond the realm of possibility that the non-standard physics responsible for neutrino masses also generates neutrino magnetic moments, thereby enabling the decay $\gamma \to \nu + \bar{\nu}$.

We obtain for the decay rate of a photon with energy E well above threshold: $\Gamma \simeq (1/6\pi)\lambda^2 \mu_B^2\, \delta_{\gamma\nu}^2 E^3$, where the magnetic moment of the neutrino is $\lambda\mu_B$, where $\mu_B \equiv e\hbar/2m_e$ is the Bohr magneton. Because the magnetic interaction is dimension five (rather than four, like the electric), the decay rate is *quadratic* (rather than linear) in $\delta_{\gamma\nu}$. Thus the range of an energetic photon is cosmological even if the neutrino magnetic moment is (implausibly) set equal to its experimental upper bound [2] $\lambda = 2 \times 10^{-10}$. Consequently, no strong bound on $\delta_{\gamma\nu}$ can be deduced from observations of energetic cosmic-ray photons.

Vacuum Čerenkov Radiation? Suppose $\delta_{a\gamma} > 0$ for a charged particle a with mass m_a. If its energy E exceeds $E_{a\gamma} = m_a/\sqrt{\delta_{a\gamma}}$, the particle travels faster than light and radiates by a process resembling Čerenkov radiation, but occuring in the vacuum and at all accessible photon momenta. Elsewhere [1] we discuss how the energy of a superluminal particle rapidly degrades to the threshold energy. Thus, the mere fact that primary cosmic-ray protons have been observed with energies up to 2×10^{20} eV (and perhaps beyond) yields a strong constraint: $\delta_{p\gamma} < 3 \times 10^{-23}$. *This bound is over 20 times stronger than the best limit based on searches for atomic anisotropies due to laboratory motion relative to a preferred frame.* [3]. We obtain weaker bounds for other charged particles:

$$\delta_{e\gamma} < 2.5 \times 10^{-13}\,[\text{TeV}/E_e]^2, \quad \delta_{\mu\gamma} < 10^{-8}\,[\text{TeV}/E_\mu]^2, \quad \delta_{\pi^\pm\gamma} < 2 \times 10^{-8}\,[\text{TeV}/E_\pi]^2,$$

where E_i is the maximum energy at which the correponding particle is observed or deduced to propagate.

Proton Beta Decay? Consider the beta decays $n \to p + e^- + \bar{\nu}$ and $p \to n + e^+ + \nu$. The former process is allowed at low energy, the latter not. We examine the special case: $c_p > c_n > c_e$ and $c_\nu \geq c_e$. The second constraint implies that $E_{\min}$, for both processes, is attained with no energy carried by the neutrino (regarded as massless). We also assume $\delta_{pn} m_p + \delta_{en} m_e > 0$. If $\delta_{pn} > 0$ and $\delta_{en} < 0$ are comparable in magnitude, $E_{\min}(p)$ will have a maximum because $m_p \delta_{pn} + m_e \delta_{en} > 0$. Following Exercise I, we find:

$$Y_{\max} \simeq (\delta_{pn} - \delta_{en})(m_e^2/\delta_{pn} - m_p^2/\delta_{en}) \quad \text{and} \quad \hat{p} \simeq (\delta_{pn} - \delta_{en})(m_p^2/\delta_{en}^2 - m_e^2/\delta_{pn}^2),$$

and obtain the following approximate results:

- Neutron beta decay is allowed if its energy E is not betweeen $E_1 \simeq \sqrt{(m_n^2 - m_p^2)/\delta_{pn}}$ and $E_2 \simeq m_n(\sqrt{\delta_{pn} - \delta_{en}} + \sqrt{\delta_{pn}})/\delta_{pn}$, where $E_2 \gg E_1$. Otherwise, neutrons cannot beta decay.

- Proton beta decay is allowed if its energy exceeds E_3, where $E_3 \simeq E_1$. Otherwise, protons cannot beta decay.

At energies exceeding $E_1 \simeq E_3 \simeq 2.4 \times 10^{16} [10^{10}/\sqrt{\delta_{pn}}]$ eV (and below E_2), neutrons are stable and protons not. Departures from Lorentz invariance small enough to have evaded detection may allow the highest energy cosmic rays to include neutrons, but forbid protons! These ultra high energy cosmic neutrons are not significantly deflected by galactic magnetic fields and should indicate the direction of their source.

Escape from the GZK Bound? Greisen, and independently Zatsepin and Kuzmin [4] showed that the propagation of ultra-high energy cosmic-ray nucleons over cosmological distances is affected by scattering off photons of the cosmic background radiation. The photopion production proceess,

$$p + \gamma(\text{CBR}) \longrightarrow p + \pi \qquad (3)$$

becomes possible for nucleon energies exceeding E_1, where:

$$E_1 = \frac{m(2M + m)}{4\omega} \simeq 3.1 \times 10^{11} \text{ GeV} \left[\frac{\omega_0}{\omega}\right],$$

with M the nucleon mass, m the pion mass, ω the target photon energy, and $\omega_0 = 2.35 \times 10^{-4}$ eV (corresponding to the CBR temperature of 2.73 K). Because of its large cross section, the special case resonance formation is especially important:

$$p + \gamma(\text{CBR}) \longrightarrow \Delta, \qquad (4)$$

where $\Delta(1232$ MeV$)$ is the first pion-nucleon resonance. For this process, the relevant nucleon energy is:

$$E_2 = \frac{M'^2 - M^2}{4\omega} \simeq 6.8 \times 10^{11} \text{ GeV} \left[\frac{\omega_0}{\omega}\right],$$

with M' the Δ mass.

These processes result in the so-called GKZ bound (an energy somewhat below 10^{11} GeV), above which extragalactic cosmic ray nucleons cannot reach us. Recent observations of events beyond the GKZ bound presents an intriguing puzzle, and have led several authors to go beyond the standard model seek an explanation [5].

Here we point out that departures from Lorentz invariance tiny enough to have evaded detection can be sufficient to undo the GZK bound. Suppose that the maximal attainable velocity (MAV) of the pion exceeds that of the nucleon: $\delta \equiv c_\pi^2 - c_p^2 > 0$. A purely kinematic analysis shows that $E_1(\delta)$ is monotone increasing and diverges as:

$$\delta \longrightarrow \delta_{\text{crit}} = \frac{4\omega^2}{m^2} \simeq 1.1 \times 10^{-23} \left[\frac{\omega}{\omega_0}\right]^2.$$

For $\delta \geq \delta_{\text{crit}}$, the photoproduction process is forbidden for all nucleon energies.

An even smaller Lorentz violating effect is sufficient to prevent the formation reaction 4. Let $\delta' \equiv c_\Delta^2 - c_p^2 > 0$, where c_Δ is the MAV of the Δ resonance. We find that $E_2(\delta')$ is monotone increasing and diverges as:

$$\delta' \longrightarrow \delta'_{\text{crit}} = \frac{4\omega^2}{M'^2 - M^2} \simeq 3.5 \times 10^{-25} \left[\frac{\omega}{\omega_0}\right]^2.$$

For $\delta' \geq \delta'_{\text{crit}}$, the formation reaction is forbidden at all nucleon energies.

Should future cosmic-ray experiments confirm the presently suggestive evidence of a failure of the GZK bound, we would be compelled to take seriously the possible existence of departures from exact Lorentz symmetry.

IV OTHER CONSEQUENCES OF BROKEN LORENTZ SYMMETRY

We discuss dynamical consequences of Lorentz non-invariance, beginning with a brief mention of TCP-violation due to the terms $A \cdot B/l$ and $\bar{\psi}(a\gamma_0 + b\gamma_5)\psi$ for QED (and analogous terms for the standard model). The former term involves the vector potential A, but makes a gauge-invariant contribution to the action yielding a velocity difference between photons of different helicity. This leads to 'vacuum faraday rotation' of polarized light or radio signals from sources at cosmological distances. Because no such effect is observed, a lower bound on l comparable to the horizon size is obtained [6].

As a dimensionless parameter appropriate to particle physics, the limit on the $A \cdot B$ term becomes $1/(lm_e) < 10^{-38}$, which is far smaller than any Lorentz-violating effect so far considered. Seemingly, searches for TCP-violation due to the fermionic terms $a\gamma_0$ and $b\gamma_0\gamma_5$ are bound to fail, because these terms might be expected to generate, through radiative corrections, an effective and empirically excluded $A \cdot B$ term. However, they do not![2] The analysis of TCP-violation due to the a and b terms and their generalizations should be examined, and their potentially observable effects searched for.[3]

Muon Decay and Muon Colliders Besides the kinematic effects so far discussed, there are dynamical consequences of flavor-changing Lorentz non-invariant perturbations. Consider the electrodymamic interactions among the charged leptons. In general, their mass matrix is a hermitean 3×3 matrix. However, this matrix may be diagonalized by a transformation leaving the electromagnetic couplings unchanged. Thus, for QED (or for the minimal standard model), such decays as $\mu \to e + \gamma$ are forbidden by what may be regarded as an accidental symmetry.

The situation is different if Lorentz symmetry is violated. Here we consider the effect of the TCP-invariant terms, those giving rise to Hermitean 3×3 'maximal velocity matrices' for each of the lepton heelicity states. Clearly, the transformation diagonalizing the lepton masses need not diagonalize these velocity matrices: the accidental symmetry is broken and the forbidden decays are permitted.

For simplicity, consider the system of two charged leptons, the muon and the electron. In the basis in which their masses are diagonal, the relevant (flavor-changing) portion of the Lorentz non-invariant Lagrangian includes the term:

$$\delta\mathcal{L} = \tfrac{1}{2}\bar{e}\,\vec{\gamma}\cdot(\vec{p}-e\vec{A})\,(\delta v_l \sin 2\theta_l\, P^- + \delta v_r \sin 2\theta_r\, P^+)\,\mu + \text{h.c.},$$

where $P^\pm = \tfrac{1}{2}(1 \pm \gamma_5)$, the angles $\theta_{l,r}$ define the velocity eigenstates in the high energy limit, $\delta v_{l,r}$ are the velocity differences between these states for each helicity, and a complex phase (irrelevant for the present purpose) is omitted. Treating $\delta\mathcal{L}$ as a perturbation, we compute the decay rate of each muon helicity state for the unseen decay mode $\mu^- \to e^- + \gamma$.[4] With the neglect of the electron mass, and the notation $\epsilon_{l,r} \equiv (\delta v_{l,r} \sin 2\theta_{l,r})^2$, we obtain:

$$\Gamma_l = \frac{\alpha}{16\,EP}\left[\epsilon_r(X+Y) + \epsilon_l(X-Y)\right], \tag{5}$$

and ditto with l and r interchanged. In Eq. 5, E and P denote the muon energy and momentum, and:

$$X = \frac{2}{15}P\left(138\,E^4 - 121\,E^2 M^2 + 13\,M^4\right),$$
$$Y = \frac{4}{15}EP^2\left(67\,E^2 - 27\,M^2\right). \tag{6}$$

---

[2] Sidney Coleman, private communication.
[3] A proposed experiment [8] could establish the limit $b/m_e < 4 \times 10^{-18}$ for the electron-positron system.
[4] We are deeply indebted to Mark Wise for his assistence in carrying out this calculation.

If the muon decays at rest, this result becomes:

$$\Gamma_l = \Gamma_r = \frac{\alpha}{4}[\epsilon_l + \epsilon_r] M.$$

The current upper limit on the branching ratio for $\mu \to e + \gamma$ at rest yields $\epsilon_l + \epsilon_r < 7 \times 10^{-26}$.

If the decaying muon is ultrarelativistic, we find:

$$\Gamma_l = \frac{\alpha}{30}[68\,\epsilon_r + \epsilon_l]\gamma^3 M,$$

where $\gamma = E/M \gg 1$. If the performance of a conjectural muon-muon collider is not to be compromised, we must have $\epsilon_l + \epsilon_r < 4 \times 10^{-33}$. Current tests of the validity of special relativity do not guarantee that a muon collider with TeV beam energies will operate as designed. However, searches for an anomalous energy dependence of the muon lifetime (such as may be performed at MACRO or SNO) may provide the necessary constraint [9].

Velocity Oscillations of Neutrinos

Neutrino physics can set better limits on some of these parameters. Again, we limit ourselves to two lepton families. The leptonic charged weak current, $\mu_l \gamma_\lambda \nu_\mu + \bar{e}_l \gamma_\lambda \nu_e$, is given in the basis where e and μ masses are diagonal. The neutrino mass eigenstates (corresponding to masses m_1 and m_2) are:

$$\nu_{m_1} = \cos\theta_m \nu_\mu + e^{i\eta}\sin\theta_m \nu_e \quad \text{and} \quad \cos\theta_m \nu_e - e^{i\eta}\sin\theta_m \nu_\mu.$$

The irremovable phase $e^{i\eta}$ is probably not observable in practice [10]. Flavor-changing Lorentz non-invariance leads to another basis in which the neutrino velocity eigenstates (with MAV difference δv_l) are:

$$\nu_{v_1} = \cos\theta_l \nu_\mu + e^{i\eta'}\sin\theta_l \nu_e \quad \text{and} \quad \cos\theta_l \nu_e - e^{i\eta'}\sin\theta_l \nu_\mu,$$

where the parameters θ_l and δv_l are as previously defined.

Neutrino oscillations result from an interplay between mass and velocity effects. The probability for ν_μ to become ν_e (or vice versa) is:

$$P = \sin^2 2\Theta \, \sin^2\{\Delta R/4E\}, \tag{7}$$

where R is the pathlength and we put $\hbar = c = 1$. The mixing angle Θ and phase factor Δ are determined by:

$$\Delta \sin 2\Theta = \left|\delta m^2 \sin 2\theta_m + 2E^2 \delta v_l \sin 2\theta_l\right|,$$
$$\Delta \cos 2\Theta = \left|\delta m^2 \cos 2\theta_m + 2e^{i(\eta-\eta')}E^2 \delta v_l \cos 2\theta_l\right|. \tag{8}$$

The result depends non-trivially on E, and as well, on δm^2, δv_l, $\cos\theta_m$, and $\cos\theta_l$. An additional phase, $\eta - \eta'$, appears because velocity and mass eigenstates are related by a complex unitary transformation. Eq. 7 reduces to the familar formula for mass mixing at small E and to pure 'velocity mixing' at large E. If all three neutrinos suffer both mass and velocity mixing, the analysis is far more complicated.

Neutrinos can experience velocity oscillations even if their masses are too small to produce detectable effects in the laboratory. Indeed, velocity oscillations can be responsible for alleged solar and atmospheric neutrino oscillations [11]. Neutrino experiments at high energy, especially those with long baselines, will provide the most sensitive tests for the existence of flavor-changing departures from strict Lorentz invariance [or the equivalence principle. [12].

We conclude with a second-order speculation: that (1) there be a second 'sterile' photon, and (2) the velocity eigenstates be linear combinations of the two states. If the mixing is large, cosmological considerations show that the velocity difference must be tiny: $\delta v/c \lesssim \sim 10^{-32} c$ [13]. For small mixing, the cosmological constraint is much less severe and astrophysical searches for photon oscillations may be feasible [14].

REFERENCES

1. See: S. Coleman and S.L. Glashow, Phys. Lett. B405, 249 (1997) and ms in preparation.
2. A.I. Derbin *et al.*, Phys. At. Nucl. 57, 222 (1994).
3. S. K. Lamoreaux, J. P. Jacobs, B. R. Heckel, F. J. Raab, and E. N. Fortson, Phys. Rev. Lett. 57, 3125 (1986).
4. K. Greisen, Phys. Rev. Lett. 16, 748 (1966) ; G.T. Zatsepin and V.A. Kuzmin, JETP Lett. 41, 78 (1966).
5. See for example: V. Berezinsky, M. Kachelreiß, and A. Vilenkin, Phys. Rev. Lett. 79, 4302 (1997) .
6. M. Carroll, G.B. Field and R. Jackiw, Phys. Rev. D40, 1231 (1990); M. Goldhaber and V. Trimble, J. Astrophys. Astr. 17, 17 (1996) .
7. Sidney Coleman, private communication.
8. R. Bluhm, V.A. Kostalecký and N. Russell, Indiana Univ. preprint IUHET 358.
9. B. Barish, private communication.
10. J. Schechter and J.W.F. Valle, Phys. Rev. D23, 66 (1981).
11. *E.g.*, S.L. Glashow, *et al.*, Phys. Rev. D56, 2433 (1997) .
12. M. Gasperini, Phys. Rev. D38, 2635 (1988); A. Halprin and C.N. Leung, Phys. Rev. Lett. 67, 1833 (1991) .
13. S. L. Glashow, Astro-ph/9803202 (to appear in Physics Letters B).
14. F. Vannucci, private communication.

Gravity Couplings in the Standard Model: CPT Nonconservation

Lay Nam Chang[1] & Chopin Soo[2]

[1] *Physics Department, Virginia Tech*
Blacksburg, Virginia 24061-0435[1]
[2] *Center for Theoretical Science, National Tsinghua University*
Hsinchu, Republic of China[2]

Abstract. Chiral asymmetric couplings are a basic feature of the standard model. We show that gravity couplings which manifestly preserve this feature are necessarily complex, and that as a result parity nonconservation can take place in the gravity sector. Implications for breakdown of CPT symmetry are also discussed.

In general relativity, gravity couplings to matter are prescribed by the equivalence principle. The idea is that locally gravity effects can be eliminated by a proper choice of coordinate system, and that gravity effects show up as being due to the metric tensor and derivatives thereof. These couplings are invariably real, and further, they preserve whatever symmetries the system possesses in the absence of gravity couplings. The prescription works well for macroscopic systems, and at the microscopic level, for tensor fields.

The prescription involves the introduction of a connection coefficient and its attendant covariant derivative. The connection should be compatible with the equivalence principle, and this requirement translates into the restriction that

$$0 = D_\mu g_{\nu\lambda} \tag{1}$$
$$= \partial_\mu g_{\nu\lambda} - \Gamma^\rho{}_{\nu\mu} g_{\rho\lambda} - \Gamma^\rho{}_{\lambda\mu} g_{\nu\rho} \tag{2}$$

The restriction can be inverted and the connection coefficient shown to be the familiar Levi-Civita connection [1].

However, in the presence of spinor fields, the prescription needs some refinement. The reason is that spinor fields carry representations that are defined relative to the

---

[1] e-mail:laynam@vt.edu
[2] e-mail:cpsoo@phys.nthu.edu.tw

local Lorentz frame, and one needs to be able to compare these representations in neighboring frames. One needs a connection relating local Lorentz transformations.

Gravity couplings to fermions test the consistency of the standard model and is therefore worthy of attention. There is no reason to suspect that there would be inconsistencies in the presence of gravity, but it is prudent to check that that is so. In addition, there is a good deal of interest these days to look for and study extra-galactic sources of neutrinos. It is quite likely that these sources could be regions of intense gravitational fields, and the interplay between the standard model dynamics and gravity would then be of importance.

Let $\psi(x)$ be a spinor field. One defines the covariant derivative on this field as

$$D_\mu \psi = \partial_\mu \psi + \frac{1}{2} \sigma^{AB} A_{AB\mu} \psi \tag{3}$$

where σ^{AB} are the generators of the local Lorentz algebra, with A, B labelling the directions of the local Lorentz frame. For Dirac fields,

$$\sigma^{AB} = \frac{1}{4}\left[\gamma^A, \gamma^B\right] \tag{4}$$

The connection coefficient $A_{AB\mu}$ are referred to as the spin connection field.

We will need direction cosine fields to relate the directions of the local Lorentz frame with those of the coordinate system. These are referred to as the vierbein fields e^A_μ and their inverses E^μ_A:

$$e^A_\mu e^B_\nu \eta_{AB} = g_{\mu\nu} \tag{5}$$
$$E^\mu_A E^\nu_B \eta^{AB} = g^{\mu\nu} \tag{6}$$
$$E^\mu_A e^B_\mu = \delta^B{}_A \tag{7}$$

The indices A, B are raised and lowered by the usual Lorentz metric η_{AB} which defines the local light-cone.

We may convert any tensor field V^λ into a vector Lorentz field by $V^\lambda e^A_\lambda$. In order to be compatible with the equivalence principle which works on tensor field, we must now generalize the covariantly constant condition on the metric tensor Eqn.[2] by

$$0 = \nabla_\mu e^A_\nu \tag{8}$$
$$= \partial_\mu e^A_\nu - \Gamma^\rho{}_{\nu\mu} e^A_\rho + A^A{}_{B\mu} e^B_\nu \tag{9}$$

with a similar condition involving E^μ_A. These relations serve to relate the two sets of connections. As a result of the spin connection, the tensor connections Γ need no longer take the Levi-Civita form and can be anti-symmetric in the lower indices. Indeed,

$$\Gamma^\alpha_{[\mu\nu]} = \frac{1}{2}(\Gamma^\alpha_{\mu\nu} - \Gamma^\alpha_{\nu\mu}) \tag{10}$$

is related to the torsion

$$T_A = \frac{1}{2}T_{A\mu\nu}dx^\mu \wedge dx^\nu = de_A + A_{AB} \wedge e^B \tag{11}$$

by

$$T_{A\mu\nu} = 2\Gamma^\alpha_{[\nu\mu]}e_{\alpha A}. \tag{12}$$

In the presence of torsion, the Dirac matrices $\gamma^\mu = E^\mu_A \gamma^A$ satisfy

$$\partial_\mu \gamma^\mu + (\partial_\mu \ln e)\gamma^\mu + \frac{1}{2}A_{\mu AB}[\sigma^{AB}, \gamma^\mu] = 2\Gamma^\nu_{[\nu\mu]}\gamma^\mu \tag{13}$$

with $\partial_\mu \ln e = \Gamma^\nu_{\nu\mu}$. However, if the torsion vanishes, then

$$D_\mu e \gamma^\mu f = e \gamma^\mu D_\mu f. \tag{14}$$

The Lagrangian for the spinor fields can be written as

$$S^- = -\int_M d^4x \, e \overline{\Psi} i \not{D} \Psi, \tag{15}$$

where $i\not{D} = \gamma^\mu(i\partial_\mu + W_{\mu a}T^a + \frac{i}{2}A_{\mu AB}\sigma^{AB})$, and e denotes the determinant of the vierbein. $W_{\mu a}$ is the internal gauge connection while $A_{\mu AB}$ is the spin connection.

Is this action real? The action so written is not Hermitian. We may add to the action the Hermitian conjugate. In the absence of internal gauge fields, the result can be expressed in Majorana form:

$$S_{Majorana} = \frac{1}{2}(S^- + (S^-)^\dagger) = -\int_M d^4x \, e \overline{\Psi}_M i \not{D} \Psi_M, \tag{16}$$

with

$$\Psi_M = \frac{1}{\sqrt{2}}(\Psi_L + C_4 \overline{\Psi}_L^T)$$

$$\overline{\Psi}_M = \frac{1}{\sqrt{2}}(\overline{\Psi}_L + \Psi_L^T C_4)$$

$$\tag{17}$$

being the Majorana spinors. The quantity C_4 denotes the charge conjugation matrix that raises and lowers spinor indices.

What happens when there are internal gauge couplings? It turns out that for real and pseudoreal representations of the internal gauge group, it is still possible to write the action in a Majorana form [2]. For complex representations, only the gravity part alone can be so expressed.

Fermions in the standard model have chirally asymmetric gauge couplings, and parity is not conserved. The fermions also belong to complex internal gauge representations. In the above, what that means is that we need to include a left-handed projection operator on the spinor field. The net result of this inclusion is that only the part of the spin connection field that is anti-self-dual in the Lorentz index AB will be coupled to the left-handed spinor field [2]. The self-dual part of the spin connection field, which is the complex conjugate of the anti-self-dual part, is coupled to right-handed spinor fields.

Now in the absence of gravity, a chiral field can be relabelled as a conjugate field of the opposite chirality. So for example, a left-handed electron field is equivalent in its diagonal couplings to other matter to a right-handed positron field, which will be coupled in like fashion to the same matter. All couplings are identical in form when we do this relabelling. The action can be written in two equivalent ways. This is the statement of CPT invariance, and from this perspective, it emerges from the form of local couplings dictated purely by gauge symmetry considerations. In the presence of gravity, however, the couplings of these two fields are to different, albeit relatively complex conjugate, sets of spin connection field. In the presence of gravity therefore, CPT invariance has to be imposed, and doesn't emerge from local couplings dictated by purely gauge symmetry considerations. Once imposed, CPT demands that the action contains terms symmetric in the complex conjugate spin connection fields. The symmetry also implies that parity is conserved by gravity couplings.

Is there any reason why the two sets of spin connnection fields should both appear? In fact, as Ashtekar showed [3], the complex anti-self-dual connections can be thought of as gauge potentials for the complex group SO(3,C), which is isomorphic to the Lorentz group. Further, the Einstein field equations can be expressed entirely in terms of these gauge fields, without reference to the complex conjugate self-dual connections. In coupling to spinor matter, only left-handed fields need be used. The resulting equations of motion can be shown to possess both gauge invariance and diffeomorphism invariance [2]. They are also identical to those resulting from an action that respects CPT.

The action in Eqn.[15] contains only the anti-self-dual connection field. It is therefore complex. The resulting equation of motion is

$$ie\gamma^\mu D_\mu \Psi_L = 0 \qquad (18)$$

On the other hand, we may consider the adjoint of of the action:

$$(S^-)^\dagger = \int_M d^4x e \left[i(\partial_\mu \overline{\Psi}_L)\gamma^\mu \Psi_L - \overline{\Psi}_L(\frac{i}{2}A_{\mu AB}\sigma^{AB} + W_{\mu a}T^a)\gamma^\mu \Psi_L \right] \qquad (19)$$

which contains the complex conjugate connections. The internal gauge currents are automatically hermitian,

$$\left\{\overline{\Psi}_L \gamma^\mu T^a \Psi_L\right\}^\dagger = \left\{\overline{\Psi}_L \gamma^\mu T^a \Psi_L\right\} \qquad (20)$$

$$= \left\{\overline{\Psi}_{cR} \gamma^\mu T_c^a \Psi_{cR}\right\} \qquad (21)$$

provided T^a is Hermitian. The fields Ψ_{cR} belong to the complex conjugate representation of the internal gauge group, and T_c^a is the corresponding matrix representation. The gauge connection is the same for both representations, and the coupling respects CPT invariance.

On the other hand, the conjugate currents obey:

$$\{\overline{\Psi}_L \gamma^\mu \sigma^{AB} \Psi_L\}^\dagger = \{\overline{\Psi}_L \sigma^{AB} \gamma^\mu \Psi_L\} \tag{22}$$

$$= \{\overline{\Psi}_R \gamma^\mu \sigma^{AB} \Psi_R\} \tag{23}$$

and, from the perspective provided by Ashtekar, are coupled to the self-dual gauge connections, which are relevant to an entirely different SO(3,C) group. To preserve CPT invariance, we need to add the two actions Eqns.[15, 19]. It can be checked that the equations which follow from the combined action, which will now be Hermitian, appear as

$$ie\gamma^\mu (D_\mu + B_\mu)\Psi_L = 0 \tag{24}$$

where $\Gamma^\nu_{[\nu\mu]} \equiv B_\mu$ is the related to the trace of the torsion field. The two spinor equations differ by the presence of this field.

Now there is a relationship between this field and the spinor fields, which follow from the equations of motion governing the classical gravity fields. This relationship is part of the equations of constraint found by Ashtekar [3]. The form can be deduced from an action given by Samuel-Jacobson-Smolin [4]. When adjoined to Eqn.[15], we get

$$S^- + S^-_{SJS} = \frac{1}{16\pi G}\int_M e^A \wedge e^B \wedge *F_{AB} - \frac{2\lambda}{16\pi G}\int_M (*1) + \frac{1}{2}(S^- + (S^-)^\dagger)$$
$$- \frac{i}{2}\int_M d\left\{\frac{1}{3!}(\epsilon_{ABCD}\overline{\Psi}_L\gamma^A\Psi_L e^B \wedge e^C \wedge e^D) - \frac{1}{8\pi G}e^A \wedge T_A\right\}$$
$$- \frac{i}{16\pi G}\int_M \Theta_A \wedge \Theta^A \tag{25}$$

where

$$\Theta_A = T_A + (2\pi G)\epsilon_{ABCD}\overline{\Psi}_L \gamma^B \Psi_L e^C \wedge e^D, \tag{26}$$

and $*$ denotes the Hodge duality operator. The Samuel-Jacobson-Smolin action[3] which contains only $A^-_{AB} = \frac{1}{2}(-iA_{AB} + \frac{1}{2}\epsilon_{AC}{}^{CD}A_{CD})$, rather than the full spin connection A_{AB}, and the vierbein is

$$S^-_{SJS} = \frac{i}{8\pi G}\int_M \Sigma^{-AB} \wedge F^-_{AB} - \frac{i\lambda}{3(16\pi G)}\int_M \Sigma^{-AB} \wedge \Sigma^-_{AB}$$

---

[3] The cosmological constant term is included here for completeness.

$$= \frac{1}{16\pi G}\int_M \{e^A \wedge e^B \wedge *F_{AB} - 2\lambda(*1)\} + \frac{i}{16\pi G}\int_M \{d(e^A \wedge T_A) - T_A \wedge T^A\} \tag{27}$$

with anti-self-dual two-forms

$$F^-_{AB} = dA^-_{AB} + A^-_{AC} \wedge A^{-C}{}_B$$
$$= \frac{1}{2}(-iF_{AB} + \frac{1}{2}\epsilon_{AB}{}^{CD}F_{CD}),$$
$$\Sigma^-_{AB} = \frac{1}{2}(-ie_A \wedge e_B + \frac{1}{2}\epsilon_{AB}{}^{CD}e_C \wedge e_D). \tag{28}$$

F_{AB} is the curvature of the full spin connection.

The remarkable thing about this combined action is that on its extremum, the torsion trace B_μ vanishes, and so the two sets of spinor field equations derived the action Eqn.[15] and its Hermitized form agree at the classical level. The two actions are therefore completely equivalent at the tree level. Differences will arise at the quantum level.

We can gain a better appreciation of how these differences arise by looking at the imaginary part of Eqn.[15]:

$$i\text{Im}(S^-) = -\frac{i}{2}\int_M d^4x e \nabla_\mu(\overline{\Psi}_L \gamma^\mu \Psi_L). \tag{29}$$

The covariant derivative is dependent upon the torsion trace field, and at the same time, it is also given by the singlet axial current. This current receives an anomaly due to quantum fermion loop corrections. The two actions therefore are expected to generate different amplitudes in the presence of torsion, and in the presence of the singlet anomaly.

We may summarize the differences in the two actions in the following way. The complex Weyl action Eqn.[15] manifests chiral asymmetry, a defining property of the standard model, and does not respect CPT invariance. (For previous discussions on CPT non-invariance, see [5,6].) On the other hand, the spin current, defined as the source of the anti-self-dual Ashtekar field, $A_{\mu A,B}$, contains all of the generators of the Lorentz group, here identified as the complex SO(3,C) group. The action is manifestly locally Lorentz symmetric. By contrast, the Hermitized action respects CPT invariance, contains both spin connections, and the spin current, defined in a similar way, is now Hermitian. It contains only the rotation part of the Lorentz group. The boosts are contained in the "orbital" part of the angular momentum curren, defined as the first moment of the energy-momentum tensor. The action does not exhibit in an obvious way the Lorentz gauge symmetry.

Although equivalent at the classical level, the two actions define distinct quantum theories. Fundamentally, the distinction arises from the complexity of one relative to the other. Within our present discussion, the imaginary parts of the Weyl action are seen to be controlled by torsion terms and the anomalous singlet axial current to which it is coupled. Therefore, any possible source of a breakdown of CPT due to the chiral couplings will only occur in manifolds with torsion, or for which the singlet anomaly is present. Or both. Perturbative amplitudes about flat space will fully respect CPT symmetry. Consequences of CPT violations under these circumstances will be reported elsewhere.

We conclude by discussing briefly the issue of unitarity for scattering amplitudes that follow from complex actions. Unitarity is preserved for amplitudes generated perturbatively from the Hermitized action. Is unitarity preserved for the Weyl action? This question is not so easy to answer within the context of the most general space-time manifolds. Our experience with the equivalence of the two sets of spinor equations from the two actions indicate that apparent disparities can be taken account by considering gravity dynamics. Ultimately, therefore, the answer to the question of unitarity must await a more complete understanding of quantum gravity fluctuations. On the other hand, as in the case for CPT nonconservation, any possible source of a breakdown of unitarity can come about only through torsion and the singlet axial anomaly. As a result, simple scattering against background fields which are solutions of the usual Einstein field equations will be fully unitary.[4]

This research was supported in part by the US Department of Energy under Grant No. DE-FG05-92ER40709-A005.

REFERENCES

1. S. Weinberg, *Gravitation and Cosmology*, John Wiley.
2. L. N. Chang and C. Soo, Phys. Rev. D **53**, 5682 (1996); C. Soo and L. N. Chang, hep-th/9702171.
3. A. Ashtekar, Phys. Rev. Lett. **57**, 2244 (1986); Phys. Rev. **D36**, 1587(1986); *Lectures on nonperturbative canonical gravity*, (World Scientific, Singapore, 1991) and references therein.
4. J. Samuel, Pramāna J. Phys. **28**, L429(1987); Class. Quantum Grav. **5**, L123 (1988); T. Jacobson and L. Smolin, Phys. Lett. **B196**, 39 (1987), Class. Quantum Grav. **5**, 583 (1988).
5. V. A. Kostelecky and R. Potting, Nucl. Phys. B**359**, 545 (1991); V. A. Kostelecky, R. Potting and S. Samuel, in *Proceedings of the 1991 Joint International Lepton-Photon Symposium and Europhysics Conference in High Energy Physics* eds. S. Hegarty et al. (World Scientific, Singapore, 1992).
6. D. N. Page, Phys. Rev Lett. **44**, 301 (1980); R. M. Wald, Phys. Rev. D **21**, 2742 (1980); L. Alvarez-Gaume and C. Gomez, Comm. Math. Phys. **89**, 235 (1983); J.

---

[4] The issue of unitarity is of course further complicated by regions in space-time cloaked by event horizons, which we do not discuss here.

Ellis, J, S. Hagelin, D. V. Nanopoulos and M. Srednicki, Nucl. Phys. B**251**, 381 (1984); J. Ellis, N. E. Mavromatos and D. V. Nanopoulos, Phys. Lett. B**293**, 142 (1992); for discussions on precision tests of CPT-violation, see for instance, P. Huet and M . E. Peskin, Nucl. Phys. B **434**, 3 (1995); J. Ellis, J. Lopez, N. Mavromatos, and D. Nanopoulos, Phys. Rev. D **53**, 3486 (1996); V. A. Kostelecky, *Testing CPT symmetry*, hep-ph/9709263.

Experimental Tests of CP, T and CPT Symmetries using K^0 and $\bar{K}^0$

D. Zavrtanik

School of Environmental Sciences, Nova Gorica and
J. Stefan Institute, Ljubljana, Slovenia
(danilo.zavrtanik@ses-ng.si)

for the CPLEAR Collaboration

A. Angelopoulos[1], A. Apostolakis[1], E. Aslanides[11], G. Backenstoss[2], P. Bargassa[13], O. Behnke [17], A. Benelli [9], V. Bertin[11], F. Blanc[7,13], P. Bloch[4], P. Carlson[15], M. Carroll[9], E. Cawley[9], M.B. Chertok[3], M. Danielsson[15], M. Dejardin[14], J. Derre[14], A. Ealet[11], C. Eleftheriadis[16], L. Faravel [7], P. Fassnacht[11], W. Fetscher[17], M. Fidecaro[4], A. Filipčič[10], D. Francis[3], J. Fry[9], E. Gabathuler[9], R. Gamet[9], H.- J. Gerber[17], A. Go[4], A. Haselden[9], P.J. Hayman[9], F. Henry-Couannier[11], R.W. Hollander[6], K. Jon-And[15], P.-R. Kettle[13], P. Kokkas[4], R. Kreuger[6], R. Le Gac[11], F. Leimgruber[2], I. Mandić[10], N. Manthos[8], G. Marel[14], M. Mikuž[10], J. Miller[3], F. Montanet[11], A. Muller[14], T. Nakada[13], B. Pagels [17], I. Papadopoulos[16], P. Pavlopoulos[2], G. Polivka[2], R. Rickenbach[2], B.L. Roberts[3], T. Ruf[4], M. Schäfer[17], L.A. Schaller[7], T. Schietinger[2], A. Schopper[4], L. Tauscher[2], C. Thibault[12], F. Touchard[11], C. Touramanis[9], C.W.E. Van Eijk[6], S. Vlachos[2], P. Weber[17], M. Wolter[17], D. Zavrtanik[10] and D. Zimmerman[3]

[1] *University of Athens, Greece,* [2] *University of Basle, Switzerland,* [3] *Boston University, USA,* [4] *CERN, Geneva, Switzerland,* [5] *LIP and University of Coimbra, Portugal,* [6] *Delft University of Technology, Netherlands,* [7] *University of Fribourg, Switzerland,* [8] *University of Ioannina, Greece,* [9] *University of Liverpool, UK,* [10] *J. Stefan Inst. and Phys. Dep., University of Ljubljana, Slovenia,* [11] *CPPM, IN2P3-CNRS et Université d'Aix-Marseille II, France,* [12] *CSNSM, IN2P3-CNRS, Orsay, France,* [13] *Paul Scherrer Institut(PSI), Villigen, Switzerland,* [14] *CEA, DSM/DAPNIA, CE-Saclay, France,* [15] *Royal Institute of Technology, Sweden,* [16] *University of Thessaloniki, Greece,* [17] *ETH-IPP Zürich, Switzerland*

Abstract. The CPLEAR experiment at CERN measured the CP and CPT violation parameters and determined in a direct way the T violation. The results allow the determination of the CPT violation parameters in the neutral kaon mixing with a precision better than a few 10^{-4}. The mass equality between K^0 and $\bar{K}^0$ is tested down to the level of 10^{-19} GeV. In addition, physics on a scale close to the Planck mass is probed for the first time.

Introduction

According to our present knowledge of weak interactions, the discrete symmetries C, P and T are not exact symmetries in our Universe. The same applies to the combined symmetry CP. This fact has been experimentally well established and well accomodated in the Standard Model, but its origin is not yet understood. Within the framework of a local field theory, of Lorentz invariance and of the usual spin-statistics requirements, the triple product of C, P and T symmetries represents an exact symmetry expressed by the CPT theorem [1]. CPT invariance, being a natural consequence of local quantum field theory, warranties the equality of lifetimes and masses of particles and antiparticles. It is conceivable, however, that a small violation of CPT symmetry could occur in extensions of non-local theories. Thus, it is imperative to check CPT invariance wherever possible.

At present, the neutral kaon system remains the most precise laboratory for measuring the totality of the parameters which describe, in the most general way, the exactness of discrete symmetries. CPLEAR has succesfully developed a new experimental approach [2] to measure the relevant CP, T and CPT parameters in all main decay modes of the neutral kaon with the same apparatus.

CP, T and CPT symmetry violations can be studied by comparing decay properties of particles and antiparticles. The symmetric production of neutral kaons with known strangeness through the annihilation of low energy antiprotons at LEAR allows CPLEAR to compare K^0 and $\bar{K}^0$ decay rates. In addition, by measuring the strangeness of the neutral kaon when it decays semileptonically, CPLEAR measures the time evolution of the neutral kaon strangeness.

Neutral Kaon System

Although the neutral kaon phenomenology is described in many text-books and reviews [3], we present here a brief outline of the basic formalism for completness.

The time evolution of the $K^0/\bar{K}^0$ system can be described by a 2×2 Hamiltonian matrix $\hat{H}$. Because the weak interaction allows $K^0 \leftrightarrow \bar{K}^0$ transitions and kaon decays, the matrix is non-diagonal and complex [4]

$$\hat{H} = M - \frac{i}{2}\Gamma \qquad (1)$$

where M is the mass and Γ the decay matrix, both of them Hermitian if unitarity holds.

The smallness of the weak interaction allows for a perturbation expansion

$$M_{\alpha\beta} = <\alpha|H|\beta> + \sum_n \mathcal{P}\frac{<\alpha|H|n><n|H|\beta>}{m_K - m_n} \qquad (2)$$

$$\Gamma_{\alpha\beta} = 2\pi \sum_n <\alpha|H|n><n|H|\beta> \delta(m_K - m_n) \qquad (3)$$

where $\mathcal{P}$ denotes the principal value. The sum in n goes over all possible states while the delta function in eq. (3) confines it to real states only.

K_L and K_S are the eigenvectors of the $\hat{H}$ matrix with the following eigenvalues:

$$\Lambda_{L,S} = m_{L,S} - \frac{i}{2}\Gamma_{L,S} \qquad (4)$$

where $m_{L,S}$ and $\Gamma_{L,S}$ are the mass and decay width of the K_L and K_S, respectively. K_L and K_S have different lifetimes ($\tau_S = (89.27 \pm 0.09)\,\text{ps}; \tau_L = (51.7 \pm 0.4)\,\text{ns} \sim 580\,\tau_S$) and masses ($\Delta m = m_L - m_S = (530.4 \pm 1.4) \times 10^7\,\hbar/\text{s} = (3.491 \pm 0.009) \times 10^{-12}\,\text{MeV}$) [5].

Applying C, P and T operators to eqs. (2) and (3) one obtains that, if $\hat{H}$ in invariant under $\mathcal{T}$, $\mathcal{CPT}$ and $\mathcal{CP}$ transformations, the following conditions must be satisfied:

$\mathcal{T}$ invariance: $\qquad |\Lambda_{12}| = |\Lambda_{21}|$
$\mathcal{CPT}$ invariance: $\qquad \Lambda_{11} = \Lambda_{22}$
$\mathcal{CP}$ invariance: $\qquad |\Lambda_{12}| = |\Lambda_{21}|$ and $\Lambda_{11} = \Lambda_{22}$

Each of the hermitian M and Γ matrix depends on 2 real and 1 complex parameter. It is convenient to use as parameters the 4 physical quantities m_L, m_S, Γ_L, Γ_S and the 2 complex parameters:

$$\epsilon_T = sin\phi_{SW} \frac{|\Lambda_{12}|^2 - |\Lambda_{21}|^2}{\Delta\Gamma\Delta m} e^{i\phi_{SW}} \qquad (5)$$

$$\delta = cos\phi_{SW} \frac{(M_{22} - M_{11}) - \frac{i}{2}(\Gamma_{22} - \Gamma_{11})}{\Delta\Gamma} e^{i(\phi_{SW} + \pi/2)} \qquad (6)$$

with $\Delta\Gamma = \Gamma_S - \Gamma_L$, and $\Delta m = m_L - m_S$ and $\phi_{SW} = \text{atan}(2\Delta m/\Delta\Gamma)$. ϕ_{SW} is often reffered as the superweak phase and is very close to 45^0. From the definition of ϵ_T and δ, it appears very clearly that a non-zero value of ϵ_T implies $\mathcal{CP}$ and $\mathcal{T}$ violation, while a non-zero value of δ implies $\mathcal{CP}$ and $\mathcal{CPT}$ violation.

The time evolution of an initialy pure strangeness state can be written as:

$$|K^0(\tau)> = \frac{1}{2}[(1 - 2\delta)|K^0> + (1 - 2\epsilon_T)e^{-i\phi_\Gamma}|\bar{K}^0>]e^{-i\lambda_S\tau} +$$
$$+ \frac{1}{2}[(1 + 2\delta)|K^0> - (1 - 2\epsilon_T)e^{-i\phi_\Gamma}|\bar{K}^0>]e^{-i\lambda_L\tau} \qquad (7)$$

$$|\bar{K}^0(\tau)> = \frac{1}{2}[(1 + 2\delta)|\bar{K}^0> + (1 + 2\epsilon_T)e^{-i\phi_\Gamma}|K^0>]e^{-i\lambda_S\tau} +$$
$$+ \frac{1}{2}[(1 - 2\delta)|\bar{K}^0> - (1 + 2\epsilon_T)e^{-i\phi_\Gamma}|K^0>]e^{-i\lambda_L\tau} \qquad (8)$$

where ϕ_Γ is the phase of Γ_{12}.

CPLEAR Method

The CPLEAR experiment measures time-dependent decay asymmetries of the following form:

$$A_f(t) = \frac{R_{\overline{K}^0 \to \bar{f}}(t) - R_{K^0 \to f}(t)}{R_{\overline{K}^0 \to \bar{f}}(t) + R_{K^0 \to f}(t)} \qquad (9)$$

where $R_{\overline{K}^0 \to \bar{f}}(t)$ and $R_{K^0 \to f}(t)$ are the decay rates of neutral kaons that were produced as $\overline{K}^0$ and K^0, and are decaying into the charge conjugate final states $\bar{f}$ and f. The neutral kaons are produced by $p\bar{p}$ annihilation at rest, $p\bar{p} \to K^\pm \pi^\mp K^0(\overline{K}^0)$. The strangeness of the neutral kaon is tagged by the accompanying charged kaon. In the case of semileptonic decays, the strangeness of the neutral kaons may also be tagged at the decay time ($\Delta S = \Delta Q$ rule). The CPLEAR detector allows different final states to be selected. In the case of hadronic decays, CPLEAR measures the rates of $\pi^+\pi^-$, $\pi^0\pi^0$, $\pi^+\pi^-\pi^0$ and $\pi^0\pi^0\pi^0$ decays, where $f = \bar{f}$. In the case of semileptonic decays, CPLEAR measures $\pi^\pm e^\mp \nu(\bar{\nu})$ decay rates, where K^0 or $\overline{K}^0$ are the final states ($\Delta S = \Delta Q$ rule). The use of the asymmetries defined in Eq. (9) minimizes systematic errors as most of the acceptances cancel.

Experimental setup

Initial K^0 and $\overline{K}^0$ are produced concurrently in the CPLEAR experiment via the annihilations of antiprotons at rest into $K^-\pi^+K^0$ and $K^+\pi^-\overline{K}^0$. Both reactions occur at a branching ratio of $\approx 2 \times 10^{-3}$. The strangeness of the neutral kaon is tagged by the charge of the accompanying charged kaon. The antiprotons are delivered by low-energy antiproton ring LEAR at CERN and are stopped in a gaseous hydrogen target in the centre of the CPLEAR detector.

The experimental apparatus [6] is shown in Fig. 1. The detector consists of tracking devices (2 Proportional Chambers, 6 Drift Chambers, 2 layers of Streamer Tubes) defining the fiducial decay volume of the neutral kaons. The material in the decay region up to the streamer tubes is minimized by using a gas target with mylar-kevlar walls and an innovative low-mass chamber construction, thus reducing regeneration effects of neutral kaons. A 32-segment scintillator-Čerenkov-scintillator provides particle identification (kaons/pions/electrons). Photons are detected by a 18-layer fine-grain high-gain tube / lead sampling calorimeter. The whole apparatus is embedded in the magnetic field of a solenoid. An online event reconstructiom is performed by a sophisticated multilevel trigger system allowing the detector to operate at a $\bar{p}$ rate of 1 MHz.

In total, about 10^8 K^0 and $\overline{K}^0$ decays were reconstructed. The results refer to the analysis of the complete data-set i.e. of 70 M $K^0, \overline{K}^0 \to \pi^+\pi^-$ decays with $\tau > 1\,\tau_S$, 0.5 M $K^0, \overline{K}^0 \to \pi^+\pi^-\pi^0$ decays, 2 M $K^0, \overline{K}^0 \to \pi^0\pi^0$ decays, 17×10^3 $K^0, \overline{K}^0 \to \pi^0\pi^0\pi^0$ decays and 1.8 M $K^0, \overline{K}^0 \to e\pi\nu$ decays.

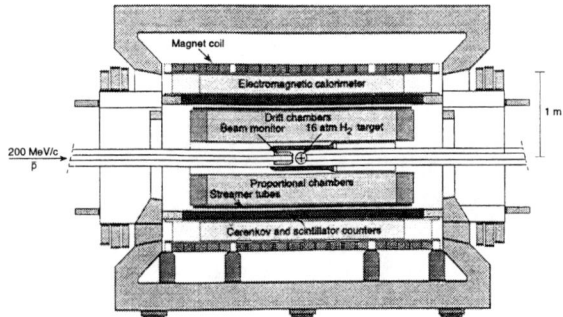

FIGURE 1. View of the CPLEAR detector.

Corrections

Although $\overline{K}^0$ and K^0 are symmetrically produced, corrections have to be applied to construct the asymmetries from the measured decay rates. This is done on an event by event basis, in the same way for all the decay channels.

The tagging efficiencies of $\overline{K}^0$ and K^0 differ due to different interactions of charged kaons and pions with the detector material. This leads to a relative normalization which depends on the kinematics and topology of the primary $K^\pm\pi^\mp$ pair.

Corrections due to the regeneration of neutral kaons in the detector material have to be applied. Because of the lack of experimental data on regeneration amplitudes, the systematic error of some CP violating parameters (for example ϕ_{+-}) would be dominated by regeneration effects. A dedicated run was then devoted to the measurement of the regeneration amplitudes. Details and results are given elsewhere [7].

Background contributions and decay time resolution are obtained from the Monte Carlo simulation and the asymmetries are modified accordingly.

An additional normalization has to be applied to semileptonic decays and is due to the difference in relative detection efficiency of the π^+e^- and π^-e^+ pairs. The normalization factor is determined from pionic annihilation and photon conversion data.

Results

Hadronic decays

From the rate asymmetries as defined in Eq.(9), with $f = 2\pi, 3\pi$, we can determine the CP violating parameters η_{+-}, η_{+-0}, η_{00} and η_{000}. As an example of the accuracy achieved by the CPLEAR experiment, we present the asymmetry A_{+-} (Fig. 2) which resulted in the determination of $|\eta_{+-}|$ and ϕ_{+-}. Details of the analysis and the sources of systematic errors are described in Ref. [8] for the $\pi^+\pi^-$ decays, in Ref. [9] for the $\pi^0\pi^0$ decays, in Ref. [11] for the $\pi^0\pi^0\pi^0$ decays, and in Ref. [10] for the $\pi^+\pi^-\pi^0$ decays. Results on CP violation parameters in hadronic

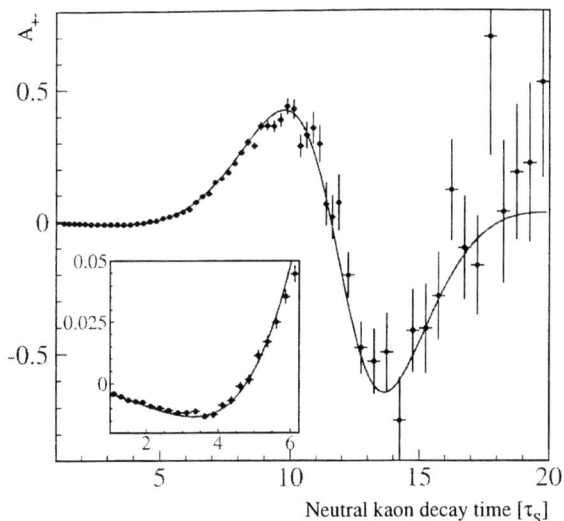

FIGURE 2. $\pi^+\pi^-$ time-dependent asymmetry. The region between 0 and $6\tau_S$ is blown up in the insert.

decays of neutral kaon decays are summarized in Table 1. These measurements provide the most precise determination of ϕ_{+-} and the best limits on η_{+-0} and η_{000}.

Parameter	CPLEAR results
$|\eta_{+-}|$	$(2.254 \pm 0.034) \times 10^{-3}$
ϕ_{+-}	$43.6^0 \pm 0.6^0 \pm 0.4^0_{\Delta m}$
$|\eta_{00}|$	$(2.47 \pm 0.39) \times 10^{-3}$
ϕ_{00}	$42.0^0 \pm 5.9^0$
$Re(\eta_{+-0})$	$(-2 \pm 8) \times 10^{-3}$
$Im(\eta_{+-0})$	$(-2 \pm 9) \times 10^{-3}$
$Re(\eta_{000})$	0.18 ± 0.15
$Im(\eta_{000})$	0.15 ± 0.20

TABLE 1. Results on CP violation parameters in hadronic decays of neutral kaons.

Semileptonic decays

Among the highlights of the CPLEAR experiment are certainly the direct tests of T and CPT invariance.

The CPLEAR experiment can simultaneously measure the four decay rates:

$$R_{\overline{K}^0 \to \pi^+ \ell^- \bar{\nu}}, \quad R_{\overline{K}^0 \to \pi^- \ell^+ \nu}, \quad R_{K^0 \to \pi^+ \ell^- \bar{\nu}} \text{ and } R_{K^0 \to \pi^- \ell^+ \nu}.$$

With $f = \overline{K}^0$ and $\bar{f} = K^0$ in Eq. (9), forming the asymmetry A_T, we directly compare the rate of a neutral kaon produced as a $\overline{K}^0$ and decaying as a K^0 with the rate of a kaon produced as a K^0 and decaying as a $\overline{K}^0$. Any asymmetry is an evidence of T violation. For decay times much larger than K_S lifetime, the A_T equals to approximately $4\mathcal{R}(\epsilon_T)$.

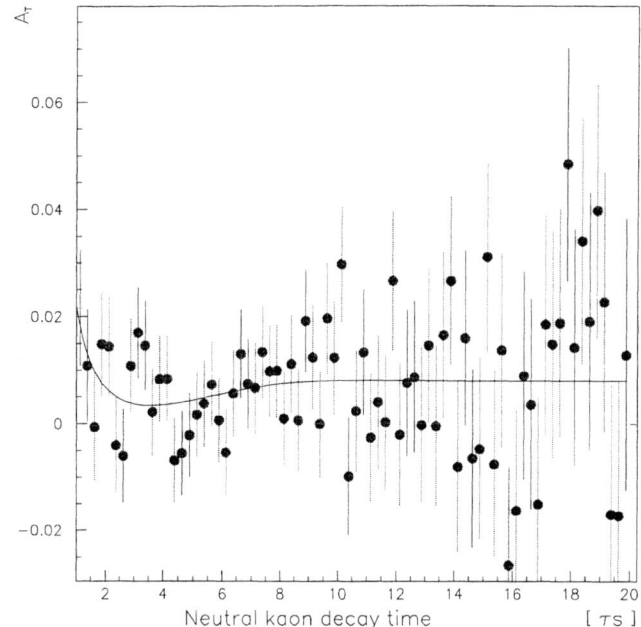

FIGURE 3. Time-dependent asymmetry A_T.

The preliminary result of our fit on the full data set (Fig. 3) yields:

$$A_T = (8.0 \pm 1.7_{stat} \pm 1.0_{syst}) \times 10^{-3}$$

which represents the first direct measurement of T violation with approximately 4σ significance.

Similarly, with $f = K^0$ and $\bar{f} = \overline{K}^0$ we get from Eq. (9) the asymmetry A_{CPT}. The sum of A_T and A_{CPT} properly normalized is equal to $8\mathcal{R}(\delta)$. The value obtained is:

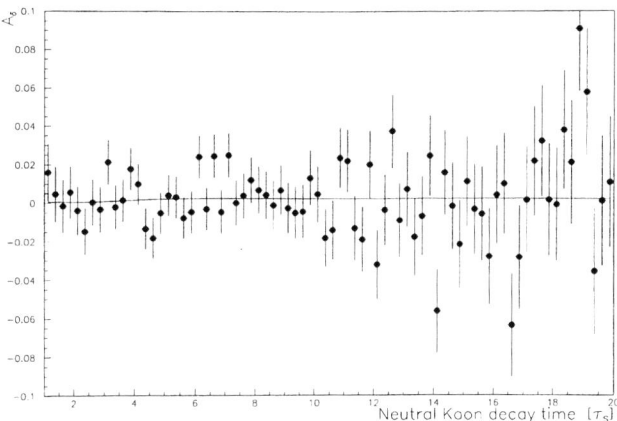

FIGURE 4. Time-dependent asymmetry $A_T + A_{CPT}$.

$$\mathcal{R}(\delta) = (2.96 \pm 3.34) \times 10^{-4}.$$

The result represents a major improvement in direct measurement of the CPT invariance and has to be compared with the value of $\mathcal{R}(\delta) = (180 \pm 200) \times 10^{-4}$ obtained by previous experiments.

An indirect test of CPT

Using the fact that K_S and K_L are eigenstates of the Wigner-Weisskopf hamiltonian and unitarity, one can derive the Bell-Steinberger relation

$$\mathcal{R}(\epsilon) + i\mathcal{I}(\delta) = \frac{\Gamma + i\Delta m}{2\Gamma^2 + \Delta m^2} \sum A_{L,f}^* A_{S,f} \qquad (10)$$

where the sum has to be taken over all final states, and $\Gamma = \frac{1}{2}(\Gamma_S + \Gamma_L)$. The products of amplitudes are understood to be integrated over phase space and summed over spin when necessary. Eq. (10) can be solved for $\mathcal{I}(\delta)$. The Bell-Steinberger relation yields:

$$\mathcal{I}(\delta) = (0.0 \pm 1.9) \times 10^{-5}$$

Parameters not measured by CPLEAR where taken from PDG [5].

Assuming CPT conservation in all decays of the neutral kaons, i.e. assuming CPT violation only in the mixing of K^0 and $\bar{K}^0$, one can transform the value of $\mathcal{I}(\delta)$ into a limit on the mass difference of K^0 and $\bar{K}^0$ [12]:

$$|m_{\bar{K}^0} - m_{K^0}| = (0.1 \pm 2.5) \times 10^{-19} \text{ GeV}/c^2,$$

only a factor of 3 away from the inverse Planck mass.

Test of CPT symmetry and QM

The relation of CPT symmetry to the existence of the 'time arrow' through causality and gravity is fundamental. According to Hawking [13], quantum gravity suggests that quantum field theory should be modified in such a way that pure quantum states evolve into mixed states. This necessarily entails a violation of CPT. Such modification of the quantum-mechanics description of the neutral-kaon system induces three new parameters α, β and γ (see Ref. [14]) describing the loss of quantum coherence in the observed system. By fitting the CPLEAR data on $\pi^+\pi^-$ and $\pi e \nu$ decays [15], we constrain the parameters to the following values:

$$\alpha = (-0.5 \pm 2.8) \times 10^{-17} \text{ GeV},$$
$$\beta = (2.5 \pm 2.3) \times 10^{-19} \text{ GeV},$$
$$\gamma = (1.1 \pm 2.5) \times 10^{-21} \text{ GeV}$$

Our results are close to $m_K^2/m_{Planck} \sim 2 \times 10^{-20}$ GeV which is the presumed order of magnitude for these quantities.

Summary

With its strangeness-tagging capacity in the production and decay of neutral kaons, the CPLEAR experiment measured all the relevant CP, T and CPT violation parameters with an unprecedented accuracy. T violation was observed for the first time at the level of 4σ. The measured limit on the CPT violation parameter $\mathcal{R}(\delta)$ is two orders of magnitude better than previously measured. The limit obtained on $\mathcal{I}(\delta)$ is one order of magnitude better than the limit for $\mathcal{R}(\delta)$. This probes the K^0 and $\bar{K}^0$ mass difference to the order of 10^{-19} GeV.

In addition, the CPLEAR experiment has demonstrated the impressive sensitivity of the neutral kaon system to physics beyond the Standard model.

REFERENCES

1. J.S. Bell, Proc. Royal Soc. A231 (1955) 479;
   G.L. Lüders, Ann. Phys. 2 (1957) 1;
   R. Jost, Helv. Phys. Acta 30 (1957) 409.
2. E. Gabathuler and P. Pavlopoulos, Strong and Weak CP violation at LEAR, Proc. Workshop on Physics at LEAR with Low Energy Cooled Antiprotons, eds. U. Gastaldi and R. Klapisch, Plenum, New York (1982) 747.
3. T.D. Lee and C.S. Wu, Ann. Rev. Nucl Sci. 16 (1996) 511.
4. V. Weisskopf and E. Wigner, Z. Phys. C63 (1930) 54.
5. Review of Particle Properties, Phys. Rev. D54 (1996) 1.
6. R. Adler et al., CPLEAR Collaboration, Nucl. Instr. Methods A379 (1996) 76.
7. A. Angelopoulos, CPLEAR Collaboration, Phys. Lett. B 413 (1997) 422.
8. R. Adler et al., CPLEAR Collaboration, Phys. Lett. B363 (1995) 243.

9. R. Adler et al., CPLEAR Collaboration, Z. Phys. C70 (1996) 211.
10. R. Adler et al., CPLEAR Collaboration, Phys. Lett. B407 (1997) 193.
11. A. Angelopoulos et al., CPLEAR Collaboration, Search for CP violation in the decay of tagged $\bar{K}^0$ and K^0 to $\pi^0\pi^0\pi^0$, Phys. Lett. B. (1998), in print.
12. P. Pavlopoulos, CPLEAR Collaboration, Proc. Workshop on K physics, ed. L. Iconomidou-Fayard, Edition Frontières, Gif-sur-Yvette, France (1997) 307.
13. S. Hawking, Comm. Math. Phys. 87 (1982) 395.
14. J. Ellis, J.S. Hagelin, D.V. Nanopoulos and M. Srednicki, Nucl. Phys. B241 (1984) 381;
    J. Ellis, N.E. Mavromatos and D.V. Nanopoulos, Phys. Lett. B293 (1992) 37.
15. R. Adler et al., CPLEAR Collaboration, Phys. Lett. B364 (1995) 239.

Lepton Flavour Violation Experiments – Some Recent Developments

Klaus P. Jungmann

Physikalisches Institut, Universität Heidelberg
Philosophenweg 12, D-69129 Heidelberg, Germany

Abstract. Dedicated experiments searching for lepton flavour violation can be performed very sensitively using K-decays and μ-decays as well as neutrinoless double β-decay and muonium to antimuonium conversion. Although there is no confirmed signal reported yet, stringent limits for parameters in speculative extensions to the standard model can be set. Some models could recently be ruled out.

I INTRODUCTION

All confirmed experimental data acquired to date indicate the conservation of lepton numbers. This fact can be described by several different empirical laws [1–5], some of which follow additive and some obey multiplicative, parity-like, schemes. Experiments have given no indication yet for favouring any of them. The standard model states for every lepton flavour a separate additively conserved quantum number. However, such lepton numbers have no status, unless their conservation can be associated with a local gauge invariance [6]. Mixings between different generations are well known in the quark sector and the Cabbibo-Kobayashi-Maskawa matrix [7] relates the weak quark eigenstates with their mass eigenstates. A familiar example are the K^0-$\overline{K^0}$ oscillations. At present we are left puzzled why leptons do not show any similar mixing. Recent experimental hints for neutrino oscillations, which have a potential for changing this situation, are not covered here (see e.g. [8]).

Many extensions to the standard model have been proposed and are presently discussed which try to explain further some of its not well understood features like e.g. parity violation in weak interaction or particle mass spectra. They are put by hand into this remarkable theoretical framework which appears to serve as an extremely robust description of all confirmed particle physics. Lepton flavor violation (LFV) appears naturally in such models which include Left-Right-Symmetry, Supersymmetry, Technicolor, Grand Unification, String Theories, Compositeness, and many others. They continue to stimulate experimental searches in a large range of energies. With some low energy experiments new physics can be probed at mass scales far beyond the reach of present accelerators or such planned for the future.

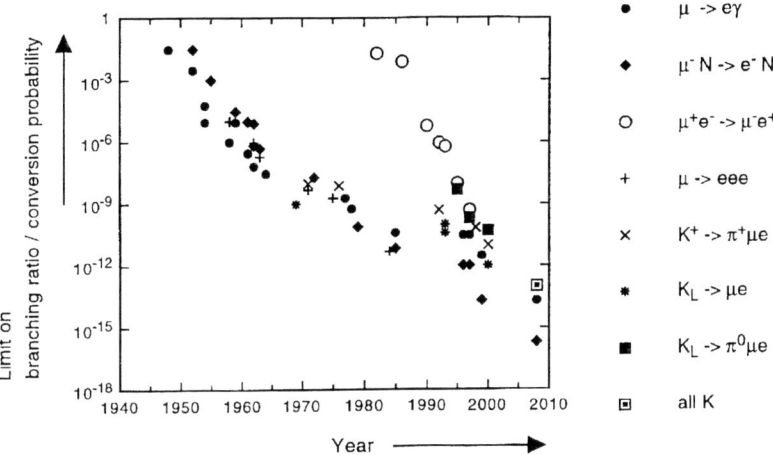

FIGURE 1. Dedicated searches for lepton number violating processes involving muons (μ) and kaons (K). Recent K experiments and $\mu^+e^- - \mu^-e^+$ conversion exhibit the most significant gains in sensitivity. Points in 1998 and beyond are projections of possibilities by the respective experimenters.

Highest sensitivity has generally been reached in dedicated search experiments particularly on Kaons (K) and muons (μ) (Table 1), where also a high discovery potential for new physics exists [9], as well as in non accelerator experiments searching for neutrinoless double β-decay. The decays of heavier objects created in high energy collisions, however, can be observed less accurately. The progress in the K and μ (see sec. IV and V) field is indicated in Fig.1 which shows more than 10 decades of improvement since the first experiments in the late 1940's. The highest recent gain in sensitivity is for muonium (M=μ^+e^-) to antimuonium ($\overline{\text{M}}=\mu^-e^+$) conversion due to a new, yet unused signature (see sec. V C).

TABLE 1. Recently obtained upper limits on lepton number violating processes (90% C.L.).

decay		limit		decay		limit	
Z^0	$\to \mu e$	$2.5 \cdot 10^{-6}$	[10]	K_L	$\to \mu e$	$2 \cdot 10^{-11}$	[14]
Z^0	$\to \tau e$	$7.3 \cdot 10^{-6}$	[10]	K_L	$\to \pi^0 \mu e$	$3.2 \cdot 10^{-9}$	[16]
Z^0	$\to \tau \mu$	$1.0 \cdot 10^{-5}$	[10]	K^+	$\to \pi^+ \mu e$	$4 \cdot 10^{-11}$	[17]
D^0	$\to \mu e$	$1.9 \cdot 10^{-5}$	[11]	μ^+	$\to e + \nu_\mu \overline{\nu}_e \mu e$	$2.5 \cdot 10^{-3}$	[18]
D^0	$\to \pi^0 \mu e$	$8.6 \cdot 10^{-5}$	[11]	μ	$\to eee$	$1 \cdot 10^{-12}$	[19]
D^0	$\to \Phi \mu e$	$3.4 \cdot 10^{-5}$	[11]	μ	$\to e\gamma$	$3.8 \cdot 10^{-11}$	[20]
B^0	$\to \mu e$	$5.9 \cdot 10^{-6}$	[12]	μ^-Ti	$\to e^-$Ti	$6.1 \cdot 10^{-13}$	[21]
B^0	$\to \tau e$	$5.3 \cdot 10^{-4}$	[12]	μ^-Ti	$\to e^+$Ca	$1.7 \cdot 10^{-12}$	[22]
B^0	$\to \tau \mu$	$8.3 \cdot 10^{-4}$	[12]	μ^+e^-	$\to \mu^-e^+$	$G_{M\overline{M}} < 3 \cdot 10^{-3} G_F$	[23]
B^0	$\to$ K μe	$1.8 \cdot 10^{-5}$	[12]	^{76}Ge	$\to ^{76}$Se $\ e^-e^-$	$T_{1/2} > 1.2 \cdot 10^{25} y$	[24]
τ	$\to e\gamma$	$2.7 \cdot 10^{-6}$	[13]			$m_{\nu_e}(Maj.) < 0.45 eV$	[24]
τ	$\to \mu \gamma$	$3.0 \cdot 10^{-6}$	[13]				

II NEUTRINOLESS DOUBLE β-DECAY

A β-decay of a nucleus involving two electrons and no neutrinos would violate electronic lepton flavour by two units. It has been suggested in many models, particularly such involving neutrinos of Majorana type. It is being searched for in many experiments (see Table 2) using ^{48}Ca, ^{76}Ge, ^{82}Se, ^{100}Mo, ^{116}Cd, ^{130}Te and ^{136}Xe. Among those the Heidelberg-Moscow Germanium experiment provides the most stringent half life limit of $T_{1/2} \leq 1.2 \cdot 10^{25}$ y [24]. It uses most advantageously isotopically enriched material with 86% ^{76}Ge as a semiconductor detector to watch its own nuclei decay. It is situated in a clean and carefully against background radiation shielded environment in the Gran Sasso underground laboratory. The measures include purging with purified nitrogen as well as using copper material in the cooling system in the vicinity of the actual detector which was selected for low intrinsic radiation. Remaining background counts were further suppressed by pulse shape analysis. The result achieved in 31.8 kg years with the 11.5 kg detector can be used to impose an upper limit on the electron neutrino Majorana mass of 0.45 eV, which is well below the electron neutrino mass limit of 3.9 eV established in model independent general direct searches using tritium decay [25].

With 1 ton enriched ^{76}Ge distributed in 288 individual detectors, as suggested by the GENIUS proposal, one could expect in 10 years running time a limit of $T_{1/2} < 6 \cdot 10^{28}$ y corresponding to a Majorana neutrino mass limit of below 6 meV/c^2 [24]. It is a particularly nice feature of most experiments searching for neutrinoless double β-decay that they can also contribute to sensitive searches for cold dark matter, particularly weakly interacting massive particles (WIMPS) in mass regions above ≈ 20 GeV/c^2.

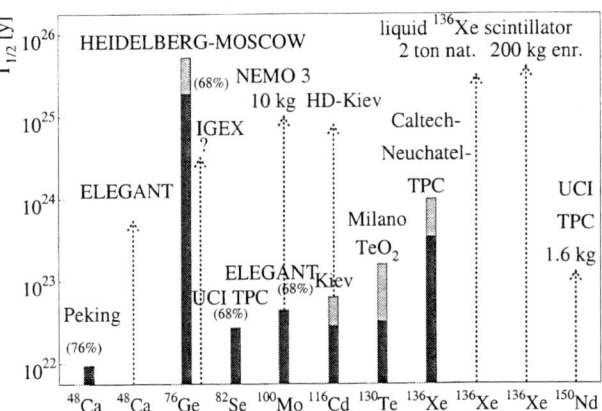

FIGURE 2. Experiments searching for neutrinoless double β-decay. The most sensitive ones use enriched ^{68}Ge detectors. The dark areas represent the current status and the lighter color indicates near future possibilities. The dashed arrows are long term future plans. Among the most ambiguous projects ranks a 1 ton Ge detector which could be used to gain two orders of magnitude in sensitivity (from ref. [24]).

III EXPERIMENTS AT ELECTRON-POSITRON COLLIDERS - Z^0 AND $W^\pm$ BOSONS AND τ LEPTONS

The general purpose detectors installed at the large high energy electron-positron colliders provide the possibility heavy elementary particles and gauge bosons like the τ lepton or the $W^\pm$ and Z^0 bosons to be observed for rare decays and particularly for lepton number violating effects. Their high mass offers for each particle a large number of different possible purely leptonic and semileptonic decay channels. Z^0 bosons were produced in large quantities at the LEP storage ring of CERN and the Stanford Linear Collider. With the LEP200 upgrade a significant number of $W^\pm$ bosons became available. For τ's the CLEO detector at the Cornell CESR facility provided a significant amount of the available data particularly on neutrinoless τ decays [13,26] as well as on B^0 and D^0 decays [11,12].

The sensitivity of all analyses for lepton number violating (LNV) decays have a principle limit set by statistics. For a particular decay channel further restrictions arise from finite acceptances for the final state particles which explain the course differences in the upper bounds reported for the different channels although starting from the same initial state (Table 1). The limits on branching ratios are in general much higher than the ones obtained in dedicated experiments on K's and μ's. For τ's one expects in the near future a sensitivity not better than 10^{-7} for any decay mode.

However, such bounds are of great value for discriminating theoretical models where mass scaling runs with a high power of the mass ratios. In the framework of superstring models, for example, the decay $l \to e\gamma$, where l stands for μ or τ, scales with the fifth power of the lepton mass m_l. In this particular case the upper limit of $2.7 \cdot 10^{-6}$ for $\tau \to e\gamma$ can compete with the present $3.8 \cdot 10^{-11}$ limit on $\mu \to e\gamma$ due to the $1.3 \cdot 10^6$ enhancement factor from the mass ratio $m_\tau/m_\mu \approx 16.8$. However in general the mass scaling is expected to be less dramatic.

IV EXPERIMENTS ON KAONS

With the availability of intense Kaon sources at the Fermi National Accelerator Laboratory (FNAL) and the Brookhaven National Laboratory (BNL) and with novel experimental techniques developed to cope with high data rates the search for LFV K-decays has gained a lot of interest. Here the experiments BNL-871 searching for $K^+ \to \pi^+ e\mu$, BNL-865 searching for $K_L \to \mu e$ and the Fermilab KTeV effort FNAL-799II investigating $K_L \to \pi^0 \mu e$ promise significant improvements (see Table 2), where the LFV decays are searched along with measurements on very rare K decay channels. Among the new physics that could be revealed are new heavy gauge bosons with masses up to order 50-200 TeV/c^2, far beyond the reach of even any planned accelerator [27]. At the projected Japanese Hadron Facility (JHF) one could expect significant improvements beyond the present status.

TABLE 2. Three presently ongoing searches for lepton flavour violating K decays.

	past limit	present limit (1998)	anticipated limit of ongoing experiment	future possibility
$K^+ \to \pi^+ e\mu$	$2 \cdot 10^{-10}$ BNL-777	$4 \cdot 10^{-11}$	$\approx 3 \cdot 10^{-12}$ BNL-865	10^{-13}
$K_L \to \mu e$	$3 \cdot 10^{-11}$ BNL-791	$3 \cdot 10^{-12}$	$\approx 8 \cdot 10^{-13}$ BNL-871	10^{-13}
$K_L \to \pi^0 \mu e$	$3.2 \cdot 10^{-9}$ FNAL-799		$\approx 1 \cdot 10^{-11}$ FNAL799II	10^{-13}

V EXPERIMENTS INVOLVING MUONS

The decay $\mu \to e\gamma$ was the first being searched for shortly after the muon's nature as a heavy electron-like particle became apparent. Searches for rare and forbidden muon decays have been among the most precise experiments in physics since and have always been of special interest in the context of unified gauge theories, as they can provide accurate tests of speculative models and because of the achievable experimental precision they may be able to discriminate between such [28]. Recently forbidden muon decays have attracted attention, when their possible sensitivity to effects arising in minimal supersymmetry (SUSY) were discussed in theoretical studies [29]. It was pointed out that for values of $tan\beta$ (the ratio of the vacuum expectation values of the two Higgs fields involved) which exceed about 3, the branching ratio should be above $\approx 10^{-14}$ for a decay $\mu \to e\gamma$ and above $\approx 10^{-16}$ for $\mu \to e$ conversion on a Ti nucleus, almost independent of all other parameters in the model. This has stimulated a letter of intent to the Paul Scherrer Institute (PSI), Switzerland, and a proposal to BNL to search for the respective processes.

In the field of searching for SUSY effects in low energy experiments rare decay experiments are in some competition with the just started new precision measurement [30] of the muon magnetic anomaly a_μ where the contribution from SUSY is of order $a_\mu(SUSY) = 140 \cdot 10^{-11} tan\beta * (100 GeV/\tilde{m})^2$ with $\tilde{m}$ the mass of the lightest SUSY particle (see [31]). The measurement goal is $\Delta a_\mu(exp) = 40 \cdot 10^{-11}$ and should be reached around the year 2001.

A $\mu \to e\gamma$ decay

The signature of a $\mu \to e\gamma$ event is a 52.8 MeV positron emitted back to back with a 52.8 MeV photon. The MEGA experiment at the late Los Alamos Meson Physics Facility (LAMPF) consisted of a magnetic spectrometer to observe the charged final state particle and three pair spectrometers for detecting the photon through its e^+e^- pair creation in lead converters. Random coincidences at high rates are reported as major background. Data taking is completed and 16% of the data could be analyzed leading to an upper limit on the branching ratio of

$3.8 \cdot 10^{-11}$ [20] which slightly improves the value of $4.9 \cdot 10^{-11}$ established in a crystal box detector also at LAMPF [32].

At PSI new efforts are being discussed to reach a sensitivity of about $5 \cdot 10^{-14}$ for this decay mode within the next couple of years. The suggested instruments include solutions like a large solid angle magnetic spectrometer for the e^+ surrounded by a crystal calorimeter for the γ, or liquid Xe calorimeters for the γ and others [33].

It should be noted that the tightest bounds on bileptonic gauge bosons, which are common to many speculative standard model extensions, come from $\mu \to e\gamma$, if flavour democracy is assumed [34].

B $\mu \to e$ conversion

Many constraints for speculative models arise from the present experimental bound on the conversion process $\mu + Z \to e + Z$ (Table 1), which is the tightest for all studied LNV decays. The variety of theoretically possible processes that can be tested includes, e.g. supersymmetric loop graphs, heavy neutrinos, leptoquarks, compositeness, Higgs bosons and heavy Z' bosons with anomalous couplings. Generally it is more sensitive to new Physics than $\mu \to e\gamma$ in a wide class of models where the process is generated at the one loop level [35].

The process needs to involve a nucleus to assure elementary conservation laws. If the nucleus is left in its ground state, a conversion event is signaled through the release of a 105 MeV electron, which is uniquely distinguishable from normal muon decay electrons ranging up to 53 MeV. Among the physically relevant intrinsic background processes is μ decay in the atomic orbit after a muonic atom has been formed, which can release much higher energetic electrons, and radiative muon capture, where the photokinematic end point can be close to the signal electron energy.

The ongoing SINDRUM II experiment uses the worldwide brightest continuous muon channel π E5 at PSI. Their new results limit tyhe branching ratios $\mu^- \text{Ti} \to e^+ \text{Ca}^{gs}$ to below $1.7 \cdot 10^{-12}$ for the Ca nucleus in the ground state [22], $\mu^- \text{Ti} \to e^+ \text{Ca}^{GDR}$ to below $3.6 \cdot 10^{-11}$ leaving Ca with giant dipole resonance excitation [22], and $\mu^- \text{Ti} \to e^- \text{Ti}$ to below $6.1 \cdot 10^{-13}$ for Ti in the ground state [21]. For the ground state processes the nucleons interact coherently which enhances the possible effect. In order to boost accuracy in the near future the SINDRUM II collaboration wants to take advantage in the gain of muon flux through a $\pi - \mu$ converter, a novel superconducting device in the beam line which collects π's and releases only μ's with very low π contamination. The latter point is essential as π's are a source of potential background due to nuclear reactions. The projections of the collaboration for the achievable limit in the coherent $\mu^- \text{Ti} \to e^- \text{Ti}$ case are in the 10^{-14} region.

The new Muon Electron Conversion (MECO) experiment proposed at BNL [36] (see Fig.3) is very close in its design to a proposal by Lobashev and collaborators for the Moscow Meson Factory. The setup consists of a target station for π/μ production which uses a proton beam from the AGS accelerator, an S-shaped transport

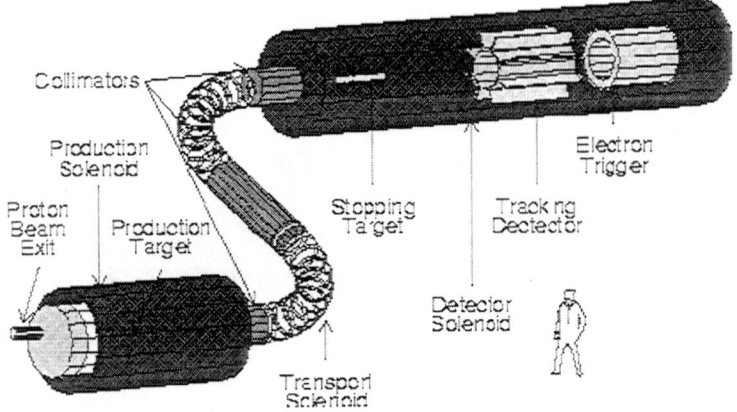

FIGURE 3. The MECO experiment planned at BNL (see [36]).

and purification section and a detector the basic idea of which is to let electrons from normal muon decay pass without being seen and to observe only the 105 MeV signal electrons. The goal is the 10^{-16} level in sensitivity, which will stringetly test supersymmetric models; there is an anticipated ultimate capability for 10^{-18} [15].

C $\mu^+ e^- \to \mu^- e^+$ conversion

The hydrogen-like muonium atom consists of two leptons from different generations. The close confinement of the bound state offers excellent opportunities to explore precisely fundamental electron-muon interactions [37,38]. Since the effect of all known fundamental forces in this system are calculable very well mainly in the framework of quantum electrodynamics (QED), it renders the possibility to search sensitively for yet unknown interactions between both particles.

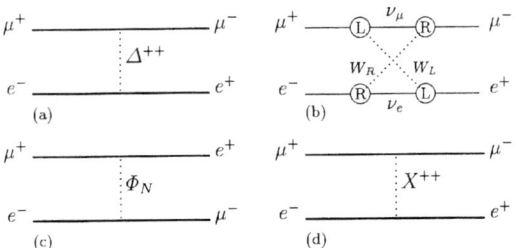

FIGURE 4. Muonium-antimuonium conversion in theories beyond the standard model. The interaction could be mediated by (a) a doubly charged Higgs boson Δ^{++} [39], (b) heavy Majorana neutrinos [40], (c) a neutral scalar Φ_N [41], e.g. a supersymmetric τ-sneutrino $\tilde{\nu}_\tau$ [42,6], or (d) a dileptonic gauge boson X^{++} [43].

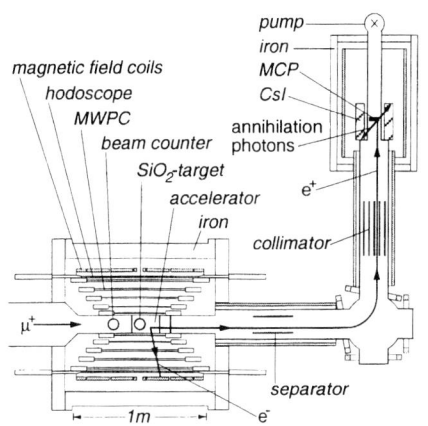

FIGURE 5. Top view of the MACS (Muonium - Antimuonium - Conversion - Spectrometer) apparatus at PSI to search for $M - \overline{M}$ - conversion [44].

An M-$\overline{\rm M}$-conversion would violate additive lepton family number conservation and is discussed in many speculative theories (see Fig. 4). It would be an analogy in the lepton sector to K^0-$\overline{K^0}$ oscillations.

The setup at PSI (Fig. 5) [44] is designed to employ the signature developed in a predecessor experiment at LAMPF, which requires the coincident identification of both particles forming the antiatom in its decay [45,46]. Muonium atoms in vacuum with thermal velocities, which are produced from a SiO_2 powder target, are observed for antimuonium decays. Energetic electrons from the decay of the μ^- in the antiatom can be observed in a magnetic spectrometer at 0.1 T magnetic field consisting of five concentric multiwire proportional chambers and a 64 fold segmented hodoscope. The positron in the atomic shell of the antiatom is left behind after the decay with 13.5 eV average kinetic energy [47]. It can be accelerated to 7 keV in a two stage electrostatic device and guided in a magnetic transport system onto a position sensitive microchannel plate detector (MCP). Annihilation radiation can be observed in a 12 fold segmented pure CsI calorimeter around it.

The relevant measurements were performed during in total 6 month distributed over 4 years during which $5.7 \cdot 10^{10}$ muonium atoms were in the interaction region. One event fell within a 99% confidence interval of all relevant distributions (Fig. 6). The expected background due to accidental coincidences is 1.7(2) events. Thus an upper limit on the conversion probability of $P_{M\overline{M}} \leq 8.2 \cdot 10^{-11}/S_B$ (90% C.L.) was found, where S_B accounts for the interaction type dependent suppression of the conversion in the magnetic field of the detector due to the removal of degeneracy between corresponding levels in M and $\overline{\rm M}$. The reduction is strongest for $(V\pm A)\times(V\pm A)$, where S_B=0.35 [48,49]. This yields for the traditionally quoted upper limit on the coupling constant in effective four fermion interaction $G_{M\overline{M}} \leq 3.0 \cdot 10^{-3} G_F (90\% C.L.)$ with G_F the weak interaction Fermi constant.

This new result, which exceeds bounds from previous experiments [45,50] by a factor of 2500 and the one from an early stage of the experiment [44] by 35, has

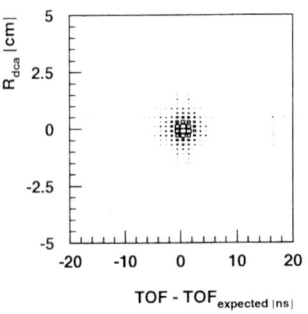
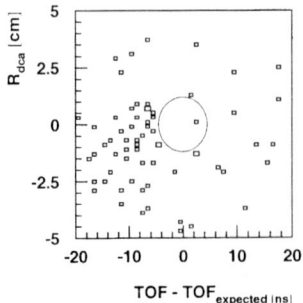

FIGURE 6. Time of flight (TOF) and vertex quality for a muonium measurement (left) and the same for all data of the final 4 month search for antimuonium (right). One event falls into the indicated 3 standard deviations area.

some impact on speculative models. A certain Z_8 model is ruled out with more than 4 generations of particles where masses could be generated radiatively with heavy lepton seeding [51].

A new lower limit of $m_{X^{\pm\pm}} \geq 2.6$ TeV/c² $*g_{3l}$ (95% C.L.) on the masses of flavour diagonal bileptonic gauge bosons in GUT models is extracted which lies well beyond the value derived from direct searches, measurements of the muon magnetic anomaly or high energy Bhabha scattering [43,34]. Here g_{3l} is of order 1 and depends on the details of the underlying symmetry. For 331 models this translates into $m_{X^{\pm\pm}} \geq 850$ GeV/c² which excludes their minimal Higgs version in which an upper bound of 600 GeV/c² has been extracted from an analysis of electroweak parameters [52,53]. The 331 models may still be viable in some extended form involving a Higgs octet [54]. In the framework of R-parity violating supersymmetry [42,6] the bound on the coupling parameters could be lowered by a factor of 15 to $|\lambda_{132}\lambda^*_{231}| \leq 3*10^{-4}$ for assumed superpartner masses of 100 GeV/c². Further the achieved level of sensitivity allows to narrow slightly the interval of allowed heavy muon neutrino masses in minimal left-right symmetry [39] (where a lower bound on $G_{M\overline{M}}$ exists, if muon neutrinos are heavier than 35 keV) to ≈ 40 keV/c² up to the present experimental bound at 170 keV/c².

In minimal left right symmetric models, in which $M\overline{M}$ conversion is allowed, the process is intimately connected to the lepton family number violating muon decay $\mu^+ \to e^+ + \nu_\mu + \overline{\nu}_e$. With the limit achieved in this experiment this decay is not an option for explaining the excess neutrino counts in the LSND neutrino experiment at Los Alamos [55,56].

The consequences for atomic physics of muonium are such that the expected level splitting in the ground state due to $M - \overline{M}$ interaction is below 1.5 Hz/$\sqrt{S_B}$ reassuring the validity of fundamental constants determined in muonium spectroscopy.

A future $M - \overline{M}$ experiment could take particularly advantage of high intense pulsed beams. In contrast to other LNV muon decays, the conversion through its nature as particle - antiparticle oscillation, has a time evolution in which the probability for finding $\overline{M}$ in the ensemble remaining after muon decay increases quadratically in time, giving the signal an advantage growing in time over major exponentially decaying background [46].

VI LONG TERM FUTURE POSSIBILITIES

It appears that the availability of particles limits the ability to find very rare processes or to impose significantly improved limits in continuation of the search program of dedicated experiments. Therefore any measure to boost the respective particle fluxes is a very important step forward. The $\pi - \mu$ converter at PSI or the dedicated tailored muon production of the planned MECO experiment at BNL are examples of novel attempts to overcome this problem. In principle, we need significantly more intense accelerators, such as they are presently discussed at various places. In the intermediate future the Japanese Hadron Facility (JHF) or a possible European Spallation Source (ESS) are important options. Also the discussed Oak Ridge neutron spallation source could in principle accommodate intense muon beams. The most promising facility would be, however, a muon collider [57], the front end of which could provide muon rates 5-6 orders of magnitude higher than present beams (see Table 3).

TABLE 3. Muon fluxes of some existing and future facilities, Rutherford Appleton Laboratory (RAL), Japanese Hadron Facility (JHF), European Spallation Source (ESS), Muon collider (MC).

	RAL(μ^+)	PSI(μ^+)	PSI(μ^-)	JHF(μ^+)	ESS(μ^+)	MC (μ^+, μ^-)
Intensity (μ/s)	3×10^6	3×10^8	1×10^8	4.5×10^7	4.5×10^7	7.5×10^{13}
Momentum bite Δ pm/p[%]	10	10	10	10	10	5-10
Spot size (cm $\times$ cm)	1.2$\times$2.0	3.3$\times$2.0	3.3$\times$2.0	1.5$\times$2.0	1.5$\times$2.0	few$\times$few
Pulse structure	82 ns	50 MHz	50 MHz	300 ns	300 ns	50 ps
	50 Hz	continuous	continuous	50 Hz	50 Hz	15 Hz

It was noted already in the early 60ies that, e.g. the process $e^-e^- \rightarrow \mu^-\mu^-$ is closely related to muonium-antimuonium conversion [58]. Indeed such scattering experiments were carried out at the Princeton-Stanford storage rings at Stanford yielding the at the time best limit on the coupling constant $G_{M\overline{M}}$ [59]. Today, similar proposals have been made for scattering of high energy e^- on e^-, e^- on e^+, μ^- on μ^- and μ^- on μ^+ [60–62]. They were mainly discussed in connection with bileponic gauge bosons. Even a lower limit for the cross section of the process $e^-e^- \rightarrow \mu^-\mu^-$ was found, provided the sum of the light neutrino masses exceeds ≈ 90 eV [62]. Pronounced resonances have been predicted particularly for such experiments at the Next Linear Collider or the high energy end of a muon collider.

Although lepton flavour conservation remains a mystery and searches for its violation were not blessed with a successful observation yet, both the theoretical and experimental work in this connection have led to a deeper understanding of particle interactions. One particular value of the experiments are their continuos contributions towards guiding theoretical developments by excluding various speculative models.

ACKNOWLEDGMENTS

It is a pleasure to thank the organizers of the first tropical workshop for creating the atmosphere for a wonderful and stimulating conference and for their great hospitality and support. The author is grateful to M. Cooper, W. Molzon, A.v.d. Schaaf, M. Zeller for discussions and updates, respectively latest results from their experiments.

REFERENCES

1. Y.B. Zeldovitch, Dan. SSR **86**, 505 (1952)
2. B. Pontecorvo, Sov.Phys.-JETP **37**, 1751 (1959) and Sov. Phys. JETP **6**, 381 (1958)
3. N. Cabbibo and R. Gatto, Phys.Rev.Lett. **5**, 114 (1960); N. Cabbibo, Nuovo Cim. **19**,612 (1961)
4. E.J. Konopinski and H.M. Mahmoud, Phys.Rev.**92**, 1045 (1953)
5. G. Feinberg and S. Weinberg, Phys.Rev.Lett. **6**, 381 (1961)
6. A. Halprin and A. Masiero, Phys.Rev.D**48**, 2987 (1993)
7. M. Kobayashi and T. Maskawa, Prog. Theor. Phys. **49**, 652 (1973)
8. J. Stone, this volume
9. R.N. Mohapatra, Prog.Part.Nucl.Phys. **31**, 39 (1993)
10. O. Adriani et al. Phys. Lett. B **316**, 427 (1993); L. Bugge et al., in: Proc. Europhysics Conference on High-energy Physics, Brussels, J. Lemonne et al. (eds.), World Scientific, Singapore (1996); P. Abreu et al., Z. Phys. C **73**, 243 (1997)
11. A. Freyberger et al., Phys.Rev.Lett.**76**, 3065 (1996)
12. R. Ammar et al., Phys.Rev.D **49**, 5701 (1994)
13. K. Edwards et al., Phys.Rev.D **55**, 3919 (1997)
14. W. Molzon, JHF98 workshop, Tsukuba, (1998); see also ref. [15]
15. T. Kirk, JHF98 workshop, Tsukuba (1998)
16. J. Belz, Proc. Intersections between Particle and Nuclear Physics, 6th conf, T.W. Donnelly (ed.), AIP Press, New York, p.763 (1997); R. Ray, JHF98 workshop, Tsukuba (1998)
17. M. Zeller, priv. com.;see also ref. [15] and S. Eilerts, loc. cit. [16], p. 779 (1997)
18. K. Eitel, doctoral thesis, University of Karlsruhe (1995)
19. W. Bertl et al., Nucl.Phys. B**260**, 1 (1985)
20. M.D. Cooper et al., loc. cit. [16], p. 34 (1997)
21. S. Eggli et al., publication in preparation (1998)
22. J. Kaulard et al., submitted for publication (1998)
23. V. Meyer et al., loc. cit. [16], p. 429 (1997)

24. H.V. Klapdor-Kleingrothaus and M. Hirsch, Z.Phys.A**359**, 361 (1997); H.V. Klapdor-Kleingrothaus, Proc. Beyond the Desert Conference, Institute of Physics Publishing, Bristol, p.485 (1998)
25. V.M. Lobashev, in: Proc. Neutrino 96, World Scientific, Singapore(1997)
26. D. Bliss et al., Phys.Rev.D **57**,5903 (1998)
27. S.H. Kettell, hep-ex/9801016 (1998) and references therein
28. T.S. Kosmas, G.K. Leontaris, J.D. Vergados, Prog.Part.Nucl.Phys. **33**, 397 (1994)
29. R. Barbieri, L. Hall and A. Strumia, Nucl.Phys. B**445**, 219 (1995)
30. J.P. Miller et al, loc. cit. [16], p. 792 (1997)
31. U. Chattopadhyay and P. Nath, Phys.Rev. D**53**, 1648 (1996)
32. T. Bolton et al., Phys. Rev D**38**, 2077 (1988)
33. A. v.d. Schaaf et al., Letter of Intent to PSI, R98-05.0 (1998)
34. F. Cuypers and S. Davidson, Eur.Phys.J. **C2**, 503 (1998)
35. M. Raidal and A. Santamaria, hep-ph/9710389 (1997)
36. W. Molzon et al., Proposal to BNL E-940 (1997)
37. V.W. Hughes and G. zu Putlitz, in: *Quantum Electrodynamics*, World Scientific, Singapore, T. Kinoshita (ed.), p. 822 (1990)
38. K. Jungmann, in: *Atomic Physics 14* (New York: AIP Press), D. Wineland et al. (ed.), p. 102 (1994)
39. P. Herczeg and R.N. Mohapatra, Phys.Rev.Lett. **69**, 2475 (1992)
40. A. Halprin, Phys.Rev.Lett. **48**, 1313 (1982)
41. W.S. Hou and G.G. Wong, Phys.Rev. D**53** 1537 (1996)
42. R.N. Mohapatra, Z.Phys. C**56**, S117 (1992)
43. H. Fujii et al., Phys.Rev. D **49** 559 (1994)
44. R. Abela et al., Phys.Rev.Lett. **77** 1951 (1996)
45. B.E. Matthias et al., Phys.Rev.Lett. **66**, 2716 (1991)
46. L. Willmann and K. Jungmann, Lecture Notes in Physics, Vol. 499, (1997)
47. L. Chatterjee et al., Phys. Rev. D**46**, 46 (1992)
48. K. Horrikawa and K. Sasaki, Phys. Rev. D**53**, 560 (1996)
49. G.G. Wong and W.S. Hou, Phys.Lett.B**357**, 145 (1995)
50. V.A. Gordeev et al, JETP Lett. **59**, 589 (1994)
51. G.G. Wong and W.S. Hou, Phys.Rev.D**50**, R2962 (1994)
52. P.Frampton , Phys.Rev.Lett**69**, 1889 (1994); see also: hep-ph/97112821 (1997)
53. P.Frampton and S. Harada, hep-ph/9711448 (1997))
54. P. Frampton, priv. comm. (1998)
55. P. Herczeg, Conference "Beyond the Desert 97", Castle Ringberg (1997)
56. Athanassopoulos C et al. Phys.Rev. C54 2685 (1996)
57. R.B. Palmer and J.C. Gallardo, physics/9802002 (1998); R.B. Palmer, physics/9802005 (1998)
58. S. Glashow, Phys.Rev.Lett. **6**, 196 (1961)
59. W.C. Barber et al, Phys.Rev.Lett. **22**, 902 (1969)
60. P.Frampton , Phys.Rev. D **45**, 4240 (1992)
61. W.S. Hou, Nucl. Phys. B**51A**, 40 (1996)
62. M.Raidal, Phys.Rev.D**57**, 2013 (1998)

Spontaneous CP Violation

Paul H. Frampton

Department of Physics and Astronomy
University of North Carolina
Chapel Hill, NC 27599-3255, USA

Abstract. In this talk I begin with some general discussion of the history of CP violation, then move on to aspects of the aspon model including the production of new particles at LHC, implications for B decay, generalized Cabibbo mixing and a reevaluation of kaon CP violation. Finally there is a summary.

I HISTORY

The parity operation is a symmetry of Newton's Laws provided we assume a strong form of the Third Law: Action and Reaction are equal and opposite and directed along the line of centers. For quantum mechanics, Parity was introduced by Wigner in 1927 [1]. The violation of P was first entertained by Lee and Yang in 1956 [2]; it was quickly verified by Madame Wu [3] and others [4].

Time reversal T is an invariance of Newton's Laws. In quantum mechanics T was introduced as the now-familiar anti-unitary operator by Wigner [5]. [T violation was studied in classical statistical mechanics earlier by Boltzmann and Panlevé, but T violation in microscopic laws was not seriously questioned until 1964.]

The operation of charge conjugation (C) could hardly be conceived of before the Dirac equation [6] in 1928 predicted the e^+, discovered in 1932. The C invariance of quantum electrodynamics was first discussed by Kramers [7] in 1937.

The invariance under CPT was proven for quantum field theory in 1954 by Luders [8] under the weak assumptions of lorentz invariance and the spin-statistics connection.

After Lee and Yang, but before P violation was discovered, Landau [9] suggested that CP is an exact symmetry.

In [10] CP violation was discovered in the decay of neutral kaons. The longer-lived CP eigenstate K_L was observed to decay 0.2% of the time into $\pi\pi$, disallowed if CP is exact. The CP violation is characterized by the parameter ϵ. Since CP violation has never been seen outside of the kaon system, ϵ is the only accurately measured (to within 1%) CP violation parameter.

In a remarkable paper containing an all-time favorite idea in particle theory, in 1966 Sakharov [11] proposed that the baryon number of the universe arose due to a combination of three ingredients: (1) B violating interactions. (2) Thermodynamic disequilibrium. (3) CP Violation.

When GUTs became popular, Yoshimura [12] and others illustrated this idea. More recently baryogenesis at the electroweak phase transition is discussed based on the same three ingredients.

In 1973, Kobayashi and Maskawa(KM) [13] proposed their mechanism for CP violation assuming, with great foresight, three fermion generations. The issue now is whether KM is the full explanation of the observed CP violation.

In 1976 't Hooft [14] emphasised the strong CP problem that a parameter θ in QCD must be fine-tuned to $\bar{\theta} < 10^{-9}$ to avoid conflicting with the upper limit on the neutron electric dipole moment.

In the decade of the 1980s, the areas of weak CP violation and strong CP proceeded along largely separate tracks.

Having mentioned time-honored classics of the subject of CP, in the rest of the talk I shall specialize to six recent papers on a specific CP model - the aspon model - published: two [15,16] in 1991, one [17] in 1992, one [18] in 1994, one [19] in 1997, and finally one in 1998 [20].

II ASPON MODEL

Because QCD has a possible term involving $\bar{\theta}$ in its lagrangian, there is the potential for unacceptably large CP violation. One approach which is much less motivated now than twenty years ago is to introduce a color-anomalous U(1); a second is to assume the up quark is massless, although this clashes with successes of chiral perturbation theory. The third direction, exemplified by the aspon model is to assume CP is a symmetry of the fundamental theory and to arrange that θ is zero at tree level, remaining sufficiently small from radiative corrections.

In the aspon model the gauge group of the standard model is extended to $SU(3) \times SU(2) \times U(1) \times U(1)_{new}$. The new charge Q_{new} is not carried by any of the fields of the SM. One additional doublet of Dirac quarks (U, D) with charge $Q_{new} = 1$ is introduced, together with two complex singlet scalars $\chi^\alpha, \alpha = 1, 2$.

The χ^α acquire VEVs with a non-zero relative phase, spontaneously breaking both the gauged $U(1)_{new}$ and CP. The gauge boson of $U(1)_{new}$ becomes massive by the Higgs mechanism and is called the "aspon".

The Yukawa couplings with χ involve the right-handed U and D but not the left-handed counterparts. As a result there are zeros [21] in the 4×4 quark mass matrices such that although there are complex entries the determinant is real. Hence $\bar{\theta} = 0$ at tree level.

Such a mass matrix is diagonalized by a bi-unitary transformation which is conveniently expanded in the small parameters $x_i = F_i/M$ where F_i are the off-diagonal elements and M is the Dirac mass. We may regard the x_i as independent of the family number i and simply write $|x_i| = x$. It turns out that x is constrained to lie in the window $3 \times 10^{-5} < x^2 < 10^{-3}$ by the constraints of $\bar{\theta}$ and of CP violation.

A FCNC

Since we have introduced right-handed doublets, a first concern is with the size of the induced Flavor-Changing Neutral Currents (FCNC). It turns out that these are more than adequately suppressed.

B $\bar{\theta}$

At one loop level $\bar{\theta}$ acquires a non-zero value and this leads to an upper limit on the product (λx^2) where λ is the coefficient of the quartic coupling $|\phi|^2|\chi|^2$ between the standard Higgs ϕ and the χ fields.

C Weak CP Violation

Fitting to the CP violation parameter ϵ and to the allowed range for $Re(\epsilon'/\epsilon)$ gives an upper limit on the symmetry breaking scale for $U(1)_{new}$ of about 2TeV. One thus predicts that, assuming the gauge coupling is not much larger than the others of the standard model, the new particles Q and A lie well below 1TeV. This fits ones intuition that if the new states are too heavy the diagrams contributing to CP violation in the kaon system will be too small.

III PRODUCTION OF A AND Q AT LHC

Production of $\bar{Q}Q$ is dominated by gluon fusion diagrams just like $\bar{t}t$ production. The aspon A can be bremsstrahlunged from a heavy quark. Detailed calculations show that the cross-section for aspon production is a few picobarns corresponding to a few tens of thousands of events per year at LHC.

Of special interest is the decay width of A which depends sensitively on the A mass relative to the Q mass M. For the most suppressed decay, when $M(A) < M(Q)$, the decay width can be as small as 1KeV which is striking for a particle weighing several hundred GeV!

IV B DECAY

The KM mechanism can be nicely checked from the unitarity triangle formed by the complex numbers in the equation:

$$V_{ub}^* V_{ud} + V_{tb}^* V_{td} + V_{cb}^* V_{cd} = 0 \tag{1}$$

with corresponding angles α, β, and γ. Using the expansion of the CKM matrix as a power series in the Cabibbo angle [22], it is profitable to define the ratios:

$$R_b = \left| \frac{V_{ud}^* V_{ub}}{V_{cd}^* V_{cb}} \right| \qquad (2)$$

and

$$R_t = \left| \frac{V_{cd}^* V_{tb}}{V_{cd}^* V_{cb}} \right| \qquad (3)$$

Clearly if the angle β, for example, is a significant value, well away from zero or π (as would follow if the KM mechanism is the full explanation of the CP violation in kaon decay), the $R_b + R_t > 1$.

It is well-known [23] how to establish the angle β from the expected data on B decay coming from the B Factories under construction at SLAC and KEL Laboratories.

A CP Asymmetries in B Decay

In the aspon model the 3×3 mixing matrix for the light quarks is a real orthogonal one up to corrections of order x^2. This means that the CP asymmetries of B decay are predicted to be at least three orders of magnitude smaller than predicted by the KM mechanism.

In a general way, we may say that the KM mechanism is special in that the CP violation in B decay is enhanced by a factor $(m_t/m_c)^2 \sim 10^4$ relative to that in K decay. In most alternative models of CP violation such as the apon model, there is no reason to expect this enhancement.

A clear prediction of the aspon model is that, to within less than 0.1%, $R_b + R_t = 1$. An unbiased study of the present data shows that this is well within the present range.

V KAON SYSTEM REEVALUATED

The value of $|\epsilon_K| = 2.26 \times 10^{-3}$ implies (from aspon exchange) that

$$\kappa/x^2 = 2.8 \times 10^3 GeV \qquad (4)$$

which, given the range for x^2, implies that $29 TeV > \kappa > 870 GeV$ from which the aspon mass is expected in the range 260GeV to 8.7TeV.

Contributions to $Re(\epsilon'/\epsilon)$ come from tree diagrams and penguin diagrams. A careful comparison to the standard model gives a suppression of at least two orders of magnitude. Consequently, observation of a value above 10^{-4} would exclude this model.

VI SUMMARY

The main attractions of the aspon model are that it solves the strong CP problem, accommodates weak CP violation, and makes testable predictions. A reader who wishes to know more may consult the References listed.

ACKNOWLEDGMENTS

I am delighted to visit San Juan, thanks to Jose F. Nieves and the organizers, and especially to visit with my 1989 PhD student, Marcelo Ubriaco.

This work was supported in part by the US Department of Energy under Grant No. DE-FG05-85ER-40219.

REFERENCES

1. E.P. Wigner, Physik **43**, 624 (1927)
2. T.D. Lee and C.N. Yang, Phys. Rev. **104**, 254 (1956).
3. C.S. Wu et al., Phys. Rev. **105**, 1413 (1957).
4. L. Lederman et al., Phys. Rev. **105**, 1415 (1957).
   V. Telegdi et al., ibid. **105**, 1681 (1957).
5. E.P. Wigner, Gott. Nachr. **546** (1932).
6. P.A.M. Dirac, Proc. Roy. Soc. **A117**, 610 (1928).
7. H.A. Kramers, Proc.Acad. Amst. **40**, 814 (1937).
8. G. Luders, Kg. Dansk. Vidersk. Selsk. Mat.-Fys. Medd. **28**, No.5 (1954).
9. L. D. Landau, Nucl. Phys. **3**, 127 (1957).
10. J.H. Christensen et al., Phys. Rev. Lett. **13**, 138 (1964).
11. A.D. Sakharov, JETP Letters **5**, 24 (1967).
12. M. Yoshimura, Phys. Rev. Lett. **41**, 281 (1978).
13. M. Kobayashi and T. Maskawa, Prog. Theor. Phys. **49**, 652 (1973).
14. G.'t Hooft, Phys. Rev. Lett. **37**, 8 (1976).
15. P.H. Frampton and T.W.Kephart, Phys. Rev. Lett. **66**, 1666 (1991).
16. P.H. Frampton and D. Ng, Phys. Rev. **D43**, 3034 (1991).
17. P.H. Frampton et al., Phys. Rev. Lett. **68**, 2129 (1992).
18. A.W. Ackley et al., Phys. Rev. **D50**, 3560 (1994).
19. P.H. Frampton and S.L. Glashow, Phys. Rev. **D55**, 1691 (1997).
20. P.H. Frampton and M. Harada, UNC-Chapel Hill IFP-757-UNC (1998); *hep-ph/9803416*.
21. A.Nelson, Phys. Lett. **136B**, 387 (1984);
    S.M. Barr, Phys. Rev. **D30**, 11005 (1984).
22. L. Wolfenstein, Phys. Rev. Lett. **51**, 1945 (1983).
23. M. Neubert, Int. J. Mod. Phys. **A11**, 4173 (1996).

New Physics Models and CP Violation Experiments

Dennis Silverman

Department of Physics and Astronomy
University of California, Irvine
Irvine, CA 92697-4575

I INTRODUCTION

In order to decide whether the B Factory is significantly testing the standard CKM model of CP violation, and to determine the "reach" of the B factory in ruling out "new physics" models, we need several alternate but full physical models for comparison. The flavor changing neutral current (FCNC) model resulting from extra weak iso-singlet down quarks, as in E_6 with one for each generation, has been very effective in causing new effects, but not being ruled out by present experiments [1–3]. In this presentation, the predictions are extended to the full range of angular experimental results. The constraints are updated for new x_s bounds, and for new data on R_b, the D0 bound on $K \to \mu\mu X$, and the BNL results on $K^+ \to \pi^+ \nu \bar{\nu}$. With Herng Tony Yao [4] we also analyze the Left-Right Symmetric Model (LRSM).

II FCNC IN Z^0 COUPLINGS FROM EXTRA ISO-SINGLET DOWN QUARKS

In models with extra down quarks, we assume that only the next heaviest would contribute significant mixing, and use only a 4×4 down quark mixing matrix. This gives a matrix with six angles and three phases, in which the leading contributions in each element are

$$V = \begin{array}{c} u \\ c \\ t \\ 4 \end{array}\begin{pmatrix} \overset{d}{c_{34}} & \overset{s}{s_{12}c_{34}} & \overset{b}{s_{13}e^{-i\delta_{13}}} & \overset{D}{s_{14}e^{-i\delta_{14}}} \\ -s_{12} & 1 & s_{23} & s_{24}e^{-i\delta_{24}} \\ (s_{12}s_{23} - s_{13}e^{i\delta_{13}}) & -s_{23} & 1 & s_{34} \\ V_{4d} & V_{4s} & V_{4b} & V_{44} \end{pmatrix} \quad (1)$$

where

$$V_{4d}^* = -s_{14}e^{-i\delta_{14}} + s_{24}e^{-i\delta_{24}}s_{12} - s_{34}(s_{12}s_{23} - s_{13}e^{-i\delta_{13}}), \qquad (2)$$

$$V_{4s}^* = -s_{24}e^{-i\delta_{24}} - s_{14}e^{-i\delta_{14}}s_{12} + s_{34}(s_{23} + s_{12}s_{13}e^{-i\delta_{13}}), \qquad (3)$$

$$V_{4b}^* = -s_{34} - s_{24}e^{-i\delta_{24}}s_{23} - s_{14}e^{-i\delta_{14}}s_{13}e^{i\delta_{13}}. \qquad (4)$$

The FCNCs are given by the mixings to the fourth down quark by $-U_{ij} \equiv V_{4i}^* V_{4j}$ for $i \neq j$. The FCNC couplings of the down quarks to the Z are then given by:

$$\mathcal{L}_{FCNC}^Z = -\frac{e}{2\sin\theta_W \cos\theta_W} U_{ij}\bar{d}_{iL}\gamma^\mu d_{jL} Z_\mu. \qquad (5)$$

There will be constraints on three FCNC amplitudes: $-U_{ds} = V_{4d}^* V_{4s}$, $-U_{sb} = V_{4s}^* V_{4b}$, and $-U_{bd} = V_{4b}^* V_{4d}$ The diagonal neutral current couplings are reduced in strength by the amplitudes into the FCNC, becoming

$$\mathcal{L}_{NC}^Z = -\frac{e}{2\sin\theta_W \cos\theta_W} \sum_i (1 - |V_{4i}|^2)\bar{d}_{iL}\gamma^\mu d_{iL} Z_\mu. \qquad (6)$$

$B_d - \bar{B}_d$ Mixing

The FCNC can act twice with amplitude U_{bd} via an intermediate Z^0 to convert a B_d to a $\bar{B}_d$ charge congugate meson as shown in the figure, to interfere with the standard model box diagram

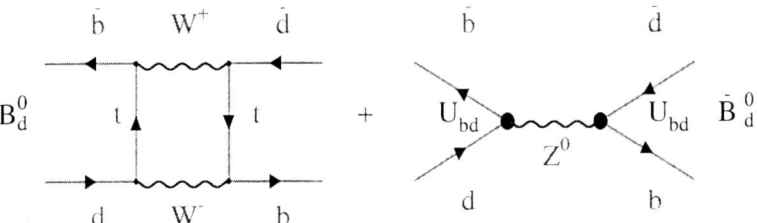

This leads to a contribution to mixing in x_d

$$x_d = \frac{2G_F}{3\sqrt{2}} B_B f_B^2 m_B \eta_B \tau_B \left| U_{std-db}^2 + (U_{db})^2 \right| \qquad (7)$$

where $x_d = \Delta m_{B_d}/\Gamma_{B_d} = \tau_{B_d}\Delta m_{B_d} = \tau_{B_d} 2|M_{12}|$, and we have normalized the standard model amplitude for $B_d - \bar{B}_d$ mixing so that it adds directly to the U_{db}^2 term

$$U_{std-db}^2 \equiv \frac{\alpha}{4\pi \sin^2\theta_W} y_t f_2(y_t)(V_{td}^* V_{tb})^2. \qquad (8)$$

Experimentally, $x_d = 0.73 \pm 0.05$, and theoretically, we use $\sqrt{B_B} f_B = 0.200 \pm 0.040$ GeV. To leading order in s_{34}, $s_{24}e^{-i\delta_{24}}$, and $s_{14}e^{-i\delta_{14}}$, we have

$$U_{db} = -s_{34}(s_{34}V_{td}^* + s_{14}e^{-i\delta_{14}} - s_{24}e^{-i\delta_{24}}s_{12}). \qquad (9)$$

The first term has the phase δ_{13} as in the standard model, but the other two phases can also account for CP violation.

Bounds on FCNC CP Violation in the K Meson System

The FCNC U_{sd} can act twice via an intermediate Z^0 to mix K^0 to $\bar{K}^0$ giving a CP violation in ϵ

$$|\epsilon| = \frac{G_F f_K^2 B_K m_K}{12 \Delta m_K} \left| \text{Im}(U_{std-ds}^2) + \text{Im}(U_{ds}^2) \right|, \tag{10}$$

where

$$U_{std-ds}^2 = \frac{\alpha}{4\pi \sin^2 \theta_W} (\eta_2 y_t f_2(y_t)(V_{td}^* V_{ts})^2 + \cdots). \tag{11}$$

Experimentally, $|\epsilon| = 2.27 \times 10^{-3}$. BNL E-787, analyzing their 1995 data, has now detected one event in $K^+ \to \pi^+ \nu \bar{\nu}$, giving a branching ratio BR($K^+ \to \pi^+ \nu \bar{\nu}$) = $4.2^{+9.7}_{-3.5} \times 10^{-10}$. The standard model range is $(0.6 - 1.5) \times 10^{-10}$. The new result replaces their older bound of 2.4×10^{-9} at 90% CL. The FCNC can contribute to this process with a U_{ds} coupling to a $Z^0 \to \nu \bar{\nu}$. To cancel out the hadronic interactions, this is scaled to a ratio

$$\frac{\text{BR}(K^+ \to \pi^+ \nu \bar{\nu})}{\text{BR}(K^+ \to \pi^0 e^+ \nu)} = \frac{1}{2} \frac{|U_{ds}|^2}{|V_{us}|^2}. \tag{12}$$

Bound From $Z^0 \to b\bar{b}$

The weak isovector part of $Z^0 \to b\bar{b}$ is reduced by $(1 - |V_{4b}|^2)$ This gives

$$\frac{\Gamma_{b\bar{b}}^{std+FCNC}}{\Gamma_{b\bar{b}}^{std}} = (1 - 2.29|V_{4b}|^2). \tag{13}$$

Present data are from Aleph five tag ($R_b = 0.2158 \pm 0.0014$), and LEP plus SLD ($R_b = 0.2170 \pm 0.0009$), while theory gives $R_b = 0.2158 \pm 0.0003$, leading to a difference from LEP plus SLD of 0.0012 ± 0.0009. Since the $|V_{4b}|$ effect is to decrease R_b, the conservative approach is to interpret the upward discrepancy as a fluctuation of statistical or systematic origin, and to still use the standard model result as the comparison point for setting the limit on $|V_{4b}|$. With the above interpretation, the errors on the experiment are expanded to equal the deviation of 0.0012.

CKM, FCNC, and B Factory Experiments

The table below shows the experimental constraints used in the four types of analyses and the matrix elements to which they are related.

- **CKM Experiments (9) for Standard Model - Present Data**

- **Charged Current (5):** V_{ud}, V_{us}, V_{cd}, $|V_{cb}| = 0.039 \pm 0.004$, and $|V_{ub}/V_{cb}| = 0.080 \pm 0.020$.
- **Neutral Current (4):**

  x_d U_{db}
  $|\epsilon|$ $\text{Im}(U_{ds}^2)$
  $K_L \to \mu\mu$ $\text{Re}(U_{ds})$
  x_s bound U_{sb}

- **Exclusively FCNC Experiments for Four Down Quark Model (4)**

  $B \to \mu\mu X$ UA1 and D0 $|U_{db}|, |U_{sb}|$
  $Z^0 \to b\bar{b}$ (R_b) $|V_{4b}|$
  $K^+ \to \pi^+ \nu\bar\nu$ $|U_{ds}|$

- **B Factory Projected Experiments (2) for SM and Four Down Quarks**

  $\sin(2\beta)$ U_{db}^2
  $\sin(2\alpha)$ U_{db}^2

CP Violating Asymmetries at the *B* Factory

In the four down quark mixing model, $\sin(2\alpha)$ and $\sin(2\beta)$ are now generally defined as the measured B factory asymmetries, but they do not correspond to unitarity triangle or quadrangle angles. They are the CP violating decay asymmetries defined here as

$$\sin(2\beta) \equiv A_{B_d^0 \to \Psi K_s^0} = \text{Im}\left[\frac{(U_{std-db}^2 + U_{db}^2)(V_{cb}^* V_{cs})}{|U_{std-db}^2 + U_{db}^2|(V_{cb}^* V_{cs})^*}\right], \tag{14}$$

and

$$\sin(2\alpha) \equiv -A_{B_d^0 \to \pi^+\pi^-} = -\text{Im}\left[\frac{(U_{std-db}^2 + U_{db}^2)(V_{ub}^* V_{ud})}{|U_{std-db}^2 + U_{db}^2|(V_{ub}^* V_{ud})^*}\right]. \tag{15}$$

In $\bar{b} \to \bar{u} + c + \bar{s}$ processes, another CP violating asymmetry can be measured. The B_s oscillation amplitude is almost real, allowing phases from decay amplitudes to be measured alone. The phase is

$$\phi_{qf} = -(2\phi(B_q \to \bar{B}_q) + \phi(\bar{B}_q \to f) - \phi(B_q \to f)) \tag{16}$$

Two of the four processes, out of those for B_s and $\bar{B}_s$ decays into final states f and $\bar{f}$, must be measured together. The largest branching ratios for $\sin(\gamma)$ processes is in $B_s, \bar{B}_s \to D_s^\pm + K^\mp$, where the BR $\approx 2 \times 10^{-4}$. Here, the $\bar{B}_s$ decay to the final state involves V_{ub}^*, but the conjugate process for B_s does not involve $b \to u$ and is thus almost real. Since $V_{ub} = s_{13}e^{-i\gamma}$, and B_s mixing is almost real, the final asymmetry which is the imaginary part, only involves $\sin(\gamma)$, not $\sin(2\gamma)$. In the four down quark model

$$\sin(\gamma) = \text{Im}\left(\frac{(U^2_{std-bs} + U^2_{bs})(V^*_{ub}V_{cs})}{|U^2_{std-bs} + U^2_{bs}||V^*_{ub}V_{cs}|}\right). \tag{17}$$

Since this imaginary part can be both positive and negative, we now include the full range $-1 \leq \sin(\gamma) \leq 1$ motivated by Wolfenstein's approach of finding the full range of predictions, rather than perturbations around the SM result.

In studying the frequency of B_s oscillations, x_s, we exclude the theoretical uncertainty on $B_{B_s}f^2_{B_s}$, and take the ratio of x_s to x_d, which is better calculated, and in which we have also included the FCNC with Z^0 exchange

$$x_s = 1.35 \ x_d \frac{|U^2_{std-bs} + U^2_{bs}|}{|U^2_{std-db} + U^2_{db}|} \tag{18}$$

We now include the amplitude method analysis of LEP with SLD to assign a $\Delta\chi^2$ for each x_s calculated in the parameter grid.

We also calculate the B_s decay asymmetry, A_{B_s}. In the standard model, B_s mixing involves $(V^*_{ts}V_{tb})^2$ which is almost exactly real. Thus no CP violating phase develops in the most likely B_s decays involving the real amplitude for $b \to \bar{c}+c+\bar{s}$. This occurs in the decays $B_s \to J/\Psi\phi$, $B_s \to D^+_s D^-_s$, and $B_s \to J/\Psi K_S$. In the four down quark model, the CP violating decay asymmetry is

$$A_{B_s} = \text{Im}\left(\frac{(U^2_{std-bs} + U^2_{bs})(V^*_{cb}V_{cs})}{|U^2_{std-bs} + U^2_{bs}|(V_{cb}V^*_{cs})}\right). \tag{19}$$

Since this is an imaginary part of an $e^{i\Phi}$, we again include the full range $-1 \leq A_{B_s} \leq 1$.

Comprehensive Maximum Likelihood Analysis: Experiments and Degrees of Freedom

The following summary gives the parameters, experiments and degrees of freedom for the four modes of analyses. In the analyses of the Standard Model, the CKM matrix has four parameters (3 angles and 1 phase). With nine present experiments, this gives five degrees of freedom. Adding the B factory with pairs of two experiments in their respective plots, gives seven degrees of freedom. In the four down quark model, the 4×4 mixing matrix has nine parameters (6 angles and 3 phases). With the 13 present experiments analyzed, we have four degrees of freedom. For the analysis with two B factory experiments added, we have six degrees of freedom.

$(\sin(2\alpha), \sin(2\beta))$ *Plots*

For the standard model, the results of this plot for present data (not shown) gives $\sin(2\beta) \approx 0.65 \pm 0.15$. and $\sin(2\alpha)$ between -0.7 to 0.8 at 1-σ.

The four down quark model $(\sin(2\alpha), \sin(2\beta))$ plot with present limits need not be presented because the entire plane is allowed at 1-σ. Thus, even a rough, early measurement of $\sin(2\beta)$ might give a surprising result.

The Decay Asymmetry from B_s Mixing, A_{B_s}

This is almost zero in the standard model, i.e. $A_{B_s} \leq 0.025$. In the four down quark model (not shown), $A_{B_s} = \text{Im}(e^{i\Phi})$ is extended here to negative values, and appears roughly symmetric about 0. $|A_{B_s}| \leq 0.4$ at 1-σ is possibly detectable in the four down quark model.

(ρ, η) Plots

Standard Model Plots

The (ρ, η) plots (broad contours), are shown with projected B-Factory cases (small contours). The contours are labeled: 1-σ (solid), 2-σ (dashed), and 3-σ (dotted). Cases of B factory results at $\sin(2\alpha) = 1, \quad 0, \quad -1$ are shown from left to right, respectively, with $\sin(2\beta) = 0.65$.

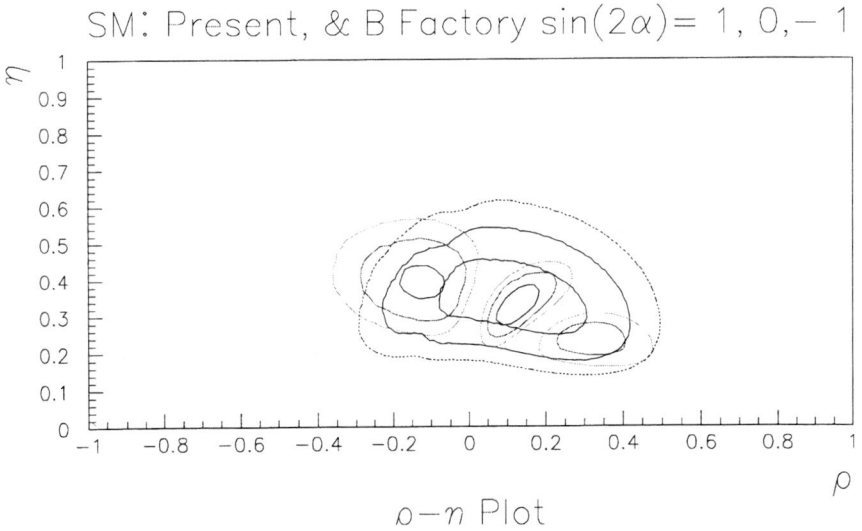

Four Down Quark Model: (ρ, η) Plots

Here, the plotted ρ and η are taken as the coordinates of V_{ub}^* in $\rho + i\eta = V_{ub}^*/|V_{cd}V_{cb}|$. The plots are now extended to negative $\eta = \text{Im}(V_{ub}^*/|V_{cd}V_{cb}|)$. There is now a unitarity quadrangle, $V_{ub}^*V_{ud} + V_{cb}^*V_{cd} + V_{tb}^*V_{td} - U_{bd} = 0$, where the last term has limits $|U_{bd}/V_{cb}^*V_{cd}| \leq 0.1$.

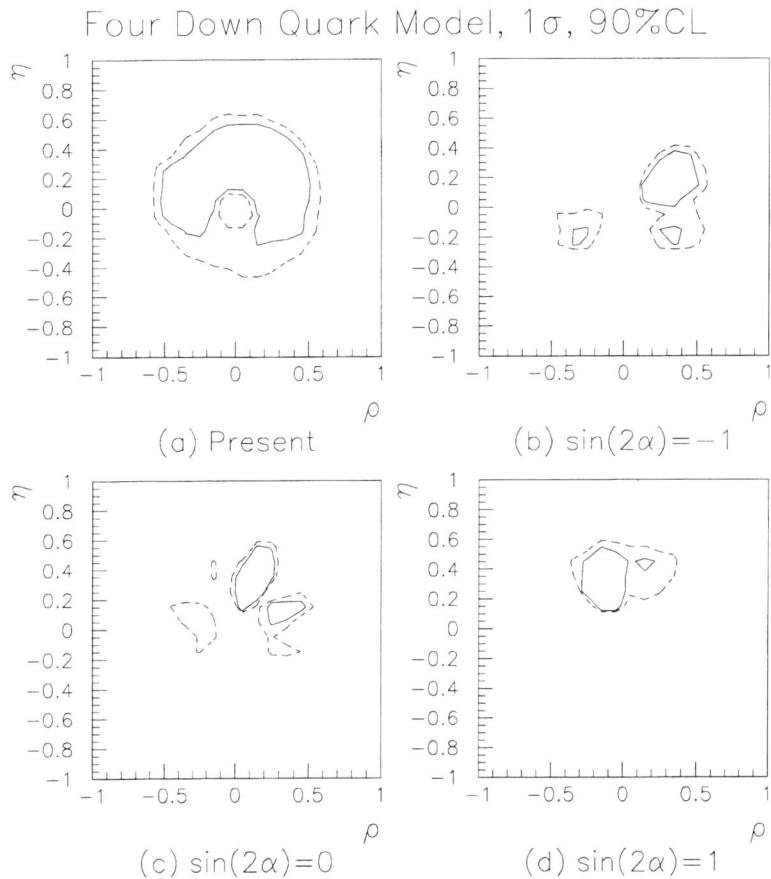

Four Down Quark Model, 1σ, 90%CL

(a) Present (b) sin(2α)=−1 (c) sin(2α)=0 (d) sin(2α)=1

The partial circles in $\delta = \delta_{13}$ ($V_{ub}^* = s_{13}e^{i\delta_{13}}$) are due to δ_{14} or δ_{24} being the source of the observed CP violation in ϵ, so that δ_{13} is unconstrained. The multiple regions are now extended into the full (ρ, η) plane.

$(x_s, \sin(\gamma))$ Plots

The standard model $(x_s, \sin(\gamma))$ plots with the present data analyzed (not shown) gives $0.75 \leq \sin(\gamma) \leq 1$ and $14 \leq x_s \leq 29$ at 1-σ. With the three B factory cases well separated in $\sin(2\alpha)$, we find than an x_s measurement can strongly constrain $\sin(2\alpha)$.

For the four down quark model $(x_s, \sin(\gamma))$ plots, we have extended the $\sin(\gamma)$ range to $-1 \leq \sin(\gamma) \leq 1$ for completeness. Here, almost all $\sin(\gamma)$ is allowed at 1-σ, as opposed to the limited standard model range. The x_s allowed region is 12 to 50 at 1-σ, larger than in the standard model.

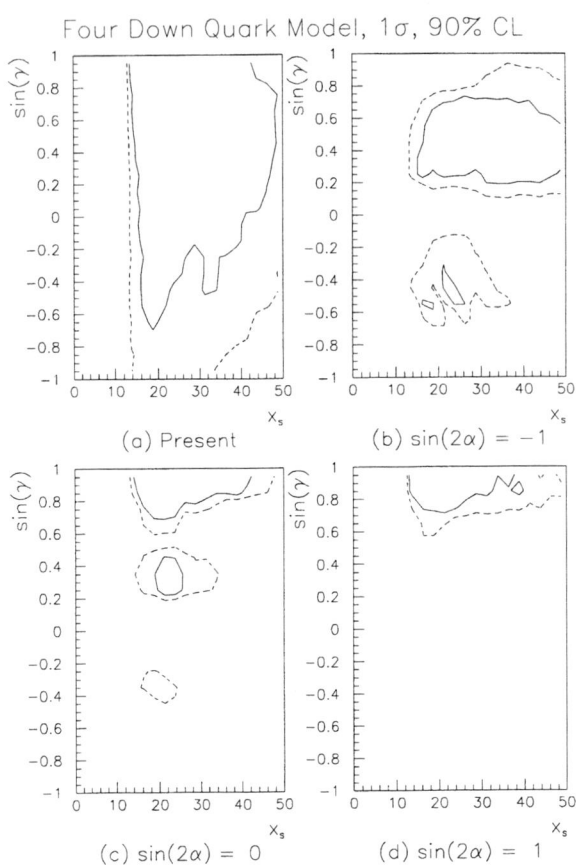

III LEFT-RIGHT SYMMETRIC MODEL ANALYSIS BY D. SILVERMAN AND HERNG TONY YAO

Another contrasting model to compare to the standard model is the $SU(2)_L \times SU(2)_R \times U(1)_{B-L}$ model with two new heavy charged $W_R^\pm$ and one new Z_0' gauge bosons. In this case, new box diagrams arise with one left handed and one right handed W's exchanged. The right handed 3×3 quark mixing matrix, V^R, can have six new phases and three new mixing angles The lefthanded quark mixing matrix V^L is that of the standard model. For our purpose of showing the effect of a new phase in the CP violation B physics sectors, we take two models that only have one right-handed phase [4].

CP Violation in K Mesons in the LRSM

Langacker and Sankar [5] have made a detailed analysis on W_R mass limits, and conclude that the lower limit of the W_R mass of 1.6 TeV for $V^L = V^R$ can be reduced by taking either of the forms of V^R below, which decouple the t and c quarks from either the d quarks (Case B_s) or the s quarks (Case B_d). The two cases are labeled by the B_q in which they show an effect. ϵ_K is given by the Hamiltonian $H(\Delta S = 2)$ which arises from the box diagrams mediated by two W_L, two W_R, or a W_L-W_R pair. The two W_R part H^{RR} gives no contribution. The contribution to ϵ_K from H^{LR} only comes from the surviving terms which involve the exchange of one u quark. Due to the smallness of $\epsilon_K = (2.28 \pm 0.02) \times 10^{-3}$, in this model one adjusts the parameters in V^R so that no contribution to ϵ_K will come from H^{LR}. This leaves the models in the form:

$$V^R_{B_s} = \begin{matrix} u \\ c \\ t \end{matrix} \begin{pmatrix} \overset{d}{1} & \overset{s}{0} & \overset{b}{0} \\ 0 & c & se^{i\sigma} \\ 0 & se^{-i\beta} & -ce^{i(\sigma-\beta)} \end{pmatrix} \text{ (Case } B_s\text{):} \quad V^R_{B_d} = \begin{matrix} u \\ c \\ t \end{matrix} \begin{pmatrix} \overset{d}{0} & \overset{s}{1} & \overset{b}{0} \\ 0 & 0 & e^{i\sigma} \\ 1 & 0 & 0 \end{pmatrix} \text{ (Case } B_d\text{).}$$

(20)

Here $s = \sin\theta$ and $c = \cos\theta$ ($0 \leq \theta \leq 90°$), and β is the angle of the unitarity triangle at the $\rho = 1$ vertex. We have eliminated an overall phase which does not appear in the processes which we consider here. Thus in the models there is only one variable phase, and Case B_s also has a variable angle.

$B^0 - \bar{B}^0$ Mixing

Mixing occurs from box diagrams in $M_{12} = M_{12}^{LL} + M_{12}^{RR} + M_{12}^{LR}$, corresponding to the diagrams in which two W_L, two W_R and a W_L-W_R pair are exchanged.

In $B_d - \bar{B}_d$ mixing, effects are only present in Case B_d, where (c,t) exchange causes $M_{12}^{LR} \sim M_{12}^{LL}$ at $M_{W_R} = 5$ TeV, but is $\leq 10^{-2} M_{12}^{LL}$ at 10 TeV.

For $B_s - \bar{B}_s$ mixing, effects are only present in Case B_s, where (t,t) exchange causes $M_{12}^{LR} \sim M_{12}^{LL}$ for $M_{W_R} = 5$ TeV, but is much smaller than M_{12}^{LL} at $M_{W_R} = 10$ TeV.

χ^2 Plots for (ρ, η), $(\sin(2\alpha), \sin(2\beta))$, and $(x_s, \sin(\gamma))$ in the LRSM

- **CKM Experiments (5) for SM - Present Data**
  - **Charged Current:** λ is fixed, $|V_{cb}| = 0.039 \pm 0.004$, and $|V_{ub}/V_{cb}| = 0.080 \pm 0.016$.
  - **Neutral Current:** x_d, $|\epsilon|$, and x_s bounds.
- **B Factory Experiments (2):** $\sin(2\beta)$ and $\sin(2\alpha)$.
- **Left Right Symmetric Model:** (for various choices of W_R)

With B factory experiments included, we have $5 + 2 = 7$ experiments.
Case B_s: 6 parameters (3 angles and 2 phases), or 2 degrees of freedom
Case B_d: 4 parameters (2 angles and 2 phases), or 3 degrees of freedom.

Experimental results predicted with M_{12} including the LRSM

$$\sin(2\beta) \equiv \mathrm{Im}\left(\frac{M_{12}^*}{|M_{12}|}\frac{V_{cb}^* V_{cs}}{V_{cb}V_{cs}^*}\right), \quad \sin(2\alpha) \equiv \mathrm{Im}\left(\frac{M_{12}^*}{|M_{12}|}\frac{V_{ub}^* V_{ud}}{V_{ub}V_{ud}^*}\right)$$

$$\sin(\gamma) \equiv \mathrm{Im}\left(\frac{M_{12}^{B_s}}{|M_{12}^{B_s}|}\frac{V_{ub}^* V_{cs}}{|V_{ub}V_{cs}|}\right), \quad x_s = 1.35 x_d \frac{|M_{12}^{B_s}|}{|M_{12}^B|}.$$

LRSM Plots for Case B_s (Case I)

LRSM (x_s, A_{B_s}) Plots for Case B_s

Whereas A_{B_s} is almost null in the SM, in the LRSM for $M_{W_R} \simeq 1$ TeV, the plots (not shown) saturate at $|A_{B_s}| \leq 1$. Even at larger $M_{W_R} \simeq 5$ TeV, A_{B_s} could be as large as 0.4.

LRSM $(x_s, \sin(\gamma))$ Plots for Case B_s

The entire $\sin(\gamma)$ range is allowed for the smaller W_R values. Larger x_s values are allowed for intermediate W_R values.

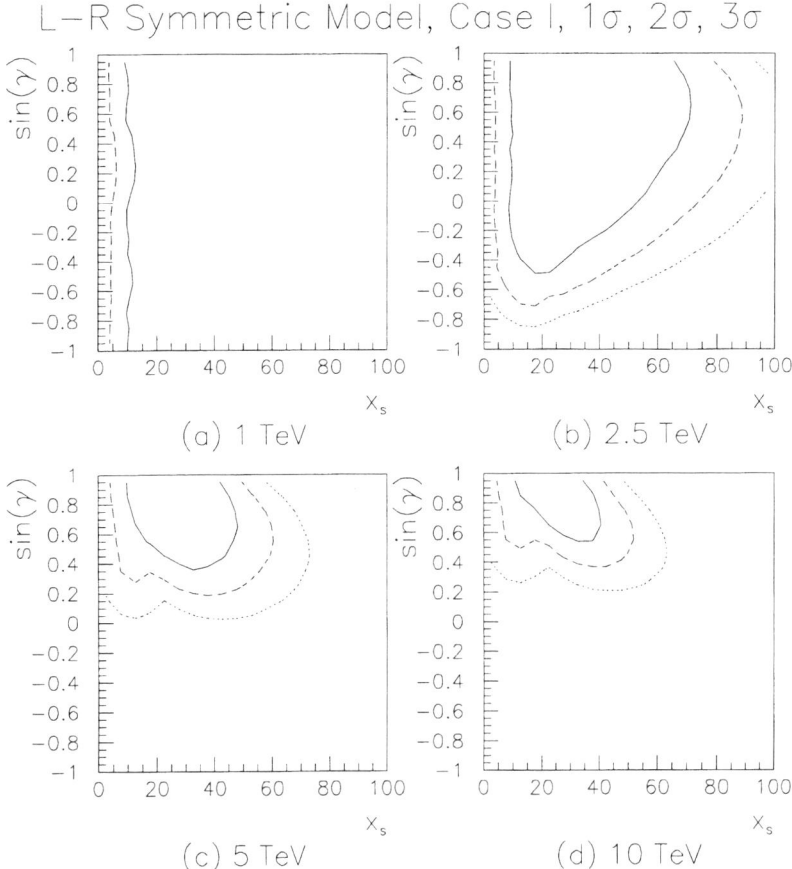

LRSM Plots for Case B_d (Case II)

LRSM (ρ, η) Plots for Case B_d

For $1.0 \leq M_{W_R} \leq 10$ TeV the plots (not shown) have larger regions than in the SM, but enclosing the SM results.

LRSM $(x_s, \sin(\gamma))$ Plots for Case B_d

Lower $-0.1 \leq \sin(\gamma) \leq 1$ is allowed than in the SM for $M_{W_R} \leq 2.5$ TeV (not shown).

LRSM $(\sin(2\alpha), \sin(2\beta))$ Plots for Case B_d

Different preferred 1-σ regions from those in the standard model exist between $M_{W_R} = 1.5$ TeV and $M_{W_R} = 5$ TeV.

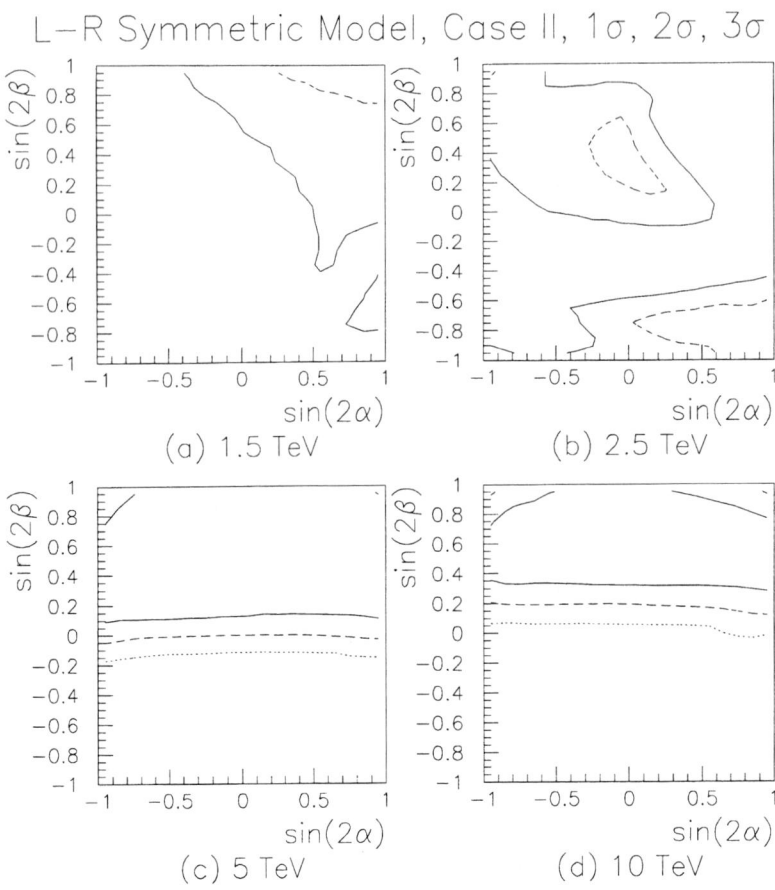

(a) 1.5 TeV (b) 2.5 TeV (c) 5 TeV (d) 10 TeV

This research was supported in part by the U.S. Department of Energy under Contract No. DE-FG0391ER40679.

REFERENCES

1. W.-S. Choong and D. Silverman, Phys. Rev. **D49**, 2322 (1994).
2. D. Silverman, Int. Journal of Modern Physics **A13**, 2253 (1995), hep/9504387.
3. General formulation and review of isosinglet up and down quark models: V. Barger, M.S. Berger, and R.J.N. Phillips, Phys. Rev. D **52**, 1663 (1995) and on hep-ph/9503204.
4. W_R Effects on CP Aysmmetries in B Meson Decays, D. Silverman and H. Yao, UC Irvine TR-97-5, LBNL-40272 Preprint, hep-ph/9706359.
5. P. Langacker and S. Sankar, Phys. Rev. **D40**, 1569 (1989).

SUPERSYMMETRY AND NEW PHYSICS

Pre-LHC SUSY Searches: an Overview

A. Masiero

SISSA, Via Beirut 2-4, 34013 Trieste, Italy
and
INFN, sez. di Trieste, Padriciano 99, 34012 Trieste, Italy

L. Silvestrini

Technische Universität München, Physik Department
D-85748 Garching, Germany

Abstract. We discuss the prospects for searches of low-energy supersymmetry in the time interval separating us from the advent of LHC. In this period of time "indirect" searches may play a very relevant role. We refer to manifestations of supersymmetry in flavour changing neutral current and CP violating phenomena and to signals of the lightest supersymmetric particle in searches of dark matter. In the first part of the talk we critically review the status of the minimal supersymmetric model to discuss the chances that direct and indirect supersymmetric searches may have before the LHC start. In the second part we point out what we consider to be the most promising grounds where departures from the standard model prediction may signal the presence of new physics, possibly of supersymmetric nature. We argue that the often invoked complementarity of direct and indirect searches of low-energy supersymmetry is becoming even more true in the pre-LHC era.

I INTRODUCTION

It is not rare to hear the following gloomy forecast: if no supersymmetric signal is seen at LEP, then we have nothing else to do but wait for LHC. We do not agree with this statement. Apart from the fact that even for direct searches one should take into account the relevant potentialities of Tevatron in the LEP-LHC time interval, one should not neglect that indirect searches for new physics signals are going to be flourishing before 2005. We refer to processes exploring flavour physics (with or without CP violation) where new particles can play an active role being exchanged in the loop contributions and to several new astroparticle observations which may constitute privileged places to obtain information on physics beyond the Standard Model (SM).

We wish to present here a brief overview (which is necessarily biased by our theoretical prejudices) of what we consider most promising in this effort of looking for indirect signals of low-energy Supersymmetry (SUSY) before the LHC advent. First we will review the status and prospects for direct SUSY searches, then we will discuss the role that SUSY may play in Flavour Changing Neutral Current (FCNC) and CP violating phenomena. Finally we will briefly comment on searches for the lightest SUSY particle in experiments looking for Dark Matter (DM).

II STATUS OF THE MSSM

It is known that, even asking for the minimal content of superfields which are necessary to supersymmetrize the SM and imposing R parity, one is still left with more than 100 free parameters most of which are in the flavour sector. It is also true that very large portions on this enormous SUSY parameter space are already ruled out by the present phenomenology (in particular FCNC and CP constraints). If one wants to reduce the number of free parameters one has to make assumptions on what lies well beyond low-energy SUSY, in particular on the quite unknown issue of the origin of SUSY breaking. The two most popular drastic reductions of free SUSY parameters are provided by minimal supergravity (SUGRA) [1] (with the further assumption of unification of gauge couplings and gaugino masses at some grand unification scale) and by the models of Gauge-Mediated SUSY Breaking (GMSB) [2]- [4]. In minimal SUGRA and the minimal version of GMSB we have only 3 or 4 parameters in addition to those of the SM and so we can become much more predictive.

In the context of the minimal supergravity model (with electroweak radiative breaking), we ask the following relevant questions for direct SUSY searches: i) given the present experimental lower bounds on the masses of SUSY particles, how much room have we got in the SUSY parameter space to explore, or, in other words, when should we give up with SUSY if searches are fruitless? ii) is there any experimental signature of low-energy SUSY which is independent from the choice of the SUSY parameters in particular of the soft breaking sector? iii) are the electroweak precision tests telling us something relevant on low-energy SUSY?

  i) SUSY must be a low-energy symmetry if it has to deal with the issue of the gauge hierarchy problem. This fact is usually translated into the statement that SUSY particle masses should not be significantly larger than $O(1\ \text{TeV})$ given that SUSY breaking should not exceed this energy scale to realize a suitable "protection" of the mass of the scalar Higgs responsible for the electroweak breaking. Actually one may try to be more quantitative [5]. First one relates the Z mass to the value of the 4 parameters of the minimal SUGRA run from the large scale, at which the soft breaking terms originate, down to the electroweak scale. Then one establishes a degree of naturalness corresponding to the amount of fine tuning of the initial SUSY parameters which is needed to reproduce the correct Z mass for increasing values of the low-energy SUSY masses. For instance it is clear that to have all SUSY particles with a mass of $O(1\ \text{TeV})$ would require a severe fine tuning of the boundary conditions.

As for all naturalness criteria, also in this case there is a large amount of subjectivity, but one message emerges quite clearly: already now, in particular with the lower bound on chargino masses exceeding 90 GeV, we are entering an area of parameter space where a certain degree of fine tuning is needed. Hence we are already at the stage where we may expect "naturally" to find SUSY particles. Moreover such naturalness analyses confirm that LHC represents kind of "definitive" machine for SUSY direct searches: if no SUSY particle is discovered at LHC, the degree of fine tuning becomes so severe that it is hard to still defend the idea of low-energy SUSY. Finally, an important comment on the degree at which different SUSY masses are constrained by such naturalness criteria: due to the large difference in the Yukawa couplings of the third (heaviest) generation with respect to the first two generations, it turns out that only the sbottoms and stops are required not to be very heavy, whilst squarks of the first two generations can be quite heavy (say tens of TeV) without severely affecting the correct electroweak breaking. This observations may play a relevant role in tackling the FCNC problem in SUSY (see below).

ii) If one allows the SUSY parameters to take larger and larger values, all the SUSY particles become heavier and heavier with only one remarkable exception. In the Higgs mass spectrum of SUSY models the lightest scalar always remains light. The mass of the light CP-even neutral Higgs in the MSSM is calculable at tree level in terms of two SUSY parameters of the Higgs potential. At this level it is smaller than the mass of the Z. When radiative corrections are included, the mass of the light Higgs becomes a function also of the other SUSY parameters and its upper bound increases significantly [6]. However, even varying the MSSM parameters as much as one wishes, it is not possible to exceed $130 - 135$ GeV for its mass. Indeed, taking $m_t = 175$ GeV and for a stop lighter than 1 TeV one obtains that the upper bound on the lightest Higgs is 125 GeV allowing for "maximal" mixing in the top squark sector (the bound decreases for smaller stop mixing).

It is not easy to significantly evade the above upper bound on the mass of the lightest Higgs even if one gives up the minimality of the SUSY model. For instance, if one adds a singlet to the two Higgs doublets (i.e., one goes to the so called Next-to-Minimal SUSY Standard Model, NMSSM), then a new parameter shows up in the scalar potential: the coupling of the singlet with the two doublets. If one imposes that all couplings remain perturbative up to the Planck scale, then the consequent upper bound on this new coupling implies that the lightest Higgs should not be heavier than 150 GeV or so [7].

Obviously having a possibly "exotic" Higgs below 150 GeV does not necessarily mean that it can be seen at LHC. While the lightest Higgs of the MSSM seems to be detectable at LHC, there may still be some significant loopholes for searches of the light Higgs in the NMSSM context.

iii) It is known that the MSSM is a decoupling theory. In the limit where we send the SUSY parameters to infinity all SUSY masses become infinite, with only the lightest Higgs remaining light and coinciding with the usual SM Higgs. In this limit we would recover the SM. It turns out that as far as electroweak precision tests are concerned, the decoupling of the MSSM is quite fast: already for SUSY masses above 200 − 300 GeV the effects due to the exchange of SUSY particles in radiative corrections to the electroweak observables become negligible. Notice that this is not true if, instead of electroweak precision tests, we consider FCNC and CP tests. In this latter case, the decoupling may be much slower with squarks and gluinos of 1 TeV still providing sizeable contributions in loop diagrams to some rare processes.

Obviously the SM fit of electroweak precision data is now so good that there is no point in trying to improve it by the addition of the several degrees of freedom represented by the SUSY particles. The situation was different a couple of years ago when the discrepancy between the SM prediction and the data in the decay of the Z into a b quark pair resulted in a SM fit which could be significantly improved. Now the goal of the game has changed: one looks for regions of the SUSY parameter space where (some) SUSY masses are sufficiently small so that virtual SUSY contributions to electroweak observables are sizeable [8]. Some of these regions may cause unbearably high departures from the SM predictions and hence they can be ruled out. In this way it is possible to exclude some (limited) portions of the MSSM parameter space which would be otherwise allowed by the limits on SUSY parameters coming from direct searches of SUSY particles.

Finally, we make a comment related to the prediction of one low-energy parameter (the electroweak angle or, as it is the case nowadays, the value of the strong coupling at the Z mass scale) when one asks for the unification of the gauge coupling constants in the MSSM. The value predicted for $\alpha_S(m_Z)$ in the MSSM is a couple of standard deviations higher than the experimental value. We do not consider this as a problem for the MSSM. Indeed, high energy thresholds generated from the masses of superheavy GUT particles may conceivably produce corrections able to account for such discrepancy. Taking into account the uncertainties in the dynamics at the GUT scale, we consider the argument of unification of couplings as a support to the existence of low-energy SUSY.

Before starting our discussion of indirect searches of SUSY, let us emphasise that direct production and detection of SUSY particles remain the only way to definitely prove the existence of low-energy SUSY. However it is true that if LEP II is not going to find a SUSY signal and unless some surprise possibly comes from Tevatron, we will have to wait almost ten years to obtain an answer from such direct searches. In view of this fact and of what we said in this section we think that indirect searches of SUSY in the pre-LHC era deserve a very special attention.

III   FCNC AND SUSY

The generation of fermion masses and mixings ("flavour problem") gives rise to a first and important distinction among theories of new physics beyond the electroweak standard model.

One may conceive a kind of new physics which is completely "flavour blind", i.e. new interactions which have nothing to do with the flavour structure. To provide an example of such a situation, consider a scheme where flavour arises at a very large scale (for instance the Planck mass) while new physics is represented by a supersymmetric extension of the SM with supersymmetry broken at a much lower scale and with the SUSY breaking transmitted to the observable sector by flavour-blind gauge interactions [2]-[4]. In this case one may think that new physics does not cause any major change to the original flavour structure of the SM, namely that the pattern of fermion masses and mixings is compatible with the numerous and demanding tests of flavour changing neutral currents.

Alternatively, one can conceive a new physics which is entangled with the flavour problem. As an example consider a technicolour scheme where fermion masses and mixings arise through the exchange of new gauge bosons which mix together ordinary and technifermions. Here we expect (correctly enough) new physics to have potential problems in accommodating the usual fermion spectrum with the adequate suppression of FCNC. As another example of new physics which is not flavour blind, take a more conventional SUSY model which is derived from a spontaneously broken N=1 supergravity and where the SUSY breaking information is conveyed to the ordinary sector of the theory through gravitational interactions. In this case we may expect that the scale at which flavour arises and the scale of SUSY breaking are not so different and possibly the mechanism itself of SUSY breaking and transmission is flavour-dependent. Under these circumstances we may expect a potential flavour problem to arise, namely that SUSY contributions to FCNC processes are too large.

The potentiality of probing SUSY in FCNC phenomena was readily realized when the era of SUSY phenomenology started in the early 80's [9]. In particular, the major implication that the scalar partners of quarks of the same electric charge but belonging to different generations had to share a remarkably high mass degeneracy was emphasised.

Throughout the large amount of work in this last decade it became clearer and clearer that generically talking of the implications of low-energy SUSY on FCNC may be rather misleading. In minimal SUGRA FCNC contributions can be computed in terms of a very limited set of unknown new SUSY parameters. Remarkably enough, this minimal model succeeds to pass all the set of FCNC tests unscathed. To be sure, it is possible to severely constrain the SUSY parameter space, for instance using $b \to s\gamma$, in a way which is complementary to what is achieved by direct SUSY searches at colliders.

However, the MSSM is by no means equivalent to low-energy SUSY. A first sharp distinction concerns the mechanism of SUSY breaking and transmission to the observable sector which is chosen. As we mentioned above, in models with gauge-mediated SUSY breaking (GMSB models [2]- [4]) it may be possible to avoid the FCNC threat "ab initio" (notice that this is not an automatic feature of this class of models, but it depends on the specific choice of the sector which transmits the SUSY breaking information, the so-called messenger sector). The other more "canonical" class of SUSY theories that was mentioned above has gravitational messengers and a very large scale at which SUSY breaking occurs. In this talk we will focus only on this class of gravity-mediated SUSY breaking models. Even sticking to this more limited choice we have a variety of options with very different implications for the flavour problem.

First, there exists an interesting large class of SUSY realizations where the customary R-parity (which is invoked to suppress proton decay) is replaced by other discrete symmetries which allow either baryon or lepton violating terms in the superpotential. But, even sticking to the more orthodox view of imposing R-parity, we are still left with a large variety of extensions of the MSSM at low energy. The point is that low-energy SUSY "feels" the new physics at the superlarge scale at which supergravity (i.e., local supersymmetry) broke down. In this last couple of years we have witnessed an increasing interest in supergravity realizations without the so-called flavour universality of the terms which break SUSY explicitly. Another class of low-energy SUSY realizations which differ from the MSSM in the FCNC sector is obtained from SUSY-GUT's. The interactions involving superheavy particles in the energy range between the GUT and the Planck scale bear important implications for the amount and kind of FCNC that we expect at low energy.

Given a specific SUSY model it is in principle possible to make a full computation of all the FCNC phenomena in that context. However, given the variety of options for low-energy SUSY (even confining ourselves here to models with R matter parity), it is important to have a way to extract from the whole host of FCNC processes a set of upper limits on quantities which can be readily computed in any chosen SUSY frame.

The best model-independent parameterisation of FCNC effects is the so-called mass insertion approximation [10]. It concerns the most peculiar source of FCNC SUSY contributions that do not arise from the mere supersymmetrization of the FCNC in the SM. They originate from the FC couplings of gluinos and neutralinos to fermions and sfermions [11]. One chooses a basis for the fermion and sfermion states where all the couplings of these particles to neutral gauginos are flavour diagonal, while the FC is exhibited by the non-diagonality of the sfermion propagators. Denoting by Δ the off-diagonal terms in the sfermion mass matrices (i.e. the mass terms relating sfermion of the same electric charge, but different flavour), the sfermion propagators can be expanded as a series in terms of $\delta = \Delta/\tilde{m}^2$ where $\tilde{m}$ is the average sfermion mass. As long as Δ is significantly smaller than $\tilde{m}^2$, we can just take the first term of this expansion and, then, the experimental information concerning FCNC and CP violating phenomena translates into upper bounds on these δ's [12]- [14].

Obviously the above mass insertion method presents the major advantage that

x	$\sqrt{\left|\mathrm{Re}\left(\delta^d_{12}\right)^2_{LL}\right|}$	$\sqrt{\left|\mathrm{Re}\left(\delta^d_{12}\right)^2_{LR}\right|}$	$\sqrt{\left|\mathrm{Re}\left(\delta^d_{12}\right)_{LL}\left(\delta^d_{12}\right)_{RR}\right|}$
0.3	1.9×10^{-2}	7.9×10^{-3}	2.5×10^{-3}
1.0	4.0×10^{-2}	4.4×10^{-3}	2.8×10^{-3}
4.0	9.3×10^{-2}	5.3×10^{-3}	4.0×10^{-3}
x	$\sqrt{\left|\mathrm{Re}\left(\delta^d_{13}\right)^2_{LL}\right|}$	$\sqrt{\left|\mathrm{Re}\left(\delta^d_{13}\right)^2_{LR}\right|}$	$\sqrt{\left|\mathrm{Re}\left(\delta^d_{13}\right)_{LL}\left(\delta^d_{13}\right)_{RR}\right|}$
0.3	4.6×10^{-2}	5.6×10^{-2}	1.6×10^{-2}
1.0	9.8×10^{-2}	3.3×10^{-2}	1.8×10^{-2}
4.0	2.3×10^{-1}	3.6×10^{-2}	2.5×10^{-2}
x	$\sqrt{\left|\mathrm{Re}\left(\delta^u_{12}\right)^2_{LL}\right|}$	$\sqrt{\left|\mathrm{Re}\left(\delta^u_{12}\right)^2_{LR}\right|}$	$\sqrt{\left|\mathrm{Re}\left(\delta^u_{12}\right)_{LL}\left(\delta^u_{12}\right)_{RR}\right|}$
0.3	4.7×10^{-2}	6.3×10^{-2}	1.6×10^{-2}
1.0	1.0×10^{-1}	3.1×10^{-2}	1.7×10^{-2}
4.0	2.4×10^{-1}	3.5×10^{-2}	2.5×10^{-2}

TABLE 1. Limits on $\mathrm{Re}\left(\delta_{ij}\right)_{AB}\left(\delta_{ij}\right)_{CD}$, with $A, B, C, D = (L, R)$, for an average squark mass $m_{\tilde{q}} = 500\,\mathrm{GeV}$ and for different values of $x = m_{\tilde{g}}^2/m_{\tilde{q}}^2$. For different values of $m_{\tilde{q}}$, the limits can be obtained multiplying the ones in the table by $m_{\tilde{q}}(\mathrm{GeV})/500$.

one does not need the full diagonalisation of the sfermion mass matrices to perform a test of the SUSY model under consideration in the FCNC sector. It is enough to compute ratios of the off-diagonal over the diagonal entries of the sfermion mass matrices and compare the results with the general bounds on the δ's that we provide here from all available experimental information.

There exist four different Δ mass insertions connecting flavours i and j along a sfermion propagator: $(\Delta_{ij})_{LL}$, $(\Delta_{ij})_{RR}$, $(\Delta_{ij})_{LR}$ and $(\Delta_{ij})_{RL}$. The indices L and R refer to the helicity of the fermion partners. Instead of the dimensional quantities Δ it is more useful to provide bounds making use of dimensionless quantities, δ, that are obtained dividing the mass insertions by an average sfermion mass.

Let us first consider CP-conserving $\Delta F = 2$ processes. The amplitudes for gluino-mediated contributions to $\Delta F = 2$ transitions in the mass-insertion approximation have been computed in refs. [13,14]. Imposing that the contribution to $K - \bar{K}$, $D - \bar{D}$ and $B_d - \bar{B}_d$ mixing proportional to each single δ parameter does not exceed the experimental value, we obtain the constraints on the δ's reported in table 1, barring accidental cancellations [14] (for a QCD-improved computation of the constraints coming from $K - \bar{K}$ mixing, see ref. [15]).

We then consider the process $b \to s\gamma$. This decay requires a helicity flip. In the presence of a $\left(\delta^d_{23}\right)_{LR}$ mass insertion we can realize this flip in the gluino running in the loop. On the contrary, the $\left(\delta^d_{23}\right)_{LL}$ insertion requires the helicity flip to occur in the external b-quark line. Hence we expect a stronger bound on the $\left(\delta_{23}^d\right)_{LR}$ quantity. Indeed, this is what happens: $\left(\delta_{23}^d\right)_{LL}$ is essentially not bounded, while $\left(\delta_{23}^d\right)_{LR}$ is limited to be $< 10^{-3} - 10^{-2}$ according to the average squark and gluino masses [14].

Given the upper bound on $\left(\delta_{23}^d\right)_{LR}$ from $b \to s\gamma$, it turns out that the quantity x_s of the $B_s - \bar{B}_s$ mixing receives contributions from this kind of mass insertions which are very tiny. The only chance to obtain large values of x_s is if $\left(\delta_{23}^d\right)_{LL}$ is large, say of $O(1)$. In that case x_s can easily jump up to values of $O(10^2)$ or even larger.

Then, imposing the bounds in table 1, we can obtain the largest possible value for BR($b \to d\gamma$) through gluino exchange. As expected, the $\left(\delta_{13}^d\right)_{LL}$ insertion leads to very small values of this BR of $O(10^{-7})$ or so, whilst the $\left(\delta_{13}^d\right)_{LR}$ insertion allows for BR($b \to d\gamma$) ranging from few times 10^{-4} up to few times 10^{-3} for decreasing values of $x = m_{\tilde{g}}^2/m_{\tilde{q}}^2$. In the SM we expect BR($b \to d\gamma$) to be typically $10 - 20$ times smaller than BR($b \to s\gamma$), i.e. BR($b \to d\gamma$) = $(1.7 \pm 0.85) \times 10^{-5}$. Hence a large enhancement in the SUSY case is conceivable if $\left(\delta_{13}^d\right)_{LR}$ is in the 10^{-2} range. Notice that in the MSSM we expect $\left(\delta_{13}^d\right)_{LR} < m_b^2/m_{\tilde{q}}^2 \times V_{td} < 10^{-6}$, hence with no hope at all of a sizeable contribution to $b \to d\gamma$.

An analysis similar to the one of $b \to s\gamma$ decays can be performed in the leptonic sector where the masses $m_{\tilde{q}}$ and $m_{\tilde{g}}$ are replaced by the average slepton mass $m_{\tilde{l}}$ and the photino mass $m_{\tilde{\gamma}}$ respectively. The most stringent bound concerns the transition $\mu \to e\gamma$ with $\left(\delta_{12}^l\right)_{LR} < 10^{-6}$ for slepton and photino masses of $O(100$ GeV) [14].

IV CP AND SUSY

The situation concerning CP violation in the MSSM case with $\Phi_A = \Phi_B = 0$ and exact universality in the soft-breaking sector can be summarised in the following way: the MSSM does not lead to any significant deviation from the SM expectation for CP-violating phenomena as d_N^e, ε, ε' and CP violation in B physics; the only exception to this statement concerns a small portion of the MSSM parameter space where a very light $\tilde{t}$ ($m_{\tilde{t}} < 100$ GeV) and χ^+ ($m_\chi \sim 90$ GeV) are present. In this latter particular situation sizeable SUSY contributions to ε_K are possible and, consequently, major restrictions in the $\rho - \eta$ plane can be inferred (see, for instance, ref. [16]). Obviously, CP violation in B physics becomes a crucial test for this MSSM case with very light $\tilde{t}$ and χ^+. Interestingly enough, such low values of SUSY masses are at the border of the detectability region at LEP II.

We now turn to CP violation in the model-independent approach that we are proposing here. For a detailed discussion we refer the reader to our general study [14]. Here we just summarise the situation in the following three points:

i) ϵ provides bounds on the imaginary parts of the quantities whose real part was limited by the K mass difference which are roughly one order of magnitude more severe than the corresponding ones derived from Δm_K.

ii) The nature of the SUSY contribution to CP violation is generally superweak, since the constraints from ε are always stronger (in the left-left sector) or at least equal (in the left-right sector) to the ones coming from ε'/ε.

iii) the experimental bound on the electric dipole moment of the neutron imposes very stringent limits on $\mathrm{Im}\left(\delta_{11}^d\right)_{LR}$ (of $O(10^{-6})$ for an average squark and gluino mass of 500 GeV.) In conclusion, although technically it is conceivable that some SUSY extension may provide a sizable contribution to ε'/ε, it is rather difficult to imagine how to reconcile a relatively large value of $\mathrm{Im}\left(\delta_{12}^d\right)_{LR}$ with the very strong constraint on the flavour-conserving $\mathrm{Im}\left(\delta_{11}^d\right)_{LR}$ from d_N^e.

We now move to the next frontier for testing the unitarity triangle in general and in particular CP violation in the SM and its SUSY extensions: B physics. We have seen above that the transitions between 1st and 2nd generation in the down sector put severe constraints on $\mathrm{Re}\,\delta_{12}^d$ and $\mathrm{Im}\,\delta_{12}^d$ quantities. To be sure, the bounds derived from ε and ε' are stronger than the corresponding bounds from ΔM_K. If the same pattern repeats itself in the transition between 3rd and 1st or 3rd and 2nd generation in the down sector we may expect that the constraints inferred from B_d–$\bar{B}_d$ oscillations or $b \to s\gamma$ do not prevent conspicuous new contributions also in CP violating processes in B physics. We are going to see below that this is indeed the case ad we will argue that measurements of CP asymmetries in several B-decay channels may allow to disentangle SM and SUSY contributions to the CP decay phase.

New physics can modify the SM predictions on CP asymmetries in B decays by changing the phase of the B_d–$\bar{B}_d$ mixing and the phase and absolute value of the decay amplitude. The general SUSY extension of the SM that we discuss here affects both these quantities.

The crucial question is then: where and how can one possibly distinguish SUSY contributions to CP violation in B decays [17]?

In terms of the decay amplitude A, the CP asymmetry reads

$$\mathcal{A}(t) = \frac{(1-|\lambda|^2)\cos(\Delta M_d t) - 2\mathrm{Im}\lambda \sin(\Delta M_d t)}{1+|\lambda|^2} \tag{1}$$

with $\lambda = e^{-2i\phi^M}\bar{A}/A$. In order to be able to discuss the results model-independently, we have labeled as ϕ^M the generic mixing phase. The ideal case occurs when one decay amplitude only appears in (or dominates) a decay process: the CP violating asymmetry is then determined by the total phase $\phi^T = \phi^M + \phi^D$, where ϕ^D is the weak phase of the decay. This ideal situation is spoiled by the presence of several interfering amplitudes.

Incl.	Excl.	ϕ^D_{SM}	r_{SM}	ϕ^D_{SUSY}	r_{250}	r_{500}
$b \to c\bar{c}s$	$B \to J/\psi K_S$	0	–	ϕ_{23}	0.03 – 0.1	0.008 – 0.04
$b \to s\bar{s}s$	$B \to \phi K_S$	0	–	ϕ_{23}	0.4 – 0.7	0.09 – 0.2
$b \to u\bar{u}s$		P 0				
	$B \to \pi^0 K_S$		0.01 – 0.08	ϕ_{23}	0.4 – 0.7	0.09 – 0.2
$b \to d\bar{d}s$		T γ				
$b \to c\bar{u}d$		0				
	$B \to D^0_{CP}\pi^0$		0.02	–	–	–
$b \to u\bar{c}d$		γ				
	$B \to D^+ D^-$	T 0	0.03 – 0.3		0.007 – 0.02	0.002 – 0.006
$b \to c\bar{c}d$				ϕ_{13}		
	$B \to J/\psi \pi^0$	P β	0.04 – 0.3		0.007 – 0.03	0.002 – 0.008
	$B \to \phi \pi^0$	P β	–		0.06 – 0.1	0.01 – 0.03
$b \to s\bar{s}d$				ϕ_{13}		
	$B \to K^0 \bar{K}^0$	u-P γ	0 – 0.07		0.08 – 0.2	0.02 – 0.06
$b \to u\bar{u}d$	$B \to \pi^+\pi^-$	T γ	0.09 – 0.9	ϕ_{13}	0.02 – 0.8	0.005 – 0.2
$b \to d\bar{d}d$	$B \to \pi^0\pi^0$	P β	0.6 – 6	ϕ_{12}	0.06 – 0.4	0.02 – 0.1
	$B \to K^+ K^-$	T γ	0.2 – 0.4		0.04 – 0.1	0.01 – 0.03
$b\bar{d} \to q\bar{q}$				ϕ_{13}		
	$B \to D^0 \bar{D}^0$	P β	only β		0.01 – 0.03	0.003 – 0.006

TABLE 2. CP phases for B decays. ϕ^D_{SM} denotes the decay phase in the SM; T and P denote Tree and Penguin, respectively; for each channel, when two amplitudes with different weak phases are present, one is given in the first row, the other in the last one and the ratio of the two in the r_{SM} column. ϕ^D_{SUSY} denotes the phase of the SUSY amplitude, and the ratio of the SUSY to SM contributions is given in the r_{250} and r_{500} columns for the corresponding SUSY masses.

We summarise the results in table 2 which is taken from the recent analysis of ref. [18]. We refer the interested reader to our work [18] for all the details of how our computation in the SM and in SUSY is carried out. Φ^D_{SM} denotes the decay phase in the SM; for each channel, when two amplitudes with different weak phases are present, we indicate the SM phase of the Penguin (P) and Tree-level (T) decay amplitudes. For $B \to K_S \pi^0$ the penguin contributions (with a vanishing phase) dominate over the tree-level amplitude because the latter is Cabibbo suppressed. For the channel $b \to s\bar{s}d$ only penguin operators or penguin contractions of current-current operators contribute. The phase γ is present in the penguin contractions of the $(\bar{b}u)(\bar{u}d)$ operator, denoted as u-P γ in table 2. $\bar{b}d \to \bar{q}q$ indicates processes occurring via annihilation diagrams which can be measured from the last two channels of table 2. In the case $B \to K^+ K^-$ both current-current and penguin operators contribute. In $B \to D^0 \bar{D}^0$ the contributions from the $(\bar{b}u)(\bar{u}d)$ and the $(\bar{b}c)(\bar{c}d)$ current-current operators (proportional to the phase γ) tend to cancel out.

SUSY contributes to the decay amplitudes with phases induced by δ_{13} and δ_{23} which we denote as ϕ_{13} and ϕ_{23}. The ratios of A_{SUSY}/A_{SM} for SUSY masses of 250 and 500 GeV are reported in the r_{250} and r_{500} columns of table 2.

We now draw some conclusions from the results of table 2. In the SM, the first six decays measure directly the mixing phase β, up to corrections which, in most of the cases, are expected to be small. These corrections, due to the presence of two amplitudes contributing with different phases, produce uncertainties of $\sim 10\%$ in $B \to K_S\pi^0$, and of $\sim 30\%$ in $B \to D^+D^-$ and $B \to J/\psi\pi^0$. In spite of the uncertainties, however, there are cases where the SUSY contribution gives rise to significant changes. For example, for SUSY masses of O(250) GeV, SUSY corrections can shift the measured value of the sine of the phase in $B \to \phi K_S$ and in $B \to K_S\pi^0$ decays by an amount of about 70%. For these decays SUSY effects are sizeable even for masses of 500 GeV. In $B \to J/\psi K_S$ and $B \to \phi\pi^0$ decays, SUSY effects are only about 10% but SM uncertainties are negligible. In $B \to K^0\bar{K}^0$ the larger effect, $\sim 20\%$, is partially covered by the indetermination of about 10% already existing in the SM. Moreover the rate for this channel is expected to be rather small. In $B \to D^+D^-$ and $B \to K^+K^-$, SUSY effects are completely obscured by the errors in the estimates of the SM amplitudes. In $B^0 \to D^0_{CP}\pi^0$ the asymmetry is sensitive to the mixing angle ϕ_M only because the decay amplitude is unaffected by SUSY. This result can be used in connection with $B^0 \to K_s\pi^0$, since a difference in the measure of the phase is a manifestation of SUSY effects.

Turning to $B \to \pi\pi$ decays, both the uncertainties in the SM and the SUSY contributions are very large. Here we witness the presence of three independent amplitudes with different phases and of comparable size. The observation of SUSY effects in the $\pi^0\pi^0$ case is hopeless. The possibility of separating SM and SUSY contributions by using the isospin analysis remains an open possibility which deserves further investigation. For a thorough discussion of the SM uncertainties in $B \to \pi\pi$ see ref. [19].

In conclusion, our analysis shows that measurements of CP asymmetries in several channels may allow the extraction of the CP mixing phase and to disentangle SM and SUSY contributions to the CP decay phase. The golden-plated decays in this respect are $B \to \phi K_S$ and $B \to K_S\pi^0$ channels. The size of the SUSY effects is clearly controlled by the the non-diagonal SUSY mass insertions δ_{ij}, which for illustration we have assumed to have the maximal value compatible with the present experimental limits on B^0_d–$\bar{B}^0_d$ mixing.

V DM AND SUSY: A BRIEF COMMENT

We have strong indications that ordinary matter (baryons) is insufficient to provide the large amount of non-shining matter which has been experimentally proven to exist in galactic halos and at the level of clusters of galaxies [20]. In a sense, this might constitute the "largest" indication of new physics beyond the SM. This statement holds true even after the recent stunning developments in the search for non-shining baryonic objects. In September 1993 the discovery of massive dark objects ("machos") was announced. After five years of intensive analysis it is now clear that in any case machos cannot account for the whole dark matter of the galactic halos.

It was widely expected that some amount of non-shining baryonic matter could exist given that the contribution of luminous baryons to the energy density of the Universe $\Omega = \rho/\rho_{cr}$ ($\rho_{cr} = 3H_0^2/8\pi G$ where G is the gravitational constant and H_0 the Hubble constant) is less than 1%, while from nucleosynthesis we infer $\Omega_{baryon} = \rho_{baryon}/\rho_{cr} = (0.06 \pm 0.02)h_{50}^{-2}$, where $h_{50} = H_0/50$ Km/s Mpc. On the other hand, we have direct indications that Ω should be at least 20% which means that baryons can represent not more than half of the entire energy density of the Universe [20].

We could make these considerations on the insufficiency of the SM to obtain a large enough Ω more dramatic if we accept the theoretical input that the Universe underwent some inflationary era which produced $\Omega = 1$. In that case, at least 90% of the whole energy density of the Universe should be provided by some new physics beyond the SM.

Before discussing possible particle physics candidates, it should be kept in mind that DM is not only called for to provide a major contribution to Ω, but also it has to provide a suitable gravitational driving force for the primordial density fluctuations to evolve into the large-scale structures (galaxies, clusters and superclusters of galaxies) that we observe today [20]. Here we encounter the major difficulties when dealing with the two "traditional" sources of DM: Cold (CDM) and Hot (HDM) DM.

Light neutrinos in the eV range are the most typical example of HDM, being their decoupling temperature of O(1 MeV). On the other hand, the Lightest Supersymmetric Particle (LSP) in the tens of GeV range is a typical CDM candidate. Taking the LSP to be the lightest neutralino, one obtains that when it decouples it is already non-relativistic, being its decoupling temperature typically one order of magnitude below its mass.

Both HDM and CDM have some difficulty to correctly reproduce the experimental spectrum related to the distribution of structures at different scales. The conflict is more violent in the case of pure HDM. Neutrinos of few eV's tend to produce too many superlarge structures. The opposite problem arises with pure CDM: we obtain too much power in the spectrum at low mass scales (galactic scales).

A general feature is that some amount of CDM should be present in any case. A possibility which has been envisaged is that after all the whole Ω could be much smaller than one, say 20% or so and then entirely due to CDM. However, if one keeps on demanding the presence of an inflationary epoch, then it seems unnatural to have Ω so different from unity (although lately some variants of inflationary schemes leading to Ω smaller than one have been proposed). Another possibility is that CDM provides its 20% to Ω, while all the rest to reach the unity value is given by a nonvanishing cosmological constant.

Finally, the possibility which encounters quite some interest is the so-called Mixed Dark Matter (MDM) [21], where a wise cocktail of HDM and CDM is present. An obvious realization of a MDM scheme is a variant of the MSSM where neutrinos get a mass of few eV's. In that case the lightest neutralino (which is taken to be the LSP) plays the role of CDM and the light neutrino(s) that of HDM. With an appropriate choice of the parameters it is possible to obtain contributions to Ω from the CDM and HDM in the desired range.

In the MSSM with R parity the lightest SUSY particle (LSP) is absolutely stable. For several reasons the lightest neutralino is the favourite candidate to be the LSP fulfilling the role of CDM [22].

The neutralinos are the eigenvectors of the mass matrix of the four neutral fermions partners of the W_3, B, H_1^0 and H_2^0. There are four parameters entering this matrix: M_1, M_2, μ and $\tan \beta$. The first two parameters denote the coefficients of the SUSY breaking mass terms $\tilde{B}\tilde{B}$ and $\tilde{W}_3\tilde{W}_3$ respectively. μ is the coupling of the $H_1 H_2$ term in the superpotential. Finally $\tan \beta$ denotes the ratio of the VEV's of the H_2 and H_1 scalar fields.

In general M_1 and M_2 are two independent parameters, but if one assumes that grand unification takes place, then at the grand unification scale $M_1 = M_2 = M_3$, where M_3 is the gluino mass at that scale. Then at M_W one obtains:

$$M_1 = \frac{5}{3} \tan^2 \theta_w M_2 \simeq \frac{M_2}{2}, \qquad M_2 = \frac{g_2^2}{g_3^2} m_{\tilde{g}} \simeq \frac{m_{\tilde{g}}}{3}, \qquad (2)$$

where g_2 and g_3 are the SU(2) and SU(3) gauge coupling constants, respectively.

The above relation between M_1 and M_2 reduces to three the number of independent parameters which determine the lightest neutralino composition and mass: $\tan \beta, \mu$ and M_2. Hence, for fixed values of $\tan \beta$ one can study the neutralino spectrum in the (μ, M_2) plane. The major experimental inputs to exclude regions in this plane are the request that the lightest chargino be heavier than M_Z and the limits on the invisible width of the Z hence limiting the possible decays $Z \to \chi\chi,$-$\chi\chi'$.

Let us focus now on the role played by χ as a source of CDM. χ is kept in thermal equilibrium through its electroweak interactions not only for $T > m_\chi$, but even when T is below m_χ. However for $T < m_\chi$ the number of χ's rapidly decrease because of the appearance of the typical Boltzmann suppression factor $\exp(-m_\chi/T)$. When T is roughly $m_\chi/20$ the number of χ diminished so much that they do not interact any longer, i.e. they decouple. Hence the contribution to Ω_{CDM} of χ is determined by two parameters: m_χ and the temperature at which χ decouples (T_D). T_D fixes the number of $\chi's$ which survive. As for the determination of T_D itself, one has to compute the χ annihilation rate and compare it with the cosmic expansion rate [23].

Several annihilation channels are possible with the exchange of different SUSY or ordinary particles, $\tilde{f}, H, Z$, etc. Obviously the relative importance of the channels depends on the composition of χ. For instance, if χ is a pure gaugino, then the $\tilde{f}$ exchange represents the dominant annihilation mode.

Quantitatively [24], it turns out that if χ results from a large mixing of the gaugino ($\tilde{W}_3$ and $\tilde{B}$) and Higgsino ($\tilde{H}_1^0$ and $\tilde{H}_2^0$) components, then the annihilation is too efficient to allow the surviving χ to provide Ω large enough. Typically in this case $\Omega < 10^{-2}$ and hence χ is not a good CDM candidate. On the contrary, if χ is either almost a pure Higgsino or a pure gaugino then it can give a conspicuous contribution to Ω

In the case χ is mainly a gaugino (say at least at the 90% level) what is decisive to establish the annihilation rate is the mass of $\tilde{f}$. If sfermions are light the χ annihilation rate is fast and the Ω_χ is negligible. On the other hand, if $\tilde{f}$ (and hence $\tilde{l}$, in particular) is heavier than 150 GeV, the annihilation rate of χ is sufficiently suppressed so that Ω_χ can be in the right ballpark for Ω_{CDM}. In fact if all the $\tilde{f}'s$ are heavy, say above 500 GeV and for $m_\chi << m_{\tilde{f}}$, then the suppression of the annihilation rate can become even too efficient yielding Ω_χ unacceptably large.

In the minimal SUSY standard model there are five new parameters in addition to those already present in the non–SUSY case. Imposing the electroweak radiative breaking further reduces this number to four. Finally, in simple supergravity realizations the soft parameters A and B are related. Hence we end up with only three new, independent parameters. One can use the constraint that the relic χ abundance provides a correct Ω_{CDM} to restrict the allowed area in this 3–dimensional space. Or, at least, one can eliminate points of this space which would lead to $\Omega_\chi > 1$, hence overclosing the Universe. For χ masses up to 150 GeV it is possible to find sizable regions in the SUSY parameter space where Ω_χ acquires interesting values for the DM problem. A detailed discussion on this point is beyond the scope of this talk. The interested reader can find a thorough analysis in the review of Ref. [22] and the original papers therein quoted.

There exist two ways to search for the existence of relic neutralinos. First we have direct detection: neutralinos interact with matter both through coherent and spin dependent effects. Only coherent effects are currently accessible to direct detection. The sensitivity of the direct detection experiment has reached now an area of the SUSY parameter space of the MSSM which is of great interest for neutralinos in the 50 GeV - 200 GeV range.

The indirect detection is based on the search for signals coming from pair annihilation of neutralinos. Such annihilation may occur inside celestial bodies (Earth, Sun, etc.) where neutralinos may be gravitationally captured. The signal is then a flux of muon neutrinos which can be detected as up-going muons in a neutrino telescope. Another possibility is that the neutralino annihilation occurs in the galactic halo. In this case the signal consists of photon, positron and antiproton fluxes. They can be observed by detectors placed on balloons or satellites. The computation of these fluxes is strongly affected by the composition of the lightest neutralino. In any case also these indirect searches for relic neutralinos are now probing interesting areas of the MSSM parameter space.

A very different prospect for DM occurs in the GMSB schemes. In this case the gravitino mass ($m_{3/2}$) loses its role of fixing the typical size of soft breaking terms and we expect it to be much smaller than what we have in models with a hidden sector. Indeed, given the well-known relation [1] between $m_{3/2}$ and the scale of

SUSY breaking $\sqrt{F}$, i.e. $m_{3/2} = O(F/M)$, where M is the reduced Planck scale, we expect $m_{3/2}$ in the keV range for a scale $\sqrt{F}$ of $O(10^6$ GeV) that has been proposed in models with low-energy SUSY breaking in a visible sector.

A gravitino of that mass behaves as a Warm Dark Matter (WDM) particle, that is, a particle whose free streaming scale involves a mass comparable to that of a galaxy, $\sim 10^{11-12} M_\odot$.

However, critical density models with pure WDM are known to suffer for serious troubles [25]. Indeed, a WDM scenario behaves much like CDM on scales above λ_{FS}. Therefore, we expect in the light gravitino scenario that the level of cosmological density fluctuations on the scale of galaxy clusters ($\sim 10\,h^{-1}$ Mpc) to be almost the same as in CDM. As a consequence, the resulting number density of galaxy clusters is predicted to be much larger than what observed.

We have recently considered different variants of a light gravitino DM dominated model. It seems that in all cases there exist difficulties to account correctly for cosmic straucture formation. This provides severe cosmological constraints on the GMSB models [26].

In conclusion SUGRA models with R parity offer the best candidate for CDM. It is remarkable that as a by-product of the MSSM we obtain a lightest neutralino which can provide the correct amount of DM in a wide area of the SUSY parametr space. Even more interesting, we are now experimentally approaching the level of sensitivity which is needed to explore (directly or indirectly) large portions of this area of parameter space. The complementarity of this exploration to that performed by using FCNC and CP tests and direct collider SUSY searches looks promising.

ACKNOWLEDGEMENTS

We are grateful to our "FCNC collaborators" M. Ciuchini, E. Franco, F. Gabbiani, E. Gabrielli and G. Martinelli and our "DM collaborators" S. Borgani, E. Pierpaoli and M. Yamaguchi who contributed to most of our recent production on the subject which was reported in these talks. A.M. thanks the organizers for the stimulating settling in which the workshop and the symposium took place. The work of A.M. was partly supported by the TMR project "Beyond the Standard Model" contract number ERBFMRX CT96 0090. L.S. acknowledges the support of the German Bundesministerium für Bildung und Forschung under contract 06 TM 874 and DFG Project Li 519/2-2.

REFERENCES

1. For a phenomenologically oriented review, see:
   P. Fayet and S. Ferrara, *Phys. Rep.* **32C** (1977) 249;
   H.P. Nilles, *Phys. Rep.* **110** (1984) 1.
   H.E. Haber and G.L Kane, *Phys. Rep.* **117** (1987) 1;

For spontaneously broken N=1 supergravity, see:
E. Cremmer, S. Ferrara, L. Girardello and A. Van Proeyen, *Nucl. Phys.* **B 212** (1983) 413 and references therein.
P. Nath, R. Arnowitt and A.H. Chamseddine, *Applied N=1 Supergravity* (World Scientific, Singapore, 1984);
A.G. Lahanas and D.V. Nanopoulos, *Phys. Rep.* **145** (1987) 1.
2. M. Dine, W. Fischler and M. Srednicki, *Nucl. Phys.* **B 189** (1981) 575;
S. Dimopoulos and S. Raby, *Nucl. Phys.* **B 192** (1981) 353;
M. Dine and W. Fischler, *Phys. Lett.* **B 110** (1982) 227;
M. Dine and M. Srednicki, *Nucl. Phys.* **B 202** (1982) 238;
M. Dine and W. Fischler, *Nucl. Phys.* **B 204** (1982) 346;
L. Alvarez-Gaumé, M. Claudson and M. Wise, *Nucl. Phys.* **B 207** (1982) 96;
C. Nappi and B. Ovrut, *Phys. Lett.* **B 113** (1982) 175;
S. Dimopoulos and S. Raby, *Nucl. Phys.* **B 219** (1983) 479.
3. A. Nelson and M. Dine, *Phys. Rev.* **D 48** (1993) 1277;
M. Dine, A. E. Nelson and Y. Shirman, *Phys. Rev.* **D 51** (1995) 1362;
M. Dine, A. Nelson, Y. Nir and Y. Shirman, *Phys. Rev.* **D 53** (1996) 2658.
4. E. Poppitz and S. Trivedi, *Phys. Rev.* **D 55** (1997) 5508;
N. Arkani-Hamed, J. March-Russel and H. Murayama, *Nucl. Phys.* **B 509** (1998) 3;
H. Murayama, *Phys. Rev. Lett.* **79** (1997) 18;
S. Dimopoulos, G. Dvali, G. Giudice and R. Rattazzi, *Nucl. Phys.* **B 510** (1998) 12 ;
S. Dimopoulos, G. Dvali and R. Rattazzi, *Phys. Lett.* **B 413** (1997) 336;
M. Luty, *Phys. Lett.* **B 413** (1997) 71;
T. Hotta, K.-I. Izawa and T. Yanagida, *Phys. Rev.* **D 55** (1997) 415;
N. Haba, N. Maru and T. Matsuoka, *Nucl. Phys.* **B 497** (1997) 31;
L. Randall, *Nucl. Phys.* **B 495** (1997) 37;
Y. Shadmi, *Phys. Lett.* **B 405** (1997) 99;
N. Haba, N. Maru and T. Matsuoka, *Phys. Rev.* **D 56** (1997) 4207;
C. Csaki, L. Randall and W. Skiba, *Phys. Rev.* **D 57** (1998) 383;
Y. Shirman, *Phys. Lett.* **B 417** (1998) 281;
For a complete review, see: G. Giudice and R. Rattazzi, hep-ph/9801271.
5. R. Barbieri and G. Giudice, *Nucl. Phys.* **B 306** (1988) 63;
G. Anderson and D. Castano, *Phys. Rev.* **D 52** (1995) 1693;
P.H. Chankowski, J. Ellis and S. Pokorski, *Phys. Lett.* **B 423** (1998) 327;
R. Barbieri and A. Strumia, hep-ph/9801353.
6. H. Haber and R. Hempfling, *Phys. Rev. Lett.* **66** (1991) 1815;
J. Ellis, G. Ridolfi and F. Zwirner, *Phys. Lett.* **B 257** (1991) 83;
Y. Okada, M. Yamaguchi and T. Yanagida, *Prog. Theor. Phys.* **85** (1991) 1.
7. S. Dimopoulos and S. Thomas, *Nucl. Phys.* **B 465** (1996) 23.
8. P.H. Chankowski and S. Pokorski, hep-ph/9707497;
D. Pierce and J. Erler, hep-ph/9708374.
9. J. Ellis and D.V. Nanopoulos, *Phys. Lett.* **B 110** (1982) 44;
R. Barbieri and R. Gatto, *Phys. Lett.* **B 110** (1982) 211.

10. L.J. Hall, V.A. Kostelecky and S. Raby, *Nucl. Phys.* **B 267** (1986) 415.
11. M.J. Duncan, *Nucl. Phys.* **B 221** (1983) 285;
    J.F. Donoghue, H.P. Nilles and D. Wyler, *Phys. Lett.* **B 128** (1983) 55;
    A. Bouquet, J. Kaplan and C.A. Savoy, *Phys. Lett.* **B 148** (1984) 69.
12. F. Gabbiani and A. Masiero, *Nucl. Phys.* **B 322** (1989) 235;
    J.S. Hagelin, S. Kelley and T. Tanaka, *Nucl. Phys.* **B 415** (1994) 293.
13. E. Gabrielli, A. Masiero and L. Silvestrini, *Phys. Lett.* **B 374** (1996) 80.
14. F. Gabbiani, E. Gabrielli, A. Masiero and L. Silvestrini, *Nucl. Phys.* **B 477** (1996) 321.
15. J.A. Bagger, K.T. Matchev and R.-J. Zhang, *Phys. Lett.* **B 412** (1997) 77.
16. M. Misiak, S. Pokorski and J. Rosiek, hep-ph/9703442.
17. For some recent discussions tackling this question, see:
    N. Deshpande, B. Dutta and S. Oh, *Phys. Rev. Lett.* **77** (1996) 4499;
    J. Silva and L. Wolfenstein, *Phys. Rev.* **D 53** (1997) 5331;
    A. Cohen, D. Kaplan, F. Leipentre and A. Nelson, *Phys. Rev. Lett.* **78** (1997) 2300;
    Y. Grossman and M. Worah, *Phys. Lett.* **B 395** (1997) 241;
    Y. Grossman, Y. Nir and R. Rattazzi, hep-ph/9701231;
    M. Ciuchini, E. Franco, G. Martinelli, A. Masiero and L. Silvestrini, *Phys. Rev. Lett.* **79** (1997) 978;
    Y. Grossman, Y. Nir and M. Worah, *Phys. Lett.* **B 407** (1997) 307;
    R. Barbieri and A. Strumia, *Nucl. Phys.* **B 508** (1997) 3.
18. M. Ciuchini, E. Franco, G. Martinelli, A. Masiero and L. Silvestrini, in ref. [17].
19. M. Ciuchini, E. Franco, G. Martinelli and L. Silvestrini, *Nucl. Phys.* **B 501** (1997) 271.
20. For an introduction to the DM problem, see, for instance:
    R. Kolb and S. Turner, *The Early universe* (addison-Wesley, New York, N.Y.) 1990;
    Dark Matter, ed. by M. Srednicki (North-Holland, Amsterdam) 1989;
    J. Primack, D. Seckel and B. Sadoulet, *Ann. Rev. Nucl. Part. Sci.* **38** (1988) 751.
21. Q. Shafi and F.W. Stecker, *Phys. Lett. B* **53** (1984) 1292;
    S.A. Bonometto and R. Valdarnini, *Astroph. J.* **299** (1985) L71;
    . S. Achilli, F. Occhionero and R. Scaramella, *Astroph. J.* **299** (1985) L77;
    J.A. Holtzman, *Astroph. J. Suppl.* **71** (1981) 1;
    A.N.Taylor and M. Rowan-Robinson, *Nature* **359** (1992) 396;
    J.A. Holtzman and J. Primack, *Astroph. J.* **396** (1992) 113;
    D. Pogosyan and A. Starobinski, *Astroph. J.* **447** (1995) 465;
    A. Klypin, J. Holtzman, J. Primack and E. Regos, *Astroph. J.* **415** (1993) 1.
22. G. Jungman, M. Kamionkowski and K. Griest, *Phys. Rep.* **267** (1996) 195, and references therein.
23. J. Ellis, J.S. Hagelin, D.V. Nanopoulos, K. Olive and M. Srednicki, *Nucl. Phys.* **B 238** (1984) 453.
24. A. Bottino, F. Donato, N. Fornengo and S. Scopel, hep-ph/9710295.
25. S. Colombi, S. Dodelson and L.M. Widrow, astro-ph/9505029.
26. E. Pierpaoli, S. Borgani, A. Masiero and M. Yamaguchi, *Phys. Rev.* **D 57** (1998) 2089.

R-Symmetry in MSSM and Beyond

Q. Shafi

Bartol Research Institute
University of Delaware
Newark, DE 19716, USA

Abstract. This talk explores the role that $U(1)$ (or some discrete) R-symmetries can play in the construction of realistic supersymmetric models. By exploiting $U(1)_R$, the first part attempts to obtain a more robust version of MSSM in which neutrinos are massive and the resolution of the μ and strong CP problems is intimately related. The second part goes beyond the MSSM gauge group by considering $SU(3)_c \times SU(2)_L \times SU(2)_R \times U(1)_{B-L}$, supplemented by $U(1)_R$. The parameter $\tan\beta \simeq m_t/m_b$, which allows one to predict the top quark mass in the right ball park, and also explain why the 'standard model' higgs h° has so far not been found at LEPII. Its tree level mass is M_{Z° and, after radiative corrections, one expects $m_{h^\circ} \sim 110 \pm 10$ GeV. This particular scheme also resolves the μ problem of MSSM and enables one to realize an inflationary epoch with the spectral index of density fluctuations very close to unity ($n \approx 0.98$).

Although quite compelling, the minimal supersymmetric standard model (MSSM) fails to address a number of important challenges. For instance, to explain the apparent stability of the proton, it must be assumed that the dimensionless coefficients accompanying dimension five operators are of order 10^{-8} or less. The strong CP and μ problems loom large in the background, and the observed baryon asymmetry of the universe (BAU), it appears, cannot be explained within the MSSM framework. Last, but by no means least, there is increasing evidence for non-zero neutrino masses from a variety of experiments.

In this part of my talk, based on work done with George Lazarides [1], I will present a resolution of the problems listed above without departing from the $SU(3)_c \times SU(2)_L \times U(1)_Y$ framework of MSSM. We first observe that the MSSM superpotential possesses a global $U(1)$ R-symmetry [2] in which Z_2 matter parity is embedded. We show how neutrino masses can be incorporated while preserving a (redefined) R-symmetry. When extended to higher orders, this symmetry ensures the appearance of global $U(1)_B$, thereby guaranteeing proton stability. In the case where right handed neutrinos are included, the BAU can arise via leptogenesis. The approach followed here also provides the framework for an elegant resolution of the strong CP and μ problems of MSSM, with the R-symmetry once again playing an essential role.

The MSSM superpotential W contains the renormalizable terms (with no distinction between the generations)

$$H^{(1)}QU^c,\ H^{(2)}QD^c,\ H^{(2)}LE^c,\ H^{(1)}H^{(2)} . \tag{1}$$

Here $H^{(1)}$, $H^{(2)}$ are the two higgs superfields, Q denotes the $SU(2)_L$ doublet quark superfields, U^c and D^c are the $SU(2)_L$ singlet quark superfields, while L (E^c) stands for the $SU(2)_L$ doublet (singlet) lepton superfields. A Z_2 matter parity (Z_2^{mp}) under which only the 'matter' superfields change sign ensures the absence of terms such as QD^cL and $U^cD^cD^c$ which lead to rapid proton decay.

W in Eq.(1) possesses three global symmetries, namely, $U(1)_B$ (baryon number), $U(1)_L$ (lepton number) and $U(1)_R$. The 'global' R charges (normalized such that W carries two units) of the various superfields are as follows:

$$H^{(1)}(1),\ H^{(2)}(1),\ Q(1/2),\ U^c(1/2),\ D^c(1/2),\ L(1/2),\ E^c(1/2) . \tag{2}$$

The introduction of the right handed neutrino superfields, ν^c, gives rise, consistent with Z_2^{mp}, to two additional renormalizable superpotential couplings

$$H^{(1)}L\nu^c ,\ M^R\nu^c\nu^c , \tag{3}$$

where M^R is the Majorana mass matrix of the superheavy right handed neutrinos. The first term in Eq.(3) fixes the quantum numbers of the right handed neutrinos, namely, $B(\nu^c) = 0$, $L(\nu^c) = -1$, $R(\nu^c) = 1/2$. The second term violates both $U(1)_L$ and $U(1)_R$, but the combination $R' = R - (1/2) L$ is now the new R-symmetry of W. In addition, the Z_2^{lp} (lepton parity) subgroup of $U(1)_L$, under which only the lepton superfields L, E^c change sign, remains unbroken. Consequently, the global symmetries of the renormalizable superpotential containing all the couplings in Eqs.(1) and (3) are $U(1)_{R'}$, $U(1)_B$ and Z_2^{lp}. With the couplings in Eq.(3), the observed neutrinos acquire masses via the see-saw mechanism.

Note that both $U(1)_B$ and lepton parity are automatically implied by $U(1)_{R'}$. Moreover, this remains true even if non-renormalizable terms are included in the superpotential. Indeed, by extending the $U(1)_{R'}$ symmetry to higher order terms, we will first show that $U(1)_B$ follows as a consequence. To see this, note that $U(1)_{R'}$ contains Z_2^{bp} (baryon parity) under which only the color triplet, antitriplet $(3, \bar{3})$ superfields change sign. This means that superpotential couplings containing, in addition to color singlet and $(3 \cdot \bar{3})^m$ ($m \geq 0$) factors, the $U(1)_B$ violating combinations $(3 \cdot 3 \cdot 3)^n$ or $(\bar{3} \cdot \bar{3} \cdot \bar{3})^n$ with $n =$ odd ≥ 1 are not allowed. Similarly, analogous couplings but with $n =$ even ≥ 2 are also not allowed since their R' charge exceeds two units and cannot be compensated. In particular, the troublesome dimension five operators $QQQL$ and $U^cU^cD^cE^c$ are eliminated.

One can next show that $U(1)_{R'}$ implies Z_2^{lp} to all orders. Because of $U(1)_B$, the quark superfields must appear in 'blocks' $QU^c(1)$ and $QD^c(1)$, where the parenthesis indicates the R' charge. The other non-leptonic 'blocks' are $H^{(1)}(1)$ and $H^{(2)}(1)$. The leptonic superfields are $L(0)$, $E^c(1)$, $\nu^c(1)$. To violate lepton parity, we need an odd number of lepton superfields. Therefore, we should consider: i) odd number of L's together, by $U(1)_{R'}$ symmetry, with two non-leptonic blocks belonging to the four types described above; ii) even number of L's and a single E^c or ν^c, together with one non-leptonic block; iii) odd number of L's with two out of the E^c's and ν^c's. In all three cases, one ends up with an odd number of $SU(2)_L$ doublets which is not gauge invariant.

In summary, both Z_2^{lp} and $U(1)_B$ are present in the scheme to all orders as mere consequences of the $U(1)_{R'}$ symmetry and remain exact, although $U(1)_{R'}$ is explicitly broken to its maximal non-R-subgroup Z'_4 (which includes Z_2^{bp}) by the supersymmetry breaking terms.

We now present an alternative scheme for neutrino masses (familiar from Grand Unified Theories) which was recently considered within the non-supersymmetric standard model framework in Ref. [3]. Introduce, in MSSM, an $SU(2)_L$ triplet pair T, $\bar{T}$, with hypercharges +1, -1 and renormalizable superpotential couplings

$$TLL, \ \bar{T}H^{(1)}H^{(1)}, \ MT\bar{T}, \tag{4}$$

such that $B(T) = 3DB(\bar{T}) = 3D0$, $L(T) = 3D - 2$, $L(\bar{T}) = 3D0$, $R(T) = 3D1$, $R(\bar{T}) = 3D0$, from the first two couplings, and M is some superheavy scale (taken real and positive by suitable phase redefinitions of T, $\bar{T}$). The mass term in Eq.(4) breaks $U(1)_R$ and $U(1)_L$ but the superpotential defined by the terms in Eqs.(1) and (4) possesses a redefined R-symmetry generated by $R'' = 3DR - (1/2)\,L$. The R'' charges of the various superfields are:

$$H^{(1)}(1), \ H^{(2)}(1), \ Q(1/2), \ U^c(1/2), \ D^c(1/2), \ L(0), \ E^c(1), \ T(2), \ \bar{T}(0)\,. \tag{5}$$

Both $U(1)_B$ and lepton parity remain unbroken in this case too. Finally, as with the $U(1)_{R'}$ symmetry, one can readily show that $U(1)_{R''}$ implies conservation of B and Z_2^{lp} to all orders despite its explicit breaking to its maximal non-R-subgroup Z''_4 (including Z_2^{bp}) by the supersymmetry breaking terms in the visible sector.

The scalar component of T acquires a vacuum expectation value (vev) $\sim M_W^2/M$ ($\ll M_W$), with the electroweak breaking playing an essential role. This is due to the fact that the two last terms in Eq.(4), after electroweak breaking, give rise to a term linear with respect to T in the scalar potential. $\langle T \rangle$ leaves $U(1)_B$, Z_2^{lp} and the Z''_2 subgroup of $U(1)_{R''}$ unbroken and generates non-zero neutrino masses. The low energy spectrum is given by the MSSM since T, $\bar{T}$ are superheavy.

The coexistence of all the couplings in Eqs.(1), (3) and (4) provides us with a scheme where the light neutrino masses acquire contributions from the see-saw mechanism as well as the triplet vev. In this 'combined' case, W possesses a $U(1)$ R-symmetry $U(1)_{\hat{R}}$ which coincides with $U(1)_{R'}$ or $U(1)_{R''}$ when restricted to the superfields where these symmetries are defined. (Note that $U(1)_{R'}$ and $U(1)_{R''}$ become identical when restricted to the MSSM superfields.) This R-symmetry implies $U(1)_B$ and Z_2^{lp} to all orders. Finally, notice that B and Z_2^{lp} conservation is a consequence of the 'redefined' R-symmetries $U(1)_{R'}$, $U(1)_{R''}$ or $U(1)_{\hat{R}}$, in the 'combined' case, and not of the original $U(1)_R$ which allows couplings like $QQQL$ and $U^c U^c D^c E^c$.

The two mechanisms considered above for generating masses for the neutrinos have an additional far reaching consequence. This has to do with the generation of the observed BAU. The basic idea [4] is to generate an initial lepton asymmetry which is partially transformed through the non-perturbative electroweak sphaleron effects, that 'actively' violate $B+L$ at energies above M_W, to the observed BAU. Actually, this is the only way to generate baryons in the present scheme, since baryon number is otherwise exactly conserved. This mechanism has been discussed in detail [4] for the case where the lepton asymmetry is created by a decaying massive Majorana neutrino (say from the ν^c superfields) and exploits the couplings given in Eq.(3). If $T, \bar{T}$ with the couplings given in Eq.(4) are also present, additional diagrams must be considered.

The complete set of double-cut diagrams for leptogenesis from a decaying fermionic ν^c, which is the relevant case for inflationary models where the inflaton predominantly decays to a fermionic right handed neutrino, is displayed in Fig.1. The resulting lepton asymmetry, in this case, can be estimated [5] to be

$$\frac{n_L}{s} \approx \frac{3}{16\pi} \frac{T_r}{m_{infl}} M_i^R \frac{\text{Im}(M^D\, m^\dagger\, \tilde{M}^D)_{ii}}{|\langle H^{(1)}\rangle|^2 (M^D\, M^{D\dagger})_{ii}} . \quad (6)$$

Here T_r is the 'reheat' temperature, m_{infl} the inflaton mass, M^D the neutrino 'Dirac' mass matrix in the basis where M^R is diagonal with positive entries, and M_i^R is the mass of the decaying ν_i^c. Also

$$m \approx -\alpha\, t\, \frac{\langle H^{(1)}\rangle^2}{M} - \tilde{M}^D \frac{1}{M^R} M^D \quad (7)$$

is the light neutrino mass matrix in the same basis, with t (a complex symmetric matrix) and α being the coefficients of the first and second terms in Eq.(4) respectively. It should be noted that this estimate holds provided that M_i^R is much smaller than the mass of the other ν^c 's and the mass M of the triplets. Eq.(6) gives [5] the bound

$$\left|\frac{n_L}{s}\right| \lesssim \frac{3}{16\pi} \frac{T_r}{m_{infl}} \frac{M_i^R\, m_{\nu_\tau}}{|\langle H^{(1)}\rangle|^2} , \quad (8)$$

which, for $T_r \approx 10^9$ GeV (consistent with the gravitino constraint), $m_{infl} \approx 3 \times 10^{13}$ GeV, $M_i^R \approx 10^{10}$ GeV (see Ref. [5]), $|\langle H^{(1)}\rangle| \approx 174$ GeV, and $m_{\nu_\tau} \approx 5$ eV (providing the hot dark matter of the universe), gives $|n_L/s| \lesssim 3\times 10^{-9}$. This is large enough to account for the observed BAU. It is important though to ensure that the lepton asymmetry is not erased by lepton number violating $2 \to 2$ scatterings at all temperatures between T_r and 100 GeV. This requirement gives [6] $m_{\nu_\tau} \lesssim 10$ eV.

In non-inflationary (and perhaps some inflationary) models, leptogenesis from the decay of bosonic ν^c 's as well as bosonic and fermionic $T, \bar{T}$'s may be present

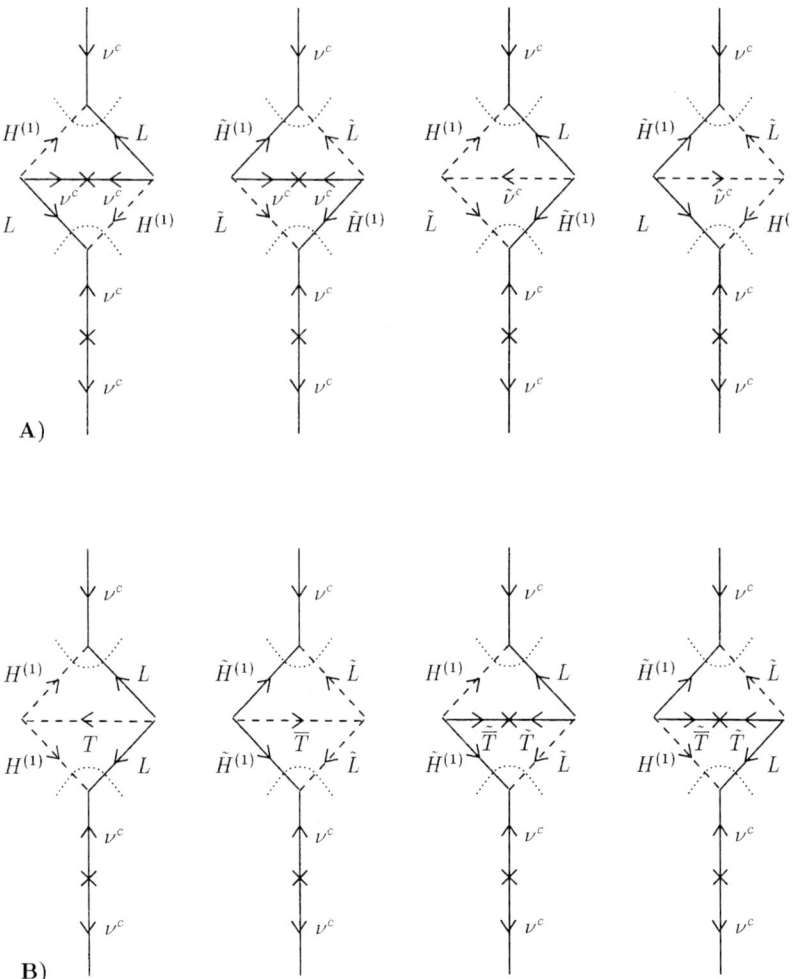

FIGURE 1. Double-cut diagrams for leptogenesis via the decay of fermionic ν^c 's. The diagrams A (B) correspond to the exchange of ν^c (T, $\bar{T}$) superfields. Continuous (dashed) lines represent fermions (bosons), while tildes denote the supersymmetric partners.

too. Most of the relevant double-cut diagrams can be obtained from the ones in Fig.1 by breaking up the ν^c, T, $\bar{T}$ internal lines and joining the external ν^c lines. The only extra diagram is a diagram of the same type with bosonic ν^c external lines and a fermionic T, $\bar{T}$ internal line. Diagrams of the type in Fig.1 with no ν^c internal or external lines cannot be constructed. Thus, efficient leptogenesis can occur only in the presence of ν^c 's.

We have seen how $U(1)_B$ arises as a consequence of requiring the superpotential W (including higher order terms) to respect a $U(1)$ R-symmetry. Among other things, this explains why the proton is so stable. However, the learned reader may be concerned that requiring the non-renormalizable terms in the superpotential to respect a continuous R-symmetry may not be a reasonable thing to do. Indeed, one may wonder if continuous global symmetries such as $U(1)_{\tilde{R}}$ or the Peccei-Quinn $(U(1)_{PQ})$ symmetry, rather than being imposed, can arise in some more 'natural' manner. One way how this may occur was pointed out in Ref. [7]. Here, discrete (including R-) symmetries that typically arise after compactification could effectively behave as if they are continuous. Furthermore, such 'continuous' symmetries can be very useful in resolving problems other than the one of proton stability. To see this, let us now address the strong CP and μ problems of MSSM. It has been noted by earlier authors [8] that a continuous $U(1)$ R-symmetry can be relevant for the solution of the μ problem. By invoking $U(1)_{PQ}$ and combining it with the $U(1)$ R-symmetry above, we will provide a resolution of both the strong CP and μ problems, with the $U(1)$ R-symmetry playing an essential role in controlling the structure of the terms that are permitted at the non-renormalizable level.

It has been recognized for some time that, within the supergravity extension of MSSM, the existence of D- and F-flat directions in field space can generate an intermediate scale M_I which, in the simplest case, is given by $M_I \sim \sqrt{m_{3/2}M_P} \sim 10^{11} - 10^{12}$ GeV, where $m_{3/2} \sim 1$ TeV is the supersymmetry breaking scale and $M_P = 3D1.22 \times 10^{19}$ GeV is the Planck mass. It seems 'natural' to try and identify M_I with the symmetry breaking scale f_a of $U(1)_{PQ}$, such that $\mu \sim m_{3/2} \sim f_a^2/M_P$ [9]. We will now see how this idea, which simultaneously resolves the strong CP and μ problems, can be elegantly realized in the presence of the $U(1)$ R-symmetry.

We supplement the MSSM spectrum with a pair of superfields N, $\bar{N}$ whose vevs will break $U(1)_{PQ}$ at an intermediate scale. W contains [9] the following terms:

$$H^{(1)}QU^c,\ H^{(2)}QD^c,\ H^{(2)}LE^c,\ N^2H^{(1)}H^{(2)},\ N^2\bar{N}^2\ . \tag{9}$$

The global symmetries of this superpotential are $U(1)_B$, $U(1)_L$ (N, $\bar{N}$ being neutral under both), an anomalous Peccei-Quinn symmetry $U(1)_{PQ}$, and a non-anomalous R-symmetry $U(1)_{\tilde{R}}$. The PQ and $\tilde{R}$ charges are

$$\begin{aligned} PQ:\quad & H^{(1)}(1),\ H^{(2)}(1),\ Q(-1/2),\ U^c(-1/2),\ D^c(-1/2),\\ & L(-1/2),\ E^c(-1/2),\ N(-1),\ \bar{N}(1)\ ;\\ \tilde{R}:\quad & H^{(1)}(0),\ H^{(2)}(0),\ Q(1),\ U^c(1),\ D^c(1),\ L(1),\ E^c(1),\ N(1),\ \bar{N}(0)\ . \end{aligned} \tag{10}$$

The quartic terms in Eq.(9) carry a coefficient proportional to M_P^{-1}. $U(1)_{\tilde{R}}$ ensures that undesirable terms such as $N\bar{N}$, which spoil the flat direction, are absent.

After taking the supersymmetry breaking terms into account, one finds [9] that, for suitable choice of parameters, a solution with $|\langle N \rangle| = 3D |\langle \bar{N} \rangle| \sim \sqrt{m_{3/2} M_P}$ is preferred over the one with $\langle N \rangle = 3D \langle \bar{N} \rangle = 3D0$. To see this, let us consider the relevant part of the scalar potential

$$(m_{3/2}^2 + \lambda^2 |N\bar{N}|^2/M_P^2)(|N|^2 + |\bar{N}|^2) + (A m_{3/2} \lambda N^2 \bar{N}^2/(2M_P) + h.c) , \qquad (11)$$

where $\lambda/(2M_P)$ is the coefficient of the last term in Eq.(9) and A the dimensionless coefficient of the corresponding supersymmetry breaking term (λ is taken positive by phase redefinitions). This potential can be rewritten as

$$(m_{3/2}^2 + \lambda^2 |N\bar{N}|^2/M_P^2)[(|N| - |\bar{N}|)^2 + 2|N||\bar{N}|]$$
$$+ |A| m_{3/2} \lambda (|N\bar{N}|^2/M_P) \cos(\epsilon + 2\theta + 2\bar{\theta}) , \qquad (12)$$

where ϵ, θ, $\bar{\theta}$ are the phases of A, N, $\bar{N}$ respectively. Minimization of the potential then requires $|N| = 3D|\bar{N}|$, $\epsilon + 2\theta + 2\bar{\theta} = 3D\pi$ and the potential takes the form

$$2|N|^2 m_{3/2}^2 \left(\lambda^2 \frac{|N|^4}{m_{3/2}^2 M_P^2} - |A|\lambda \frac{|N|^2}{2 m_{3/2} M_P} + 1 \right) . \qquad (13)$$

For $|A| > 4$, the minimum is at $|\langle \bar{N} \rangle| = 3D|\langle N \rangle|$ with

$$|\langle N \rangle| = 3D(m_{3/2} M_P)^{\frac{1}{2}} \left(\frac{|A| + (|A|^2 - 12)^{\frac{1}{2}}}{6\lambda} \right)^{\frac{1}{2}} . \qquad (14)$$

Note that $\langle N \rangle$, $\langle \bar{N} \rangle$ together break $U(1)_{\tilde{R}} \times U(1)_{PQ}$ down to Z_2^{mp}. Substitution of these vevs in Eq.(9) shows that the μ parameter of MSSM is of order $m_{3/2}$ as desired.

We can extend this discussion to include either massive ν^c's or the $SU(2)_L$ triplets T, $\bar{T}$. In the ν^c case, the new terms in W are $H^{(1)} L \nu^c$ and $N \nu^c \nu^c$. The first term yields $B(\nu^c) = 3D0$, $L(\nu^c) = 3D - 1$, $PQ(\nu^c) = 3D - 1/2$, $\tilde{R}(\nu^c) = 3D1$. The second term leaves $U(1)_B$ unbroken but breaks $U(1)_L$, $U(1)_{PQ}$ and $U(1)_{\tilde{R}}$ to a 'redefined' Peccei-Quinn symmetry $U(1)_{PQ'}$ and a 'redefined' R-symmetry $U(1)_{\tilde{R}'}$ with $PQ' = 3DPQ - L$ and $\tilde{R}' = 3D\tilde{R} + PQ - (1/2)L$. Thus, the strong CP problem can be resolved. Note that $U(1)_{\tilde{R}'}$ coincides with $U(1)_{R'}$ when restricted to the superfields where $U(1)_{R'}$ is defined. Moreover, just as $U(1)_{R'}$, the R-symmetry $U(1)_{\tilde{R}'}$ contains Z_2 baryon parity as a subgroup and implies $U(1)_B$ to all orders. Z_2^{mp} is contained in $U(1)_{PQ'}$ and, thus, Z_2^{lp} is also present but not as an automatic consequence of $U(1)_{\tilde{R}'}$ in this case.

A similar discussion applies if the triplets T, $\bar{T}$ are introduced in the scheme of Eq.(9). The new superpotential terms are TLL, $\bar{T} H^{(1)} H^{(1)}$ and $NT\bar{T}$. The first two terms give $B(T) = 3DB(\bar{T}) = 3D0$, $L(T) = 3D - 2$, $L(\bar{T}) = 3D0$, $PQ(T) = 3D1$, $PQ(\bar{T}) = 3D - 2$, $\tilde{R}(T) = 3D0$ and $\tilde{R}(\bar{T}) = 3D2$. The last term leaves unbroken the symmetries $U(1)_B$, $U(1)_{PQ''}$ and $U(1)_{\tilde{R}''}$ with $PQ'' = 3DPQ - L$ and $\tilde{R}'' = 3D\tilde{R} + PQ - (1/2)L$. $U(1)_{\tilde{R}''}$ is an extension of $U(1)_{R''}$, contains Z_2^{bp} and implies $U(1)_B$ to all orders.

Finally, it should be pointed out that ν^c and T, $\bar{T}$ can coexist with all the couplings mentioned being present. In this 'combined' case, W possesses a $U(1)$ Peccei-Quinn and R-symmetry, $(U(1)_{\bar{PQ}}, U(1)_{\bar{\tilde{R}}})$, which coincides with $U(1)_{PQ'}$ ($U(1)_{\tilde{R}'}$) or $U(1)_{PQ''}$ ($U(1)_{\tilde{R}''}$) when restricted to the superfields where these symmetries are defined. The $U(1)_{\bar{\tilde{R}}}$ symmetry implies $U(1)_B$ to all orders.

To summarize the discussion so far, we have considered MSSM and certain extensions, with Z_2 matter parity embedded in a $U(1)$ R-symmetry. Neutrino masses, consistent with a redefined $U(1)$ R-symmetry, can be introduced in two ways. By requiring the higher order superpotential couplings to respect this redefined R-symmetry, one can i) explain proton stability to be a consequence of an automatic $U(1)_B$, and ii) show that the BAU can arise via a primordial leptogenesis provided right handed neutrinos are present. Finally, simultaneous resolutions of the strong CP and μ problems, with $\mu \sim f_a^2/M_P$, can be elegantly accommodated in this scheme.

There is, however, at least one crucial undetermined parameter in MSSM which strongly suggests that an extension of the gauge symmetry $SU(3)_c \times SU(2)_L \times U(1)$ is warranted. This parameter is $\tan\beta$, the ratio of the vevs of the 'up' and 'down' higgs doublets. A knowledge of $\tan\beta$ can, under suitable circumstances, shed light on a host of questions that are left unanswered in MSSM. These include:

i. Nature of the lightest sparticle (LSP);

ii. Hierarchy of sparticle masses;

iii. Higgs spectroscopy.

In 1991 [10] it was shown that within the framework of minimal supersymmetric SO(10), $\tan\beta \simeq m_t/m_b$ which, after taking the asymptotic relation $m_b^{(0)} = 3Dm_\tau^{(0)}$ into account, gave a prediction for the top quark mass which is in good agreement with its subsequent discovery at Fermilab. For the higgs sector it was soon pointed out [11] that this prediction of $\tan\beta$ would lead to the conclusion that LEPII may narrowly 'miss' discovering the 'Weinberg-Salam' higgs that is contained in MSSM.

Since we are taking a 'bottom-up' approach in this talk, let us consider a relatively modest extension of MSSM which preserves the above prediction of $\tan\beta (\simeq m_t/m_b)$. Namely, the two higgs doublets of MSSM will be 'unified' within the gauge group $G_{LR} = 3DSU(2)_L \times SU(2)_R \times U(1)_{B-L}$ which has a long and distinguished history [12,13]. Here I will follow an approach that was developed in collaboration with Gia Dvali and George Lazarides [14]. It has the following distinguishing features:

i. The μ problem of MSSM is resolved and a suitable $B\mu$ term is generated;

ii. Hybrid inflation is readily possible, and the spectral index of density fluctuations $n \simeq 0.98$. Moreover, the magnitude of density fluctuations can be taken from experiments to fix the symmetry breaking scale of $SU(2)_R \times U(1)_{B-L} \to U(1)_Y$. It turns out to be just shy of 10^{16} GeV;

iii. With $\tan\beta \simeq m_t/m_b$, the 'Weinberg-Salam' higgs mass is around 110 ± 10 GeV (its tree-level mass is M_{Z^0}).

It is worth noting that a $U(1)$ R-symmetry (or some suitable discrete symmetry) plays an essential role in this analysis.

Within the framework of $G_{LR}(\equiv SU(2)_L \times SU(2)_R \times U(1)_{B-L})$, our resolution of the μ problem is based on the observation that whenever a large vev of some higgs field is triggered through its couplings with a gauge singlet field S, the supersymmetry breaking effects shift the vev of S from zero by an amount of the order of the low energy supersymmetry breaking scale (which can be parameterized by the gravitino mass $m_{3/2} \sim 100$ GeV – 1 TeV in theories with gravity-mediated supersymmetry breaking). This allows one to induce the μ term via the coupling $Sh^{(1)}h^{(2)}$ in the superpotential, where $h^{(1)}, h^{(2)}$ denote the MSSM higgs doublets. As we will see below, this shift of the vev of S from zero is quite insensitive to the value of the large scale M. A priori, any symmetry group broken at M is suitable for generating the μ term. Our choice of $SU(2)_R \times U(1)_{B-L}$ is partially motivated by the fact that it also can lead to 'hot' dark matter in neutrinos with mass in the eV range. A global $U(1)$ R-symmetry plays an essential role in our analysis. Its unbroken Z_2 subgroup acts as 'matter parity', which implies a stable LSP. The R-symmetry also gives rise, analogous to the MSSM case discussed earlier, to an accidental $U(1)_B$, including higher order terms in the superpotential, which leads to a stable proton. The model predicts a spectral index of primordial density fluctuations that is extremely close to unity and which is consistent with a 'cold' plus 'hot' dark matter scenario of structure formation [15].

Let us first describe the main features of the G_{LR} symmetric model which solve the μ problem. In fact, the solution can work in any scheme with gravity-mediated supersymmetry breaking provided there is an additional symmetry group factor broken at some scale $\gg m_{3/2}$ (or even simply a superheavy vev). The $SU(2)_R \times U(1)_{B-L}$ group is broken by a pair of $SU(2)_R$ doublet chiral superfields $l^c, \bar{l}^c$ which acquire a vev $M \gg m_{3/2}$. In order to achieve this breaking by a renormalizable superpotential, we will need a gauge singlet chiral superfield S. This singlet plays a crucial three-fold role: 1) it triggers $SU(2)_R$ breaking; 2) it generates the μ and $B\mu$ terms of MSSM after supersymmetry breaking; and 3) it leads to hybrid inflation [16]. To see this, we first ignore the matter fields of the model and consider the superpotential

$$W = 3DS(\kappa l^c \bar{l}^c + \lambda h^2 - \kappa M^2), \qquad (15)$$

where the chiral superfield $h = 3D(h^{(1)}, h^{(2)})$ belongs to a bidoublet $(2, 2)$ representation of $SU(2)_L \times SU(2)_R \times U(1)_{B-L}$, and h^2 denotes the unique bilinear invariant

$\epsilon^{ij}h_i^{(1)}h_j^{(2)}$. Note that, through a suitable redefinition of the superfields, the parameters κ, λ and M can be made real and positive. Also note that W in Eq.(15) has the most general renormalizable form invariant under the gauge group and a continuous $U(1)$ R-symmetry under which S carries the same charge as W, while $h, l^c \bar{l}^c$ are neutral. Clearly, a vev of S will generate a μ term with $\mu = 3D\lambda < S >$. Moreover, the vev of its F component, $F_S = 3D\frac{\partial W}{\partial S}$, together with the soft trilinear gravity-mediated terms will generate a bilinear soft term, $B\mu h^{(1)} \epsilon h^{(2)}$, in the scalar potential.

To understand how the μ problem is solved, let us analyze the minimum of the scalar potential. In the unbroken global supersymmetry limit, the vacuum is at

$$S = 3D0, \quad \kappa l^c \bar{l}^c + \lambda h^2 = 3D\kappa M^2, \quad l^c = 3De^{i\phi}\bar{l}^{c*} \quad h_i^{(1)} = 3De^{i\theta}\epsilon_{ij}h^{(2)j*}, \quad (16)$$

where the last two conditions arise from the requirement of D flatness. We see that there is a flat direction (with two real dimensions) at generic points of which both $SU(2)_L$ and $SU(2)_R$ are spontaneously broken. The supersymmetry breaking will lift this degeneracy. Clearly, the desirable vacuum is the one with $h^{(1),(2)} = 3D0$ and $l^c \bar{l}^c = 3DM^2$ (up to higher order corrections). Whether this indeed is the case depends on the parameters of the model and the cosmological history (see below). Let us investigate the theory about this supersymmetric minimum. The $SU(2)_R$ is broken and all states in $l^c, \bar{l}^c$ obtain masses $\sim M$ either through the superhiggs effect or through their mixing with the S superfield. The masses of S and of the higgs components of $l^c, \bar{l}^c$ are $m_{infl} = 3Dm_S = 3Dm_{l^c} = 3D\sqrt{2}\kappa M$. The only massless (up to supersymmetry breaking corrections) non-gauge degrees of freedom so far are the two higgs doublets in h. Their 'masslessness' can be simply understood from their pseudogoldstone nature: they are zero modes of the vacuum flat direction. Gravity-mediated soft terms lift the degeneracy along the flat direction, the doublets acquire masses $\sim m_{3/2}$, and the μ term is generated by a shift ($\sim m_{3/2}$) of the vev of S from zero. Such an automatic generation of the $\mu^2 \sim B\mu(\sim m_{3/2}^2)$ terms after supersymmetry breaking is a generic feature of the models in which the higgs doublets are pseudogoldstone particles.

To make the connection more transparent, let us take $\lambda = 3D\kappa$ for a moment. In this limit, the superpotential in Eq.(15) has an accidental $U(4)$ symmetry under which the $(l^c, h^{(1)})$ and $(\bar{l}^c, \epsilon h^{(2)})$ states transform as the fundamental and antifundamental representations respectively. Of course, the G_{LR} gauge interactions and the Yukawa couplings break the $U(4)$ symmetry explicitly, but this breaking cannot affect the vacuum degeneracy as long as supersymmetry is unbroken. Thus, the degeneracy of the vacuum manifold is as if we had an exact $U(4)$ symmetry broken to $U(3)$ by the vevs of $l^c, \bar{l}^c$. This breaking produces seven would-be goldstone superfields, three of which are the true 'goldstones' that are absorbed by the massive $SU(2)_R \times U(1)_{B-L}/U(1)_Y$ gauge superfields. The remaining four are the physical states $h^{(1)}, h^{(2)}$. They are 'pseudogoldstones' and obtain masses only after supersymmetry breaking. At tree level, however, one combination, $h^{(1)} + \epsilon h^{(2)*}$, must be exactly massless as a result of the $U(4)$ symmetry and only gets a mass by radiative corrections. This implies the generic relation $\mu^2 + m_{3/2}^2 = 3D - B\mu$ in the $U(4)$ symmetric case. For $\lambda \neq \kappa$, the $U(4)$ symmetry is explicitly broken in the superpotential and the above relation holds only approximately.

To see how all this works in detail, we write the full low-energy scalar potential including the soft terms (for simplicity, we discuss the case with a canonical Kähler metric, but the results stay essentially intact even in the general case). We have

$$V = 3D|\kappa l^c \bar{l}^c + \lambda h^2 - \kappa M^2|^2 + (m_{3/2}^2 + \kappa^2|\bar{l}^c|^2 + \kappa^2|l^c|^2 + \lambda^2|h|^2)|S|^2 + m_{3/2}^2(|\bar{l}^c|^2$$
$$+ |l^c|^2 + |h|^2) + \left(Am_{3/2}S(\kappa l^c \bar{l}^c + \lambda h^2) - (A-2)m_{3/2}S\kappa M^2 + \text{h.c.}\right). \quad (17)$$

Since S and the $l^c, \bar{l}^c$ fields are heavy in the vanishing $m_{3/2}$ limit, their vevs cannot be shifted significantly by the tiny soft supersymmetry breaking effects. For a leading order estimate of the vev of S, we can substitute in V the supersymmetric vevs of the $SU(2)_R$ doublets. We see that S gets a destabilizing tadpole term $\simeq 2\kappa m_{3/2}M^2 S + \text{h.c.}$, and taking account of the term $\simeq 2\kappa^2 M^2|S|^2$, the resulting vev of S is $\simeq -m_{3/2}/\kappa$. The vev of S will generate a μ term with $\mu = 3D\lambda <S> = 3D - m_{3/2}\lambda/\kappa$, whereas the vev of its F-component together with the soft trilinear gravity-mediated terms will generate a $B\mu$ term in the scalar potential with

$$B\mu = 3D\lambda \left(F_S^* + m_{3/2}SA\right). \quad (18)$$

The magnitude of $B\mu$ is readily found using the equation of motion of l^c:

$$\kappa \bar{l}^c(\kappa l^c \bar{l}^c + \lambda h^2 - \kappa M^2)^* + (\kappa^2|S|^2 + m_{3/2}^2)l^{c*} + Am_{3/2}S\kappa \bar{l}^c = 3D0. \quad (19)$$

One obtains $B\mu \simeq -2\lambda m_{3/2}^2/\kappa$. The above leading order estimates can be confirmed by an explicit minimization of V using the iterative series $l^c = 3D\bar{l}^c = 3DM(1 + \sum_{n\geq 1} c_n(m_{3/2}/M)^n)$ and $S = 3D - (m_{3/2}/\kappa)(1 + \sum_{n\geq 1} d_n(m_{3/2}/M)^n)$.

So far we were expanding the theory about the 'good' minimum at $h = 3D0, l^c = 3D\bar{l}^c = 3DM$, assuming that this is the prefered ground state of the system. Now we must check the self-consistency of this assumption. One can show that, in the case of minimal Kähler potential, both the 'good' ($h = 3D0, l^c, \bar{l}^c \neq 0$) and the 'bad' ($h \neq 0, l^c, \bar{l}^c = 3D0$) points are local minima of the potential for all values of the parameters. In fact, they are the only minima for $\kappa \neq \lambda$. For $\kappa = 3D\lambda$, the degeneracy of the vacuum is not totally lifted and these points lie on a flat direction, with one real dimension. A simple way to see that the 'good' stationary point is never unstable is to observe that the electroweak higgs mass squared matrix has no negative eigenvalues. This is due to the fact that the diagonal elements of this matrix are equal to $\lambda^2|S|^2 + m_{3/2}^2$, while its off diagonal elements are $B\mu = 3D(-\lambda/\kappa)(\kappa^2|S|^2 + m_{3/2}^2)$, as one can deduce from Eqs.(18) and (19). The determinant of this matrix is then equal to $(\lambda^2/\kappa^2 - 1)(\lambda^2\kappa^2|S|^4 - m_{3/2}^4)$ which, replacing S by its leading order vev, becomes equal to $(\lambda^2/\kappa^2 - 1)^2 m_{3/2}^4 \geq 0$. This together with the obvious positivity of its trace imply that this matrix has no negative eigenvalues for all values of the parameters. This statement remains true even if higher order corrections to the vev of S are included. As it turns out, for any choice of the parameters, the 'good' and 'bad' minima of V are degenerate to leading order (up to corrections of order $m_{3/2}^4$ in the energy difference). If we allow for a non-minimal Kähler potential, the degeneracy between these minima can be lifted. In fact, we can find ranges of parameters where the 'bad' point ceases even to be a local minimum. For example, by replacing $(A-2)$ in Eq.(17) by $(A-2\alpha)$, we can have the 'good' point as the unique local minimum of the potential provided $|\alpha|^2 \geq \lambda/\kappa > 1$ or $|\alpha|^2 \leq \lambda/\kappa < 1$. As we will see, the cosmological evolution prefers the first choice.

We are now ready to introduce the matter fields of the model as well. The superpotential takes the following form:

$$W = 3DS(\kappa l^c \bar{l}^c + \lambda h^2 - \kappa M^2) + g_{ab}^q h Q_a Q_b^c + g_{ab}^l h l_a l_b^c + h_{ab}\frac{\bar{l}^c \bar{l}^c l_a^c l_b^c}{M_P}, \qquad (20)$$

where Q_a, Q_a^c, l_a, l_a^c are chiral quark, antiquark and lepton superfields respectively and $a, b = 3D1, 2, 3$ are the family indices. The last term gives superheavy masses to the right handed neutrinos which, through the seesaw mechanism, can lead to a 'tau' neutrino mass in the eV range. The superpotential in Eq.(20) has the most general form (up to quartic terms) allowed by the gauge group and a $U(1)$ R-symmetry under which the superfields have the following charges: $R_S = 3D1$, $R_{l^c} = 3D - R_{\bar{l}^c} = 3DR_h = 3D0, R_{Q_a} = 3DR_{Q_a^c} = 3DR_{l_a} = 3DR_{l_a^c} = 3D1/2$. Note that the Z_2 subgroup of this R-symmetry remains unbroken and indeed is the conventional MSSM 'matter parity'.

The $U(1)$ R-symmetry, in contrast to 'matter parity', also eliminates the dimension five operators responsible for proton decay. If one assumes that the R-symmetry is an exact symmetry then, in fact, it eliminates the baryon number violating operators from the superpotential to all orders. Note that even if the R-symmetry is not explicitly broken by Planck scale suppressed operators in the superpotential it must, in any case, be broken by the hidden sector superfields which break supersymmetry spontaneously and also produce a non-vanishing vev of the superpotential (gravitino mass). So, in general, we can expect some higher dimensional baryon number violating terms coming from a nonminimal Kähler potential. These are, however, adequately suppressed. The proton is essentially stable in the present scheme.

Next let us show that this model has a built-in inflationary trajectory in the field space along which the F_S term takes a constant value as in the supersymmetric hybrid inflationary scenario [16]. The relevant trajectory in the field space is parameterized by S. The key point here is that S has no self-interactions and appears in the superpotential only linearly. At a generic point of this trajectory with $|S| > \max\left(M, M\sqrt{\kappa/\lambda}\right)$, all the gauge non-singlet higgs fields obtain masses of order $|S|$ and, therefore, they decouple. The massless degrees of freedom along this trajectory are: the singlet S, the massless G_{LR} gauge supermultiplet, and the massless matter superfields. The effective low energy superpotential, which is obtained after integrating out the heavy superfields, can be readily constructed by simply using holomorphy and symmetry arguments. This superpotential is linear in S, namely

$$W_{infl} = 3D - \kappa S M^2. \tag{21}$$

Were it not for the mass scale M in the superpotential, the trajectory parameterized by S would simply correspond to a supersymmetry preserving vacuum direction remaining flat to all orders in perturbation theory. The F_S term, however, lifts this flat direction so that, at tree level, it takes an asymptotically constant value for arbitrarily large $|S|$. As a result, the trajectory of interest is represented by a massless degree of freedom S, whose vev sets the mass scale for the heavy particles and provides us with a constant tree level vacuum energy density

$$V_{\text{tree}} = 3D\kappa^2 M^4, \tag{22}$$

which is responsible for inflation. The above result can be easily rederived by an explicit solution of the equations of motion along the inflationary trajectory. To this end, we can explicitly minimize all the D and F terms for large values of $|S|$. It is easy to check that, for

$$|S| > \max\left(M, M\sqrt{\kappa/\lambda}\right), \tag{23}$$

all the other vevs vanish and, therefore, a nonzero contribution to the potential comes purely from the constant $F_S = 3D - \kappa M^2$ term. Whenever the condition in Eq.(23) is violated, the $l^c, \bar{l}^c$ components become tachyonic and compensate the F_S term. We see that, if $\kappa > \lambda$, h will become tachyonic earlier and the system will evolve towards the 'wrong' minimum. Thus, we prefer $\kappa < \lambda$. The system rapidly approaches the supersymmetric vacuum and oscillates about it. At tree level, the potential along the inflationary trajectory is exactly flat. Radiative corrections, however, create a logarithmic slope that drives the inflaton toward the minimum. The origin of these corrections can be understood in the following way. As we have shown, the value of $|S|$ sets the mass scale for the heavy particles along the inflationary trajectory. Thus, we can think of the low energy theories at different points on this trajectory as being a single theory at different energy scales. This gives rise to a wave function renormalization of the S field through loops involving the $\bar{l}^c, l^c, h$ particles. Since their mass is set by the value of $|S|$, a nontrivial dependence on this value arises providing an effective one-loop potential for the inflaton field. For large field strengths or, in other words, for masses of the particles in the loop suitably larger than M, this potential assumes the following form

$$V_{\inf} \simeq \kappa^2 M^4 \left[1 + \frac{\kappa^2 + \lambda^2}{8\pi^2} \ln\left(\frac{SS^*}{\Lambda^2}\right)\right]. \tag{24}$$

This simply is an asymptotic form of the one-loop corrected effective potential with

$$\delta V_{one-loop} = 3D \frac{1}{64\pi^2} \sum_i (-1)^{F_i} M_i^4 \ln(\frac{M_i^2}{\Lambda^2}). \tag{25}$$

The contribution to Eq.(25) comes from the $\bar{l}^c, l^c$, and h supermultiplets, since they receive at tree level a non-supersymmetric contribution to the masses of their scalar components from the F_S term. All other states have either no mass splitting due to a vanishing coupling with S or have no inflaton dependent mass (these are the gauge and matter fields).

Inflation can end when the condition in Eq.(23) breaks down, thereby signaling that some of the fields become tachyonic and the system moves towards the global minimum. This is indeed the case provided the slow roll conditions are not violated before this instability occurs. As we have argued above, for $\lambda > \kappa$, the system is destabilized towards the $SU(2)_R$ breaking vacuum and oscillates about it. These $S, l^c, \bar{l}^c$ oscillations will create, among other particles, right handed neutrinos. The subsequent decay of these neutrinos gives rise to a primordial lepton number [4]. The observed baryon asymmetry of the universe can then be obtained by partial conversion of this lepton asymmetry through non-perturbative sphaleron effects. This process, in a certain range of the parameter space, can lead to a successful baryogenesis via leptogenesis [4]. The gravitino constraint on the 'reheat' temperature can also be simultaneously satisfied and the spectral index of primordial density fluctuations turns out to be extremely close to unity. The details are involved and will not be discussed here, but one fact is worth noting. The $SU(2)_R \times U(1)_{B-L}$ breaking scale turns out to be about an order of magnitude lower than the MSSM unification scale and, thus, the right handed neutrino masses are restricted to be smaller than about 10^{13} GeV. This implies that at least the 'tau' neutrino mass can be in the eV range.

The model discussed above, although based on $SU(2)_L \times SU(2)_R \times U(1)_{B-L}$ gauge group, is L-R asymmetric in the higgs sector. This is acceptable since, in reality, the above scheme must be a low energy remnant of a more fundamental theory in which the L-R discrete symmetry is spontaneously broken at a scale $M_{LR} > M$.

In summary, we have presented a supersymmetric model based on the $SU(3)_c \times SU(2)_L \times SU(2)_R \times U(1)_{B-L}$ gauge group where the μ and $B\mu$ terms are automatically generated after supersymmetry breaking. Moreover, the model gives rise to hybrid inflation. The observed baryon asymmetry of the universe can be successfully generated through primordial leptogenesis, while the 'hot' component of the dark matter in the universe, in light of the recent Super K results on atmospheric neutrinos, consists of at least two nearly 'mass' degenerate neutrinos in the eV range.

Acknowledgements

This workshop could not have taken place without the leadership provided by my friend Jose F. Nieves, ably supported by Maru Rodriguezas well as many members of the University of Puerto Rico (Rio Pedras). I would like to thank my collaborators George Lazarides, Gia Dvali and Robert Schaefer. I acknowledge support from DOE under contract number DE-FG02-91ER40626, and from NATO under Grabt Number CRG-970149. Finally, without Sherri Evans' great typing skills this manuscript could not have been finished.

REFERENCES

1. G. Lazarides and Q. Shafi, Bartol preprint BA-98-11 (1998).
2. For another approach to MSSM with R-symmetry, see L. Hall and L. Randall, Nucl. Phys. **B352** (1991) 289.
3. E. Ma and U. Sarkar, hep-ph/9802445.
4. M. Fukugita and T. Yanagida, Phys. Lett. **B174** (1986) 45; G. Lazarides and Q. Shafi, Phys. Lett. **B258** (1991) 305; G. Lazarides, C. Panagiotakopoulos and Q. Shafi, Phys. Lett. **B315** (1993) 325.
5. G. Lazarides, R.K. Schaefer and Q. Shafi, Phys. Rev. **D56** (1997) 1324; G. Lazarides, hep-ph/9802415 (to appear in the proccedings of the 6th BCSPIN School).
6. L. Ibáñez and F. Quevedo, Phys. Lett. **B283** (1992) 261.
7. G. Lazarides, C. Panagiotakopoulos and Q. Shafi, Phys. Rev. Lett. **65** (1986) 432.
8. For a recent discussion and additional references, see H. Nilles and N. Polonsky, Nucl. Phys. **B484** (1997) 33.
9. G. Dvali, private communication.
10. B. Ananthanarayan, G. Lazarides and Q. Shafi, Phys. Rev. **D44** (1991) 1613.
11. Q. Shafi and B. Ananthanarayan, 1991 ICTP Summer School Proceedings, World Scientific Publishers, Singapore.
12. J.C. Pati and A. Salam, Phys. Rev. **D10** (1974) 275; R.N. Mohapatra and J.C. Pati, Phys. Rev. **D11** (1975) 566; G. Senjanovic and R.N. Mohapatra, Phys. Rev. **D12** (1975) 1502; G. Senjanovic, Nucl. Phys. **B153** (1979) 334; M. Magg, Q. Shafi and C. Wetterich, Phys. Lett. **B87** (1979) 227.
13. See, e.g., M. Cvetic and J.C. Pati, Phys. Lett. **B135** (1984) 57; R. Kuchimanchi and R.N. Mohapatra, Phys. Rev. **D48** (1993) 4352; Phys. Rev. Lett. **75** (1995) 3989; C.S. Aulakh, K. Benakli and G. Senjanovic, hep-ph/9703434; C.S. Aulakh, A. Melfo and G. Senjanovic, hep-ph/9707256=2E
14. G. Dvali, G. Lazarides and Q. Shafi, hep-ph/9710314.
15. Q. Shafi and F.W. Stecker, Phys. Rev. Lett. **53** (1984) 1292; for a recent review and other references, see Q. Shafi and R.K. Schaefer, hep-ph/9612428.
16. G. Dvali, Q. Shafi and R.K. Schaefer, Phys. Rev. Lett. **73** (1994) 1886; G. Dvali, Phys. Lett. **B355** (1995) 78; Phys. Lett. **B387** (1996) 471; G. Lazarides, R.K. Schaefer and Q. Shafi, Phys. Rev. **D56** (1997) 1324.

Searches for SUSY Particles at LEP2

Sylvie Braibant*
on behalf of the LEP Collaborations

*CERN, EP Division, CH-1211 Geneva 23, Switzerland

Abstract. Since November 1995, LEP has been operated at centre-of-mass energies up to $\sqrt{s} = 183$ GeV. These are the highest energies ever attained at an e^+e^- collider providing an unique opportunity to search for new particles. Searches for supersymmetric particles have been performed using the data samples collected at these centre-of-mass energies. Searches for R-parity violating decays of supersymmetric particles and searches for long-lived heavy charged particles are also described. Model independent limits on the production cross-sections and exclusion plots in the context of the Minimal Supersymmetric Standard Model and in the context of gauge-mediated supersymmetry breaking theories are presented.

INTRODUCTION

Since November 1995, the LEP e^+e^- collider has been operated at centre-of-mass energies well above the Z^0 peak: first at $\sqrt{s} = 130$-136 GeV (LEP 1.5) and then at $\sqrt{s} = 161$ GeV, 172 GeV and 183 GeV (LEP 2). These are the highest energies ever attained in e^+e^- collisions. Each of the four LEP experiments, ALEPH, DELPHI, L3 and OPAL collected integrated luminosities of about 10 pb^{-1} at LEP1.5. At LEP2, about 10 pb^{-1} at 161 GeV, about 10 pb^{-1} at 172 GeV and about 60 pb^{-1} at 183 GeV have been collected per experiment. This provided an opportunity to search for the existence of new particles in a new energy range. Searches for several types of new particles have been performed. In the following, searches for particles predicted by supersymmetric theories are reviewed. In particular, searches for charginos, neutralinos, scalar top and bottom quarks (stop and sbottom quarks) and charged scalar leptons (sleptons) are presented. Searches for R-parity violating decays of these particles and long-lived heavy charged particles are also reviewed.

In supersymmetric (SUSY) models [1], each particle is accompanied by a supersymmetric partner whose spin differs by half a unit. SUSY models require a minimum of two Higgs doublets to generate the masses of bosons and fermions. The fermionic partners of the $W^\pm$ and of the charged Higgs bosons, $H^\pm$, mix to form two mass eigenstates, the charginos $\tilde{\chi}^\pm_{1,2}$. The partners of the γ, of the Z^0 and of the neutral Higgs bosons mix to form four mass eigenstates, the neutralinos

$\tilde{\chi}_i^0$ ($i = 1, 4$ in increasing mass order). A new multiplicative quantum, R-parity ($\equiv (-1)^{2S+3B+L}$), is introduced with value $+1$ for the Standard Model particles and -1 for the SUSY particles. It is often assumed that R-parity is conserved and that the lightest neutralino, $\tilde{\chi}_1^0$, is the lightest supersymmetric particle (LSP). R-parity conservation implies that SUSY particles would always be pair produced and always decay, through cascade decays, to ordinary particles and $\tilde{\chi}_1^0$. Moreover, the $\tilde{\chi}_1^0$ is stable and escapes detection due to its weakly interacting nature. Therefore, a characteristic signature of all events containing SUSY particles is missing energy and momentum carried away by the neutralinos. If R-parity is violated, the sparticles can decay to Standard Model particles and any sparticle could be the LSP. Moreover, the LSP may be unstable and directly decay to Standard Model particles. Supersymmetric particles could also then be singly produced.

In the framework of the Minimal Supersymmetric extension of the Standard Model (MSSM) [2], under the assumption of a common scalar mass, m_0, at the Grand Unification (GUT) scale, all sparticle masses and couplings are completely determined by m_0 and a set of four parameters: M_2, the SU(2) gaugino mass parameter at electroweak scales[1]; μ, the mixing parameter of the two Higgs doublets; $\tan\beta = v_2/v_1$, the ratio of the vacuum expectation values for the two Higgs doublets and A the supersymmetric trilinear coupling.

CHARGINOS

The charginos, $\tilde{\chi}_{1,2}^\pm$, could be pair produced at LEP either through γ or Z^0 exchange in the s-channel or through sneutrino exchange in the t-channel. The production cross-section could be fairly large and therefore the search for charginos is one of the most appealing SUSY searches. The $\tilde{\chi}_1^\pm$ can decay into a $\tilde{\chi}_1^0$ and an ordinary lepton : $\tilde{\chi}_1^\pm \to \tilde{\chi}_1^0 l^\pm \nu$ (leptonic decay), or into a neutralino and a quark pair : $\tilde{\chi}_1^\pm \to \tilde{\chi}_1^0 q \bar{q}'$ (hadronic decay) through a virtual W*, slepton or scalar quark emission. The experimental signature for $\tilde{\chi}_1^\pm$ production is therefore : a) two acoplanar leptons, b) one lepton plus jets or c) multi-jets; all these topologies share the characteristic of a large missing energy carried away by the neutralinos. The most challenging search is when the mass difference between the $\tilde{\chi}_1^\pm$ and $\tilde{\chi}_1^0$ ($\Delta m = m_{\tilde{\chi}_1^\pm} - m_{\tilde{\chi}_1^0}$) is small, because the event visible energy also becomes very small. Detailed searches for charginos in the small Δm region have been performed. In the case of a very small mass difference ($\Delta m \leq 0.1$ GeV), the charginos would be quasi-stable charged particles. The searches are based on the specific ionisation loss measurement, dE/dx, provided by the central detectors of each experiment. The DELPHI experiment also uses the particle identification provided by their RICH detectors. No evidence for long-lived chargino production has been observed by the four LEP collaborations. In most of these searches, candidates have been selected by the analyses but their number is compatible with the expected background from

---

[1] We assume that M_1, the U(1) gaugino mass at electroweak scales, is related to M_2 by the usual gauge unification condition: $M_1 = \frac{5}{3} \tan^2 \theta_W M_2$.

Standard Model processes. Therefore, 95% C.L. upper limits on the quasi-stable $\tilde{\chi}_1^\pm$ pair production cross-section have been computed. The left plot [3] in Figure 1 shows the pair production cross-section upper limit at the 95% C.L.; this limit varies from 0.05 to 0.19 pb in the mass range between 45 and 89.5 GeV. Within the MSSM framework, long-lived charginos with a mass smaller than 89.5 GeV have been excluded at the 95% C.L. for every choice of the MSSM parameters and assuming a heavy sneutrino ($m_{\tilde{\nu}} > 500$ GeV).

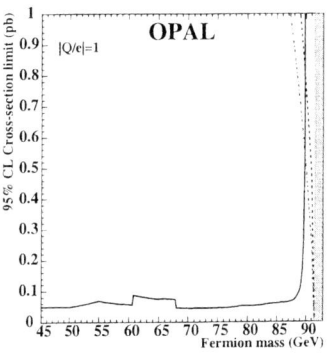
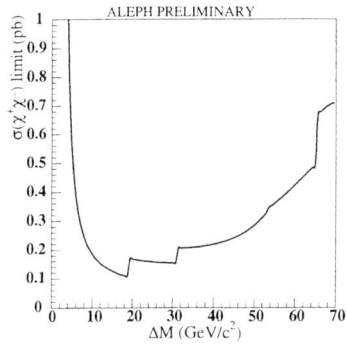

FIGURE 1. 95% C.L. upper limit on the long-lived $\tilde{\chi}_1^\pm$ pair production cross-section at $\sqrt{s} \simeq$ 183 GeV from the OPAL collaboration (left) and 95% C.L. upper limit on the $\tilde{\chi}_1^\pm$ pair production cross-section at $\sqrt{s} \simeq$ 183 GeV from the ALEPH collaboration as a function of Δm for a 91 GeV $\tilde{\chi}_1^\pm$ (right).

For a small mass difference ($0.1 \leq \Delta m \leq 0.5$ GeV), the DELPHI Collaboration has searched for events with kinks in their central tracking detectors. Charginos with a mass difference of $0.5 \leq \Delta m \leq 3.0$ GeV would yield a signal topology with very small visible energy. The sensitivity to this topology is increased by searching for events with initial state radiative photons.

No evidence for chargino production has been observed by the four LEP collaborations in any of the Δm regions. Therefore, 95% C.L. upper limits on the $\tilde{\chi}_1^\pm$ pair production cross-section have been computed. The right plot [4] in Figure 1 shows the pair production cross-section upper limit as a function of Δm for a 91 GeV $\tilde{\chi}_1^\pm$, i.e. close to the kinematic limit. Figure 2 shows exclusion plots [5] in the (μ, M_2) plane in the MSSM framework, for two values of $\tan \beta$ =1.5, 35. The 95% C.L. lower limits on the $\tilde{\chi}_1^\pm$ mass obtained by the four LEP collaborations for the case of a heavy sneutrino and $\Delta m \geq 10$ GeV are close to the kinematic limit.

NEUTRALINOS

The neutralinos could be pair produced through s-channel virtual Z^0 exchange or t-channel scalar electron exchange. Since the $\tilde{\chi}_1^0$ is experimentally *invisible*,

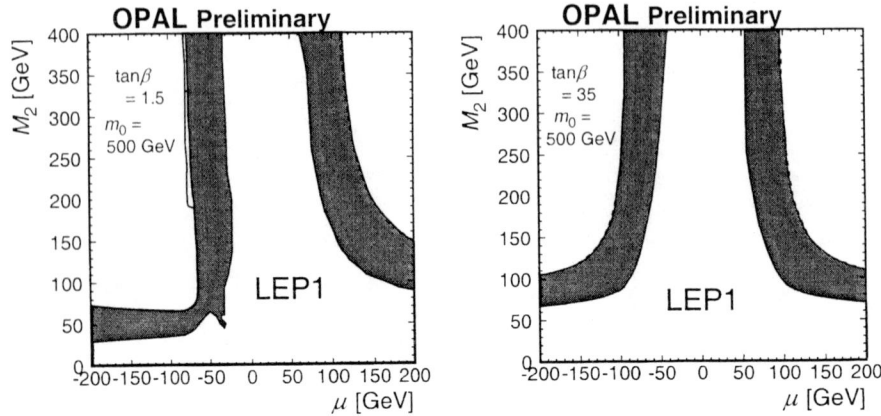

FIGURE 2. Exclusion regions at the 95% C.L. in the (μ, M_2) plane for two values of $\tan\beta = 1.5$ (left), 35 (right) and for $m_0 \leq 500$ GeV from the OPAL collaboration. The light shaded areas show the LEP1 excluded regions and the dark shaded areas show the additional exclusion region using the data from $\sqrt{s} \simeq 183$ GeV.

the only way to look for $\tilde{\chi}_1^0 \tilde{\chi}_1^0$ production is through the $e^+e^- \to \tilde{\chi}_1^0 \tilde{\chi}_1^0 \gamma$ process. Alternatively one can look directly for the production of a $\tilde{\chi}_2^0 \tilde{\chi}_1^0$ pair. The $\tilde{\chi}_2^0$ could then decay into $\tilde{\chi}_1^0 l^+ l^-$, $\tilde{\chi}_1^0 \nu \bar{\nu}$ or $\tilde{\chi}_1^0 q\bar{q}$, through a virtual Z^0, Higgs boson, slepton or squark exchange. The event topologies are similar to those studied for $\tilde{\chi}_1^+ \tilde{\chi}_1^-$ events.

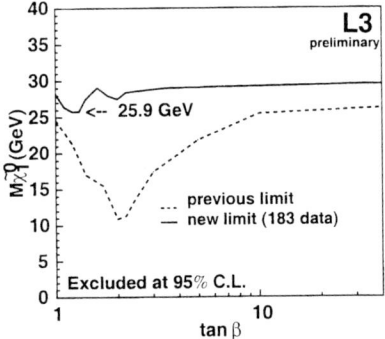

FIGURE 3. 95% C.L. limits on the $\tilde{\chi}_1^0$ (LSP) mass as a function of $\tan\beta$ and for any value of m_0 from the L3 Collaboration.

No evidence of $\tilde{\chi}_2^0 \tilde{\chi}_1^0$ production has been observed by the four LEP collaborations. Therefore, 95% C.L. upper limits on the $\tilde{\chi}_2^0 \tilde{\chi}_1^0$ pair production cross-section have been computed by all the LEP experiments. From these studies and also using results from the slepton searches, a $\tilde{\chi}_1^0$ has been excluded at the 95% C.L. with a mass smaller than 30.1 GeV for $m_0 > 200$ GeV by the ALEPH Collaboration, smaller than 28.1 GeV for $m_0 > 1$ TeV by the DELPHI Collaboration, smaller than 30.1 GeV for $m_0 > 200$ GeV by the L3 Collaboration and smaller than 30.1 GeV for $m_0 > 500$ GeV by the OPAL Collaboration. A 95% C.L. lower mass limit of $m_{\tilde{\chi}_1^0} > 25.9$ GeV has been derived by the L3 Collaboration for any value of $\tan\beta$ and any value of m_0 [6], as shown in Figure 3.

SCALAR QUARKS

Because of the possible large mass splitting by left-right mixing, the lowest mass eigenstate of the stop quark, $\tilde{t}_1 = \tilde{t}_L \cos\theta_{mix} + \tilde{t}_R \sin\theta_{mix}$, could be the lightest charged supersymmetric particle. The stop quark pair production cross-section depends on the stop mass, $m_{\tilde{t}_1}$, and the mixing angle θ_{mix}. If the stop quark is assumed to be lighter than every other charged sparticle, the dominant decay mode is the two-body flavour changing decay $\tilde{t}_1 \to c + \tilde{\chi}_1^0$. The event topologies would therefore be two acoplanar jets with missing energy.

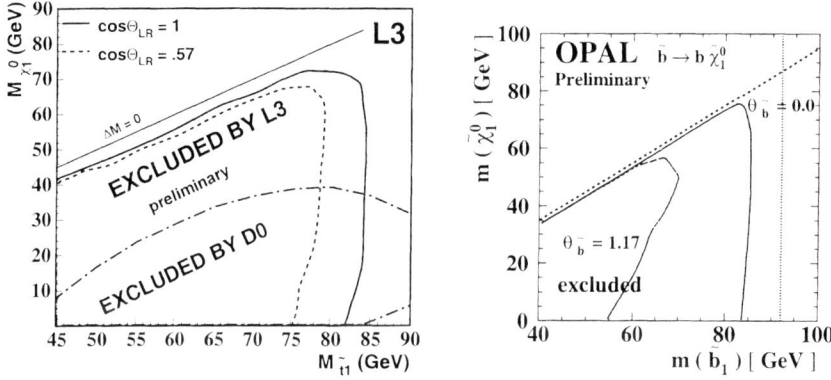

FIGURE 4. 95% C.L. exclusion regions in the $m_{\tilde{t}_1} - m_{\tilde{\chi}_1^0}$ plane for pair produced stop quarks followed by a two-body decay (left) and in the $m_{\tilde{b}_1} - m_{\tilde{\chi}_1^0}$ plane for pair produced sbottom quarks (right).

The three-body decay, $\tilde{t}_1 \to b + \tilde{\chi}_1^{*+}$, followed by $\tilde{\chi}_1^{*+} \to \tilde{\nu} + \ell^+$, is dominant if it is kinematically accessible and for a $\tilde{\chi}_1^{\pm}$ lighter than about 200 GeV. In this case, the event topologies would be two acoplanar jets and an acoplanar pair of leptons with missing energy. No evidence for $\tilde{t}_1\bar{\tilde{t}}_1$ production has been observed by any of the LEP collaboration. The left plot in Figure 4 shows the excluded regions for the $\tilde{t}_1$ mass as a function of the $\tilde{\chi}_1^0$ mass for the two-body decay. The region excluded by the D0 experiment [7] is also shown. Excluded regions as a function of the $\tilde{\nu}$ mass have been obtained for the three-body decay of the stop quark.

TABLE 1. 95% C.L. lower $\tilde{t}_1$ and $\tilde{b}_1$ masses.

Lower $\tilde{t}_1$ mass (GeV) for the two-body decays		
ALEPH	74	$10 < \Delta m < 40$ GeV
DELPHI	70	Massless $\tilde{\chi}_1^0$
L3	75	$\Delta m > 10$ GeV
OPAL	81	$\Delta m > 10$ GeV
Lower $\tilde{t}_1$ mass (GeV) for the three-body decays		
ALEPH	82	$\Delta m > 10$ GeV
OPAL	79	$\Delta m > 10$ GeV
Lower $\tilde{b}_1$ mass (GeV) for the two-body decays		
ALEPH	79	$\Delta m > 10$ GeV
OPAL	84	$\Delta m > 10$ GeV

For large $\tan \beta$, there could be a large mixing also between the right- and left-handed $\tilde{b}$ quarks, resulting in two states, $\tilde{b}_1$ and $\tilde{b}_2$. The lowest lying state, $\tilde{b}_1$, would decay primarily to $\tilde{b}_1 \to b + \tilde{\chi}_1^0$, leading to a topology identical to that of the pair production of stop quarks followed by a two-body decay. No evidence for $\tilde{b}_1 \bar{\tilde{b}}_1$ production has been observed by any of the LEP collaborations. The right plot in Figure 4 shows the excluded regions for the $\tilde{b}_1$ mass as a function of the $\tilde{\chi}_1^0$ mass. Similar exclusion regions have been obtained by the four LEP Collaborations. The numerical mass bounds [8] are listed in Table 1 for $\theta_{mix} = 56^0$ for the stop quark and for a purely left-handed $\tilde{b}_1$ ($\theta_{mix} = 0^0$) quark.

SLEPTONS

In SUSY theories, each lepton has two scalar partners, the right and left-handed sleptons, denoted $\tilde{\ell}_R$ and $\tilde{\ell}_L$, according to their helicity states. Sleptons could be pair produced through s-channel Z^0 or γ exchange. Selectrons can also be produced through t-channel neutralino exchange. The selectron cross-section is enhanced by the t-channel contribution compared to the smuon and stau production cross-sections. The dominant slepton decay mode is : $\tilde{\ell}^{\pm} \to \ell^{\pm} + \tilde{\chi}_1^0$, leading to event topologies of two acoplanar leptons and missing energy.

No evidence for pair produced slepton events has been observed and slepton lower mass limits at the 95% C.L. have been derived. Exclusion regions for right-handed selectron and smuon pair production are shown in Figure 5. Similar plots are obtained for the stau pair production. The numerical mass bounds [9] are listed in Table 2 for right-handed sleptons within the framework of the MSSM.

A search for heavy long-lived sleptons has been performed similarly to that for long-lived charginos. No evidence for the production of such heavy quasi-stable sleptons has been observed and a 95% C.L. upper limit on the production cross-section of 0.05 to 0.1 pb has been derived for masses between 45 and 89.5 GeV [3]. Within the framework of the MSSM, 95% C.L. lower limits of 82.5 GeV and of

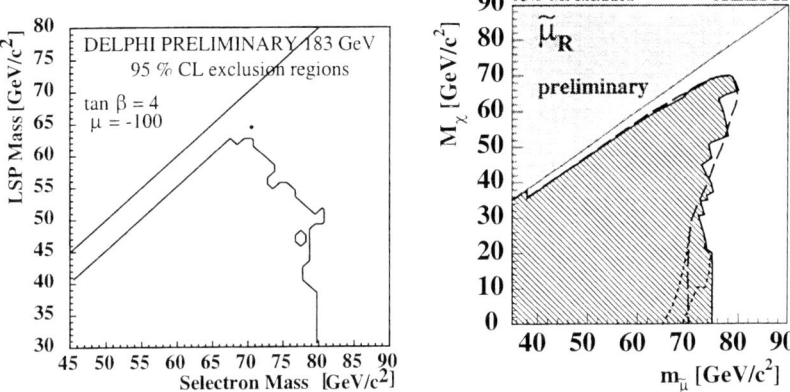

FIGURE 5. 95% C.L. lower mass limits for right-handed selectron (left) and smuon (right) pair production as a function of the $\tilde{\chi}_1^0$ mass. For the selectron case, $\tan\beta = 4$ and $\mu = -100$ GeV have been used.

TABLE 2. 95% C.L. lower masses of right-handed sleptons.

		Lower $\tilde{e}_R$ mass (GeV)
ALEPH	81	$\Delta m > 10$ GeV, $\mu = -200$ GeV, $\tan\beta = 2$
DELPHI	77	$m_{\tilde{\chi}_1^0} < 50$ GeV, $\mu = -100$ GeV, $\tan\beta = 4$
L3	80	$\Delta m > 15$ GeV, $\mu = -200$ GeV, $\tan\beta = 1.41$
OPAL	78	$\Delta m > 2$ GeV, $\mu < -200$ GeV, $\tan\beta = 1.5$
		Lower $\tilde{\mu}_R$ mass (GeV)
ALEPH	71	$\Delta m > 5$ GeV
DELPHI	66	$m_{\tilde{\chi}_1^0} < 65$ GeV
		$\Delta m > 5$ GeV
L3	69	$\Delta m > 15$ GeV
OPAL	65	$\Delta m > 3$ GeV
		Lower $\tilde{\tau}_R$ mass (GeV)
ALEPH	65	$\Delta m > 15$ GeV
DELPHI	60	$m_{\tilde{\chi}_1^0} < 35$ GeV
		$\Delta m > 10$ GeV
OPAL	66	$\Delta m > 20$ GeV

83.5 GeV on the masses of the quasi-stable right- and left-handed smuons have been derived by the OPAL Collaboration [3].

SUSY WITH R-parity VIOLATION

If R-parity is violated, the sparticles can decay directly to Standard Model particles and any of them is a credible candidate for being the LSP. The supersymmetric particles can also be singly produced. As a matter of fact, for most of the analyses, the event signature is different from the R-parity conserving model, where the LSP escapes detection and takes away a large fraction of the total momentum. With the MSSM particle content, explicit breaking of R-parity conserving interactions are parametrised in a gauge-invariant super-potential that includes the following Yukawa coupling terms:

$$W_{RPV} = \lambda_{ijk} L_i L_j \overline{E}_k + \lambda'_{ijk} L_i Q_j \overline{D}_k + \lambda''_{ijk} \overline{U}_i \overline{D}_j \overline{D}_k$$

where i, j, k are the generation indices of the superfields L, Q, E, D and U. L and Q are lepton and quark left-handed doublets respectively. $\overline{E}, \overline{D}$ and $\overline{U}$ are right-handed singlet charge-conjugate superfields for charged leptons, down- and up-type quark, respectively. λ_{ijk} are antisymmetric under the interchange of the first two indices, $\lambda_{ijk} = - \lambda_{jik}$ and are non-vanishing only if $i \neq j$, so that at least two different generations are coupled in the purely leptonic vertices. λ''_{ijk} is antisymmetric under the interchange of the last two indices. The λ and λ' couplings both violate the lepton number (L) and the λ'' couplings violate the baryon number (B). There are 9 λ couplings for the triple lepton vertices, 27 λ' for the lepton-quark-quark vertices and 9 λ'' couplings for the triple quark vertices. There are therefore a total of 45 new couplings.

Recently, supersymmetric models with R-parity violation (RPV) have attracted a considerable theoretical and phenomenological interest. Because it is still an open question, it is of major importance to consider the phenomenology of possible R-parity violating scenarios. Not only, the branching ratios of some of the R-parity violating decay modes can be comparable or even larger than the R-parity conserving mode. This is for instance the case for the scalar top quark (stop) decay modes to the third generation fermions. The simultaneous presence of the couplings λ'' (B-violating) and λ' (L-violating) is forbidden since it would allow fast squark mediated proton decay at the tree level. This experimental non-observation of the proton decay places strong bounds on the product of these two couplings, i.e., $\lambda' \times \lambda'' < 10^{-24}$.

Due to the large amount of results available in this field, only a few representative results from each Collaboration are shown here.

No evidence for chargino pair production has been observed. Therefore, 95% C.L. upper limits on the $\tilde{\chi}_1^\pm$ pair production cross-sections with R-parity violating decays via one dominant Yukawa-like λ coupling have been computed. OPAL has been able to exclude cross-sections larger than 0.2 pb at the 95% C.L. [10]. The

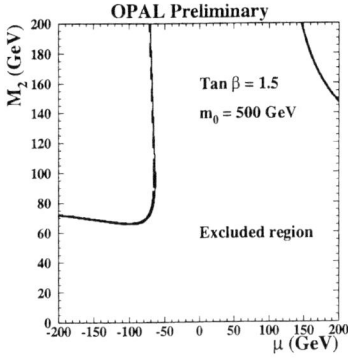
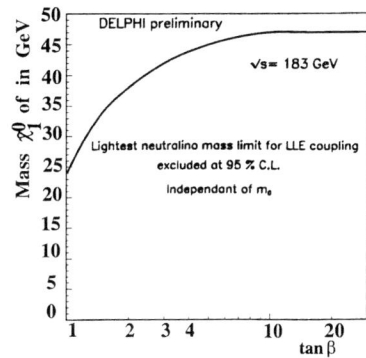

FIGURE 6. Exclusion region at the 95% C.L. in the (μ, M_2) plane for $\tan\beta = 1.5$ and $m_0 = 500$ GeV from the OPAL Collaboration (left) and $\tilde{\chi}_1^0$ mass limit for a non-zero λ coupling as a function of $\tan\beta$ for any value of m_0 from the DELPHI Collaboration (right).

left plot in Figure 6 shows the exclusion plot in the (μ, M_2) plane for $\tan\beta = 1.5$ and $m_0 = 500$ GeV. The DELPHI experiment has searched for particular R-parity violating decays of the $\tilde{\chi}_1^0$ induced by a non-zero λ coupling. As can be seen in the right plot of Figure 6, a neutralino with a mass smaller than 24 GeV has been excluded at the 95% C.L. [11].

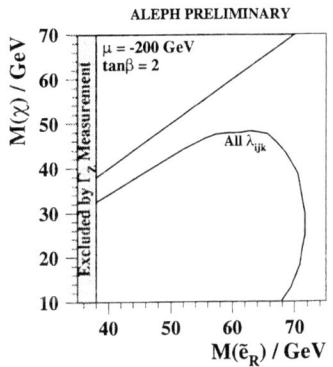
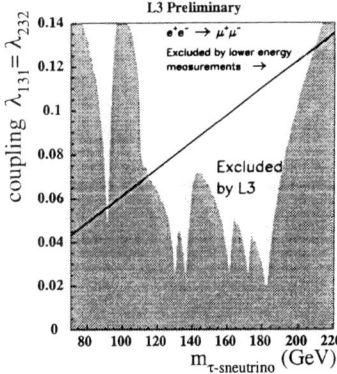

FIGURE 7. 95% C.L. exclusion region for right-handed selectrons with R-parity violating decays via one dominant Yukawa-like λ' coupling within the framework of the MSSM (left) from the ALEPH Collaboration and 95% C.L. exclusion region on coupling $\lambda_{131} = \lambda_{232}$ as a function of the sneutrino mass (right) from the L3 Collaboration.

The ALEPH Collaboration has performed a complete search [12] for pair produced sleptons, sneutrinos, stop and sbottom quarks with R-parity violating decays via one dominant Yukawa-like λ' coupling using the data collected at centre-of-mass energies of 130 GeV to 172 GeV. The left plot in Figure 7 shows the exclusion region for right-handed selectrons within the framework of the MSSM.

The production cross-section of two leptons could be affected by the presence of new heavy particles which would contribute to the measured cross-section via the $s-$ and/or $t-$ channel. For instance, the presence of a $\tilde{\nu}_\tau$ with both couplings $\lambda_{131} = \lambda_{232}$ non-zero would give rise to a modified $\mu^+\mu^-$ cross-section due to a $s-$channel $\tilde{\nu}_\tau$ exchange. The 95% C.L. exclusion region on coupling $\lambda_{131} = \lambda_{232}$ as a function of the sneutrino mass is shown in the right plot [13] of Figure 7.

GAUGE MEDIATED SUSY

It is typically assumed that SUSY is broken in some "hidden" sector of new particles, and is "communicated" (or mediated) to the "visible" sector of Standard Model and SUSY particles by one of the known interactions. Two scenarios for this mediation have been widely investigated: gravity and gauge mediation [14]. In gravity-mediated SUSY breaking, the hidden sector is necessarily at an energy scale where the gravitational coupling to the visible sector is non-negligible, *i.e.* greater than approximately 10^{11} GeV. In gauge-mediated SUSY breaking (GMSB), the hidden sector can lie at much lower energies, down to about 10^4 GeV. In most current GMSB theoretical work, it is assumed that this hidden sector is coupled to a messenger sector, which in turn couples to the visible sector through normal SM gauge interactions. The advantage of GMSB is that flavour changing neutral currents cannot be induced by SUSY breaking, because the normal gauge interactions are flavour blind. A feature which distinguishes gravity- from gauge-mediated models is the mass of the gravitino, $\tilde{G}$. In gravity-mediated models, $\tilde{G}$ is usually too heavy to have a significant effect on SUSY phenomenology, while in GMSB models, the $\tilde{G}$ is typically very light (< 1 GeV) and is the lightest SUSY particle, the LSP. While $\tilde{G}$ is a spin 3/2 particle, only its spin 1/2 component (which has "absorbed" the goldstino associated with spontaneous SUSY breaking via the "superhiggs" mechanism) interacts with weak, rather than gravitational, strength interactions, and contributes to phenomenology.

The charginos, $\tilde{\chi}^\pm_{1,2}$, could be pair produced and then decay into a $\tilde{\chi}^0_1$ and a virtual W^* with the subsequent decay of $\tilde{\chi}^0_1 \to \gamma\tilde{G}$. The topologies would be two energetic photons plus jets and/or isolated leptons plus missing energy. Alternatively, one can look directly for the production of a $\tilde{\chi}^0_1\tilde{\chi}^0_1$ pair. The $\tilde{\chi}^0_1$ could then decay into $\gamma\tilde{G}$. For the pair production of $\tilde{\chi}^0_1$, the topologies would be 2 energetic photons and missing energy carried away by the gravitinos. The left plot in Figure 8 shows the 95% C.L. upper limits [15] on the $\tilde{\chi}^0_1$ pair production cross-sections as a function of the $\tilde{\chi}^0_1$ decay length for two $\tilde{\chi}^0_1$ masses in this GMSB model. The right plot instead shows the exclusion region [16] compared to the region allowed by the SUSY interpretation of the CDF event $e^+e^-\gamma\gamma$ with missing energy in the selectron scenario [17]. In this scenario, pair produced selectrons with the subsequent decay $\tilde{e} \to e\tilde{\chi}^0_1$ followed by $\tilde{\chi}^0_1 \to \gamma\tilde{G}$ would yield such a topology.

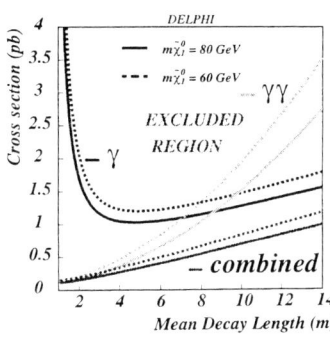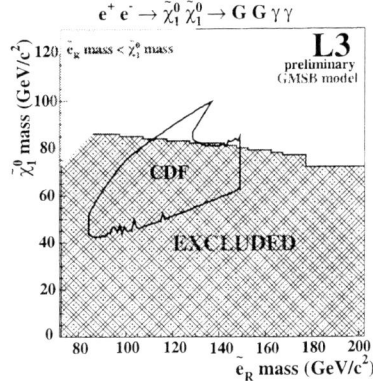

FIGURE 8. 95% C.L. upper limits on the $\tilde{\chi}_1^0$ pair production cross-sections as a function of the $\tilde{\chi}_1^0$ decay length for two $\tilde{\chi}_1^0$ masses (left) and exclusion region compared to the region allowed by the SUSY interpretation of the CDF event in the selectron scenario (right).

CONCLUSIONS

Since November 1995, the LEP collider has been operated at centre-of-mass energies up to 183 GeV. Many searches for physics beyond the Standard Model have been performed using these data. The four LEP experiments have searched for SUSY particles predicted by different theoretical scenarios. Some candidates have been selected by the analyses but their number is compatible with the expected background from Standard Model processes. So far, no excess of events in any of the presented topologies has been observed and no evidence of these new particles has been found. Upper limits have been derived which greatly improve the limits obtained previously at LEP, running at the Z^0 peak. These searches will be further improved with the higher energies and larger data samples that LEP will collect in the next years.

REFERENCES

1. Gol'fand Y. and Likhtam E., *JETP Lett.* **13**, 323 (1971);
   Volkov D. and Akulov V., *Phys. Lett.* **B46**, 109 (1973);
   Wess J. and Zumino B., *Nucl. Phys.* **B70**, 39 (1974).
2. Nilles H. P., *Phys. Rep.* **110**, 1 (1984);
   Haber H. E. and Kane G. L., *Phys. Rep.* **117**, 74 (1985).
3. OPAL Collaboration, *"Search for Stable and Long-Lived Massive Charged Particles in e^+e^- Collisions at $\sqrt{s} = 130\text{-}183$ GeV,"* CERN-EP/98-039, March 1998.
4. ALEPH Collaboration, *"Search for Charginos and Neutralinos at 183 GeV for Large Slepton Masses,"* ALEPH 98-016, CONF 98-006, *Contribution to the 1998 Winter Conferences.*

5. OPAL Collaboration, *"Search for Chargino and Neutralino Production at $\sqrt{s} = 181$-184 GeV at LEP,"* OPAL Physics Note, PN339, March 1998.
6. L3 SUSY Group, *"Search for Scalar Leptons, Charginos and Neutralinos in e^+e^- Collisions at $\sqrt{s} = 183$ GeV,"* L3 Note 2231, March 1998, *Contribution to the 1998 Winter Conferences.*
7. D0 Collaboration, Abachi F. et al., *Phys. Rev. Lett.* **76**, 2222 (1996).
8. ALEPH Collaboration, *"Scalar Quark Searches at $\sqrt{s} = 181$-184 GeV in e^+e^- Collisions,"* ALEPH 98-015, CONF 98-005, *Contribution to the 1998 Winter Conferences;*
   DELPHI Collaboration, *Private Communication,* De Boer W.;
   L3 SUSY Group, *"Search for Stop and Sbottom Decays with the L3 Detector at $\sqrt{s} = 183$ GeV,"* L3 Note 2238, March 1998, *Contribution to the 1998 Winter Conferences;*
   OPAL Collaboration, *"Search for Scalar Top and Scalar Bottom Quarks at $\sqrt{s} = 181$-184 GeV in e^+e^- Collisions,"* OPAL Physics Note, PN337, March 1998.
9. ALEPH Collaboration, *"Search for Sleptons in e^+e^- Collisions at centre-of-mass energies up to 184 GeV,"* ALEPH 98-014, CONF 98-004, *Contribution to the 1998 Winter Conferences;*
   DELPHI Collaboration, *Private Communication,* De Boer W.;
   L3 SUSY Group, *"Search for Scalar Leptons, Charginos and Neutralinos in e^+e^- Collisions at $\sqrt{s} = 183$ GeV,"* L3 Note 2231, March 1998, *Contribution to the 1998 Winter Conferences;*
   OPAL Collaboration, *"Search for Anomalous Production of Acoplanar Di-lepton Events at $\sqrt{s} = 161, 172$ and 183 GeV,"* OPAL Physics Note, PN341, March 1998.
10. OPAL Collaboration, *"An Update on the Searches for R-parity Violating Decays of Charginos and Stop Quarks at $\sqrt{s} \simeq 183$ GeV,"* OPAL Physics Note, PN335, March 1998.
11. DELPHI Collaboration, *Private Communication,* De Boer W.
12. ALEPH Collaboration, *"Search for Supersymmetry with a Dominant R-parity Violating $LQ\overline{D}$ Coupling in e^+e^- Collisions at centre-of-mass energies of 130 GeV to 172 GeV,"* ALEPH 98-017, CONF 98-007, *Contribution to the 1998 Winter Conferences.*
13. L3 Fermion-pair Group, *"Search for R-parity Breaking Sneutrino Exchange at LEP,"* L3 Note 2135, March 1998, *Contribution to the 1998 Winter Conferences.*
14. OPAL Collaboration, *"Searches for Charginos and Neutralinos in Models with Gauge-Mediated SUSY breaking in e^+e^- Collisions at $\sqrt{s} = 183$ GeV,"* OPAL Physics Note, PN332, March 1998.
15. DELPHI Collaboration, *Private Communication,* De Boer W.
16. L3 SUSY Group, *"Single and Multi-Photon Events with Missing Energy in e^+e^- Collisions at $\sqrt{s} = 183$ GeV,"* L3 Note 2130, March 1998, *Contribution to the 1998 Winter Conferences.*
17. Ellis J. et al., *Phys. Lett.* **B394**, 354 (1997).

Searches for Leptoquarks at Fermilab

David Hedin

for the CDF and DØ Collaborations

Northern Illinois University, DeKalb, Illinois 60115

Abstract.
We report on searches for leptoquarks produced in $p\bar{p}$ collisions at $\sqrt{s} = 1.8$ TeV using the CDF and DØ detectors at Fermilab. Mass limits for scalar leptoquarks are presented for all three possible generations.

Leptoquarks (LQ) are bosons predicted in many extensions to the Standard Model [1]. They carry both lepton and color quantum numbers, couple to leptons and quarks, and decay via $LQ \to l+q$. To satisfy experimental constraints on flavor changing neutral currents, leptoquarks of mass accessible to current collider experiments are constrained to couple to only one generation of leptons and quarks [2]. Therefore, only leptoquarks which couple within a single generation are considered here.

This paper summarizes the results of searches for leptoquarks produced in $p\bar{p}$ collisions at $\sqrt{s} = 1.8$ TeV. We assume that leptoquarks are produced in pairs by strong interactions, such as $p\bar{p} \to g \to LQ \, \overline{LQ} + X$. This process dominates over other production mechanisms which depend on the unknown leptoquark-lepton-quark coupling λ under the standard condition $\lambda \leq \sqrt{4\pi\alpha_{EM}}$. The production cross section is the same for all three generations of leptoquarks for a fixed leptoquark mass (M_{LQ}). Cross section limits as a function of M_{LQ} from data collected by the CDF and DØ detectors have been set for each LQ generation. Limits on the masses of scalar leptoquarks are determined utilizing the NLO theoretical cross section [3], and are given in this paper.

Searches are made for leptoquark decay into either charged leptons or neutrinos. We define β as the branching fraction into charged leptons, and $B = 1 - \beta$ as the branching fraction into neutrinos. Most models of leptoquarks give $\beta = 0, 1/2$ or 1 and we will quote mass limits for those branching fraction values.

I. First Generation Leptoquarks

TABLE 1. Analysis summary for the dielectron channel.

	CDF	DØ
Luminosity	110 pb^{-1}	123 pb^{-1}
Event Selection		
Two electrons $E_T >$	25 GeV	20 GeV
Two jets $E_T >$	30,15 GeV	15 GeV
m_{ee} cut if	$76 < m_{ee} < 106$ GeV/c^2	$82 < m_{ee} < 100$ GeV/c^2
correlated E_T	$E_{Tj1} + E_{Tj2} > 70$ GeV	$S_T > 350$ GeV
	$E_{Te1} + E_{Te2} > 70$ GeV	
	$M_{ej1} \approx M_{ej2}$	

CDF and DØ have searched for first generation leptoquarks through their decays to one or two isolated high E_T electrons plus jets [4]. The event selection for the dielectron channel (summarized in Table 1) requires two or more jets in addition to two electrons. An invariant mass cut on the electron pair reduces the Z boson background. Remaining Drell-Yan events are removed by utilizing the leptoquark decay kinematics. The two body decay of two heavy leptoquarks will produce high E_T electrons and jets, with the invariant mass of two electron-jet combinations being approximately that of the leptoquark's mass. CDF requires the E_T sum of each of the electron pair and the jet pair be greater than 70 GeV. Also, the two invariant mass combinations which are closest to each other are required to be within the resolution error of the leptoquark's assumed mass. For assumed masses greater than 180 GeV/c^2, no events remain in the data. DØ defines S_T as the scalar sum of the E_T of the electrons and jets, and a requirement that $S_T > 350$ GeV gives an estimated background of 0.41 ± 0.06 events, with no events in the data passing the selection criteria. Correcting for acceptance gives a 95% C.L. limit on the productrion cross section of 0.10 pb for CDF and 0.076 pb for DØ ($M_{LQ} = 200$ GeV/c^2). The corresponding limits on the mass of a first generation scalar leptoquark are 213 GeV/c^2 and 225 GeV/c^2, assuming $\beta = 1$.

Searches are also made for the final state with one leptoquark decaying into an electron plus jet while the other decays into a neutrino plus jet. The event selection for this single electron channel (summarized in Table 2) requires two or more jets in addition to the electron. Missing transverse energy ($\not{E}_T$) greater than 35 GeV (CDF) or 30 GeV (DØ) is used to identify the neutrino. A cut on the transverse mass M_T of the electron and $\not{E}_T$ reduces the W boson background. Remaining W and top quark events are removed using similar kinematic quantities as in the dielectron analysis. CDF requires the E_T sum of the jet pair be greater than 80 GeV, and that the electron-jet invariant mass and neutrino-jet transverse mass be within the resolution error of the leptoquark's assumed mass. Depending on the assumed mass, either 0 or 1 events remain in the data. DØ uses a neural network algorithm to define a contour in the S_T versus $(M_{ej} - M_{LQ})/M_{LQ}$ plane, and no events remain in the data following a cut in this plane. Mass limits are obtained assuming $\beta = 0.5$ and are 180 GeV/c^2 for CDF and 175 GeV/c^2 for DØ. DØ has combined the dielectron and single electron channels raising the limit for $\beta = 0.5$

TABLE 2. Analysis summary for the single electron channel.

	CDF	DØ
Luminosity	110 pb^{-1}	115 pb^{-1}
Event Selection		
One electron $E_T >$	20 GeV	20 GeV
Two jets $E_T >$	30,15 GeV	20 GeV
$\not{E}_T >$	35 GeV	30 GeV
$M_{Te\nu} >$	120 GeV	110 GeV
correlated E_T	$E_{Tj1} + E_{Tj2} > 80$ GeV	S_T vs $(M_{ej} - M_{LQ})/M_{LQ}$
	$M_{ej1} \approx M_{T\nu j2}$	contour

to 204 GeV/c^2.

DØ has also searched in the channel where both leptoquarks decay into neutrinos. Cuts on $\not{E}_T > 40$ GeV, two or more jets with $E_T > 30$ GeV, and on the relative jet and $\not{E}_T$ directions yield 3 events in a 7.4 pb^{-1} data sample. The background is estimated to be 3.5 ± 1.2 events from W and Z bosons. A 95% C.L. limit of $M_{LQ} > 79$ GeV/c^2 is obtained for $\beta = 0$.

II. Second Generation Leptoquarks

CDF and DØ have searched for second generation leptoquarks through their decays to two isolated high p_T muons plus jets [5]. The event selection (summarized in Table 3) requires two or more jets in addition to two muons. In order to reduce the Z boson background, CDF performs an invariant mass cut on the muon pair while DØ makes a χ^2 cut on a fit to a Z decay. Each experiment also requires $m_{\mu\mu} > 10$ GeV/c^2 to remove low mass resonances. Remaining Drell-Yan events are removed by utilizing the leptoquark decay kinematics. CDF again requires that the two muon-jet invariant mass combinations which are closest to each other be within the resolution error of the leptoquark's mass. Depending on the assumed leptoquark mass, either 0 or 1 events remain in the data. DØ defines H_T as the scalar sum of the E_T of the jets. Requiring $H_T > 100$ GeV plus making a cut that the maximum gap in ϕ between a muon and its nearest neighboring jet is lesss than π eliminates all remaining events. Correcting for acceptances, and comparing the resultant cross section limits to theory gives limits on the mass of a second generation scalar leptoquark of 195 GeV/c^2 for CDF and 184 GeV/c^2 for DØ, assuming $\beta = 1$.

III. Third Generation Leptoquarks

CDF has searched for charge 2/3 or 4/3 third generation leptoquarks decaying into a tau plus a b quark giving a $\tau^+\tau^- jj$ final state [6]. The branching fraction β is 1 as the decay into top quarks is kinematically suppressed. One τ is required to decay leptonically and is identified by requiring an isolated e or μ with $p_T > 20$ GeV/c and $\not{E}_T$ within 50° of the lepton's direction in the transverse plane. The other τ is identified by requiring 1 or 3 charged tracks within a 10° cone and no

TABLE 3. Analysis summary for the dimuon channel.

	CDF	DØ
Luminosity	110 pb^{-1}	95 pb^{-1}
Event Selection		
Two muons $E_T >$	30,20 GeV	15 GeV
Two jets $E_T >$	30,15 GeV	15 GeV
Z cut if	$76 < m_{\mu\mu} < 106$ GeV/c^2	Z fit χ^2
$m_{\mu\mu}$ cut if	$m_{\mu\mu} < 10$ GeV/c^2	$m_{\mu\mu} < 10$ GeV/c^2
correlated E_T	$M_{\mu j1} \approx M_{\mu j2}$	$H_T > 100$ GeV
		max. $\phi(\mu\text{-jet})$ gap $< \pi$

other tracks above 1 GeV/c between the 10° and 30° cones. Two or more jets with $E_T > 20$ GeV are required in the event. No explicit b-tagging is made. To aid in eliminating Z bosons, events with the invariant mass of the lepton and leading tau-jet track between 70 and 100 GeV/c^2 are removed.

For an integrated luminosity of 110 pb^{-1}, one event is left, consistent with an expected background of $2.4^{+1.2}_{-0.6}$ events (primarily from $Z \to \tau\tau$). A 95% C.L. cross section limit of 10 pb for $M_{LQ} = 100$ GeV/c^2 is obtained yielding a mass limit of 99 GeV/c^2 for charge 2/3 or 4/3 third generation scalar leptoquarks.

DØ has searched for charge 1/3 third generation leptoquarks decaying into a tau neutrino plus a b quark giving a $\nu\bar{\nu}b\bar{b}$ final state [7]. The decay of a b quark is indicated by the presence of a muon associated with a jet. Three triggers were used with the greatest sensitivity coming from a single muon ($p_T^\mu > 3.5$ GeV/c) plus jet trigger with an integrated luminosity of 19.5 pb^{-1}. Either two muons, each associated with its own jet, or a single jet-associated muon was required. If the event had only one muon-jet combination, an additional jet with $E_T^j > 25$ GeV was required. Requiring $\not{E}_T > 35$ GeV and the azimuthal angular separation between the missing energy and the nearest jet be greater than 0.7 radians removed QCD multijet events. Backgrounds from W boson decays were removed by cuts on muon-jet correlations whereas top events were minimized by cuts on the sum of jet E_T.

Following all selection criteria, two events remained in the data. The total background is estimated to be 2.5±0.6 events, mostly from $t\bar{t}$ and $W \to \mu\nu$ sources. The 95% C.L. upper limits on scalar leptoquark pair production as a function of leptoquark mass were determined, and a limit of 12.7 pb was obtained for $M_{LQ} = 100$ GeV/c^2. This yielded a mass limit of 94 GeV/c^2 for charge 1/3 third generation scalar leptoquarks (with $B = 1$ since $M_{LQ} < m_t$).

We thank the staffs at Fermilab and collaborating institutions for their contributions. This work was supported by the Department of Energy and National Science Foundation (U.S.A.); Commissariat à L'Energie Atomique (France); State Committee for Science and Technology and Ministry for Atomic Energy (Russia); CAPES and CNPq (Brazil); Departments of Atomic Energy and Science and Education (India); Colciencias (Colombia); CONACyT (Mexico); Istituto Nazionale di Fisica Nucleare (Italy); Ministry of Education, Science and Culture (Japan); Natural Sciences and Engineering Research Council (Canada); National Science Council (China); Ministry of Education and KOSEF (Korea), and CONICET and UBACyT (Argentina).

REFERENCES

1. J.L. Hewett and T.G. Rizzo, Phys. Rep. **183**, 193 (1989), and references therein. Also, J.L. Hewett and T. Rizzo, PRD **56**, 5709 (1997).
2. M. Leurer, Phys. Rev. D **49**, 333 (1994).
3. M. Krämer, T. Plehn, and P.M. Zerwas, Phys. Rev. Lett. **79**, 341 (1997).
4. DØ Collaboration, B. Abbott et al., Phys. Rev. Lett. **79**, 4321 (1997); CDF Collaboration, F. Abe et al., Phys. Rev. Lett. **79**, 4327 (1997); DØ Collaboration, B. Abbott et al., Phys. Rev. Lett. **80**, 2051 (1998).
5. CDF Collaboration, H.S. Kambara, Fermilab-Conf-97/225-E.
6. CDF Collaboration, F. Abe et al., Phys. Rev. Lett. **78**, 2906 (1997).
7. DØ Collaboration, B. Abbott et al., Fermilab-Pub-98/081-E, to be published in Phys. Rev. Lett.

Searches for SUSY at the Tevatron

Andrei Nomerotski[1]

University of Florida, Gainsville, FL 32611, USA

Abstract. This paper presents recent results on Supersymmetry searches in various channels from the CDF and D0 experiments at the Tevatron.

I SUPERSYMMETRY

The Standard Model (SM) very successfully describes the existing experimental information. But it has several problems from the theoretical point of view. The SM Lagrangian contains terms that are divergent unless exceptional fine tuning takes place [1]. In addition it fails to provide a dark matter candidate.

SUperSYmmetric (SUSY) model [2] introduces a new type of symmetry, between bosons and fermions. It assigns to each of the known particles a superpartner with a spin different by $\frac{1}{2}$. Table I lists ordinary particles and corresponding to them superparticles as predicted by the Minimal Supersymmetric Standard Model (MSSM) [2]. Note that the MSSM requires two Higgs doublets and that the superpartners of gauge and Higgs bosons do mix forming four neutralinos and two charginos.

SUSY solves the problem of fine-tuning and provides a dark matter candidate. The price for this is a large number of free parameters in the model. This number can be greatly reduced if we demand a grand unification [3]. In the supergravity inspired approach, called SUGRA, only 5 independent parameters survive : M_0, $M_{1/2}$, A_t, $\tan\beta$ and $sgn(\mu)$, where M_0 is the common boson mass, $M_{1/2}$ is the common fermion mass at the GUT scale, A_t is the trilinear coupling, $\tan\beta$ is the ratio of the vacuum expectation values of the two Higgs doublets and μ is the Higgs mass parameter.

If R parity is conserved (i.e. decay products of a supersymmetric particle include at least one supersymmetric particle), then the Lightest Supersymmetric Particle (LSP) must be stable, and neutral for cosmological reasons. From the experimental point of view, a stable neutral LSP escapes the detector undetected manifesting

---

[1] representing the CDF and D0 collaborations

Particle			Sparticle		
Fermion (spin 1/2)	Lepton	ℓ	Sfermion (spin 0)	Slepton	$\tilde{\ell}_L, \tilde{\ell}_R$ (only $\tilde{\nu}_L$)
	Quark	q		Squark	$\tilde{q}_L, \tilde{q}_R$ ($\tilde{b}_{1,2}, \tilde{t}_{1,2}$)
(spin 1)	Gluon	g	(spin 1/2)	Gluino	$\tilde{g}$
Gauge Boson (spin 1)	Photon	γ	Gaugino (spin 1/2)	Photino	$\tilde{\gamma}$
	Z boson	Z		Zino	$\tilde{Z}$
Higgs Boson (spin 0)	light Higgs	h	Higgsino (spin 1/2)	Higgsino	$\tilde{h}$ → Neutralino $\tilde{\chi}^0_{1,2,3,4}$
	heavy Higgs	H		Higgsino	$\tilde{H}$
	Pseudoscalar Higgs	A		Higgsino	$\tilde{A}$
Gauge Boson (spin 1)	W boson	$W^\pm$	Gaugino (spin 1/2)	Wino	$\tilde{W}^\pm$ → Chargino $\tilde{\chi}^\pm_{1,2}$
Higgs Boson (spin 0)	Charged Higgs	$H^\pm$	Higgsino (spin 1/2)	Higgsino	$\tilde{H}^\pm$
(spin 2)	Graviton	G	(spin 3/2)	Gravitino	$\tilde{G}$

itself as missing transverse energy ($\not{E}_T$). $\not{E}_T$ is a very important attribute of all SUSY signatures discussed in this paper.

Here we present the results on searches for SUSY particles performed recently with the CDF and D0 detectors which operate at the Tevatron $p\bar{p}$ collider at Fermilab. The center of mass energy of the $p\bar{p}$ collisions is 1.8 TeV. The paper summarizes the results of several analyses : $\not{E}_T$ and jets from D0, trilepton analysis from CDF and searches for SUSY with γ signatures from D0 and CDF.

II SQUARK AND GLUINO SEARCHES

A classical signature for SUSY is the multiple jets and $\not{E}_T$ where jets are produced in direct and cascade decays of squarks and gluinos.

$$p\bar{p} \to (\tilde{q}\tilde{q}, \tilde{g}\tilde{g}, \tilde{q}\tilde{g}), \quad \tilde{g} \to q\bar{q}\tilde{\chi}^0_i, \quad \tilde{g} \to q\bar{q}'\tilde{\chi}^\pm_i, \quad \tilde{q} \to q\tilde{\chi}^0_i, \quad \tilde{q} \to q'\tilde{\chi}^\pm_i.$$

The final event signature depends on the decay channels of the charginos and neutralinos and involve $\not{E}_T$ and multiple jets. Here and below the $\tilde{\chi}^0_1$ is assumed to be the LSP and R parity is assumed to be conserved. Therefore the $\tilde{\chi}^0_1$ is stable and carries away energy producing the $\not{E}_T$ signature.

The D0 experiment searched for events with multiple jets and $\not{E}_T$ using the data sample of 79 pb^{-1} [4]. Three or more jets with $E_T > 25$ GeV were required where at least one of the jets is central with $E_T > 115$ GeV. Events with a large $\not{E}_T$ were selected where the $\not{E}_T$ was uncorrelated with the direction of jets. Events with isolated leptons (e/μ) were rejected.

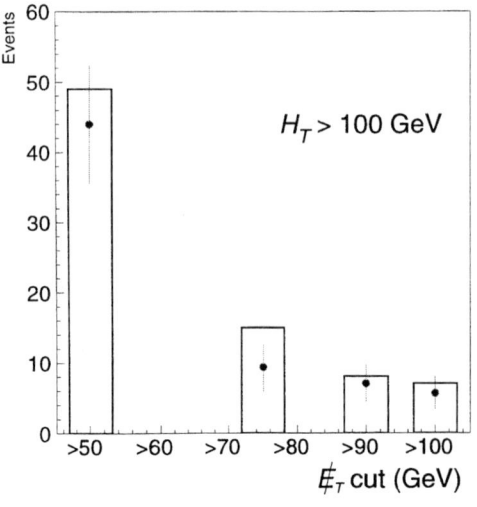 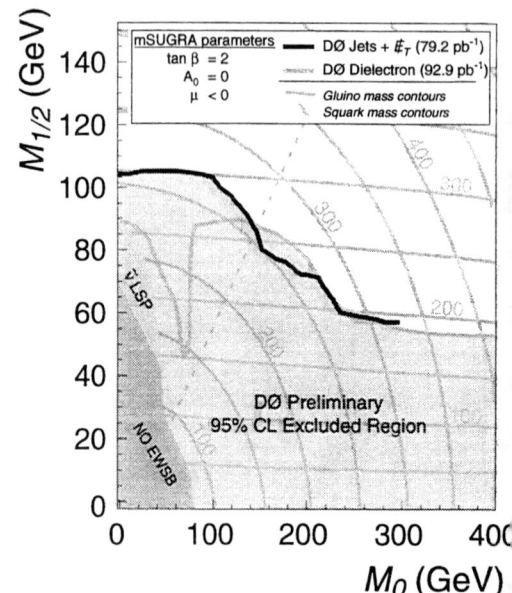

FIGURE 1. a) Comparison of the number of events observed by D0 with expected background for different values of the $\not\!\!E_T$ cut. b) D0 exclusion region in the $M_{1/2}$ and M_0 parameter space in minimal SUGRA.

The main backgrounds in this analysis come from W/Z + jets processes, top quark production and QCD multijet production in which the $\not\!\!E_T$ originates from a mismeasurement. The variable H_T (sum of E_T of the second, third, etc. jets) was found to be efficient in background suppression. Final $\not\!\!E_T$ and H_T cuts were optimized in each M_0 and $M_{1/2}$ point in the minimal SUGRA parameter space ($\not\!\!E_T \geq$ 50 - 150 GeV; $H_T \geq$ 100 - 250 GeV).

Figure 1a compares the number of observed events with expected background for different values of the $\not\!\!E_T$ cut. No excess of data has been found. This is interpreted as an exclusion region in the $M_{1/2}$ and M_0 parameter space (see Figure 1b).

III SUSY IN THE TRILEPTON CHANNEL

Superpartners of gauge and Higgs bosons, charginos and neutralinos, are predicted to be the lightest supersymmetric particles in many versions of MSSM and SUGRA. If they are produced in $p\bar{p}$ collisions their decay gives a distinct trilepton + $\not\!\!E_T$ signature [5]:

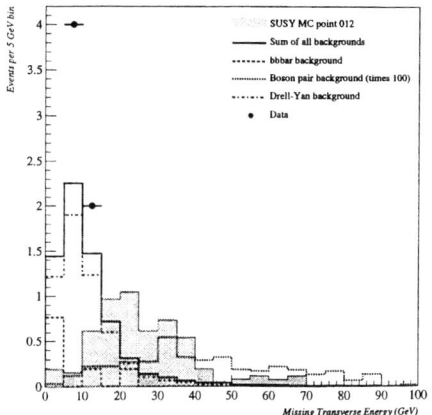 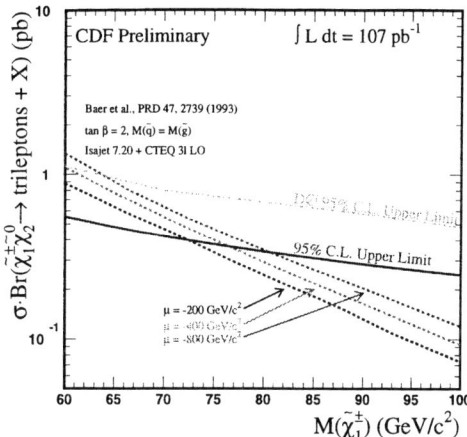

FIGURE 2. a) distribution of $\not{E}_T$ in the data, main backgrounds and one of the signal points before the final $\not{E}_T$ cut of 15 GeV at CDF. b) CDF 95% C.L. limit on the production cross section times Branching Ratio as a function of chargino mass. Overlayed are the theoretical predictions for different values of μ. Models predicting cross section that is higher than the 95% C.L. limit are excluded.

$$p\bar{p} \to \tilde{\chi}_1^+ \tilde{\chi}_2^0, \ \tilde{\chi}_1^+ \to l\nu\tilde{\chi}_1^0, \ \tilde{\chi}_2^0 \to ll\tilde{\chi}_1^0.$$

This process is considered to be the golden signature of SUSY because of the low SM backgrounds. CDF has searched the data sample of 107 pb^{-1} for events with 3 isolated leptons and $\not{E}_T > 15\ GeV$ and found none [6]. The expected SM background is 1.2 ± 0.2 events. This result was used to set limits on models that predict production of more than 3.2 events. Figure 2a shows the distribution of $\not{E}_T$ for the data, for main backgrounds and for one of the signal points before the final $\not{E}_T$ cut. Figure 2b shows the CDF 95% C.L. limit on the production cross section times Branching Ratio as a function of chargino mass. Overlayed are the theoretical predictions for different values of μ.

IV PHOTON ENRICHED SUSY

SUSY models predict a wide variety of signatures. After CDF recorded the famous $ee\gamma\gamma\not{E}_T$ event [7] several scenarios with photon signatures were suggested to explain it.

$E_T^\gamma > 12$ GeV Threshold		
Signature (Object)	Obs.	Expected
$\not{E}_T > 35$ GeV, $|\Delta\phi_{\not{E}_T - \text{jet}}| > 10°$	1	0.5 ± 0.1
$N_{\text{jet}} \geq 4$, $E_T^{\text{jet}} > 10$ GeV, $|\eta^{\text{jet}}| < 2.0$	2	1.6 ± 0.4
Central e or μ, $E_T^{e \text{ or } \mu} > 25$ GeV	3	0.3 ± 0.1
Central τ, $E_T^\tau > 25$ GeV	1	0.2 ± 0.1
b-tag, $E_T^b > 25$ GeV	2	1.3 ± 0.7
Central γ, $E_T^{\gamma 3} > 25$ GeV	0	0.1 ± 0.1
$E_T^\gamma > 25$ GeV Threshold		
Object	Obs.	Exp.
$\not{E}_T > 25$ GeV, $|\Delta\phi_{\not{E}_T - \text{jet}}| > 10°$	2	0.5 ± 0.1
$N_{\text{Jet}} \geq 3$, $E_T^{\text{Jet}} > 10$ GeV, $|\eta^{\text{Jet}}| < 2.0$	0	1.7 ± 1.5
Central e or μ, $E_T^{e \text{ or } \mu} > 25$ GeV	1	0.1 ± 0.1
Central τ, $E_T^\tau > 25$ GeV	0	0.03 ± 0.03
b-tag, $E_T^b > 25$ GeV	0	0.1 ± 0.1
Central γ, $E_T^{\gamma 3} > 25$ GeV	0	0.01 ± 0.01

TABLE 1. Number of observed and expected $\gamma\gamma$ events with additional objects at CDF in 85 pb^{-1}

A Light Gravitino LSP

In the framework of gauge mediated models in the MSSM the gravitino could be the LSP [10]. For most of the parameter space within these models, the lightest neutralino decays to a photon and gravitino: $\tilde{\chi}_1^0 \to \gamma \tilde{G}$. Therefore, any pair produced sparticles will yield a pair of photons and $\not{E}_T$.

B Search for events with $\gamma\gamma + \not{E}_T$

CDF has performed systematic searches for events with 2 photons and any additional object, $\gamma\gamma X$ [7]. The results of these searches are summarized in Table 1 for two photon energy thresholds - 12 and 25 GeV. The number of observed events is in agreement with that expected from SM sources.

In the light gravitino scenario, M_2 (SU(2) group mass parameter) is a parameter which together with $\tan\beta$ and the sign of μ controls the gaugino masses. The masses of the lightest chargino and the second lightest neutralino are approximately equal to M_2. For most of parameter space pair production of sparticles (dominated by gaugino pair production) eventually leads to final states containing $\gamma\gamma + \not{E}_T$. The results of the counting experiment in the $\gamma\gamma \not{E}_T$ channel with a $\not{E}_T$ cut at 35 GeV were used to set limits on the light gravitino models. 1 event passes all cuts with expected background of 0.5 ± 0.1 events.

Figure 3a) shows the distribution of $\not{E}_T$ before the final $\not{E}_T$ cut in the CDF $\gamma\gamma \not{E}_T$ sample. In Figure 3b) we present the excluded region of M_2 vs $\tan\beta$ plane

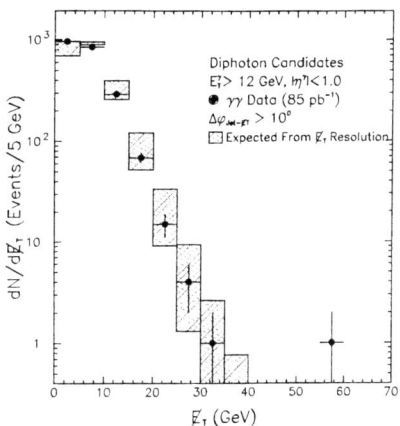 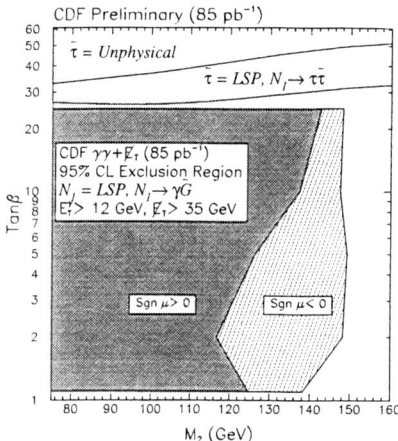

FIGURE 3. a) Distribution of $\not{E}_T$ before the final $\not{E}_T$ cut in the CDF $\gamma\gamma\not{E}_T$ sample. b) CDF excluded region of M_2 vs $\tan\beta$ plane for the positive and negative signs of μ.

for the positive and negative signs of μ.
See also [8] for $\gamma\gamma + \not{E}_T$ results from D0.

C Higgsino LSP

Another way to produce photon enriched signatures is to have a photino-like $\tilde{\chi}_2^0$ and Higgsino-like $\tilde{\chi}_1^0$. In this class of models $\tilde{\chi}_2^0 \to \tilde{\chi}_1^0 + \gamma$ yielding one or more photons in the final state.

D Search for events with $\gamma + \not{E}_T$

Processes with production of Higgsino LSP

$$p\bar{p} \to (\tilde{q}, \tilde{g}, \tilde{\chi}_2^0) \to \tilde{\chi}_2^0 + X \to \gamma\tilde{\chi}_1^0 + X$$

will result in $\gamma + \not{E}_T$ events with multijets [9].

D0 analysed 99 pb^{-1} of data looking for a photon with E$_T$ > 20 GeV, two or more hadronic jets with E$_T$ > 20 GeV and $\not{E}_T$ > 25 GeV [11]. 378 events were observed in the data. The number of data events agrees well with the expected background. The background is dominated by fakes (fake $\not{E}_T$ and a real or fake γ). The contribution from W/Z production is small.

The event selection was optimized in the $\not{E}_T$, H$_T$ plane ($\not{E}_T$ > 45 GeV, H$_T$ > 220 GeV). After final cuts 5 events have been found in data with 8.1 ± 5.8 background events expected. For $m(\tilde{q}) = m(\tilde{g}) = 300$ GeV/c^2 the efficiency for the signal predicted by the Higgsino LSP model is 21.5% and 11.3 events are expected. Figure 4a

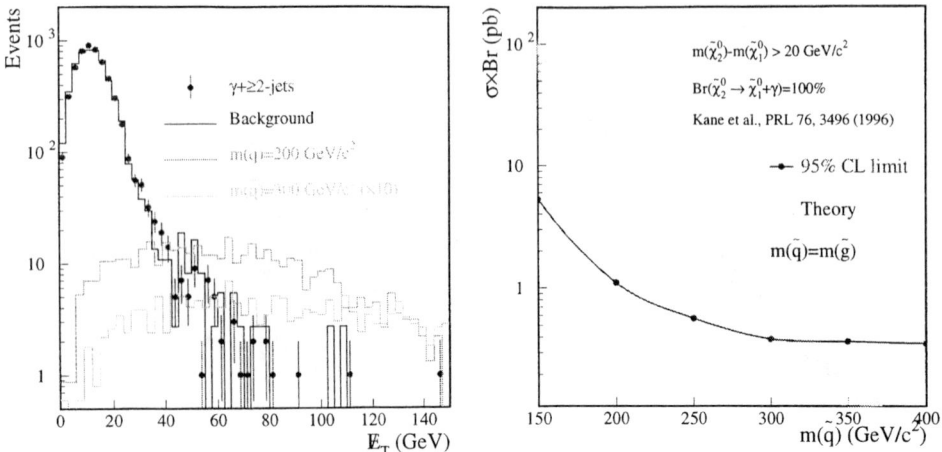

FIGURE 4. a) Distribution of $\not{E}_T$ before the final cuts. b) D0 95% CL limit on the cross section with overlayed theoretical prediction.

shows the distribution of $\not{E}_T$ before the final cuts. Figure 4b shows the 95% CL limit on the cross section with overlayed theoretical prediction.

E Search for events with $\gamma + b + \not{E}_T$

In association with the light stop (top superpartner) hypothesis the Higgsino LSP scenario yields a photon plus heavy flavor plus $\not{E}_T$ signature [12]:

$$p\bar{p} \to \tilde{\chi}_2^0 \tilde{\chi}_1^+, \ \tilde{\chi}_2^0 \to \gamma \tilde{\chi}_1^0, \ \tilde{\chi}_1^+ \to b\tilde{t}, \ \tilde{t} \to c\tilde{\chi}_1^0.$$

CDF has searched for events with an isolated photon $E_T > 25$ GeV and a b-tagged jet. The $\not{E}_T$ spectrum of these events is shown in Figure 5a. Overlayed are the background estimate and the expected signal distributions. For $\not{E}_T > 40$ GeV we find 2 events. Without performing background subtraction we set the 95% C.L. limit at 6.46 event level. We can exclude a region in squark and gluino mass plane (see Figure 5b) for the processes where squarks and gluinos cascade down to charginos and neutralinos.

V CONCLUSION

Searches for Supersymmetric signatures at the Tevatron are consistent with SM expectations. As a result a significant area of the parameter space of various SUSY models has been excluded.

The CDF and D0 collaborations are looking forward to Run II with the Main Injector and upgraded detectors. A factor of 20 increase in luminosity will allow us

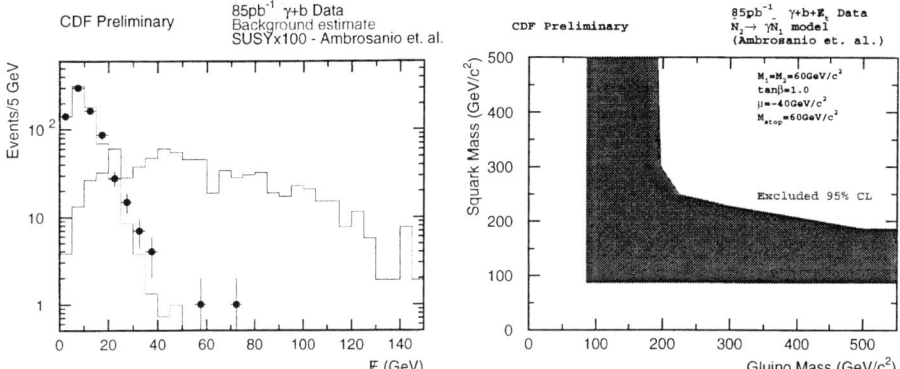

FIGURE 5. a) $\not{E}_T$ spectrum of the γ b events. Overlayed are the background estimate and the expected signal distribution. b) Excluded by CDF region in squark and gluino mass plane.

to probe deeper the Supersymmetric parameter space and either discover or rule out a wider class of models.

VI ACKNOWLEDGEMENTS

We thank the Fermilab staff and the technical staffs of the participating institutions for their vital contributions. This work was supported by the U.S. Department of Energy and National Science Foundation; the Istituto Nazionale di Fisica Nucleare (Italy); the Ministry of Education, Science and Culture (Japan); the Natural Sciences and Engineering Research Council (Canada); the National Science Council (China); the A. P. Sloan Foundation; the Swiss National Science Foundation; the Commissariat a L'Energie (France); the State Committee for Science and Technology and Ministry for Atomic Energy (Russia); CAPES and CNPq (Brasil); the Department of Atomic Energy and Science and Education (India); Colciencias (Colombia); CONACyT (Mexico); the Ministry of Education and KOSEF (Korea); and CONICET and UBACyT (Argentina).

REFERENCES

1. See for example: L. Susskind, *Phys. Rev.* D **20**, 2619 (1979).
2. For reviews of SuSy see H. P. Niles, *Phys. Rep.* **110**, 1 (1984); H. E. Haber and G. L. Kane, *Phys. Rep.* **117**, 75 (1985).
3. See for example: R. Arnowitt and P. Nath, *Phys. Rev. Lett.* **69**, 725 (1992); G. L. Kane, *Phys. Rev.* D **49**, 6173 (1994).
4. S. Abachi *et al*, FERMILAB-CONF-97/357-E.
5. H. Baer *et al*, *Phys. Rev.* D **48**, 2978 (1993).

6. F. Abe *et al*, FERMILAB-PUB-98/084-E. Submitted to Phys. Rev. Lett. March 3, 1998.
7. F. Abe *et al*, FERMILAB-PUB-98/024-E. Submitted to Phys. Rev. Lett. January 16, 1998.
8. S. Abachi *et al*, *Phys. Rev. Lett.* **80**, 442 (1998).
9. G. L. Kane *et al*, *Phys. Rev. Lett.* **76**, 3498 (1996).
10. See for example: P. Fayet, *Phys. Rev.* B **70**, 461 (1977); P. Fayet, *Phys. Rev.* B **84**, 416 (1979); R. Casalbuoni *et al*, *Phys. Rev.* B **215**, 313 (1988); S. Dimopoulos *et al*, *Phys. Rev.* D **54**, 3283 (1996); S. Dimopoulos *et al*, *Nucl. Phys. Proc. Suppl.* A **52**, 38 (1997).
11. S. Abachi *et al*, FERMILAB-CONF-98/174-E.
12. S. Ambrosanio *et al*, *Phys. Rev.* D **54**, 5395 (1996); S. Mrenna and G. Kane *Phys. Rev. Lett.* **77**, 3502 (1996).

Higgs Boson Searches at LEP2

Ulrich Schwickerath[1]

Institut für Experimentelle Kernphysik, Engesserstraße 7, 76128 Karlsruhe, Germany

Abstract. In this talk preliminary results of Higgs boson searches at LEP2 up to centre-of-mass energies of 183 GeV have been summarised. The main topics are searches for the Minimal Standard Model (MSM) Higgs boson and searches for neutral Higgs bosons of the Minimal Supersymmetric Standard Model (MSSM). No indication for Higgs production has been found. The experiments have published their Higgs search results for $\sqrt{s}$=133 GeV up to $\sqrt{s}$=172 GeV. A preliminary combination of these results yields a mass limit on the Standard Model Higgs boson of 77.5 GeV/c^2 at 95% CL. During the winter conferences 1998 first results of the 1997 data taking at $\sqrt{s}$=183 GeV have been presented. Individual limits of the experiments on the MSM Higgs boson mass are around 85 GeV/c^2, while benchmark mass limits on the light scalar and the pseudoscalar Higgs bosons within the MSSM are between 70 GeV/c^2 and 75 GeV/c^2. Other searches cover the charged Higgs boson production and Higgs decays into invisible final states.

I INTRODUCTION

Precision tests of the Minimal Standard Model [1] show an excellent agreement of MSM predictions for electroweak observables with electroweak precision data. However, electroweak symmetry breaking requires (at least) one scalar Higgs boson to explain the origin of the particle masses by the Higgs mechanism. Up to now, no experimental evidence for Higgs boson production has been found. Although the MSM makes no direct prediction of the Higgs boson mass, predictions of electroweak data depend on the Higgs mass via radiative corrections. At one loop level, the gauge boson masses aquire loop corrections including top, bottom and Higgs particles, as shown in fig. 1 [2]. While top-corrections are proportional to M_t^2, Higgs corrections yield only a logarithmic dependence. However, with the precise electroweak data it is possible to restrict the range of the Higgs boson mass. Precision tests of the MSM including all available preliminary electroweak precision data [2] favour a light Higgs boson. The MSM predictions of the electroweak observables [2] have been calculated with the ZFITTER 5.1 package [3], including two loop electroweak corrections. From the χ^2 of the fit shown in fig. 2, an upper limit of 212 GeV/c^2 at 95% CL is derived. The preferred Higgs boson mass turns out

---

[1] e-mail: Ulrich.Schwickerath@Physik.Uni-Karlsruhe.de

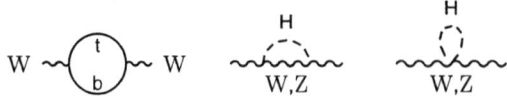

FIGURE 1. Radiative corrections to the gauge boson masses including top and Higgs particles in the MSM.

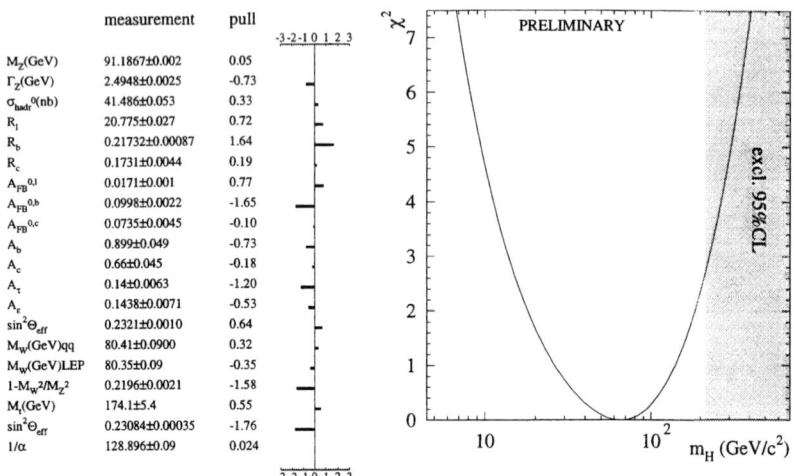

FIGURE 2. Results of fits of M_t, M_H, α_s and M_Z to electroweak precision data within the MSM as a function of the Higgs mass M_H. α was allowed to vary within its error in the fit. From the fits a 95% upper limit on the Higgs boson mass of 215 GeV can be derived.

to be $M_H = 65^{+74}_{-39}$ GeV/c^2, and the $\chi^2/d.o.f.$ of the fit is 16.0/15, corresponding to a fit probability of 38%. The measurement of the neutrino-nucleon neutral to charged currents ratio from [4] has a small dependence on the top and Higgs mass. This has been taken into accoount as in ref. [2]. Measurements, MSM predictions and pulls are given in fig 2. The pulls are defined as the difference between the measurement and the prediction, divided by the error.

Higgs boson searches are also important to test physics beyond the MSM. Supersymmetry has been discussed as a possible extension because it allows to solve many problems of the MSM simultanously. The Standard Model is contained in the limit of a large SUSY masses. As each MSM particle has a massive supersymmetric partner in these models, at least two Higgs doublets are needed, so that additional Higgs particles are predicted by the theory. The MSSM with two Higgs doublets has been shown to give an equally good description of precision data [8]. It has two

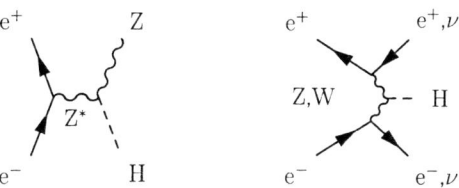

FIGURE 3. Diagrams for the production of the MSM Higgs boson in electron-positron annihilations at LEP2, Higgsstrahlung (left) and W, Z fusion (right).

scalar, one pseudoscalar and two charged Higgs bosons. At tree level, the masses of the scalar Higgs bosons $M_{h,H}$ and the pseudoscalar Higgs boson M_A are related by

$$M_{h,H}^2 = \frac{1}{2}\left[M_A^2 + M_Z^2 \mp \sqrt{(M_A^2 + M_Z^2)^2 - 4M_Z^2 M_A^2 \cos^2 2\beta}\right], \quad (1)$$

where $\tan\beta$ is the ratio of the vacuum expectation values v_2 and v_1. Eq. 1 implies an upper limit on the light scalar Higgs mass of $M_h < M_Z$ [5]. Radiative corrections are large and cannot be neglected. At one loop level the increases to 150 GeV, but second order corrections reduce this limit significantly [7].

For large M_A the light scalar Higgs boson behaves like the MSM Higgs, but the theoretical limit on the light Higgs boson mass remains. Thus, improved limits on the light scalar Higgs boson mass by direct searches at LEP2 limit the allowed parameter space of such models.

II MSM HIGGS BOSON PRODUCTION AND DECAYS

The most important process for the production of the Standard model Higgs boson at LEP is the Higgsstrahlung process in which a Higgs boson is produced in association with a Z boson. The Higgs can also be produced by fusion of Z or W bosons, as indicated in fig. 3, giving an additional small contribution to leptonic and missing energy final states [6].

At LEP the Higgs boson would either decays immediately into a pair of fermions or through triangular graphs into gluons or photons as indicated in fig. 4. As the coupling of the Higgs to fermions is proportional to the fermion mass[2], decays into heavy fermions dominate. For Higgs masses accessible at LEP2 energies, decays into gluons or photons are rare, and the Higgs decays mainly into b- quarks or τ- leptons. Table 1 lists the branching fractions for a Higgs mass of 85 GeV/c^2, estimated with the HZHA generator [5]. About 84% of the Higgs bosons decay into b-quarks. Fig. 4 shows the fractions of the resulting final states if the Higgs boson decays into a $b\bar{b}$ pair or a $\tau\tau$ pair for $M_H=85$ GeV. The fully hadronic channel, in

---

[2]) for quarks the running mass at scale M_H must be used

MSM Higgs M_H=85 GeV decays $H \to$	branching fraction (%)
$b\bar{b}$	84.1
$c\bar{c}$	3.8
$\tau^+\tau^-$	7.8
gg	4.1
$\gamma\gamma$	0.1
other	0.1

TABLE 1. Higgs branching fractions

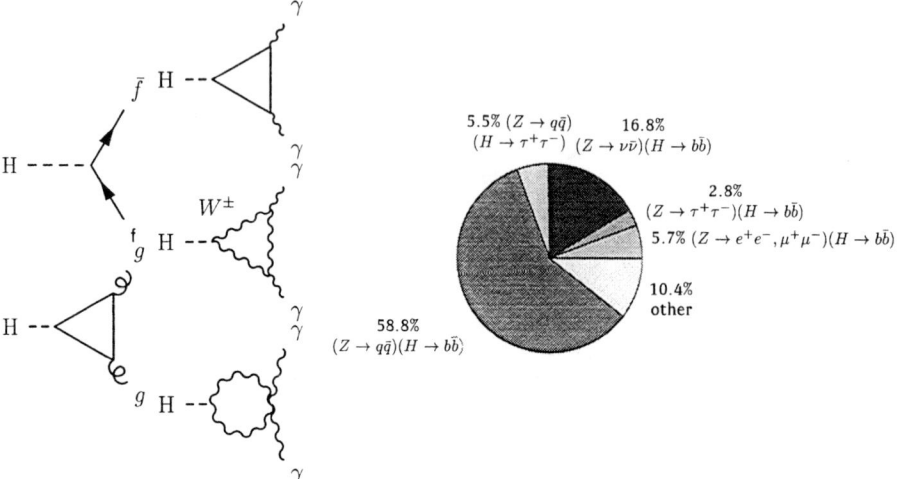

FIGURE 4. Diagrams contributing to MSM Higgs decays (left) and MSM Higgs boson decay channels for M_H=85 GeV (right).

which the Z decays into any allowed pair of quarks, is by far the most important one with a fraction of almost 60%. The topology of this kind of events is indicated in fig. 5. They are characterized by a large number of final state particles which typically form four jets. Furthermore, the events have a high b-content which allows to distinguish in particular from WW background.

In around 20% of the cases the Z decays into a neutrino pair, leaving a characteristic signature with two b-jets and a missing mass close to M_Z in the detector. This channel is called the missing energy channel. It's topology is indicated in fig. 5. The four-jet and the missing energy channel cover 75% of the Higgs final states. However, final states with leptonic decays of the Z are also experimentally accessible. Such events give final states with two isolated leptons having an invariant mass close to the Z-mass and two b-jets from the Higgs decay. The topology is indicated in fig. 5.

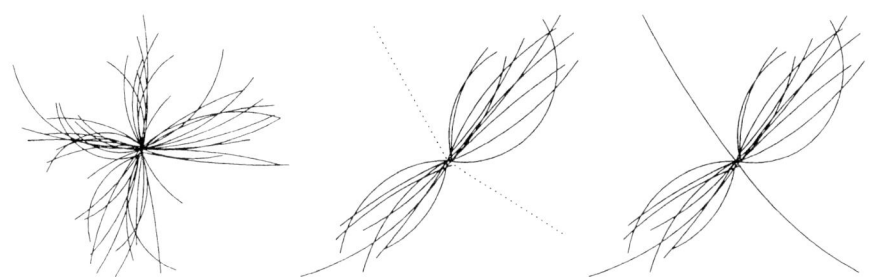

FIGURE 5. Characteristic signatures for Higgs production. The four-jet fully hadronic channel (left hand side), the missing energy channel (middle) and the leptonic channel (right hand side). In addition, all final states have large b-contents from the Higgs - boson decaying into b-quarks.

III SEARCH STRATEGIES

The search strategy depends on the characteristics of the particular final state. The following types of final states are considered:

- the four-jet channel ($hZ \rightarrow q\bar{q}q\bar{q}$),

- the missing energy channel ($hZ \rightarrow q\bar{q}\nu\bar{\nu}$),

- the leptonic channels ($hZ \rightarrow q\bar{q}l^+l^-$, l=$e,\mu$)

- final states containing τ leptons, ($hZ \rightarrow q\bar{q}\tau^+\tau^-$ and $hZ \rightarrow \tau^+\tau^-q\bar{q}$)

Due to the small production cross section of the signal a high signal selection efficiency and an efficient background rejection are required. As typically 84% of the Higgs bosons decay involve b-quarks, b-tagging is very important in Higgs searches. The b-tagging is mainly based on the long lifetime of b-hadrons which allows them to travel a significant distance before they decay. The decay leaves characteristic secondary vertices. This information is combined with additional information from the decay kinematics and allows an efficient tagging of b-quarks. The LEP experiments have optimised their analyses for Higgs boson decays into b-quarks. The invariant mass of the Z decay products can also be used for classification. In the leptonic channels and in the missing energy channel, the invariante mass of the lepton pair or the missing mass should be close to M_Z, respectively. In the four-jet final state, the pairing of the jets is not unique. With four jets in the final state there are six possibilities for the Z pair. However, the invariant mass of the jet pairs and jet b-tagging can be used to determine the pairing with the highest probability to come from the Z decay. The choice of the correct pairing is also important for a good mass reconstruction, because the candidate mass is used as additional information for setting limits.

Important backgrounds come from two and four fermion processes. QCD ($e^+e^- \rightarrow q\bar{q}(\gamma)$) has typically a two orders of magnitude larger cross section than

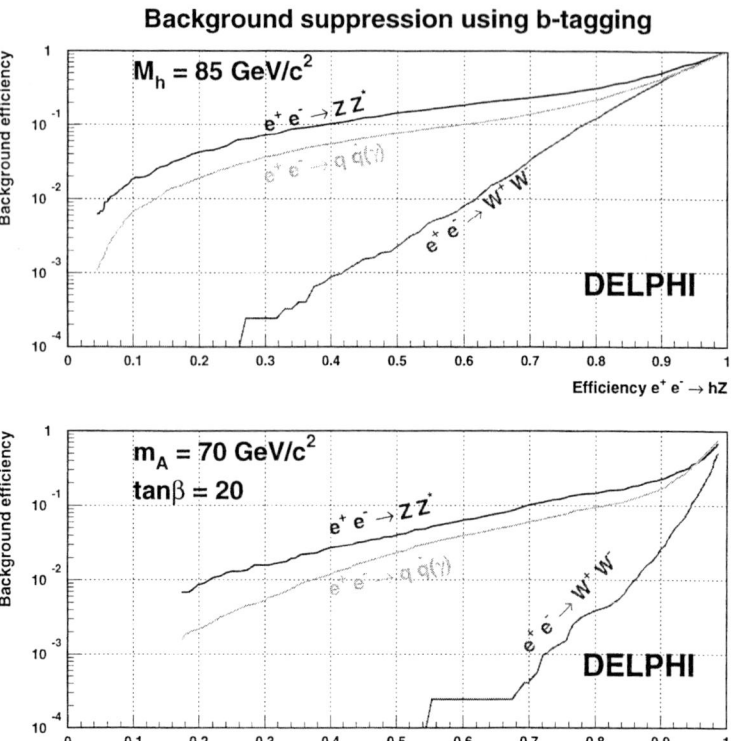

FIGURE 6. Background selection efficiency as a function of the hZ (upper plot) and hA (lower plot) selection efficiency obtained by b-tagging.

the signal processes. About two third of these events are radiative returns to the Z. Most of the QCD background can be reduced by preselection cuts on carefully chosen event shape variables, depending on the final state. WW and ZZ^* events can look very signal like. WW events are mainly rejected by b-tagging. It also reduces the ZZ^* and QCD background. For centre-of-mass energies above the ZZ threshold, an irreducible background from ZZ production arrises for Higgs boson masses close to M_Z.

IV LEP2 DATA TAKING

The world largest e^+e^- collider LEP is located at CERN close to Geneva in Switzerland. It has a circumference of 26.7 km. The four experiments ALEPH, DELPHI, L3 and OPAL and started operating in 1989. In the first phase LEP operated at the Z peak. The second phase of LEP started in 1995 with increased centre-

Year	Energy $\sqrt{s}$[GeV]	approx. integrated Luminosity (pb^{-1}/experiment.)
1995	130-136	6
1996	161,172	10
1997	183,130-136	55, 6
1998	188 GeV-189 GeV	goal:150

TABLE 2. LEP2 running periods and approximate collected luminosities per experiment

of-mass energies of 130 GeV and 136 GeV. Each experiment accumulated about 3 pb^{-1} at both energies. In the following year the energy was increased and each experiment accumulated around 10 pb^{-1} at the WW threshold of $\sqrt{s}$ =161 GeV and at 172 GeV. The largest amount of data was accumulated in 1997 when LEP operated at a centre-of-mass energy of about 183 GeV. Each experiment accumulated around 55 pb^{-1}. The analysis of these data is not finalized yet, but first preliminary results have been presented at the Winter conference 1998 in Moriond. For 1998, a centre-of-mass energy of 189 GeV is forseen. Data taking started on 25 Mai 1998. An integrated luminosity of about 150 pb^{-1} per experiment should be possible. The LEP2 running periods are summarised in table 2.

V   COMBINED RESULTS UP TO 172 GEV

Current limits are based on the combination of published results up to centre-of-mass energies of 172 GeV. Four different methods have been considered to perform this limit. Method A and B are based on the probability or likelihood function for the signal+background hypothesis and differ by the definition of the test statistics. Method C is a fractional event counting method. These methods take into account the mass distribution of the events. With method D it is possible to combine existing limits. More details can be found in reference 3. Both the expected and the observed limits have been calculated. They are summarised in table 3. All methods give compatible results. Finally, the most conservative number has been chosen. MSM Higgs boson masses below 77.5 GeV/c^2 can be excluded at 95% CL.

VI   PRELIMINARY RESULTS FROM 1997 DATA TAKING

Different techniques have been applied by the experiments to separate signal from background [11–15]. In a first step events are preselected keeping a high signal selection efficiency. For the classification of the remaining events different techniques are used. Most of the experiments made independent analyses with

	Statistical method							
	A		B		C		D	
Experiment	exp.	obs.	exp.	obs.	exp.	obs.	exp.	obs.
ALEPH	68.5	69.6	68.8	69.6	68.6	69.6	68.5	69.6
DELPHI	65.4	65.9	65.3	65.9	65.1	65.5	-	-
L3	66.1	69.4	65.7	69.3	65.0	69.2	-	-
OPAL	65.9	69.0	65.6	68.6	65.3	68.9	-	-
LEP	75.8	77.5	76.0	77.8	75.6	77.7	75.7	77.9

TABLE 3. Expected (left) and observed (right) combined limits on M_H up to $\sqrt{s}$=172 GeV for different combination methods. The most conservative choice yields a limit of 77.5 GeV. The method proposed by ALEPH (D) does not recalculate individual limits, therefore there are no entries for DELPHI, L3 and OPAL in the last column.

different methods to cross-check their results. In the four-jet and missing energy channel ALEPH combined the results of a sequentiel cuts and a Neural network analysis [12]. In the τ channels they give the results of the Neural Network analysis only. Other leptonic channels are covered by cut based analyses. DELPHI prefers sequential analyses in the four-jet and the leptonic final states [13]. The missing energy channel analysis is based on an iterated second order discriminant analysis. The L3 collaboration [14] also uses Neural Networks to discriminate signal and background in the four-jet, the missing energy and the electron and muon channels. No cut on the neural network output is done in the four-jet and missing energy channels, but the full shape is used to obtain the limit. OPAL [15] calculates likelihood ratios in the four-jet, the missing energy and the electron and muon channels. L3, DELPHI and OPAL presented their results of cut based analyses in the τ channels. The search results in the individual channels are summarised in table 4. Table 4 also summarises the number of expected and the number of observed events. The number of obsevered events is in good agreement with the expected background from MSM processes. No indication for Higgs production is found. Individual limits are listed in table 5. The combined limit is expected to be around 90 GeV [11].

VII NEUTRAL MSSM HIGGS BOSON SEARCHES

A Scalar MSSM Higgs Bosons

The scalar Higgs bosons of the MSSM can be produced in a Higgsstrahlungs process in association with a Z boson. This process is similar to the MSM Higgs boson production, see fig. 3, but the couplings of the light and heavy scalar Higgs to the Z are reduced by a factor $\sin(\beta - \alpha)$ or $\cos(\beta - \alpha)$, respectively. α is the

Channel	Background	Data	Signal eff. (%) M_h=80 GeV	Signal eff. (%) M_h=85 GeV
ALEPH				
Hq$\bar{q}$	4.45	4		39.2
H$\nu\bar{\nu}$	0.41	0		32.3
H(e^+e^-, $\mu^+\mu^-$)	2.0	3		76.2
q$\bar{q}\tau^+\tau^-$	0.17	0		19.7
(H→$\tau^+\tau^-$)q$\bar{q}$	0.16	0		13.7
Total	7.19	7		
DELPHI				
Hq$\bar{q}$	5.34	4		40.1
H$\nu\bar{\nu}$	0.58	1		30.1
He^+e^-	0.66	1		39.9
H$\mu^+\mu^-$	0.44	2		62.4
(H→q$\bar{q}$)$\tau^+\tau^-$	0.41	0		20.0
(H→$\tau^+\tau^-$)q$\bar{q}$	0.74	1		20.0
Total	8.2	9		
L3				
(H→q$\bar{q}$)q$\bar{q}$	315*	321*	65*	66*
(H→q$\bar{q}$)$\nu\bar{\nu}$	50.4*	56*	76*	75*
(H→q$\bar{q}$)e^+e^-	5.4	6	78	80
(H→q$\bar{q}$)$\mu^+\mu^-$	1.4	2	44	45
(H→q$\bar{q}$)$\tau^+\tau^-$	0.7	0	15	14
(H→$\tau^+\tau^-$)q$\bar{q}$	1.7	1	19	17
Total	10.6	11		
OPAL				
(H→b$\bar{b}$)q$\bar{q}$	6.06	6	38.2	
(H→b$\bar{b}$)$\nu\bar{\nu}$	1.35	0	40.9	
He^+e^- H$\mu^+\mu^-$	0.72	1	58.0, 61.6	
(H → q$\bar{q}$)$\tau^+\tau^-$, (H → $\tau^+\tau^-$)q$\bar{q}$	3.30	1	25.2	
Total	11.4	8		
LEP	37.4	37		

TABLE 4. Preliminary results of MSM Higgs boson searches at 183 GeV. For L3, the number of events entering in the Neural Network analysis are given, indicated with a star. For the total number of events only the most significant candidates are counted.

Experiment.	luminosity pb^{-1}	expected limit GeV/c^2	observed limit GeV/c^2
ALEPH	57	85.5	87.9
DELPHI	53.95	86.3	84.4
L3	55	86.7	87.6
OPAL	55.3	86.5	84.2

TABLE 5. Preliminary limits on M_H including the 1997 data taking at 183 GeV

Higgs mixing angle. Couplings are different for up and down type fermions and depend on α. Taking this into account, the MSM Higgs search results can be also applied for the light scalar MSSM Higgs.

B Associated Production of Neutral Higgs Bosons

In addition, a scalar and the pseudoscalar Higgs boson can be produced in association if kinematically allowed. This process is indicated in fig. 7. The coupling to the Z is proportional to $\cos(\beta - \alpha)$. Typical decay channel fractions are indicated in fig. 7. The most important search topology is the fully hadronic channel, $hA \to b\bar{b}b\bar{b}$. The semileptonic final states $hA \to b\bar{b}\tau^+\tau^-$ and $\tau^+\tau^- b\bar{b}$ also play a role. Due to the large b-content in the fully hadronic channel, b-tagging variables are the most important discriminating quantities, see also fig. 6.

The ALEPH analysis is based on the combination of a Neural Network and a cut based analysis, DELPHI uses a Neural Network for the four-jet channel and cut selections in the τ channels. L3 makes a three step analysis: After a preselection, the events are passed through a series of automatically tuned cuts [9]. Then a discriminating variable is calculated which is used in the confidence level calculation. OPAL uses a likelihood analysis in the four-jet channel.

In some regions of the parameter space of the MSSM, the light scalar Higgs has more than twice the mass of the pseudoscalar Higgs boson. In these regions, the scalar Higgs may decay into a pair of pseudoscalar Higgs bosons, giving final states with six b-quarks. Due to the different topology, the sensitivity of the standard four-jet analyses can be significantly reduced which is the case for very light pseudoscalar Higgs boson masses.

C MSSM Benchmark Limits

The Higgs sector of the model can be parameterized by M_A and $\tan \beta$. As already explained, radiative corrections are large and depend on the parameters of the model. The superpartners of left and right handed fermions mix to give the physical sfermion states. As the mixing is proportional to the SM particle

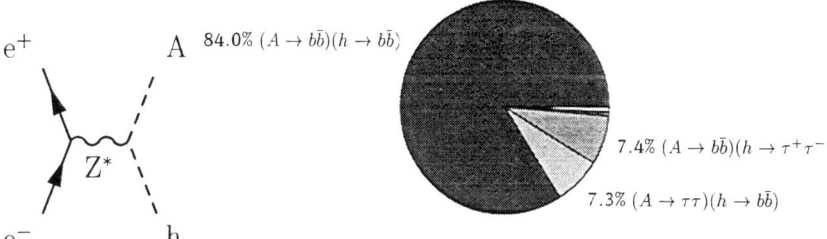

FIGURE 7. Associated production of two Higgs bosons (left) and fractions of final states (right) of the pseudoscalar Higgs boson in the MSSM. Decay fractions are given for $M_h = M_A = 70$ GeV, $\tan\beta = 20$.

Channel	ALEPH			DELPHI			L3				OPAL		
	bkg	obs	ϵ_{hA} M_A=75	bkg	obs	ϵ_{hA} M_A=70	bkg	obs	ϵ_{hA} M_A=70	ϵ_{hZ} M_h=85	bkg	obs	ϵ_{hA} M_A=70
$b\bar{b}b\bar{b}$	2.39	2	60.3	1.4	0	55	380.5	376	78	84	6.3	7	50.9
$b\bar{b}\tau^+\tau^-$	0.07	0	28.6	0.45	0	22.6	14.0	17	32	27	-	-	-
$\tau^+\tau^-q\bar{q}$	-	-	-	-	-	-	-	-	-	30	-	-	-
total	2.46	2		1.9	0		-	-			6.3	7	

TABLE 6. Preliminary results on the search for associated production of a scalar and a pseudoscalar Higgs boson, $e^+e^- \to hA$. Efficiencies are given in per cent and masses in GeV/c^2. L3 numbers are given before the last step of their analysis, see text. The L3 analysis for $e^+e^- \to hZ$ differs from their MSM analysis, thus the results of this analysis have been added in this table.

preliminary benchmark MSSM mass limits				
	ALEPH	DELPHI	L3	OPAL
M_h (GeV/c^2)	72.2	74.4	70.7	70.0
M_A (GeV/c^2)	76.1	75.2	71.0	70.5

TABLE 7. Preliminary MSSM benchmark limits on M_h and M_A

mass, stop mixing must be taken into account. The experiments use the following benchmarks as proposed in [5] to convert the search results in the MSSM into mass limits:

- Top mass $M_t = 175$ GeV
- common squark mass scale $M_{sq} = 1000$ GeV
- gaugino mass $M_g = 1000$ GeV
- consider three stop mixing scenarios:
    - minimal mixing, $A_t = 0$ and $\mu = -100$ GeV
    - typical mixing, $A_t = M_{sq}$ and $\mu = 1000$ GeV
    - maximal mixing, $A_t = \sqrt{6}M_{sq}$ and $\mu = -100$ GeV

Benchmark limits are obtained in the $\tan\beta$-M_A and the $\tan\beta$-M_h planes. Figure 8 shows preliminary results of the DELPHI collaboration. The search results for lower centre-of-mass energies are included in this figure. In the no-mixing scenario a lower limit of $\tan\beta = 1.7$ can be derived. The results are summarised in table 7.

More sophisticated scans have been performed based on 1996 data [16]. Similar scans have been performed by the OPAL collaboration [15]. In the latter analysis unification of sfermion masses and gaugino masses is assumed at a large scale M_{GUT}. The corresponding unified masses M_0 and $M_{1/2}$ are varied up to 1 TeV and the top-quark mass is varied between 165 GeV and 185 GeV. As a consequence, the theoretically excluded regions are smaller, and no limit can be set on $\tan\beta$. In addition, a non-excluded island appears in the region where $M_h > 2M_A$, and the limits in table 7 are significantly reduced. For $M_A = M_h$ the results are unchanged.

VIII CHARGED HIGGS BOSON PRODUCTION

There is also the possibility of associated production of two charged Higgs bosons in the general two doublet model. The final states which have been investigated are listed in table 8. Each Higgs may decay into a quark or a lepton pair. As no b-quarks are produced, the b-tagging can only be applied as a veto. Thus, these events are much more difficult to distinguish from possible background processes, and the limits are much weaker than the limits on neutral Higgs bosons.

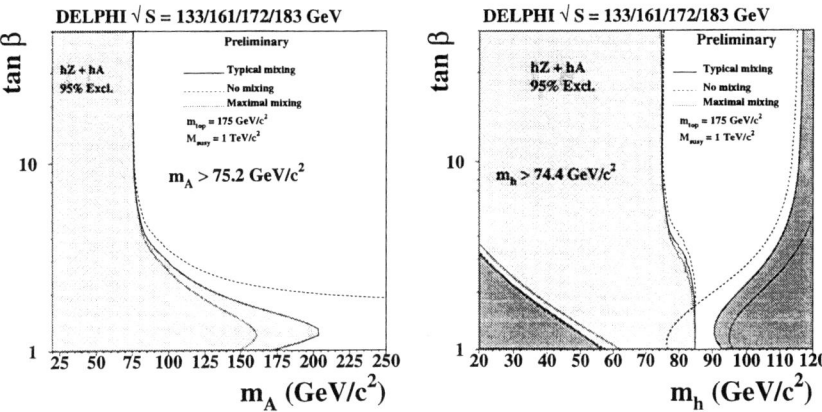

FIGURE 8. Preliminary MSSM results from the DELPHI collaboration for the benchmark parameter settings. The dark shaded areas are theoretically inaccessible.

channel	signature
leptonic	τ pair
($\tau\nu\tau\nu$)	high missing momentum
semileptonic	one τ, missing momentum
($\tau\nu c\bar{s}$)	two jets
hadronic	four-jets,
($c\bar{s}c\bar{s}$)	pairing with similar mass

TABLE 8. Final states of the process $e^+e^- \to H^+H^-$

Preliminary 95% CL mass limits are:

ALEPH	DELPHI	OPAL
52 GeV/c^2	56.7 GeV/c^2	52 GeV/c^2

IX   INVISIBLE HIGGS DECAYS

In some models the Higgs boson can also decay into invisible final states like Majorons or neutralinos [6]. In this case, only the decay products of the Z boson are visible in the detector, leaving a characteristic signature with visible mass close to M_Z and large missing energy. Under the assumption that the branching ratio for h,A→ invisible is 100%, ALEPH and L3 give mass limits of 69.6 GeV/c^2 and 71.3 GeV/c^2, respectively.

X SUMMARY

Much progress on Higgs boson mass limits has been obtained due to the increase of the centre-of-mass energy at LEP2. The combination of MSM Higgs boson limits yields $M_H >77.5$ GeV/c^2 at 95% CL. The analysis of around 55 pb^{-1}/experiment of data taken in 1997 at $\sqrt{s} \approx 183$ GeV is almost finalised. Preliminary individual 95% CL limits are around $M_H >85$ GeV/c^2. After combination a limit around 90 GeV/c^2 can be expected. Preliminary individual benchmark limits on neutral MSSM Higgs bosons vary between 70 GeV/c^2 and 76 GeV/c^2. Charged Higgs boson limits in the general two doublet model are between 52 GeV/c^2 and 56 GeV/c^2. ALEPH and L3 have also given preliminary limits on invisible Higgs decay channels which are around 70 GeV/c^2.

ACKNOWLEDGEMENT

I want to thank the organisers of the workshop for the pleasant and creative atmosphere and the LEP collaborations for providing material for the talk.

REFERENCES

1. The LEP Collaborations ALEPH, DELPHI, L3, OPAL and the LEP Electroweak Working Group, CERN-PPE/97-154 (1997).
2. C. Paus, Talk given on LEPC - meeting 31.3.1998, Geneva; LEPEWWG 98-01 internal note; C. Paus, priv. com.
3. D. Bardin et al., Z. Phys. **C44** (1989) 493; Comp. Phys. Comm. **59** (1990) 303; Nucl. Phys. **B351**(1991) 1; Phys. Lett. **B255** (1991) 290 and CERN-TH 6443/92 (May 1992).
4. NuTeV Collaboration, K. McFarland, talk presented at the XXXIIIth Rencontres de Moriond, Les Arcs, France, 15-21 March, 1998.
5. G. Altarelli, T.Sjöstrand, F.Zwirner, Physics at LEP2, Vol. 2, CERN 96-01.
6. J.F.Gunion, H.E. Haber, G.Kane, S.Dawson, The Higgs Hunter's Guide, Frontiers in Physics Series (Volume No. 80), Addison-Wesley Publishing Company.
7. S. Heinemeyer, W. Hollik, G. Weiglein, KA-TP-06-1998, hep-ph/9806250 and references therein.
8. U. Schwickerath, WIN97, Capri, June 1997, to appear in the proceedings; W. de Boer, A. Dabelstein, W. Hollik, W. Mösle, U. Schwickerath, Z.Phys.**C75 627-640**,1997; W. de Boer et al., hep-ph/9603350; W. de Boer et al., Z. Phys. C67 , (1995) 647-664; W. de Boer et al., hep-ph/9603346 and references therein.
9. ALEPH Collaboration, R. Barate et al., Phys. Lett.**B412** (1997) 155.; DELPHI Collaboration, P. Abreu et al., Eur. Phys. Journ. C2 (1998) 1.; L3 Collaboration, M. Acciarri et al., Phys. Lett.**B411** (1997) 373.; OPAL Collaboration, K. Ackerstaff et al., Eur. Phys. Journ. C1 (1998) 425.

10. The LEP working group for Higgs boson searches; ALEPH, DELPHI, L3 and OPAL Collaborations, Lower bound for the Standard Model Higgs boson mass from combining the results of the four LEP experiments, **CERN-EP/98-046**.
11. S. de Jong, talk given at the XXXIIInd Rencontres de Moriond, Les Arcs, 14-21 March, 1998; V. Ruhlman-Kleider, talk given at the XXXIIInd Rencontres de Moriond, Les Arcs, 14-21 March, 1998.
12. J. Carr, Talk given at LEPC meeting for the ALEPH Collaboration, 31.3.1998, Geneva; ALEPH Collaboration, R. Barate et al., ALEPH 98-029 CONF (CONFerence) 98-017.
13. K. Mönig, Talk given at LEPC meeting for the DELPHI Collaboration, 31.3.1998, Geneva.
14. J. Mnich, Talk given at LEPC meeting for the L3 Collaboration, 31.3.1998, Geneva; L3 Collaboration, M. Acciarri et al., **CERN-EP/98-52**; L3 Collaboration, M. Acciarri et al., **CERN-EP/98-72**.
15. M. Thomson, Talk given at LEPC meeting for the OPAL Collaboration, 31.3.1998, Geneva; OPAL Collaboration, K. Ackerstaff et al., **CERN-EP/98-029**.
16. A. Sopczak, Presented at 1st International Workshop on Nonaccelerator New Physics (NANP 97), Dubna, 7-11 Jul 1997, EKP-KA-97-14, hep-ph/9712283.

Standard Model Tests at very high Q^2 at HERA

G. Eckerlin
on behalf of the H1 and ZEUS Collaborations

DESY, Hamburg
Notkestr. 85, 22607 Hamburg, Germany

Abstract. With the e-p collider HERA at DESY tests of the Standard Model at highest available Q^2 have been performed. Recent results on the cross section for deep inelastic e-p scattering up to $Q^2 = 40000 \text{GeV}^2$ are presented. An analysis of events with high missing transverse momentum p_t and an isolated lepton is discussed. Limits on the production of leptoquarks, excited fermions and on Contact Interactions are given.

INTRODUCTION

With the e-p collider HERA at DESY electron proton scattering processes [1] at a maximum center of mass energy $\sqrt{s} = 300 \text{GeV}$ offer the possibility to study deep inelastic scattering (DIS) in charged and neutral current processes up to a momentum transfer of $Q^2 = 90000 \text{GeV}^2$.

With the large data sample from 1997 both collaborations H1 and ZEUS have refined their analyses and measured cross sections up to $Q^2 = 40000 \text{GeV}^2$. This extends previous measurements of fixed target experiments by 2 orders of magnitude in Q^2. A new domain in DIS has been explored and searches for possible deviations from the Standard Model (like Leptoquarks or Contact Interactions) have been performed. With increased luminosity new limits on excited fermions have been derived extending beyond the LEP mass range. A detailed description of the H1 and ZEUS experiments can be found in [1] and [2] respectively.

DIS

At HERA deep inelastic scattering of 27.5GeV positrons on 820GeV protons is studied and cross sections are measured over more than 4 orders of magnitude both

---

[1] Although HERA has run with positrons instead of electrons from 1994 until 1997 throughout the paper the name electron is used for e^+ as well.

in Q^2 and x. The momentum transfer $Q^2 = -q^2 = -(k-k')^2$ is derived from k and k' the four vectors of the initial and final lepton respectively. In addition further Lorentz-invariants can be defined:
the Bjorken scaling variable:

$$x = \frac{Q^2}{2Pq} \qquad (1)$$

and the fractional energy transfer to the proton in its rest frame:

$$y = \frac{Pq}{Pk}, \qquad (2)$$

where P is the 4-momentum of the incoming proton. In the quark parton model the mass in the electron quark system is given by $M = \sqrt{sx}$, with s being the ep invariant mass squared. At a given s two of the Lorentz-invariants are needed to fully characterize the kinematics of the ep scattering.

In the data collected during 1994 - 1996 both H1 and ZEUS reported an access of DIS events at highest Q^2 and high y [3,4]. With the data recorded in 1997 the statistics available is more than doubled and an extensive analysis of the cross sections were performed.

Neutral Current Cross Sections

Neutral current DIS mediated by the neutral bosons γ or Z show a scattered electron and a hadronic final state. The events are expected to be balanced in transverse momentum. The kinematics can be derived from 4 measurable quantities, the energy E_e and the polar angle θ_e of the scattered electron, as well as the energy E_h and the angle γ_h of the hadronic system [5]. The H1 analysis [3] uses the electron quantities to derive the kinematics (electron method), whereas ZEUS [4] uses the two measured angles (double angle method). Both experiments used the alternative method as a cross check.

The differential cross section is given by :

$$\frac{d^2\sigma}{dx dQ^2} = \frac{2\pi\alpha^2}{xQ^4}[Y_+ F_2(x,Q^2) - y^2 F_L(x,Q^2) - Y_- x F_3(x,Q^2)] \qquad (3)$$

with $Y_\pm = 1 \pm (1-y^2)$.

Fig.1 shows the single differential cross section $d\sigma/dQ^2$ measured by the ZEUS collaboration using the data from 1994 to 1997. All points agree with the Standard Model (SM) predictions within error, with the exception of the highest bin ($Q^2 > 35000 \text{GeV}^2$) which contains two events where 0.29 are expected.

The H1 collaboration measured the double differential cross section. The results are presented in the form of the reduced cross section

$$\sigma(e^+p) = \frac{xQ^4}{2\pi\alpha^2} \frac{1}{Y_+} \frac{d^2\sigma}{dxdQ^2} \tag{4}$$

in figure 2 for 6 different x ranges compared with the standard model expectation based on the MRSH evolution equations [6]. Also shown are the low Q^2 data from SLAC [7], BCDMS [8] and NMC [9]. The low Q^2 data join well the high Q^2 cross section. Both the standard model expectation based on MRSH parton distributions as well as a next to leading order H1 fit describe the observed cross section over 4 orders of magnitude for all ranges in x. Only for $x = 0.45$ and $Q^2 > 15000 \text{GeV}^2$ the data points are above the expectation, however with less significance than for the 1994 - 1996 data alone. Table 1 gives the number of observed events seen by H1 above $Q^2 = 15000 \text{GeV}^2$ for $0.1 < y < 0.9$, comparing the results from the pre 1997 data with the full statistics up to 1997.

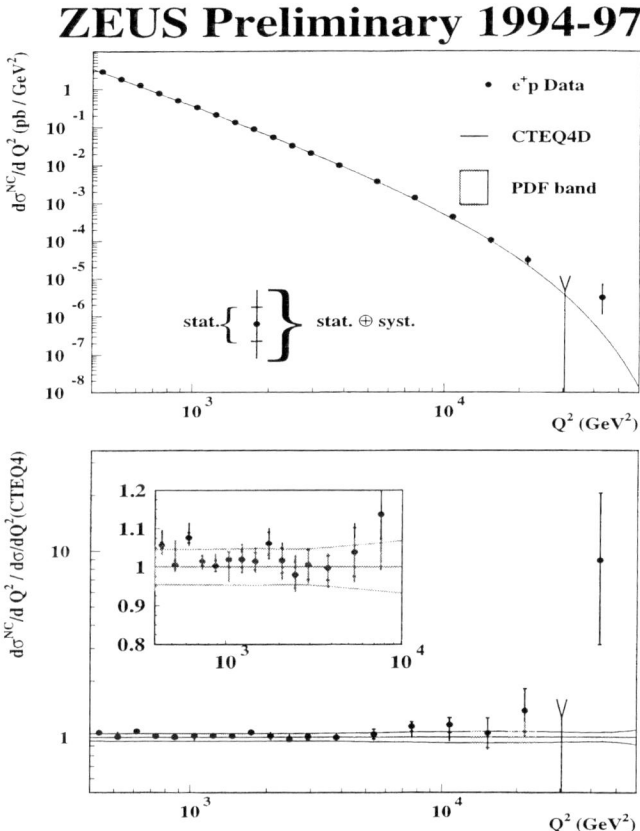

FIGURE 1. The e^+p neutral current cross section $\frac{d\sigma}{dQ^2}$ measured by ZEUS (upper figure) and the ratio data/SM (lower figure) is shown. The Standard Model expectation shown is based on the CTEQ4D parametrisation of the parton density function.

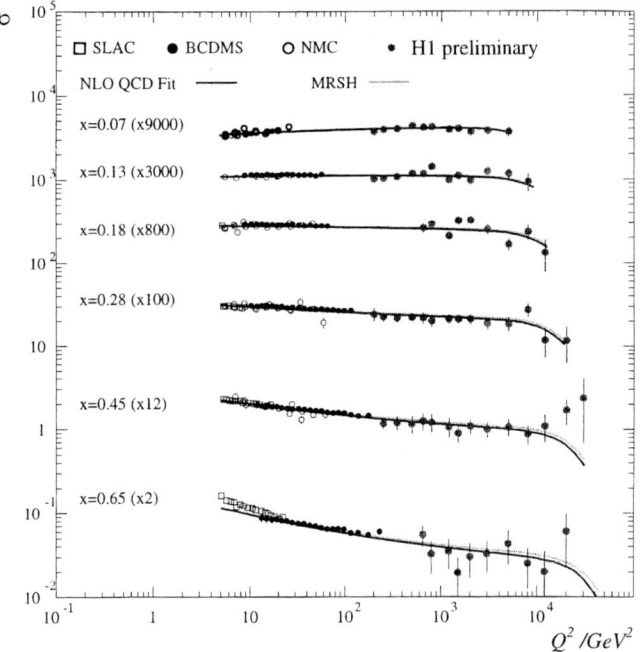

FIGURE 2. The e^+p neutral current reduced cross section (see text) measured by H1 is shown together with low Q^2 data from SLAC, BCDMS and NMC. The MRSH based Standard Model expectation and a next to leading order QCD Fit from H1 are shown as curves.

TABLE 1. Number of events above $Q^2 > 15000 \text{GeV}^2$ measured by H1 and the Standard Model expectation in comparison for the data up to 1996 and the full data sample

NC Events with Q^2 > 15000GeV^2	1994 - 1996 data ($14pb^{-1}$)	1994 - 1997 data ($37pb^{-1}$)
N observed	12	22
N expected	4.71 ± 0.76	14.8 ± 2.13
Probability	6×10^{-3}	5.9×10^{-2}

Limits on Leptoquarks

H1 analysed the neutral current data sample in terms of Leptoquarks by comparing the event rate at various M bins with the Standard Model expectations, interpreting a possible deviation from the SM as a signal for Leptoquarks. Scalar leptoquarks can be produced as s-channel resonances at HERA, leading to a flat cross section $d\sigma/dy$ whereas the t-channel exchange in standard DIS leads to a $1/y^2$ dependence. By choosing a proper cutoff value y_{cut} for the data to be analysed, the signal to background ratio can be optimised. Figure 3 shows the mass spectrum

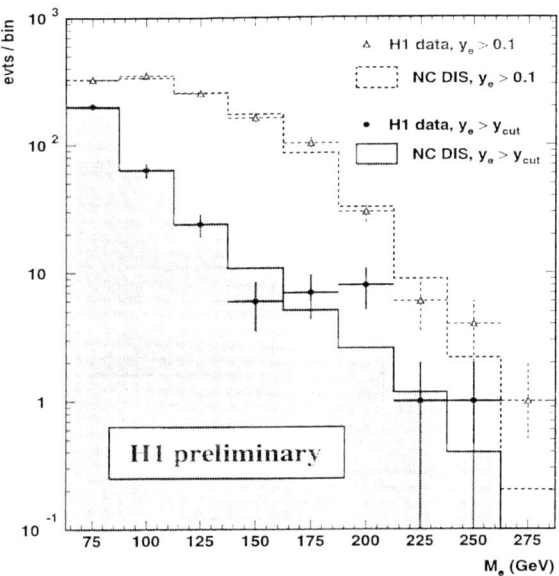

FIGURE 3. The Mass spectrum of the measured neutral current DIS events by H1 shown as triangles for the full data sample and as circles for the data with y_{cut} applied. y_{cut} varies from 0.6 at M=60GeV to 0.2 at M=250GeV. The Standard Model expectation is shown as a histogram.

for all NC data as well as for data with a mass dependent y_{cut} [10] applied. The standard model NC DIS expectation is shown for comparison. The excess around $M_e = 200$GeV seen for the data with $y > y_{cut}$ is mainly due to the events already published in [3] and is in this analysis considered to be a fluctuation. The analysis has three parameters, the unknown coupling λ of the leptoquark to the e-q system, the Mass of the leptoquark M and the branching fraction β of the leptoquark decaying into eq. To derive limits for leptoquarks two different approaches are used. Either the coupling λ is assumed to have a given value and the limit is derived in the $M - \beta$ plane, or β is fixed, taken from a specific leptoquark model (here Buchmüller, Rückel, Wyler [11]) and the limits are derived in the $M - \lambda$ plane. The limits for fixed λ are shown in figure 4. For $\beta = 1$ leptoquarks are excluded

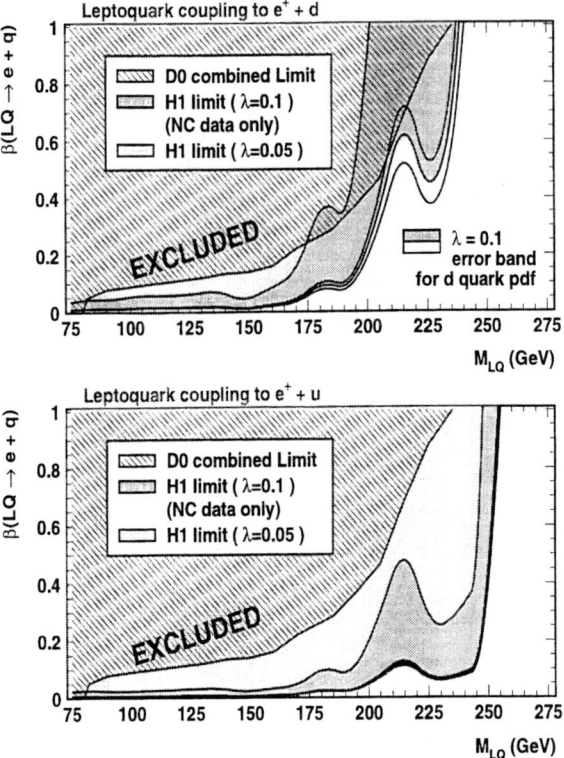

FIGURE 4. H1 limits on leptoquarks in the $\beta - M_{LQ}$ plane for 2 different λ values. The results for the leptoquarks coupling to the d-quark (upper figure) and for the u-quark (lower figure) are shown. The combined D0 limits are shown for comparison.

up to 250GeV at 95% CL, comparable to previous results from D0 [12]. But for $\beta \ll 1$ there is still a high discovery potential for leptoquarks even below 200GeV.

Contact Interaction

The ZEUS collaboration has analysed the NC-DIS data in terms of possible signs for Contact Interactions. Contact Interactions lead to additional terms in the Lagrangian as shown in equation 5.

$$L_{NC} = L_{SM} \\
+ \eta_{LL}(\bar{e}_L\gamma^\mu e_L)(\bar{q}_L\gamma^\mu q_L) + \eta_{LR}(\bar{e}_L\gamma^\mu e_L)(\bar{q}_R\gamma^\mu q_R) \\
+ \eta_{RL}(\bar{e}_R\gamma^\mu e_R)(\bar{q}_L\gamma^\mu q_L) + \eta_{RR}(\bar{e}_R\gamma^\mu e_R)(\bar{q}_R\gamma^\mu q_R)$$

(5)

TABLE 2. Parameter combinations considered in the Contact Interaction analysis by ZEUS

Label	η_{LL}^u	η_{LR}^u	η_{RL}^u	η_{RR}^u	η_{LL}^d	η_{LR}^d	η_{RL}^d	η_{RR}^d
VV	+a	+a	+a	+a	+a	+a	+a	+a
AA	+a	-a	-a	+a	+a	-a	-a	+a
VA	+a	-a	+a	-a	+a	-a	+a	-a
X1	+a	-a	0	0	+a	-a	0	0
X2	+a	0	+a	0	+a	0	+a	0
X3	+a	0	0	+a	+a	0	0	+a
X4	0	+a	+a	0	0	+a	+a	0
X5	0	+a	0	+a	0	+a	0	+a
X6	0	0	+a	-a	0	0	+a	-a
U5	0	+a	0	+a	0	0	0	0
U1	+a	0	-a	0	0	0	0	0
U4	0	+a	+a	0	0	0	0	0

with a = -1 or +1

TABLE 3. Limits on Λ for Contact Interactions from the ZEUS analysis

Label	a	Fit of Λ ZEUS	Limit on Λ ZEUS	Limit on Λ CDF	Limit on Λ OPAL	Limit on Λ Aleph
VV	+1	∞	4.9	3.5	5.8	3.6
VV	-1	> 10	4.6	5.2	5.8	4.6
AA	+1	2.5	2.0	3.8	5.6	5.0
AA	-1	> 10	4.0	4.8	2.5	3.3
X3	+1	∞	2.8	-	4.3	3.7
X3	-1	> 10	1.5	-	3.4	3.2
X4	+1	∞	4.5	-	4.3	2.7
X4	-1	> 10	4.1	-	5.4	4.3
U4	+1	∞	4.6	-	-	2.1
U4	-1	> 10	4.4	-	-	2.6

Here only the vector terms are taken into account and the convention $\eta = a4\pi/\Lambda^2$ with $a = \pm 1$ is used. A binned as well as an unbinned likelihood analysis has been performed considering the following constraints :

- From atomic parity violation results : $\eta_{LL}^q + \eta_{LR}^q - \eta_{RL}^q - \eta_{RR}^q = 0$
- SU(2) symmetry implies $\eta_{iL}^u = \eta_{iL}^d$
- At high x the cross section at HERA is dominated by up quarks

The possible combinations of the parameters considered are shown in table 2. The result of the fit is summarised in table 3. The ZEUS results are competitive with previously published results from the LEP and Tevatron experiments [13]. The upper Limit of the coupling Λ is in the range of 2-5 TeV.

Charged Current Cross Sections

Charged current DIS processes mediated by W bosons have due to the undetected neutrino ν only a hadronic system visible in the final state and hence an imbalance in the detected transverse momentum. The kinematics can only be derived from the measured energy and angle of the hadronic system [14].

In leading order QCD the e^+p cross section is related to the quark parton densities:

$$\frac{d^2\sigma_{CC}}{dxdQ^2} = \frac{G_F^2}{2\pi}\left(\frac{1}{1+Q^2/m_W^2}\right)^2(\bar{u}+\bar{c}+(1-y)^2(d+s)). \qquad (6)$$

H1 has measured the double differential reduced cross section defined similar to the neutral current case by

$$\sigma_{CC} = \frac{2\pi}{G_F^2}\left[1+\frac{Q^2}{M_W^2}\right]^2\frac{d^2\sigma_{CC}}{dxdQ^2}. \qquad (7)$$

Figure 5 shows the measured cross section in 6 different ranges of Q^2 up to $Q^2 = 16000 \text{GeV}^2$ compared to the MRSH based Standard Model prediction.

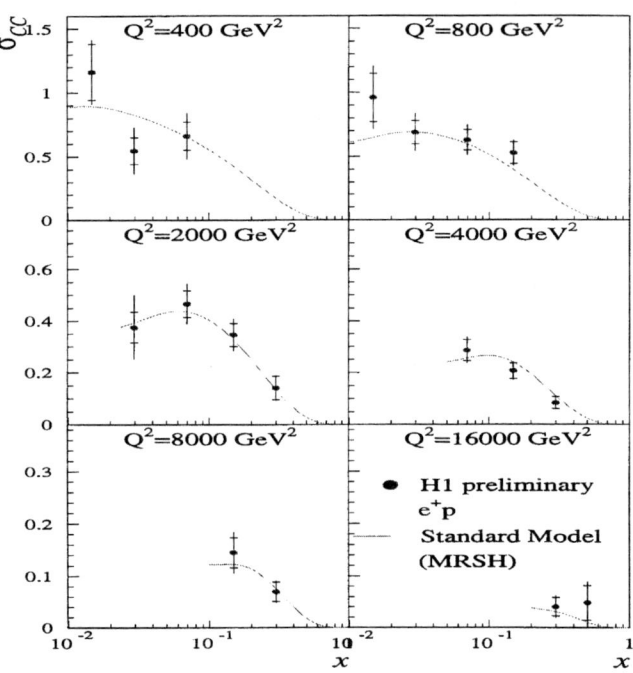

FIGURE 5. Charged current reduced cross section (see text) measured by H1. The Standard Model expectation using the MRSH parametrisation is shown as dotted line.

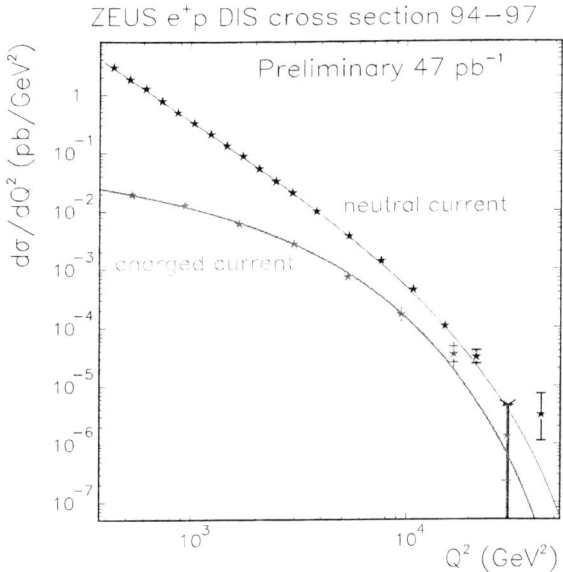

FIGURE 6. Charged and neutral Current differential cross sections measured by ZEUS. The Standard Model expectation is shown as full line

The single differential cross section $\frac{d\sigma}{dQ^2}$ measured by ZEUS for both the neutral current as well as the charged current is shown in figure 6. Both experiments see, for charged current DIS, good agreement with the standard model expectation up to highest Q^2.

EVENTS WITH ISOLATED LEPTON AND LARGE MISSING P_T

The H1 collaboration has observed events with an isolated lepton and large missing p_t and has searched for possible explanations of these events [15]. Within the Standard Model, processes with this signature are expected from W production with subsequent semileptonic decay [17]. The lowest order diagrams for these processes are shown in figure 7.

A dedicated analysis has been performed by using the charged current event selection [16] including the following criteria (for details see [15]):

- $P_{trans}(Calorimetry) > 25 \text{GeV}$
- requiring a well measured central track with $p_t > 10 \text{GeV}$
- reconstructed vertex in the interaction region
- topological and timing filters against cosmic and halo muons

- rejection of badly measured neutral current events

leading to 124 events.

For this data sample the isolation of high p_t tracks with respect to jets or other tracks in the event has been investigated. The isolation is measured by the distance of a track in the $r - \phi$-plane to its nearest neighbor track (D_{track}) and to the nearest jet (D_{jet}). Figure 8 shows the data sample in the D_{track}-D_{jet}-plane, after the above mentioned cuts. Six tracks are clearly displaced in the D_{track}-D_{jet}-plane, all of them fulfill the H1 lepton criteria [15]. For five events an isolated μ and for one event an isolated e^- is detected.

To further study the kinematics a Monte Carlo simulation of the Standard Model W-production was performed with 500 times the statistics used in the data sample. Figure 9 shows the generated events in the plane of hadronic transverse momentum, P_T^X, and transverse mass M_T, together with the six isolated tracks found in the data for the e^- and the μ candidates separately, here M_T is calculated out of

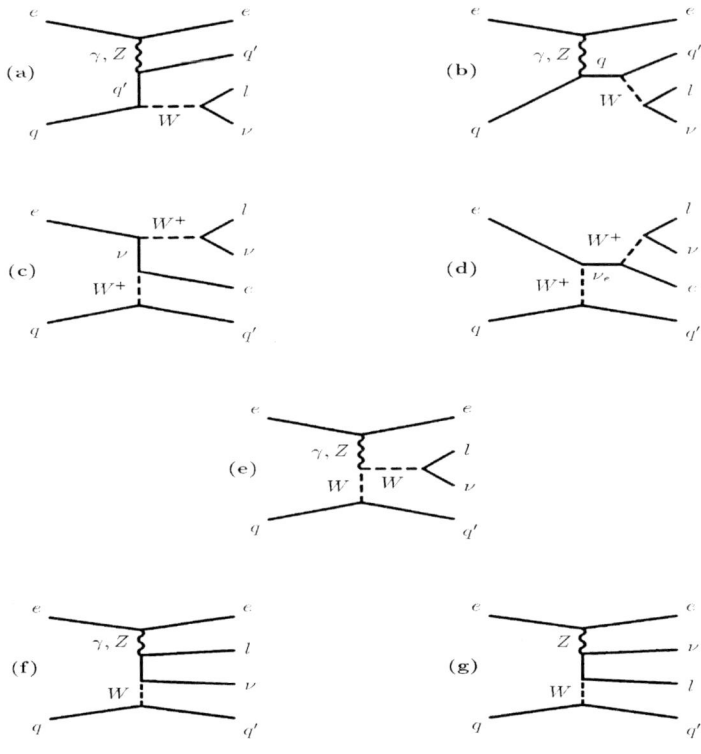

FIGURE 7. Parton level diagrams for the process $eq \to eql\nu$.

TABLE 4. Number of events with an isolated high p_t lepton observed by H1

H1	N_{obs}	$N_{exp}(W)$	background
e^-	1	1.65 ± 0.47	<0.14
$\mu^\pm$	5	0.53 ± 0.11	<0.28

TABLE 5. Number of events with an isolated high p_t lepton measured by ZEUS

ZEUS	N_{obs}	$N_{exp}(W)$	background
e^+	4	2.22 ± 0.2	1.24
$\mu^\pm$	0	0.46 ± 0.02	0.84

the transverse momentum of the lepton and the neutrino. The missing transverse energy is hereby attributed to a hypothetical neutrino. Due to the kinematics of the process one would expect P_T^X to be low and to observe a Jacobean Peak around the W-mass, which is clearly seen for the Monte Carlo events. Table 4 summarises the results. The expectation from standard model W-production as well as the expected background is shown.

ZEUS did a similar analysis and Monte Carlo study, the result is shown in figure 10 for the e^+ case and summarised in Table 5.

Whereas H1 sees 3 muon events in a region which is not preferred by the standard model W-production, ZEUS does not see any excess above Standard Model expec-

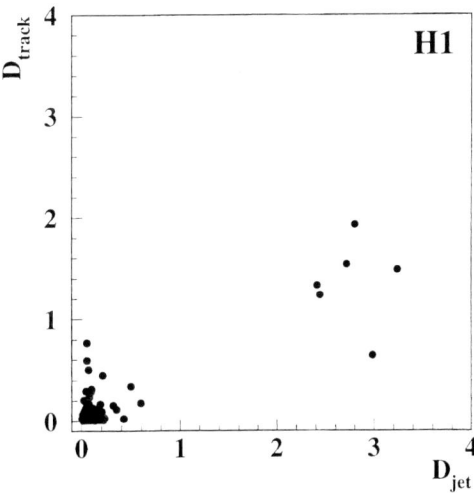

FIGURE 8. Correlation between the distances D_{jet} and D_{track} (see text) to the closest hadronic jet and track, for all high p_t tracks.

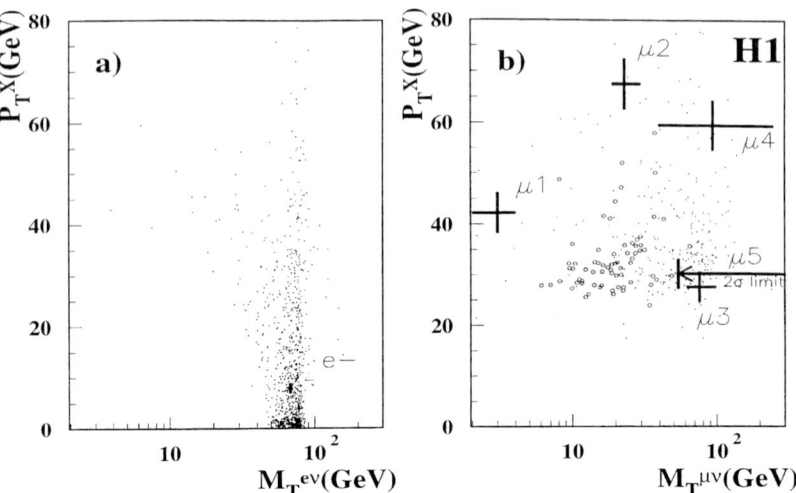

FIGURE 9. Distribution of the events in the P_T^X-M_T plane measured by H1. Figure a shows the electron channel figure b the muon channel. The 1-σ uncertainty on the measured kinematic parameters are indicated by the crosses. (For event $\mu 5$ the 2σ lower limit is shown.) The dominant SM contributions (dots for W production, open circles for $\gamma\gamma$ processes in the μ channel) are shown for an accumulated luminosity which is a factor 500 higher than in the data. For the μ channel no significant contribution is expected below $P_T^X = 25$ GeV because of a cut in the event selection.

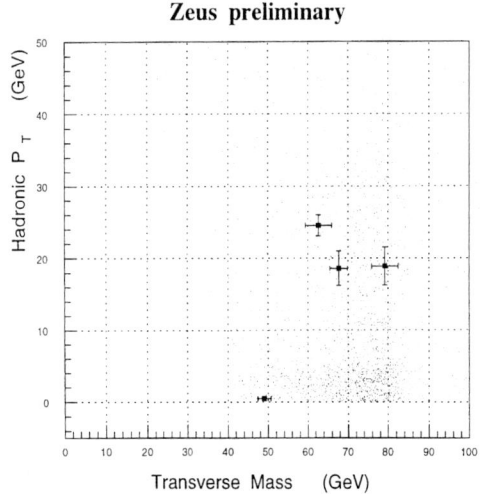

FIGURE 10. Distribution of e^+ events in the $p_t^{hadronic}$ M_T plane observed by ZEUS, compared with a Monte Carlo study of W production with $L_{MC} = 1000 * L_{data}$

tation. Further studies with higher statistics will certainly clarify the situation.

LIMITS ON EXCITED FERMIONS

With the full statistics of 1994 - 1997 data, H1 did a search for excited fermions and improved production limits considerably compared to previous published HERA results [19]. The processes considered, the event selection and the observed events are listed in table 6.

TABLE 6. Excited lepton channels analysed by H1

Channel	Selection	Data	Background	Efficiency
$e^* \to e\gamma$	$2EM\,clusters$	223	239± 7	85%
$e^* \to eZ \to ee$	$3EM\,clusters$	3	1.4± 0.3	78%
$e^* \to eZ \to \nu\bar{\nu}$	$1e + P_t^{miss}$	1	3.6± 0.7	70%
$e^* \to eZ \to q\bar{q}$	$2jets + 1e$	38	48± 3	41%
$e^* \to \nu W \to e\nu$	$1e + P_t^{miss}$	1	3.6± 0.7	70%
$e^* \to \nu W \to q\bar{q}$	$2jets + P_t^{miss}$	3	3.8± 0.5	40%
$\nu^* \to \nu\gamma$	$1phonton + P_t^{miss}$	0	1.3± 0.8	38%
$\nu^* \to \nu Z \to ee$	$2e + P_t^{miss}$	0	0.38± 0.2	40%
$\nu^* \to \nu Z \to q\bar{q}$	$2jets + P_t^{miss}$	3	3.8± 0.5	40%
$\nu^* \to eW \to e\nu$	$2e + P_t^{miss}$	0	0.38± 0.2	40%
$\nu^* \to eW \to q\bar{q}$	$2jets + 1e$	38	48 ± 3	41%

As an example the invariant mass spectrum for the $e\gamma$ channel is shown in figure 11, compared with the Standard Model expectation. From the derived cross section,

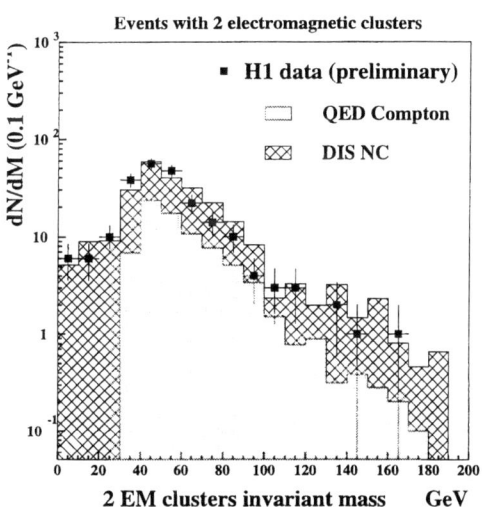

FIGURE 11. Invariant mass of 2 electro magnetic clusters compared with the QED compton and DIS neutral current simulations

limits on the characteristic coupling f/Λ [18] are obtained. The limits for the excited electron channels are shown as an example in figure 12. The limits extend

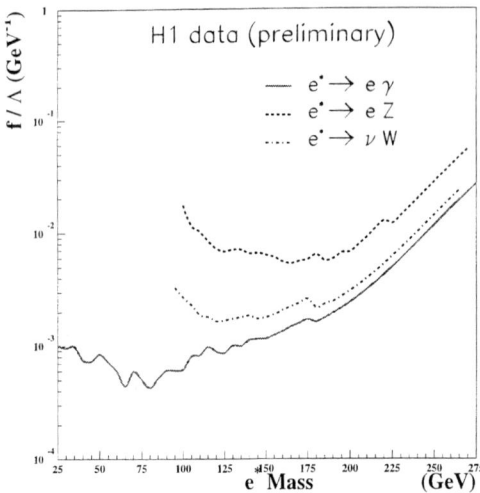

FIGURE 12. Upper limits at 95% confidence level on the coupling f/Λ as a function of the mass for the excited electron

well above the LEP mass range and exclude excited electrons up to 275GeV for $f/\lambda > 3*10^{-2} \text{GeV}^{-1}$.

SUMMARY AND OUTLOOK

Using the full data sample 1994 - 1997 the H1 and ZEUS collaborations measured the deep inelastic scattering cross sections up to $Q^2 = 40000 \text{GeV}^2$ extending measurements from previous fixed target experiments by two orders of magnitude. The results from H1 and ZEUS in deep inelastic scattering are well described by the standard model over more than 4 orders of magnitude in Q^2.

With the data available a new domain for Standard Model tests is accessible and searches for new phenomena have been performed by both experiments, pushing the limits for leptoquarks and contact interaction well above the 200GeV mass range. A slight access of events with very high Q^2 at high y remains to be clarified with improved statistics in coming years.

H1 has seen 6 events with an isolated high p_t lepton and large missing p_t, 3 of them cannot easily be explained by Standard Model W processes.

Searches for excited leptons have been performed and previously published limits from HERA have been improved by a factor of 2.

In the 1998/1999 data taking period HERA will go back to electron operation and the proton beam energy will be raised from 820GeV to 920GeV. An expected $50 nb^{-1}$

of 27.5GeV e^- on 920GeV p should help in obtaining improved measurements in the above mentioned topics. In the year 2000 HERA will considerably upgrade its capability and a factor of five more luminosity is expected from then on, providing $1 fb^{-1}$ to each experiment by the year 2005.

REFERENCES

1. H1 Collaboration, I.Abt et al., Nucl. Instr. and Meth. A386(1997) 310 and 348.
2. ZEUS Collaboration, The ZEUS Detector, Status Report 1993, DESY 1993.
3. H1 Collaboration, C.Adloff et al., Z. Phys. C74(1997) 191.
4. ZEUS Collaboration, J.Breitweg et al., Z. Phys. C74(1997) 207.
5. S.Bentvelsen, J.Engelen, P.Kooijman, Physics at HERA Hamburg 1991, Proceedings of the Workshop, Volume 1, 23, eds. W.Buchmüller and G.Ingelman.
6. A.D.Martin, W.J.Stirling and R.G.Roberts, Phys. Rev. D50(1994) 6734 and Phys. Rev. D51(1995) 4756.
7. L.W.Whitlow et al., Phys. Lett. B282 (1992) 475;
   L.W. Whitlow, preprint SLAC-357 (1990).
8. BCDMS Collaboration, A.C.Benvenuti et al., Phys. Lett. B223(1989) 485; CERN preprint CERN-EP/89-06.
9. NMC Collaboration, M.Arneodo et al., Phys. Lett. B 364(1995) 107.
10. H1 Collaboration, I.Abt et al., Nucl. Phys. B396(1993) 3; ibid, T.Ahmed et al. Z. Phys. C64(1994) 545; ibid, T.Ahmed et al., Phys. Lett. B369(1996) 173.
11. W.Buchmüller, R.Rückel and D.Wyler, Phys. Lett. B 191(1987) 442.
12. D0 Collaboration, B.Abbott et al., Phys. Rev. Lett. 79(1997) 4321;
    D0 Collaboration, B.Abbott et al., Phys. Rev. Lett. 80(1998) 2051.
13. CDF Collaboration, F.Abe et al., Phys. Rev. Lett. 79(1997) 2198;
    OPAL Collaboration, K.Ackerstaff et al., CERN-PPE/97-101(1997);
    ALEPH Collaboration, ALEPH 98-021(1998) except for U4 values from contributed paper #602 to EPS HEP Conference 1997, Jerusalem.
14. F.Jacquet and A.Blondel, Proceedings of the Study for an ep Facility in Europe, DESY 79/48(1979) 391, ed. U.Amaldi.
15. H1 Collaboration, C.Adloff et al., DESY 98-063 (subm. to Euro Phys. C).
16. H1 Collaboration, C.Adloff et al., Z. Phys. C67(1995) 565, Phys. Lett. B379(1996) 319.
17. U. Baur, J.A.M.Vermaseren, D.Zeppenfeld, Nucl. Phys. B375(1992) 3.
18. K.Hagiwara, S.Komamiya and D.Zeppemfeld, Z. Phys. C29(1985) 115.
19. H1 Collaboration, C.Adloff et al., Nucl. Phys. B483(1997) 44;
    ZEUS Collaboration, J.Breitweg et al., Z. Phys. C76(1997) 631.

W PHYSICS AND STANDARD MODEL TESTS

Standard Model tests and new physics at LEP

Riccardo Faccini

INFN Rome and University of California S. Diego

Abstract. The stringent tests of the Standard Model performed with LEP data taken between 1989 and 1995 severely constrain the possible sources of new physics beyond it. The most recent measurements of the Standard Model parameters obtained with more than 20 million Z boson decays are summarized and interpreted from this point of view. The data collected by LEP at higher energies allow to investigate the direct production of new particles. The status of the searches for new particles, other than Higgs bosons and supersymmetric partners is reported.

I INTRODUCTION

In the years between 1989 and 1995, the four LEP experiments collected overall more than 20 million Z bosons. About 100 pb^{-1} were collected per experiment at center of mass energy $\sqrt{s} \sim M_Z$, and 40pb^{-1} approximately two GeV above and below. Since November 1995 the beam energy has been increased. The center of mass energy passed the W pair-production threshold in 1996 and the Z pair one in 1997. The integrated luminosities collected per experiment are approximately 6, 11, 12 and 55 pb^{-1} at $\sqrt{s}$=130-140, 161, 172 and 183 GeV respectively(LEP2).

The analysis of LEP1 data is almost finalized and it yields a very precise measurement of the Standard Model parameters. The level of accuracy is such that the comparison of the results with the predicted radiative corrections severely limits the existence of new physics.

The LEP2 project is still going on and each upgrade in energy opens new accessible phase space for the production of new particles. Among the possible extensions of the Standard Model that the LEP experiments are able to probe, this article reviews:

- those predicted by <u>compositeness models</u>.
  The possibility to explain the mass hierarchy of the existing particles as a spectrum of particles composed of smaller constituent has appealed physicists for a long time [1]. Most of the models developed have the additional advantage that dynamic symmetry breaking can occur without requiring the existence of the Higgs boson, the major missing piece of the Standard Model [2].
  As far as new particles are concerned, these models predict the existence of

excited states of the known particles that decay in the stable ones emitting a gauge boson. The binding forces (TechniColour, TC) are typically strong enough to give contributions to the electroweak radiative corrections and indeed these models are probed at LEP1. In addition particles carrying both leptonic and barionic number are predicted (leptoquarks [3]).

- models predicting <u>more than four families</u>. The next sequential family can either have a neutral leptonic component (N) lighter than the charged one (C) ore vice-versa. Since the existence of other massless N particles is excluded, in this scenario the mixing between families is very likely so that C and N particles can also decay into ordinary leptons.

- generic <u>contact interaction</u> models. New interactions mediated by bosons with masses well above the center of mass energies can be approximated by four-fermion vertices, resulting therefore in deviations from the Standard Model cross sections and asymmetries of the fermion pair-production ($e^+e^- \to f\bar{f}$).

II ELECTROWEAK TESTS AT LEP1

The measurements of the properties of the Z boson decays can be divided in two sets, the cross sections and the asymmetries. Preliminary results based on the full data set collected at LEP1 are discussed here.

The cross section measurements of the processes $e^+e^- \to Z \to f\bar{f}$ are related to the couplings of the Z boson to the fermion f, the Z mass and total width via

$$\sigma_{f\bar{f}}^0 = \frac{12\pi \Gamma_{ee} \Gamma_{ff}}{M_Z^2 \Gamma_Z^2} \qquad \Gamma_{ff} \propto (\bar{g}_{V,f}^2 + \bar{g}_{A,f}^2)$$

where $\sigma_{f\bar{f}}^0$ are the peak cross sections, Γ are the decay widths and $g_{A,V}$ are the axial and vector couplings. Measuring cross sections as a function of the center of mass energy close to $\sqrt{s}=M_Z$ allows the determination of M_Z Γ_Z and the individual Γ_{ff}. To this aim good measurements of the integrated luminosity collected, which has an error smaller than 0.1% since 1994, and of the center of mass energy are both crucial. The current leakage into LEP of some current of the close-by railway (the so called "train effect") has been understood [4] so that the present error are reduced below all the other experimental systematics. With this improvement the present measurement of the mass and the width of the Z are:

$$M_Z = 91186.7 \pm 2.0 \text{ MeV} \qquad \Gamma_Z = 2494.8 \pm 2.5 \text{ MeV}$$

Among the cross section measurements the most sensitive to new physics are those involving heavy quarks since in most of the models the coupling of the new particles is proportional to the masses. While on partial statistics there were indications of deviations of Γ_{cc} and Γ_{bb} from the predictions the analysis of the full data samples shows no significant discrepancy (see figure 1a).

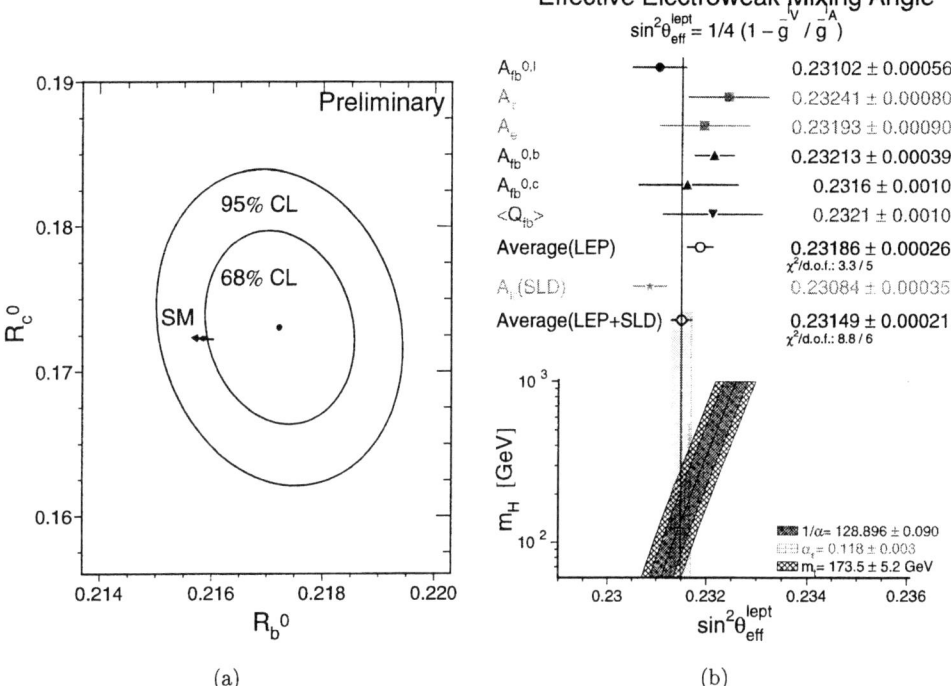

FIGURE 1. a) Measurement of the Z→b$\bar{b}$ and Z→c$\bar{c}$ branching ratios. b) Comparison of the measurements of $\overline{\sin^2 \theta_w}$ at LEP and SLC.

Several asymmetries can be defined, depending on how the two states whose rates are compared are defined: the direction of flight of the out-coming fermion with respect to the electron beam (forward-backward asymmetry A_{FB}), the polarization of the out-coming fermion (polarization asymmetry), the polarization of the beams (left-right asymmetry A_{LR}) and any combination of them. LEP is able to measure only the forward-backward asymmetry for the leptons, for the b and the c quarks and for generic hadronic events (Q_{FB}) and the tau polarization asymmetry. The Stanford Linear Collider is capable of measuring in addition the left right asymmetries of leptons, b and c quarks and hadrons. All these measurements allow to extract

$$A_f = \frac{\overline{g}_{A,f}\overline{g}_{V,f}}{\overline{g}_{A,f}^2 + \overline{g}_{V,f}^2} = \frac{1 - 4q_f\overline{\sin^2 \theta_w}}{1 + (1 - 4q_f\overline{\sin^2 \theta_w})^2}$$

where $\overline{\sin^2 \theta_w}$ is the effective weak mixing angle. In the Standard Model all the asymmetry measurements are therefore related to one single parameter and this sets a very tight consistency check, as shown in figure 1b. The agreement is extremely better than it was during the 1997 Winter conferences, mainly due to three new, very precise measurements: an update on higher statistics of measurement of the left-right asymmetry at SLC [7], a new measurement of the b forward-backward asymmetry [5] and of the tau polarization [6].

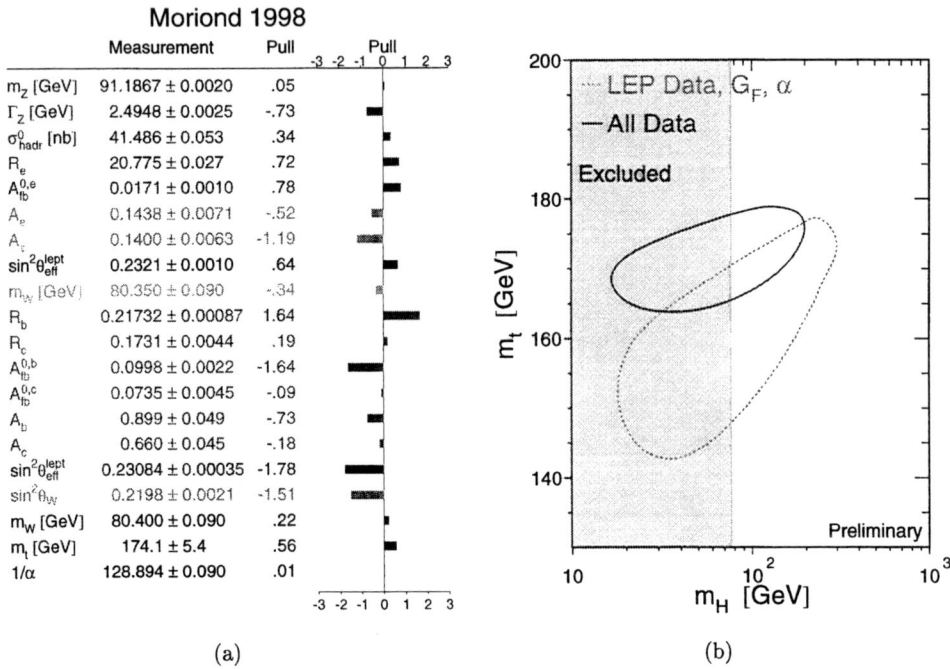

FIGURE 2. a) Results of the fit of LEP and SLD data to the Standard Model and corresponding pulls. b) Top mass versus Higgs mass 68% C.L. contours.

The extremely precise measurements of the Standard Model parameters extracted with the Standard Model fits [7] are shown in figure 2a. The agreement with the predictions is overall very good. The accuracy achieved allows both the study of the radiative corrections and to extract indications of possible new physics. The radiative corrections are particular sensitive to the top quark mass because of the symmetry breaking needed in order to have a very heavy top quark. On the contrary the sensitivity to the Higgs mass scales only as the logarithm of the mass itself. The constraints on M_t and M_H are reported in figure 2b. The top mass constraint is in perfect agreement with the direct measurement at Fermilab. The LEP data, assuming the minimal Standard Model symmetry breaking mechanism, tend to favour a low mass Higgs (below about 200 GeV at 95% C.L.) [8].

Further interpretations of the measured radiative corrections were given [9] showing actual evidence of deviations from the Born approximation, in agreement with the predictions of the Standard Model. Nevertheless, the LEP average of $A_{FB}(b\bar{b})$ and the SLC A_{LR} are still about 2.5 standard deviations apart, which has led to many theoretical speculations. In particular, releasing the assumption of universality for the b quark and fitting for its right and left couplings to the Z boson one obtains [10]

$$g^b_l = -0.4167 \pm 0.0027 \qquad g^b_r = 0.100 \pm 0.010$$

to be compared with the predictions $g^b_l = -0.4208$ and $g^b_r = 0.0774$. The disagreement is somehow more pronounced in the right coupling.

III SEARCH FOR NEW PHYSICS AT LEP2

Due to the low annihilation cross section of electrons and positrons at energies above the Z resonance, the statistics collected at LEP2 is limited and most of the searches are performed in a direct way, looking for signatures differing from the ones predicted in the Standard Model. In this article the attention is concentrated to the searches that are more related to the measurements of the cross sections of fermion pair-production $e^+e^- \to f\bar{f}$. These measurements are important since they provide tests of the Electroweak theory at higher energies. This allows the evaluation of the interference term between the annihilation into a Z and a photon, crucial for the interpretation of the LEP1 data in a Standard Model independent framework [7]. Figure 3a shows the evolution of the expected cross section with the center-of-mass energy and, as an illustration, the results of the ALEPH collaboration. At $\sqrt{s} > M_Z$ a good fraction of the events tend to radiate an energetic initial state photon in order to produce an on-shell Z boson. From the point of view of the search for new physics, these events have little interest since they do not access kinematic domains different from LEP1 ones. Because of this effect the cross section are usually also reported after requiring the initial state radiation to be small ("non-radiative events").

A Contact interactions and Leptoquarks

A new interaction between the known particles mediated by very massive particles would show up in the effective Lagrangian as contact interactions $e^+e^- \to f\bar{f}$ [11]:

$$\mathcal{L}_{ci} = \frac{g^2}{(1+\delta_{fe})\Lambda^2} \sum_{i,j=L,R} \eta_{ij} (\bar{e}_i \gamma^\mu e_i)(\bar{f}_j \gamma_\mu f_j) \qquad (1)$$

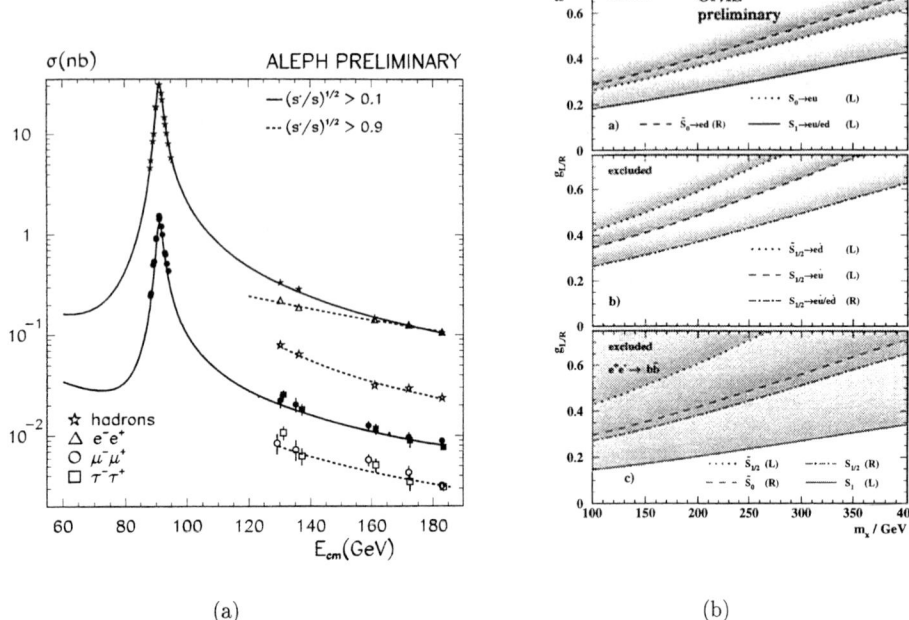

FIGURE 3. a) Fermion pair-production cross section as a function of the center of mass energy (E_{cm}). The reduced cross-section of the "non-radiative events" ($\sqrt{\frac{s'}{s}} > 0.9$) is also reported. b) Excluded scalar leptoquark couplings as a function of the leptoquark mass hypothesis in case of spin zero. Similar exclusions are obtained for vector leptoquarks.

where Λ is the scale of new physics and η_{ij} are the relative couplings of the chiral states. Atomic parity violation measurement strongly favour the hypothesis $|\eta_{ij}| = 1$, and this is therefore assumed hereafter. The models taken into consideration differ for the assumed η_{ij}, as shown in table 1. The possibility that the new interaction couples only with the leptons, the up-type quarks or the down-type quarks have been considered separately [12]. As shown in figure 3a no evidence for deviations is observed and this translates into limits at 95% confidence level on the compositeness scale Λ. The two solutions, corresponding to a simultaneous change of sign of all the η_{ij}, are considered. The regions excluded by at least one LEP experiment, in case Standard Model like couplings are assumed, are reported in the table.

If the hypothesis of high mass of the intermediate boson is released the fermion pair-production studies are sensitive also to the presence of leptoquarks. Com-

Model	LL	RR	LR	RL	VV	AA	V0	A0
η_{LL}	±1	0	0	0	±1	±1	±1	0
η_{RR}	0	±1	0	0	±1	±1	±1	0
η_{LR}	0	0	±1	0	±1	∓1	0	±1
η_{RL}	0	0	0	±1	±1	∓1	0	±1
Λ_+ (TeV) <	6.1	5.9	6.4	6.9	11.8	8.4	10.1	9.6
Λ_- (TeV) <	5.6	5.1	6.1	6.1	9.7	9.5	7.5	7.9

TABLE 1. Models of contact interaction considered. The parameters η_{ij} (i,j = L, R) define to which helicity amplitudes the contact interaction contributes. The excluded compositeness scales at 95% C.L. by any of the LEP experiments are also reported in the assumption that the relative coupling of the new interaction between leptons and hadrons is the same as in the Standard Model.

positeness model trying to explain the excess of high Q^2 events at Hera in 1996 [13] suggested that deviations should also be seen in the $e^+e^- \to q\bar{q}$ cross section at LEP2 [14]. With this interpretation the exchange of a leptoquark in the t-channel would give an enhancement in the hadronic cross section. The contribution to the Lagrangian would then be the same as $\mathcal{L}_{ci}$ having replaced Λ with the propagator. Limits can therefore be set on the coupling for a given mass hypothesis (see figure 3). It can be noted that the region $M_X \sim 200$ GeV where the Hera signal was more likely to be, is partially covered by these searches.

B Sequential Heavy Leptons

Another possible extension of the Standard Model is the existence of more than three lepton families. Several scenarios are possible depending on the masses of the additional neutral (M_N) and charged (M_C) leptons.

In case $M_C > M_N$, the neutral lepton would be stable and masses $M_N < \frac{M_Z}{2}$ are already excluded at 95% C.L. by LEP1 data. Decays $Z \to N\bar{N}$ would in fact increase the invisible width of the Z boson [7]

$$\Gamma_{n.p.} = \Gamma_Z - \Gamma_{had} - \sum_l \Gamma_l < 2.9 \text{MeV} \qquad @95\% \text{ C.L.}$$

where the sum runs over all charged and neutral leptons, or as an increase of the number of events with only a single photon from initial state radiation. The single photon cross section measurement at LEP1 can be converted in a measurement of the numer of families. The most precise measurement is $N_\nu = 2.98 \pm 0.10$ [15].

If the neutral lepton is massive the main search channel of the new family is the pair-production of the charged lepton that decays into the neutral one and a W boson. The visible energy decreases as the two masses become closer. Ultimately if the two masses are close enough the charged lepton lifetime becomes long enough that it is stable in the detector. This scenario is studied in detail because the signatures involved are similar to the chargino production in some supersymmetric models.

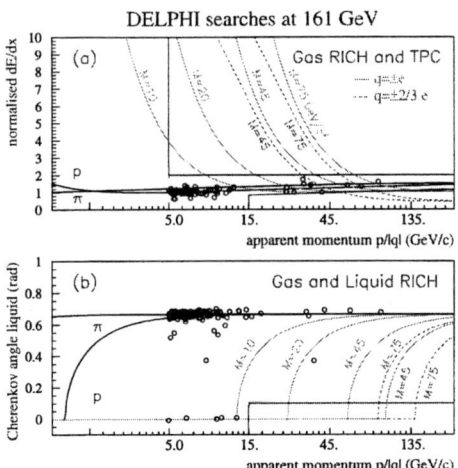

FIGURE 4. Mass separation power of the ionization in the TPC and of the Chereckov angle in the RICH of the Delphi experiment.

Heavy charged stable particles would be mainly identified in the tracking chambers and in the Cherenkov detectors given their low speed (see figure 4a). No evidence for signal is reported and production cross sections larger than about 0.2 pb are excluded for $M_C < 87$ GeV at 95% C.L. [16].

In case $M_N > M_C$ the main signature is the production of a pair of W bosons and two additional charged leptons (N→lW). Since $M_N \neq 0$, the mixing among leptons is natural and different searches are performed depending on which family the fourth one couples most. The analysis also depends on whether the neutral leptons are Majorana or Dirac type objects: on one side the production cross section of the processes $e^+e^- \to N\bar{N}$ and $e^+e^- \to NN$ would be different, on the other, in case of Majorana objects the two leptons produced in the events could have equal signs. The LEP2 data do not show evidence of such events and the existence of Dirac (Majorana) neutral heavy lepton is excluded at 95% C.L. for $M_N < $ 86.3(72.5), 87.7(74.9) and 75.4 (63.6) GeV assuming they couple exclusively to the three families respectively.

C Compositeness and excited particles

Among the predictions of the composite models [1], the existence of new leptons that are excited bound states of the same constituents as the ordinary ones can be tested at LEP2. In electron-positron annihilation the excited leptons can be either pair-produced ($e^+e^- \rightarrow l^*\bar{l}^*$) or produced in pair with an ordinary lepton of the same generation($e^+e^- l^*\bar{l}$). In the first case the coupling of the photon or the Z boson with the lepton pair is assumed to be equal to the one with ordinary matter and the production cross section is a function of the mass of the excited lepton only. The production cross section of the single production depends also on the strength of the interaction between ordinary and excited leptons:

$$\mathcal{L}_{\text{eff}} = \frac{\lambda}{4m_{l^*}} \overline{\Psi}_l^* \sigma^{\mu\nu}(gf\vec{\tau}\vec{W}_{\mu\nu} + g'f'YB_{\mu\nu})\Psi_l \qquad (2)$$

Finally the existence of the excited leptons of the first generation could also give contribution to the $e^+e^- \rightarrow VV$ ($V=\gamma$,W and Z) production via the t-channel.

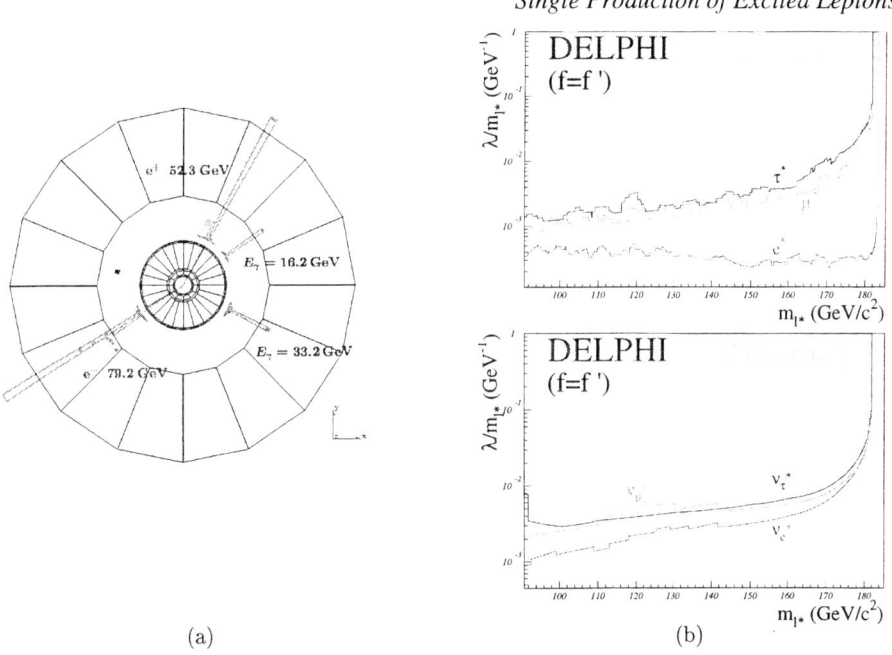

FIGURE 5. a) $e^+e^- \rightarrow e^{+*}e^{-*}$ candidate in the L3 detector. b) 95% C.L. excluded couplings of the excited leptons with the ordinary ones as a function of their mass.

The direct searches for these particles have been performed covering a large number of signatures [17]. The excited leptons can in fact decay in an ordinary lepton of the same family and one of the vector bosons. The signatures where photons are produced are very clear: two leptons and two photons. Figure 5a shows a candidate that survives all the selection criteria apart from the requirement that the two electron-photon invariant masses are equal.

As the accessible mass range increases, the decay into W and Z bosons, which in turn decay into a pair of fermions, becomes more likely. No significant deviations from the Standard Model predictions are observed and limits on the production cross sections are set at 95% C.L.. In the case of the pair-production signatures this is directly converted into limits on the mass. The preliminary limits from the data at $\sqrt{s}$=183 GeV are 91.0 GeV on the masses of the charged excited leptons. In the case of the neutral excited leptons the hypothesis on their dipole moment affects the final result. In the most pessimistic hypothesis the limit is set at 88.1 GeV. From the single production the upper limits can be set on the couplings f and f' as a function of the mass of the excited lepton as shown in figure 5b under the assumption f = f'. The contribution of the t-channel in the search for the first generation leptons is clearly visible.

IV SUMMARY AND CONCLUSIONS

The very successful operation of the LEP machine both at the Z peak and above the W pair-production threshold has allowed very precise tests of the Standard Model. No deviations from the predictions of the theory was found and actually the indications of possible new physics reported on smaller samples of data have proved to be statistical fluctuation: the measurements of R_b and R_c are now in agreement with the Standard Model, the χ^2 of the measurements of $\sin^2\theta_w$ at LEP and SLC is 8.8 with 6 degrees of freedom and the excess of events in the four jets channel (described in [18]) was not reproduced any more. Assuming the minimal Standard Model symmetry breaking mechanism the data are consistent with a light Higgs, M_H < 200 GeV at 95% C.L., i.e. within reach of the present or next generation accelarators.

Direct searches at LEP2 of leptoquark, heavy leptons and sequential leptons and excited leptons have not found any evidence of signal for most of the kinematically allowed parameter space. The data taking at $\sqrt{s}$=189 GeV in 1998 and $\sqrt{s}$=200 GeV 1999 and 2000 will open a window on new domains.

REFERENCES

1. J.C. Pati, A. Salam, Phys. Rev. **D10**(1974)275
   K.Hagiwara, S.Komamiya and D.Zeppenfeld, Z.Phys. **C29**(1985)115 and references therein.
2. B. Schrempp and F. Schrempp, Nuc. Phys. **B242**(1984)203 and reference therein.

3. J.L. Hewett and T.G. Rizzo, Phys. Rep. **183**(1989)193.
4. The LEP Energy Working Group, "Calibration of centre-of-mass energies at LEP1 for precise measurements of Z properties", CERN-EP/98-40, CERN-SL/98-12. Submitted to EPJ C.
5. R.Barate et al., ALEPH Collaboration, CERN-EP/98-038, submitted to Phys. Lett. B.
6. M.Acciarri et al., L3 Collaboration, CERN-EP/98-028, submitted to Phys. Lett. B.
7. A Combination of Preliminary Electroweak Measurements and Constraints on the Standard Model, CERN-PPE/97-154 .
8. C. Paus, "Combined Electroweak Results from Lep and other Accelerators"', LEPC meeting CERN 31st March 1998.
9. G.Altarelli et al., Int. Journal of Modern Physics **A13**(1998) 1031.
10. J. Field, HEP-PH/9801403.
11. E.Eichten, K. Lane, M.Peskin, Phys. Rev. Lett. **50**(1981) 811
12. M.Acciarri et al., L3 Collaboration, CERN-EP/98-031, submitted to Phys. Lett. B
    K. Ackerstaff et al., Eur. Phys. J. **C2** (1998)441.
13. C. Adloff et al., H1 Collaboration, Z. Phys. **C74** (1997) 191
    J. Breitweg et al., Zeus Collaboration, Z. Phys. **C74** (1997) 207.
14. J.Kalinowski et al., Z. Phys. **C74** (1997) 595.
15. M.Acciarri et al., L3 Collaboration, CERN-EP/98-025, submitted to Phys. Lett. B.
16. R.Barate et al., ALEPH Collaboration,Phys. Lett. **B405**(1997) 379
    P.Abreu et al., DELPHI Collaboration, Phys. Lett. **B396**(1997) 315
    M.Acciarri et al., L3 Collaboration, Phys. Lett. **B415**(1997) 299
    K. Ackerstaff et al., CERN-EP/98-039, submitted to Phys. Lett. B.
17. R.Barate et al., ALEPH Collaboration, CERN-EP/98-022, submitted to EPJ C
    P.Abreu et al., DELPHI Collaboration, Phys. Lett. **B393**(1997) 245
    M.Acciarri et al., L3 Collaboration, Phys. Lett. **B401**(1997) 139
    K. Ackerstaff et al., CERN-EP/98-039, EPJ **C1**(1998)45.
18. R.Barate et al., ALEPH Collaboration, CERN-EP/97-156, submitted to Phys. Lett. B.

The Measurement of the Mass of the W Boson from the Tevatron

Randy Thurman-Keup[1]

Argonne National Lab / HEP-362
9700 South Cass Avenue
Argonne, IL 60439

Abstract. This paper presents measurements of the mass of the W vector boson from the CDF and DØ experiments using data collected at $\sqrt{s} = 1.8$ TeV during the 1994-1995 data taking run. CDF finds a preliminary mass of $M_W = 80.43 \pm 0.16$ GeV and DØ measures a mass of $M_W = 80.44 \pm 0.12$ GeV.

I INTRODUCTION

During the 1994-1995 collider run of the Fermilab Tevatron, the CDF and DØ experiments collected data corresponding to integrated luminosities of 90 pb^{-1} and 82 pb^{-1} respectively. From this data, CDF extracted $\sim 21,000$ $W \to \mu\nu$ events while DØ obtained $\sim 28,000$ $W \to e\nu$ events. These W events were used to make measurements[2] of the mass of the W boson [1].

A Motivation

When going from tree level in the Standard Model (SM) to next-to-leading order, the relation between the mass of the W boson and the other SM parameters gets modified as follows:

$$M_W^2 = \left(\frac{\pi \alpha_{EM}(0)}{\sqrt{2} G_F \sin^2 \theta_W} \right) \longrightarrow \left(\frac{\pi \alpha_{EM}(M_Z^2)}{\sqrt{2} G_F \sin^2 \theta_W} \right) \frac{1}{1 - \Delta r}, \qquad (1)$$

where Δr embodies the loop corrections to the W propogator [2]. The dominant contributions to the corrections come from the top quark which introduces a quadratic dependence on its mass, and the higgs boson which produces a logarithmic dependence on its mass. These dependencies allow one to probe for the

---

[1] representing the CDF and DØ collaborations
[2] The CDF measurement is only a preliminary measurement.

higgs given a precision measurement of the W mass. In addition, Standard Model extensions and/or replacements produce their own corrections and here again, a precision measurement may be used to uncover these theories.

B Tevatron Environment

The majority of W events at the Tevatron are produced from s channel quark-antiquark interactions. The W's of interest in this measurement are the ones that decay to muon-neutrino or electron-neutrino pairs, since these are the cleanest decays. Unfortunately, the fact that the neutrino is undetected, prevents us from measuring the invariant mass of the W. One could infer the neutrino 3-momentum by requiring momentum conservation in the event if it weren't for the fact that the energy of the incident quarks is not known. Unlike electron-positron colliders, the quarks are bound inside the (anti)protons and their momentum is governed by parton distribution functions (PDFs). Fortunately we can still enforce momentum conservation in the plane transverse to the beam direction since the since the incident quarks have essentially no transverse momentum.

This leaves us with two possible quantities from which to extract the mass: the transverse momentum of the lepton, p_T^ℓ, or the transverse mass of the lepton-neutrino pair, $M_T = \sqrt{2 p_T^e p_T^\nu (1 - \cos \Delta\phi)}$. The transverse mass is analogous to the invariant mass except that only components of momentum transverse to the beam are used. The transverse mass is less sensitive than p_T^ℓ to the transverse momentum of the W[3] and is upwardly bounded by M_W thereby still providing sensitivity to the mass. It is, however, more sensitive to the energy resolution of the calorimeter (see Sec. III A) since it depends on the inferred momentum of the neutrino. Which method is ultimately chosen depends on the relative precision of the measurement of the transverse momentum distribution of the W (increased systematic uncertainty in p_T) when compared to the calorimeter resolution (increased statistical uncertainty in M_T). In the present case, M_T wins.

C Event Selection

The signature for W events is a high momentum muon or electron ($p_T >$ 25 GeV)[4], and large missing energy from the undetected neutrino ($\not{E}_t > 25$ GeV). This missing energy is the transverse momentum needed to balance the visible transverse momentum in the event. The visible momentum is the sum of the muon or electron momentum and a vector sum of the energy in the calorimeter towers of the detector (not including the energy from the lepton of course; see Sec. III A). The lepton must be central ($|\eta| < 1$) and satisfy various quality cuts. For DØ, these

---

[3] Corrections are of $\mathcal{O}(\beta^2)$ compared with $\mathcal{O}(\beta)$ for p_T^ℓ.
[4] Throughout this paper, $\hbar = c = 1$; thus mass, momentum, and energy all have units of eV.

cuts require the electron to be isolated, to have a calorimeter shower shape consistent with Monte Carlo expectations, and to have a track pointing at the calorimeter cluster. For CDF, the muon is required to have deposited energy in the calorimeter consistent with a minimum ionizing particle, to have a track pointing at the hits in the muon chambers, and to *not* be consistent with a cosmic ray. There is also a requirement that the vector sum of the energy in the calorimeter (not including the contribution from the lepton), $\vec{u}$, be less than 15 GeV for DØ and less than 20 GeV for CDF. This serves to further reduce the background from QCD inspired processes and results in cleaner events. After these requirements are placed on the data sample, CDF has 21,000 W's remaining and DØ has 28,000 W's left.

There are other datasets that are used in this analysis, chief among them being $Z \to \ell^+\ell^-$. The requirements on Z events are similar to W's for one lepton and typically loosened somewhat for the other lepton resulting in 2200 Z's for DØ and 1400 Z's for CDF.

II LEPTON MOMENTUM CALIBRATION

CDF and DØ take similar approaches to calibrating the energy scale for the lepton. Both involve comparing measured mass resonances with known mass values. In the case of CDF, $J/\psi \to \mu\mu$ decays are used to set the momentum scale of the tracking chamber; at DØ, $Z \to ee$, $\pi^0 \to \gamma\gamma$, and $J/\psi \to ee$ are used to set the energy scale of the calorimeter.

A CDF Momentum Scale

The momentum scale of the tracking chamber / magnetic field is set using the invariant mass distribution of $\sim 250,000$ $J/\psi \to \mu\mu$ events (Fig. 1), which is fit using a simulated lineshape. The data are corrected for magnetic field variations over the course of data taking and the momenta are corrected for energy lost in the material before the tracking chamber. The J/ψ simulation includes QED radiative contributions and both prompt and B decay sources of J/ψ's. These two effects can be seen in Figure 1. The latter is important because the tracks from the muons are constrained to originate from the beamline which introduces a systematic bias in tracks that did not originate from the beamline. This beam constraint is applied to J/ψ events to uncover any possible unknown biases that may affect the W mass measurement.

Fitting the lineshape to the J/ψ data results in a mass of 3096.2 ± 1.5 MeV. This translates to a momentum scale of 0.99977 ± 0.00048 leading to an uncertainty of 40 MeV on the W mass. The dominant uncertainties in the momentum scale are the dE/dx energy loss correction and the extrapolation of the momentum scale from the J/ψ mass to the W mass.

Muon Energy Loss – The muons from J/ψ decays traverse material before entering the tracking volume and thus lose energy. The amount of energy they lose

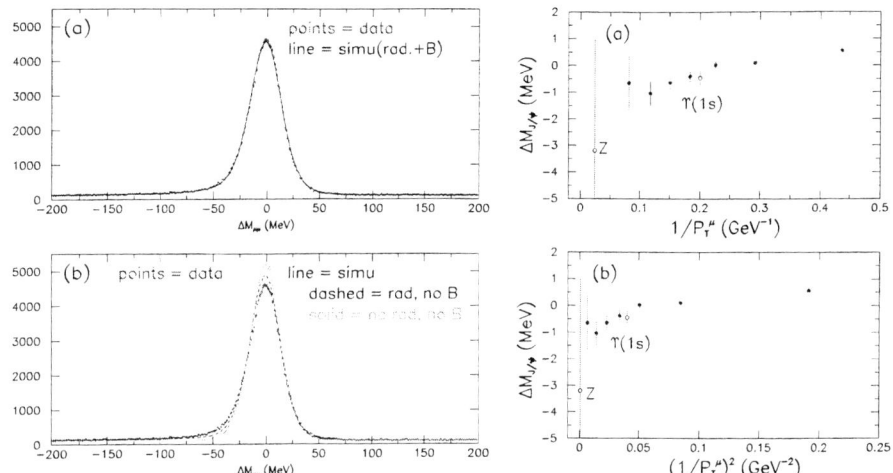

FIGURE 1. Left: a) Invariant mass distribution of $J/\psi \to \mu\mu$ events. b) The effect of adding various features to the simulation. Right: J/ψ mass as a function of a) inverse p_T^μ and b) inverse $(p_T^\mu)^2$.

depends on the amount and type of material. The amount of material can be obtained using photon conversions to electron-positron pairs. These conversions are used to map out the material in the inner detector, and when combined with knowledge about the composition of the various structures provide the necessary corrections. The uncertainty of 1.0 MeV produced in the J/ψ mass is due to uncertainties in the material types in the various regions and to residual variation in the J/ψ mass with region.

Extrapolation to the W – The momentum scale is obtained using muons from J/ψ decays which have an average momentum of 4 GeV. This is a long way from the typical 38 GeV momenta of W decay muons. Fortunately the relevant quantity is not momentum but inverse momentum, which is proportional to the curvature of the track. This is what the tracking chamber measures and where deviations from expected behavior should occur. Figure 1 shows the variation of J/ψ mass with inverse momenta. The advantage is that the distance in inverse momenta from the J/ψ to the W is shorter than the spread in the J/ψ data making for an effortless extrapolation. Since the extrapolated difference is small and since wrong dE/dx corrections, for example, can fake a variation here, it is taken as an uncertainty rather than a correction. This uncertainty, if expressed in terms of the J/ψ mass, is also 1.0 MeV.

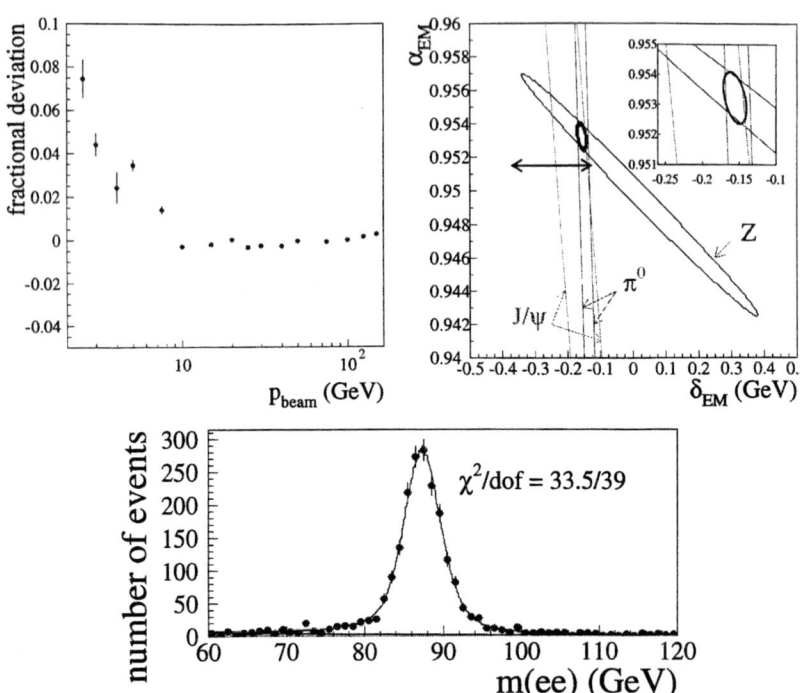

FIGURE 2. Top Left: Deviations from the expected response for the DØ electromagnetic calorimeter for electrons as a function of the electron energy. Top Right: The contours in the scale parameter's plane for $J/\psi \to ee$, $\pi^0 \to \gamma\gamma$, and $Z \to ee$. The thick contour is all three combined including only the statistical component. The double arrow indicates the systematic uncertainty in δ due to deviations in the testbeam results at low energy (top left plot). Bottom: Invariant mass distribution of electron pairs near the Z compared with simulation where the simulation is adjusted for the energy scale to allow comparison with data.

B DØ Energy Scale

The DØ calorimeter is a Uranium/Liquid Argon sampling calorimeter which, since the LAr has unit gain, is extremely stable over time. This enables electron testbeam data to be used to obtain a simple functional form for the energy scale (Fig. 2). The testbeam results indicate that the calorimeter response is very linear and only deviates from the expected response below 10 GeV. Thus a linear function is used to relate the true energy to the measured energy, $E_{meas} = \alpha_{EM} E_{true} + \delta_{EM}$. The deviations from this linear behavior are treated as systematic uncertainties.

Like CDF, DØ determines the constants in the above equation by comparing mass resonances with known masses. Three resonances are used: $J/\psi \to ee$, $\pi^0 \to \gamma\gamma$, and $Z \to ee$.

$J/\psi \to ee$ **Decays** – The relation between the measured J/ψ mass peak and the

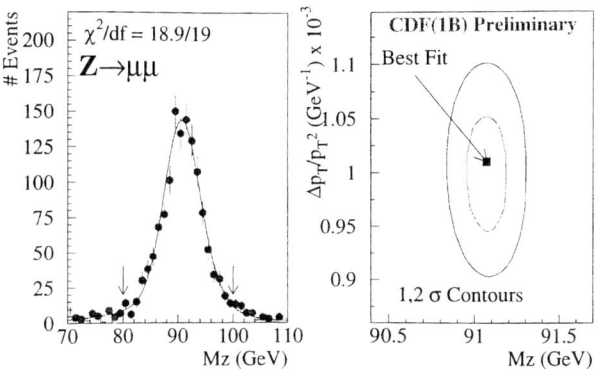

FIGURE 3. Left: Invariant mass distribution of $Z \to \mu\mu$ events from CDF. The width of the peak is a convolution of the linewidth of the Z and the momentum resolution. Right: Contours in momentum resolution and mass with best fit point.

J/ψ mass is determined from simulation to be $m(ee) = \alpha_{EM} m_{J/\psi} + 0.56 \delta_{EM}$. This equation defines a contour in the α_{EM}-δ_{EM} plane (Fig. 2).

$\pi^0 \to \gamma\gamma$ Decays – The decay $\pi^0 \to \gamma\gamma$ is detected by identifying the electrons resulting from the conversions of the two photons[5]. Because the calorimeter clusters from the 4 electrons overlap, a quantity called the symmetric mass is constructed which is equal to the invariant mass if the energy of the two photons is equal. As with the $J/\psi \to ee$ decays, a similar equation relating the symmetric mass to the actual mass can be obtained and again results in a contour in the α_{EM}-δ_{EM} plane (Fig. 2).

$Z \to ee$ Decays – Both the previous resonances are low masses and as such do a good job of constraining δ_{EM} but do a poor job of constraining α_{EM}. To constrain α_{EM} the $Z \to ee$ resonance is used. The Z has the added benefit that the decay electrons are not monochromatic thus providing another albeit weak measurement of δ_{EM}. Again, a contour in α_{EM}-δ_{EM} is found and together with the J/ψ and π^0 provides a tight constraint on α_{EM} and δ_{EM} (Fig. 2). The best fit returns $\delta_{EM} = -0.16^{+0.03}_{-0.21}$ GeV and $\alpha_{EM} = 0.9533 \pm 0.0008$ where the uncertainties include systematic contributions.

C Energy and Momentum Resolutions

The lepton energy or momentum resolution is obtained by fitting for the width of either the $Z \to ee$ (DØ; Fig. 2) or $Z \to \mu\mu$ (CDF; Fig. 3) distributions.

The electron energy resolution for DØ is parameterized as

$$\frac{\sigma_E}{E} = \frac{13.5\%}{\sqrt{E \sin\theta}} \oplus \kappa \oplus \frac{n}{E} \qquad (2)$$

---

[5] Each of the two conversion pairs appears as a doubly ionized track in the tracking chamber

where the first term (stochastic) is determined from testbeam data and the last term (n/E) is the contribution from other energy in the event and is determined from calorimeter towers near the electron. The constant term, κ, is measured from the width of the Z and is found to be $(1.15^{+0.27}_{-0.36})\%$.

The momentum resolution for CDF is parameterized as $\sigma_{1/p_T} = \kappa \cdot (1/p_T)$ since the tracking chamber measures the curvature of a track which is proportional to $1/p_T$. The constant is extracted from the width of the Z and is $(0.101 \pm 0.005)\%$.

III RECOIL MEASUREMENT AND CALIBRATION

A Recoil Measurement

There are only two quantities to measure in every W event: the lepton p_T, and the transverse recoil momentum, $\vec{u}$, from the p_T of the W. Together they can be used to infer the neutrino p_T. The recoil momentum is defined as $\vec{u} = \sum_i (E_i \cdot \sin\theta_i)\hat{\phi}_i$ where the sum runs over calorimeter towers and θ and $\hat{\phi}_i$ are the polar angle and azimuthal unit vector of the tower containing energy E. The towers containing energy deposited by the lepton are removed from the sum; however, the removed towers also contain small contributions from the recoil which must be accounted for. The CDF measurement replaces the removed towers with an average recoil event energy determined from nearby towers. The DØ measurement on the other hand duplicates the removal in the simulated data and also corrects the simulated electron energy to account for this small recoil contamination.

B Recoil Calibration

The calibration of the recoil measurement is obtained from Z data where the recoil measurement can be compared to the p_T of the Z measured with the leptons. There is a slight complication in that $\vec{u}$ contains not only the recoil energy but also the energy from the (anti)proton breakup plus any overlapping $\bar{p}p$ interaction. Thus the calibration also includes a minimum bias component.

DØ Calibration — The parameterization of $\vec{u}$ is given by

$$\vec{u} = \left[R_{rec}\, \vec{q}_T + s_{rec}\sqrt{R_{rec}\, q_T}\, \hat{q}_t \right] - \Delta \vec{u} + \alpha_{mb} \vec{E}_T^{mb} \tag{3}$$

$$R_{rec} = \alpha_{rec} + \beta_{rec} \log(q_T) \tag{4}$$

where R_{rec} represents the recoil response, s_{rec} is the response resolution, Δu is the small correction for the tower removals mentioned in Section III A, and the last term is a literal minimum bias event weighted by α_{mb}.

The form of R_{rec} is obtained from a Herwig-Geant $Z \to ee$ simulation (Fig. 4) and does a good job of describing the DØ jet response. Z data is used to constrain the parameters α_{rec} and β_{rec} and the values obtained ($\alpha_{rec} = 0.693 \pm 0.060$ and

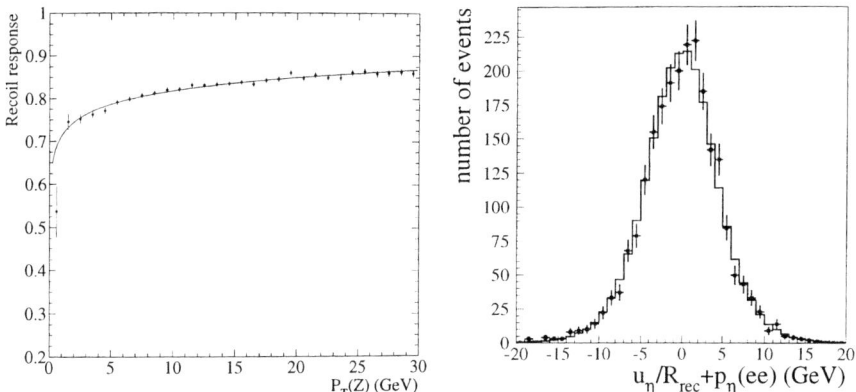

FIGURE 4. Left: Simulated detector response to recoil from Z's in Herwig-Geant $Z \to ee$ simulation. Right: Difference between p_η and u_η after being corrected by the response function. The points are the data and the histogram is the simulation. This distribution should have a mean of zero and a width characteristic of the recoil resolution. (η is the angular bisector of the two electrons and is the axis with the least sensitivity to the electron energy resolution.)

$\beta_{rec} = 0.040 \pm 0.021$) agree with the Herwig-Geant values of 0.713 and 0.046. The parameter s_{rec} is the recoil part of the resolution parameterization and is also obtained from $Z \to ee$ events along with the parameter α_{mb} representing the non-recoil part of the resolution. This non-recoil part is modeled by a minimum bias event (chosen such that the luminosity distribution is the same as in the Z data) multiplied by a weight, α_{mb}, which is constrained by the Z data[6]. Comparisons of Monte Carlo $Z \to ee$ events with data in Figure 4 show good agreement in both the mean of the distribution (response) and the width (resolution).

CDF Calibration — The CDF parameterization of $\vec{u}$ is

$$\vec{u} = R_{rec}\, \vec{q}_T + S\, \vec{\sigma}_{mb} \qquad (5)$$

$$R_{rec} = \alpha_{rec} + \beta_{rec}\, e^{-\lambda q_T} \qquad (6)$$

where, as with DØ, R_{rec} is the recoil response and, unlike DØ, all the resolution is contained in the second term.

The form of R_{rec} is obtained from Z data and is plotted in Figure 5. The parameters are additionally constrained by W data distributions to improve the uncertainty in the recoil response.

The resolution term handles both resolution from the minimum bias contribution and from the recoil response resolution. This works because for most W's, the recoil tends to look like the minimum bias contribution and thus one resolution is a fairly good approximation. The starting point for the resolution term is minimum bias data in which energy fluctuations are parameterized in terms of total energy in the

---

[6] α_{mb} must be adjusted for W data which have a different luminosity distribution.

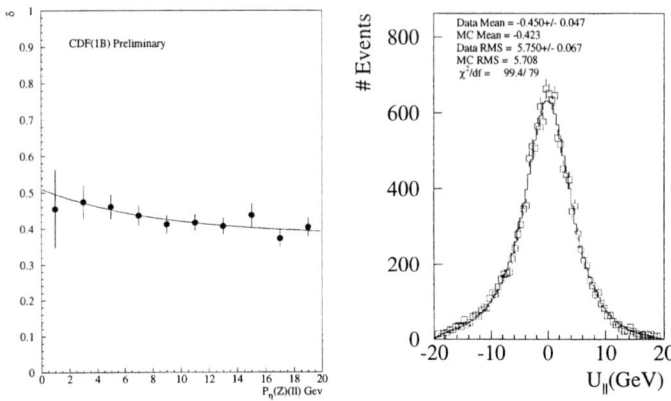

FIGURE 5. Left: Plot of $1-R_{rec}$ vs. p_η for Z events. The points are both $Z \to ee$ and $Z \to \mu\mu$ events and the solid line is the fit to these points. Right: Comparison of $u_\parallel$ from W data and simulation. The variable $u_\parallel$ is the projection of $\vec{u}$ onto the lepton momentum axis. This quantity should have a negative mean since the lepton is boosted in the direction of the W (opposite $\vec{u}$).

event which itself is a function of the luminosity. This is weighted by S which is constrained from $|\vec{u}|$ distributions in the W data.

After obtaining the parameters for the recoil model, one can compare the simulation with W data. Figure 5 is one such comparison of $u_\parallel$ in data and simulation from CDF. Comparisons from DØ show similar agreement.

IV   MONTE CARLO SIMULATION

The simulations used by both CDF and DØ are similar in form. They start with a tree level calculation including parton distribution functions (PDF's). In an effort to separate out the various effects of the PDF's, DØ parameterizes the Q^2 effect of the PDF's as $e^{-\beta Q}/Q$ where β is determined from Monte Carlo studies. The choice of which PDF to use is somewhat debatable. CDF attempts to constrain the allowed range of PDF's using W decay charge asymmetry data where the asymmetry is a function of the u to d ratio. The W lineshape is also dependent on this u-d ratio and thus the asymmetry can be used to set limits on the allowed PDF's. Unfortunately, the range of current PDF's does not fill the allowed asymmetry space making it difficult to use this method. DØ has chosen a small set of recent PDF's and taken the variation as a systematic uncertainty.

NLO QCD contributions are incorporated using a calculation [3] which matches the $\mathcal{O}(\alpha_s^2)$ large q_T perturbative result with a small q_T soft gluon resummation. There are 3 parameters ($g1$, $g2$, $g3$) plus Λ_{QCD} in the most recent incarnation of this calculation. CDF fixes $g3$ to the Ladinsky-Yuan value and constrains $g1$ and $g2$ with Z data (Λ_{QCD} is varied by varying PDF's). DØ sets $g1$ and $g3$ to the

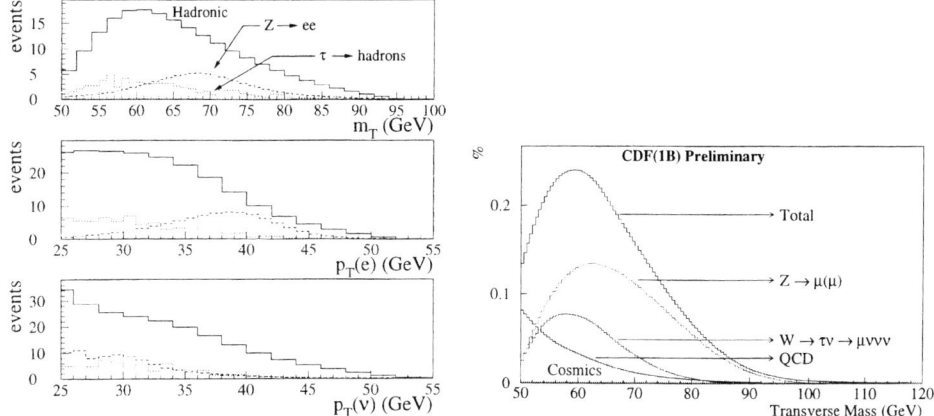

FIGURE 6. Left: DØ background shapes for transverse mass and lepton and neutrino p_T distributions. The hadronic background is $(1.3\pm0.2)\%$, the Z background is $(0.42\pm0.08)\%$, and the $\tau \to hadrons$ is 0.24%. There is also a $W \to \tau\nu \to e\nu\nu\nu$ which is generated simultaneously with the $W \to e\nu$ events and contributes 1.6% events. Right: CDF background shapes for the transverse mass distribution. The dominant background is $Z \to \mu\mu$ where one muon is not detected in the tracking chamber and amounts to $(3.6\pm0.5)\%$. The other backgrounds are $W \to \tau\nu \to \mu\nu\nu\nu$ at 0.8%, QCD at $(0.4\pm0.2)\%$, and Cosmic Rays which add $(0.1\pm0.05)\%$.

Ladinsky-Yuan values and constrains $g2$ with Z data while allowing Λ_{QCD} to vary within reasonable bounds.

The decay model includes QED radiative effects using a calculation by Berends and Kleiss [4]. The results were checked by both CDF and DØ using two-photon generators [5].

Backgrounds are included in the simulated lineshape and are detailed in Figure 6.

V RESULTS

The mass of the W is extracted using a likelihood fit of the Monte Carlo lineshapes to the data. The most precise value is obtained from fits to the transverse mass (Fig. 7).

The results for CDF and DØ are

$$M_W^{D\emptyset} = 80.44 \pm 0.12 \text{GeV} \tag{7}$$
$$M_W^{CDF} = 80.43 \pm 0.16 \text{GeV}. \tag{8}$$

The uncertainties in these measurements are documented in Table 1. Combining these numbers with previous measurements by CDF and DØ and with measurements made by the LEP II experiments and NuTeV leads to a world average of

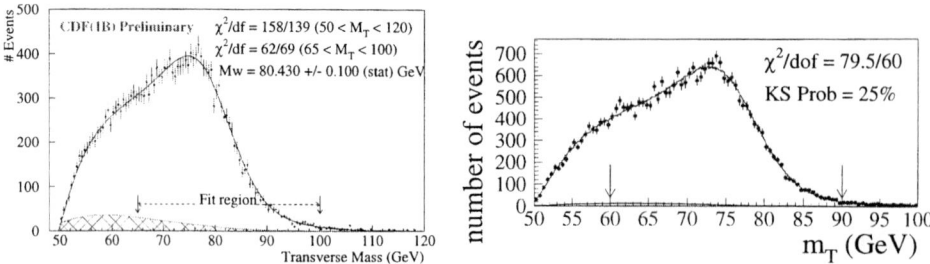

FIGURE 7. Left: CDF transverse mass distribution compared with simulation. The fit region is 65 to 100 GeV. Right: Same plot for DØ with a fit region of 60 to 90 GeV.

Uncertainty	CDF (MeV)	DØ
Statistical	100	70 } 95(stat)
Momentum / Energy Scale	40	65
Calorimeter Linearity	–	20
Lepton Resolution	25	20
Recoil Modeling	*90*	40
Input W p_T and PDF's	50	25
Radiative Decays	20	15
Higher Order Corrections	20	–
Backgrounds	25	10
Lepton Angle Calibration	–	30
Fitting	10	–
Miscellaneous	20	15
Systematics Total	115	70
Total (MeV)	**155**	**120**

TABLE 1. Uncertainties in the W mass measurement for both CDF and DØ. It is suspected that the 90 MeV recoil modeling uncertainty for CDF is due in part to deficiencies in the model and should decrease in the final analysis. The energy scale uncertainty for DØ is dominated by the statistics of the Z data.

$$M_W = 80.420 \pm 0.055 \text{GeV}. \qquad (9)$$

These measurements are listed in Figure 8 together with the current best indirect determination of M_W. Also in Figure 8 is a graphical representation of the state of the art in precision electroweak testing. The data point is the world average from above.

A The Future

CDF is currently working on adding the electron decay channel to the mass measurement while improving the muon result. The expectations are that the

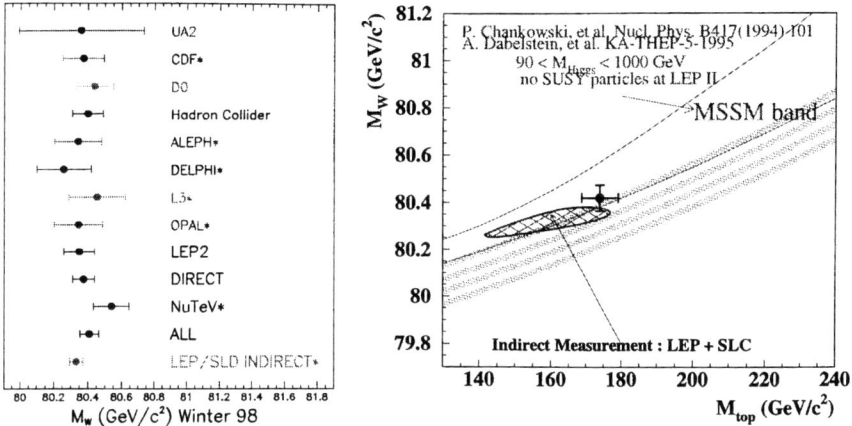

FIGURE 8. Left: W mass measurements from around the world. The CDF and DØ points are combined numbers from all their respective data taking runs. Right: W mass versus top mass for various higgs masses and SUSY.

CDF uncertainty can be reduced to a little under 100 MeV. DØ is working on increasing the pseudorapidity range of accepted electrons by including those that traverse the endcap calorimeter. Here too, the aim is to reach 100 MeV.

Both experiments are also upgrading the detectors in preparation for Run 2 of the Tevatron which will bring a factor of 20 more data. With this larger dataset and a lot of perseverence on the part of the experimenters, the uncertainty is expected to reach ~ 40 MeV. Even farther in the future, TeV33 is supposed to supply another factor of 10 in data and a corresponding decrease in the W mass uncertainty, perhaps as low as 20-30 MeV.

REFERENCES

1. B. Abbott *et al.*, (DØ Collaboration), Phys. Rev. Lett. **80**, 3008 (1998), and submitted to Phys. Rev. D. More information on both the CDF and DØ measurements can be found at http://www-cdf.fnal.gov/physics/ewk/ewk.html and http://www-d0.fnal.gov/public/wz/ewk_public.html.
2. W. Hollik and W. Marciano, *Precision Tests of the Standard Electroweak Model*, ed. by P. Langacker (World Scientific, Singapore, 1994).
3. G.A. Ladinsky and C.P. Yuan, Phys. Rev. D **50**, 4239 (1994); P.B. Arnold and R.P. Kauffman, Nucl. Phys. B **349**, 381 (1991).
4. F.A. Berends and R. Kleiss, Z. Phys. C **27**, 365 (1985); F.A. Berends *et al.*, Z. Phys. C **27**, 155 (1985).
5. CDF used PHOTOS and DØ used a calculation by U. Baur *et al.*, Phys. Rev. D **56**, 140 (1997).

W couplings measurements at LEP

Marco Verzocchi

*Fakultät für Physik,
Albert–Ludwigs Universität
Herman–Herder-Straße 3
D-79104 Freiburg i. Br.
Germany*

Abstract. I will review measurements performed by the LEP experiments on properties of the W bosons: the production cross section, decay branching ratios and couplings at vertices involving three electroweak gauge bosons.

INTRODUCTION

The LEP collider has been running since 1996 at centre–of–mass energies above the threshold for pair–production of W bosons, allowing precision measurements of the properties of the W bosons competitive with those performed at hadron colliders [1]. This review describes measurements of the couplings of the W with both fermions and neutral electroweak bosons. This sector of the Standard Model ($\mathcal{SM}$) is essentially untested, in contrast to precision reached in the measurements of couplings between fermions and neutral vector bosons at LEP and SLC [2]. The availability of pure samples of W bosons allows for the first time direct measurements of their decay branching ratios into hadrons and into leptons. It also allows one to test an important consequence of the non–Abelian nature of the $SU(2) \otimes U(1)$ gauge group of the $\mathcal{SM}$, the coupling between gauge bosons, which in turn is related to the mechanism provoking the spontaneous breaking of the gauge symmetry. Evidence for anomalous couplings would signal new physics beyond the $\mathcal{SM}$. This review will describe the analysis techniques and preliminary results obtained from the data collected in 1997 (≈ 57 pb^{-1} for each experiment at centre–of–mass energies around $\sqrt{s}=183$ GeV). Where relevant the results are combined with those based on the 1996 data samples (≈ 10 and ≈ 11 pb^{-1} collected by each experiment at $\sqrt{s}=161$ and 172 GeV respectively) [3-6]. Measurements of the W boson mass performed at LEP are reviewed elsewhere in these proceedings [7].

CROSS SECTION AND BRANCHING RATIO MEASUREMENTS

Pair production of W bosons in e^+e^- collisions proceeds through three doubly resonating diagrams (also known as CC03 diagrams): two s–channel annihilation diagrams $e^+e^- \to \gamma/Z^0 \to W^+W^-$ and one t–channel ν_e exchange diagram. Precise theoretical predictions including higher order corrections are available for this CC03 cross section. Given that the W bosons are short lived what is measured experimentally is the cross section for producing four fermions in the final state. A large number of diagrams contribute to this cross section, which is known in most cases only numerically: depending on the invariant mass of the fermions the diagrams leading to one final state can be classified as doubly, singly or non–resonating diagrams. Final states which involve the decay of both Ws into the same fermion–antifermion pair can be produced also through diagrams which involve only the exchange of neutral electroweak bosons. The CC03 cross section is obtained from the experimentally measured one applying correction factors determined from Monte Carlo simulations.

The $e^+e^- \to W^+W^-$ final states are classified as leptonic ($W^+W^- \to \ell^-\bar{\nu}_\ell \ell'^+\nu_{\ell'}$), semileptonic ($W^+W^- \to q\bar{q}'\ell\bar{\nu}_\ell$) or hadronic ($W^+W^- \to q\bar{q}'q\bar{q}'$), with expected branching ratios of 10.5%, 43.9% and 45.6% respectively. In the following, the criteria used to select events for the cross section measurements are reviewed. The same procedures are generally used also for selecting the events used in the measurements of the W mass and of the triple gauge boson couplings.

Leptonic final states

Fully leptonic WW events are characterised by the presence of two energetic leptons or low multiplicity jets produced by the decay of τs and large missing momentum due to the neutrinos. The event selections are generally based on the event topology and require two low multiplicity acoplanar jets (here a jet may also be a single leptonic track) and large missing transverse momentum. Events where a high energy lepton pair is produced trough γ/Z^0 annihilation are vetoed. High selection efficiencies are reached applying loose lepton identification criteria, although this leads to some cross–contamination from the τ decay channels, which must be then corrected for. The average efficiency is for all experiment around 70%, with purities in excess of 90%, and samples of 60–70 events have been selected by each experiment at $\sqrt{s}$=183 GeV, as shown in table 1. These are only average values, the efficiencies and purities for channels with τs in the final state being generally lower. Backgrounds are dominated by lepton pair production in the collisions of pairs of virtual photons emitted from the beams and by 4–fermion events of the ZZ type. The cross sections measured by the 4 LEP experiments at $\sqrt{s}$=183 GeV assuming for the W branching ratios the $\mathcal{SM}$ values are given in table 1.

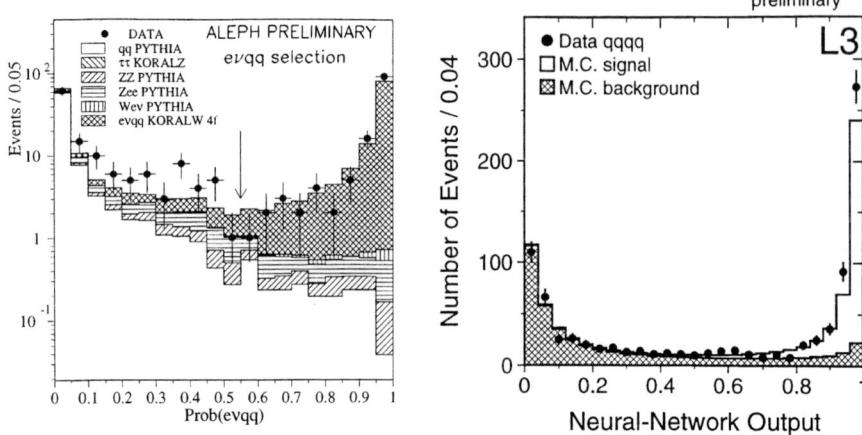

FIGURE 1. Right: Probability distribution for the events selected for the $q\bar{q}'e\bar{\nu}_e$ channel by the ALEPH collaboration. The arrow shows the cut above which events are kept. Left: Distribution of the output of the neural network used by the L3 collaboration in the analysis of the $e^+e^- \to qqqq$ events, comparing the signal and background Monte Carlo to the data. The cross section for $W^+W^- \to q\bar{q}'q\bar{q}'$ is obtained from a fit to the distribution.

Semileptonic final states

These events are characterised by the presence of a high energy lepton, two jets and missing energy. The selection typically begins with the identification of the lepton as the highest momentum isolated track. Then the rest of the event is forced into two jets. Vetoes are applied on events with isolated photons which may fake the signature of an electron, and on four fermion events with two charged leptons in the final states. The signal is then isolated by requiring high missing transverse momentum and placing cuts on the energy and isolation of the lepton. These variables may also be combined in a probability, as shown in an example from the ALEPH experiment in figure 1. The selection of the $W^+W^- \to q\bar{q}'\tau\bar{\nu}_\tau$ channel proceeds in a similar way: the τ decay product are identified as a low multiplicity narrow jet. Tighter cuts on the selection variables are then applied, resulting usually in a lower efficiency for this channel. Purities in excess of 90% for efficiencies around 80% are obtained, and samples of around 300 events were collected during the 1997 run. The background is dominated by $e^+e^- \to q\bar{q}(\gamma)$ events and four-fermion final states which are not reached through the CC03 process. The cross sections obtained for this channel, assuming the Standard Model values for the branching ratios are given in table 1.

TABLE 1. Results on WW production at $\sqrt{s}$=183 GeV.

	Average efficiency	purity	Selected events	Cross section
$W^+W^- \to \ell^-\bar{\nu}_\ell\ell'^+\nu_{\ell'}$				
ALEPH	71%	89%	60	$1.34 \pm 0.20 \pm 0.05$ pb
DELPHI	60%	83%	59	$1.64 \pm 0.26 \pm 0.09$ pb
L3	56%	86%	54	$1.53 \pm 0.24 \pm 0.05$ pb
OPAL	77%	94%	78	$1.69 \pm 0.20 \pm 0.06$ pb
$W^+W^- \to q\bar{q}'\ell\bar{\nu}_\ell$				
ALEPH	80%	95%	322	$6.79 \pm 0.40 \pm 0.14$ pb
DELPHI	73%	89%	288	$7.04 \pm 0.47 \pm 0.28$ pb
L3	78%	93%	311	$6.82 \pm 0.42 \pm 0.11$ pb
OPAL	86%	90%	362	$6.65 \pm 0.39 \pm 0.13$ pb
$W^+W^- \to q\bar{q}'q\bar{q}'$				
ALEPH	82%	84%	423	$7.44 \pm 0.41 \pm 0.30$ pb
DELPHI	82%	76%	391	$7.33 \pm 0.48 \pm 0.29$ pb
L3	88%	80%	475	$8.25 \pm 0.45 \pm 0.25$ pb
OPAL	84%	80%	433	$7.15 \pm 0.43 \pm 0.31$ pb
CC03 cross section				
ALEPH				$15.51 \pm 0.61 \pm 0.36$ pb
DELPHI				$16.01 \pm 0.71 \pm 0.43$ pb
L3				$16.66 \pm 0.66 \pm 0.30$ pb
OPAL				$15.52 \pm 0.62 \pm 0.35$ pb

Hadronic final states

This channel is characterised by final states which have four well separated hadronic jets. The missing energy is only due to radiation from the initial state and is therefore small. Events are selected by requiring large visible energy and invariant mass. Multijet production in $e^+e^- \to q\bar{q}(\gamma)$ collisions is characterised by smaller jet energies and jet–jet angles with respect to the $W^+W^- \to q\bar{q}'q\bar{q}'$ final states. This information is usually combined with global event properties (sphericity, the y_{34} jet parameter, ...) and in some cases also with mass related informations into neural networks or likelihoods. The neural network output used by the L3 collaboration is shown in figure 1. The results of the selections applied to the data collected at $\sqrt{s}$=183 GeV are reported in table 1.

Cross section and branching ratio fits

Each collaboration performs several fits to obtain the CC03 cross section and the W decay branching ratios. In all cases a maximum likelihood fit to the number of observed events is performed. The number of expected events takes into account the presence of the background, the "cross talk" between the different decay channels of the W and the different branching ratios. The CC03 cross section is always one of the free parameters of the fits. In addition the leptonic branching ratios can be fitted for, with or without the additional assumption of lepton universality. Results from all the experiments are then combined by the LEP Electroweak Working Group, to obtain the best possible values. All the results confirm the hypothesis of lepton universality and yield for the leptonic branching ratio an average value of 10.46 ± 0.26%, in agreement with the $\mathcal{SM}$ prediction of 10.8%. The hadronic branching ratio is determined to be 68.6 ± 0.8% also in agreement with the $\mathcal{SM}$ prediction of 67.5%. The CC03 cross sections measured by the four LEP experiments at $\sqrt{s}$=183 GeV [8] are given in table 1. The different measurements of W pair production performed at LEP are shown in figure 2 as a function of the centre–of–mass energy and compared with the expectations assuming a W mass of 80.35 GeV/c^2. Models without γWW or Z^0WW vertices are excluded at a high confidence level from these measurements.

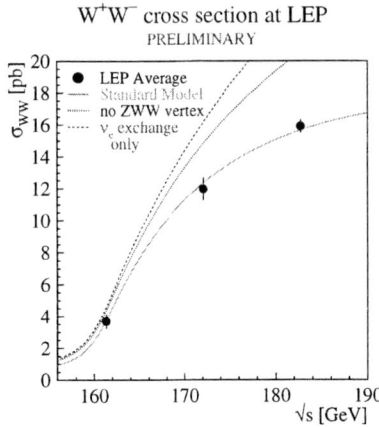

FIGURE 2. The W^+W^- cross section at LEP.

TRIPLE GAUGE BOSON VERTICES

The most general Lagrangian which describes a triple gauge boson vertex (γWW or Z^0WW) depends linearly on seven terms [9], giving a total of 14 unmeasured parameters [1]. Of these six violate $\mathcal{CP}$ and two violate $\mathcal{C}$ and $\mathcal{P}$ separately. Current LEP2 statistics are insufficient to measure all fourteen couplings simultaneously and therefore additional theoretical constraints are applied. If only terms corresponding

---

[1] The $\mathcal{SM}$ predicts also vertices with four gauge bosons. However these are not accessible experimentally with the current centre–of–mass energies and luminosities.

to operators with dimension smaller than 6 and which do not violate $\mathcal{C}$, $\mathcal{P}$ and $\mathcal{CP}$ are considered and electromagnetic gauge invariance is invoked, only 5 terms remain, and their values are predicted in the $\mathcal{SM}$ to be:

$$\kappa_\gamma = \kappa_Z = g_1^Z = 1, \; \lambda_\gamma = \lambda_Z = 0.$$

These couplings can be related to a multipole expansion of the electric and magnetic charge of the W boson, and of their weak charge equivalents. If, in addition, constraints from lower energy measurements are taken into account (these are mainly the *oblique* corrections to the Z propagator), the set of anomalous couplings to be investigated can be restricted to three parameters:

$$\alpha_{W\phi} = \Delta g_1^Z \cos^2 \theta_W,$$
$$\alpha_{B\phi} = \Delta \kappa_\gamma - \Delta g_1^Z \cos^2 \theta_W,$$
$$\alpha_W = \lambda_\gamma.$$

This corresponds to the set of most loosely constrained couplings if one assumes that $\lambda_\gamma = \lambda_Z$ and $\Delta \kappa_Z = -\Delta \kappa_\gamma \tan^2 \theta_W$ [2].

The presence of anomalous couplings affects the relative contribution of each of the two W helicity states to the total cross section as a function of the W–pair scattering angle, defined as the polar angle of the W^- relative to the electron beam. As a consequence both the total and the differential cross section for W production have a bilinear dependence on the anomalous couplings. The weak interaction in the decay of the W boson provides the polarimeter which allows to measure the helicities of the two W bosons.

Experimental strategies

If the finite W width and initial state radiation (ISR) are both neglected, the final state (the W–pair scattering angle and the helicities of the two W bosons) is fully determined through the measurements of the five angles shown in figure 3, the momenta of the two Ws being determined through energy conservation. This approximation remains good even though the two W momenta are only loosely constrained, due to the non–negligible W width and the presence of ISR.

The measurement of the full set of five angles is, however, not possible experimentally since the neutrinos (which are at least two in a $W \to \tau$ decay) are undetected and the charges of the quarks cannot be measured on a event by event basis. Therefore, depending on the decay channels of the two W bosons some variables remain unmeasured or possess some ambiguity. In addition, effects of the detector resolution and of background have to be taken into account. The event selections are in some cases tightened with respect to those used in the cross section measurements to ensure that the kinematical quantities are properly measured

---

[2] Here Δ refers to the difference from the $\mathcal{SM}$ value.

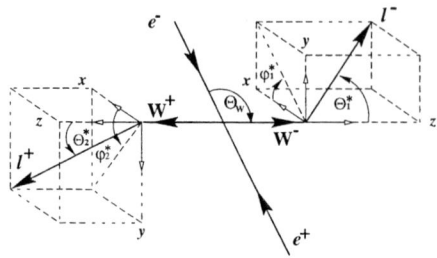

FIGURE 3. The 5 angles determining the W–pair kinematic variables.

(this applies mostly for the τ channel) or that the jet assignment to different Ws is correct. Resolution on the decay angles is improved for channels involving at least one hadronic decay of the W with a kinematical fit which typically uses the constraint of equal W masses.

The channel which gives most informations is $W^+W^- \to q\bar{q}'\ell\bar{\nu}_\ell$: here three of the angles can be measured without ambiguity, since the lepton charge provides the sign of the W scattering angle. A twofold ambiguity remains for the angle of the quarks and jet charge techniques do not help to resolve it on a event by event basis. This is even more problematic for the $W^+W^- \to q\bar{q}'q\bar{q}'$ channel. Jet charges techniques can be applied to distinguish the charges of the two bosons provided the correct combination of jets is chosen out of the three possible ones. Although in this case the difference of charge to be measured is largest (2 units), the sign of the scattering angle can be reliably measured only in approximately 70% of the events. The reliability of the jet charge method can be tested on the data themselves using $W^+W^- \to q\bar{q}'\ell\bar{\nu}_\ell$ decays. The difference of statistical power between the $W^+W^- \to q\bar{q}'\ell\bar{\nu}_\ell$ and $W^+W^- \to q\bar{q}'q\bar{q}'$ channels can be seen in figure 4. All LEP collaborations have so far investigated these two channels so far [10–12].

The OPAL collaboration has investigated also the $W^+W^- \to \ell^-\bar{\nu}_\ell\ell'^+\nu_{\ell'}$ channel. If the energy of the leptons is determined correctly, all 5 angles can be measured in these events (up to a twofold ambiguity for three of them) giving sensitivity to the correlation between the helicities of the two W bosons. Therefore the decay channels with τs cannot be used, reducing the useful branching ratio. Despite the low statistics this channel helps particularly in improving the limits on α_W [12].

Fit methods

The couplings are obtained using both the information from the total cross section and from the differential one. The differential distributions are fitted using different methods: unbinned and binned maximum likelihood methods and the *optimal observables* method [13], which in the end all give similar results and are thus used to cross–check the results.

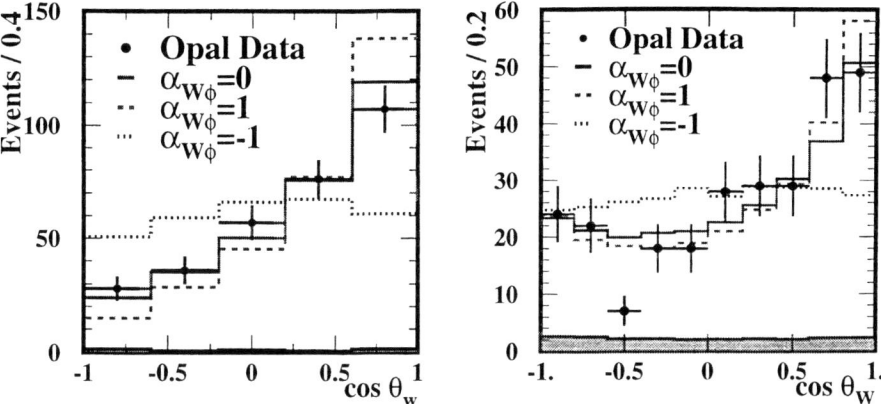

FIGURE 4. Angular distribution of the $W^+W^- \to q\bar{q}'\ell\bar{\nu}_\ell$ (right) and $W^+W^- \to q\bar{q}'q\bar{q}'$ (left) events selected by the OPAL collaboration. Data are shown as dots and the prediction of the $\mathcal{SM}$ Monte Carlo as a full line. The bump at $\cos\theta_W = -1$ in the hadronic case is due to wrong measurements of the W charge. The hatched area shows the background from non–WW events. The dashed and dotted lines show the expected distribution for different values of the anomalous couplings.

In the unbinned maximum likelihood method, the differential distribution is parametrised as a function of the anomalous couplings by a simple analytic function which does not take into account the width of the W boson and ISR. These have to be folded in the fitting function together with the detector resolution and acceptance, a rather complicated procedure. The advantage of this method is that it uses all the information present in the data, and does not require large Monte Carlo samples. In the binned maximum likelihood method all physical and detector effects are taken into account, at the cost of a loss of statistical precision, performing a fit of the data to distributions obtained from large Monte Carlo samples which have different values of the couplings.

In the *optimal observables* method the differential distributions are not directly fit, but use to form a function $\mathcal{O}_\alpha$ which is essentially the derivative of the cross section with respect to the anomalous couplings. The direction in the 5–dimensional space which is most sensitive to the couplings is picked up by this optimal observable. The value of the couplings can then be obtained from the mean of the distribution of $\mathcal{O}_\alpha$, or from a likelihood fit to the 1–dimensional distribution. For small values of the couplings this method retains the full power of an unbinned maximum likelihood fit, with the advantage of being much simpler.

Monte Carlo samples generated at different values of the couplings are used to check for biases in the event selection and in the fit procedure. Since the available data samples are small and the detector resolution is not gaussian, simulated events are used to calibrate the errors. The fit results are presented as likelihood curves which are then added when combining results from different channels and from different experiments. For the preliminary results shown below only 1–dimensional fits are given, setting the remaining couplings to their $\mathcal{SM}$ values. However, multi-dimensional fits where all the couplings are free parameters of the fit at the same time have been performed. Data from different experiments have been combined following the procedure outlined in [14], taking into account correlations of systematic errors.

Other channels

Additional information on the anomalous couplings can be obtained measuring other reactions. The measurement of the total and differential cross section for the process $e^+e^- \to \nu\bar{\nu}\gamma$ is directly sensitive to the couplings at the γWW vertex. The $\nu_e\bar{\nu}_e\gamma$ final state can be produced through the WW $\to \gamma$ fusion process, where the two virtual W bosons are emitted from the electron and positron in the initial state. The signature of this channel is that of a single energetic photon in the detector and the main background, the $e^+e^- \to \gamma Z^0$ process can be vetoed with cuts on the photon energy. The ALEPH collaboration has studied this channel and obtained the following limits on the couplings: $|\Delta\kappa_\gamma| \leq 2.2$ and $|\lambda_\gamma| \leq 3.2$ [15].

Another process which is sensitive to the anomalous couplings is the single W production. The basic process is the γW $\to$ W fusion, and as in the previous case the two virtual bosons are emitted from the electron and positron in the initial state. This reaction leads to the same final states as for W–pair production, when one of the Ws decays into an electron, but can be isolated through kinematical cuts. The electron which emits the virtual photon is generally scattered at small angle and remains unobserved. The signature of the events is therefore high missing transverse momentum from the neutrino(s) plus two acoplanar hadronic jets or a lepton from the decay of the W. All experiments have investigated this reaction, although only ALEPH [16] and L3 [17] have presented updated results using the 1997 data:

$$-2.60 \leq \Delta\kappa_\gamma \leq 0.50 \qquad -1.60 \leq \lambda_\gamma \leq 1.60 \qquad \text{ALEPH}$$
$$-0.86 \leq \Delta\kappa_\gamma \leq 0.38 \qquad -1.60 \leq \lambda_\gamma \leq 1.60 \qquad \text{L3}.$$

The interpretation of the single W production cross section in terms only of the γWW couplings of the single W production is complicated by the presence of a large background from $W^+W^- \to q\bar{q}'\tau\bar{\nu}_\tau$ events (which may lead to some double counting of the events), and by some residual sensitivity to the couplings in the Z^0WW vertex. For this reason the result of L3 is not included in the combined limits shown below.

Results

Measurements of the three anomalous couplings performed by the four LEP experiments are summarised and compared with recently published measurements from the DØ collaboration [18] [3] in table 2 and figure 5. At 183 GeV, ALEPH data entering in this combination come at present only from the analysis of the γWW coupling in the single photon final state, and single W production channels. All results are compatible with the $\mathcal{SM}$ values. The combined results of the LEP experiments reach a precision comparable with that obtained by DØ for $\alpha_{B\phi}$ and α_W and are better for a factor 2 for $\alpha_{W\phi}$, since $\alpha_{W\phi}$ is measured in hadronic collisions only in the processes $p\bar{p} \to$ WZ, WW for which only a few events are observed. On the other side, limits on κ_γ and λ_γ independent from the couplings at the Z^0WW vertex obtained at Tevatron are typically better than those obtained at LEP. Combined limits using both the LEP preliminary results and the DØ published have become available after the conference and are reported in table 2 for completeness.

TABLE 2. Results on triple gauge boson couplings.

	Measurements of anomalous couplings		
	$\alpha_{W\phi}$	$\alpha_{B\phi}$	α_W
ALEPH	$-0.14^{+0.27}_{-0.25}$	$0.28^{+0.80}_{-0.94}$	$-0.04^{+0.52}_{-0.47}$
DELPHI	$0.01^{+0.11}_{-0.08}$	$0.31^{+0.76}_{-0.51}$	$-0.11^{+0.16}_{-0.16}$
L3	$-0.12^{+0.11}_{-0.10}$	$-0.43^{+0.33}_{-0.27}$	$-0.25^{+0.23}_{-0.16}$
OPAL	$-0.03^{+0.13}_{-0.13}$	$0.25^{+0.61}_{-0.51}$	$-0.05^{+0.23}_{-0.21}$
LEP	$-0.05^{+0.06}_{-0.06}$	$-0.04^{+0.33}_{-0.24}$	$-0.09^{+0.13}_{-0.12}$
DØ	$0.11^{+0.16}_{-0.16}$	$-0.08^{+0.34}_{-0.34}$	$0.00^{+0.10}_{-0.10}$
LEP + DØ	$-0.03^{+0.06}_{-0.06}$	$-0.05^{+0.22}_{-0.20}$	$-0.03^{+0.08}_{-0.08}$
	95% confidence levels		
	$\alpha_{W\phi}$	$\alpha_{B\phi}$	α_W
LEP	$[-0.12, +0.13]$	$[-0.44, +0.95]$	$[-0.21, +0.27]$
DØ	$[-0.22, +0.44]$	$[-0.77, +0.58]$	$[-0.20, +0.20]$
LEP + DØ	$[-0.14, +0.10]$	$[-0.42, +0.43]$	$[-0.18, +0.13]$

CONCLUSIONS

The first two years of LEP2 running have allowed direct measurements of the W branching ratios and determinations of the triple gauge boson couplings with a precision comparable to that obtained at hadron colliders. No deviation from the

---

[3] In the DØ analyses the anomalous couplings are modified by dipole form factors with a scale $\Lambda = 2$ TeV.

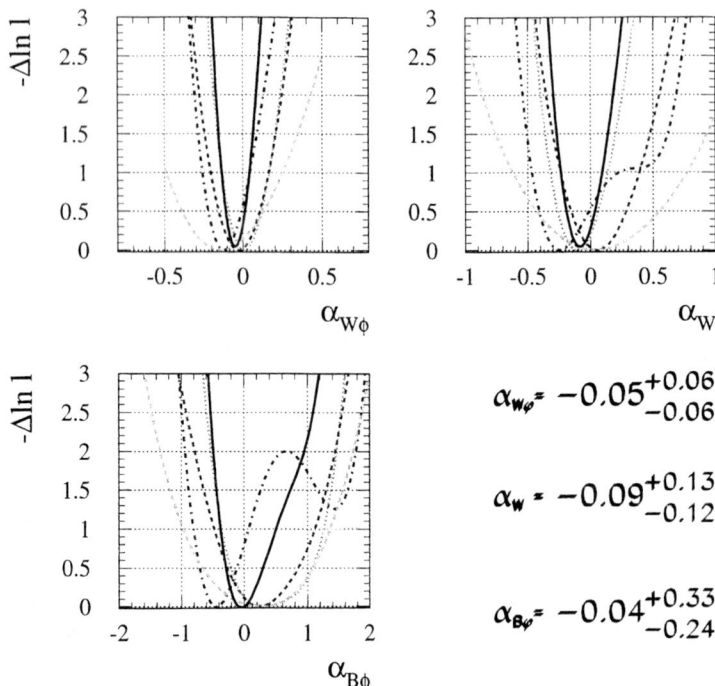

FIGURE 5. LEP combined log$\mathcal{L}$ curves as a function of the anomalous couplings obtained from the combination of the curves from the individual experiments which are shown as dotted lines. The minimum value has been subtracted in all cases.

$\mathcal{SM}$ has been observed so far. With three years of data still to come at higher luminosities the measurements of the triple gauge boson couplings will enter a region yet unexplored. Much work will have to be done on the experimental side to use all possible channels to investigate the anomalous couplings and to keep under control systematic errors.

ACKNOWLEDGEMENTS

I should like to thank Gideon Bella, Dave Charlton, Clara Matteuzzi, Eric Lançon, Peter Molnar and Mark Thomson for their help in compiling results and some useful discussions. I should also like to thank the organisers of the Latin American Symposium for providing such an enjoyable conference.

REFERENCES

1. H.C. Valdez, *these proceedings*,
   R.M. Thurman, *these proceedings*.
2. R. Faccini, *these proceedings*.
3. ALEPH Collaboration, R. Barate *et al.*, *Phys. Lett.* **B422** (1998) 369, *Phys. Lett.* **B415** (1997) 435, *Phys. Lett.* **B401** (1997) 347.
4. DELPHI Collaboration, P. Abreu *et al.*, *Phys. Lett.* **B423** (1998) 194, CERN–PPE/97–160, to be published in Eur. Phys. J. **C**, *Phys. Lett.* **B397** (1997) 1158.
5. L3 Collaboration, M. Acciarri *et al.*, *Phys. Lett.* **B413** (1997) 176, *Phys. Lett.* **B407** (1997) 419, *Phys. Lett.* **B403** (1997) 168, *Phys. Lett.* **B398** (1997) 223.
6. OPAL Collaboration, K. Ackerstaff *et al.*, CERN–PPE/97–125, to be published in *Eur. Phys. J.* **C**, *Eur. Phys. J.* **C1** (1998) 425, *Phys. Lett.* **B397** (1997) 147.
7. R. Edgecock, *these proceedings*.
8. ALEPH Collaboration, ALEPH–CONF 98–019 (march 1998), DELPHI Collaboration, DELPHI] 98–20 CONF 120 (march 1998), L3 Collaboration, L3 Note 2236 (march 1998), OPAL Collaboration, OPAL Physics Note PN331 (march 1998).
9. Physics at LEP2, Edited by G. Altarelli, T. Sjostrand and F. Zwirner, CERN 96-01 (1996), vol. 1, p. 525.
10. DELPHI Collaboration, DELPHI 98–21 CONF 121 (march 1998).
11. L3 Collaboration, L3 Note 2236 (march 1998).
12. OPAL Collaboration, OPAL Physics Note PN329, january 1998, OPAL Physics Note PN336 (march 1998).
13. M. Diehl and O. Nachtmann, *Zeit. Phys.* **C62** (1994) 397, C.G. Papadopoulos, *Phys. Lett.* **B386** (1996) 442, G.K. Fanourakis, D. Fassouliotis and S.E. Tsamarias, DEMO-97/09, hep-ex/9711015.
14. The LEP–TGC Combination group, Internal Note LEPEWWG/TGC/97–01 (august 1997).
15. ALEPH Collaboration, ALEPH–CONF 98–012 (march 1998).
16. ALEPH Collaboration, ALEPH–CONF 98–023 (march 1998).
17. L3 Collaboration, L3 Note 2239 (march 1998).
18. DØ Collaboration, D. Abbot *et al.*, FERMILAB-PUB-98/094-E, submitted to *Phys. Rev.* **D**.
19. LEPEWWG/TGC/98-01, DØ Note 3437, A combination of preliminary measurements of triple gauge boson coupling parameters measured by the LEP and DØ experiments (may 1998).

Measurement of $|V_{cs}|$ with DELPHI Experiment

DELPHI Collaboration
Boštjan Golob

Faculty of Mathematics and Physics, University of Ljubljana, SI-1000, Ljubljana, Slovenia, and J. Stefan Institute, SI-1000, Ljubljana, Slovenia

Abstract. Pair production of charged weak bosons $W^\pm$ at LEP2 collider can be exploited to measure the absolute value of the V_{cs} element of Cabbibo-Kobayashi-Maskawa matrix. The value can be most accurately extracted from the measured hadronic branching ratio of $W^\pm$ bosons. An independent method to obtain the $|V_{cs}|$ value consists of tagging the flavour of primary quarks in jets, produced in $W^\pm$ decays. Using both methods on the data collected with DELPHI experiment during 1996 and 1997 runs, we obtained $|V_{cs}| = 0.99 \pm 0.06(\text{stat.}) \pm 0.04(\text{syst.})$. Combined result of $|V_{cs}|$ measurements with four LEP experiments enables a test of CKM matrix unitarity.

I INTRODUCTION

In the Standard Model of electroweak interaction, with $SU(2) \times U(1)$ as the gauge group, the quark mass eigenstates are not the same as the weak eigenstates. For six quarks, the two bases are related by a unitary 3×3 Cabbibo-Kobayashi-Maskawa (CKM) matrix [?] [?]. Table 1 shows absolute values of CKM elements [?] obtained from direct measurements of individual elements through the quoted processes, without implementation of unitarity constraints.

Apart from the elements describing the top quark decays, $|V_{cs}|$ is measured with the worst precision. The quoted error of the measurement is completely dominated

Table 1. Absolute values of CKM elements, together with the process enabling the most accurate direct measurement of individual element.

$\|V_{ud}\| = 0.9736 \pm 0.0010$	$\|V_{us}\| = 0.2205 \pm 0.0018$	$\|V_{ub}\| = 0.0033 \pm 0.0009$
nuclear β decay/muon decay	$K \to \pi e \nu$	$b \to u \ell \nu$
$\|V_{cd}\| = 0.224 \pm 0.016$	$\|V_{cs}\| = 1.01 \pm 0.18$	$\|V_{cb}\| = 0.041 \pm 0.003$
ν_μ induced c production on N	$D \to \overline{K} e^+ \nu_e$	$B \to \overline{D}^* \ell^+ \nu_\ell$
$\|V_{td}\| = ?$	$\|V_{ts}\| = 0.05 \pm 0.02$	$\|V_{tb}\| = ?$
B mixing	$b \to s \gamma$	B mixing

by the theoretical uncertainty in calculation of a D meson form factor. A precise knowledge of $|V_{cs}|$ value is important since it represents one of fundamental parameters of the Standard Model and as such offers a test of the theory through the unitarity of CKM matrix.

Hadronic decays of charged weak bosons $W^{\pm}$, produced in e^+e^- collisions at LEP2, offer new measurement methods of this quantity. In decays $W^{\pm} \to q_1 \bar{q}_2$ the coupling of $W^{\pm}$ to the quarks is proportional to the appropriate CKM matrix element $|V_{q_1 q_2}|$. Hence from the measured rate of $W^+ \to c\bar{s}$ [1] decays one can extract the value of $|V_{cs}|$.

The value of $|V_{cs}|$ first of all reflects in the ratio of hadronic to all $W^{\pm}$ decays. By measuring the hadronic branching ratio of W's and by setting other parameters of the Standard Model to the presently known values, one can therefore determine $|V_{cs}|$.

An additional piece of information can be obtained by tagging the flavour of primary quarks in jets, produced in hadronic $W^{\pm}$ decays. Although the flavour of primary quarks is heavily veiled in the process of hadronisation, some properties of jets can still reveal the information about the jet origins. Since W^+ bosons decay dominantly into $c\bar{s}$ and $u\bar{d}$ quark pairs this method is based on the ability of separating c and s jets from u and d jets. It represents an independent direct measurement of $c\bar{s}$ production in W^+ decays. The method for the flavour tagging of jets described bellow would be of use also in measurements of Triple Gauge Couplings [?], where ambiguities on the production angle of quarks in $W^{\pm}$ decays are limiting the accuracy of measurements.

II $|V_{cs}|$ FROM $W^{\pm}$ HADRONIC BRANCHING RATIO

Figure 1 shows the expected number of produced W-pair events as a function of the $|V_{cs}|/|V_{cs}|_0$ ratio, where $|V_{cs}|_0 = 0.974$ is the value obtained by imposing unitarity constraint to the CKM matrix [?]. While the total number of events is almost insensitive to the value of $|V_{cs}|$, numbers of fully hadronic events, where both W's decay into a pair of quarks, mixed events, with one $W^{\pm}$ decaying into a quark and the other into a lepton pair, and fully leptonic events with both W's decaying into leptons, are sensitive to the changes of $|V_{cs}|$. By measuring the total cross-section for $W^{\pm}$ pair production $\sigma(W^+W^-)$ and partial cross-sections for the three different classes of events, which depend on the hadronic branching ratio $\mathcal{B}$ of charged weak bosons as

$$\sigma(4j) = \mathcal{B}^2 \, \sigma(W^+W^-)$$
$$\sigma(2j\ell\bar{\nu}_\ell) = 2 \, \mathcal{B} \, (1 - \mathcal{B}) \, \sigma(W^+W^-)$$
$$\sigma(2\ell) = (1 - \mathcal{B})^2 \, \sigma(W^+W^-) \, ,$$

one can determine the hadronic branching ratio $\mathcal{B}$.

---

[1] Throughout the paper references to a specific charge state are meant to imply the charge conjugate states as well, unless explicitly stated otherwise.

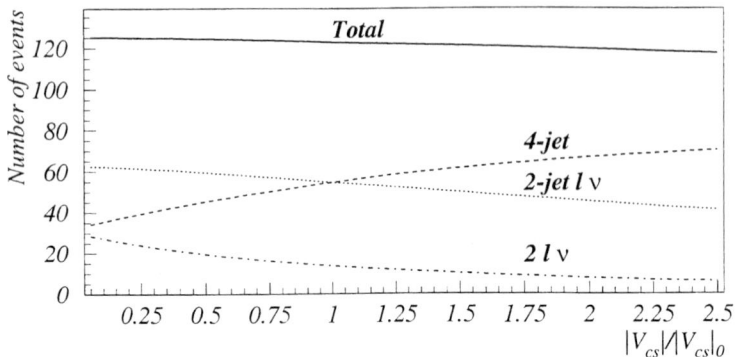

Figure 1. Expected number of produced W-pair events in e^+e^- collisions at 172 GeV centre-of-mass energy with an integrated luminosity of 10 pb^{-1}, with respect to the $|V_{cs}|/|V_{cs}|_0$ ratio. The total number of events is shown together with events in three different decay channels: hadronic events, with both W's decaying into a quark pair, mixed events, with one $W^\pm$ decaying into a quark and the other into a lepton pair, and leptonic events with both W's decaying into leptons. The curves were obtained in the leading-order approximation.

The hadronic branching ratio

$$\mathcal{B} = \frac{C_{\alpha_s} \sum_{i>j} |V_{ij}|^2}{1 + C_{\alpha_s} \sum_{i>j} |V_{ij}|^2} , \qquad (1)$$

where C_{α_s} includes higher order corrections in α_s, depends on the sum of squares of all CKM matrix elements apart of those, involving the top quark. From the measured value of $\mathcal{B}$ one can thus extract the value of $|V_{cs}|$ by means of equation 1.

The data sample collected by DELPHI in 1996 and 1997 corresponds to an integrated luminosity of around 10 pb^{-1} at the average centre-of-mass energy of 161 GeV, around 10 pb^{-1} at the energy of 172 GeV and approximately 50 pb^{-1} at the energy of 183 GeV. From 885 selected W-pair event candidates, the hadronic branching ratio was found to be [?] [?] [?]

$$\mathcal{B} = 0.675 \pm 0.015(\text{stat.}) \pm 0.009(\text{syst.}) . \qquad (2)$$

By assuming all necessary parameters of the Standard Model to have presently known values, quoted in [?], this result can be converted into the value

$$|V_{cs}| = 0.98 \pm 0.07(\text{stat.}) \pm 0.04(\text{syst.}) . \qquad (3)$$

III $|V_{cs}|$ FROM FLAVOUR TAGGING OF JETS

A Selection of $W^\pm$ events

An additional information about the $|V_{cs}|$ can be obtained by tagging the flavour of primary quarks in hadronic jets, produced in $W^\pm$ decays. The first step in flavour tagging is the selection of W-pair event candidates. For this purpose a probabilistic approach was used. After a loose preselection several kinematical properties of events were combined into a single variable, separating the W-pair signal from the background. The discriminating power of these variables steams from the topological properties of W-pair events, as well as from the fact that the most severe background, coming from the $q\bar{q}$ creation in e^+e^- annihilations, is frequently accompanied by a photon radiated from the initial state [?]. This results in a smaller effective CMS energy. Jets from $q\bar{q}(\gamma)$ events are also expected to be distributed less uniformly in space than jets from W-pair decays, hence the signal events can be separated from the background using angular distributions of $W^\pm$ decay products. In mixed events with leptons one can also profit from the identification of electrons and muons with electromagnetic calorimeter and muon chambers of DELPHI [?]. In order to improve the momentum and angle resolution, a kinematically constrained fit was performed [?], imposing the 4-momentum conservation and nominal W-mass to dijets or a dijet and lepton-missing momentum combination.

Relying on the appropriate simulated distributions of discriminating variables one can calculate the probability for an event with a particular value of such variable to be either a signal or a background event. By combining probabilities calculated from different discriminating variables a single separator was constructed. Distributions of separator P_{4j} for hadronic and P_{2j} for mixed events with $e^\pm$, $\mu^\pm$ or $\tau^\pm$ candidates, are shown in fig.2.

Cuts indicated by arrows in fig.2 were defined by maximising the product of efficiency and purity in each channel. Only events to the right of the arrows were used for further flavour tagging of jets. 500 hadronic and 340 mixed events were selected with the expected number of background events, arising mainly from $e^+e^- \to q\bar{q}(\gamma)$, around 150 in hadronic and 30 in mixed channel.

B Flavour tag

As already mentioned in the introduction, tagging of jet flavour is based on the ability for the separation of $c\bar{s}$ from $u\bar{d}$ dijets. For that purpose one can fully exploit DELPHI Vertex and Ring Imaging Cherenkov detectors [?]. The flavour tagging was based on the following discriminating properties of individual quark flavours:

- if the primary quark in a jet is a c or an s quark, the jet is likely to contain a high momentum charged kaon. Such a kaon is composed of a primary s quark or of an s quark from the dominant $c \to s$ decay chain. On the other

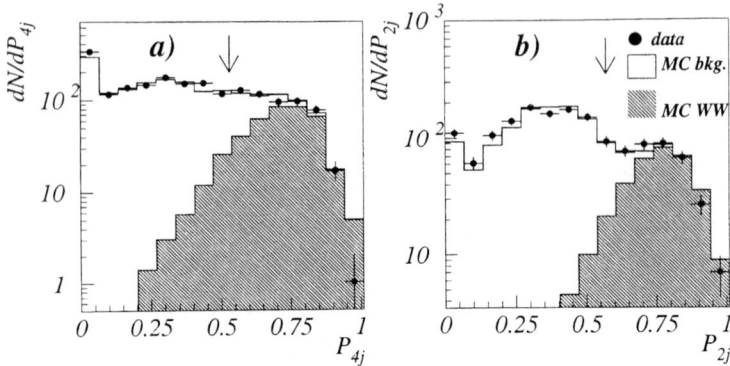

Figure 2. Distributions of separator for isolation of W-pair events in a) hadronic and b) mixed events with $e^{\pm}$, $\mu^{\pm}$ or $\tau^{\pm}$ candidates. Events to the right of arrows were selected as W-pair candidates. Data are represented by dots, empty histrogram shows contribution of the background and hatched histogram distribution of the signal.

hand the presence of a high momentum charged pion in a jet is an indication of a u or a d primary quark. Fig.3 shows the expected momentum spectra of the leading particles in u, d, c and s jets when they were identified as $K^{\pm}$ or $\pi^{\pm}$. From the simulated distributions a momentum dependent probability for a particular jet flavour was calculated for each jet of a W-pair candidate event, by considering only the highest momentum identified particle in a jet.

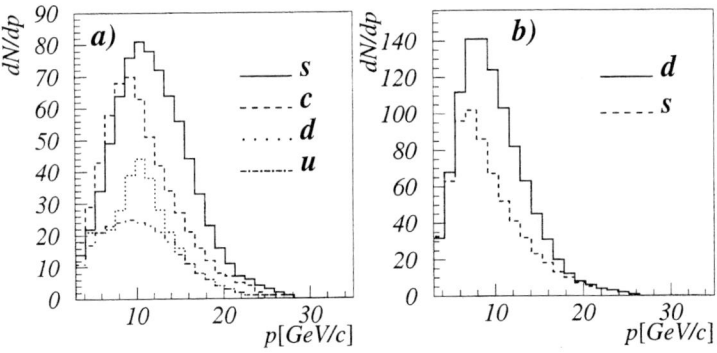

Figure 3. a) Expected momentum spectrum of identified charged kaons in simulated u, d, c and s jets when the kaon is the leading particle in a jet. b) Same for identified charged pions in d and s jets.

- Due to a finite lifetime and large mass of hadrons containing c quarks, decay products of charmed hadrons have on average larger impact parameters than particles arising from the primary $W^\pm$ decay vertex. A lifetime tag [?] used for b quark tagging at the Z^0 peak can thus also be used to separate c jets from light quark jets in $W^\pm$ decays. From the measured impact parameters of tracks in a jet one can calculate the probability that all tracks originate from a primary vertex. This probability is larger for light quark jets than for c jets, as shown in fig.4.

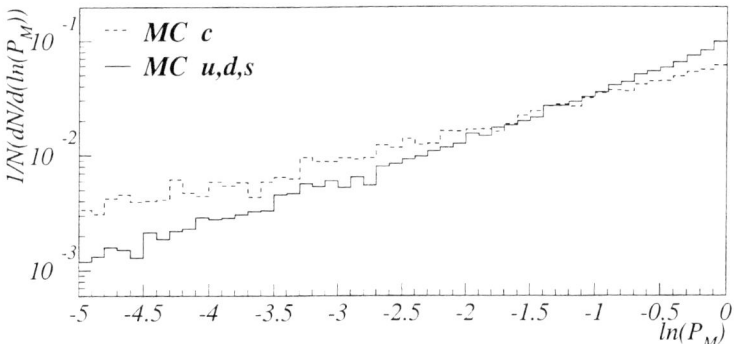

Figure 4. Simulated probability distribution for all tracks in c (dashed histogram) and light quark (full histogram) jets to originate from the primary vertex.

- On average, positively charged weak bosons fly more in the forward direction, determined by the incoming e^- beam, than negatively charged ones. Distribution of simulated polar angle measured from the e^- direction is shown in fig.5 a). In fully hadronic events the fitted direction of W's can be converted into a probability for a certain charge assignment of weak bosons, while in mixed events the charge of W's is determined from the charge of the lepton candidate. As a consequence of the V-A structure of $W^\pm$ decays, in the boson rest-frame uplike quarks and anti-quarks fly more along the $W^\pm$ momentum than downlike quarks and anti-quarks (fig.5 b)). Combining both angular informations one can deduce a probability for a certain jet to be either an u-like, d-like, $\bar{u}$-like or a $\bar{d}$-like jet.

- As a minor contribution to the flavour tagging also a search for identified muons, neutral kaons and Λ-baryons was performed. While muons indicate a semileptonic decay of a c quark, K^0's and Λ's in jets are a signature of a primary s quark.

The above signatures were combined into probabilities $\mathcal{P}_{\text{jet}}(q)$ for each jet to be a u, d, c or an s jet. The final separator of $W^+ \to c\bar{s}$ decays was constructed as

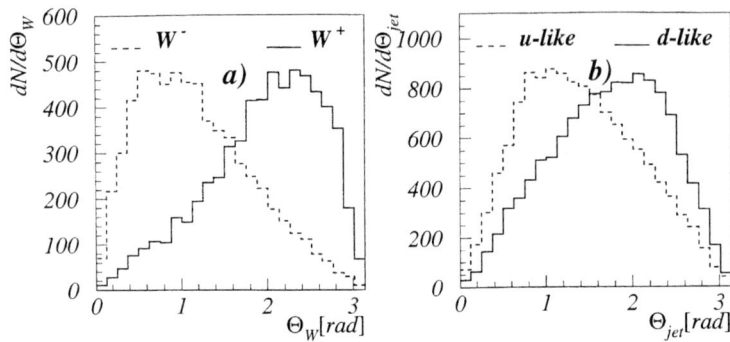

Figure 5. a) Simulated distribution of fitted $W^\pm$ polar angle, measured from the direction of the incoming e^- beam. b) Angle between a jet direction of uplike and downlike quarks and momentum of the corresponding $W^\pm$ in the rest-frame of W's.

$$\mathcal{P}_{cs} = \frac{\mathcal{P}_1(c)\mathcal{P}_2(\bar{s}) + \mathcal{P}_1(\bar{s})\mathcal{P}_2(c) + \mathcal{P}_1(\bar{c})\mathcal{P}_2(s) + \mathcal{P}_1(s)\mathcal{P}_2(\bar{c})}{K_{nor}}$$

$$K_{nor} = \mathcal{P}_1(c)\mathcal{P}_2(\bar{s}) + \mathcal{P}_1(\bar{s})\mathcal{P}_2(c) + \mathcal{P}_1(\bar{c})\mathcal{P}_2(s) + \mathcal{P}_1(s)\mathcal{P}_2(\bar{c}) +$$
$$+ \mathcal{P}_1(u)\mathcal{P}_2(\bar{d}) + \mathcal{P}_1(\bar{d})\mathcal{P}_2(u) + \mathcal{P}_1(\bar{u})\mathcal{P}_2(d) + \mathcal{P}_1(d)\mathcal{P}_2(\bar{u}) .$$

In the above equation only $c\bar{s}$ and $u\bar{d}$ combinations were taken into account while the Cabbibo suppressed decays $W^+ \to c\bar{b}$ ($u\bar{s}$, $c\bar{d}$) were considered as a minor contribution to the $u\bar{d}$ final state.

Simulated distributions of $\mathcal{P}_{cs}$ for $W^\pm$ decays into $c\bar{s}$ and $u\bar{d}$ quark pairs are shown in fig.6.

C Results

The expected distributions of $\mathcal{P}_{cs}$ separator for different $W^\pm$ decays (fig.6) and for the background were fitted to the distribution measured on the data. A likelihood function was constructed as a multinomial distribution

$$\mathcal{L} = \frac{N!}{r_1! r_2! \ldots r_k!} p_1^{r_1} p_2^{r_2} \ldots p_k^{r_k} , \qquad (4)$$

where N is the total number of observed jet pairs, r_i is the number of jet pairs in i-th bin of $\mathcal{P}_{cs}$ distribution and p_i is the probability for a jet pair to fall into the i-th bin of distribution. p_i's depend on the value of $|V_{cs}|$, e.g. for the fully hadronic channel

$$p_i(\text{had.}) = p_i(\text{MC; bkg.}) + p_i(\text{MC; 0 } cs \text{ dijets}) + \qquad (5)$$

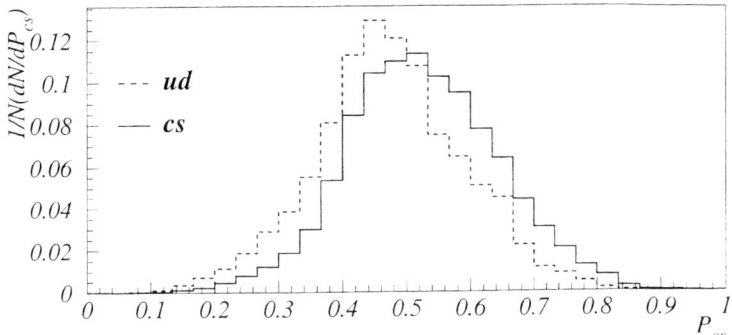

Figure 6. Expected normalised distribution of the $\mathcal{P}_{cs}$ separator for $W^+ \to c\bar{s}$ (solid line) and $W^+ \to u\bar{d}$ (dashed line) decays.

$$+ \frac{|V_{cs}|^2}{|V_{cs}|_0^2} p_i(\text{MC; 1 } cs \text{ dijet}) + \frac{|V_{cs}|^4}{|V_{cs}|_0^4} p_i(\text{MC; 2 } cs \text{ dijets})$$

with $|V_{cs}|_0$ being the value used for the MC sample generation.

$\mathcal{P}_{cs}$ distribution was fitted simultaneously for the hadronic and mixed channel. Results of the fit are presented in fig.7. The best fit is shown together with the

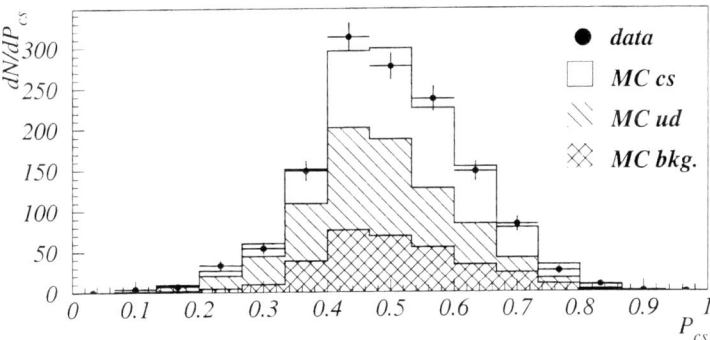

Figure 7. Measured $\mathcal{P}_{cs}$ distribution (dots) with best fit superimposed (histogram). Contributions of $W^+ \to c\bar{s}$ decays, $W^+ \to u\bar{d}$ decays together with the Cabbibo suppressed decays and background are shown separately with empty, singly and doubly hatched histogram, respectively.

measured distribution and expected contributions of $c\bar{s}$ dijets, sum of $u\bar{d}$ dijets and Cabbibo suppressed decays, and background. The result of the fit is

$$|V_{cs}| = 1.01^{+0.12}_{-0.10} \qquad (6)$$

with statistical error only.

Different sources of possible systematic errors were studied and are evaluated in table 2. The largest contributions arise from a slight disagreements between data and MC distributions of lifetime tag probability P_M and momentum spectra of identified $\pi^\pm$ and $K^\pm$. Systematic uncertainties due to the P_M distribution were estimated using Z^0 decays collected during short runs at lower energies in 1996 and 1997. By cutting at different values of lifetime tag probability in one hemisphere, samples of events with different fractions of $b\bar{b}$, $c\bar{c}$ and light quark pairs were obtained. Real and simulated P_M distributions in the second hemisphere were then compared in order to obtain corrections for the MC distribution. The corrections were applied to simulated W-pair events. However, due to a limited statistics of recorded Z^0 decays, corrections are known with a finite accuracy. These errors were propagated through the fit to give a systematic error of ± 0.079 on $|V_{cs}|$. A similar procedure was followed for evaluation of systematic error arising from differences in real and simulated momentum spectra of identified charged kaons and pions. By cutting at different values of P_M and requiring identified kaons and pions in one hemisphere, c, s or u and d enhanced samples were obtained. From a comparison of $K^\pm$ and $\pi^\pm$ momentum distributions in the other hemisphere a systematic shift of ± 0.058 on the $|V_{cs}|$ value was obtained. Including systematic

Table 2. List of systematic errors on $|V_{cs}|$ obtained from the measured hadronic branching ratio of $W^\pm$, from the flavour tagging of jets, and combination of both methods.

Source	$\mathcal{B}$	flavour tag	Combined
Eff. calcul.	±0.027	-	±0.018
Bkg. normal.	±0.028	±0.003	±0.019
"CC03" corr.	±0.015	<0.001	±0.011
$|V_{cd}|$	±0.004	<0.001	±0.003
α_s	±0.001	<0.001	±0.001
Lifetime tag	-	±0.079	±0.024
$K^\pm$, $\pi^\pm$ spectra	-	±0.058	±0.018
MC statistics	-	±0.031	±0.010
Total	±0.04	±0.10	±0.04

errors in tagging of jet flavours one obtains the value

$$|V_{cs}| = 1.01^{+0.12}_{-0.10}(\text{stat.}) \pm 0.10(\text{syst.}) \qquad (7)$$

which can also be expressed as a direct measurement of $c\bar{s}$ production in W^+ decays:

$$r_{c\bar{s}} = \frac{\Gamma(W^+ \to c\bar{s})}{\Gamma(W^+ \to \text{hadrons})} = 0.49 \pm 0.07 \quad . \qquad (8)$$

QCD AND τ PHYSICS

QCD at the Tevatron:
W, Z, and Direct-γ Production

Dylan P. Casey
for the DØ and CDF Collaborations

Department of Physics and Astronomy
Michigan State University, East Lansing, Michigan 48824

Abstract. We present measurements W, Z, and direct-γ production made by the DØ and CDF collaborations at the Tevatron with $\sqrt{s} = 1.8$ TeV. All of the measurements are consistent with calculations of perturbative QCD.

The study of the properties of W, Z, and direct-γ production in $\bar{p}p$ collisions allows for precision testing of perturbative QCD (pQCD) at high momentum transfers. Among the advantages that make the W and Z, in particular, excellent laboratories for studying pQCD, are the presence of well-measured leptons, low backgrounds, and a well-defined event vertex. Similarly, direct-γ production has the benefit of a well-measured photon in the final state. All these features combine to provide higher-precision studies of the strong interaction than can generally be obtained from studying events with only jet final states.

In these proceedings, we present results from the two collider detectors at the Tevatron, DØ and CDF. From DØ, we present the results of measurements of the transverse momentum (p_T) distributions of the W and Z. From CDF, we present the results of a measurement of cross section for W production as a function of jet multiplicity, as well as the ratio of $W+ \geq 1$ jet production to the inclusive W cross section as a function of the minimum E_T of the jet (E_T^{min}). Lastly, we present measurements from both DØ and CDF of the cross section for direct-γ production, as a function of the E_T of the photon.

I W AND Z p_T

With the W and Z p_T distributions, we can test next-to-leading-order (NLO) pQCD predictions in the region where $p_T \sim M_{VB}$, where M_{VB} is the mass of the vector boson. We can also evaluate the resummation techniques used to solve pQCD in the low-p_T region where standard pQCD fails. Besides being of interest on its own merits, detailed knowledge of vector boson production is required when making

Uncertainty	m_T method (MeV)	$p_T(e)$ method (MeV)
p_T model for the W	10	50
Total W production and decay model	30	75
Total detector systematics	60	50
Total statistical uncertainty	95	108
Total uncertainty	115	140

TABLE 1. Comparison of some of the sources of uncertainty in measuring the mass of the W boson.

other measurements of processes with leptons in the final state, e.g., backgrounds for top, Higgs, and diboson production. In particular, detailed knowledge of vector boson production at low-p_T, where the cross section is largest, is important for making a precise measurement of the mass of the W boson. With the next round of data from the Tevatron (coinciding with the Main Injector turning on), it is expected that the uncertainty on the mass of the W can be reduced to ∼40 GeV. The preferred method for extracting M_W will be via the $p_T(e)$ distribution, rather than the transverse mass distribution, because the $p_T(e)$ measurement will not be systematically limited by the reconstruction of the total missing transverse energy ($\not{E}_T$). However, as can be seen in Table 1 the $p_T(e)$ measurement has a significantly larger contribution from uncertainties in the vector boson production than the transverse mass measurement [1].

In the high-p_T region, where $p_T^2 \approx Q^2$, the differential cross section is expected to be well-described by standard, "fixed-order", perturbation theory, in which the cross section is expanded in terms of the strong coupling constant (α_s) [2]. In this case, each power of α_s corresponds to the radiation of a single gluon or quark into the final state. However, as $p_T \to 0$, this fixed-order calculation of the cross section diverges due to the presence of terms that go as $\log^n(Q^2/p_T^2)$. Physically, the failure of fixed-order perturbation theory at low-p_T is due to soft-gluon radiation from the initial partons being an important contributor to the transverse momentum of the vector boson. This difficulty in performing the perturbative calculation can be remedied by resumming the perturbation series in terms of the large logarithms, rather than strictly in terms of powers of α_s. The resulting calculation is an "all-orders" calculation, i.e., each piece in the new sum contains terms to all-orders in α_s. However, the largest terms dropped in the latest calculation are $O(\alpha_s^2)$, so it is considered to be accurate to $O(\alpha_s^2)$.

Formally, the calculation is carried out in impact-parameter space (b-space), rather than in transverse momentum space, and following relation describes the differential cross section [3] [4] [5] [6] [7] [8] :

$$\frac{d\sigma}{dp_T^2 dy} \sim \int_0^\infty d^2 b e^{i\vec{p}_T \cdot \vec{b}} W(b,Q) + Y \qquad (1)$$

where

$$W(b,Q) \sim exp(-S(b,Q)) \qquad (2)$$

$$S(b,Q) = \int_{b/b_o}^{q^2} \frac{d\mu^2}{\mu^2} [\ln(\frac{Q^2}{\mu^2})A(\alpha_s(\mu)) + B(\alpha_s(\mu))] \qquad (3)$$

$S(b,Q)$ is called the "Sudakov" factor. It is via the exponentiation of $S(b,Q)$ that the sum over the logarithms is implemented. The Y term in Eq.1 is a correction from fixed-order perturbation theory.

The resummed calculation becomes undefined at high b values (corresponding to very low p_T), where $b \geq 1/\Lambda_{QCD}$, and perturbation theory is expected to fail in general. A parameterization applicable to the low-p_T region is introduced in order to account for the non-perturbative effects. Formally, this is accomplished by cutting off the integral in Eq.1 at some value b_{max} and replacing $W(b,Q)$ with $W(b_*,Q)e^{-S_{NP}(b,Q)}$, where $b_* = b/\sqrt{1+b/b_{max}}$ and S_{NP} is the non-perturbative function being introduced. It has been shown that the non-perturbative function has the following universal form [3]:

$$S_{NP}(b,Q) = h_1(b,x_A) + h_1(b,x_B) + h_2(b)\ln(\frac{Q}{2Q_o}) \qquad (4)$$

where x_A and x_B are the momentum fraction of the incoming partons; Q_o is an arbitrary momentum scale; and the functions $h_1(b,x)$ and $h_2(b)$ must be determined from experiment.

The two suggested parameterizations for the non-perturbative function are [4] [5],

$$S_{NP}^{DWS}(b,Q) = g_1 b^2 + g_2 b^2 \ln(\frac{Q}{2Q_o}) \qquad (5)$$

$$S_{NP}^{LY}(b,Q) = g_1 b^2 + g_2 b^2 \ln(\frac{Q}{2Q_o}) + g_3 b \ln(100 x_A x_B) \qquad (6)$$

where x_A and x_B are the momentum fractions of the incoming partons, b is fourier conjugate to the transverse momentum (impact parameter), Q_o is an arbitrary momentum scale. The parameters g_1, g_2, and g_3 cannot be predicted by QCD, and, therefore, must be measured to provide guidance to phenomenology.

Fits to Drell-Yan data were performed in order to obtain values for the g parameters. Using Eq. 5, with the Duke and Owens [9] parton distribution functions (PDFs), $b_{max} = 0.5$ GeV^{-1}, and $Q_o = 2$ GeV, Davies, Weber, and Stirling, obtained the values $g_1 = 0.15$ GeV^2 and $g_2 = 0.4$ GeV^2 [4]. Using Eq. 6, with the CTEQ2M PDFs, $b_{max} = 0.5 GeV^{-1}$, and $Q_o = 1.6$ GeV, Ladinsky and Yuan, obtained the values $g_1 = 0.11^{+0.04}_{-0.03}$ GeV^2, $g_2 = 0.4^{+0.1}_{-0.2}$ GeV^2, $g_3 = -1.5^{+0.1}_{-0.2}$ GeV^{-1} [5].

The data for the measurements of the W and Z p_T distributions were taken with the DØ detector, which has been described elsewhere in detail. [10] The Ws have been selected from the 1993-94 run of the Tevatron, corresponding an integrated luminosity, $\int \mathcal{L} dt = 12$ pb^{-1}. The Zs have been selected from the 1994-96 data corresponding to an integrated luminosity, $\int \mathcal{L} dt = 108.5$ pb^{-1}.

In identifying both W and Z events, only the electron decay channel is used, picking the one(s) with the highest transverse momentum to reconstruct the vector boson. The electrons are required to be of good quality, i.e., well-isolated clusters in the calorimeter, with a large electromagnetic fraction (> 95%). The following kinematic and fiducial requirements are imposed for the W selection: total missing transverse energy, which corresponds to the transverse energy of the undetected neutrino, $p_T(\nu) > 25$ GeV, and one electron in the central region ($|\eta_e| < 1.1$) with $p_T(e) > 25$ GeV. For the Z the following requirements are imposed: 2 electrons with $p_T(e) > 25$ GeV, where at least one is in the central region and the other may be central or forward ($1.5 < |\eta_e| < 2.5$). In the case of the Z, the invariant mass of the dielectron system is also required be near the true Z mass ($75 < M_{ee} < 105$ GeV). After the event selection criteria, there are 7132 W candidates and 6407 Z candidates.

The geometrical acceptance as a function of p_T for both the W and Z samples is determined using Monte Carlo (MC) with a parameterized detector simulation. The variation in the event selection efficiency as a function of p_T is determined using a full detector Monte Carlo simulation. The absolute normalization of the efficiency is determined using Z events in which one electron is tagged for event identification and the second, unbiased electron is used to determine the efficiency. The corrections due to acceptance and efficiency variations as a function of p_T are about 5%.

The background for both the Z and W data samples is dominated by QCD dijet and photon-jet production. The shape and normalization of the background was determined directly from data. The total background for the Z sample is $\sim 2\%$ for Zs in which both electrons are central and $\sim 7\%$ when one electron is central and the other is forward. The total amount of background in the W sample is $\sim 4\%$. Figure 1 shows the background fraction from QCD processes as a function of p_T for both the W and Z samples. Backgrounds to $W \to e\nu$ production from τ production via $W \to \tau\nu$ and $Z \to \tau\tau$ are estimated from MC to be $\sim 2\%$. Backgrounds to $Z \to ee$ production from $Z \to \tau\tau$, top production, etc. have all been estimated with Monte Carlo and are negligible.

A parameterized description of the DØ detector has been used to smear the theoretical predictions in order to account for detector effects and compare to the measured p_T distributions. The results for the W are shown in Fig.2. The results for the Z are shown in Figs. 3 and 4. Figures 2 and 3 compare the measured results to the resummed calculation of the vector boson p_T. Only the Z p_T is sufficiently well-measured to discriminate between the two parameterizations shown. Figure 4 compares the measurement to NLO standard perturbation theory over a larger range of p_T. One can observe the divergence of the theory from the data at low-p_T, corraborating the need for the resummation calculation in that region. As can been seen the the corresponding (Data-Theory)/Theory ratio, there is good agreement with pQCD over a wide range of transverse momenta in both the Z and W distributions.

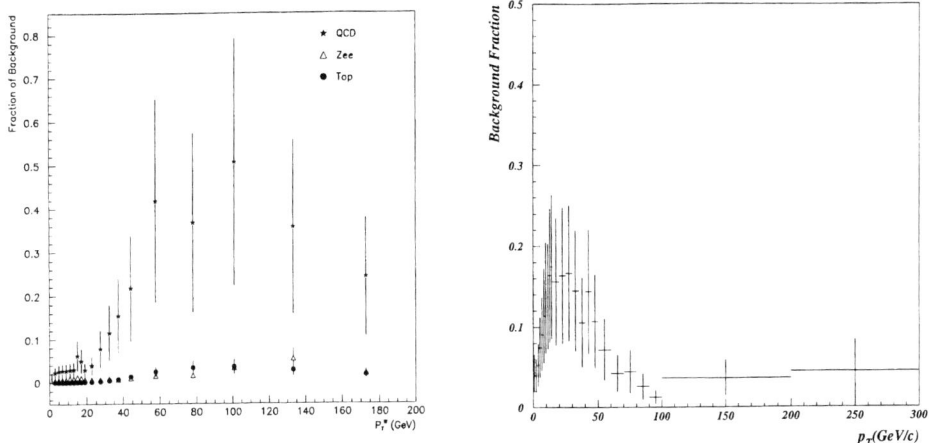

FIGURE 1. Left: The background fraction to the $W \to e\nu$ signal from QCD, $Z \to ee$ and top production as a function of the p_T of the W. Right: The background fraction from QCD processes to the $Z \to ee$ signal as a function of the p_T of the Z.

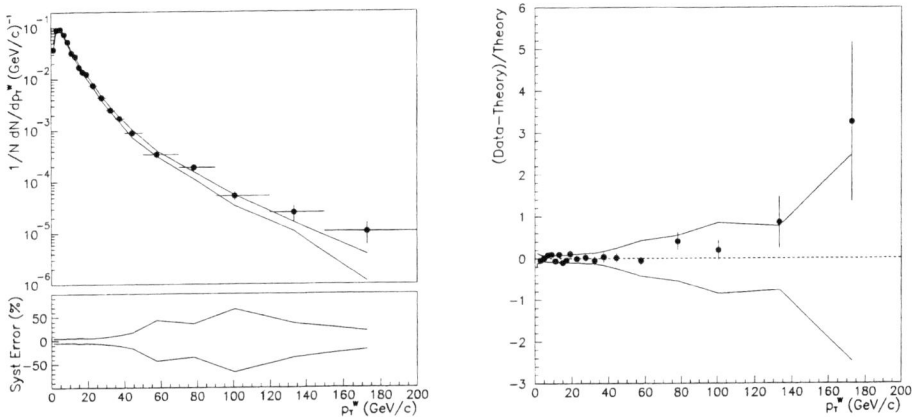

FIGURE 2. Left: DØ data (solid points) with statistical uncertainty shown compared to the theoretical prediction, which is smeared with detector resolutions. The theoretical prediction shown uses the Davies-Weber-Stirling form for the non-perturbative function, with parameter values as discussed in the text. Data and theory are independently area normalized. The band shown in the lower portion of the plot shows the fractional systematic uncertainty on the data. Right: The ratio (Data-Theory)/Theory shown as a function of the p_T of the W. The error bars indicate the statistical uncertainty and the band indicates the systematic uncertainty.

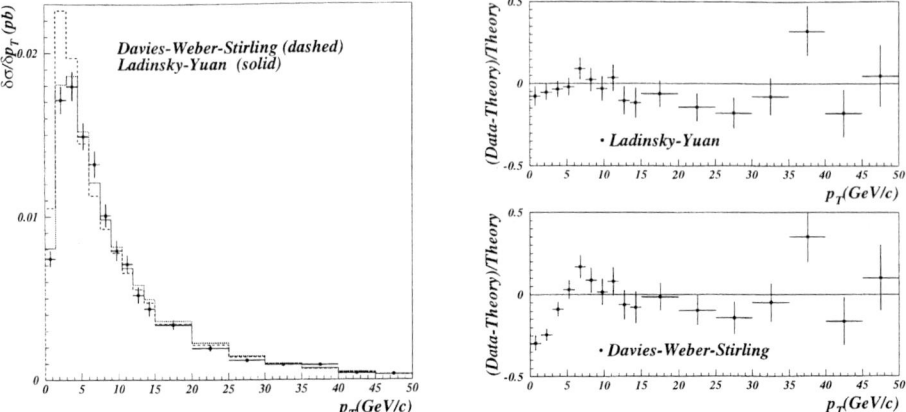

FIGURE 3. Left: DØ data (solid points) with total uncertainty shown compared to two theoretical predictions, which are smeared with detector resolutions. The theoretical predictions shown are those of Davies-Weber-Stirling and Ladinsky-Yuan, with parameter values as discussed in the text. The data is normalized to the DØ measured inclusive Z production cross section (preliminary); the theory is normalized to the data. Right: The ratio (Data-Theory)/Theory as a function of p_T of the Z shown for the two parameterizations of the non-perturbative function in the theory.

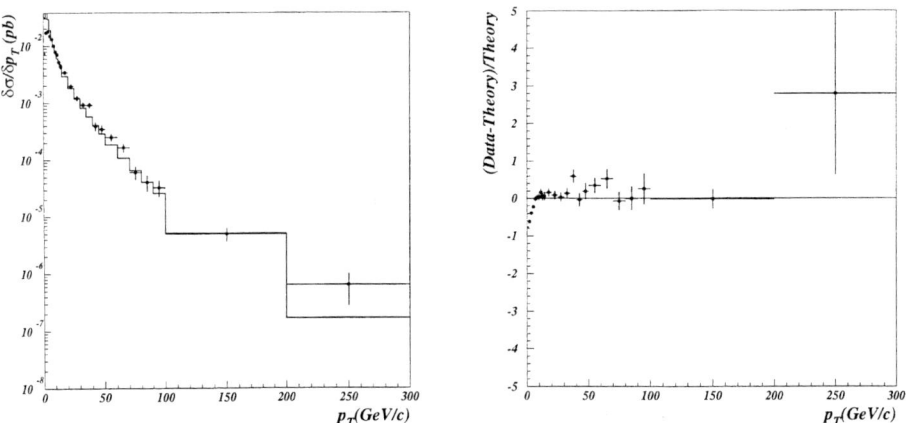

FIGURE 4. Left: DØ data (solid points) with total uncertainty shown compared to the NLO perturbative prediction, which was smeared with detector resolutions. The data is normalized to the DØ measured inclusive Z production cross section (preliminary); the theory is normalized to its own prediction. Right: The ratio (Data-Theory)/Theory as a function of p_T of the Z shows the known need for resummation in the very low-p_T region.

II W+JET PRODUCTION

CDF measures the production cross section for $W \to e\nu$ as a function of jet multiplicity, $\sigma_n = \sigma(W+ \geq n\ jets)$, $n = 1...4$, as well as the ratio, $R_{10} = \sigma_1/\sigma_0$ as a function of minimum E_T of the jet (E_T^{min}). The data are selected from the 1992-1995 runs at the Tevatron and correspond to an integrated luminosity of 108 pb^{-1}. The electrons are restricted to be in the central rapidity region ($|\eta_3| < 1.1$), with $E_T \geq 20$ GeV, and are required to satisfy tight selection criteria [12]. The missing E_T ($\not{E}_T$) is required to be > 30 GeV. The jets are reconstructed using a fixed cone algorithm with a cone-radius of 0.4 in η-ϕ and restricted to the rapidity region $|\eta_{jet}| < 2.4$. Additionally, the jets are required to be well-separated from the electrons, with the minimum electron-jet separation allowed $\delta R = 0.52$ in η-ϕ. In the measurements of σ_n, the jet is required to have $E_T \geq 15$ GeV.

The backgrounds to the sample are determined as a function of jet multiplicity. The dominant background is QCD multijet production in which the jet is incorrectly reconstructed as an electron and large $\not{E}_T$ results from shower fluctuations and mismeasuring the jet energy. The multijet background is measured from the data by removing the electron isolation and $\not{E}_T$ requirement in the event selection [13]. The estimated background level varies from $\approx 2.9\%$ to $\approx 27\%$ for the $n = 0$ to $n = 4$ cases. The background due to other processes which contain an electron in the final state, e.g., top quark and diboson production, are estimated using the VECBOS MC program [14] and a simulation of the CDF detector response. The levels of contamination in the signal from these processes vary between $\approx 0.1\%$ to $\approx 17\%$ for the $n = 0$ to the $n = 4$ cases. For the measurement of R_{10}, the above backgrounds are determined in the $W+ \geq 1$ jetcase as a function of the E_T^{min}. The total background increases from $(22 \pm 5\%)$ at $E_T^{min} = 15$ GeV to $(44 \pm 13\%)$ at $E_T^{min} = 95$ GeV. The overall background for the inclusive W sample is $(5.9 \pm 1.2\%)$. Figure 5(left) shows the various background levels as a function of E_T^{min} of the jet in the $W+ \geq 1$ jet sample.

The acceptance due to kinematic and fiducial requirements in the event selection is determined for each jet E_T^{min} using VECBOS and a CDF detector simulation. The acceptance for $W+ \geq 1$ jet events increases with jet E_T^{min} from 24% to 36%. The acceptance for the inclusive W sample is $(23.9 \pm 0.5\%)$. The overall detection efficiency includes the trigger efficiency, the electron identification efficiency, and the electron-jet overlap efficiency. The electron-jet overlap efficiency accounts for losses due to the electron-jet separation requirement. The electron identification and electron-jet overlap efficiencies are determined from the data using $Z \to ee$ events. The combined acceptance and detection efficiency for the inclusive W sample is $(19.5 \pm 0.5\%)$. The combine acceptance and detection efficiency for the $W+ \geq 1$ jet sample ranges from $(19 \pm 1\%)$ at jet $E_T^{min} = 15$ GeV to $(25 \pm 3\%)$ at jet $E_T^{min} = 98$ GeV. Figure 5 (right) shows a plot of combine acceptance and detection efficiency as a function of jet E_T^{min} for $W+ \geq 1$ jet events.

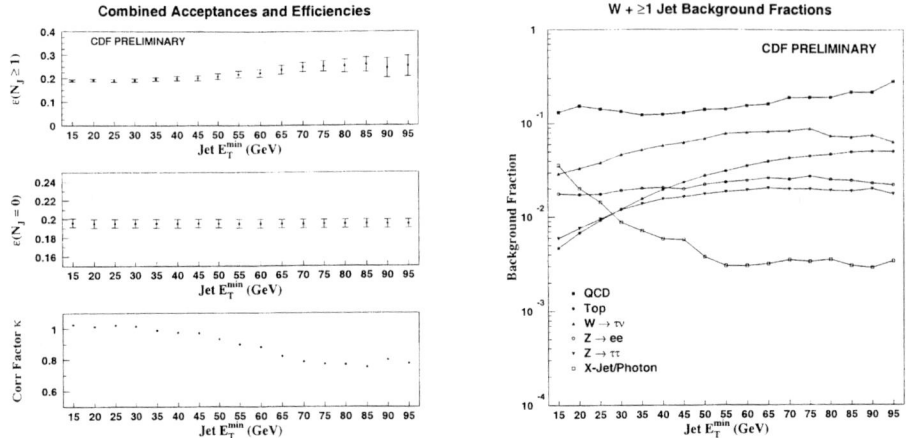

FIGURE 5. Leftt: Combined acceptances and efficiencies for $W+ \geq 1$ jet (top), exclusive W production (middle), and the correction factor applied to the measured ratio in order to account for the effects (bottom). Right: Background fraction to $W+ \geq 1$ jet jet production for several processes shown as a function of E_T^{min}.

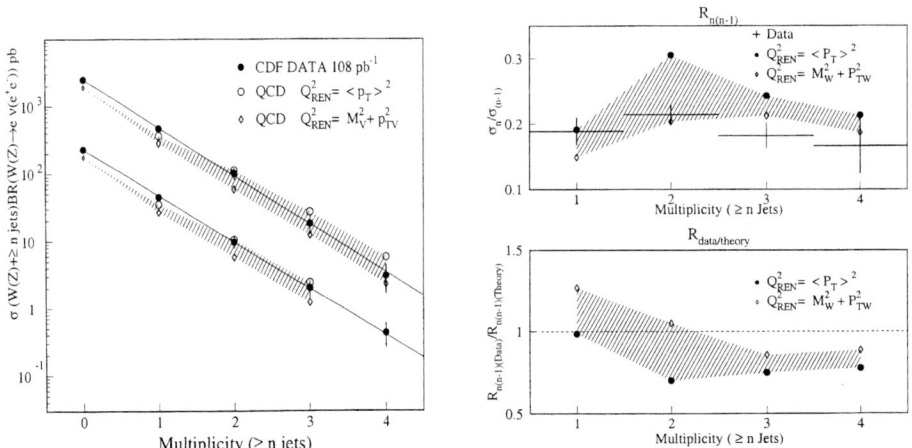

FIGURE 6. Left: CDFs $\sigma_n(W+ \geq 1\ \text{jets})$ as a function of jet multiplicity. The result for $\sigma_n(Z+ \geq n\ \text{jets})$ is also shown. Right: Ratios σ_n/σ_{n-1} for $n = 1-4$ of the W cross sections shown on the right compared to LO QCD (top) along with the ratio to theory (bottom).

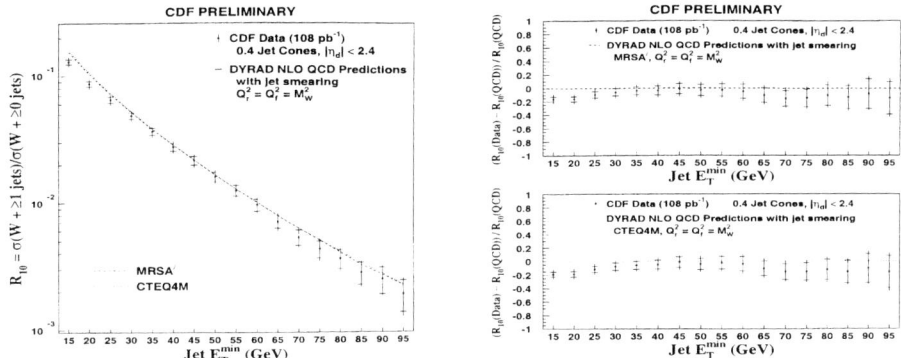

FIGURE 7. Left: R_{10} as measured by CDF (solid dots) as a function of E_T^{min} compared to the MRSA' and CTEQ4M parton distribution functions. Right: The (Data-Theory)/Theory ratio for the data shown on the right, showing excellent agreement between the data and the prediction.

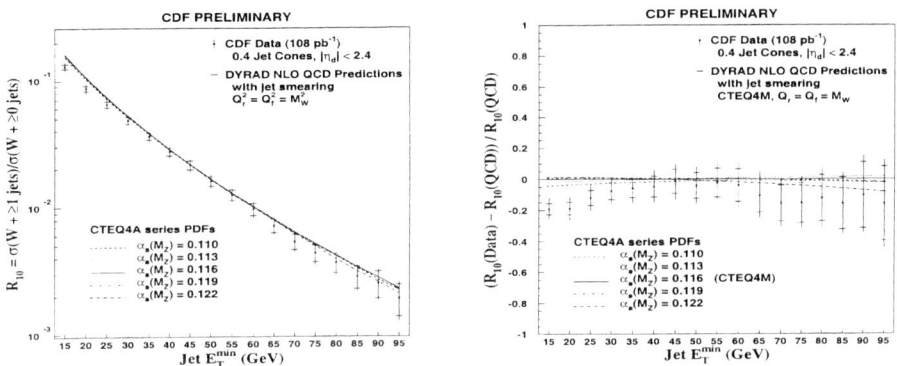

FIGURE 8. Left: R_{10} as measured by CDF (solid dots) as a function of E_T^{min} of the jet compared to the CTEQ4A parton distribution functions using different values of the strong coupling (α_s). Right: The ratio (Data-Theory)/Theory for the data on the right shows little sensitivity to the value of the strong coupling.

The measurement of σ_n [15] is compared to predictions from leading order (LO) QCD matrix element calculations [16] by including gluon radiation and hadronic fragmentation using the HERWIG shower simulation program [17]. The events are then introduced into the CDF full detector simulation, with the resulting jets identified and selected as they are in the data, allowing a comparison between the theoretical predictions and the data. Figure II shows the CDF data compared to the QCD prediction for two choices of renormalization and factorization scale, $Q^2_{REN,FAC}$. The published $Z+ \geq n$ Jets cross section is also shown [15]. One observes that the LO calculation is quite sensitive to the renormalization scale.

Also shown in the figure are the ratios σ_n/σ_{n-1} for $n = 1 - 4$.

The measurement of R_{10} is compared to NLO perturbative QCD predictions obtained from the DYRAD MC program [19]. Figure7 shows the measured R_{10} compared to the theoretical prediction. Figure8 shows the (Data-Theory)/Theory ratio for the same result. There is excellent agreement over the full range of E_T^{min}, except for the lowest thresholds ($E_T^{min} > 25$ GeV), which is interpreted as an indication of the need for resummation in the calculation of the cross sections. As can be seen in the figures, the variation in the prediction over different PDFs and different values of the strong coupling constant (α_s) is very small compared to the measurement uncertainties.

III DIRECT-γ PRODUCTION

Direct-γ production also provides a means for precision testing pQCD. Since the lowest order production processes are all proportional to $\alpha_s\alpha$, the smallness of the electromagnetic coupling assures that the perturbative series converges more rapidly than is the case with all jets in the final state. The invariant cross section [20] is given by

$$E\frac{d^3\sigma}{dp^3} = \sum_{a,b} x_a G(x_a, Q^2) x_b G(x_b, Q^2) \frac{d\hat{\sigma}}{d\hat{t}}(ab \to jet\ \gamma) = \frac{d^3\sigma}{dE_T^2 d\eta_\gamma d\eta_{jet}} \quad (7)$$

where the first expressions show the factorization of the cross section into a partonic cross section and PDFs. The last expression shows the variables that are directly measured in the detector: transverse energy (E_T) and pseudo-rapidity(η). At leading order, direct-γ production in $\bar{p}p$ collisions occur via quark-gluon scattering and quark-quark annihilation. Quark-gluon scattering dominates the cross section for $E_T < 100$ GeV (at $\sqrt{s} = 1.8$ TeV), prompting one to consider such an interaction to be a good laboratory for measuring the gluonic structure of the proton. However, due to the difficulties of measuring low-E_T objects at high η, the region accessible to the Tevatron is limited to ($10^{-2} < x < 10^{-1}$) for photon measurements and in this region the variations in PDFs are less than 10% [21].

In the detector, direct-γ production is characterized by the presence of a single, highly-electromagnetic cluster in the calorimeter and at least one jet which balances the total momentum in the event. In the measurements being presented, the cross section is measured as a function of the E_T of the photon for a specific range of rapidity ($d^2\sigma/dE_Td\eta$). CDF has measured the differential cross section in the central region, ($|\eta| < 0.9$), and DØ has measured the differential cross section in this region and the forward region, ($1.6 < |\eta| < 2.5$), as well. Both CDF and DØ require the candidate photon clusters to be highly electromagnetic and isolated in E_T. CDF requires the isolation cone around the photon to be $R = \sqrt{(\delta\eta)^2 + (\delta\phi)^2} = 0.7$ whereas DØ requires $R = 0.4$. In both cases, the measurement are corrected for the effects of geometrical acceptance and resolution smearing. Both CDF and DØ compare their measurements to NLO perturbative QCD calculations using the

CTEQ2M and CTEQ4M parton distribution functions (PDFs) with a renormalization scale $\mu = E_T^\gamma$. Figure 9 shows the (Data-Theory)/Theory ratio for DØ in the central and forward regions and also a comparison of CDF and DØ to NLO perturbative QCD in the central region.

CDF observes some discrepancy in the low-E_T region, leading to the consideration that some sort of "k_T effect", i.e., transverse motion of the partons within the proton, is enhancing the production of photons at low-E_T. However, the DØ measurement is consistent with the NLO pQCD prediction within the large systematic uncertainties and reveals little indication of an enhancement of direct-γ production within those uncertainties.

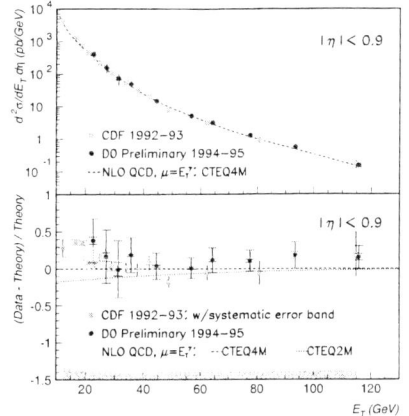
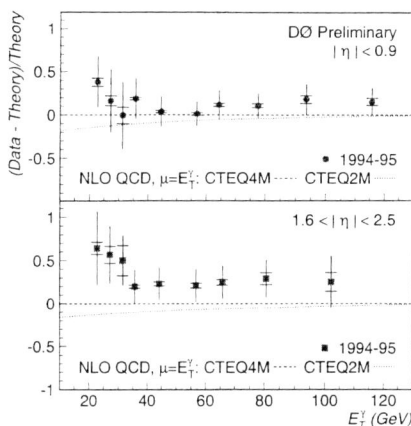

FIGURE 9. Left: Comparison of DØ data (solid circles), CDF data (solid squares) and NLO QCD (dashed line) for direct-γ production in the central region shown as a function of E_T of the photon. In the DØ result, the inner error bars reflect the statistical uncertainty and the outer ones reflect the systematic uncertainty. In the CDF result, the error bars reflect the statistical uncertainty and the band in the lower plot reflects the systematic uncertainty. Right: The ratio (Data-Theory)/Theory for the DØ measurement of the direct-γ cross section as a function of E_T of the photon for the central (top) and forward (bottom) regions. The inner error bars reflect the statistical uncertainty and the outer error bars reflect the systematic uncertainty.

IV   CONCLUSIONS

Using approximately 100 pb^{-1} of data, both of the CDF and DØ collaborations have made precise measurements of vector boson production properties and the direct-γ production cross section. The measured W and Z transverse momentum distributions from DØ agree well the theoretical predictions and the uncertainties present in the Z measurement are small enough to distinguish between two models of vector boson production. The $W+ \geq 1$ jet cross sections measured by CDF agree well with LO perturbative QCD, provided that one selects a suitable renormalization scale. The ratio of the production cross section for $W+ \geq 1$ jet to the inclusive W cross section (R_{10}), measured as a function of the minimum E_T of the jet (E_T^{min}), agrees very well with NLO pQCD predictions, except at the lowest values of E_T^{min}, where it is expected that resummation techniques must be included when calculating the cross sections.

V ACKNOWLEDGEMENTS

We thank the staffs at Fermilab and the acknowledge support from the Department of Energy, the DØ and CDF collaborating institutions and their funding agencies in this work.

REFERENCES

1. B.Abbott et al, Phys. Rev. Lett. **80**, 3000 (1998). B.Abbott et al, submitted to Phys. Rev. D.
2. P.B. Arnold and M.H. Reno, Nucl. Phys. **B319**, 37 (1989); R.J. Gonsalves, J. Pawlowski, and C-F. Wai, Phys. Rev. D **40** 2245 (1989).
3. J.C. Collins, D.E. Soper, G.Sterman, Nucl. Phys. **B250**, 199 (1985).
4. C.T.H. Davies, B.R. Weber, W.J. Stirling, Nucl. Phys. **B256**, 413 (1985).
5. G.A. Ladinsky and C.-P. Yuan, Phys. Rev. **D50**, 4239 (1994). C. Balazs and C.-P. Yuan, Phys. Rev. D**56**, 5558 (1997).
6. C.T.H. Davies, W.J. Stirling, Nucl. Phys. **B244**, 337 (1984).
7. G. Altarelli, R.K. Ellis, M. Greco, and G. Martinelli, Nucl. Phys. **B246**, 12 (1984).
8. P.B. Arnold and R.P. Kaufman, Nucl. Phys. **B349**, 381 (1991).
9. D.W. Duke and J.F. Owens, Phys. Rev. D**30**, 49 (1984).
10. DØ Collaboration, S. Abachi et al., Nucl. Instr. Meth. **A338**, 185 (1994).
11. CDF Collaboration, F. Abe et al., Nucl. Instr. Meth. **A271**, 387 (1988).
12. F. Abe et al, Phys. Rev. D **44**, 29 (1991); the selection is the same as in this reference except for (i) $0.5 < E/(pc) < 2.0$ and (ii) $\chi^2 < 10$.
13. F. Abe et al, Phys. Rev. D **52**, 2624 (1995).
14. F.A. Berends, W.T. Giele, H. Kuijf, and B. Tausk, Nucl. Phys. **B 357**, 633 (1993).
15. F. Abe et al, Phys. Rev. Lett. **77**, 448 (1996).
16. F. A. Berends et al, Nucl. Phys. **B310**, 461 (1998).
17. G. Marchesini and B. Weber, Nucl. Phys. **B310**, 461 (1988).
18. F. Abe et al, Phys. Rev. Lett. **77**, 448 (1996).
19. W.T. Giele et al, Nucl. Phys. **B403**, 633 (1993).
20. J.F. Owens, Rev. Mod. Phys. **59**, 465 (1987).
21. H.L. Lai et al, Phys. Rev. D **55**, 1280 (1997).

HIGH PRECISION TESTS OF QCD AT LEP

J. Fuster Verdú

IFIC, Centro Mixto Universitat de València –CSIC, Doctor Moliner 50
E-46100 Burjassot, València, Spain

Abstract. In this report, the latest measurements involving high precision tests of QCD at LEP are reviewed. The status of the strong coupling constant, α_s, is discussed and summarized. The relative ratio of the normalized three-jet cross-section of b and light, $\ell = u, d, s$, quarks has been determined by the DELPHI collaboration in agreement with the QCD prediction including NLO radiative corrections with mass effects. In this study the running b quark mass at the M_Z scale has been measured providing the first experimental evidence of the running of the b quark mass. As a consequence of this study the flavour independence of α_s for b and ℓ quarks is established within 1% accuracy. The charged multiplicity of quark and gluon jets has been investigated for various three-jet event topologies and compared for different energy and transverse momentum-like scales. An increase of the gluon jet multiplicity with the scale was observed and this was twice as strong as the corresponding increase for quarks.

INTRODUCTION

The strong coupling constant is *the* fundamental parameter of QCD, and its determination is consequently of tremendous experimental endeavour. Here, the most recent analyses, performed by the LEP experiments to extract α_s, are presented. These measurements of α_s were performed at different center-of-mass energies, $\sqrt{s}$, and involved using different methods which analysed the hadronic event shape variables and the dependence of their mean values with respect to $\sqrt{s}$ on the basis of QCD calculations that included $\mathcal{O}(\alpha_s^2)$, next-to-leading logarithmic contributions (NNLA) and combined schemes using hadronization corrections obtained with fragmentation model generators as well as using an analytical power ansatz.

The masses of quarks are also fundamental quantities of the QCD lagrangian not predicted by the theory. The definition of the quark masses is however not unique leading to various possible scenarios. The perturbative pole mass, M_q, and the running mass, m_q, of the $\overline{MS}$ scheme are among the most currently used. At leading order, LO, the predicted expression for any observable is not able to resolve this mass ambiguity and, only when next-to-leading order, NLO, or higher order terms are included, the mass definition becomes known. This is because at orders higher than LO the renormalization scheme used as the baseline of the calculation needs to be chosen and this contains the information about the mass definition. Earlier calculations of the three-jet cross-section in e^+e^- including mass terms already existed at LO [1,2] and have been used to evaluate mass effects for the b-quark when testing the universality of α_s [3,4]. They could not however be used to evaluate the mass of the b-quark, m_b, because these calculations are ambiguous in this parameter. Recently, expressions at NLO, for the multi-jet production rate in e^+e^- are available [5–9] and, thus, they enable measuring m_b in case the flavour independence of α_s is assumed and enough experimental precision is achieved [10].

In addition to the α_s or b mass determinations there are also other interesting studies intended to test certain aspects of the theory of strong interactions, such as these which try to establish experimental evidence for differences between quark and gluon jets. In the context of perturbative QCD, gluons carry a colour charge greater than that of quarks. The Casimir factors, C_A and C_F, denote the relative gluon-gluon and quark-gluon coupling strengths and the LO predicted value for the ratio C_A/C_F of 9/4 reveals in a larger probability for gluons to radiate. Therefore, different properties in the multiplicity and the particle energy spectrum are expected when comparing gluons jets with respect to quark jets.

There are well known existing difficulties to measure all the above parameters in quantitative agreement with the predictions from perturbative QCD, since partons, quarks and gluons, are not directly observed in nature and only the stable particles, produced after the fragmentation process, are experimentally detected. However, the massive statistics and improved jet tagging techniques available at LEP presently allow overcoming these difficulties by applying restrictive selection criteria which lead to quark and gluon jet samples with high purities. The selected data samples are almost background free and small corrections to account for impurities are needed. A smaller model dependence than ever is now achieved, bringing the possibility to perform quantitative studies of quark and gluon fragmentation according to perturbative QCD.

The analyses reported in here include LEP data collected at center-of-mass energies corresponding to the Z resonance, $\sqrt{s} \approx M_Z$ GeV, and above, $\sqrt{s} \approx 133-183$ GeV. In the first section α_s measurements are presented. In the second section, the b mass determination is analyzed and, in the third section, the ratio between the gluon jet multiplicity and the quark jet multiplicity, $r = \langle N_g \rangle / \langle N_q \rangle$, is discussed.

I RECENT α_S MEASUREMENTS AT LEP

Over the past years (1995, 1996 and 1997), the RF system of LEP has been upgraded with the inclusion of new superconducting RF cavities. Thus the energy of the LEP e^- and e^+ beams has increased, giving data with center-of-mass energy of 133 [1], 161, 172 and 183 GeV. The analysis of the hadronic system produced in the

---

[1]) The true energies were 130 and 136 GeV. However, as the integrated luminosity at these two energy points was the same, the data are usually combined and presented as data collected at the

$e^+e^- \to Z/\gamma^* \to q\bar{q}$ process allows the measurement of the running of the strong coupling constant, α_s, and the test of QCD predictions. Any deviation can thus be interpreted as a sign of new physics (e.g. production of light gluinos) appearing at the new energy domain accessible at LEP: LEP II.

Theoretical predictions

It is well known that α_s is a running quantity whose value depends on the physical energy scale (μ) of the process. The beta function describes the renormalization scale dependence of the strong coupling constant:

$$\mu \frac{d\alpha_s(\mu)}{d\mu} = -\frac{\beta_0}{2\pi}\alpha_s^2(\mu) - \frac{\beta_1}{4\pi^2}\alpha_s^3(\mu) + \mathcal{O}(\alpha_s^4) \qquad (1)$$

Regarding SU(3) as the colour gauge symmetry describing the strong interactions: $\beta_0 = 11 - 2n_f/3$ and $\beta_1 = 51 - 19n_f/3$, with n_f being the number of active quark flavours. Solving equation 1, the α_s dependence with the scale goes as: $\ln(\mu^2/\Lambda^2)$, where Λ is the QCD scale.

Combination of individual results

It is always a matter of debate how the different measurements of α_s are combined, given that most of the theoretical uncertainties are correlated [11]. In this paper, the average of a set of measurements, $x_i \pm \sigma_i$, is calculated as $\langle x \rangle = \Sigma x_i w_i$ (with weights w_i inversely proportional to squares of the error). The experimental error is obtained in this way, while the theoretical uncertainty is taken in average. The total error corresponds with the quadratic sum of the experimental and theoretical uncertainties.

Measurements of α_s at LEP I

The high statistics collected at the Z peak by the four LEP experiments (over 15 million of hadronic Z decays) has allowed $\alpha_s(M_Z)$ to be measured with high accuracy[2]. All four LEP experiments have published α_s measurements with a wide range of methods. These can be summarized by: inclusive quantities as R_h and R_τ (the total hadronic decays of the Z and the τ), analysis of the global event shapes (thrust, broadness...), scaling violations [12,13], three-jet rates, etc. Another set of important measurements was the test of the α_s universality for all the quark flavours [3,4]. Also important are the measurements using $q\bar{q}\gamma$ events, where the

---

mean energy of 133 GeV.
[2]) Due to this, most of the published results on α_s from other experiments at different energy scales are extrapolated to a common reference scale, usually taken as M_Z.

boost to the center-of-mass of the hadronic system allows measurements of the running of α_s [14–16].

The global average of all LEP I measurements is: $\alpha_s(M_z) = 0.121 \pm 0.005$, where the main contribution to the error is the theoretical uncertainty.

Measurements of α_s at LEP II

The α_s measurements at LEP II have new problems with respect to the analyses performed at LEP I. The first is the small number of statistics. The second problem is the background, mainly from the radiative return to the Z events, and from $\gamma\gamma$ collisions. Another type of background comes from the W^+W^- pairs decaying into $q\bar{q}q\bar{q}$ and $q\bar{q}l\nu$, however this background is only present with LEP operating above the W^+W^- production threshold. Due to these problems, the total hadronic event sample available for each individual experiment is rather small and is summarized in Tab. 1 for each energy point.

TABLE 1. Available data at LEP per Experiment.

	LEP I	LEP I.5	LEP II		
$\sqrt{s}$ (GeV)	~91	130–136	161	172	183
Year	'89–'95	'95+'97	'96		'97
# Non Rad. Evts.	4×10^6	$3+7\times 10^2$	3×10^2	3×10^3	1.5×10^3

Global event shapes

The LEP experiments obtained new results on $\alpha_s(\sqrt{s})$ by comparing the global event shape variables with the exact $\mathcal{O}(\alpha_s^2)$ QCD calculations plus either NLLA or resummed series. The new LEP II results for α_s are summarized in figure 1.

The first measurement of α_s(133 GeV) was published by L3 [17]. In that analysis the distributions of thrust, scaled heavy jet mass and jet broadening variables were compared with resummed $\mathcal{O}(\alpha_s^2)$ QCD calculations. ALEPH [18] used the differential two jet rate with the Durham jet algorithm. In the DELPHI analysis [19] a similar set of event shape variables was used, plus the three jet rate with the Durham and the JADE algorithms. The variation of the mean values of the event shape variables as a function of the center-of-mass energy was studied using an anlytical power ansatz to account for the fragmentation corrections. In the OPAL analysis [20] the distributions of $1-T$, scaled heavy jet mass, jet broadening variables and the Durham differential two jet rate were fitted to the $\mathcal{O}(\alpha_s^2)$+NLLA

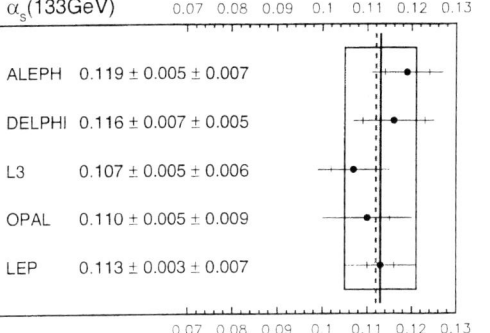
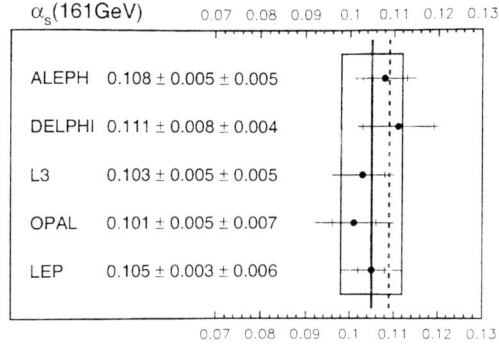
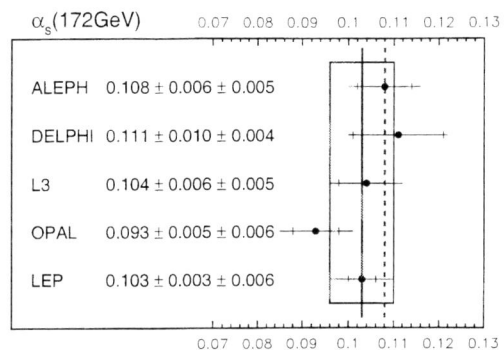

FIGURE 1. Measurements of α_s in LEP II at 133, 161 and 172 GeV. For each measurement, the errors quoted are the experimental and theoretical respectively. Dashed lines correspond to the QCD prediction assuming $\alpha_s(M_Z) = 0.118$.

QCD predictions. The same event variables were used in the analysis of the 161 GeV data, providing the first $\alpha_s(161\text{ GeV})$ LEP measurement [21].

The results shown in figures 1 were provided by the QCD representatives of each experiment. Due to the fact that at the time of the conference not all experiments had their results at $\sqrt{s} = 183$ GeV figure 1 only includes up α_s-values up to $\sqrt{s} = 172$ GeV but the individual results can be seen at figure 2

The size of the non-perturbative effects in the event shape variables decreases when increasing the energy, as shown in a DELPHI study [19,22]. This is due to the convergence of the $\mathcal{O}(\alpha_s^2)$ perturbative calculations with the non-perturbative contribution described by a power series of α_s.

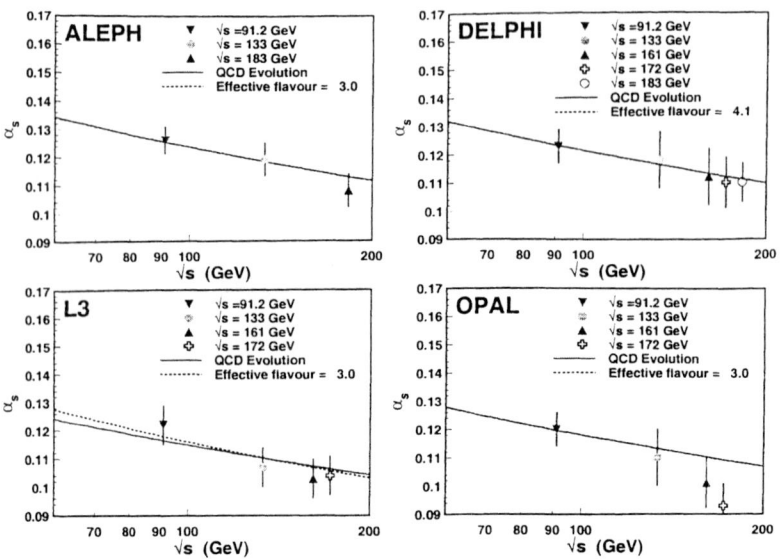

FIGURE 2. Measurements of α_s in LEP II at 133, 161, 172 and 183 GeV foreach measurement.

LEP average

In figure 3, the global LEP values for α_s are presented for each energy available at LEP. The lines corresponds to the running of α_s as obtained assuming $\alpha_s(M_Z) = 0.118 \pm 0.003$ and using the renormalization group equation. The new measurements are compatible with the expected values of α_s. Combining all values of $\alpha_s(M_Z)$ measured so far at LEP I and LEP II, the new LEP average is:

$$\alpha_s(M_Z) = 0.120 \pm 0.005$$

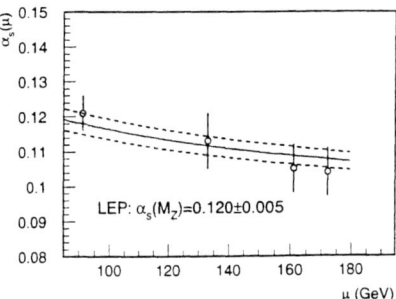

FIGURE 3. Running of the strong coupling constant: measurements at different scales in LEP (M_Z, 133, 161 and 172 GeV by order). Only total errors are displayed.

DETERMINATION OF m_b AT THE M_Z SCALE

For many observable quantities at LEP energies, $\sqrt{s} \geq M_Z$, quark mass effects usually appear in terms proportional to m_q^2/M_Z^2. This represents a ~ 0.003 correction for $m_q = m_b$ which in most of the cases can be safely neglected. This argument, for instance is true for the total hadronic cross-section [2] but cannot be applied for the differential multi-jet cross-sections that depend on the jet-resolution parameter, y_c. The reason being the new scale, $E_c = M_Z\sqrt{y_c}$, introduced in the analysis by the new variable which enhances the mass effects in the form $m_b^2/E_c^2 = (m_b^2/M_Z^2)/y_c$. At $\sqrt{s} \approx M_Z$ the three-jet production rate for b-quarks is in fact suppressed by a factor $\sim 5 - 10\%$ with respect to that of light quarks [1-3]. This difference can then be expressed as a function of m_b [2] and, therefore, used to measure its value.

The experimental observation of such effects is however difficult and delicate because the effect is after all small and furthermore the correct theoretical framework to resolve the mass definition ambiguities is needed. This means that the observable has to be calculated including mass effects at NLO. For this purpose a recent calculation [6-8] of the ratio of the normalized three-jet cross-sections between b-quarks and light uds-quarks [10]:

$$R_3^{b\ell}(y_c) = \frac{\Gamma_{3j}^{Z \to b\bar{b}g}(y_c)/\Gamma_{tot}^{Z \to b\bar{b}}}{\Gamma_{3j}^{Z \to \ell\bar{\ell}g}(y_c)/\Gamma_{tot}^{Z \to \ell\bar{\ell}}} = 1 + r_b(\mu) \cdot \left(b_I(y_c, r_b(\mu)) + \frac{\alpha_s(\mu)}{\pi} \cdot b_{II}(y_c, r_b(\mu)) \right)$$

(2)

where $\Gamma_{3j}^{Z \to q\bar{q}g}$ and $\Gamma_{tot}^{Z \to q\bar{q}}$ are the differential three-jet and total cross-sections, respectively, for the b ($q = b$) and light ($q = \ell \equiv u, d, s$) quarks. The functions b_I and b_{II} can be found in reference [7] and the parameter $r_b(\mu)$ is $m_b^2(\mu)/M_Z^2$. has been performed

The normalization in $R_3^{b\ell}$ to the total decay rates is introduced to cancel possible weak corrections depending on the top quark mass [23] and the ratio of the three-jet cross-sections between b and light uds-quarks minimizes uncertainties due to the hadronization process [10].

The overall correction to the observed value of $R_3^{b\ell-mes}(y_c)$ was about 10% (averaged over all years) from which $\sim 1\%$ was due to the hadronization process. All years' data sets after correction agreed with each other within one standard deviation of the statistical error. Therefore they were combined without further requirements. Fig. 4 shows the corrected data values of $R_3^{b\ell}(y_c)$ together with the theoretical expectations at LO ($\mathcal{O}(\alpha_s)$) from references [1,2] and at NLO ($\mathcal{O}(\alpha_s^2)$) from references [7,8] for $m_b(M_Z)$ mass values of 2.8 GeV/c^2 or a b-qaurk pole mass of $M_b = 4.6$ GeV/c^2.

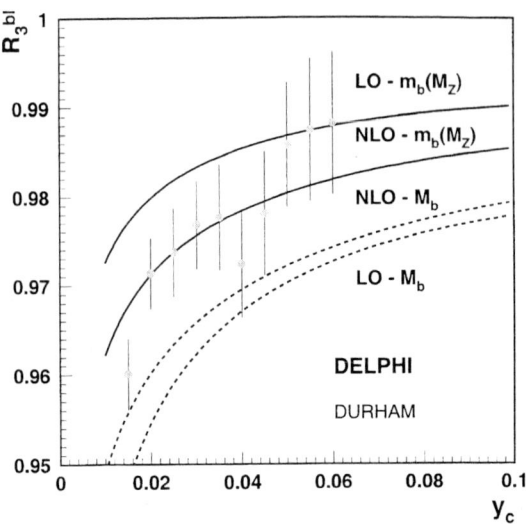

FIGURE 4. Corrected data values of $R_3^{b\ell}(y_c)$ Durham (grey points) in comparison with the LO and NLO theoretical results for this observable expressed in terms of the pole mass, M_b (dashed curves) or the running mass $m_b(M_Z)$ (solid curves).

The corrected values of $R_3^{b\ell}$ are shown in Fig. 4 as a function of y_c. All data points at different y_c values are highly correlated. The measurement of m_b was therefore based on a single point. The optimisation of the statistical error advised the use of small y_c values, but y_c had to lie in a region where the non-perturbative effects and the contributions due to final states with more than three jets were small ($y_c > 0.015$). The value $y_c = 0.02$ was chosen, though any value in the range $0.015 \leq y_c \leq 0.03$ was equally valid as their total errors are approximately the same and furthermore their $R_3^{b\ell}$ values are fully compatible, as seen in Fig. 4. The result is [10]

$$R_3^{b\ell}(0.02) = 0.971 \pm 0.005 \text{ (stat.)} \pm 0.007 \text{ (frag.)}, \tag{3}$$

where the statistical errors from the data and the simulation have been added and the fragmentation error accounts for the uncertainties arising from both the fragmentation model and the tuning parameters.

Additional theoretical uncertainties enter when the measurement of $R_3^{b\ell}$ is transformed into a determination of $m_b(M_Z)$ by means of Eq. (2). Firstly, as also happens for α_s measurements, there is an unphysical *μ-scale dependence* which needs to be quantified. In addition, there is a *mass ambiguity* arising from the fact that there are two ways of expressing Eq. (2) in terms of the running b mass at the M_Z scale [7,6]. This latter uncertainty is labelled as the *mass ambiguity* and it reflects the ambiguity of writing Eq. (2) either directly in terms of the running mass at the M_Z scale or using the pole mass as an intermediate stage, transforming it to the running mass at the pole mass scale, M_b, and then making the evolution to the M_Z scale using the renormalization group equations. Both ways are equally valid, but the contributions due to the higher order terms enter differently in the two procedures because truncated and resummed expressions are used differently. The average of the two results was taken and their difference provides an estimate of the unknown higher order contributions.

The result thus obtained for m_b at $\mu = M_Z$ was [10]

$$m_b(M_Z) = 2.67 \pm 0.25 \text{ (stat.)} \pm 0.34 \text{ (frag.)} \pm 0.27 \text{ (theo.)} \text{ GeV}/c^2,$$

where the statistical and fragmentation errors correspond to the errors expressed for $R_3^{b\ell}$ in Eq. (3) and the theoretical error includes the *mass ambiguity* uncertainty (0.25 GeV/c^2) and the variation of the *scale* in the range $0.5 \leq \mu/M_Z \leq 2$ (0.10 GeV/c^2). Evolving this result down to the b mass scale using $\alpha_s = 0.118 \pm 0.003$ would give $m_b(m_b) = 3.91 \pm 0.67$ GeV/c^2.

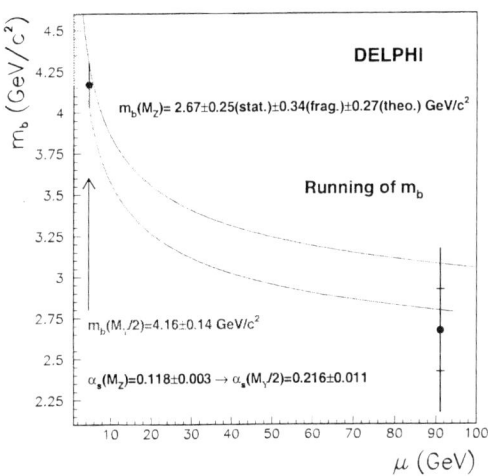

FIGURE 5. The running of $m_b(\mu)$. The $m_b(M_Z)$ value is displayed together with the statistical and total errors. The hatched area corresponds to the band associated to $m_b(\mu)$ when running the non-weighted average value of $m_b(\mu)$ at $M_\Upsilon/2$ up to the M_Z scale using $\alpha_s(M_Z) = 0.118 \pm 0.003$ and the QCD renormalization group equations.

The values of the running b mass at the M_Υ [25] and M_Z scales were compared. The measured difference between them is

$$m_b(M_Z) - m_b(M_\Upsilon/2) = -1.49 \pm 0.52 \text{ GeV}/c^2,$$

where the value of $m_b(M_\Upsilon/2) = 4.16 \pm 0.14$ GeV/c^2 has been calculated as the non-weighted average of all the results appearing in reference [25] at $\mu = M_\Upsilon/2$.

The observed change of the running b mass value from $M_\Upsilon/2$ to M_Z is an effect of 2-3 standard deviations and represents the first experimental evidence of the running property of any fermion mass. The result is in good agreement with the predicted QCD evolution, as Fig. 5 shows, and confirms that QCD radiative corrections including mass effects describe the data correctly from the $M_\Upsilon/2$ scale to the M_Z scale.

The result can also be interpreted as a test of the flavour independence of the strong coupling constant by using the relation [26,6]

$$\frac{\alpha_s^b}{\alpha_s^\ell} = R_3^{b\ell} - H(m_b(M_Z)) + 1.94\frac{\alpha_s(M_Z)}{\pi}\left(R_3^{b\ell} - H(m_b(M_Z)) - 1\right), \quad (4)$$

where α_s^b and α_s^ℓ represent the strong coupling constants for b and ℓ quarks respectively, and $H(m_b(M_Z))$ is the mass correction term of Eq. 2. At $y_c = 0.02$ the value of $H(m_b(M_Z))$ is -0.036 ± 0.005, where the error takes into account the theoretical uncertainties due to the μ scale and to the mass ambiguity as discussed above. Combining this value with that obtained for $R_3^{b\ell}$ from Eq. 3 yields

$$\frac{\alpha_s^b}{\alpha_s^\ell} = 1.007 \pm 0.005 \text{ (stat.)} \pm 0.007 \text{ (frag.)} \pm 0.005 \text{ (theo.)}, \quad (5)$$

which verifies the flavour independence of the strong coupling constant for b and light quarks.

MULTIPLICITIES OF QUARK AND GLUON JETS

Gluon and quark jets are usually selected using hadronic three-jet events. Jets are mainly reconstructed using the DURHAM algorithm although the JADE algorithm was also used [14], in particular, to observe the effects due to different angular particle acceptance of the various algorithms.

In the gluon splitting process ($g \to q\bar{q}$), the heavy quark production is strongly suppressed [27]. Gluon jets can thus be extracted from $q\bar{q}g$ events by applying b tagging techniques. The two jets which satisfy the experimental signatures of being initiated by b quarks are associated to the quark jets and the remaining one is, *by definition*, assigned to be the gluon jet without any further requirement. Algorithms for tagging b jets exploit the fact that the decay products of long lived B hadrons have large impact parameters and/or contain inclusive high momentum leptons coming from the semileptonic decays of the B hadrons. Gluon purities of 94% and 85% are achieved when using these techniques, respectively. Obviously, the quark jets belonging to these events cannot be used to represent an unbiased quark sample. Thus the quark jets whose properties are to be compared with the gluon jets must be selected from other sources which in any case should preserve the same kinematics. Two possibilities have been proposed in the current literature. One consists in selecting symmetric three-jet event configurations [14,28,29] in which one (Y) or the two (Mercedes) quark jets have similar energy to that of the gluon jet. The quark jet purities reached are $\sim$52% and $\sim$66%, for Y and for Mercedes events, respectively. In a second solution [14,28,30] radiative $q\bar{q}\gamma$ events are selected, allowing a sample of quark jets with variable energy to be collected. In this latter case, misidentification of γ's due to the π° background and radiative $\tau^+\tau^-\gamma$ contamination give rise to quark jet purities of $\sim$92%. This method gives a higher purity but unfortunately suffers from the lack of statistics.

For instance, the b-quark purity in the $b\bar{b}g$ sample reached in DELPHI analyses is $\sim$93% and for the light uds-quarks is $\sim$80%.

Results on the charged multiplicity of quark and gluon jets [28,29] using symmetric Y configurations and reconstructed with DURHAM at 24 GeV gluon jet energy, give a ratio of $r \approx 1.23 \pm 0.04$(stat.+syst.) which does not depend on the cut-off parameter (y_{cut}) selected to reconstruct jets [28]. It is significantly higher than one, which indicates that quark and gluons in fact fragment differently, but it remains far from the asymptotic lowest order expectation of $C_F/C_A = 9/4$, suggesting that higher order corrections and non-perturbative effects are very important to understand the measured value. A next-to-leading order correction [31] in MLLA (Modified Leading Log Approximation) at $O(\sqrt{\alpha_s})$ already lowers the prediction towards r values slightly below two and exhibits a small energy dependence due to the running of α_s. However this is still insufficient to explain the value of r determined by the experiments. Solutions based on the Monte Carlo method give a better approximation [14]. The parton shower option of the JETSET generator [24] which uses the Altarelli-Parisi splitting functions for the evolution of the parton shower reduces the theoretical prediction [14] for r. At parton level, at 24 GeV jet energy, the expected value is $\sim$1.4 and it is further reduced to $\sim$1.3 if the value of r is computed after the fragmentation process. In both cases there is a clear dependence of r with the jet energy [14] which can be parametrized using straight lines with slopes of $\Delta r/\Delta E = (+90 \pm 3\text{(stat.)}) \cdot 10^{-4}$ GeV^{-1} at parton level and $\Delta r/\Delta E = (+76 \pm 2\text{(stat.)}) \cdot 10^{-4}$ GeV^{-1} after fragmentation. The absolute value of r predicted at parton level is however largely affected by the choice on the Q_0 parameter (cut-off at which the parton evolution stops) but has negligible influence on its relative variation with the energy, i.e., the slope. The DELPHI analysis uses symmetric and non-symmetric three-jet event configurations with quark and gluon jets of variable energy, allowing thus all these properties and predictions to be tested. A value of $r = 1.23 \pm 0.03$ (stat.+syst.) is measured corresponding to an average jet energy of $\sim$27 GeV. The energy dependence of r is also suggested at 4σ significance level, with a fitted slope of $\Delta r/\Delta E = (+104 \pm 25\text{(stat.+syst.)}) \cdot 10^{-4}$ GeV^{-1}.

In a recent review [32] all published data from various experiments [14,28,29,34,35] were used to perform a general study of r as a function of the jet energy. At present, more data can be added to this comparison. These are the new analysed DELPHI data sample presented above and the most recent measurements of r performed by CLEO [36] and OPAL [37] at 4-7 GeV and 39 GeV

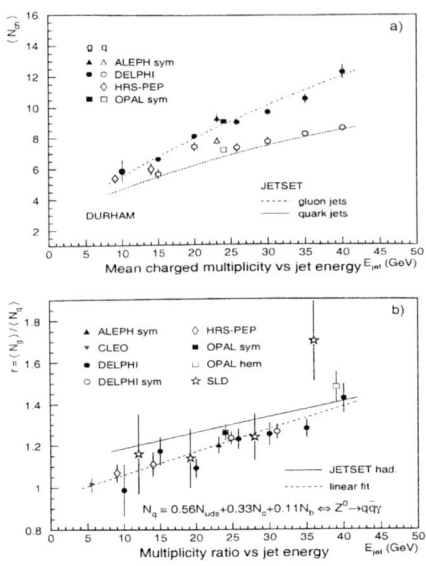

FIGURE 6. (a) Mean charged multiplicity of quark and gluon jets and (b) multiplicity ratio r as a function of the jet energy

average jet energies, respectively. The updated new DELPHI analysis incorporates two times more statistics than the previous analysis [14], therefore, significantly reduces the statistical errors. The analysis from CLEO compares the charged particle multiplicity in $\Upsilon(1S) \to gg\gamma$ decays to that observed in $e^+e^- \to q\bar{q}\gamma$ just in the continuum. This study does not rely on the Monte Carlo simulation to associate the final hadrons to the initial partons and can consequently be fairly considered as being model independent. The obtained value is $r = 1.04 \pm 0.05$ (stat.+syst.). The OPAL analysis uses a new technique [38] which selects gluon jets at ~39 GeV by dividing the events into two hemispheres. While one of these hemispheres is required to contain two tagged quark jets, the other is left untouched being regarded as the gluon jet. The result from OPAL, expressed for only light uds-quarks, is $r_{uds} = 1.55 \pm 0.07$ (stat.+syst.). As it can be observed in figure 6.b all these data agree with the predicted energy behaviour of [14,33,32] when the correction to the quark multiplicity to account for the same flavour composition is applied. In our case it is: 56% uds's, 33% c's and 11% b's. The OPAL number considering this quark mixture becomes $r = 1.48 \pm 0.07$ (stat.+syst.).

All these results thus give evidence for an energy dependence of r. The measured increase is

$$\frac{\Delta r}{\Delta E} = (+110 \pm 13 \text{ (stat.+syst.)}) \cdot 10^{-4} \text{ GeV}^{-1},$$

representing a ~8σ effect.

The measured value of r remains systematically lower than the JETSET prediction over the whole energy range, having an average value of

$$r = 1.23 \pm 0.01 \; (stat.) \pm 0.03 \; (syst.),$$

which corresponds to an average energy of ~23 GeV. This ratio can be further expressed as

$$r_{uds} = 1.30 \pm 0.01 \; (stat.) \pm 0.04 \; (syst.),$$

if r is computed only for the light uds-quarks, extracting the b and c quark contribution to the quark jet multiplicity.

The absolute value of r depends on the reconstruction jet algorithm. For both the JADE and CONE schemes different results are obtained w.r.t. the DURHAM scheme [14,28]. This is due to the combined effect of the different sensitivity of the various jet reconstruction algorithms to soft particles at large angles and of the expected different angular and energy spectra of the emitted soft gluons in the quark and gluon jets. A precise deconvolution of both effects is, at present, impossible [39]. This jet algorithm dependence of r becomes however less apparent as the jet energy increases. The results from OPAL [37], $r = 1.48 \pm 0.07$ (stat.+syst.) and those from DELPHI [14] at ~40 GeV presented in this conference, $r = 1.43 \pm 0.07$ (stat.+syst.) for DURHAM and $r = 1.52 \pm 0.11$ (stat.+syst.) for JADE, agree within errors for the various methods and algorithms used. For the low energy interval, the JADE and DURHAM jet algorithms give a different description of the gluon jet properties [14], although the DURHAM algorithm is in better agreement to those, *model independent*, results obtained by CLEO. Hence, the DURHAM jet algorithm seems to be better suited to decribe the intermediate energy region than the JADE algorithm is.

The interpretation of these results in combination with those obtained by OPAL [40] and ALEPH [41] restrict the validity of the statement that gluon and b-quark jets have similar properties to the jet energy interval around 24 GeV and cannot be applied to the whole jet energy spectrum.

The ALEPH collaboration has investigated other scale definitions to study the jet multiplicities [42]. The angular separation between jets was entered in the description by means of a transverse momentum-like scale whose definition was

$$Q_q = E_Q \sin\left(\frac{\theta_{qg}}{2}\right)$$

for the quark jet, and

$$\overline{Q}_g = \sqrt{Q_q Q_{\bar{q}}}$$

for the gluon jet.

The results showed that the mean jet multiplicities were not functions of the jet energy alone as indicated in Fig. 7 though they couldn't simultaneously describe both quark and gluon jet scale dependences by the same QCD-inspired ansatz.

Following this study DELPHI [43] was able to resolve this ambiguity and has provided a procedure based on NLO QCD calculations which could give a proper

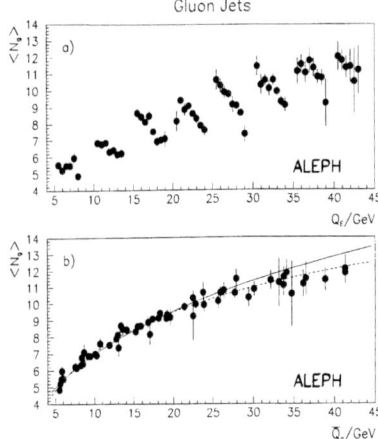

FIGURE 7. The mean gluon jet charged multiplicity as a function of a) the jet energy ($E_g = Q_E$) and b) the topological scale ($\overline{Q}_g$)

common description of the quark and gluon jet multiplicities. As the result of this study the observed increase of the gluon multiplicity with the scale was found to be twice as strong as the corresponding increase for quarks. This implied that the hadronic multiplicity could be quantitatively understood as resulting from a gluon radiation process which is proportional to the colour charges of the initial quarks and gluons whose ratio was measured to be:

$$\frac{C_A}{C_F} = 2.34 \pm 0.04 \text{ (stat)} \pm 0.11 \text{ (scale)}$$

ACKNOWLEDGEMENTS

The author acknowledges many useful discussions with colleagues from the DELPHI QCD and LEP working groups, especially to S. Martí, Roger Jones and Dominique Duchesneau for for their help in providng me with some of the material presented in this report. Also, I express my gratitude to the organizing committee for what was a splendid school in a *paradisiac* atmosphere.

REFERENCES

1. A. Ballestrero, E. Maina, S. Moretti, Phys. Lett. **B294** (1992) 425 and Nucl. Phys. **B415** (1994) 265.
2. M. Bilenky, G. Rodrigo, A. Santamaría, Nucl. Phys. **B439** (1995) 505.
3. DELPHI Coll., P. Abreu et al., Phys. Lett. **B307** (1993) 221;
   J. Chrin, Proc. 28th Rencontre de Moriond, 1993, p. 313, ed. J.T.T. Van;
   J. Valls, PhD Thesis, Universitat de València, 1994, unpublished.
4. L3 Coll., B. Adeva et al., Phys. Lett. **B263** (1991) 551;
   OPAL Coll., R. Akers et al., Z. Phys. **C65** (1995) 31;
   ALEPH Coll., D. Buskulic et al., Phys. Lett. **B355** (1995) 381;
   SLD Coll., K. Abe et al., SLAC-PUB-7573, Jan. 1997.
5. G. Rodrigo, QCD-96 Montpellier, hep-ph/9609213 and Nucl. Phys. Proc. Suppl. **54A** (1997) 60.
6. G. Rodrigo, PhD Thesis, Universitat de València, 1996, hep-ph/9703359 and ISBN:84-370-2989-9.
7. G. Rodrigo, M. Bilenky, A. Santamaría, Phys. Rev. Lett. **79** (1997) 193.
8. W. Bernreuther, A. Brandenburg, P. Uwer, Phys. Rev. Lett. **79** (1997) 189.
9. P. Nason and C. Oleari, Phys. Lett. **B407** (1997) 57.
10. DELPHI Coll., P. Abreu et al., Phys. Lett **B418** (1998) 430-442.
11. M. Schmelling, MPI-H-V39-1996, Talk given at 28th Int. Conf. on High-energy Physics (ICHEP 96), Warsaw, Poland, July 1996, hep-ex/9701002.
12. ALEPH Coll. Phys. Lett. **B357** (1995) 487
13. DELPHI Coll. CERN–PPE/96–185
14. DELPHI Coll. Z. Phys. **C70** (1996) 179
15. S. Martí i García, hep-ex/9511004
16. L3 Coll. Contribution to ICHEP 96, Warsaw, pa04–021
17. L3 Coll. Phys. Lett. **B371** (1996) 137
18. ALEPH Coll. Z. Phys. **C73** (1997) 409
19. DELPHI Coll. Z. Phys. **C73** (1997) 229
20. OPAL Coll. Z. Phys. **C72** (1996) 191
21. OPAL Coll. CERN–PPE/97–015
22. K. Hamacher. Proceedings of QCD 96 conference, Montpellier.
23. A.A. Akhundov, D.Y. Bardin and T. Riemann, Nucl. Phys. **B276** (1986) 1;
    J. Bernabéu, A. Pich and A. Santamaría, Phys. Lett.**B200** (1988) 569.

24. T. Sjöstrand, Comp. Phys. Comm. **39** (1986) 346, Comp. Phys. Comm. bf82 (1994) 74.
25. C.A. Dominguez, G.R. Gluckman, N. Paver, Phys. Lett. **B293** (1992) 197 and hep-ph/9410362;
    S. Narison, Phys. Lett. **B341** (1994) 73;
    M. Neubert, Phys. Rep. **245** (1994) 259 and hep-ph/9404296;
    S. Titard, F. J. Yndurain, Phys. Rev **D49** (1994) 6007;
    M. Jamin, A. Pich, IFIC-97-06, FTUV-97-06, HD-THEP-96-55, hep-ph/9702276;
    M. Crisafulli, V. Giménez, G. Martinelli, C.T. Sachrajda, Nucl. Phys. **B457** (1995) 594;
    V. Giménez, G. Martinelli, C.T. Sachrajda, Nucl. Phys. Proc. Suppl. **53** (1997) 365 and Phys. Lett. **B393** (1997) 124 ;
    C.T.H. Davies et al., Phys. Rev. Lett. **73** (1994) 2654 and Phys. Lett. **B345** (1995) 42.
26. S. Catani et al., Phys. Lett. **B269** (1991) 432;
    N. Brown, W.J. Stirling, Z. Phys. **C53** (1992) 629.
27. M.H. Seymour, Nucl. Phys. **B436** (1995) 163.
28. OPAL Collab., Z. Phys. **C58** (1993) 387;
    OPAL Collab., Z. Phys. **C68** (1995) 179.
29. ALEPH collab., Phys. Lett. **B346** (1995) 389.
30. L3 EPS-HEP95/105 (1995).
31. J.B. Gaffney, A.H. Mueller Nucl. Phys. **B250** (1985) 109.
32. J. Fuster, S. Martí, Proceedings of XXXII EPS conference in Brussels, July (1995), hep-ex/9511002.
33. Z. Fodor, Phys. Lett. **B263** (1991) 305.
34. HRS Phys. Lett. **B165** (1985) 449.
35. SLD I. Ywasaki SLAC–R–95-460.
36. CLEO ICHEP96 PA04-050.
37. OPAL Collab., CERN-PPE/96-116.
38. J.W. Gary, Phys. Rev. **D49** (1994) 4503.
39. P.V. Chliapnikov et al. Phys. Lett. **B300** (1993) 183.
40. OPAL Collab., Z. Phys. **C69** (1996) 543.
41. ALEPH Collab., CERN-PPE/95-184.
42. ALEPH Collab., Z. Phys. **C76** (1997) 191.
43. DELPHI Collab., HEP-97 contributed paper 545.

Jet Production at the Tevatron

Freedy Nang [1]

Department of Physics, University of Arizona, Tucson, AZ 85721

Abstract.
We present several results on high P_T jet production and compare them to the most recent theoretical predictions using the latest parton distribution functions. The dijet invariant mass distribution, triple differential cross section, and dijet angular distribution measurements exhibit very good agreement with the theoretical predictions. The latter agreement obviates the need to invoke compositeness models. At $\sqrt{s} = 630$ GeV, the inclusive jet cross section lies $15 - 20\%$ below the theoretical predictions for both experiments. At $\sqrt{s} = 1800$ GeV, the DØ inclusive jet cross section is in agreement with the theoretical predictions over the entire jet E_T range, the CDF collaboration observes an excess in the cross section for E_T above 200 GeV.

INTRODUCTION

Two years ago, the CDF collaboration generated great excitement when it published the inclusive jet cross section at $\sqrt{s} = 1800$ GeV showing an excess in the high transverse energy (E_T) region when compared to a next-to-leading order (NLO) theory [1]. The inclusive jet cross section measurement could deviate from the theory due to a variety of reasons, the most interesting of which is quark compositeness. Other explanations for the excess include the existence of new particles and differences between parton distribution functions (PDF's). We examine various results from both the CDF and DØ experiments, starting with the inclusive jet cross section at $\sqrt{s} = 1800$ GeV and 630 GeV. We also present results on the dijet angular distribution, the dijet invariant mass analysis and the triple differential dijet cross section.

THEORY

Perturbative QCD (pQCD) allows us to calculate any jet cross section as a convolution of the parton distribution functions for the incoming particles and the parton-parton hard scattering cross section:

---

[1] for the CDF and DØ Collaborations.

$$\sigma \sim \sum_{i,j} f_i(x_1, Q^2) f_j(x_2, Q^2) \hat{\sigma}_{ij} ,\qquad(1)$$

where $x_{1(2)}$ represents the momentum fraction of the proton (anti-proton) carried by the colliding parton; $f_{i,j}$ represent the parton distribution functions for the initial valence quarks (u, $\bar{u}$, d, $\bar{d}$) or sea partons (g, u, d, s, ...), evaluated at the energy scale of the hard scattering, Q^2; and $\hat{\sigma}_{ij}$ represents the cross section for the scattering of partons i and j. The relationship between the momentum fraction and pseudorapidity (η) is given by

$$x_{1,2} = (\frac{1}{\sqrt{s}}) \sum_{i=1}^{N} E_{Ti} \exp(\pm \eta_i) ,\qquad(2)$$

where N=2 corresponds to a two parton (jet) final state (LO) and N=3 to a three parton (jet) final state (NLO). The pseudorapidity is defined as $\eta = \ln \cot(\theta/2)$, where θ is the polar angle of the outgoing jet with respect to the z-axis. The subscript 1(2) corresponds to a positive (negative) argument in the exponential. The energy scale of the hard scattering, Q^2, is usually defined as $Q^2 = E_T^2$ where E_T is defined as the transverse energy of the jet. Jets are ordered in descending order with respect to E_T, so the "leading jet" refers to the highest E_T jet in the event.

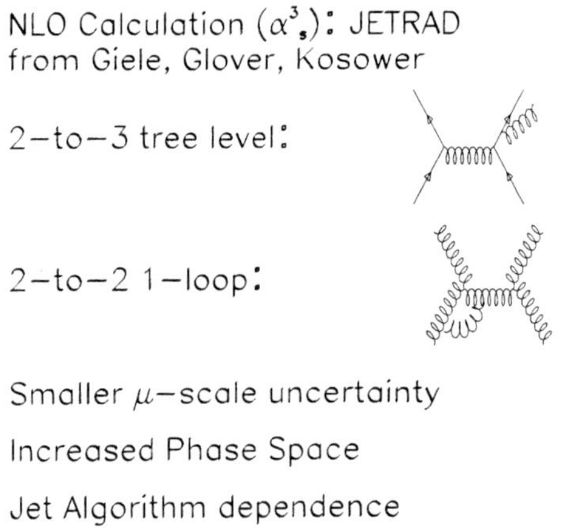

NLO Calculation (α_s^3): JETRAD
from Giele, Glover, Kosower

2-to-3 tree level:

2-to-2 1-loop:

Smaller μ-scale uncertainty

Increased Phase Space

Jet Algorithm dependence

FIGURE 1. Main features of the NLO calculation including the 2-to-3 tree level and 2-to-2 1-loop Feynman diagrams that contribute to the jet production.

Calculations of jet physics are available at NLO (order α_s^3) [2–4]. The NLO theory includes, in addition to the 2-to-2 tree level Feynman diagrams, 2-to-3 tree level and 2-to-2 one-loop diagrams as shown in Fig. 1. The NLO calculation provides a smaller uncertainty due to the renormalization/factorization scale μ [5], and increased phase space in the forward pseudorapidity regions and introduces a jet algorithm dependence. Both experiments use the Snowmass iterative cone algorithm with a cone radius of $\mathcal{R} = \sqrt{(\Delta\eta)^2 + (\Delta\phi)^2} = 0.7$ in $\eta - \phi$ space [6].

The NLO calculation packages used for the different analyses are EKS and JETRAD [3,4]. The main parameters that can be changed are the PDF's, the μ scale, and the jet cone algorithm. EKS was used with a value of $\mu = E_T^{jet}/2$, where E_T^{jet} is the E_T of the jet while JETRAD was evaluated with $\mu = E_T^{max}/2$, where E_T^{max} is the E_T of the leading jet in the event. At NLO, two partons can be $2R$ apart and still be merged into one jet where R is the jet cone radius. This criterion is used for the jet algorithm used by the CDF collaboration. The DØ collaboration restricts this separation to a value of $R_{SEP} = 1.3$, where R_{SEP} is the value in units of $\mathcal{R}$ of allowed separation [7].

INCLUSIVE JET CROSS SECTION AT $\sqrt{S} = 1800$ GEV

The inclusive jet cross section counts all the jets in the event that satisfy η and E_T criteria. The cross section is written as

$$\frac{d^2\sigma}{dE_T d\eta} = \frac{N}{\Delta E_T \Delta \eta \int \mathcal{L} dt} \qquad (3)$$

and plotted as a function of the jet E_T. N is the number of jets in the event, ΔE_T is the E_T bin size, $\Delta \eta$ is the η bin size, and $\int \mathcal{L} dt$ is the integrated luminosity.

The CDF collaboration has measured the jet cross section using the 1992-1993 data sample (run 1A) in the region $0.1 \leq |\eta| \leq 0.7$ and observed an excess above NLO calculations for jet E_T above 200 GeV [1]. The theoretical calculation is made with the EKS algorithm with the renormalization and factorization scales set at $\mu = E_T^{jet}/2$, $R_{sep} = 2.0$ and the CTEQ3M PDF [3,8]. Higher statistics data from the 1994-1995 run (run 1B) continues to exhibit an excess in the jet cross section (Fig. 2).

The DØ collaboration has measured the cross section from run 1B data sample for the regions $0.0 \leq |\eta| \leq 0.5$ and $0.1 \leq |\eta| \leq 0.7$. Figure 3 shows a residual plot for both regions depicting excellent agreement with the NLO theory for the entire E_T range. The theoretical calculation comes from JETRAD with the renormalization and factorization scales set at $\mu = E_T^{max}/2$, $R_{sep} = 1.3$, and the CTEQ3M PDF [4,8].

The data from both experiments for the region $0.1 \leq |\eta| \leq 0.7$ are shown in Fig. 4, in which the DØ data has been compared to a fit of the CDF's published result.

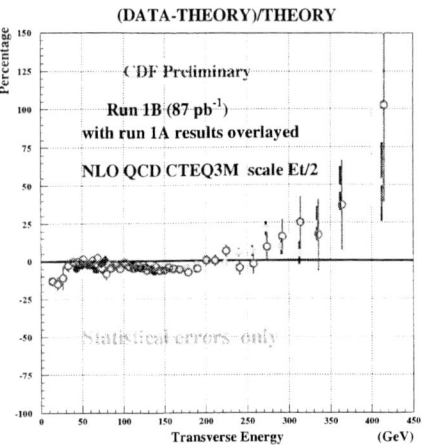

FIGURE 2. Residual plot of the inclusive jet cross section from CDF from run 1B (solid circles) and run 1A (open circles). The theory used was EKS with the CTEQ3M PDF. The error bars are statistical only.

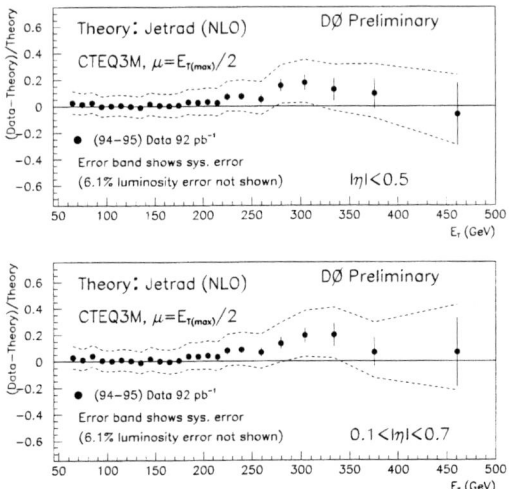

FIGURE 3. Residual plot of the inclusive jet cross section from DØ from run 1B for the regions $0.0 \leq |\eta| \leq 0.5$ (top) and $0.1 \leq |\eta| \leq 0.7$ (bottom). The theory used was JETRAD with the CTEQ3M PDF. The error bars are statistical only. The band shows $\pm 1\sigma$ systematic uncertainty.

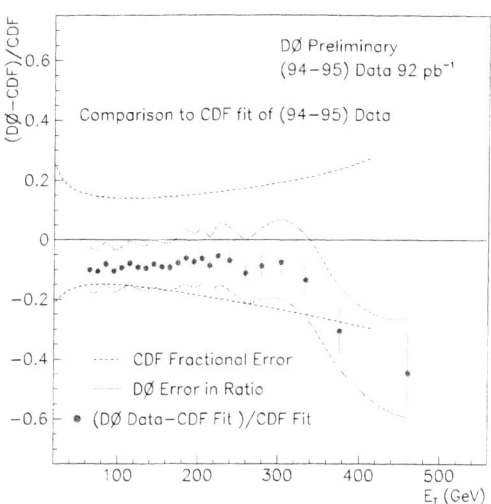

FIGURE 4. Residual plot of the inclusive jet cross section from DØ from run 1B for the region $0.1 \leq |\eta| \leq 0.7$. The reference used was a fit to the CDF's published run 1A data. The error bars are statistical only. Bands correspond to the $\pm 1\sigma$ systematic uncertainty for each experiment.

INCLUSIVE JET CROSS SECTION AT $\sqrt{S} = 630$ GEV

Both experiments have also measured the inclusive cross section at a lower center-of-mass energy, $\sqrt{s} = 630$ GeV. To allow a direct comparison with the $\sqrt{s} = 1800$ GeV data, the data are plotted as a function of the variable x_T, defined as $x_T = 2E_T/\sqrt{s}$. Figure 5 shows the ratio of the scaled cross sections at $\sqrt{s} = 630$ GeV to that at $\sqrt{s} = 1800$ GeV for both experiments, compared to the same ratio in the theory, as a function of x_T. The experiments are in good agreement with each other but the data lie 15% − 20% below the theoretical predictions.

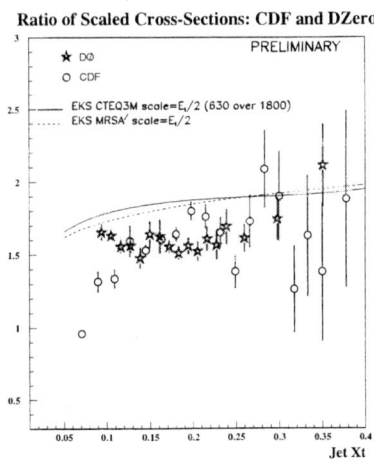

FIGURE 5. The ratio of the scaled cross sections for both CDF (open circles) and DØ (solid stars) with the same ratio from EKS using MRSA' and CTEQ3M. The error bars are statistical only. The boxes correspond to the ± 1σ systematic uncertainty for the DØ collaboration.

DIJET ANGULAR DISTRIBUTION

The dijet angular distribution, $\frac{1}{N}\frac{dN}{d\chi}$, is a good tool for determining whether any observed excess of events might be due to compositeness. Compositeness models are extensions to the Standard Model in which quarks are allowed to have substructure. One of the advantages to investigating the angular distribution is that it should be insensitive to the PDF's. One searches for dijet events and plots them as a function of χ, defined as

$$\chi = \exp(|\eta_1 - \eta_2|) = \frac{1 + \cos\theta^*}{1 - \cos\theta^*} \qquad (4)$$

for different mass bins, where $\eta_{1,2}$ represents the pseudorapidity of the two leading jets and θ^* represents the center-of-mass scattering angle. The use of χ flattens the angular distribution, facilitating comparison with theory.

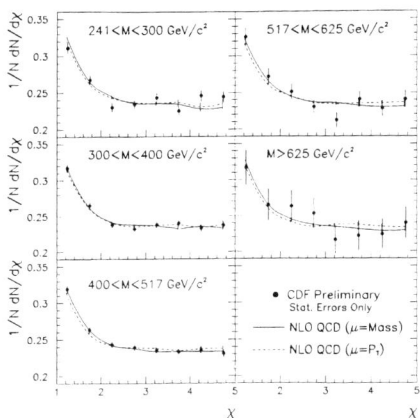

FIGURE 6. Dijet angular distribution as a function of χ (circles) for different mass bins compared to JETRAD for two different μ scales. The error bars are statistical only.

The jets are restricted to $0.1 \leq |\eta_{1,2}| \leq 2.0$ and $\chi < 5$ by the CDF collaboration [9]. Figure 6 compares data with theory and demonstrates that very little variation arises due to the different scales when looking at regions of $\chi < 5$. To determine a limit, a ratio R is defined as

$$R_\chi = \frac{Number\ of\ events\ with\ \chi < 2.5}{Number\ of\ events\ with\ \chi > 2.5} \qquad (5)$$

for each mass bin. This procedure removes correlated errors and reduces the curve to a single number. The ratio is then plotted as a function of the mass bin and compared to models with different values of contact terms, which are expressed in the form of the parameters Λ^- and Λ^+ for constructive and destructive interference terms respectively. Fig. 7 shows that the CDF data is in excellent agreement with LO and NLO QCD predictions as well as the behavior of the theoretical predictions for including different contact term values. For a model where all quarks are allowed to be composite objects, the CDF collaboration excludes at the 95% confidence level (CL) regions with $\Lambda^+ \leq 1.8$ TeV and $\Lambda^- \leq 1.6$ TeV.

The DØ search is performed for $0.0 \leq |\eta_{1,2}| \leq 3.0$, which includes χ values up to 20 when kinematically accessible [10]. Though a large range of χ introduces some sensitivity of the theoretical predictions to different renormalization/factorization scales (Fig. 8), the analysis is more sensitive to higher values of Λ. The DØ experiment defines a ratio similar to the CDF collaboration but for different ranges of χ:

$$R_\chi = \frac{Number\ of\ events\ with\ \chi < 4}{Number\ of\ events\ with\ \chi > 4} \qquad (6)$$

The DØ experiment rules out a model, in which all quarks are allowed to be composite objects, at the 95% CL regions with $\Lambda^+ \leq 2.0$ TeV.

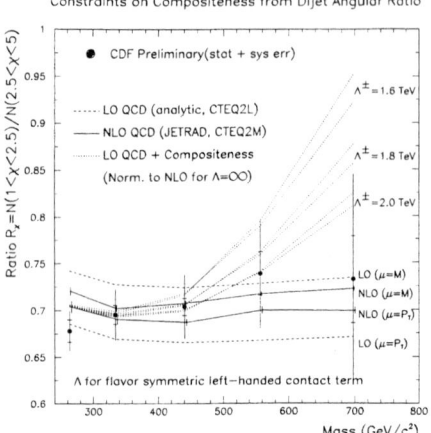

FIGURE 7. The ratio R_χ as a function of mass. The inner error bars are statistical and outer error bars indicate the sum in quadrature of statistical and systematic errors. The data exhibit excellent agreement with LO and NLO QCD. Also shown is the behavior of the theory when different contact term values are included. The higher curve corresponds to destructive interference (Λ^+).

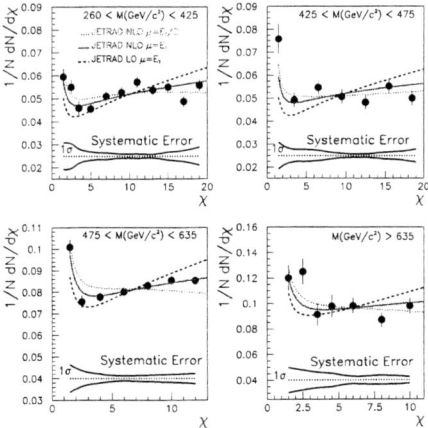

FIGURE 8. Dijet angular distribution for four different mass bins. A wider range of χ allows sensitivity to differences due to the μ scale. The inner error bars are statistical errors only. The band shows $\pm\ 1\sigma$ systematic uncertainty.

DIJET INVARIANT MASS

The existence of new particles may be identified by looking for a resonance in their hadronic decay products. The DØ analysis requires the two leading jets to be within $0.0 \leq |\eta| \leq 1.0$ and the CDF analysis requires the two leading jets to be within $0.0 \leq |\eta| \leq 2.0$ and $\cos\theta^* < 2/3$. Figures 9 and 10 show residual plots for both the CDF and DØ analyses respectively. The comparison for CDF is normalized to the first six data points while the DØ comparison is done for absolute normalization. Each result shows very good agreement with JETRAD.

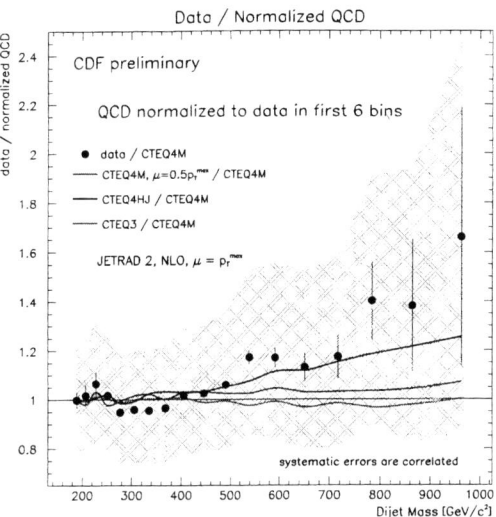

FIGURE 9. The residual plot of the dijet invariant mass with JETRAD with the CTEQ4M PDF where the theory has been normalized to the first six data points. The error bars are statistical only. The boxes show the $\pm 1\sigma$ systematic uncertainty. Also seen are residual distributions when using other PDF's or μ scales.

To reduce the systematic uncertainty, the DØ collaboration has further subdivided the sample into two η regions, and taken a ratio of the cross sections to reduce the systematic uncertainties ($0.0 \leq |\eta| \leq 0.5$ and $0.5 \leq |\eta| \leq 1.0$). The systematic uncertainty is reduced to $< 10\%$ and still shows good agreement with the theory (Fig. 11).

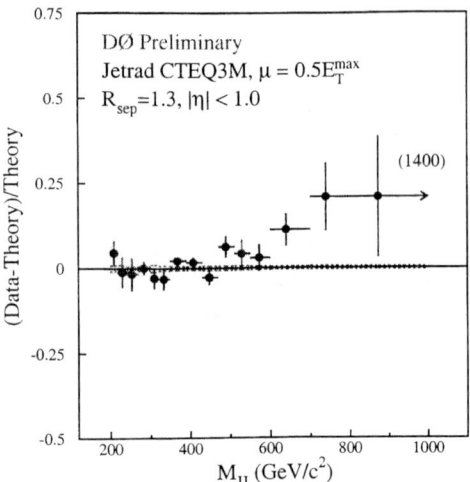

FIGURE 10. The residual plot of the dijet invariant mass with JETRAD with the CTEQ3M PDF. The error bars are statistical only. The band shows the ± 1σ systematic uncertainty of the theory. The boxes surrounding the data points show the ± 1σ systematic uncertainty.

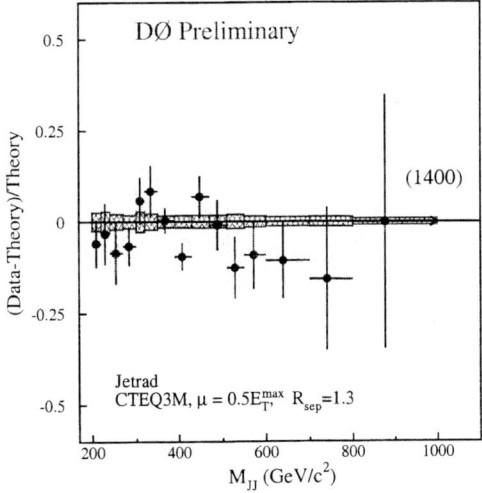

FIGURE 11. The residual plot of the ratio of the two η regions ($0.5 \leq |\eta| \leq 1.0$ over $0.0 \leq |\eta| \leq 0.5$) as a function of the invariant mass with a similar ratio from JETRAD.

TRIPLE DIFFERENTIAL DIJET CROSS SECTION

One explanation for the excess of very high E_T jets observed by the CDF collaboration is the choice of PDF used in the theory Monte Carlo generation. The CTEQ group has found that placing more gluons in the high x region does not strongly affect the fit with other experiments and has released the CTEQ4HJ PDF [11].

The triple differential dijet cross section, $d^3\sigma/dE_T d\eta_1 d\eta_2$ is ideal for the study of different PDF's because the choice of variables are sensitive to the PDF's, while being insensitive to the matrix elements. Hence, the triple differential analysis is complementary to the angular distribution analysis. The CDF collaboration requires the leading jet to be central ($0.1 \leq |\eta_1| \leq 0.7$) and plots the E_T of the leading jet for four different configurations defined by the position of the second leading jet: $0.1 \leq |\eta_2| \leq 0.7$, $0.7 \leq |\eta_2| \leq 1.4$, $1.4 \leq |\eta_2| \leq 2.1$, and $2.1 \leq |\eta_2| \leq 3.0$ as it is shown in Fig. 12.

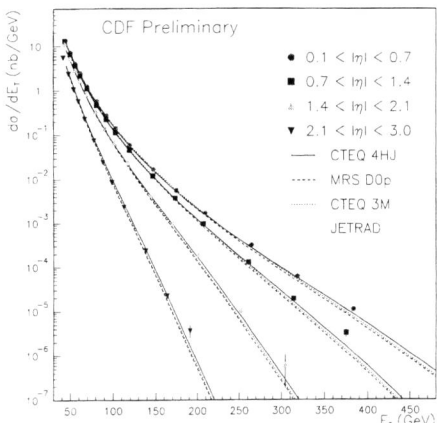

FIGURE 12. The triple differential dijet cross section for the four different η regions compared to JETRAD using different PDF's. The error bars are statistical only.

To linearize the scale, the CDF collaboration weights the cross section by different powers of E_T of the leading jet for the different pseudorapidity regions as shown in Fig. 13. The results in Fig. 13 are not sensitive enough to distinguish among the various PDF's.

CONCLUSIONS AND FUTURE PROSPECTS

A variety of high P_T jet analyses has shown that pQCD is still a very successful theoretical model. There remains a discrepancy in the $\sqrt{s} = 630$ GeV inclusive data where the theory exhibits a $15-20\%$ excess. In the $\sqrt{s} = 1800$ GeV inclusive data sample, the CDF collaboration finds an excess for $E_T > 200$ GeV while the DØ collaboration finds excellent agreement for the entire E_T range. Both experimental data sets are in good agreement for $E_T < 300$ GeV. The CTEQ4HJ PDF remains a possible candidate for

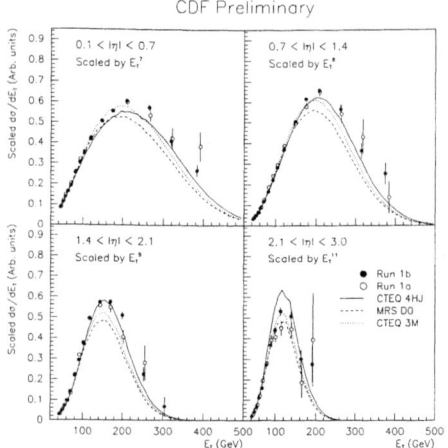

FIGURE 13. The triple differential dijet cross section for the four different η regions scaled by a region-dependent power of E_T compared to JETRAD using various PDF's. The error bars are statistical only.

explaining the excess found by the CDF collaboration. The next Tevatron run, where a factor to 20 in increased luminosity is expected and both detectors will be upgraded, will hopefully produce a definite answer to these remaining questions.

ACKNOWLEDGMENTS

We are grateful to the DØ and CDF collaborations for discussions of their data.

REFERENCES

1. F. Abe *et al.*, (CDF Collaboration), Phys. Rev. Lett. **77**, 428 (1996).
2. F. Aversa *et al.*, Nuclear Physics **B327** (1989).
3. S. Ellis, Z. Kunszt, and D. Soper, Phys. Rev. Lett. **64**,2121 (1990).
4. W. Giele, N. Glover, and D. Kosower, Phys. Rev. Lett. **B73**,2019 (1994).
5. B. Abbott *et al.*, (DØ Collaboration), Hep-Ph/9801285 (1998).
6. J. Huth *et al.*, proceedings of Research Directions for the Decade, Snowmass 1990, pp. 134-136 edited by E. L. Berger (World Scientific, Singapore 1992).
7. B. Abbott *et al.*, (DØ Collaboration), Fermilab-PUB-97/242-E (1997). Phys. Res., Sect. A, **271**, 387 (1988). Phys. Res., Sect. A, **338**, 185 (1994).
8. H. Lai *et al.*, (CTEQ Collaboration), Phys. Rev. **D51**, 4763 (1995).
9. F. Abe *et al.*, (CDF Collaboration), Phys. Rev. Lett. **78**, 4397 (1997).
10. B. Abbott *et al.*, (DØ Collaboration), Phys. Rev. Lett. **80**, 666 (1998).
11. H. Lai *et al.*, (CTEQ Collaboration), Phys. Rev. **D55**, 1280 (1997).

QCD at HERA

N. H. Brook
on behalf of the ZEUS & H1 collaborations

*Dept. of Physics & Astronomy, University of Glasgow,
Glasgow G12 8QQ, United Kingdom.*

Abstract. A review of HERA measurements of structure functions, fragmentation functions and forward jet production is presented.

DIS KINEMATICS

The event kinematics of deep inelastic scattering, DIS, are determined by the negative square of the four-momentum transfer at the lepton vertex, $Q^2 \equiv -q^2$, and the Bjorken scaling variable, $x = Q^2/2P \cdot q$, where P is the four-momentum of the proton. In the quark parton model (QPM), the interacting quark from the proton carries the four-momentum xP. The variable y, the fractional energy transfer to the proton in its rest frame, is related to x and Q^2 by $y \simeq Q^2/xs$, where $\sqrt{s}$ is the positron-proton centre of mass energy. Because the H1 and ZEUS detectors are almost hermetic the kinematic variables x and Q^2 can be reconstructed in a variety of ways using combinations of positron and hadronic system energies and angles [1].

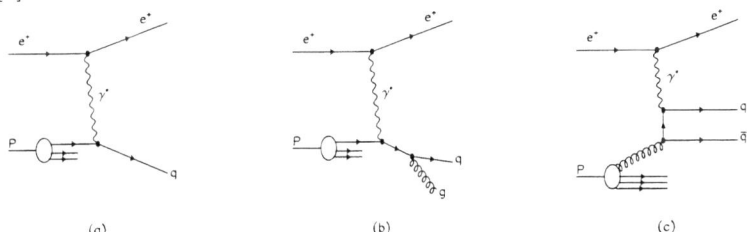

FIGURE 1. (a) QPM (b) QCDC and (c) BGF diagrams

Neutral current (NC) DIS occurs when an uncharged boson (γ, Z^0) is exchanged between the lepton and proton. In QPM there is a 1+1 parton configuration, fig. 1a, which consists of a single struck quark and the proton remnant, denoted by "+1". At HERA energies there are significant higher-order quantum chromodynamic (QCD) corrections: to leading order in the strong coupling constant, α_s, these are QCD-Compton scattering (QCDC), where a gluon is radiated by the scattered quark and boson-gluon-fusion (BGF), where the virtual boson and a gluon fuse to form a quark-antiquark pair. Both processes have 2+1 partons in the final state, as shown in fig. 1. There also exists calculations for the higher, next-to-leading (NLO) processes.

HERA provides a unique opportunity to study the scale behaviour of Quantum Chrmodynamics in a single experiment with a cleaner background environment than that obtainable at hadron colliders. The data presented here were taken in 1994 and onwards at the ep collider HERA using the H1 and ZEUS detectors. During this period HERA operated with positrons of energy $E_e = 27.5$ GeV and protons with energy 820 GeV. A detailed descriptions of the H1 and ZEUS detectors can be found in refs. [2] and [3] respectively.

STRUCTURE FUNCTIONS

Perturbative QCD (pQCD) does not predict the absolute value of the parton densities within the proton but determines how they vary from a given input. For a given initial distribution at a particular scale, Q_0^2, Altarelli-Parisi (DGLAP) evolution [4] enables the distributions at higher Q^2 to be determined. DGLAP evolution resums the leading $\log(Q^2)$ contributions associated with a chain of gluon emissions. At large enough positron-proton centre-of-mass energies there is a second large variable $1/x$ and, therefore, it is also necessary to resum the $\log(1/x)$ contributions. This is achieved by using the BFKL equation [5]. In the DGLAP parton evolution scheme the parton cascade follows a strong ordering in transverse momentum $p_{Tn}^2 \gg p_{Tn-1}^2 \gg ... \gg p_{T1}^2$, while there is only a soft (kinematical) ordering for the fractional momentum $x_n < x_{n-1} < ... < x_1$ (see figure 2.) By contrast, in the BFKL scheme the cascade follows a strong ordering in fractional momentum $x_n \ll x_{n-1} \ll ... \ll x_1$, while there is no ordering in transverse momentum.

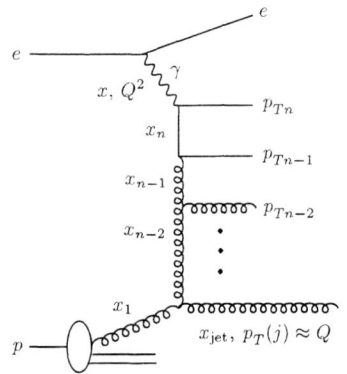

FIGURE 2. Parton ladder diagram in DIS.

At small x the dominant parton is the gluon and the description of the structure function is driven by the behaviour of the gluon. Because of gluon splitting, $g \to q\bar{q}$, pQCD suggests the small x behaviour of the sea quark and gluon distributions are strongly correlated.

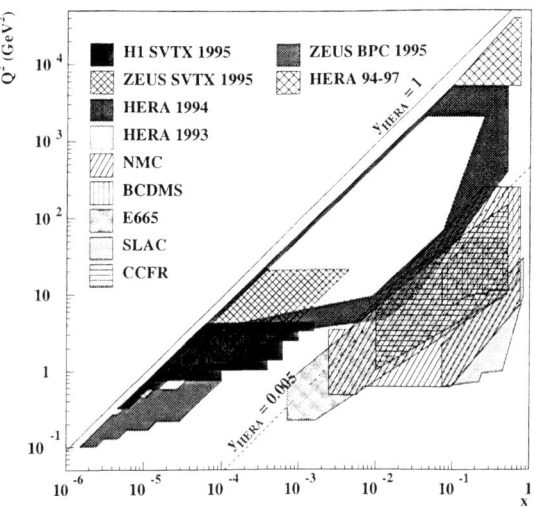

FIGURE 3. The kinematical region covered by HERA and fixed target experiments

The kinematic plane covered by HERA and the fixed target measurements is shown in fig. 3. HERA has increased the reach in Q^2 by about 2 orders of magnitude and can also probe nearly 3 orders of magnitude further down in x. The low x region is correlated with low values of Q^2. The differential NC DIS cross section is related to three structure functions:

$$\frac{d^2\sigma^{e^{\pm}p}}{dxdQ^2} = \frac{2\pi\alpha^2}{xQ^4}(Y_+ F_2(x,Q^2) - y^2 F_L(x,Q^2) \mp Y_- xF_3(x,Q^2)), \quad (1)$$

where $Y_{\pm} = 1 \pm (1-y)^2$. The structure function F_2 in QPM is just the sum of the quark densities multiplied by the appropriate electric charge; F_3 arises from the weak part of the cross section and is negligible for $Q^2 < 5000$ GeV^2, and F_L is the longitudinal structure function and only becomes important for $y > 0.6$. Hence by measuring the differential cross section at HERA one is effectively measuring the structure function F_2.

The F_2 measurements are shown in figs. 4 and 5 as a function of x and Q^2 respectively. The error bars are at the 5-10% level and the normalisation uncertainty is ~2%. There is a steep rise of F_2 with decreasing x in all Q^2 bins, fig. 4. Scaling violations in Q^2 are clearly seen in fig. 5. Both H1 and ZEUS have performed next to leading order (NLO) QCD fits [6,7] based on the DGLAP evolution equations using both HERA and fixed target data. Fig. 5 shows that these QCD fits describe

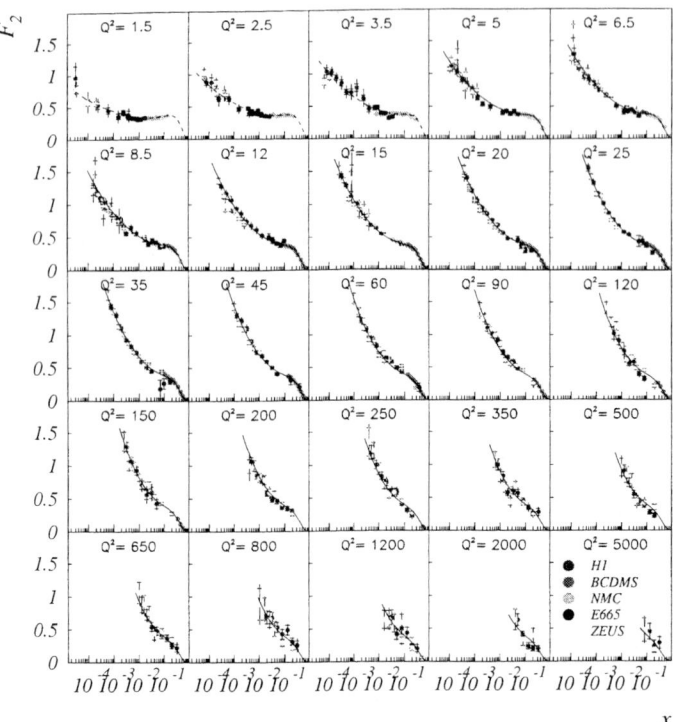

FIGURE 4. Measurement of $F_2(x, Q^2)$ as function of x.

the F_2 data well, though it should be noted that the data can also be satisfactorily described by the BFKL prediction [8].

The scaling violations from the HERA data allow an estimate of the gluon density $xg(x)$ at low values of x, whilst the fixed target data are used to constrain the high x region. The extracted gluon densities from the fits are shown in fig 6 for a fixed $Q^2 = 20$ GeV^2. The error band shows the statistical and systematic uncertainty taking into account correlations and variations in the mass of the charm quark, m_c, and the strong coupling constant, α_s. The results of the two HERA experiments are in good agreement and the extracted densities agree with the results of NMC [9] for large x. The resulting gluon distributions show a clear rise with decreasing x and have a 15% uncertainty at $x \sim 5 \times 10^{-4}$. These NLO QCD fits are also in good agreement with the global QCD analyses performed by MRS [10] and CTEQ [11], whilst the prediction from the dynamical evolution of GRV [12] is too steep for $x < 10^{-3}$.

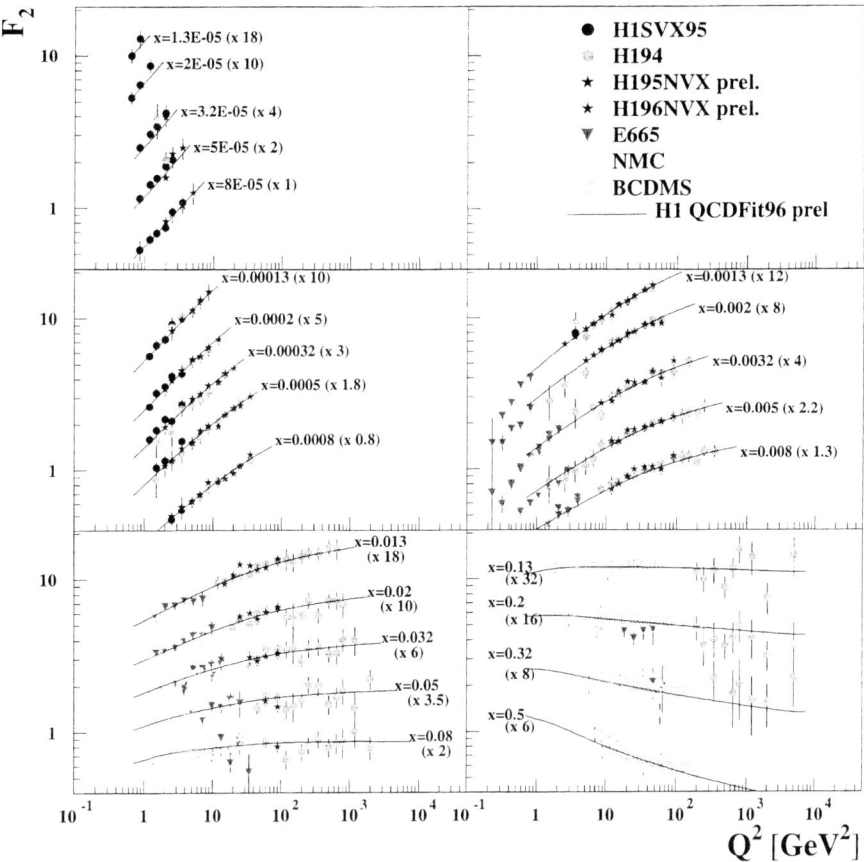

FIGURE 5. Measurement of $F_2(x, Q^2)$ as function of Q^2.

FORWARD JET PRODUCTION IN DIS

The F_2 measurements fail to distinguish between DGLAP and the BFKL approach to the QCD evolution. The hadronic final states are expected to give additional information. For events at low x, hadron production between the current jet and the proton remnant is expected to be sensitive to the effects of BFKL or DGLAP dynamics. A possible signature of BFKL dynamics is the behaviour of DIS events at low x which contain a jet that has a transverse momentum $p_T^2(j) \approx Q^2$ (so minimizing the phase space available for DGLAP evolution) and has longitudinal momentum fraction (of the proton) x_{jet}, fig 2, that is large (in order to maximise the phase space for BFKL evolution.)

In Fig. 7, recent data from H1 [13] and ZEUS [14] are compared with BFKL predictions [16] and fixed order QCD predictions as calculated with the MEPJET [15]

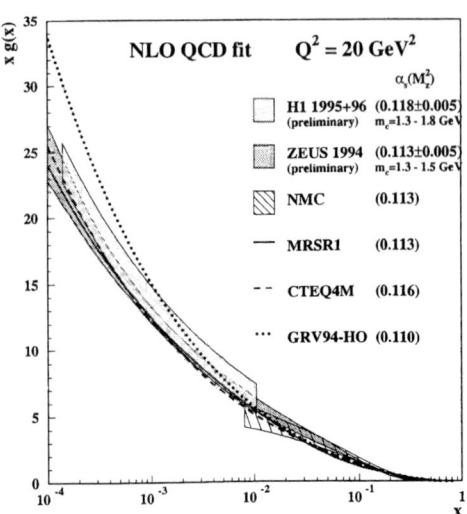

FIGURE 6. The gluon density $xg(x)$ at $Q^2 = 20$ GeV^2 extracted from NLO QCD fits by the H1, ZEUS and NMC collaborations.

program at NLO. The conditions $p_T(j) \simeq Q$ and $x_{jet} \gg x$ are satisfied in the two experiments by slightly different selection cuts. H1 selects events with a forward jet of $p_T(j) > 3.5$ GeV (in the angular region $7° < \theta(j) < 20°$) with

$$0.5 < p_T(j)^2/Q^2 < 2, \qquad x_{jet} \simeq E_{jet}/E_{proton} > 0.035 ; \qquad (2)$$

while ZEUS triggers on somewhat harder jets of $E_T(j) > 5$ GeV and $\eta(j) < 2.6$ with

$$0.5 < p_T(j)^2/Q^2 < 2, \qquad x_{jet} = p_z(j)/E_{proton} > 0.036 . \qquad (3)$$

Fig. 7 shows that both experiments observe a forward jet cross section which rises steeply with decreasing x with substantially more forward jet events than expected from NLO QCD (labelled as Born in fig. 7a.) A BFKL calculation (the stars) gives a better agreement with the data. The overall normalisation in this calculation is uncertain and the agreement may be fortuitous. Indeed, it should also be noted that both experiments observe more centrally produced dijet events than predicted by the NLO QCD calculations. The ARIADNE Monte Carlo (CDM) model describes the steeply increasing jet cross section with decreasing x. The ARIADNE model does not have a strong ordering in transverse momentum in the QCD cascade, akin to BFKL type dynamics, although it does not make explicit use of the BFKL equation. Whilst those models that adhere to the DGLAP formalism (LEPTO and HERWIG) fail to predict this large growth. Further careful investigation is necessary before claiming that BFKL is the mechanism for this enhanced forward jet production.

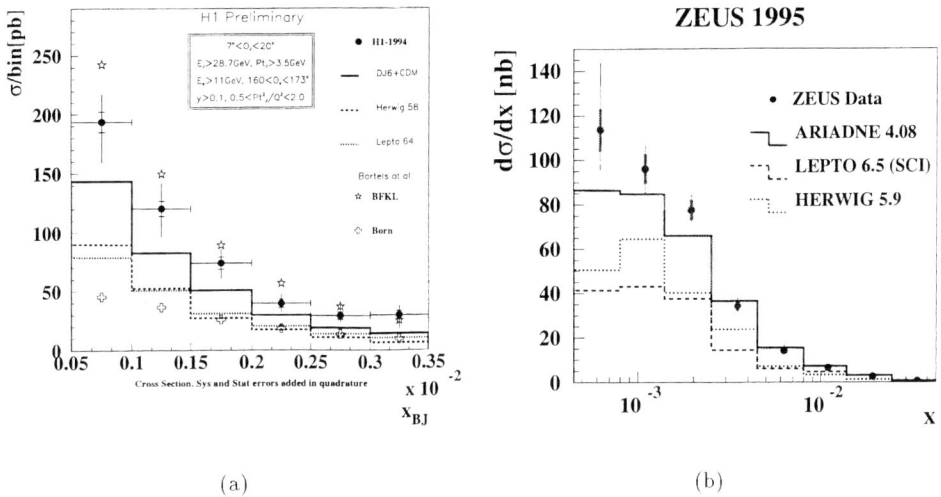

FIGURE 7. Measurement of forward jet production as function of x by (a) H1 and (b) ZEUS experiments. The errors due to the uncertainty in the jet energy scale are show in the shaded band.

FRAGMENTATION FUNCTIONS

Fragmentation functions represent the probability for a parton to fragment into a particular hadron carrying a certain fraction of the parton's energy and, like structure functions, cannot be calculated in perturbative QCD, but can be evolved from a starting distribution at a defined energy scale. If the fragmentation functions are combined with the cross sections for the inclusive production of each parton type in the given physical process, predictions can be made for the scaled momentum, x_p, spectra of final state hadrons. Small x_p fragmentation is significantly affected by the coherence (destructive interference) of soft gluons [17], whilst scaling violation of the fragmentation function at large x_p allows a measurement of α_s [18].

A natural frame in which to study the dynamics of the hadronic final state in DIS is the Breit frame [19]. In this frame the exchanged virtual boson is purely space-like with 3-momentum $\mathbf{q} = (0, 0, -Q)$, the incident quark carries momentum $Q/2$ in the positive Z direction, and the outgoing struck quark carries $Q/2$ in the negative Z direction. A final state particle has a 4-momentum p^B in this frame, and is assigned to the current region if p_Z^B is negative, and to the target frame if p_Z^B is positive. The advantage of this frame lies in the maximal separation of the outgoing parton from radiation associated with the incoming parton and the proton remnant, thus providing the optimal environment for the study of the fragmentation of the outgoing parton.

In e^+e^- annihilation the two quarks are produced with equal and opposite momenta, $\pm\sqrt{s}/2$. This can be compared with a quark struck from within the proton with outgoing momentum $-Q/2$ in the Breit frame. In the direction of the struck quark (the current fragmentation region) the particle momentum spectra, $x_p = 2p^B/Q$, are expected to have a dependence on Q similar to those observed in e^+e^- annihilation [20–22] at energy $\sqrt{s} = Q$.

In fig 8 the $\log(1/x_p)$ distributions for charged particles in the current fragmentation region of the Breit frame are shown as a function of Q^2. These distributions are approximately Gaussian in shape with mean charged multiplicity given by the integral of the distributions. As Q^2 increases the multiplicity increases and the the peak of the distributions shifts to larger values of $\log(1/x_p)$. Figure 9 shows this peak position, $\log(1/x_p)_{max}$, as a function of Q for the HERA data and of $\sqrt{s}$ for the e^+e^- data. Over the range shown the peak moves from $\simeq 1.5$ to 3.3. The HERA data points are consistent with those from TASSO and TOPAZ and a clear agreement in the rate of growth of the HERA points with the e^+e^- data at higher Q is observed.

The increase of $\log(1/x_p)_{max}$ can be approximated phenomenologically by the straight line fit $\log(1/x_p)_{max} = b\log(Q) + c$ also shown in figure 9. The values

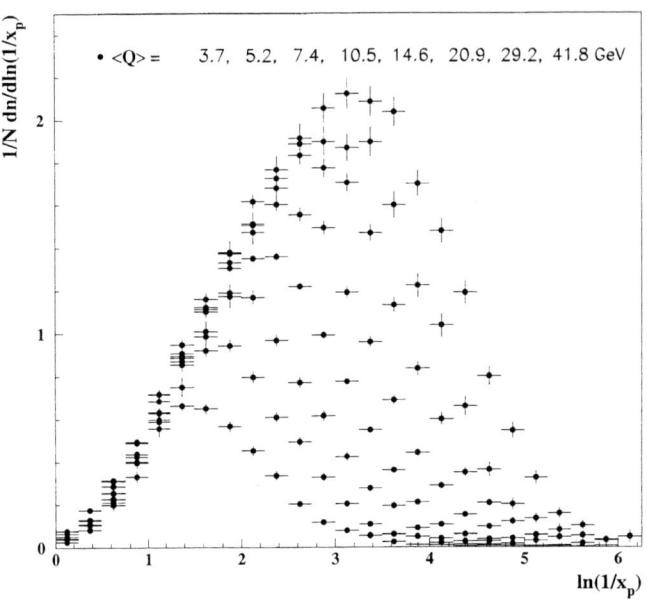

FIGURE 8. Preliminary ZEUS results of the evolution of the $1/N\,dn/d\log(1/x_p)$ distributions with Q^2. Only statistical errors are shown.

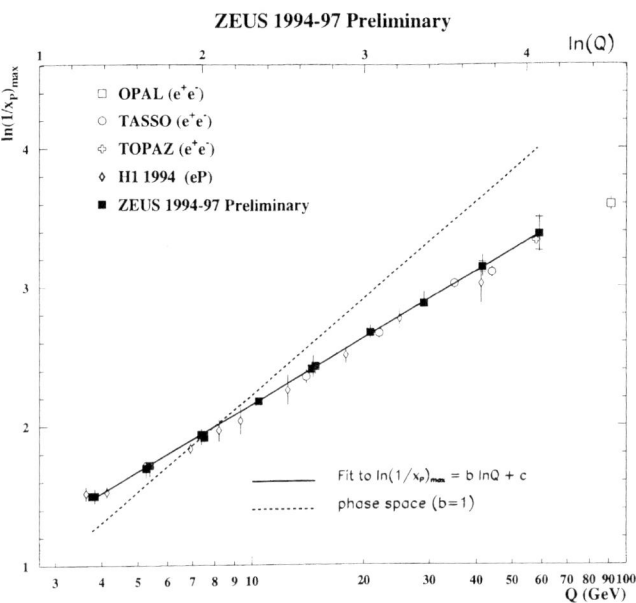

FIGURE 9. Preliminary ZEUS results of evolution of the peak position $\log(1/x_p)_{max}$ with Q^2.

obtained from the fit to the ZEUS data are $b = 0.69 \pm 0.01(\text{stat}) \pm 0.03(\text{sys})$ and $c = 0.56 \pm 0.02^{+0.08}_{-0.09}$. The gradient extracted from the OPAL and TASSO data is $b = 0.653 \pm 0.012$ (with $c = 0.653 \pm 0.047$) which is consistent with the ZEUS result. This value is consistent with that published by OPAL, $b = 0.637 \pm 0.016$, where the peak position was extracted using an alternative method [23]. A consistent value of the gradient is therefore determined in DIS and e^+e^- annihilation experiments.

Also shown is the statistical fit to the data when $b = 1$ ($c = 0.054 \pm 0.012$) which would be the case if the QCD cascade was of an incoherent nature, dominated by cylindrical phase space. The observed gradient is clearly inconsistent with $b = 1$ and therefore inconsistent with cylindrical phase space.

The inclusive charged particle distribution, $1/\sigma_{tot}\, d\sigma/dx_p$, in the current fragmentation region of the Breit frame are shown in bins of x_p and Q^2 in fig. 10. The increasingly steep fall-off, at fixed Q^2, towards higher values of x_p as Q^2 increases, shown in figure 10, corresponds to the production of more particles with a smaller fractional momentum, and is indicative of scaling violation in the fragmentation function. For $Q^2 > 80$ GeV^2 the distributions rise with Q^2 at low x_p and fall-off at high x_p and high Q^2. In figure 10 the HERA data are compared at $Q^2 = s$ to e^+e^- data [25], again divided by two to account for the production of both a q and $\bar{q}$.

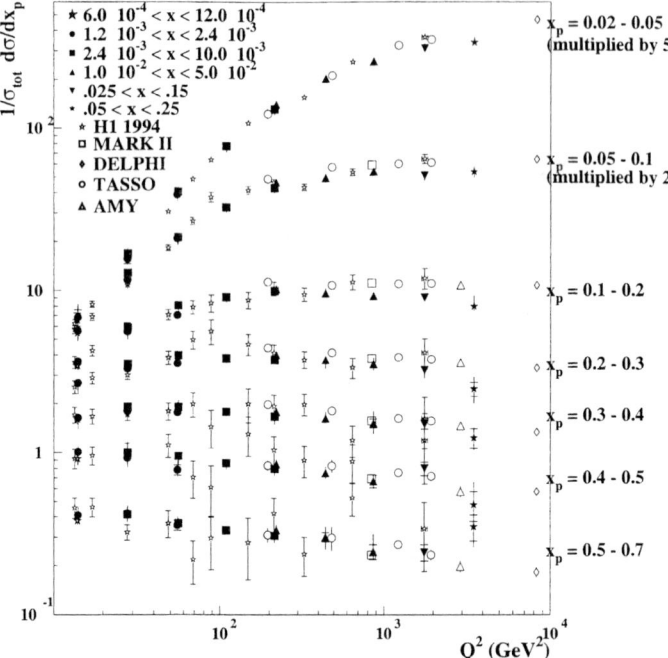

FIGURE 10. The inclusive charged particle distribution, $1/\sigma_{tot}\, d\sigma/dx_p$, in the current fragmentation region of the Breit frame as measured by H1 and ZEUS collaborations. The open points represent data from e^+e^- experiments divided by two to account for q and $\bar{q}$ production (also corrected for contributions from K_S^0 and Λ.)

In the Q^2 range shown there is good agreement between the current region of the Breit frame in DIS and the e^+e^- experiments.

SUMMARY

This review gives a brief summary of a small sample of the QCD results coming from HERA. Charged particle distributions have been studied in the current region of the Breit frame over a wide range of Q^2. These result show clear evidence for scaling violation in scaled momenta as a function of Q^2 and supports the coherent nature of QCD cascades. The observed charged particle spectra are consistent with the universality of quark fragmentation in e^+e^- and DIS.

The intriguing rise of F_2 at small x can be well described using conventional DGLAP evolution equations. The data though can also be described by the BFKL approach thus giving rise to ambiguities how to treat QCD in this small x regime. In order to resolve these ambiguities jet production in the forward direction has been studied and the cross section for such jets is seen not to be reproduced by Monte Carlo models based on DGLAP parton shower evolution.

REFERENCES

1. S. Bentvelsen, J. Engelen and P. Kooijman, Proceedings of the 1991 Workshop on Physics at HERA, DESY Vol. 1 (1992) 23.
2. H1 Collaboration, I. Abt et al., DESY preprint 93-103, *Nucl. Instr. Meth.* **386**, 310 (1997) (Vol 1) and *ibid.* p.348 (Vol 2).
3. ZEUS Collab., The ZEUS Detector, Status Report 1993, DESY 1993.
4. G. Altarelli and G. Parisi, *Nucl. Phys.* **126**, 297 (1977); V.N. Gribov and L.N. Lipatov, *Sov. J. Nucl. Phys.* **15**, 438 and 675 (1972); Yu. L. Dokshitzer, *Sov. Phys. JETP* **46**, 641 (1977).
5. E.A. Kuraev, L.N. Lipatov and V.S. Fadin, *Sov. Phys. JETP* **45**, 199 (1977); Y.Y. Balitsky and L.N. Lipatov, *Sov. J. Nucl. Phys.* **28**, 282 (1978).
6. H1 collab., Paper 262 presented at Int. Europhysics Conf. on High Energy Physics, HEP97, Jerusalem.
7. M. A. J. Botje, Proc. of 5th Int. workshop on DIS and QCD (1997), (eds. Repond and Krakauer).
8. A. J. Askew et al. *Phys. Rev.* **D47**, 3775 (1993); **D49**, 4402 (1994).
9. NMC collab., M. Arneodo et al. *Phys. Lett.* **B309**, 222 (1993).
10. A. Martin, R. Roberts and W. J. Stirling, *Phys. Lett.* **B387**, 419 (1996).
11. H. L. Lai et al., *Phys. Rev.* **D55**, 1280 (1997).
12. M. Glück, E. Reya and A. Vogt,*Z. Phys.* **C67**, 433 (1995).
13. H1 collab., paper pa03-049 submitted to 28th Int. Conf. on High Energy Physics ICHEP'96, Warsaw, Poland.
14. ZEUS collab., J.Breitweg et al. DESY preprint 98-050.
15. E. Mirkes and D. Zeppenfeld, *Phys. Lett.* **B380**, 205 (1996).
16. J. Bartels et al. *Phys. Lett.* **B384**, 300 (1996).
17. Yu. Dokshitzer, V. Khoze, A. Mueller and S. Troyan, "Basics of Perturbative QCD", Editions Frontières, Gif-sur-Yvette, France (1991).
18. G. Altarelli et al., *Nucl. Phys.* **B160**, 301 (1979); P. Nason and B. R. Webber, *Nucl. Phys.* **B421**, 473 (1994).
19. R.P. Feynman, "Photon-Hadron Interactions", Benjamin, N.Y. (1972).
20. Yu. Dokshitzer et al., *Rev. Mod. Phys.* **60**, 373 (1988).
21. A. V. Anisovich et al., *Il Nuovo Cimento* **A106**, 547 (1993).
22. K. Charchuła, *J. Phys.* **G19**, 1587 (1993).
23. OPAL Collab., M. Akrway et al., *Phys. Lett.* **B247**, 617 (1990).
24. TOPAZ collab., R. Itoh et al.,*Phys. Lett.* **B345**, 335 (1995).
25. TASSO Collab., W. Braunschweig et al.,*Z. Phys.* **C47**, 187 (1990); MARK II Collab., A. Petersen et al., *Phys. Rev.* **D37**, 1 (1988); AMY Collab., Y. K. Li et al., *Phys. Rev.* **D41**, 2675 (1990); DELPHI Collab., P. Abreu et al., *Phys. Lett.* **B311**, 408 (1993).

TAU PHYSICS FROM LEP

Doris Y. Kim

INFN Sezione Roma I, P. A. Moro 2, I-00185 Roma, Italy
The L3 Experiment, CERN, CH-1211 Geneva 23, Switzerland

Abstract. We briefly review recent analysis on the tau lepton by four LEP experiments, ALEPH, DELPHI, L3 and OPAL, based on data samples generated at the Z peak. Intrinsic properties of the tau lepton as well as applications to various physics topics are treated from an experimental point of view. The summarized analysis subjects include lifetime, leptonic branching ratios, hadronic decay structures, dipole moments, related electroweak and QCD properties, and the mass of the tau neutrino.

I INTRODUCTION

In 1975 Perl et al. [1] found a pair of new particles which decay into an electron and a muon with missing energy. Soon it was confirmed this "tau" particle decays into hadronic particles as well, generating a rich field of physics topics. Another aspect of the tau particle is its short lifetime, order of 10^{-13} second, which forces experiments to regard decay product patterns as the particle signature.

During LEP phase I, the machine was operated at the Z resonance. About 10^6 $\tau^+\tau^-$pairs were generated in the LEP experiments (ALEPH, DELPHI, L3 and OPAL) by e^+e^- collisions. The production cross-section of the $\tau^+\tau^-$pair is much smaller in the high energy range of LEP II. Our summary will be concentrated on results from the data samples produced in LEP I.

II LIFETIME AND LEPTONIC BRANCHING RATIO

According to the Standard Model(SM), decay rates of heavier leptons into lighter leptons are written as follows:

$$\Gamma(L^- \to l^- \bar{\nu}_l \nu_L) = \frac{G_L G_l m_L^5}{192\pi^3}(1+\delta_{ph})(1+\delta_W)(1+\delta_{rad}),$$

where G_L, G_l are the effective Fermi coupling constants for the mother lepton and the daughter lepton, respectively. δ_{ph} is a phase space factor calculated as a function of $(m_l/m_L)^2$, δ_W is the W propagator effect with a negligible amount and δ_{rad} is a radiative correction [4]. The uncertainty coming from the mass of the tau particle

TABLE 1. Recent measurements on the lifetime and leptonic branching ratios of the tau particle from LEP and other experiments [2].

Experiments	Lifetime [fsec]	BR($\tau^- \to e^- \bar{\nu}_e \nu_\tau$) [%]	BR($\tau^- \to \mu^- \bar{\nu}_\mu \nu_\tau$)[%]
ALEPH [6,12]	290.1 ± 1.9	17.79 ± 0.13	17.31 ± 0.12
DELPHI [7,13]	292.7 ± 2.1	18.08 ± 0.22	17.28 ± 0.13
L3 [8,14]	291.6 ± 3.4	17.67 ± 0.19	17.34 ± 0.21
OPAL [9,15]	289.2 ± 2.1	17.78 ± 0.13	17.48 ± 0.13
ARGUS [16]	a	17.30 ± 0.60	17.20 ± 0.60
CLEO [10,17]	289.0 ± 4.9	17.76 ± 0.18	17.37 ± 0.20
SLD [11]	288.1 ± 6.9	–	–
New Average	290.6 ± 1.1	17.79 ± 0.07	17.35 ± 0.06
PDG96 [3]	291.0 ± 1.5	17.83 ± 0.08	17.35 ± 0.10

[a] The ARGUS result on the lifetime is not included in the table

is also small, due to the threshold energy scan analysis by the BES collaboration [5]. Hence, comparison between Fermi coupling constants obtained by various leptonic decay rates is a benchmark check on lepton universality. In practice, $\Gamma(L^- \to l^- \bar{\nu}_l \nu_L)$ is converted from the lifetime of the mother and the corresponding branching ratio into the daughter: $\Gamma(L^- \to l^- \bar{\nu}_l \nu_L) = BR(L^- \to l^- \bar{\nu}_l \nu_L)/\tau_L$. Recent results on the lifetime and the branching ratios of the tau from LEP as well as other major experiments are shown in Table 1.

Lifetime

The decay length of tau particles in LEP experiments is $\sim 2.2mm$, due to boosting of particles at the Z resonance. The size of the beam spots perpendicular to the beam direction is well under control and high precision silicon vertex detectors reduce uncertainties in reconstruction. The lifetime is directly converted from the decay length reconstructed in 3-prong tau decays. In case of 1-prong tau decays, the lifetime is obtained indirectly from impact parameters or their combinations.

Leptonic Branching Ratio

A charged particle is identified as an electron if it has high dE/dx deposition in central tracking systems or if the matching shower pattern in electromagnetic calorimeters is sharp and prominent. A muon should behave as a minimum ionizing particle in hadronic calorimeters or should leave a track in out-most muon detectors.

Lepton Universality Test

The combined results from Table 1 are used to extract ratios between the leptonic decay widths. For convenience, the effective Fermi constant G_l is converted to a

TABLE 2. The fitted mass parameters of the a_1 resonance to $\tau \to 3\pi\nu_\tau$ decay products. KS and IMR refer to the Kühn and Santamaria and the Isgur, Morningstar and Reader models, respectively.

Exp.	M(KS) [MeV]	Γ(KS) [MeV]	M(IMR) [MeV]	Γ(IMR) [MeV]
DELPHI [21]	1255 ± 9	587 ± 34	1207 ± 9	478 ± 15
OPAL [22]	1262 ± 11	621 ± 66	1210 ± 7	457 ± 23

weak coupling constant for each lepton as $G_l = g_l^2/\sqrt{32}M_W^2$, $(l = e, \mu, \tau)$. The final numbers confirm lepton universality theory with just 0.3 % of uncertainty.

$$\frac{\Gamma(\tau^- \to \mu^- \bar{\nu}_\mu \nu_\tau)}{\Gamma(\tau^- \to e^- \bar{\nu}_e \nu_\tau)} \to \frac{g_\mu}{g_e} = 1.001 \pm 0.003 \to \mu - e \text{ universality}$$

$$\frac{\Gamma(\tau^- \to e^- \bar{\nu}_e \nu_\tau)}{\Gamma(\mu^- \to e^- \bar{\nu}_e \nu_\mu)} \to \frac{g_\tau}{g_\mu} = 1.000 \pm 0.003 \to \tau - \mu \text{ universality}$$

III HADRONIC DECAYS

The relatively high tau mass makes it possible to decay into hadronic particles. Usually the decay proceeds via intermediate resonances like ρ, K^*, a_1 and finalizes as a narrow jet in LEP detectors, consisting of $\pi^\pm$ and π^0 mostly. Decay widths predicted by the conserved vector current hypothesis, the isospin symmetry, or chiral Lagrangian theories reproduce experimental results reasonably [19]. $\pi^\pm$ are identified by dE/dx deposition in central tracking systems or by shower patterns in calorimeters. Photon showers in electromagnetic calorimeters with the matching invariant mass are identified as π^0. Characteristics of major hadronic tau decay modes as well as findings of rare decay modes are well documented in references [18,19]. We will focus our article on recent developments in a few selected modes.

Mass Parameters of the a_1 particle

The G-parity argument predicts that the axial-vector contribution dominates the $\tau \to 3\pi\nu_\tau$ decay mode with an a_1 resonance. The a_1 resonance itself decays by way of secondary resonances, so fitting mass parameters of the a_1 particle to the reconstructed mass on the 3π system becomes model dependent. A paper from the OPAL experiment [22] shows that current theoretical predictions are somewhat limited and not satisfactory when details of 3π decay structure functions are tested. It is suspected there would be either a non-resonance or an unknown resonance contribution other than the a_1. For example, the DELPHI experiment proposes [21] an $a_1'(1700)$ at the high mass tail (Figure 1). Table 2 shows the fitted mass parameters of the a_1 by DELPHI and OPAL using two theoretical models [20].

TABLE 3. Some results on kaonic tau decay modes from the ALEPH experiment published in 1997 [23].

Mode	Branching Ratio [10^{-3}]	Mode	Branching Ratio [10^{-3}]
$\bar{K}^0\pi^-\nu_\tau$	$8.55 \pm 1.17 \pm 0.66$	$K^-K^+\pi^-\nu_\tau$	$1.63 \pm 0.21 \pm 0.17$
$\bar{K}^0\pi^-\pi^0\nu_\tau$	$2.94 \pm 0.73 \pm 0.37$	$K^-\pi^+\pi^-\nu_\tau$	$2.14 \pm 0.37 \pm 0.29$
$\bar{K}^0K^-\nu_\tau$	$1.58 \pm 0.42 \pm 0.17$	$K^-K^+\pi^-\pi^0\nu_\tau$	$0.75 \pm 0.29 \pm 0.15$
$\bar{K}^0K^-\pi^0\nu_\tau$	$1.52 \pm 0.76 \pm 0.21$	$K^-\pi^+\pi^-\pi^0\nu_\tau$	$0.61 \pm 0.39 \pm 0.18$
$K^0_S K^0_L \pi^-\nu_\tau$	$1.01 \pm 0.23 \pm 0.13$	$K^-K^+K^-\nu_\tau$	< 0.19 (95 % CL)

Decays with Kaons

The strange sector of tau decay currents is suppressed by the corresponding Kobayashi-Maskawa matrix element. Recently LEP experiments made significant progress in identifying the strange sector decay modes down to branching ratios of 10^{-3}. The crucial point of the search is $K^\pm/\pi^\pm$ particle discrimination, which is handled either by measuring dE/dx deposition in central tracking systems or by utilizing TOF (Time of Flight) and RICH (Ring Image Cherenkov Counter). K_L is found by pattern recognition in hadronic calorimeters while K_S is acknowledged if there exists a secondary vertex heading from a tau decay. Some of recent results from LEP experiments are shown in Table 3.

Spectral Functions and QCD

Spectral functions are defined as distributions of $\Gamma(\tau^- \to h^-\nu_\tau)$ over the invariant mass of the daughter hadron systems [24],

$$\Gamma(\tau^- \to h^-\nu_\tau) = \frac{G_F^2}{32\pi^2 m_\tau^3} \int_0^{m_\tau^2} dq^2 (m_\tau^2 - q^2)(m_\tau^2 - 2q^2) \cdot \{[v_1(q^2) + a_1(q^2)]\cos^2\theta_c + strange\ sector\}.$$

v_1 represents the vector part where corresponding tau decays proceed via even number of pions. Figure 2 shows a nice agreement between the measured v_1 distribution from data and the predicted curve by the CVC theorem [25]. a_1 represents the axial-vector part where odd number of pions are produced by tau decays. Integration of $v_1 - a_1$ can be used to prove chiral sum rules up to m_τ^2 [25]. ¿From information extracted from the $v_1 + a_1$ distribution, $\alpha_s(m_\tau^2)$ is measured as 0.334 ± 0.022 and 0.306 ± 0.024 from the ALEPH and the CLEO experiments, respectively. Extrapolation of the numbers to the Z peak predicts $\alpha_s(m_Z^2)$ as 0.1202 ± 0.0027 (ALEPH) and 0.114 ± 0.003 (CLEO), respectively. [25,26].

IV LORENTZ STRUCTURES OF TAU DECAYS

In the previous sections we assumed that decay currents of a tau particle has the V-A structure of the electroweak force exclusively. In leptonic τ decays, the most

TABLE 4. Recent results on Michel parameters of the tau lepton from LEP and other experiments with an assumption of lepton universality [2]. The correlations between the parameters are ignored in the average numbers.

Exp.	ρ	η^a	ξ	$\delta\xi$	ξ_h
ALEPH [27]	0.751 ± 0.045	-0.04 ± 0.19	1.18 ± 0.16	0.88 ± 0.13	-1.006 ± 0.037
DELPHI [28]	0.794 ± 0.046	-0.01 ± 0.11	0.988 ± 0.074	0.682 ± 0.083	-1.006 ± 0.036
L3 [29]	0.794 ± 0.050	0.25 ± 0.20	0.94 ± 0.22	0.81 ± 0.16	-0.970 ± 0.055
OPAL [22]	–	–	–	–	-1.29 ± 0.28
ARGUS [30]	0.735 ± 0.029	0.03 ± 0.22	1.03 ± 0.11	0.62 ± 0.09	-1.017 ± 0.039
CLEO [31]	0.747 ± 0.012	-0.015 ± 0.087	1.007 ± 0.043	0.745 ± 0.027	-1.03 ± 0.07
SLD [32]	0.72 ± 0.09	$-^b$	1.05 ± 0.35	0.88 ± 0.27	-0.93 ± 0.11
Average	0.750 ± 0.010	0.01 ± 0.06	1.01 ± 0.03	0.74 ± 0.02	-1.01 ± 0.01
SM	0.75	0	1	0.75	-1

[a] Careful interpretation is needed over the results on ρ and η, since the $\tau^- \to e^- \bar{\nu}_e \nu_\tau$ decay spectrum is not sensitive to η.
[b] See [32] for the compilation result including η.

general local four fermion interaction with lepton number conservation is written as,

$$\mathcal{M} = \frac{4G_F}{\sqrt{2}} \sum_{\gamma=S,V,T} \sum_{\alpha,\beta=R,L} g^\gamma_{\alpha,\beta} (\overline{u}^\alpha_l \Gamma_\gamma v_{\nu_l})(\overline{v}^\nu_{\nu_\tau} \Gamma_\gamma u^\beta_\tau)$$

where S, V, T symbolize scalar, vector, and tensor currents and R, L represent handedness of spinors. The Standard Model claims all the coupling constants $g^\gamma_{\alpha,\beta}$ is 0 except g^V_{LL}. Binomial combinations of the coupling constants (Michel parameters) are used as parameters in experiments. For example, the energy spectrum of leptons from τ decays are written as,

$$\frac{1}{\Gamma}\frac{d\Gamma}{dx} = h_0(x) + \eta h_\eta(x) + \rho h_\rho(x) - \bar{\mathcal{P}}_\tau(\xi h_\xi(x) + \xi\delta h_{\xi\delta}(x))$$

where $\eta, \rho, \xi, \xi\delta$ are Michel parameters, $\bar{\mathcal{P}}_\tau$ is the average τ polarization and h are polynomial functions of $x = E_l/E_\tau$. Likewise in hadronic τ decays,

$$\frac{1}{\Gamma}\frac{d\Gamma}{dx} = g_0(x) + \bar{\mathcal{P}}_\tau \xi_h g_\xi(x) \; (\xi_h = \nu_\tau \; chirality)$$

Double decay spectra are used to separate $\bar{\mathcal{P}}_\tau$ from ξ and ξ_h. The sign of ξ_h is obtained from independent methods [33,22]. Table 4 shows recent results from LEP experiments as well as CLEO and ARGUS measurements. The numbers are dominated by the CLEO measurement and they do not indicate any violation of the V-A structure within errors.

V NEUTRAL CURRENTS AND TAU PRODUCTION

According to the Standard Model, only neutral weak current and photon exchange are involved in $e^+e^- \to \tau^+\tau^-$ production. $\sim 10^6$ $\tau^+\tau^-$ pairs generated

TABLE 5. Recent results from LEP experiments on the tau polarization asymmetry.

Experiments	A_e	A_τ
ALEPH [34]	$0.129 \pm 0.016 \pm 0.005$	$0.136 \pm 0.012 \pm 0.009$
DELPHI [35]	$0.140 \pm 0.013 \pm 0.003$	$0.138 \pm 0.009 \pm 0.008$
L3 [36]	$0.1678 \pm 0.0127 \pm 0.0030$	$0.1476 \pm 0.0088 \pm 0.00062$
OPAL [37]	$0.129 \pm 0.014 \pm 0.005$	$0.134 \pm 0.009 \pm 0.010$
LEP Average [38]	0.1438 ± 0.0071	0.1400 ± 0.0063

in LEP experiments make it possible to do precision tests on weak coupling constants, g_A and g_V, by measurements on $\sigma^0, A_{FB}, A_{pol}^{FB}, \mathcal{P}_\tau$, etc. In this paper, we will explain the electroweak analysis specific to the tau particle.

Tau Polarization Asymmetry

In the improved Born approximation, the (longitudinal) tau polarization asymmetry, $\mathcal{P}_\tau$, is parametrized as follows:

$$\mathcal{P}_\tau \equiv \frac{\sigma_R - \sigma_L}{\sigma_R + \sigma_L} \Rightarrow -\frac{A_\tau + 2A_e \cos\theta/(1+\cos^2\theta)}{1 + 2A_\tau A_e \cos\theta/(1+\cos^2\theta)},$$

where A_τ and A_e are a function of g_V/g_A (vector/ axial vector coupling constants) for τ and e, respectively and $\cos\theta$ is the tau production angle with respect to the incoming e^- beam (Figure 3).

$$A_l \equiv \frac{2g_{A_l}g_{V_l}}{g_{A_l}^2 + g_{V_l}^2} \Rightarrow \frac{2(1 - 4\sin^2\theta_W^{eff})}{1 + (1 - 4\sin^2\theta_W^{eff})^2}$$

Energy and decay angles of product particles from tau decays depend on the original tau spin direction. Assuming negative neutrino helicity, we get maximal spin information from the kinematic distributions. Table 5 shows the combined results from LEP experiments. With lepton universality we get $A_l = 0.1417 \pm 0.0047$, which implies $\sin^2\theta_W^{eff} = 0.2322 \pm 0.0006$.

Spin Correlation between the Tau Pair

Another test on the weak coupling constants of the tau particle is to measure correlated normal and transverse components of spin between the $\tau^+\tau^-$ pair. Though the sensitivity is not as good as that of the $\mathcal{P}_\tau$ measurement, it is important to verify the Standard Model from all the available aspects. The detailed definition of spin correlation asymmetries are explained in references [39–41].

$C_{TT} = (|g_A|^2 - |g_V|^2)/(|g_A|^2 + |g_V|^2)$ (transverse-transverse spin correlation),
$C_{TN} = -2\,Im(g_V g_A^*)/(|g_A|^2 + |g_V|^2)$ (transverse-normal spin correlation).

TABLE 6. Recent results on the spin correlation between tau pairs from LEP experiments. SM refers to the Standard Model.

Exp.	C_{TT}	C_{TN}
ALEPH [39]	$1.06 \pm 0.13 \pm 0.05$	$0.08 \pm 0.13 \pm 0.04$
DELPHI [40]	$0.87 \pm 0.20 ^{+0.10}_{-0.12}$	–
L3 [41]	$1.04 \pm 0.26 \pm 0.06$	$0.36 \pm 0.26 \pm 0.05$
SM	~ 0.989	~ -0.01

The azimuthal angle of the incoming e^- beam in a coordinate system defined by τ pair decay products is used as the analyser, whose value is independent of τ dipole moments. The result is shown in Table 6.

VI DIPOLE MOMENTS

The Standard Model with loop corrections predicts small dipole moments for the tau particle. A measurement of unexpectedly large dipole moments could be interpreted as an indication to a substructure in the tau particle. Electric dipole moments are related to the CP violating currents, where non-Standard-Model theories could generate larger contributions. Though current experiments lack the precision to verify the Standard Model on this subject, measurements on dipole moments can test and lead the way to new physics phenomena.

Anomalous Magnetic Moment

The shortness of the tau lifetime prevents us from using the spin precession method to measure the tau magnetic dipole moment. Indirect bounds from $\Gamma(Z \rightarrow \tau^+\tau^-)$ are studied in review papers where q^2 of photon is m_Z^2. A direct method has been developed to measure the magnetic dipole with radiative $Z \rightarrow \tau^+\tau^-(\gamma)$ events, where one τ particle is off-shell but q^2 of photons is 0 [42]. Anomalous terms enhance the production of photons with hight energy and large isolation angles to the taus, compared to the contribution from SM terms (Figure 4). The direct results from L3 and OPAL as well as an updated review result are shown in Table 7. Though the indirect review method gives better limits, one should keep in mind that physical interpretations would be different between two methods.

Weak (Magnetic) Dipole Moments

a_τ^w, d_τ^w at $q^2 = M_Z$ are related to a spin dependent part of the $e^+e^- \rightarrow \tau^+\tau^-$ cross-section. Invisible neutrinos in tau decay products limit information on tau flight directions and tau spin components and reduce sensitivity to the moments. Several experiments use observables based on correlated kinematic information from

TABLE 7. The limits on the anomalous magnetic moment of the tau lepton in 95 % CL. Numbers from the L3 and the OPAL experiments ($q^2 = 0$) as well as a recent review paper ($q^2 = m_Z^2$) are shown.

Experiments	a_τ	d_τ $[10^{-16} e \cdot cm]$
L3 [43]	$-0.052 < a_\tau < 0.058$	$-3.1 < d_\tau < 3.1$
OPAL [44]	$-0.068 < a_\tau < 0.065$	$-3.8 < d_\tau < 3.6$
Review [45]	$-0.004 < a_\tau < 0.006$	$-0.11 < d_\tau < 0.11$
SM	0.0011773(3)	~ 0 by P,T invariance

TABLE 8. The limits on the weak dipole moments of the tau lepton from LEP experiments in 95 % CL. a_τ^w is measured for the first time in the world.

Experiments	$Re(d_\tau^w)$ $[e \cdot cm]$	$Im(d_\tau^w)$ $[e \cdot cm]$
ALEPH [46]	$|Re(d_\tau^w)| < 1.5 \times 10^{-17}$	
DELPHI [47]	$|Re(d_\tau^w)| < 0.66 \times 10^{-17}$	$|Im(d_\tau^w)| < 2.0 \times 10^{-17}$
L3 [49]	$|Re(d_\tau^w)| < 3.0 \times 10^{-17}$	
OPAL [48]	$|Re(d_\tau^w)| < 0.56 \times 10^{-17}$	$|Im(d_\tau^w)| < 1.5 \times 10^{-17}$
SM	$\sim 3 \times 10^{-37}$	~ 0 by CPT conservation.

Experiments	$Re(a_\tau^w)$	$Im(a_\tau^w)$
L3 [49]	$|Re(a_\tau^w)| < 4.5 \times 10^{-3}$	$|Im(a_\tau^w)| < 9.9 \times 10^{-3}$
SM	-2.10×10^{-6}	0.61×10^{-6}

$\tau^+\tau^-$ pairs, which are optimized to measure d_τ^w [46-48]. Another option is to use an asymmetry constructed from single tau decay angles, which is more sensitive to a_τ^w [49]. The results in 95 % CL are shown in Table 8. There is no indication of new physics.

VII THE MASS OF THE TAU NEUTRINO

Indirect bounds on the mass of a tau neutrino from cosmology predictions and astrophysical measurements are typically less than $\mathcal{O}(1\ MeV)$. On the earth, a direct measurement of the mass is possible in hadronic tau decays, where the high invariant mass of the hadronic system limits the available phase space for energy of the tau neutrino. The classic analyzer on the mass of a tau neutrino is the endpoint of the invariant mass spectrum of the hadronic system. Together with the invariant mass, visible energy of the hadronic system or the missing momentum between the tau particle and the daughter hadronic system can be used as a variable in 2-dimensional fitting. The $\tau^- \to 3\pi^-2\pi^+(\pi^0)\nu_\tau$ decay mode is used for its high sensitivity to the neutrino mass. The $\tau^- \to 2\pi^-\pi^+(\pi^0)\nu_\tau$ mode is another good source due to its high statistics. Recent results in 95% CL from the Argus, CLEO and LEP experiments are shown in Table 9.

TABLE 9. Measurements on the mass of the tau neutrino from LEP and other experiments in 95% CL.

Experiments	Fit	m_τ	Experiments	Fit	m_τ
ARGUS [53]	1-D	$< 31\ MeV$	ALEPH [50]	2-D	$< 18.2 MeV$
CLEO [54]	1-D	$< 30 MeV$	DELPHI [51]	1-D	a
			OPAL [52]	2-D	$< 29.9 MeV$

^a The DELPHI number is preliminary and depends on a_1 decay structure models. The limits are 31 MeV(KS), 33 MeV(IMR) and 62 MeV(conservative evaluation), respectively.

VIII CONCLUSION

With a vast amount of $\tau^+\tau^-$ samples, excellent detector performances and improved analysis skills, the LEP experiments tested fairly broad area of tau physics. The measurements support the Standard Model, where the tau lepton is a point-like particle and its decay has V-A structure. Together with the ARGUS and CLEO measurements, most decay modes of the tau are found down to branching ratios of 10^{-5}. Nevertheless, some intrigues are still left. To catch up the precision in the measurements on electrons and muons, we would need contributions from next generation experiments with new samples of data and upgraded detector technology.

REFERENCES

1. Abrams et al., *Phys. Rev. Lett.* **35**, 1489 (1975).
2. W. Lohmann, private communication.
3. R.M. Barnett et al., Physical Review **D54**, 1 (1996) and 1997 off-year partial update for the 1998 edition available on the PDG WWW pages (URL: http://pdg.lbl.gov/).
4. There is a well written review article in [3].
5. BES Collab., J. Z. Bai et al., *Phys. Rev.* **D53**, 20 (1996).
6. ALEPH Collab., R. Barate et al., *Phys. Lett.* **B414**, 362 (1997)
7. A. Andreazza et al., DELPHI Note 97-87 CONF 73 (1997), submitted to HEP97.
8. M. Biasini et al., L3 Note 2127 (1997), unpublished.
9. OPAL Collab., G. Alexander et al., *Phys. Lett.* **B374**, 341 (1996).
10. CLEO Collab., Balest et al., *Phys. Lett.* **B388**, 402 (1996).
11. SLD Collab., K. Abe et al., SLAC-PUB-96-7216, ICHEP-96 PA07-064 (1996).
12. ALEPH Collab., D. Buskulic et al., *Z. Phys.* **C70**, 561 (1996).
13. K. Johansen et al., DELPHI Note 97-86 CONF 72 (1997), submitted to HEP97.
14. W. Lohmann, L3 Collab., in *Proc. of the Fourth Workshop on Tau Lepton Physics, Colorado, 1996*, Nucl. Phys. B (proc. Suppl.) 55C, 101 (1997).
15. OPAL Collab., G. Alexander et al., *Phys. Lett.* **B369**, 163 (1996); Randy Sobie, private communication on $BR(\tau^- \to \mu^- \bar{\nu}_\mu \nu_\tau)$.
16. ARGUS Collab., H. Albrecht et al., *Z. Phys.* **C53**, 367 (1992).
17. CLEO Collab., A. Anastassov et al., *Phys. Rev.* **D55**, 2559 (1997).

18. S. Gentile and M. Paul, Phys. Rep. **274** 287 (1996).
19. *Proc. of the Fourth International Workshop on Tau Lepton Physics, Colorado, 1996*, Nucl. Phys. B (proc. Suppl.) 55C, 101 (1997).
20. J. H. Kühn and A. Santamaria, *Z. Phys.* **C48**, 45 (1990); N. Isgur, C. Morningstar and C. Reader, *Phys. Rev.* **D39**, 1357 (1989); M. Feindt, *Z. Phys.* **C48**, 681 (1990).
21. DELPHI Collab., Abreu et al., CERN-EP/98-14 (1998), accepted by *Phys. Lett.* **B**.
22. OPAL Collab., K. Ackerstaff et al., *Z. Phys.* **C75**, 593 (1997)
23. ALEPH Collab., R. Barate et al., CERN-PPE/97-167 (1998), *Eur. Phys. J.* **C**;DOI 10.1007/s100529800879; ALEPH Collab., R. Barate et al., *Eur. Phys. J.* **C1**, 65 (1998).
24. Y.S. Tsai, *Phys. Rev.* **D4**, 2821 (1971).
25. ALEPH Collab., R. Barate et al., CERN-PPE/98-012 (1998), submitted to *Eur. Phys. J.* **C**; ALEPH Collab., R. Barate at al., *Z. Phys.* **C76**, 15 (1997).
26. CLEO Collab., T. Coan et al., *Phys. Lett.* **B356**, 580 (1995).
27. ALEPH Collab., D. Buskulic et al., *Phys. Lett.* **B346**, 379 (1995).
28. I. Boyko et al., DELPHI Note 98-28 CONF 124 (1998), unpublished.
29. L3 Collab., M. Acciarri et al., *Phys. Lett.* **B377**, 313 (1996).
30. ARGUS Collab., H. Albrecht et al., *Phys. Lett.* **B349**, 576 (1995); ARGUS Collab., H. Albrecht et al., DESY-97-194, hep-ex/9711022 (1997).
31. CLEO Collab., J. Alexander et al., *Phys. Rev.* **D56**, 5320 (1997) ; CLEO Collab., T. Coan et al., *Phys. Rev.* **D55**, 7291 (1997).
32. SLD Collab., K. Abe et al., *Phys. Rev. Lett.* **78**, 4691 (1997).
33. ARGUS Collab., H. Albrecht et al., *Z. Phys.* **C58**, 61 (1993); SLD Collab., K. Abe et al., *Phys. Rev. Lett.* **78**, 2075 (1997).
34. ALEPH Collab., D. Buskulic et al., *Z. Phys.* **C69**, 183 (1996).
35. S. Amato et al, DELPHI Note 96-114 CONF42(1996), submitted to ICHEP96.
36. L3 Collab., M. Acciarri et al., CERN-EP/98-26 (1998), accepted by *Phys. Lett.* **B**.
37. OPAL Collab., G. Alexander et al. *Z. Phys.* **C72**, 365 (1996).
38. *Precision Tests of the Standard Model*, a talk given in the XXXIIIrd Rencontres de Moriond: ElectroWeak Interactions and Unified Theories, Les Arcs (1998).
39. ALEPH Collab., R. Barate et al., *Phys. Lett.* **B405**, 191 (1997).
40. DELPHI Collab., Abreu et al., *Phys. Lett.* **B404**, 194 (1997).
41. R. Völkert, Ph. D. Thesis, Humboldt University Berlin, (1997), Interner Bericht, DESY-Zeuthen 97-04.
42. J. Biebel et al., *Z. Phys.* **C76**, 1997 (53); S. S. Gau et al., hep-ph/9712360 (1997); J. A. Grifols et al., *Phys. Lett.* **B255**, 1991 (611), Erratum *ibd* **B259** (1991) 512.
43. L3 Collab., M. Acciarri et al., CERN-EP/98-45 (1998), submitted to *Phys. Lett.* **B**.
44. OPAL Collab., K. Ackerstaff et al., CERN-EP/98-33 (1998), submitted to *Phys. Lett.* **B**.
45. R. Escribano and E. Massò, *Improved Bounds on the Electromagnetic Dipole Moments of the Tau Lepton*, hep-ph/9609423 (1996).
46. ALEPH Collab., D. Buskulic et al., *Phys. Lett.* **B346**, 371 (1995).
47. M.-C. Chen et al., DELPHI Note 97-70 CONF 56 (1997), submitted to HEP97.
48. OPAL Collab., K. Ackerstaff et al., *Z. Phys.* **C74**, 403 (1997).
49. L3 Collab., M. Acciarri et al., *Phys. Lett.* **B426**, 207 (1998).

50. ALEPH Collab., R. Barate et al., *Eur. Phys. J.* **C2**, 395 (1998).
51. A. Galloni et al., DELPHI Note 97-129 CONF 108 (1997), submitted to HEP97.
52. OPAL Collab., K. Ackerstaff et al., CERN-EP/98-55 (1998), submitted to *Eur. Phys. J. C*.
53. ARGUS Collab., H. Albrecht et al., *Phys. Lett.* **B292**, 221 (1992).
54. CLEO Collab., R. Ammar et al., CLNS 98/1551, CLEO 98-6 (1998), submitted to *Phys. Lett.* **B**.

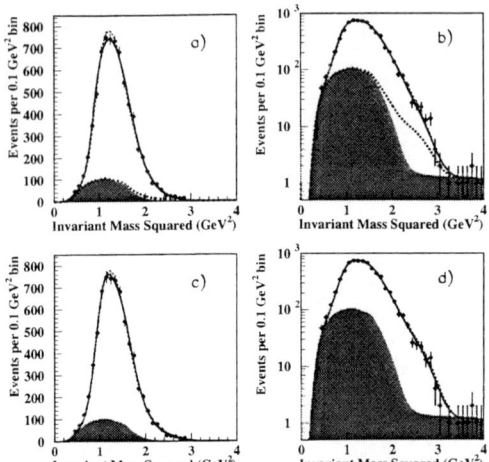

FIGURE 1. The invariant mass spectrum of the hadronic system in the $\tau \to 3\pi\nu_\tau$ decay mode, a) and c) in linear scale and b) and d) in log scale by the DELPHI experiment. The solid line is a IMR model fit while the dashed line indicates a KS model fit. In c) and d), possible $a'_1(1700)$ contribution is included in the fit by Feindt modelling[21].

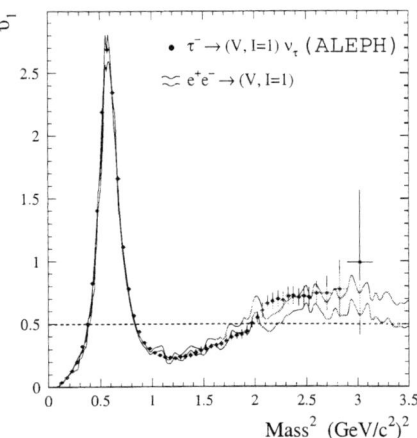

FIGURE 2. The vector spectral function generated on the data sample collected by the ALEPH experiment. The solid curve indicates the expectation by the theorem, with the shaded area representing the uncertainty. The dashed curve indicates a naive isovector quark-parton prediction [25].

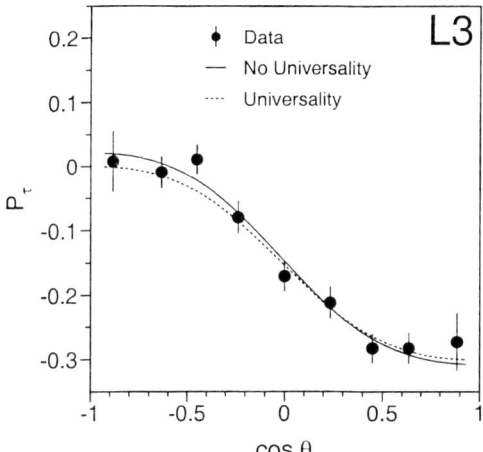

FIGURE 3. The tau polarzation asymmetry versus the tau production angle generated on the data sample collected by the L3 experiment[36].

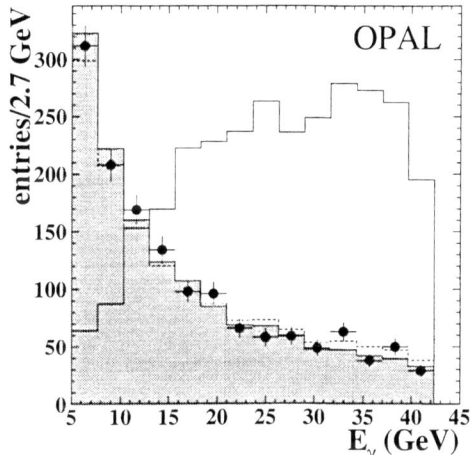

FIGURE 4. The photon energy spectrum of the radiative $Z \to \tau^+\tau^-(\gamma)$ events by the OPAL experiment. The shaded histogram represents the contribution from the normal term while the blank histogram represents the contribution from the anomalous term. The dots with error bars are the data points and the dotted curve is the fitted result. The normalizations of the histograms are arbitrary[44].

Recent Tau Results from CLEO

Richard G. Baker for the CLEO Collaboration

Cornell University, Ithaca, New York 14853

Abstract.
A data sample of 4.5×10^6 tau pairs has been analyzed to obtain a limit on the mass of the tau neutrino and also to search for CP violation in tau decay. The neutrino mass study utilizes a two-dimensional extended likelihood analysis that includes the dependence of the end-point population on m_{ν_τ}, and also, for the first time, incorporates an explicit background contribution. We use the the decays $\tau \rightarrow 5\pi\,\nu_\tau$ and $\tau \rightarrow 3\pi\,2\pi^0\,\nu_\tau$ to obtain an upper limit of 30 MeV/c^2 at 95% C.L. In addition, we have performed the first search for CP violation in tau lepton decay. CP violation could appear as an asymmetry in the angular distribution of the decay $\tau^- \rightarrow K^0\,\pi^-\,\nu_\tau$ relative to the charge conjugate decay. We define and measure an observable assymetry parameter and find no evidence for CP violation in this decay.

INTRODUCTION

The data used in these analyses were collected with the CLEO II detector [1] at the Cornell Electron Storage Ring with a center of mass energy of ~10.6 GeV. 4.5×10^6 tau pairs were produced in an integrated luminosity of 5.0 fb^{-1}. The tau pairs, created mainly via $e^+e^- \rightarrow \gamma* \rightarrow \tau^+\tau^-$, have the full beam energy and recoil back to back, modulo initial state radiation effects. The CLEO II.V detector is currently running with instantaneous peak luminosities of greater than $5 \times 10^{32}\,cm^{-2}\,s^{-1}$ and should collect an integrated luminosity of approximately 7 fb^{-1} by the end of 1998. The combined CLEO II and CLEO II.V datasets will contain more than 10^7 tau pair events.

TAU NEUTRINO MASS

Motivation

Although the tau neutrino has never been directly observed, the question of its mass is an important issue in particle physics and cosmology. While the requirement that the density of primordial relic neutrinos from the Big Bang not over-close the Universe restricts the mass of a stable neutrino to be less than ~ 100 eV/c^2 [2–6], unstable neutrinos are less restricted. Big Bang nucleosynthesis models allow for a massive neutrino in the 10 to 31 MeV/c^2 range for lifetimes in the 0.01 to 40 second interval [7][1]. This astrophysically allowed range currently overlaps the experimentally accessible bounds.

Previous Results

The lowest published upper limit on tau neutrino mass[2], by the ALEPH collaboration [9], is 24 MeV/c^2 and is obtained with a two-dimensional likelihood fit to a region near the endpoint of 25 5π (π^0) candidate events. By doubling this dataset and adding in 3π decay candidates, the ALEPH collaboration [10] has recently reported an upper limit of 18.2 MeV/c^2. The DELPHI collaboration [11] has claimed a limit of 33 MeV/c^2 in the 3π mode from a mass fit, but notes that this limit is sensitive to the contribution of a possible higher mass resonance in the 3π mass distribution. DELPHI estimates that the model dependence in the 3π mode increases the limit to 62 MeV/c^2. The OPAL collaboration [12] quotes a 35.3 MeV/c^2 limit from a two-dimensional fit to the missing momentum and missing mass in 3π decays using the event thrust axis as an estimator of the tau direction. Combining this with an earlier limit of 74 MeV/c^2 from 5π decays [13], a 29.9 MeV/c^2 limit is obtained. ARGUS [14,15] has published a 31 MeV/c^2 limit based on a one-dimensional mass fit to 20 5π events. The previous CLEO limit [16] of 32.6 MeV/c^2 was set with a fit to the mass spectrum of 60 5π events, and 53 $3\pi 2\pi^0$ events.

Method

In order to avoid a spuriously low limit from a background event, the event selection criteria are chosen to obtain a very pure event sample of $\tau \rightarrow 5\pi\,\nu_\tau$ or $3\pi\,2\pi^0\,\nu_\tau$. Details can be found in our recent publication [17]. Events are required to contain a lepton "tag" recoiling against 3 or 5 charged tracks. Very tight constraints on the tag hemisphere and standard track quality and neutral energy cuts on the signal hemisphere are optimized to minimize backgrounds from non tau events and feed-across from other tau decay modes. A data sample which satisfies the event selection criteria on the signal hemisphere, but with an invariant mass above the tau mass on the tag side, is used as an estimator of non-tau backgrounds. Monte Carlo events with full detector simulation [18] based on the event generators KORALB [19–23] and PHOTOS [24] were used to estimate tau decay backgrounds in the signal hemisphere (feed-across).

---

[1] Recently [8] it has been pointed out that these arguments do depend upon which set of astrophysical observations are used.
[2] All neutrino mass upper limits quoted herein are at the 95% confidence level.

Results

The resulting event distributions, shown in Fig. 1, contain 266 events in the 5π final state, of which 8 are above the tau mass. The $3\pi 2\pi^0$ mode contains 207 events, of which 13 are above the tau mass. Monte Carlo simulations show a reconstruction efficiency of $(3.08 \pm 0.10)\%$ for the 5π mode and $(0.43 \pm 0.02)\%$ for the $3\pi 2\pi^0$ mode. In both modes the number of events observed is consistent with the expectation from the world average branching fractions [25].

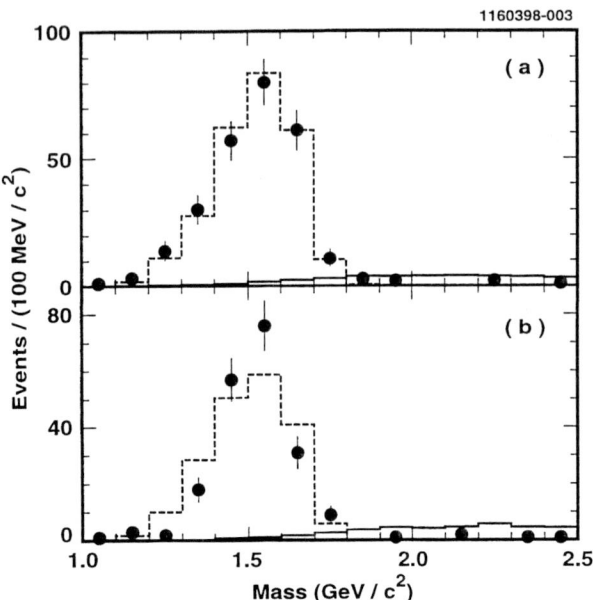

FIGURE 1. The mass distributions of the 5π (a) and $3\pi 2\pi^0$ (b) data samples. The dashed histogram represents the shape expected from Monte Carlo normalized to the number of events below the tau mass. The background estimate is shown as a solid histogram and is normalized for display purposes to five times its nominal value.

Each signal event is represented by a point in the two-dimensional plane formed by the hadronic energy scaled to the beam energy (E_X/E_B) versus hadronic mass (M_X). The sensitivity to neutrino mass is largest in the region near $M_X = m_\tau$ and $E_X/E_B = 1$. The data events are shown in Fig. 2. There are 36 5π events in the fit region, with negligible tau backgrounds and an estimated 0.3 events from non-tau backgrounds. There are 19 $3\pi 2\pi^0$ events in the fit region, with 1.0 of these expected as feed-across from other tau decay modes and 0.4 expected from non-tau backgrounds. Details of the analysis are summarized in Table 1.

We extract a measurement of the tau neutrino mass using an unbinned extended maximum likelihood technique. The procedure fits for one parameter (m_ν), taking as input the measured hadronic energy and mass of the events in the fit region

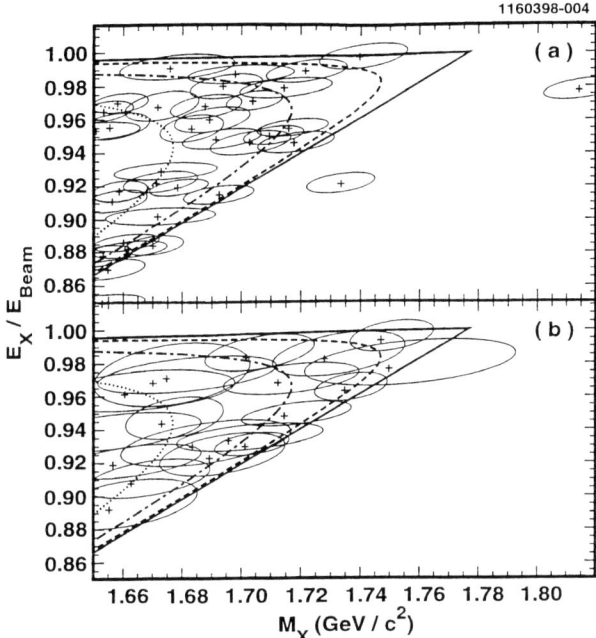

FIGURE 2. The hadronic scaled energy versus mass distribution of the 5π (a) and $3\pi 2\pi^0$ (b) data samples in the fit region. Ellipses represent 1 σ resolution errors and incorporate reconstruction induced systematic offsets. Kinematically allowed contours are drawn for neutrino masses of 0, 30, 60 and 100 MeV/c^2 as solid, dashed, dot-dashed and dotted lines respectively. Events below the kinematically allowed region are fully consistent with signal events, once initial state radiation effects are considered (as they are in the likelihood.)

shown in Fig. 2. The likelihood function, is composed of a Poisson factor, expressing the number of events expected, times the sum of a signal term and a background term. The likelihood is calculated as a function of neutrino mass using a novel

TABLE 1. Summary of signal, background, efficiency, resolution and upper limit by mode.

Mode	5π	$3\pi 2\pi^0$
Total Events	266	207
Events in Fit Region	36	19
Signal Region Purity (%)	99	93
Selection Efficiency (%)	3.08	0.43
Typical Mass Resolution (MeV/c^2)	15	25
Typical Energy Resolution (MeV)	25	50
Upper Limit @ 95% CL (MeV/c^2)	31	33

technique. Instead of using explicit parameterizations of the Monte Carlo in a likelihood convolution [26], Monte Carlo signal events are used in the evaluation of the likelihood, directly implementing the best knowledge of all physics effects including initial state radiation and detector acceptance. The only explicitly parameterized term is the detector smearing function.

The resulting extended likelihood is shown in Fig. 3. We define[3] the 95% confidence level (CL) upper limit by integrating defined likelihood above zero mass to its ninety-fifth percentile. We find 95% CL upper limits of 33, 31, and 27 MeV/c^2 for the $3\pi 2\pi^0$, 5π and combined samples. A separate analysis using only mass information yields a combined upper limit of 31 MeV/c^2.

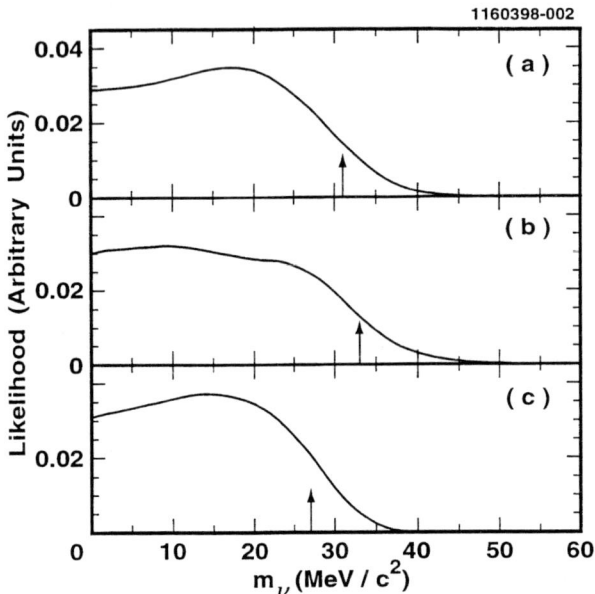

FIGURE 3. The likelihood as a function of neutrino mass for the 5π (a), $3\pi 2\pi^0$ (b) and combined (c) data samples. The 95% CL upper limits, uncorrected for systematic errors, are at 31, 33, and 27 MeV/c^2, respectively.

Systematic Error

Following the conservative prescription used by the LEP experiments [26], a linear systematic error, $\Delta(m_{95})$, is added to the limit. The largest systematic error

---

[3] The method for extracting an upper limit from a likelihood distribution at a given confidence level is not unambiguously defined; the method used here differs from that used in the analysis of Ref. [9]. Therefore comparisons of upper limits among different experiments must be done with care.

is due to uncertainties in the spectral functions. We obtain the spectral functions from a fit to a statistically independent sample of decays recoiling against pionic decays of the tau, excluding the 100 MeV/c^2 near the tau mass and including a background term. The next largest systematic error comes from the modeling of the detector smearing function. The smearing function is approximated by a sum of three two-dimensional Gaussians to model its extended tails. Other sources of error considered are uncertainty in the absolute mass and energy scales, the Monte Carlo statistics used in the likelihood evaluation and the background size. The total systematic error from all sources added in quadrature is 2.9 MeV/c^2, resulting in a final 95% C.L. upper limit on the mass of the tau neutrino of 30 MeV/c^2. Systematic errors are summarized in Table 2.

Significance

The use of the Poisson term in this analysis' likelihood is an improvement over a previous technique [9] because it reduces the variance in fitted neutrino mass, thus decreasing the probability of a spuriously low limit in the presence of a massive neutrino. The present analysis also uses measured event errors to avoid a possible bias. The smearing shape used in previous analyses [9,13] depends on repeated Monte Carlo simulation of the data points, using the measured values as input to the Monte Carlo event generator. Such a method neglects the fact that events reconstructed near the endpoint stand a larger chance of being upward fluctuations from low mass events than do events in the middle of the accepted distribution, and thus tends to underestimate the mass and energy errors associated with each data point.

The interpretation of this limit as a meaningful statement about probability is conditional upon the measured likelihood being representative of an ensemble of similar experiments. The event distribution and the observed number of events at the endpoint are consistent with our Monte Carlo estimation with zero neutrino mass, using a spectral function tuned to the data. An ensemble of Monte Carlo

TABLE 2. Systematic error summary.

Source	Error (MeV/c^2)
Spectral Function	1.9
Smearing Tails	1.6
Mass Scale	1.1
Smearing Parameters	0.7
MC Statistics	0.6
Background	0.4
Energy Scale	0.3
Total	2.9

experiments using statistics compatible with those we observe reveal that a smaller upper limit is obtained in 67% of experiments with a massless neutrino. The average expected upper limit given our sample size is found to be 25 MeV/c^2.

FIRST SEARCH FOR CP VIOLATION IN TAU DECAY

Motivation

To date CP violation has only been observed in the kaon system [27] and its origin remains unknown. In the minimal standard model (MSM) CP violation is restricted to the quark sector and cannot occur in lepton decay [28,29]. It can, however, occur in extensions to the MSM such as the three Higgs doublet model [30]. It appears that there is insufficient CP violation in the MSM to generate the apparent matter-antimatter asymmetry of the universe [5]. Searches for additional CP violation beyond the MSM may help reconcile this problem.

CP violation appears as a phase θ_{cp} in the gauge boson-fermion coupling constant, $CP:\theta_{cp} \to -\theta_{cp}$. The physical effects of such a phase are only manifest in the interference of two amplitudes with both relative CP odd phase θ_{cp} and relative CP-even phase δ (the interference term is proportional to $\cos(\delta - \theta_{cp})$). In tau lepton decay the two amplitudes could come from the MSM vector boson exchange (W) and the extended standard model scalar (Higgs) exchange. The CP-odd phase comes from the imaginary part of the complex scalar coupling constant. The CP-even phase difference is provided by the final state interaction (strong) phase that is different for s-wave scalar exchange and p-wave vector exchange and only arises in semi-leptonic decay modes with at least two final state hadrons ($\tau^- \to h_1 h_2 \nu_\tau$). The final state interaction is described by the s-wave and p-wave form factors, $F_s = |F_s|e^{i\delta_s}$ and $F_p = |F_p|e^{i\delta_p}$ respectively so that the strong phase difference is $\delta_{strong} = \delta_p - \delta_s$. The CP-violating $s-p$ wave interference term is then proportional to $|F_p||F_s|g\cos(\delta_{strong} - \theta_{cp})\cos\beta\cos\psi$, where β and ψ are physical decay angles measured in the hadronic rest frame ($\vec{p}_{h_1} + \vec{p}_{h_2} = 0$) [31,32]. The direction of the laboratory frame as viewed from the hadronic rest frame is $\vec{p}_{lab}$ and β is the angle between the direction of h_1 or h_2 and $\vec{p}_{lab}$. ψ is the angle between the tau flight direction and $\vec{p}_{lab}$. The ratio of scalar to vector coupling strength is g (i.e. g is in units of $G_F/2\sqrt{2}$). Since the sign of θ_{cp} changes for the CP-conjugate τ^- and τ^+, we define an experimentally measurable asymmetry $A_{observed}(\cos\beta\cos\psi)$ in terms of the number of events from $\tau^\pm$ decay, $N^\pm(\cos\beta\cos\psi)$, in a particular interval of $\cos\beta\cos\psi$:

$$A_{observed}(\cos\beta\cos\psi) = \frac{N^+ - N^-}{N^+ + N^-} \propto |F_p||F_s|g\sin\delta_{strong}\sin\theta_{cp}\cos\beta\cos\psi$$

Method

We select a $\tau^- \to K_S^0 h^- \nu_\tau$, $K_S^0 \to \pi^+ \pi^-$ event sample since a mass dependent Higgs-like coupling would give the largest asymmetry in this mode and with three charged tracks in the final state the decay angles are well measured. Here h^- is a charged pion or kaon. Events with 1 vs. 3 topology are selected with tight constraints on the tag hemisphere to reduce backgrounds from non-tau events. In the signal hemisphere, two of the charged tracks must be consistent with K_S^0 decay and form a vertex at least 5 mm from the primary interaction point. The invariant mass of this pair of tracks must be within 20 MeV of the known K_S^0 mass. We define a sideband region 30-100 MeV above and below the K_S^0 mass to use as a control sample. Details can be found in our recent publication [33].

Results

Figure 4 shows the invariant mass distribution after all selection criteria. Using this sample we measure the asymmetry for both signal and sideband in two intervals of $\cos\beta\cos\psi$, $A_{observed}(\cos\beta\cos\psi < 0)$ and $A_{observed}(\cos\beta\cos\psi > 0)$, given in Table 3. Both signal and sideband exhibit similar non-zero asymmetries but with low statistical significance. The measured asymmetries are insensitive to small variations in the selection criteria.

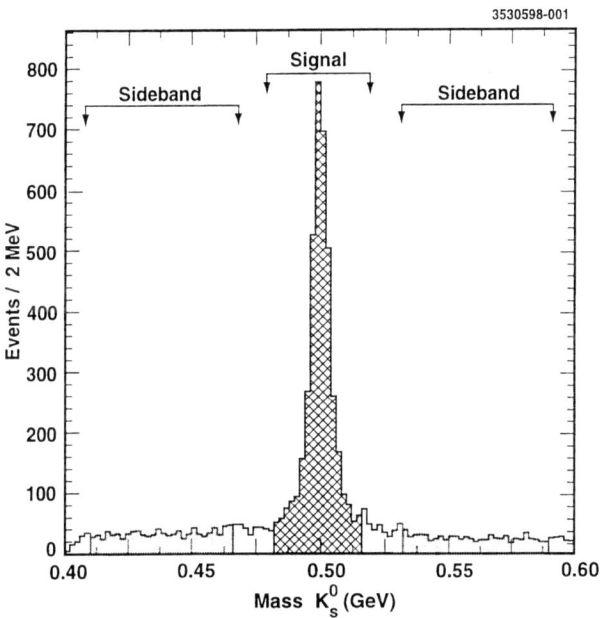

FIGURE 4. Invariant mass distribution for $K_S^0 \to \pi^- \pi^+$ in final data sample.

TABLE 3. Observed asymmetries in signal and sideband regions

	$A_{observed}(\cos\beta\cos\psi < 0)$	$A_{observed}(\cos\beta\cos\psi > 0)$
Signal	0.058 ± 0.023	0.024 ± 0.021
Sideband	0.049 ± 0.030	0.034 ± 0.033

To estimate the expected CP-violating asymmetry for a pure $\tau^- \to K_S^0 \pi^- \nu_\tau$ sample we use the KORALB Monte Carlo [19–23] to generate τ-pairs, with modifications to include a scalar Higgs coupling in addition to the standard model W boson coupling, for the signal $K_S^0 \pi^-$. We set $F_s = 1$ (i.e non-resonant decay) and F_p to be a Breit-Wigner for the $K^*(892)$ resonance so that $F_p >> F_s$ and the average strong phase difference $<\delta_{strong}>= \pi/2$. The GEANT code [18] is used to simulate detector response and assumes equal detection efficiencies for positive and negatively charged particles. We estimate $A^{K_S^0\pi^-}_{expected}(\cos\beta\cos\psi < 0) = -0.033g\sin\theta_{cp}$ and $A^{K_S^0\pi^-}_{expected}(\cos\beta\cos\psi > 0) = +0.033g\sin\theta_{cp}$ for a pure $\tau^- \to K_S^0 \pi^- \nu_\tau$ signal.

To compare this estimated asymmetry to the observed asymmetry we must take into account the effect of backgrounds since the signal region is not pure $K_S^0 \pi^-$ and also estimate the asymmetry expected from detection inefficiencies alone. Table 4 gives the estimated signal and sideband compositions by mode where the Lund Monte Carlo [34] has been used to generate the $q\bar{q}$ events.

TABLE 4. Signal and sideband mode composition. f^{mode} is the fraction of the total sample for a particular mode. α^{mode} is the approximate magnitude of asymmetry expected relative to the $\tau^- \to K_S^0 \pi^- \nu_\tau$ mode. The last column gives the dilution factor expected in the asymmetry when the measured asymmetry in the sideband control sample is subtracted from the measured asymmetry in the signal sample.

Tau Mode	α^{mode}	f^{mode}_{signal}	$f^{mode}_{sideband}$	$(f^{mode}_{signal} - f^{mode}_{sideband}) \cdot \alpha^{mode}$
$K_S^0(\pi^+\pi^-)\pi^-\nu_\tau$	1	0.525 ± 0.057	0.043 ± 0.005	0.482 ± 0.057
$K_S^0 K^-\nu_\tau$	1/20	0.124 ± 0.036	0.009 ± 0.003	0.006 ± 0.002
$a_1^- \nu_\tau$	1/80	0.106 ± 0.003	0.620 ± 0.013	-0.0064 ± 0.0002
$K_S^0 \pi^- \pi^0 \nu_\tau$	1/4	0.066 ± 0.016	0.006 ± 0.002	0.015 ± 0.004
$K_S^0 K_L^0 \pi^- \nu_\tau$	1/80	0.055 ± 0.018	0.003 ± 0.001	0.0007 ± 0.0002
$K_S^0 K^- \pi^0 \nu_\tau$	1/20	0.030 ± 0.008	0.003 ± 0.001	0.0014 ± 0.0004
$\pi^+ \pi^- \pi^- \pi^0 \nu_\tau$	1/20	0.028 ± 0.002	0.167 ± 0.007	-0.0070 ± 0.0004
$K^- \pi^+ \pi^- \nu_\tau$	1/4	0.008 ± 0.003	0.043 ± 0.007	-0.009 ± 0.002
others	0	0.012 ± 0.002	0.071 ± 0.017	0
$q\bar{q}$	0	0.044 ± 0.003	0.037 ± 0.003	0
Total	-	1.00 ± 0.07	1.00 ± 0.03	0.48 ± 0.06

The backgrounds arise from our inability to distinguish kaons and pions in the desired momentum range, lack of K_L^0 identification, particles that fall outside the fiducial region of the detector, and charged track mismeasurement. We note that the signal and sidebands are composed of different modes and it is unlikely that both samples would exhibit a similar CP-asymmetry as the strong phases, and possibly the coupling strengths are different for each mode. Also the samples exhibit an overall rate asymmetry not expected from CP-violating interference effects. However the effect of detection inefficiencies would be expected to be similar as both samples satisfy the same kinematic selection criteria.

Studies of pions from an independent $K_S^0 \to \pi^+ \pi^-$ sample indicate that at low momentum the reconstruction efficiency for π^+ is slightly greater than π^- and also the reconstruction of a K_S^0 in close proximity to a π^+ is slightly more efficient than for a π^-. The hadronic interaction of π^+ with the CLEO CsI calorimeter produces more fake electromagnetic clusters than from a π^- which may then be used as veto clusters. These effects are more pronounced at lower momentum (<1 GeV) and thus for $\cos\beta\cos\psi < 0.0$ since the pion from $\tau^- \to K_S^0 \pi^- \nu_\tau$ tends to be of lower momentum in this region. The sidebands may be used as a control sample to estimate these combined effects in our signal region in a simple empirical way providing we assume that any CP-violating effects are suppressed in the sideband modes.

The samples consist of a sum of modes, each a fraction f^{mode} of the total sample, with a possible CP-violating asymmetry suppressed by a factor α^{mode} relative to the $\tau^- \to K_S^0 \pi^- \nu_\tau$ signal mode. The suppression factor α_{mode} arises from two effects and is given for each mode in Table 4. First from the mass dependence of the Higgs coupling and second due to the dilution of the p-wave nature of the standard model final state. For example, the $\tau^- \to \pi^- \pi^+ \pi^- \nu_\tau$ mode is dominated in the standard model decay by an s-wave $\tau^- \to a_1^- \nu_\tau \to \rho^0 \pi^- \nu_\tau$ intermediate state which dilutes the $s-p$ wave interference by a factor of ≈ 4 in addition to a mass suppression of m_u/m_s relative to the $K_S^0 \pi^-$ mode.

If we assume that in the absence of any true CP violation a detector inefficiency would produce an asymmetry $A_{detector}$ common to all modes then the observed asymmetry, assuming the asymmetries are small (i.e $\ll 1$), is given by

$$A_{observed} = \Sigma_{mode} f^{mode} \alpha^{mode} A_{expected}^{K_S^0 \pi^-} + A_{detector}$$

From table 4 we see that the sideband should have negligible asymmetry with respect to the signal under the assumption of a mass dependent coupling and can be used as a control sample to subtract the detector asymmetries common to both signal and sideband.

$$A_{observed}^{signal} - A_{observed}^{sideband} = 0.48 A_{expected}^{K_S^0 \pi^-}$$

If a true CP violation exists the subtracted quantity should still exhibit significant but diluted asymmetry while detector effects should be removed. From table 3 the measured subtracted asymmetries are $A_{subtracted}(\cos\beta\cos\psi < 0) = 0.009 \pm 0.038$ and $A_{subtracted}(\cos\beta\cos\psi > 0) = -0.010 \pm 0.039$ which is consistent with no CP violation. This can be compared with a revised Monte Carlo estimate that takes into account the dilution factor of 0.48, $A(\cos\beta\cos\psi < 0) = -0.016 g \sin\theta_{cp}$,

$A(\cos\beta\cos\psi > 0) = 0.016g\sin\theta_{cp}$. To cross check our assumption of suppressed CP violation in the sidebands we measure the asymmetry in an independent high-purity high-statistics data sample of the dominant sideband mode, $\tau^- \to a_1^-\nu_\tau$, using the selection criteria of reference [35]. We find $A^{a_1}_{observed}(\cos\beta\cos\psi < 0) = -0.0013 \pm 0.0047$, $A^{a_1}_{observed}(\cos\beta\cos\psi > 0) = -0.0023 \pm 0.0047$ giving no evidence for CP violation. The higher track momentum and cluster veto thresholds combined with the absence of a K_S^0 requirement from this sample removes the contribution to the asymmetry from detection inefficiencies but a true CP-violating effect should remain. We note that by measuring the CP-violating asymmetry in the dominant sideband mode as zero our results are approximately valid for a non-mass dependent coupling but we cannot fully relax this assumption due to the difficulty of empirically isolating a sample of each background mode in which to measure the asymmetry.

In conclusion we find no evidence for CP violation in tau decay. We may compare the observed to expected asymmetries to set a constraint of $g\sin\theta_{cp} < 1.7$ at the 90 % confidence level assuming $F_s = 1$.

SUMMARY

By the end of 1998, the CLEO II and CLEO II.V detectors will have collected more than 10^7 tau pair events. The CLEO II.V detector has improved track resolution, vertex reconstruction and K^-/π^- separation relative to CLEO II. These improvements are expected to significantly reduce backgrounds in future analyses, and the ever growing size of the data sample will improve statistical precision. In 1999, the CLEO III upgrade will be installed to take advantage of yet higher luminosity anticipated at CESR, and we expect that CLEO will continue to produce important results in tau physics.

REFERENCES

1. Kubota, Y., et al., Nucl. Inst. and Meth. A **320**, 66 (1992).
2. Gershtein, S.S., and Zeldovich, Ya.B., JTEP Lett. **4**, 174 (1966).
3. Cowsik, R., and McClelland, J., Phys. Rev. Lett. **29**, 669 (1972).
4. Szalay, A.S., and Marx, G., Astron. & Astrophys. **49**, 437 (1976).
5. Kolb, E., and Turner, M., The Early Universe, New York: Addison-Wesley, 1990.
6. Peebles, P.J.E, Physical Cosmology, Princeton: Princeton University Press, 1993.
7. Kawasaki, M., et al., Nucl. Phys. B **419**, 105 (1994).
8. Rehm, J.B., Raffelt,G.G., and Weiss, A., Astronomy and Astrophysics **327**, 443 (1997).
9. Buskulic, D., et al., Phys. Lett. B **349**, 585 (1995).
10. Barate, R., et al., CERN PPE/97-138, submitted to Z. Phys. C.
11. Galloni, A., et al., A Limit on the Tau Neutrino Mass, submitted to HEP97, Jerusalem, 1997.

12. Akers, R., *et al.*, *Z. Phys. C* **72**, 231 (1996).
13. Akers, R., *et al.*, *Z. Phys. C* **65**, 183 (1995).
14. Albrecht, H., *et al.*, *Phys. Lett. B* **202**, 149 (1988).
15. Schröder, H., *et al.*, *Mod. Phys. Lett. A* **8**, 573 (1993).
16. Cinabro, D., *et al.* (The CLEO Collaboration), *Phys. Rev. Lett.* **70**, 3700 (1993).
17. Ammar, R., *et al.* (The CLEO Collaboration), CLNS 98/1551, to appear in *Phys. Lett. B* (1998).
18. Brun, R., *et al.*, *GEANT v. 3.14*, CERN Report No. CC/EE/84-1 (1987).
19. Jadach, S., and Was, Z., *Comp. Phys. Com.* **36**, 191 (1985).
20. Jadach, S., and Was, Z., *Comp. Phys. Com.* **64**, 267 (1991).
21. Jadach, S., Kühn, J.H., and Was, Z., *Comp. Phys. Com.* **64**, 275 (1991).
22. Jadach, S., Kühn, J.H., and Was, Z., *Comp. Phys. Com.* **70**, 69 (1992).
23. Jadach, S., Kühn, J.H., and Was, Z., *Comp. Phys. Com.* **76**, 361 (1993).
24. Barbiero E., van Eijk B., and Was Z., CERN-TH-5857/90 (1990).
25. Barnett, R.M., *et al.*, *Phys. Rev. D* **54** (1996).
26. Passalacqua, L., *Nucl. Phys. B* (Proc. Suppl.) **55C**, 435 (1997).
27. J.H. Christenson, J.H., Cronin, J.W., Fitch, V.L., and Turlay, R., *Phys. Rev. Lett.* **13**, 138 (1964).
28. Cabbibo, N. *Phys. Rev. Lett.* **10**, 531 (1963).
29. Kobayashi, M. and Maskawa, T., *Prog. Theor. Phys.* **49**, 652 (1973).
30. Weinberg, S., *Phys. Rev. Lett.* **37**, 657 (1976).
31. Kuhn, J.H. and Mirkes, E. *Phys. Lett. B* **398**, 407 (1997)
32. Tsai, Y.S. *Nucl. Phys. B* (Proc. Suppl.) **55C**, 293 (1997).
33. Anderson, S., *et al.* (The CLEO Collaboration), CLNS 98/1557, submitted to *Phys. Rev. Lett.* (1998).
34. Sojstrand, S.J., *LUND 7.3*, CERN-TH-6488-92 (1992).
35. Balest, R., *et al.* (The CLEO Collaboration), *Phys. Rev. Lett.* **75**, 3809 (1995).

b AND *t* PHYSICS

b Physics

Pascal Perret

*Laboratoire de Physique Corpusculaire, Université Blaise Pascal, IN2P3-CNRS
F-63177 Aubiere cedex, FRANCE.*

Abstract. A summary of the most recent and important measurements in b physics is presented. The production of beauty particles in Z decays, b quark couplings, lifetimes, B^0-$\overline{B^0}$ oscillations, semileptonic b decays and studies of the number of charm quarks produced in b decays are reviewed. Extraction of the Cabibbo-Kobayashi-Maskawa (CKM) matrix elements $|V_{td}|$, $|V_{cb}|$, $|V_{ub}|$ and implication for $|V_{ts}|$ are discussed.

I INTRODUCTION

The heavy mass of the b quark, around 5 GeV, so much greater than the strong interaction scale $\Lambda_{QCD} \sim 0.2$ GeV and the fact that it belongs to the same isospin doublet than the top quark, confere a special role to the b quark studies. Furthermore the top quark is too heavy to build hadrons and thus b hadrons are the heaviest. In that respect, b physics is a broad subject and one of the major areas of investigation of present experiments at CESR, LEP, SLC and Tevatron colliders. Experimentally b hadrons are easier to observe or to disentagle from other sources because tracks issued from their decays have higher transverse momentum and momentum due to the high mass and the hard fragmentation of the b hadrons. The lifetime of b hadrons (~ 1 ps) is relatively long and the subsequent presence of secondary vertices in detector can be used as a tag; for instance the mean decay length is 3 mm at LEP. They have also sizeable semileptonic branching ratios which allow to sign b events with the presence of leptons, cleanly identified, in decay products. Specific theoretical framework can be used for the description of the properties of b hadrons such as Heavy Quark Effective Theory (HQET) where a b hadron is considered like a hydrogen atom, Heavy Quark Symmetry (HQS) in the limit $m_b \to \infty$ or Heavy Quark Expansion (HQE) which allows expansions in $1/m_b$.

The main issues in b physics are to provide precision tests in the electroweak sector of the Standard Model (SM), to study the decay dynamics, especially the effect of strong interactions on the underlying quark decay, to understand the origin of CP violation and to measure the magnitude of the CKM matrix elements $|V_{cb}|$, $|V_{ub}|$, $|V_{td}|$ and $|V_{ts}|$, and to observe rare processes which can probe physics beyond the SM. A selection of subjects is reviewed and, due to space limitations, only a short summary is given. More extensive recent summaries can be found in [1–4]. Production in Z decays, lifetimes, B^0-$\overline{B^0}$ oscillations and decays are mainly discussed in the following. Most of the given averages are provided by LEP working groups [5–7].

In the field of beauty hadron spectroscopy, the main result is the observation and measurement of the B_c meson, by CDF at Tevatron, with a mass of 6.40±0.39±0.13 Gev/c^2 and a lifetime of $0.46^{+0.18}_{-0.16} \pm 0.05$ ps. More details can be found in the presentation of J. Troconiz, where recent results from Tevatron are covered.

II B PRODUCTION IN Z DECAYS

The relative ease with which b quarks can be separated from other quark flavours and the availability of large Z^0 event samples allow precision tests of the Standard Model to be carried out using $Z \to b\bar{b}$ decays at e^+e^- colliders. By the end of the LEP I phase (1989-1995), each of the four LEP experiments had recorded approximately 3.8 10^6 $Z \to q\bar{q}$, including nearly 0.8 10^6 $Z \to b\bar{b}$ decays, while by the end of 1997 SLD at SLC had recorded approximately 0.3 10^6 $Z \to q\bar{q}$ with polarized beams.

R_b: Due to the large mass of the b quark and the fact that it belongs to the same isospin doublet than the top quark, the Z-$b\bar{b}$ coupling is one of the most interesting windows in the search for new physics. The partial width ratio $R_b = \frac{\Gamma(Z \to b\bar{b})}{\Gamma(Z \to q\bar{q})}$ is sensitive to m_{top} via vertex corrections, while the corrections in α_s and m_H are suppressed in a first approximation. A precision measurement of R_b would therefore represent a unique test of the Standard Model and would provide significant constraints on possible new physics such as additional Higgs bosons or supersymmetry.

The first R_b measurements were done using leptons [14–17]. In inclusive charged lepton analyses, the preferred approach is to fit a two-dimensional distribution (p,p$_\perp$) for single lepton and dilepton events together, and extract simulatneously R_b with other parameters as for instance R_c or $\mathcal{B}(b \to \ell)$. Relatively large systematic errors remain due mainly to uncertainties in the modelling of the semileptonic decay of the b quark; these measurements were not precise enough to perform a stringent test. Then the informations from silicon vertex detector were used and thanks to the large statistics available, "double tagging" techniques, were applied. The double tagging technique exploits the fact that the b and $\bar{b}$ quarks are typically produced back to back, in separate hemispheres as defined by the thrust axis. A b quark tag is applied separately to each hemisphere in a sample of hadronic events and the total number of single and double-tagged events are measured. Assuming backgrounds from charm and light quark events and the correlations between the hemispheres from Monte Carlo, as well as R_c from its SM value, both R_b and the b tagging efficiency can be extracted from data. In 1995 the world average R_b showed a discrepancy with more than 3σ (dominated by systematics) from the SM, and it

TABLE 1. Summary of R_b results.

ALEPH [8]	mult	1992-95	$0.2159 \pm 0.0009 \pm 0.0011$
DELPHI [9]	mult	1994-95	$0.2166 \pm 0.0008 \pm 0.0009$
L3 [10]	mult	1994-95	$0.2179 \pm 0.0015 \pm 0.0026$
L3 [11]	shape	1991	$0.2223 \pm 0.0030 \pm 0.0064$
OPAL [12]	mult	1992-94	$0.2178 \pm 0.0014 \pm 0.0017$
SLD [13]	vtx mass	1993-97	$0.2158 \pm 0.0017 \pm 0.0014$
LEP [14-17]	leptons		$0.2227 \pm 0.0020 \pm 0.0025$
LEP + SLC	corrected for γ exchange		0.21732 ± 0.00087

was shown that charm systematics were a worry (exponential charm lifetime tail (D^+) is difficult to cut away) and so a b purity of 94 % was not enough controlled, as well as the understanding of the correlations. A new round of analyses has been performed [8–10,13]. An increased purity is achieved by exploiting b/c hadrons masses and kinematical differences, by including for instance the invariant mass of the significant tracks. The primary vertex was initially measured with all tracks of the event, including a correlation between hemispheres. Primary vertices were then reconstructed, one per hemisphere. Finally lifetime tag was used in conjunction with other tags; variables are combined in multivariate analyses (neural network for instance). With this kind of analyses, a purity greater than 98 % with an efficiency greater than 30 % has been achieved by DELPHI [9]. Measurements are summarised in table 1. The combined result [5], $R_b^0 = 0.21732 \pm 0.00087$, corresponds to a precision $\frac{\Delta R_b}{R_b} = 0.4\%$. The statistical and systematic errors are comparable, the latter receiving contributions mainly from uncertainties in gluon splitting $g \to b\bar{b}$ and $g \to c\bar{c}$, from the tracking resolution of the detector, and from hemisphere correlations.

Gluon splitting: The gluon splitting $g \to b\bar{b}$ is an important ingredient in the R_b measurement, constituting the largest single systematic uncertainty. New methods have been developed to measure this parameter by searching for b-tagged jets in 4 jet events. The 2 b-tagged jets have to form a small angle and the initial quarks are required to be in opposite hemispheres. DELPHI has measured $g \to b\bar{b}$ = $(0.21 \pm 0.11 \pm 0.09)\%$ [18] and ALEPH $(0.26 \pm 0.04 \pm 0.09)\%$ [19], the average being $(0.24 \pm 0.09)\%$.

b Asymmetry: The other electroweak quantity of interest is the forward backward charge asymmetry A_{FB}^b, obtained from measurements of the angular distribution $\frac{d\sigma}{d\cos\theta} \propto 1 + \cos^2\theta + \frac{3}{8}A_{FB}^b \cos\theta$; where θ is the angle of the outgoing b quark with respect to the initial e^- direction. The asymmetry A_{FB}^b arises from differences in the coupling strengths of the Z to left- and right-handed fermions, and is one of the most sensitive quantities to the effective electroweak mixing angle $\sin^2\theta_{\text{eff}}^{\text{lept}} = 1/4(1 - g_{V\ell}/g_{A\ell})$ [5]. To measure A_{FB}^b, one needs to select a b sample, to define the b quark direction (usually approximated by the thrust axis), and to estimate the electric charge of the quark to assign the b quark to the forward or backward

TABLE 2. Summary of A^b_{FB} results at $\sqrt{s} \approx m_Z$.

ALEPH [20]	leptons	1990-95	$0.0965 \pm 0.0044 \pm 0.0026$
DELPHI [21]	leptons	1991-94	$0.1075 \pm 0.0077 \pm 0.0031$
L3 [22]	leptons	1990-95	$0.0963 \pm 0.0065 \pm 0.0035$
OPAL [23]	leptons	1990-95	$0.0910 \pm 0.0044 \pm 0.0020$
ALEPH [24]	jet-charge	1991-95	$0.1017 \pm 0.0038 \pm 0.0032$
DELPHI [21]	jet-charge	1991-94	$0.0995 \pm 0.0072 \pm 0.0040$
L3 [25]	jet-charge	1994	$0.0855 \pm 0.0118 \pm 0.0056$
OPAL [26]	jet-charge	1991-95	$0.1004 \pm 0.0052 \pm 0.0046$
LEP + SLD	$A^{0,b}_{FB}$	winter 98	0.0998 ± 0.0022

direction. Leptons are good candidates, high p and $p_\perp$ leptons come mainly from b quarks and their electric charge allows to identify the b quark hemisphere [20-23]. Tag of Z $\to b\bar{b}$ using a lifetime/mass tag and using a momentum weighted track charge in each hemisphere to flag the b quark is also performed [24,21,25,26]. Both approaches are still statistically limited and achieve a similar overall precision. $D^{*\pm}$ or $K^\pm$ tag can also be used but are less performing [21,27]. For instance, at SLD, $b\bar{b}$ events are tagged using a mass tag, while kaons from $b \to c \to s \to K$ are used to sign the b quark direction with their charge and direction. The main measurements are summarized in table 2, and the average of the pole asymmetry is $A^{0,b}_{FB} = 0.0998 \pm 0.0022$ [5]. $A^{0,b}_{FB}$ can be expressed as a measurement of the effective angle $\theta^{\text{lept}}_{\text{eff}}$: $\sin^2 \theta^{\text{lept}}_{\text{eff}} = 0.23213 \pm 0.00039$.

III LIFETIMES

In the quark spectator model, the heavy quark decays weakly without interacting with the other light quark(s). As a result, all the hadrons containing a b quark should have the same lifetime. As in the case of the charm hadrons, non-spectator effects, such as final state interference, W exchange, weak annihilation and helicity suppression lead to significant differences in the lifetimes of beauty hadrons. In heavy quark expansion (HQE) theory, a theoretical approach based on QCD and where the decay rates of a beauty hadron are expressed as an expansion in powers of $1/m_b$, the lifetime difference of baryons and mesons depends on terms of the order of $1/m_b^2$ and higher, while the lifetime of the different B mesons depend on terms $1/m_b^3$ [29]. The following hierarchy among the various species $\tau_{\Lambda_b} < \tau_{B^0_d} \simeq \tau_{B^0_s} < \tau_{B^+}$ is expected [28], but it seems that corrections in "$\mathcal{O}(1/m_b^3)$" could be large in the ratio $\tau(B^+)/\tau(B^0)$ without model assumptions [30,31]. The experimental determination of the magnitude of these differences is needed.

To measure the proper lifetime of a B hadron, it is necessary to determine its decay length and its momentum. Several different and complementary methods have been developed to perform such measurements. Fully reconstructed beauty hadron final states are the cleanest way. These measurements [32] benefit from the precise determination of the secondary vertex, and since there are no missing particles, the momentum is well determined. Consequently these measurements have little dependence on simulation. However this technique is limited at LEP/SLC due to the available statistics. Larger samples are obtained by using the presence of a high momentum lepton to select semileptonic b decays, and by fully or partially reconstructing a charm hadron of the appropriate charge in the same jet [32–39]. The vertex resolution is still good due to the lepton, but the missing products degrade the momentum resolution. These methods suffer also from higher background due to fake contaminations and the "pollution" of B^+ and B^0 cross-contamination for instance. Another approach is based on pure topological vertexing. b decay vertices are reconstructed inclusively and the b hadron charge is determined from the total charge of the tracks associated with its vertex [40,41]. This method gives the highest statistics at the expense of a reduced purity and a greater sensitivity to the modelling simulation. Some measurements of the b lifetime over all hadron species are based on the impact parameters of tracks from b decays, generally leptons. The knowledge of the b fragmentation and of the semileptonic decay models systematically limits the accuracy of these measurements.

There are many lifetime measurements, their average [6] is given in table 3. Lifetime ratios are known experimentally close to 5% and the lifetime hierarchy among beauty hadrons is predicted correctly. No significant differences between the three B mesons lifetimes are observed and they are in good agreement with HQE predictions. However the ratio $\tau(b-\text{baryon})/\tau(B^0)$ is significantly different from unity [37,38,33,42] and smaller than usual predictions. This is correlated with a small semileptonic b-baryon semileptonic branching ratio (see below), and is the place of an intensive work.

TABLE 3. World average lifetime measurements and their ratio. Predictions are given in the last column.

$\tau(B^+)$	1.67 ± 0.04 ps	$\tau(B^+)/\tau(B^0)$	1.07 ± 0.04	1.0 - 1.1
$\tau(B^0)$	1.57 ± 0.04 ps			
$\tau(B_s)$	1.48 ± 0.06 ps	$\tau(B_s)/\tau(B^0)$	0.95 ± 0.05	0.99 - 1.01
$\tau(\Lambda_b)$	1.23 ± 0.08 ps	$\tau(\Lambda_b)/\tau(B^0)$	0.78 ± 0.06	0.9 - 1.0
$\tau(b-\text{baryon})$	1.22 ± 0.05 ps	$\tau(b-\text{baryon})/\tau(B^0)$	0.78 ± 0.04	0.9 - 1.0
$\tau(b)$	1.554 ± 0.013 ps			

IV B^0-$\overline{B^0}$ OSCILLATIONS

In the Standard Model, particle-anti-particle oscillations take place via a second order weak interaction process - box diagram - with a loop of W bosons and up-type quarks, which are dominated by top quark exchange in the case of neutral B mesons. The oscillation frequency depends on the mass difference Δm_q between the mass eigenstates. Time integrated measurements are performed, they are typically based on counting same-sign and opposite-sign lepton pairs. At LEP both neutral B meson species are produced with a rate $f_{B_d^0}$ and $f_{B_s^0}$ (see below), and the LEP average is $\overline{\chi} = f_{B_d^0}\chi_d + f_{B_s^0}\chi_s = 0.1214 \pm 0.0043$ [5], while CLEO and ARGUS, at the $\Upsilon(4s)$ where only B_d^0 mesons are produced, measured $\chi_d^{\Upsilon(4s)} = 0.156 \pm 0.024$ [43].

To measure the time dependence of the mixing, one needs to know the b flavour at production time and at decay time to define whether a mixing occured or not, as well as the B decay length and energy to reconstruct the proper decay time. Many different methods have been developed for this purpose. The final state tag is given by the charge of the decay products (lepton, $D^{*\pm}$, $D_s^\pm$ or $K^\pm$ [44]). For fully inclusive analyses based on topological vertexing, the final state tagging techniques include jet charge [45] and charge dipole methods [44]. For the initial flavour state, we can distinguish tags which exploit the B hadron decay in the opposite hemisphere using the charge of a lepton or a kaon, and those which exploit informations of the B candidate itself. These later one use the charge of a track from the primary vertex which is correlated with the production state of the B if that track is a decay product of a B^{**} state or if it is the first particle in the fragmentation chain [46,47]. The jet charge techniques work on both sides. At SLC, the beam polarization produces a sizeable forward-backward asymmetry in the Z $\to$ $b\bar{b}$ decays and provides another very interesting and effective initial state tag, based on the polar angle of the B candidate [44].

A lot of different analyses have been performed to measure Δm_d [32,44–46,49–53]. An overview is shown in figure 1. Averaging all direct Δm_d measurements from LEP, SLD and CDF, yields $0.475 \pm 0.018 \text{ps}^{-1}$ [7]. The systematic uncertainties are not negligible; they are often dominated by the sample composition, mistag probability, or b hadron lifetime contributions. Including CLEO and ARGUS measurements of χ_d [43] give the world averages [7]: $\Delta m_d^{\text{world}} = 0.466 \pm 0.018 \text{ ps}^{-1}$ and $\chi_d^{\text{world}} = 0.174 \pm 0.011$.

The B_s^0 oscillations have been the subject of many recent studies [47–49]. However, the B_s^0 mixing proceeds much faster than the B_d^0 mixing, and the time evolution has not been resolved. Only lower limits are derived, and an overview of the available sensitivities is given in figure 2. The combined 95% Confidence Level (C.L.) limit, derived from the amplitude method [54], is $\Delta m_s > 10.2 \text{ps}^{-1}$ [7].

The measurement of Δm_d and Δm_s are related, in the Standard Model, to the CKM matrix elements V_{td} and V_{ts} respectively. From Δm_d one gets $|V_{td}| = (8.8 \pm 0.2_{\Delta m_d} \mp 0.2_{m_t} \mp_{1.8th}^{1.4}) \times 10^{-3}$, with an uncertainty completely dominated by theoretical uncertainties. However, many uncertainties cancel in the frequency ratio, yelding $|V_{ts}|/|V_{td}| > 3.8$ at 95% C.L.

The B_s^0 and b baryon fractions can be extracted from branching ratio measurements. The LEP B oscillations working group estimates [7] $f_{\text{b-baryon}} = (10.6_{-2.7}^{+3.7})\%$ and $f_{B_s^0} = (10.8_{-2.9}^{+3.3})\%$. $\Delta m_d^{\text{world}}$ and χ_d^{world} can be used to improve our knowledge on

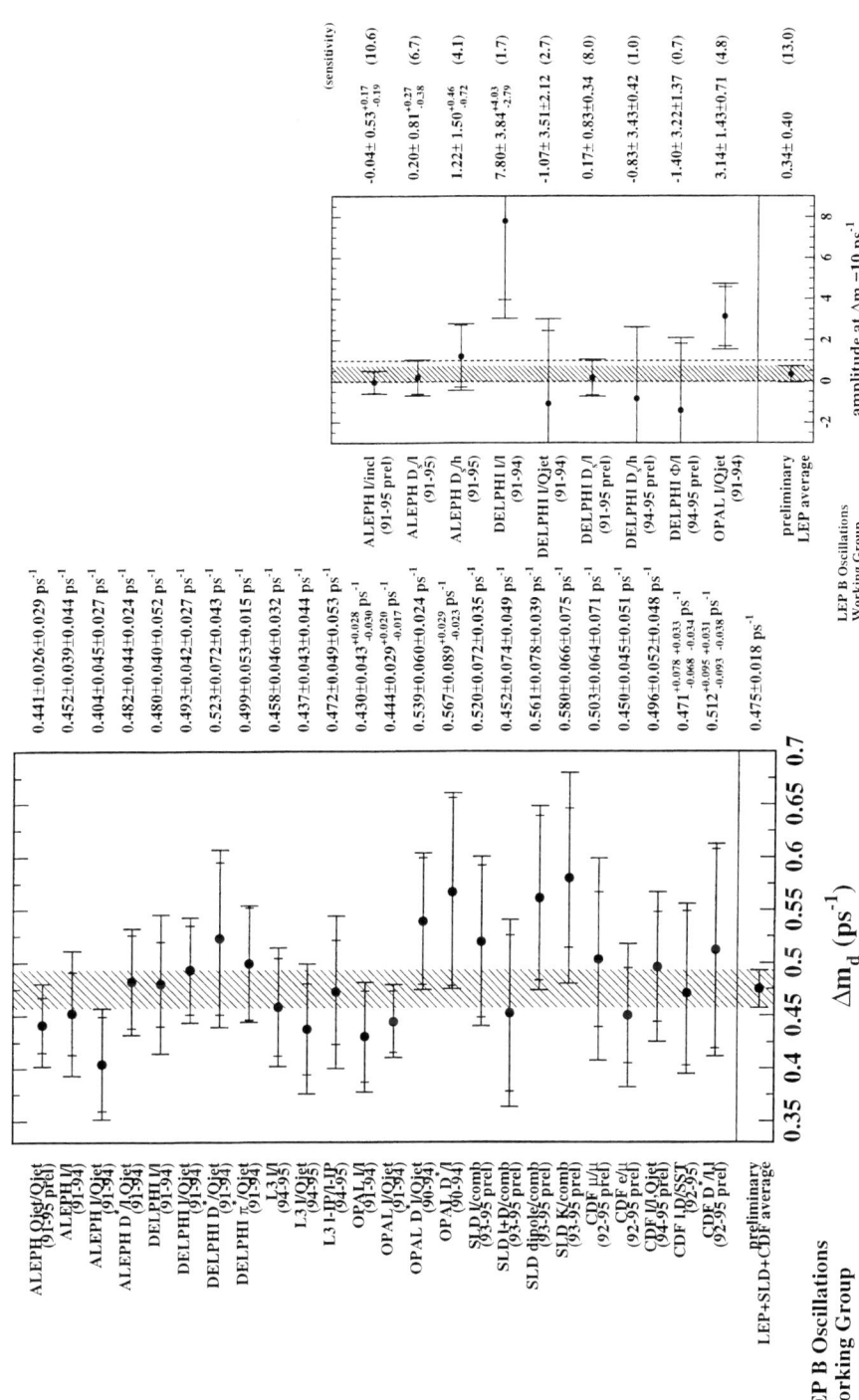

FIGURE 1. Measurements of Δm_d.

FIGURE 2. Sensitivities on Δm_s.

the fractions of weakly decaying bottom hadron in Z → $b\bar{b}$ events. If one assumes also $\chi_s = 1/2$ and $f_{B^0} = f_{B^+} = (1 - f_{B^0_s} - f_{\text{b-baryon}})/2$, another estimate of $f_{B^0_s}$ can be extracted from χ_d^{world}, from the inclusive integrated mixing rate $\bar{\chi}$ measured at LEP, from the $f_{\text{b-baryon}}$ branching ratios estimate and from the b hadron lifetimes. Combining all the informations yields $f_{B^0_s} = (10.3^{+1.6}_{-1.5})\%$, $f_{\text{b-baryon}} = (10.6^{+3.7}_{-2.7})\%$ and $f_{B^0} = f_{B^+} = (39.5^{+1.6}_{-2.0})\%$. These results, including $\Delta m_d^{\text{world}}$, have been obtained by the LEP B oscillations working group in a consistent way. There are also new measurements of $f_{\text{b-baryon}} = (10.2 \pm 0.7 \pm 2.7)\%$ from ALEPH [55] and of $f_{B^0_s} = (10.8 \pm 1.3 \pm 2.2)\%$ from DELPHI [56] (from $f_{B^0_s + B^{**}_s} = (14.4 \pm 1.7 \pm 3.0)\%$ and assuming $f_{B^0_s}/(f_{B^0_s} + f_{B^{**}_s}) = 0.25 \pm 0.05$). Including these measurements yields to $f_{B^0_s} = (10.4^{+1.4}_{-1.3})\%$, $f_{\text{b-baryon}} = (10.4^{+2.2}_{-1.9})\%$ and $f_{B^0} = f_{B^+} = (39.6^{+1.2}_{-1.4})\%$. DELPHI has also a preliminary measurement [56] of the rate of charged and neutral weak B hadrons: $\mathcal{B}(b \to X^0_B) = (57.8 \pm 0.5 \pm 1.0)\%$, $\mathcal{B}(b \to X^{\pm}_B) = (42.2 \pm 0.5 \pm 1.0)\%$.

V DECAYS

A b decay multiplicity

There are two new measurements of the mean charged multiplicity in b-hadron decays at the Z, with much smaller systematic uncertainties than previous measurements. L3 has measured $<n_b> = 4.90 \pm 0.04 \pm 0.11$ [42]; while DELPHI found $<n_b> = 4.97 \pm 0.03 \pm 0.06$ [57].

B semileptonic branching ratio

From the experimental point of vue, semileptonic branching ratios are accessible. They are relatively large and leptons have clean signatures. Furthermore all detectors have good lepton identification device. From the theoretical point of vue, despite strong interactions are quite important in these decays, they allow detailed theoretical predictions that can be tested experimentally. It is why they are among the most extensively studied decays.

At LEP, old analyses were done by performing a multi-parameter fit in the (p,$p_\perp$) lepton spectrum [14,15,64]. The 4 electroweak heavy quark flavours parameters R_b, R_c, A^b_{FB}, A^c_{FB} can all be measured simultaneously, with the following parameters: χ, $\mathcal{B}(b \to \ell)$, $\mathcal{B}(b \to c \to \ell)$, $\mathcal{B}(c \to \ell)$, b, and c fragmentation parameters. Some collaborations have separate fits for smaller sets of parameters, and restrict to a $p_\perp$ region to enrich the sample in b. ALEPH has developed new techniques, first presented in 1995 [58], to measure more accurately $\mathcal{B}(b \to \ell)$ and $\mathcal{B}(b \to c \to \ell)$ using information from the silicon vertex detector. Events are split into 2 hemispheres, and a cut on the lifetime tag probability [59] is imposed on all hemispheres to prepare a very pure sample of $Z \to b\bar{b}$ events. Typically, a purity of 96 % in b events can be achieved with an efficiency of 25 %. Then the opposite hemisphere a tagged hemisphere is used as an unbiased sample of b decays. A clear kinematic distinction allows to disentangle the $b \to \ell$ at high $p_\perp$ from $b \to c \to \ell$ at low $p_\perp$. While single leptons are sufficient to extract $\mathcal{B}(b \to \ell)$ and $\mathcal{B}(b \to c \to \ell)$, by performing a fit in the $p_\perp$ plane, the opposite-side dilepton sample, which is naturally enriched in b decays, is also used to measure at the same time the b fragmentation and the mixing parameter χ, taking advantage of the charge correlations. The charge correlations allow also to reduce the model dependence. Although the uncertainty from the semileptonic decay models is still dominant, it has been reduced nearly by a factor 2. This new analysis provides some significant improvements in systematic uncertainties thanks to the use of a very pure sample of b events, which suppresses the charm and light quark contributions, and to the fact that $\mathcal{B}(b \to \ell)$ is independent of R_b by construction.

DELPHI has performed the same kind of analysis, measuring $\mathcal{B}(b \to \ell)$, $\mathcal{B}(b \to c \to \ell)$ and χ [60]. And, OPAL has now, for this winter, an analysis of this type [64]. They use single muon sample only and a neural net to improve the discrimination between $b \to \ell$ and $b \to c \to \ell$. A measurement of $\mathcal{B}(b \to \ell)$ and $\mathcal{B}(b \to c \to \ell)$ is obtained. All LEP measurements of $\mathcal{B}(b \to \ell)$ and $\mathcal{B}(b \to c \to \ell)$ are summarized in table 4.

The new LEP average value [5] of $\mathcal{B}(b \to \ell)$ is 0.1104 ± 0.0019, while the $\Upsilon(4s)$ average is 0.1045 ± 0.0021 [2]. The discrepancy between these two numbers is 0.0059 ± 0.0028 corresponding to 2.1 σ. Furthermore, beauty baryons are produced at LEP and not at the $\Upsilon(4s)$ and their semileptonic branching ratio is smaller (this will be seen later). Consequently the LEP average is expected to be lower than the $\Upsilon(4s)$ result, contrary to the observed pattern. If we consider only the last measurements of $\mathcal{B}(b \to \ell)$, obtained with a new kind of method and corresponding to a second generation of $\mathcal{B}(b \to \ell)$ measurements [58,60,64] which are less model dependent, the Z average becomes $\mathcal{B}(b \to \ell)^Z = 0.1094 \pm 0.0030$, to be compared with $\Upsilon(4s)$ [65,66] average of $\mathcal{B}(b \to \ell)^{\Upsilon(4s)} = 0.1018 \pm 0.0040$ [2]. The discrepancy decrease to 1.5 σ.

TABLE 4. $\mathcal{B}(b \to \ell)$ and $\mathcal{B}(b \to c \to \ell)$ measurements (in %) at LEP.

	$\mathcal{B}(b \to \ell)$	$\mathcal{B}(b \to c \to \ell)$
ALEPH [14]	$11.2 \pm 0.3 \pm 0.4$	$8.8 \pm 0.3 \pm 0.8$
ALEPH [58]	$11.0 \pm 0.1 \pm 0.3$	$7.7 \pm 0.2 \pm 0.5$
DELPHI [15]	$11.3 \pm 0.5 \pm 0.7$	$7.9 \pm 0.5 \pm 1.2$
DELPHI [60]	$10.6 \pm 0.1 \pm 0.4$	$8.3 \pm 0.3 \pm 0.8$
L3 [61]	$11.4 \pm 0.5 \pm 0.4$	-
L3 [62]	$10.7 \pm 0.1 \pm 0.4$	-
OPAL [63]	$10.6 \pm 0.6 \pm 0.7$	$8.4 \pm 0.4 \pm 0.7$
OPAL(n) [64]	$10.9 \pm 0.1 \pm 0.5$	$9.9 \pm 0.3 \pm 1.3$
LEP Average	11.04 ± 0.19	8.07 ± 0.34

Historically, theoretical predictions of the semileptonic b branching ratio have been significantly larger than the measured values. Traditionally $\mathcal{B}(b \to \ell)^{\text{TH}} \geq 12.5\%$ [67], which disagrees with the experimental values. Various aspects of this problem have been scrutinized. The inclusive semileptonic b branching ratio is defined as:

$$\mathcal{B}(b \to \ell) = \frac{\Gamma_{\text{semi-leptonic}}}{\Gamma_{\text{semi-leptonic}} + \Gamma_{\text{hadronic}} + \Gamma_{\text{leptonic}}}$$

with $\Gamma_{\text{hadronic}} = \Gamma(b \to c\bar{u}d) + \Gamma(b \to c\bar{c}s) + \Gamma(b \to \text{no charm})$ and $\Gamma(b \to \text{no charm}) = \Gamma(b \to s(d)\gamma) + \Gamma(b \to s(d)g) + \Gamma(b \to u\bar{u}d)$.

Solutions of the problem consist to find a way to increase the theoretical hadronic rate. Theoretical possible solutions are that where there could be an enhancement of:

- $\Gamma(b \to c\bar{u}d)$ due to non perturbative effects. But, in the same time, these models predict $\tau_{B^+}/\tau_{B^0} \simeq 0.8$ [68] which is in desagreement with the experimental lifetime ratio of 1.07 ± 0.04 previously presented in table 3.

- $\Gamma(b \to c\bar{c}s)$ due to large higher order QCD corrections [69–71]. In the same time, these models affect the average number of charm quarks per b decay, n_c, which consequently has to be also measured experimentally (see later).

- $b \to$ no open charm, which could be a sizable fraction of $b \to c\bar{c}s$ transitions [72]. The hypothesis is that a large component of low mass $c\bar{c}$ pairs are seen as light hadrons and not as open charm. n_c would not be increased by this mechanism.

- $\Gamma(b \to \text{no charm})$ e.g. large $\mathcal{B}(b \to s\gamma)$ or $\mathcal{B}(b \to sg)$, from some sources of new physics.

charm counting : Classical charm counting experiments consist in measuring the rates of the weakly decaying charm hadrons in selected b events. The $\Upsilon(4s)$ branching ratio are from CLEO [73–75] giving $n_c^{\Upsilon(4s)} = 1.119 \pm 0.053$. LEP measurements are from ALEPH [76] $n_c = 1.230 \pm 0.036 \pm 0.038 \pm 0.053$ and from OPAL [77] $n_c = 1.061 \pm 0.045 \pm 0.060 \pm 0.037$, where the last error is due to D branching ratio, largely correlated between the experiments. A main difference between the experiments are assumptions made about the unmeasured Ξ_c contribution, which is set to 0 in the case of OPAL, whereas ALEPH estimates it to be 0.063 ± 0.021. Accepting this last estimate and including also DELPHI measurements of D^0 and D^+ rates [78], the average result is $n_c^Z = 1.202 \pm 0.067$ [2,4].

New methods have been developed to estimate n_c which can be written as: $n_c = 1 + \mathcal{B}(B \to D\bar{D}) + \mathcal{B}(B \to \text{"hidden" } c\bar{c}) - \mathcal{B}(B \to \text{no } c)$, where the "hidden" $c\bar{c}$ is the contribution of bound states (e.g. J/ψ). DELPHI has determined the fraction of b decays into 0,1 and 2 charmed hadrons thanks to an analysis of the hemisphere b tagging probability distribution in terms of Monte Carlo expectations of the three components [79]. They have measured: $\mathcal{B}(b \to 2c) = 0.136 \pm 0.042$ and

$\mathcal{B}(b \to 0\,c) = 0.033 \pm 0.021$. Substracting the hidden charm contribution of 0.026 ± 0.004 [73,80] yields a charmless B branching ratio without hidden charm of $\mathcal{B}(b \to \text{no charm}) = 0.007 \pm 0.021$, to be compared with the Standard Model expectation of 0.016 ± 0.008 [81]. Imposing this SM value they have measured $n_c = 1.147 \pm 0.041 \pm 0.008$. An upper limit at 95% CL on new physics in charmless B decays is derived: $\mathcal{B}(b \to \text{no charm})^{\text{New}} < 0.037$. In another study, correlations of identified charged kaons with inclusively reconstructed D mesons were analysed [82]. A fit of the transverse momentum spectra of same and opposite sign K(extra)K(from D) samples has given $\mathcal{B}(b \to 2\,c) = 0.170 \pm 0.035 \pm 0.032$ and $\mathcal{B}(b \to \bar{D}D_s X)/\mathcal{B}(b \to 2\,c) = 0.84 \pm 0.16 \pm 0.09$.

CLEO has selected high momentum leptons, has studied D-lepton angular and charge correlations and has looked for wrong sign D [83]. They have measured the ratio of "upper vertex" charm to "lower vertex" charm : $\mathcal{B}(B \to DX)/\mathcal{B}(B \to \bar{D}X) = 0.100 \pm 0.026 \pm 0.016$. From that the number of D's produced at the upper vertex in B decays $\mathcal{B}(B \to DX) = 0.079 \pm 0.022$ is derived, as well as $\mathcal{B}(b \to c\bar{c}s) = 0.219 \pm 0.036$ and $n_c = 1.204 \pm 0.037$.

ALEPH has performed an inclusive analysis looking for doubly-charmed B decays, $B \to D\bar{D}(X)$, where D can be either a D^0, D^+, D^{*+} or a D_s, with both charmed mesons reconstructed [84]. The following branching ratios were measured:

$$\mathcal{B}(b \to D_s D^0, D_s D^\pm(X)) = (13.1^{+2.6}_{-2.2}(\text{stat})^{+1.8}_{-1.6}(\text{syst})^{+4.4}_{-2.7}(\mathcal{B}_D))\%$$
$$\mathcal{B}(b \to D^0\bar{D}^0, D^0 D^\pm(X)) = (7.8^{+2.0}_{-1.8}(\text{stat})^{+1.7}_{-1.5}(\text{syst})^{+0.5}_{-0.4}(\mathcal{B}_D))\%$$
$$\mathcal{B}(b \to D^\pm D^\mp(X)) < 0.9\% \text{ at } 90\% \text{ C.L.}$$

providing the first evidence for doubly-charmed B decays involving no D_s production. The sum of the inclusive DD rates is:

$$(20.9^{+3.2}_{-2.8}(\text{stat})^{+2.5}_{-2.2}(\text{syst})^{+4.5}_{-2.8}(\mathcal{B}_D))\% \text{ leading to } n_c = 1.219^{+0.061}_{-0.045}.$$

An evidence for associated K_S^0 and $K^\pm$ production in the decays $B \to \bar{D}D(X)$ was also found $\mathcal{B}(B \to \bar{D}^{(*)}D^{(*)}K) = (7.1^{+2.5}_{-1.5}(\text{stat})^{+0.9}_{-0.8}(\text{syst}) \pm 0.5(\mathcal{B}_D))\%$. which showed that $B \to \bar{D}^{(*)}D^{(*)}K$ is a large part of $B \to \bar{D}D(X)$ ($\approx 70\%$).

If previous n_c measurements could show some discrepancies between results obtained at LEP and at lower energy, the agreement is now better. Combining all these measurements leads to the world average $n_c = 1.178 \pm 0.021$.

$b \to s\gamma$: The flavour changig neutral current decay $b \to s\gamma$ has been seen in both exclusive and inclusive channels. The exclusive decay $B \to K^*\gamma$ has been measured by CLEO [85] $\mathcal{B}(B \to K^*\gamma) = (4.2 \pm 0.8 \pm 0.6)\,10^{-5}$ and ALEPH has placed an upper limit on the $B_s \to \Phi\gamma$ penguin decays [86] $\mathcal{B}(B_s \to \Phi\gamma) < 29\,10^{-5}$ at 90% CL. CLEO has first observed the inclusive electromagnetic penguin decay [87] $\mathcal{B}(b \to s\gamma) = (2.32 \pm 0.57 \pm 0.35)\,10^{-4}$ while ALEPH has published the first result at LEP. The signal was isolated in lifetime-tagged $b\bar{b}$ events by the presence of a hard photon associated with a system of a high momentum and rapidity hadrons [88]. $\mathcal{B}(b \to s\gamma) = (3.11 \pm 0.80 \pm 0.72)\,10^{-4}$ was measured. The average of these two measurements is $(2.54 \pm 0.57)\,10^{-4}$, consistent with the Standard Model expectation via penguin processes $(3.76 \pm 0.30)10^{-4}$ [89].

$b \to sg$: Large rates of $b \to sg$ [90] would show up an extra sources of charged kaons, especially visible at high momentum in the B rest frame. DELPHI has looked for a high $p_\perp$ kaon, identified with their RICH or dE/dX in the TPC in b tagged events [82], and derived an upper limit $\mathcal{B}(b \to sg) < 0.05$ at 95% C.L. In their wrong sign charm paper CLEO [83] has derived also $\mathcal{B}(b \to sg) < 0.068$ at 90% C.L.

To summarize, theoretical predictions on $\mathcal{B}(b \to \ell)$ can accommodate with lower value as predicted in the framework of $1/m_Q$ expansions with higher order perturbative QCD corrections. In these models the rate $b \to c\bar{c}s$ is increased while $\mathcal{B}(b \to \ell)$ is decreased. This is in agreement with the experimental situation. Other models, which predict new physics for instance, are disfavoured.

b-baryon semileptonic branching ratio : By determining the ratio $R_{\Lambda \ell} = \mathcal{B}(\Lambda_b \to \Lambda \ell^- X)/\mathcal{B}(\Lambda_b \to \Lambda X) = 0.070 \pm 0.012 \pm 0.007$ [91], using $\Lambda - \ell$ correlations, OPAL has a measurement of the Λ_b semileptonic branching ratio. ALEPH, on this side, has a measurement of the b-baryon semileptonic branching ratio, using $p - \ell$ correlations, by determining $R_{p\ell} = \mathcal{B}(b - \text{baryon} \to p\ell X)/\mathcal{B}(b - \text{baryon} \to pX) = 0.080 \pm 0.012 \pm 0.014$ [55]. Both can be assumed to be very similar to $\mathcal{B}(b - \text{baryon} \to \ell)$. They are significantly lower than the average $\mathcal{B}(b \to \ell)$. Combining them we get $\mathcal{B}(b - \text{baryon} \to \ell) = 0.074 \pm 0.011$. This confirms that light quarks play a significant role in the decay of b-baryons, as suggested by the short b-baryon lifetime measurements presented earlier. When correlated with this short lifetime, the agreement between the ratios $\tau_{b-\text{baryon}}/\tau_{B_d^0} = 0.78 \pm 0.04$ and $\mathcal{B}(b - \text{baryon} \to \ell)/\mathcal{B}(b \to \ell) = 0.67 \pm 0.10$ is consistent with the hypothesis of a constant semileptonic decay width for all b-hadrons.

$b \to \tau \nu_\tau (X)$: The study of this channel is interesting because the decay could proceed by a W or a Higgs boson. Thus a measurement of this decay channel could be sensitive to new physics, for example supersymmetry where two Higgs doublets are introduced. The SM prediction for $\mathcal{B}(b \to \tau \nu_\tau X)$ is $(2.30 \pm 0.25)\%$ [92], and its measurement limits the ratio $\tan\beta/m_{H^\pm}$, where $\tan\beta$ is the ratio of the vacuum expectation values for the Higgs fields and $m_{H^\pm}$ is the mass of the charged Higgs boson. DELPHI has performed a new measurement [93]. First b-tagging is used to obtain a sample of $Z \to b\bar{b}$ events. Then the events are required to have large missing energy and no electron or muon candidates. The result $\mathcal{B}(b \to \tau \nu_\tau X) = (2.52 \pm 0.23 \pm 0.49)\%$ is obtained, consistent with previous measurements from ALEPH: $(2.58 \pm 0.19 \pm 0.33)\%$ [94], L3: $(1.7 \pm 0.5 \pm 1.1)\%$ [95], and OPAL: $(2.58 \pm 0.11 \pm 0.51)\%$ [97]. The LEP average is $(2.52 \pm 0.26)\%$.

The fully leptonic $b \to \tau \nu_\tau$ decay is also very interesting. The expected branching ratio in the SM is $6 \cdot 10^{-5}$, however with large uncertainty. Because of helicity conservation the rates are proportional to the square of the lepton mass. The purely leptonic decays to e and μ are expected to have the following branching ratios: $\mathcal{B}(b \to e\nu_e) = 5 \cdot 10^{-12}$ and $\mathcal{B}(b \to \mu\nu_\mu) = 2 \cdot 10^{-7}$. No signal of $B \to \tau \nu_\tau$ decays is observed and upper limits are given at 90% CL: $\mathcal{B}(b \to \tau \nu_\tau) < 1.6 \cdot 10^{-3}$ from ALEPH [94], $1.1 \cdot 10^{-3}$ from DELPHI [93] and $5.7 \cdot 10^{-4}$ from L3 [96]. From this last limit the best constraint $\tan\beta/m_{H^\pm} < 0.38$ at 90% CL is obtainbed.

C $|V_{cb}|$ measurements

There are two main approaches to determine the magnitude of the CKM matrix element $|V_{cb}|$, either from inclusive semileptonic B decays or from exclusive channels such as $B \to D^{(*)}\ell\nu$.

The first one uses the measurement of the inclusive b semileptonic branching ratio and has the advantage of great statistical power. Treating the b quark as a free particle, its semileptonic partial width is

$$\Gamma(b \to c\ell\nu) = \frac{G_F^2\, m_b^5}{192\,\pi^3}\, \Phi\, |V_{cb}|^2 \equiv \alpha\, |V_{cb}|^2 = \frac{\mathcal{B}(b \to c\ell\nu)}{\tau_b}$$

where Φ is a phase space factor. The theoretical dominant uncertainties is dominated by the knowledge of the correction due to the binding of the b quark into a hadron, and the b quark mass dependence. Recent calculations [98], using HQET in combination with the technique of Operator Product Expansion, have rather small uncertainty and show that they are now under better control. $|V_{cb}|$ is then given by:

$$|V_{cb}| = 0.0419\sqrt{\frac{\mathcal{B}(B \to X_c\ell\bar\nu)}{0.105}}\sqrt{\frac{1.55}{\tau_B}}(1 \pm 0.015 \pm 0.010 \pm 0.012)$$

The measured branching ratios need to be corrected for the $b \to u$ contribution: $\frac{\mathcal{B}(b \to u\ell\nu)}{\mathcal{B}(b \to c\ell\nu)} \simeq 2\left|\frac{V_{ub}}{V_{cb}}\right|^2 = (1.5 \pm 1.0)\%$. The value of $|V_{cb}|$ [2] is $(38.7 \pm 2.1)\,10^{-3}$ at the $\Upsilon(4s)$ and $(40.6 \pm 2.1)\,10^{-3}$ at the Z.

The second technique for measuring $|V_{cb}|$ is based on the study of exclusive channels such as $B \to D^{(*)}\ell\nu$. It has less theoretical limitations. The rate of this process is governed by $|V_{cb}|$ and Heavy Quark Symmetry provides model independent relations between the relevant weak decay form factors in the heavy quark limit. The corrections from the heavy quark symmetry breaking are calculable in the framework of HQET. The differential decay rate of $B \to D^{(*)}\ell\nu$, with respect to the boost w ($w = (m_B^2 + m_D^2 - q^2)/(2\,m_B\,m_D)$) of the $D^{(*)}$ in the B rest frame, is given by $d\Gamma(B \to D^{(*)}\ell\nu)/dw = G(w)\,|V_{cb}|^2 \mathcal{F}^2(w)$ where $G(w)$ is a known phase space function and $\mathcal{F}(w)$ is a universal hadronic form factor. $\mathcal{F}(w)$ parametrizes the effects of the strong interaction on the decay, with an unknown shape. The product $|V_{cb}|\mathcal{F}(w)$ is then extrapolated to $w = 1$, which corresponds to the maximal value of q^2, by the expansion $\mathcal{F}(w) = \mathcal{F}(1)\,[1 - \rho^2(w-1) + c(w-1)^2 + ...]$. The intercept and slope are strongly correlated, and this needs to be accounted for when averaging results of different experiments. Furthermore, the expected value of $\mathcal{F}(1)$ is not exactly unity due to correction for the finite heavy quark mass: $\mathcal{F}(1)_{D^*\ell\nu} = 0.91 \pm 0.03$ for $B \to D^*\ell\nu$ decays [99], while $\mathcal{F}(1)_{D\ell\nu} = 0.98 \pm 0.07$

for $B \to D\ell\nu$ decays [100]. The decay $B \to D^*\ell\nu$ is favoured for the measurement, as the $1/m_Q$ correction is predicted to vanish, and experimentally it has a large branching ratio and clean signal. There is a new preliminary analysis from DELPHI, using a new parametrization [101] giving a more precise expansion versus the axial form factor, $\mathcal{A}(w) = \mathcal{A}(1) [1 + \rho_A^2 \mathcal{G}(w) + ...]$. ($\mathcal{A}(1) \equiv \mathcal{F}(1)$). They got $\mathcal{A}(1)V_{cb} = (37.7 \pm 1.7(stat) \pm 1.7(syst)) \, 10^{-3}$, $\rho_A^2 = 1.36 \pm 0.17 \pm 0.14$, $\mathcal{B}(B^0 \to \ell\nu D^*) = (5.18 \pm 0.16 \pm 0.49)\%$ and $|V_{cb}| = (41.4 \pm 3.0) \, 10^{-3}$. We can hope that in the future existing data will be re-analysed using this new parametrization. The previous published measurements are presented in table 5 and the average of the V_{cb} measurements from this channel, $B \to D^*\ell\nu$, is $|V_{cb}| = (38.7 \pm 3.1) \, 10^{-3}$ [107].

TABLE 5. Experimental values of $|V_{cb}|\mathcal{F}(1)$.

	$B \to D^*\ell\nu$	$B \to D\ell\nu$
ALEPH [102]	$(32.1 \pm 1.8 \pm 1.9) \, 10^{-3}$	$(28.2 \pm 6.8 \pm 6.5) \, 10^{-3}$
ARGUS [105]	$(39.2 \pm 3.9 \pm 2.8) \, 10^{-3}$	
CLEO [106]	$(35.2 \pm 1.9 \pm 1.8) \, 10^{-3}$	$(34.2 \pm 4.4 \pm 4.9) \, 10^{-3}$
DELPHI [103]	$(36.9 \pm 2.1 \pm 2.2) \, 10^{-3}$	
OPAL [104]	$(32.6 \pm 1.7 \pm 2.2) \, 10^{-3}$	

The decay $B \to D\ell\nu$ can also be used to measure $|V_{cb}|$ in a similar manner. Here there are fewer experimental results, as it is a more challenging mode, and only ALEPH [102] and CLEO [106] have used this channel (see table 5). Combining their results gives $|V_{cb}| = (39.4 \pm 5.0) \, 10^{-3}$ [107]. The experimental values for the intercept and the slope are combined with careful attention to the correlated errors.

The excellent agreement between a wide variety of methods for extracting V_{cb} is encouraging. The world average of all V_{cb} measurements leads to $|V_{cb}| = (39.5 \pm 1.7) \, 10^{-3}$. Improvements from theoretical and experimental sides are promising.

D $|V_{ub}|$ measurements

$|V_{ub}|$ can be extracted in a same way as $|V_{cb}|$ by looking to $b \to u$ transitions instead of $b \to c$ transitions. The inclusive and exclusive approaches can also be used. Due to the fact that a b decay is dominated by the process where the b quark turns into a c quark, the experimental determinations of $|V_{ub}|$ are very much difficult than those of $|V_{cb}|$. It is also more difficult from the theoretical point of vue because the u quark in the final state is no longer heavy.

The first observations for $b \to u$ transitions were done at the $\Upsilon(4s)$ were B mesons are produced at rest [108]. These analyses have looked at the endpoint of the single lepton spectrum for leptons from B decay that are kinematically incompatible with coming from the decay of a B meson to charm meson. From the lepton excess in this corner of phase space, theoretical models are used to extrapolate the full lepton spectrum and $|V_{ub}| = (3.1 \pm 0.8)10^{-3}$ is extracted by CLEO [109]. Model uncertainties dominate the error.

ALEPH has published the first evidence for semileptonic $b \to u$ transitions in b hadrons produced at LEP [110]. ALEPH has inclusively reconstructed the hadronic system accompagning the lepton in the semileptonic B decays and has built a set of kinematic variables to discriminate between $X_u \ell \nu$ and $X_c \ell \nu$ transitions by taking advantage of the different shape properties of these final states. A neural network was used to extract the inclusive $\mathcal{B}(b \to X_u \ell \nu)$ branching ratio. They have measured $\mathcal{B}(b \to X_u \ell \nu) = (1.73 \pm 0.55 \pm 0.55)\ 10^{-3}$. An advantage of this analysis is that it integrates over the entire lepton and hadron spectrum for these decays, potentially reducing the model dependence of the result. A disadvantage is that it has to manage with the large background from $b \to c$ semileptonic decays which needs to be well understood.

The same kind of calculations than for $|V_{cb}|$ have been done to extract $|V_{ub}|$ from $\mathcal{B}(b \to X_u \ell \nu)$ [98]

$$|V_{ub}| = 0.0465 \sqrt{\frac{\mathcal{B}(B \to X_u \ell \bar{\nu})}{0.002}} \sqrt{\frac{1.55}{\tau_B}} (1 \pm 0.025_{\text{pert}} \pm 0.03_{m_b})$$

The resulting value of $|V_{ub}|$ is $(4.16 \pm 1.02)\ 10^{-3}$.

$|V_{ub}|$ has also been extracted from the exclusive decays $B \to \pi \ell \nu$ and $B \to \rho \ell \nu$ at CLEO [111]. The value is $|V_{ub}| = (3.3 \pm 0.2 \pm 0.4 \pm 0.7)\ 10^{-3}$, where the errors are statistical, experimental systematic, and theoretical model dependence respectively.

All these $|V_{ub}|$ extractions are model dependent, however in different ways. The consistency of the results is thus comforting. The world average is then $|V_{ub}| = (3.4 \pm 0.5\ )\ 10^{-3}$

VI SUMMARY AND PROSPECTS

b physics is a place of intensive work. The e^+e^- colliders and Tevatron have provided a large amount of data on the production and decay of beauty particles. In recent years the experiments at LEP, SLC and CESR have turned the studies on the b quark into precision physics. Numerous interesting new results have been provided and a lot of measurements have been performed. Among these results:

- In the electroweak sector, R_b has been measured, $R_b^0 = 0.2173 \pm 0.0009$, to an accuracy of $\sim 0.4\%$, the forward backward charge asymmetry $A_{FB}^{0,b} = 0.0998 \pm 0.0022$ leads to a measurement of $\sin^2 \theta_{\text{eff}} = 0.23213 \pm 0.00039$ with an accuracy of $\sim 0.17\%$. One of the most precise determination of this quantity.

- In bottom spectroscopy, the B_c meson has been discovered by CDF at Tevatron.

- The B^+, B^0, B^0_s and b-baryon lifetimes, and their ratio with the B^0_d lifetime, have been measured with a good precision ($\sim 5\%$). The hierarchy among b hadron lifetimes agrees with theoretical predictions. But for the b baryon lifetime, predictions are higher than measurements by $\sim 3\sigma$. The semileptonic b baryon branching ratio confirms this puzzle ...

- The B^0_d-$\overline{B^0_d}$ oscillation frequency, $\Delta m_d = 0.466 \pm 0.018\,ps^{-1}$, is also measured with similar accuracy ($\sim 4\%$); but the hadronic uncertainty limits the extracted value of $|V_{td}| = (8.8 \pm 0.2_{\Delta m_d} \mp 0.2_{m_t} \mp^{1.4}_{1.8}{}_{F\sqrt{B}}) \times 10^{-3}$ to an accuracy of $\sim 20\%$.
  The B^0_s-$\overline{B^0_s}$ oscillation is still not seen and a limit $\Delta m_s > 10.2\,ps^{-1}$ at 95% CL has been set. Nevertheless non negligible constraints on the CKM matrix with $\frac{|V_{ts}|}{|V_{td}|} > 3.8$ at 95% CL is provided.

- b semileptonic decays have been scrutinized intensively. If the situation between $\mathcal{B}(b \to \ell)$ measurements at $\Upsilon(4s)$ and Z energies is still to be clarified, the agreement being of the order of 2σ, these low $\mathcal{B}(b \to \ell)$ values agree with theoretical expectations which favour high values of the number of charm quark per b decay, as experimentally measured. A significant contribution of decays in $b \to c\bar{c}s$ transition, without D_s is observed. No hints for new physics is found in b decays.

- The CKM matrix element $|V_{cb}| = (39.5 \pm 1.7)\,10^{-3}$ has been measured to an accuracy of $\sim 4\%$. Values of $|V_{cb}|$ show good agreement in results obtained from inclusive and exclusive studies. $|V_{ub}| = (3.4 \pm 0.5)\,10^{-3}$ has also been measured but to an accuracy of $\sim 15\%$ as it is a lot more challenging. Limitations are due to models.

The accuracy of present data provides important tests of the SM and guides the developements of the understanding of the rich phenomenology of weak b decays.

No more substantial running at the Z pole is foreseen at LEP and the final analyses with optimised algorithms, or new analysis ideas, are being prepared. LEP has shown its capability for studying b physics. The large amount of data, highly efficient detectors, and good particle identification have allowed a wide range of b physics to be explored. The Z pole is very competitive to study b physics. SLC is still running and can benefits for the polarization and excellent resolution of their vertex detector to provide interesting results. CESR will continue to improve its luminosity and CLEO has upgraded the detector and will continue to play a significant role in B physics. The experiments at the e^+e^- colliders have also prepared important engineering data and developed analysis techniques that will be further exploited in the continuation of beauty physics at the next generation of b facilities.

HERA-B, at DESY, will start to collect data in 1998. Next year, in 1999, BABAR and BELLE will start to take data in the new b-factories. CDF and D0 get upgraded and will take data at much higher luminosity at the Tevatron, with perhaps B-TEV in a few years. Then the LHC experiments will enter the scene in 2005, in particular LHCb, a specific detector to study b physics at pp collider, which will be able to do many analyses with large precision.

In conclusion there is a bright future for b physics in the next decade, with new dedicated or upgraded colliders and detectors, and with some challenges. The first observation of B_s^0-$\overline{B_s^0}$ oscillations and the measurement of Δm_s, the study of rare decays and of CP violation ...

ACKNOWLEDGEMENTS

I would like to thank the ALEPH, CDF, CLEO, DELPHI, D0, L3, OPAL and SLD collaborations, and their representatives who provided results for this review; and the LEP Electroweak Heavy Flavour, B-Lifetime and B-oscillation working groups. I am also grateful to all the people who help me in preparing this review, and in particular Z. Ajaltouni, C. Ferdi, S. Monteil, V. Morenas, P. Rosnet, O. Schneider, ... I would also like to thank the organisers of this conference.

REFERENCES

1. O. Schneider, *Proceedings of the XVIII International Symposium on Lepton-Photon Interactions,Hamburg, Germany, July 1997.*
2. P. S. Drell,*Proceedings of the XVIII International Symposium on Lepton-Photon Interactions,Hamburg, Germany, July 1997.*
3. J. D. Richman, *Proceedings of the 28th International Conference on High Energy Physics, Warsaw, Poland, July 1994.*
4. M. Feindt, *Proceeding of the International Europhysics Conference on High Energy Physics, Jerusalem, August 1997.*
5. LEP electroweak heavy flavour working group. "A Combination of Preliminary Electroweak Measurements and Constraints on the Standard Model", LEPEWWG/98-01.
6. LEP B lifetimes working group.
   http://wwwcn.cern.ch/~claires/lepblife.html
7. LEP B oscillations working group. Combined results on B^0 oscillations: update for winter conferences 1998, LEPBOSC 98/1, see also http://www.cern.ch/LEPBOSC/
8. ALEPH Coll., R. Barate *et al.*, Phys. Lett. **B 401** (1997) 150; Phys. Lett. **B 401** (1997) 163.
9. DELPHI Coll., *"Measurement of the partial decay width $R_b^0 = \Gamma_{b\bar{b}}/\Gamma_{had}$ with the DELPHI detector at LEP "*, DELPHI 97-106 CONF 88, contributed paper to International Europhysics Conference on High Energy Physics (EPS-HEP-97), Jerusalem, Israel, 1997, EPS-0383. **EPS-419**.
10. L3 Coll., *"Measurement of the Z Branching Fraction into Bottom Quarks Using Double Tag Methods"*, L3 Note 2114, contributed paper to the EPS-HEP-97, Jerusalem, **EPS-489**

11. L3 Coll., O. Adriani et al, Phys. Lett. **B307** (1993) 237.
12. OPAL Collaboration, K.Ackerstaff et al., Z. Phys. **C74** (1997) 1.
13. SLD Coll., SLAC-PUB-7585, contributed paper to EPS-HEP-97, Jerusalem, **EPS-118**.
14. ALEPH Coll., D. Buskulic et al., Z. Phys. **C62** (1994) 179.
15. DELPHI Coll., P.Abreu et al., Z. Phys. **C66** (1995) 323.
16. L3 Coll., "Measurement of R_b and $BR(b \to \ell X)$ from b-quark semileptonic decays", L3 Note 1449, July 16 1993.
17. OPAL Coll., G. Alexander et al, Z. Phys. **C70** (1996) 357.
18. DELPHI coll., P. Abreu et al., Phys. Lett. **B405** (1997) 202.
19. ALEPH coll., R. Barate et al., "A Measurement of the Gluon Splitting Rate into $b\bar{b}$ Pairs in Hadronic Z Decays", contributed paper to the EPS-HEP-97, Jerusalem, **EPS97-606**.
20. ALEPH Coll., D. Buskulic et al., Phys. Lett. **B384** (1996) 414.
21. DELPHI Coll., P.Abreu et al., Z. Phys **C65** (1995) 569;
    DELPHI Coll., P.Abreu et al., Z. Phys **C66** (1995) 341;
    DELPHI Coll., " Measurement of the Forward-Backward Asymmetries of $e^+e^- \to Z \to b\bar{b}$ and $e^+e^- \to Z \to c\bar{c}$", DELPHI 95-87 PHYS 522, contributed paper to EPS-HEP-95 Brussels **eps0571**.
22. L3 Coll., "Measurement of the $e^-e^- \to Z \to b\bar{b}$ Forward-Backward Asymmetry Using Leptons", L3 Note 2112, contributed paper to the EPS-HEP-97, Jerusalem, **EPS-490**.
23. OPAL Coll., G. Alexander et al., Z. Phys. **C70** (1996) 357; OPAL Coll., R. Akers et al., "Updated Measurement of the Heavy Quark Forward-Backward Asymmetries and Average B Mixing Using Leptons in Multihadronic Events", OPAL Physics Note PN226 contributed paper to ICHEP96, Warsaw, 25-31 July 1996 **PA05-007**.
24. ALEPH coll., R. Barate et al., "Determination of A_{FB}^b using jet charge measurements in Z decays", CERN EP/98-038, (Submitted to Phys. Let. B)
25. L3 Coll., "Afb(bb) using a jet-charge technique on 1994 data", L3 Note 2129
26. OPAL Coll., R. Akers et al., Z. Phys. **C67** (1995) 365;
    OPAL Coll., K.Ackerstaff et al., Z. Phys. **C75** (1997) 385.
27. OPAL Coll., G. Alexander et al., Z. Phys. **C73** (1996) 379.
28. I. Bigi et al., B decays, ed S. Stone, 2nd Ed., World Scientific, Singapore, 1994.
29. I. I Bigi, N. G Uraltsev and A. Vainshtein, Phys. Lett **B293** (1992) 430; **B297** (1993) 477; N. Blok and M. Shifman, Nucl. Phys. **B399** (1993) 441.
30. I. I. Bigi, Nuovo Cimento **109A** (1996) 713.
31. M. Neubert, "Theory of beauty lifetimes" CERN-TH/97-148, hep-ph/9707217.
32. CDF coll. http://www-cdf.fnal.gov/physics/new/bottom/bottom.html
33. OPAL coll., "Measurements of the B_s^0 and Λ_b lifetimes" CERN-PPE/97-159.
34. DELPHI coll. "Search for $B_s^0 - \bar{B}_s^0$ oscillations and measurement of the B_s^0 lifetime" contributed paper to the EPS-HEP-97, Jerusalem, **EPS97-457**; "Search for $B_s^0 - \bar{B}_s^0$ oscillations", Phys. Lett. **B414** (1997) 382.
35. DELPHI coll., P. Abreu et al., Z. Phys. **C74** (1997) 19.
36. L3 coll., "Measurement of the B_d^0 meson lifetime using the decay $B_d^0 \to D^{*+}X\ell\nu$", contributed paper to the EPS-HEP-97, Jerusalem, **EPS97-495**.

37. ALEPH coll., R. Barate *et al.*, "Measurement of the b baryon lifetime and branching fractions in Z decays", European Physical Journal C 2 (1998) 2, 197.
38. DELPHI coll., "Determination of average b-baryon lifetime at LEP", DELPHI 97-104 CONF 86, contributed paper to the EPS-HEP-97, Jerusalem, **EPS97-454**, updated for Moriond 98.
39. OPAL coll., K. Ackerstaff *et al.* "A measurement of the B_s^0 lifetime using recontructed D_s mesons" CERN-PPE/97-095.
40. SLD coll., "Measurement of the B^+ and B^0 lifetimes using tological vertexing at SLD" SLAC-PUB-7635, contributed paper to the EPS-HEP-97, Jerusalem, **EPS97-127**.
41. SLD coll., K. Abe *et al.* Phys. Rev. Lett. **79** (1997) 590.
42. L3 coll., M. Acciari *et al.*, "Measurement of the average lifetime of B-hadrons in Z decays", Phys. Lett. **B416** (1997) 220.
43. R. M. Barnett *et al.*, Particle Data Group, Phys. Rev. D**54** (1996) 1.
44. SLD coll., "Measurement of time-dependent B_d^0-$\overline{B_d^0}$ mixing using inclusive semileptonic decays", SLAC-PUB-7228, "Measurement of time-dependent B_d^0-$\overline{B_d^0}$ mixing using topology and charge selected semileptonic B decays", SLAC-PUB-7229, "Preliminary measurements of the time dependence of B_d^0-$\overline{B_d^0}$ mixing with kaon and charge dipole tags", SLAC-PUB-7230.
45. ALEPH coll., "Inclusive lifetime and mixing measurements using topological vertexing", contributed paper to the EPS-HEP-97, Jerusalem, **EPS97-596**.
46. CDF coll., "Observation of π–B meson charge-flavour correlations and measurement of time dependent B^0-$\overline{B^0}$ mixing in p$\bar{\text{p}}$ collisions", FERMILAB-CONF-96/175-E.
47. ALEPH coll., "Study of B_s^0 oscillations using fully reconstructed D_s^+ decays", CERN PPE/97-157; "Search for B_s^0 oscillations using inclusive lepton events", contributed paper to the EPS-HEP-97, Jerusalem, **EPS97-612**.
48. DELPHI coll., W. Adam *et al.*, "Search for $B_s^0 - \bar{B}_s^0$ oscillations" Phys. Lett. **B414** (1997) 382.
49. OPAL coll. K. Ackerstaff *et al.*, Z. Phys. **C76** (1997) 401; Phys. **C76** (1997) 417.
50. ALEPH coll., D. Buskulic *et al.*, Z Phys. **C75** (1997) 397.
51. DELPHI coll., P. Abreu *et al.*, Zeit. Phys. **C76** (1997) 579.
52. L3 Coll., M. Acciarri *et al.*, "Measurement of the $B_d^0 - \bar{B}_d^0$ Oscillation Frequency", CERN-EP/98-028.
53. OPAL coll. G. Alexander *et al.*, Z. Phys. **C72** (1996) 377.
54. H.-G. Moser and A. Roussarie Nucl. Inst. Meth. **A384**(1997) 491.
55. ALEPH coll., R. Barate *et al.*, " A measurement of the semileptonic branching ratio BR(b-baryon $\to p l \nu$X) and a study of inclusive $\pi^\pm$, $K^\pm$, $(p,\bar{p})$ production in Z decays, CERN PPE/97-153, (Submitted to European Physical Journal C)
56. DELPHI coll., "A direct measurement of the branching fractions of the b-quark into strange, neutral and charged B-mesons", contributed paper to the EPS-HEP-97, Jerusalem, **EPS97-451**.
57. DELPHI coll., P. Abreu *et al.*, "Measurement of the charged particle multiplicity of weakly decaying B-hadrons", CERN-EP/98-34. (Submitted to Phys. Lett. B).
58. ALEPH Coll.,*Contributed paper to International Europhysics Conference on High Energy Physics, Brussels,1995*, EPS95 **eps0404**.

P. Perret, "Measurement of heavy quark electroweak properties in Z decays using inclusive leptons at LEP", Proceeding of the 1995 International Europhysics Conference on High Energy Physics, Brussels, edited by J. Lemonne et al. (World Scientific, New Jersey, 1996).

59. ALEPH Coll., D. Buskulic et al., Phys. Lett. **B 313** (1993) 535.
60. DELPHI coll., "Measurement of the semileptonic b branching ratios and $\bar{\chi}_b$ from inclusive leptons in Z decays", DELPHI 97-118 CONF 100, contributed paper to the EPS-HEP-97, Jerusalem, **EPS97-415**.
61. *L3 Results on R_b and $BR(b \to \ell)$ for the Glasgow Conference* L3 Note 1625 and references therein.
62. L3 Collab., M. Acciarri et al., Z. Phys. **C71** (1996) 379.
63. OPAL Coll., R. Akers et al., Z. Phys. **C60** (1993) 199.
64. OPAL Coll., "Measurement of the semileptonic branching fration of inclusive b hadrons", OPAL Physics Note PN334 (10 March 1998).
65. ARGUS coll., H. Albrecht *et al.*, Phys. Lett. **B318** (1993) 397.
66. CLEO coll., B. Barish *et al.*, Phys. Rev. Lett. **76** (1996) 1570.
67. I. I. Bigi it et al., Phys. Lett. **B 323** (1994) 408.
68. K. Honscheid it et al., Z. Phys. **C 63** (1994) 117.
69. E. Bagan it et al., Phys. Lett. **B 351** (1995) 546.
70. E. Bagan it et al., Phys. Lett. **B 342** (1995) 362; erratum-*ibid.* **B 374** (1996) 363.
71. M. B. Voloshin, Phys. Rev. **D 51** (1995) 3948.
72. I. Dunietz it et al. FERMILAB-PUB-96/421-T.
73. CLEO coll., L. Gibbons *et al.*, Phys. Rev. **D 54** (1997) 3783.
74. T. Browder and K. Honscheid, Prog. in Part. and Nucl. Phys. **35**, 81 (1995).
75. T. Browder, *Proceedings of the 7th International Symposium on Heavy Flavor Physics, Santa Barbara, California, July 1997.*
76. ALEPH Coll., D. Buskulic et al., Phys. Lett. **B 388** (1996) 648.
77. OPAL Coll., G. Alexander *et al.*, Z. Phys. **C72** (1996) 191.
78. DELPHI coll., P. Abreu *et al.*, Z. Phys. **C59** (1993) 533. Contributed paper to ICHEP'96 pa01-058.
79. DELPHI coll., P. Abreu *et al.*, "Measurement of the inclusive charmless and double-charm B branching ratios" CERN-EP/98-07. (Submitted to Phys. Lett. B).
80. G. Buchalla *et al.* Phys. Lett. **B364** (1995) 188.
81. A. Lenz *et al.* Phys. Rev. **D56** (1997) 7228.
82. DELPHI coll., "First Measurement of the Inclusive Charmless B Decay Branching Ratio and Determination of the $b \to c\bar{c}s$ Rate", DELPHI 97-80 CONF 66, submitted to HEP 97 conference, Jerusalem, Aug. 1997, EPS97-448.
83. CLEO coll., S. Glenn *et al.*, contributed paper to the EPS-HEP-97, Jerusalem, **EPS-0383**.
84. ALEPH coll., R. Barate *et al.*, "Observation of doubly-charmed B decays at LEP.", CERN EP/98-037, (Submitted to European Physical Journal C)
85. CLEO coll., R. Ammar *et al.*, Phys. Rev. Lett. **71** (1993) 674; CLEO CONF 96-05.
86. A. M. Litke in *Proceedings of the 27th International Conference on High Energy Physics, Glasgow, Scotland, July 1994.* ALEPH coll., R. Barate *et al.*,
87. CLEO coll., M. S. Alam *et al.*, Phys. Rev. Lett. **74** (1995) 2885.

88. ALEPH coll., R. Barate et al., "A measurement of the inclusive $b \to s\gamma$ branching ratio", CERN EP/98-044, (Submitted to Phys. Lett. B).
89. A. Ali, "Theory of rare B decays", DESY 97-192, (*to be published in the Proceedings of the 7th International Symposium on Heavy Flavor Physics, Santa Barbara, California, July 1997.*)
90. A. L. Kagan and J. Rathsman, "Hints for Enhanced $b \to sg$ from Charm and Kaon Counting", hep-ph/9701300.
91. OPAL Coll., K. Ackerstaff et al., Z. Phys. **C74** (1997) 423
92. A. Falk et al., Phys. Lett. **B326** (1994) 145.
93. DELPHI coll., DELPHI note submitted to XXXI Rencontres de Moriond 1998.
94. ALEPH Coll., D. Buskulic et al., Phys. Lett. **B343** (1995) 444.
95. L3 coll., M. Acciari et al., Phys. Lett. **B332** (1994) 201.
96. L3 coll., M. Acciari et al., Phys. Lett. **B396** (1997) 327.
97. OPAL Coll., "Measurement of the Branching Ratios $b \to \tau^- \bar{\nu}_\tau X$ and $b \to D^{*+}\tau^- \bar{\nu}_\tau(X)$", OPAL Physics Note PN 209.
98. I. Bigi, M. Shifman and N. Uraltsev, "Aspects of Heavy Quark Theory", TPI-MINN-97/02-T hep-ph/9703290.
99. M. Shifman, N. Uraltsev and A. Vainshtein, Phys. Rev. **D51** (1995) 2217;
    A. F. Falk and M. Neubert, Phys. Rev. **D47** (1993) 1965.
100. M. Neubert, Phys. Lett. **B338** (1994) 84.
101. I. Caprini, L. Lellouch and M. Neubert, "Dispersive Bounds on the Shape of $\overline{B} \to D^* l \bar{\nu}$ Form Factors", CERN-TH-97-091, (*Submitted to Nucl. Phys., B*).
102. ALEPH coll., D. Buskulic et al., Phys. Lett. **B395** (1997) 373.
103. DELPHI coll., P. Abreu et al., Z. Phys. **C71** (1996) 539.
104. OPAL Coll., K. Ackerstaff et al., Phys. Lett. **B395** (1997) 128.
105. ARGUS coll., H. Albrecht et al., Phys. Rept.. **276** (1996) 223.
106. CLEO coll., B. Barish et al., Phys. Rev. **D 51** (1995) 1014.
107. L. K. Gibbons, *Proceedings of the 28th International Conference on High Energy Physics, Warsaw, Poland, July 1996.*
108. CLEO coll., R. Fulton et al., Phys. Rev. Lett. **64** (1990) 16;
    J. Bartelt et al., Phys. Rev. Lett. **71** (1993) 4111.
    ARGUS coll., H. Albrecht et al., Phys. Lett. **B234** (1990) 409;
    H. Albrecht et al., Phys. Lett. **B255** (1991) 297.
109. K. Berkelman, Nucl. Phys. B (proc. Suppl.) 66 (1998) 447.
110. ALEPH coll., R. Barate et al., "Determination of $|V_{ub}|$ from the Measurement of the Inclusive Charmless Semileptonic Branching Ratio of b hadrons", CERN EP/98-067, (Submitted to Euro. Phys. Jour. C).
111. CLEO coll., J. Alexander et al., Phys. Rev. Lett. **77** (1996) 5000.

Tevatron Results on the Top Quark

Natalia Sotnikova (for the DØ and CDF collaborations)

*Nuclear Physics Institute, Moscow State University,
119899 Moscow, Russia*

Abstract. The CDF and DØ collaborations have collected over 110 pb^{-1} of data each. From these data samples, both experiments have measured the top quark mass and pair production cross section in a variety of decay channels, including the dilepton, lepton plus jets and all jets channels. The combined top quark mass from both experiments is $m_t = 173.5 \pm 5.2$ GeV/c^2. Additionally, upper limits have been obtained for certain rare or nonstandard (FCNC and charged Higgs) top quark decays.

I INTRODUCTION

The top quark was predicted by the Standard Model of particle physics, and observed in 1995 by the CDF and DØ experiments at the Fermilab Tevatron. Since the discovery of the top quark [1,2], the DØ and CDF collaborations have continued exploring the top sector by refining their measurements of the $t\bar{t}$ production cross section ($\sigma_{t\bar{t}}$) and top mass (m_t) and by initiating other activities such as the measurement of V_{tb}, the search for the top quark rare decays. The results presented here are based on the full Tevatron run I (1992-1996) data sets of 110 and 125 pb^{-1} for CDF and DØ respectively.

At the Tevatron energy of $\sqrt{s} = 1.8$ TeV the dominant top quark production mechanism is $t\bar{t}$ pair production via $q\bar{q}$ annihilation ($\sim 90\%$) or gg fusion ($\sim 10\%$), with the QCD predicted cross section of ~ 5 pb [3]. Top quarks should also be produced singly in the electroweak processes from a Wtb vertex, but the single top quark production cross section is predicted to be 2 times lower than the $t\bar{t}$ one. Analyses are underway to extract this signal at both CDF and DØ.

According to the Standard Model, top quark undergo the weak decay, almost exclusively into Wb ($|V_{tb}| \approx 1$). The W-boson will subsequently decay into fermion pairs, either $W \to \ell\nu$ or $W \to q\bar{q}$, where ℓ denotes a charged lepton and $q\bar{q}$ denotes a light-quark pair, $u\bar{d}$ or $c\bar{s}$. The final states are therefore classified by the decay of the W bosons and these primary channels are the *dilepton* (both W's decay leptonically), the *lepton+jets* (one W decays leptonically, one W decays hadronically), and the *all jets* (both W's decay hadronically). As for τ channels, since $W \to \tau\nu_\tau \to e/\mu\nu_{e/\mu}\nu_\tau$ decays are essentially indistinguishable from $W \to$

$e/\mu\nu_{e/\mu}$ decays, both experiments include such events as a part of the leptonic W decays in the dilepton or lepton+jets channels as appropriate. Other channels are the $\ell\tau$ (e or μ plus a hadronically decaying τ) dilepton channel (CDF) and the $e\nu$ channel (DØ).

II CROSS SECTION MEASUREMENTS

The $\sigma(p\bar{p} \to t\bar{t})$ measurement provides a test of QCD predictions at the high-mass scale associated with the top quark ($Q^2 \approx m_t^2$). A deviation from these predictions could indicate non-standard production or decay modes. The $t\bar{t}$ production cross section is determined as a function of m_t from the number of events observed in each of the counting experiments to obtain independent measures, which are then combined to obtain the best estimate of $\sigma_{t\bar{t}}$ using a maximum-likelihood method.

For each channel $\sigma_{t\bar{t}} = (N - B)/(\epsilon_{total} * \mathcal{L})$, where N is the number of observed events, B is the number of expected background, $\mathcal{L}$ is the integrated luminosity, and ϵ_{total} is the total acceptance consisting of the geometrical and kinematic acceptances including the branching fraction, and the trigger efficiency.

Although the CDF and DØ analyses differ somewhat in detail, their principal features are similar enough. The initial phase of the experimental search consists of the event selection and the estimation of expected number of events from background sources.

A Dilepton Channel

The searches in the dilepton channel are optimized for the decay chain
$$t\bar{t} \to W^+bW^-\bar{b} \to \ell^+\nu b \ell^- \bar{\nu} \bar{b},$$
where ℓ represents either an electron or a muon. This gives the characteristic signature of two isolated, high transverse momentum leptons, missing transverse energy from the two neutrinos, and at least two hadronic jets from the fragmentation and decay of the b-quarks and any initial-state gluons.

$ee, e\mu$, and $\mu\mu$ channels

Experimentally, these channels have the cleanest signal, but because of the small branching ratio ($\sim 5\%$ for $ee + e\mu + \mu\mu$ channels), the expected number of events is very small. The selection criteria are summarized in Table 1. These cuts reduce the contribution from $Z \to \ell\ell$, W^+W^- processes, Drell-Yan events, multijets production and the instrumental fakes. DØ makes a cut on H_T, which is defined as the scalar sum of the transverse energies (E_T) of the jets (for the $\mu\mu$ channels) or H_T^e (the sum of the E_T's of the leading electron and the jets, for the ee and $e\mu$ channels).

CDF observes 9 events (7 $e\mu$, 1 ee, 1 $\mu\mu$) passing all cuts with the expected background of 2.4 ± 0.5 and the expected top yield of 6.3 ± 0.6 events (at $\sigma_{t\bar{t}} = 7.5$ pb) [4]. DØ observes 5 events (3 $e\mu$, 1 ee, 1 $\mu\mu$) passing all cuts with the expected background of 1.4 ± 0.4 and the expected top yield of 4.1 ± 0.7 events (at $m_t = 170$ GeV/c^2) [5].

Cut	DØ	CDF
$N(\ell)$	2	2
$p_T^\ell \geq$	15, 20 (ee)	20
$|\eta|_\ell \leq$	2.5(e), 1.7(μ)	1.05
$\not{E}_T \geq$	20($e\mu$), 25(ee)	25, 50 if $\Delta\Phi < 20°$
$N(j) \geq$	2	2
$E_T^j \geq$	20	10
$|\eta|_j \leq$	2.5	2.0
$|M_{\ell\ell} - M_Z| \geq$	12 if $\not{E}_T <40$, $\mu\mu \neq Z$	15
$H_T^e \geq$	120(ee,$e\mu$)	-
$H_T \geq$	100($\mu\mu$)	-

TABLE 1. Dilepton channels $ee, e\mu, \mu\mu$: event selection.

$e\nu$ channel

To increase acceptance for $t\bar{t}$ production, the DØ collaboration has searched for so called $e\nu$ channel, which requires an isolated high E_T electron, $\not{E}_T > 50$ GeV, transverse mass of $e\nu$, $M_T^{e\nu} > 115$ GeV/c^2, and two or more jets with $E_T > 30$ GeV. The $e\nu$ channel contains top signal mainly from dileptons and e+jets top decays which fail the standard kinematic selection. DØ observes 4 events passing these cuts with an expected background of 1.2±0.4 and an expected top yield of 1.7±0.5 events (at m_t= 170 GeV/c^2) [5].

τ-decay channel

The CDF collaboration has searched for dilepton events from $t\bar{t}$ production with one electron (muon) and one hadronically decaying τ lepton from the decay $t\bar{t} \to (\ell\nu_\ell)(\tau\nu_\tau)b\bar{b}, (\ell = e, \mu)$ [6]. The analysis starts with events containing a high-p_T electron or muon ($p_T \geq 20$ GeV/c, $|\eta| \leq 1.0$), two jets with $E_T^{\text{jet}} \geq 10$ GeV, $|\eta^{\text{jet}}| \leq 2$, "$\not{E}_T$ significance" $> 3\text{GeV}^{1/2}$, and $H_T \geq 180$ GeV (where $H_T \equiv E_T^e/p_T^\mu + p_T^\tau + \not{E}_T + \Sigma_{\text{jets}} E_T$ for this channel). The first technique searches then the one-pronged hadronic τ-lepton decays by searching for isolated single tracks that are inconsistent with being either an electron or a muon. The second analysis searches for either one or three-prong τ-decays and relies on identifying τ-leptons by their calorimeter signature. Background events come from $Z \to \tau\tau$ decays and events containing jets that fake the τ signature.

Using the track (calorimeter) based τ id, CDF observes 4(4) events with an expected background of 1.3 ± 0.3 (2.5 ± 0.4) and an expected top yield of 0.7 ± 0.3 (1.1 ± 0.4). The same four events are observed in both analyses and three of these candidate events contain b-tags.

Table 2 shows the $t\bar{t}$ production cross section for the different dilepton channels.

Channels	DØ at 172 GeV/c^2	CDF at 175 GeV/c^2
ee, $e\mu$, $\mu\mu$	5.1 ± 3.3 pb	$8.2^{+4.4}_{-3.4}$ pb
$e\nu$	9.6 ± 7.5 pb	–
$\ell\tau$	–	$10.2^{+16.3}_{-10.2}$ pb

TABLE 2. $\sigma_{t\bar{t}}$ cross section for dilepton channels.

B Lepton+Jets Channels

The searches in the lepton + jets channel are optimized for the decay chain
$$t\bar{t} \to W^+ b W^- \bar{b} \to \ell^+ \nu b q \bar{q} \bar{b},$$
where ℓ represents an electron (muon) and $q\bar{q}$ is a light quark pair from the decay of the second W boson. The characteristic signature is, therefore, one isolated, high-p_T lepton, missing energy from the neutrino, and four jets from the fragmentation of the $q\bar{q}$ pair and the two b-quarks. The lepton+jets channels are categorized according to how the rejection against the W+jets background is achieved: (1) a topological tag used only by DØ, (2) a displaced vertex tag used only by CDF, and (3) a soft-lepton-tag (SLT) used by both DØ and CDF. Approach (1) exploits the differences between the topology of the signal and that of the background. Approaches (2) and (3) attempt to identify whether or not there are any b-quarks in the event (backgrounds will only rarely contain b-quarks).

Topological Tagging

Topological selection of $t\bar{t}$ events relies on the fact that the jets from heavy top quark decay will be both more centrally produced and more energetic than the jets from W + multijet background. The event selection details are shown in Table 3. In this analysis, the variables that provide significant discrimination between $t\bar{t}$ and background events are H_T; the aplanarity $\mathcal{A}$ computed using the W boson and jet momenta in the laboratory frame; and E_T^L, the scalar sum of the lepton E_T and $\not{E}_T$. H_T measures the transverse activity in the event and a large value of H_T indicates on the massive object decay. The aplanarity $\mathcal{A}$ measures the activity transverse to the plane of maximum activity and is proportional to the lowest eigenvalue of the momentum tensor for the observed objects. The $t\bar{t}$ events tend to be more spherical than the W + multijet background. A requirement on the pseudorapidity η_W of the W boson which decays leptonically is also imposed to obtain better agreement between background control samples from data and the W+jets Monte Carlo samples. Figure 1 shows plots of the two kinematic variables $\mathcal{A}$ and H_T, after imposing all cuts except those on the variables plotted for the DØ ℓ+jets data sample, $t\bar{t}$ MC, and the two background sources (QCD multijet and W+4 jets events).

The cuts indicated by the dashed lines provide good separation between the

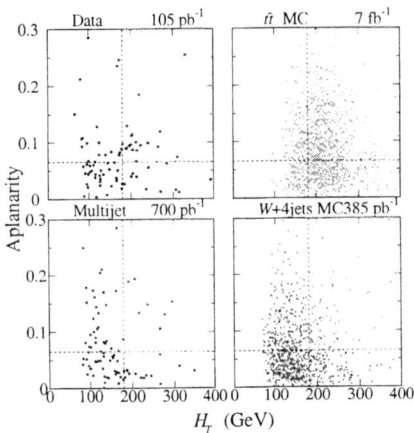

FIGURE 1. Distribution of $\mathcal{A}$ vs. H_T for DØ lepton+jets data events compared to those expected for higher luminosity samples of top MC ($m_t = 170$ GeV/c^2), QCD multijet, and W+jets backgrounds. The dashed lines represent the threshold values for the selection.

expected signal and backgrounds. DØ observes 19 events passing this selection with an expected background of 8.7 ± 1.7 and an expected top yield of 14.1 ± 3.1 events (at $m_t = 170$ GeV/c^2).

b-quark tagging techniques

The first method is used only by CDF and relies on the long lifetime (≈ 1.6 ps) of B hadrons and the large p_T of the b-quark. This combination gives rise to a B hadron decay vertex, which can be separated from the primary interaction point of the proton and antiproton by using the precision Silicon Vertex detector (SVX). For $m_t \sim 175$ GeV/c^2, the average separation is ~ 0.5 cm. The efficiency for tagging at least one b-quark in a $t\bar{t}$ event with ≥ 3 jets is determined to be $41 \pm 4\%$. The full selection criteria are given in Table 3. On the 109 pb^{-1} CDF observes 34 events passing the 3 jet requirement and SVX cut, which have a total of 42 tagged b-jets, with an expected background of 9.2 ± 1.5 events and an expected top yield of 31.0 ± 5.9 (at $\sigma_{t\bar{t}} = 7.5$ pb) [7].

Both CDF and DØ have performed searches for additional non-isolated low-p_T leptons from the semileptonic b-decay. Since 44% of top events will contain a $b \to \mu$ or a $b \to c \to \mu$ (and likewise for e), this is a very useful method for improving the signal to background ratio. The selection criteria are also shown in Table 3; both the experiments have a tagging efficiency of 20% on Monte Carlo top events. DØ observes 11 events with a combined background of 2.4 ± 0.5 and an expected top yield of 5.9 ± 0.8 ($m_t = 170$ GeV/c^2) [5], and CDF observes 44 b tags in 40 events (there are 11 events which are common to both the SVX and soft-lepton-tag set) with a predicted background of 24.3 ± 3.5 and a expected top yield of 14.3 ± 2.7.

Cut	DØ(topological)	DØ(SLT)	CDF (SLT)	CDF (SVX)
$N(\ell)$	1	1	1	1
$p_T^\ell \geq$	20	20	20	20
$\|\eta\|_\ell \leq$	2.0(e), 1.7(μ)	2.0(e), 1.7(μ)	1.05	1.05
$\displaystyle{\not}E_T \geq$	25(e), 20(μ)	20	20	20
$N(j) \geq$	4	3	3	3
$E_T^j \geq$	15	20	15	15
$\|\eta\|_j \leq$	2.0	2.0	2.0	20
$\|\eta_W\| \leq$	2.0	-	-	-
Shape	$H_T \geq 180, \mathcal{A} \geq 0.065, E_T^L \geq 60$	$H_T \geq 110, \mathcal{A} \geq 0.04$	-	-
SLT	veto μ tag events	$p_T^\mu \geq 4, \Delta R(\mu, j) \leq 0.5$	$2 \leq p_T^\ell \leq 20$	-
SVX	-	-	-	yes
Energies (momenta) are measures in GeV (GeV/c)			

TABLE 3. DØ and CDF: topological, ℓ+jets soft-lepton-tag(SLT) and displaced-vertex(SVX) selection.

Table 4 shows the $t\bar{t}$ production cross section for the lepton+jets channel.

DØ at 172 GeV/c^2		CDF at 175 GeV/c^2	
topological	SLT	SLT	SVX
4.1 ± 2.1 pb	8.3 ± 3.6 pb	$9.2^{+4.3}_{-3.6}$ pb	$6.2^{+2.1}_{-1.7}$ pb

TABLE 4. $\sigma_{t\bar{t}}$ for lepton+jets channel.

C All Jets Channel

The signature for the production of $t\bar{t}$ events in the all-jets channel is six or more high-p_T jets with no significant missing transverse energy:
$$t\bar{t} \to W^+ b W^- \bar{b} \to q\bar{q} b q\bar{q} \bar{b},$$
where $q\bar{q}$ pairs are light quark pairs from hadronic decays of the two W bosons. This channel has a largest branching fraction ($\sim$ 44 %) but there is a very large background from QCD multijet production that is kinematically quite similar to the $t\bar{t}$ signal. A significant background reduction is achieved by requiring the presence of a b-tagged jet as well as applying cuts on the event shape and kinematics.

Both the experiments have presented preliminary results from the analysis of this channel. CDF requires for the 6-jet events that the total scalar transverse energy in the event ΣE_T(jets) $\geq$ 300 GeV, that ΣE_T(jets)$/\sqrt{s} \geq 0.75$, and that there be at least one displaced vertex b tag. In addition, further cuts based on event topology (e.g. aplanarity, minimum separation between the jets) are applied [8]. For a data sample of 109 pb[1], 187 SVX tagged events are observed with an expected background of 151±5.9 events. Double SVX tagged analysis yields 157 tagged events with an expected background of 122.7±13.4 events. All jets combined result for the cross section extracted from these analyses is found to be $\sigma_{t\bar{t}}= 10.1^{+4.5}_{-3.6}$ pb, at $m_t = 175$ GeV/c^2.

DØ employs a Neural Network (NN) analysis with 13 input variables on events with six jets and a b-tagged jet [9]. These inputs are based on sophisticated topological variables such as centrality, sphericity, aplanarity, and variants of H_T. One of the inputs is the jet width which attempts to discriminate between quark jets and gluon jets, as $t\bar{t}$ events have predominantly quark jets as opposed to the background multijet process. From a simultaneous fit of the data to the expected signal and background, DØ measures the cross section to be $\sigma_{t\bar{t}}= 7.1 \pm 2.8 \pm 1.5$ pb.

The measured $t\bar{t}$ production cross section are summarazed in Fig.2 together with the current range of theoretical expectations [3], [15].

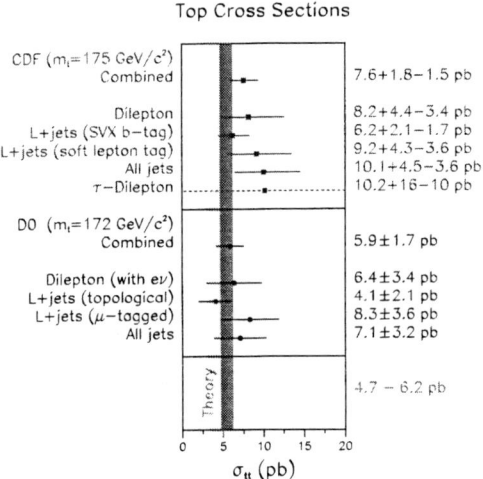

FIGURE 2. Summary of DØ and CDF cross section results and theoretical expectations.

III MASS MEASUREMENTS

A Lepton+Jets mass measurements

Both CDF and DØ have made several determinations of the top quark mass by using a variety of fitting techniques and data samples. The most precise of these are based on a two-constraint fit, which reconstructs the t and $\bar{t}$ in a subset of the lepton + jet events having four or more hadronic jets [10], [11]. The analysis starts with the events required to have one charged lepton (e or μ), four or more jets, and missing E_T, and when it is available, it makes use of b-tag information. Since the discovery period, both experiments have modified the event selection used in the mass analyses to minimize any mass biases inherent in the selection.

DØ introduced four mass insensitive variables [11], which individually provide some separation between signal and background; they are then combined into a multivariant discriminant ($0 \leq \mathcal{D} \leq 1$) which can be considered as a measure of the top probability and gives excellent separation between signal and background with essentially no correlation with top mass. DØ uses two methods (LB and NN) to obtain the discriminant values $\mathcal{D}$. For the LB ("low bias") method, the discriminant is constructed from a log likelihood function ($\mathcal{L}$) based on the relative densities of the signal and background samples as a function of these four variables. Alternatively, the NN ("Neural Network") method inputs these four variables into an artificial neural network and outputs the discriminant. The data, expected signal, and background samples are then binned in the ($\mathcal{D}, m_{\text{fit}}$) plane and a fit is made of the signal + background models to the data. The number of events per bin are shown in Fig. 3 for events passing and failing the LB cut. The likelihood values as a function of m_t for both methods are also shown. Combining the two results of the fits, taking into account the correlations (88%), DØ determines the top quark mass in the lepton+jets channels to be $m_t = 173.3 \pm 5.6(\text{stat}) \pm 5.5(\text{sys})$ GeV/c^2.

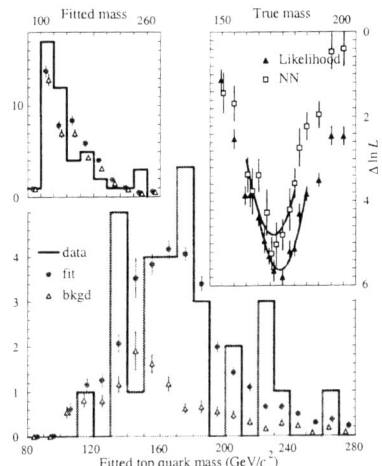
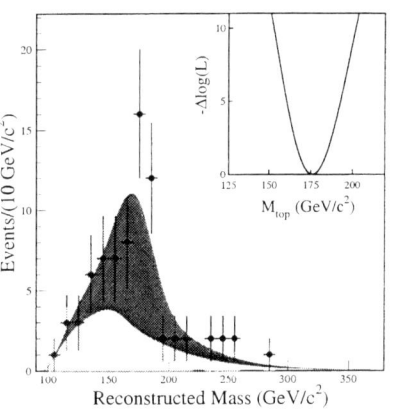

FIGURE 3. Events per bin versus m_{fit} for DØ events (a) passing or (b) failing the LB cut. (c) Log likelihood L versus true top quark mass m_t.

FIGURE 4. Reconstructed mass distributions for the CDF lepton+jets candidate events. The inset shows Log likelihood L versus true top quark mass m_t.

CDF selects lepton+jets events with four or more jets and classifies them into four independent subcategories based on the b-tagging information: (1) events with two SVX b-tagged jets, (2) events with a single SVX b-tagged jet, (3) events with a soft lepton tag, and (4) untagged events. All samples are analyzed separately as the signal to background ratio and the mass resolutions for them are different, and each sample is fitted to Monte Carlo of top signal and background model to extract the likelihood values as a function of m_t. Since the sub-samples are statistically independent, the likelihood functions are combined into one result (Fig. 4), which yields: $m_t = 175.9 \pm 4.8(\text{stat}) \pm 4.9(\text{sys})$ GeV/c^2 [10].

B Dilepton mass measurements

There are two leptonic W decays in the dilepton channel, and due to the presence of two unmeasured neutrinos, a constrained fit is not possible for dilepton events. Two different strategies are used by the experiments to get mass estimators other than reconstructed mass.

CDF has performed two analyses using different kinematic variables. One analysis uses the fact that the b-jets in top events are harder than those from background and therefore fits the observed jet energy spectra in the dilepton events. A maximum likelihood fit to the jet energy spectra of dilepton events leads to $m_t = 159 \pm 23(\text{stat}) \pm 17(\text{sys})$ GeV/c^2. A second analysis uses the correlation of the lepton-b invariant mass ($m_{\ell b}$) with the top quark to determine the mass. From this method CDF obtains a mass of $m_t = 163 \pm 20(\text{stat}) \pm 9(\text{sys})$ GeV/c^2 [12].

In the DØ analysis the events are reconstructed for every assumed top quark mass [13] and weight is computed which characterizes how likely the reconstructed final state occurs in $t\bar{t}$ decay for the assumed mass. Two algorithms are used to determine the weight: the matrix element weighting (MWT) method which uses the proton structure functions and the probability density function for the energy of the charged lepton in the rest frame of the top quark (an extension of Ref. [14]) and the neutrino weighting method (νWT) which assigns the weight based on the available phase space for the neutrinos, consistent with the measured $\not{E}_T$. The results for the two analyses are in excellent agreement : $m_t = 168.1 \pm 12.4(\text{stat})$ GeV/c^2 (MWT) and $m_t = 169.9 \pm 14.8(\text{stat})$ GeV/c^2 (νWT) with a systematic error of 3.6 GeV/c^2. By combining the two results, taking into account the correlations (77%), DØ determines the top quark mass in the dilepton channels to be $m_t = 168.4 \pm 12.3(\text{stat}) \pm 3.6(\text{sys})$ GeV/c^2.

C All Jets mass measurements

CDF has obtained a top quark mass measurement in the all jets channel [8]. To determine the top quark mass, full kinematic reconstruction based on a 3C fit to the top quark pair hypothesis, is applied to the sample of events with 6 or more jets, one or more tags with further kinematic requirements. The event sample consists of 136 events with an expected background of 108 ± 9, estimated from the tag probability. A maximum likelihood method applied to extract the top quark mass gives the result: $m_t = 186 \pm 10(\text{stat}) \pm 12(\text{sys})$ GeV/c^2.

The top mass results for both the experiments are summarized in Table 5.

In Figure 5, the top quark pair production cross section is shown at the measured mass, for both the experiments and compared to the theoretical predictions listed in Ref. [3], [15]. The Standard Model links the mass of the W boson to the masses of the top quark and the Higgs boson through radiative corrections. Figure 6

Experiment	Method	m_t (GeV/c^2)
CDF	ℓ + jets	$175.9 \pm 4.8 \pm 4.9$
	$\ell\ell$ (E_b)	$159 \pm 23 \pm 11$
	$\ell\ell$ ($m_{\ell b}$)	$163 \pm 20 \pm 9$
	all jets	$186 \pm 10 \pm 12$
DØ	ℓ + jets	$173.3 \pm 5.6 \pm 5.5$
	$\ell\ell$	$168.4 \pm 12.3 \pm 3.6$
	$\ell\ell, \ell + jets$	172.1 ± 7.1
CDF + DØ	$\ell + jets$	174.8 ± 5.5
Tevatron Average		173.5 ± 5.2

TABLE 5. Top quark mass results. If two errors are specified, the first is statistical and the second is systematic.

compares the measured W boson and top quark masses to the SM prediction for a range of Higgs boson mass. The measurements and the SM are in good agreement given the present measurement accuracies.

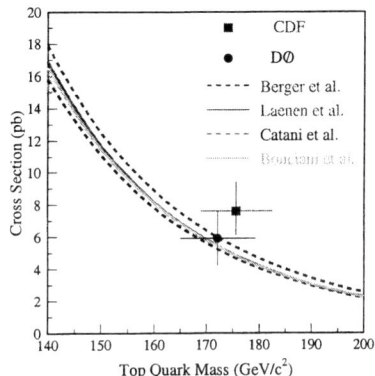

FIGURE 5. Comparison of DØ and CDF measurements of m_t and $\sigma_{t\bar{t}}$ with theoretical calculations.

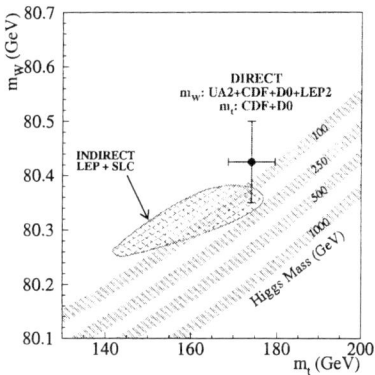

FIGURE 6. m_W vs. m_t in the SM compared to DØ, CDF, and LEP measurements.

IV OTHER TOP QUARK RESULTS

Search for the rare decays

Within the Standard Model, Flavor Changing Neutral Current (FCNC) decays of a top quark are suppressed at the level of 10^{-10} to 10^{-12}. Any appearances of FCNC decays would signal the appearance of physics beyond the Standard Model.

CDF has searched $t\bar{t}$ events with one top decaying in the standard fashion ($t \to Wb$) and the other top decaying to a rare mode [16] $t \to c\gamma$, $t \to u\gamma$ and $t \to Zq$.

The photon FCNC decay search uses both the lepton plus photon and photon plus jets (with SVX b-tag) signatures. One $\mu\gamma$ event is observed with an expected background of less than half of an event. The non-background-subtracted upper limit on the photon branching ratio of the top quark is $BR(t \to q\gamma) < 3.2\%$ at 95% confidence level.

In the Z FCNC decay search, the Z was required to decay leptonically, and the other t quark was required to decay into 3 jets. One $\mu\mu$ event was observed with an expected background of 1.2 events, giving an upper limit on the Z branching ratio of $BR(t \to qZ) < 33\%$ at 95% CL, consistent with Standard Model expectations.

CDF has also made a direct search for top decaying to charged Higgs using the decay chain $t \to H^+ b$, $H^+ \to \tau\nu$, $\tau \to$ hadrons [17]. The branching ratios of top to charged Higgs, and charged Higgs to τ depend on the parameter $\tan\beta$, which is the ratio of the vacuum expectation values for the two Higgs doublets. These branching ratios approach one, and maximum sensitivity is achieved, for large $\tan\beta$. The result is expressed as an exclusion region in the m_{H^+} vs. $\tan\beta$ plane. For $\tan\beta = \infty$, the excluded region is $m_{H^+} < 147(158)$ GeV/c^2 at 95% CL assuming $m_t = 175$ GeV/c^2 and $\sigma_{t\bar{t}} = 5.0$ (7.5) pb.

Measurement of V_{tb}

CDF has analyzed the lepton + jets and dilepton samples to measure the ratio of events with 0, 1, and 2 b-jets to extract $B = BR(t \to Wb)/BR(t \to Wq)$ ratio. B is varied to maximize likelihood of the observed number of events, taking into account the expected background, acceptance, and tagging efficiency. The B is measured to be 0.99 ± 0.29 and $B > 0.58$ at 95% confidence level. In a three-generation Standard Model, assuming unitary,

$$B = \frac{|V_{tb}|^2}{|V_{td}|^2 + |V_{ts}|^2 + |V_{tb}|^2} = |V_{tb}|^2$$

and this yields that $|V_{tb}| = 0.99 \pm 0.15$ and $|V_{tb}| > 0.76$ (95% CL) [18].

V FUTURE PROSPECTS

One of the goals of particle physics in the next decade is to perform detailed experimental studies of the properties of the top quark. For the next collider run (Run II), both the accelerator and the detectors will undergo substantial upgrades, which will allow the extension of the current studies of top quark production.

For Run II (begins late 1999), the Tevatron will undergo both energy and luminosity upgrades: it is expected to reach 2 fb^{-1} per experiment and to operate at a center-of-mass energy of 2.0 TeV. The $\sigma_{t\bar{t}}$ cross section grows then by 35 % and there is also 40% increase in single $t/\bar{t}$ cross section; it yields about 30 times current sample of top events. This will enable to measure the top quark mass to $\Delta M_{top} < 4$ GeV/c^2, and the $\sigma_{t\bar{t}}$ production cross section to $\Delta\sigma/\sigma \sim 9$ %, and perform detailed survey of top quark properties, including rare decay modes, top width and Wtb coupling.

ACKNOWLEDGMENTS

We would like to thank the organizers of "SILAFAE '98" for a most enjoyable stay here. And we also thank the Fermilab Accelerator, Computing, and Research Divisions, and the support staffs at the collaborating institutions for their contributions to this work. We also acknowledge the support of the U.S. Department of Energy, the U.S. National Science Foundation, the Commissariat à L'Energie Atomique in France, the Ministry for Atomic Energy and the Ministry of Science and Technology Policy in Russia, CNPq in Brazil, the Departments of Atomic Energy and Science and Education in India, Colciencias in Colombia, CONACyT in Mexico, the Ministry of Education, Ministry of Education and KOSEF in Korea, the Italian Istituo Nazionale di Fisica Nucleare, the Ministry of Science, CONICET and UBACyT (Argentina), Culture and Education of Japan, and the A.P. Sloan Foundation.

REFERENCES

1. CDF Collaboration, F. Abe et al., Phys. Rev. Lett., **74**, 2626 (1995).
2. DØ Collaboration, S. Abachi et al., Phys. Rev. Lett., **74**, 2632 (1995).
3. E. Laenen, J. Smith, and W. van Neerven, Phys. Lett. **321B**, 254 (1994); E. Berger and H. Contopanagos, Phys. Rev. D **54**, 3085 (1996). S. Catani, M.L. Mangano, P. Nason, and L. Trentadue, Phys. Lett. **378B**, 329 (1996).
4. CDF Collaboration, F. Abe et al, Fermilab-Pub-97/304-E.
5. DØ Collaboration, S. Abachi et al., Phys. Rev. Lett., **79**, 1203 (1997).
6. CDF Collaboration, F. Abe et al., Phys. Rev. Lett. 79, 3585 (1997)
7. CDF Collaboration, F. Abe et al., Fermilab-Pub-97/286-E.
8. CDF Collaboration, F. Abe et al., Fermilab-Pub-97/075-E.
9. N. Amos for the DØ and CDF Collaborations, presented at XXXIIInd Rencontres de Moriond: QCD and High Energy Hadronic Interactions; Les Arc, France, 1998
10. CDF Collaboration, F. Abe et al., Fermilab-Pub-97/284-E.
11. DØ Collaboration, S. Abachi et al., Phys. Rev. Lett., **79**, 1197 (1997); Fermilab-Pub-98/031-E.
12. CDF Collaboration, F. Abe et al., Fermilab-Pub-97/304-E.
13. DØ Collaboration, B. Abbott et al., Fermilab-Pub-97/172-E.
14. R.H. Dalitz and G.R Goldstein, Phys. Rev. **D45**, 1531 (1992), OUTP-92-07P; K. Kondo, Journal of the Physical Society of Japan **57**, 4126 (1988) and **60**, 836 (1991).
15. E. Berger and H. Contopanagos, Phys. Lett. **361B**, 115 (1995).
16. CDF Collaboration, F. Abe et al., Phys. Rev. Lett. **80**, 2525 (1998).
17. CDF Collaboration, F. Abe et al., Phys. Rev. Lett. **79**, 357 (1997).
18. K. Takikawa for the CDF Collaboration, Fermilab-Conf-98/054-E.

B PHYSICS AT THE TEVATRON

JORGE F. DE TROCONIZ [1]

Dpto. de Física Teórica
Universidad Autónoma de Madrid,
E28049 Cantoblanco, Spain

Abstract. Precision B-physics results from the CDF and D0 Collaborations based on data collected during the Tevatron 1992-96 run are presented. In particular we discuss the measurement of the B_s meson lifetime, B_c meson observation, and $B^0 - \bar{B}^0$ mixing results obtained using time-evolution analyses. Prospects for the next Tevatron run, starting in 1999, are also reported.

INTRODUCTION

Precision B physics became possible at hadron colliders since the CDF Collaboration installed a silicon vertex detector (SVX) in 1992. The total b production cross section at the Fermilab Tevatron is about 30 μb [1] in the rapidity region $|y| < 1$. For a typical instantaneous luminosity of 10^{31} cm^{-2}·sec^{-1} the corresponding production rate is 300 Hz. However the backgrounds are also large: the b cross section is three orders of magnitude smaller than the total inelastic cross section. Clearly the trigger performance is of vital importance. All B triggers at hadron colliders are based on leptons. The CDF and D0 Collaborations have collected 100 pb^{-1} of data during the 1992-96 run. Currently, all B physics results from the Tevatron are very competitive to the ones from the LEP and SLC experiments [2].

In the following we shall report the latest CDF results on the B_s meson lifetime measurement, B_c meson observation, and time-dependent $B^0 - \bar{B}^0$ mixing.

B_S LIFETIME MEASUREMENT

The B hadron lifetimes are sensitive to the details of the decay mechanism beyond the spectator model. Unlike the D^+/D^0 case, B decay models predict very small differences between the B_u and the B_d lifetimes (5-10%) [3,4]. Although there is some controversy among theorists about the precise size of this effect, there is agreement that the expected difference between the B^0 and B_s lifetimes is less than about 1%.

---

[1] Representing the CDF and D0 Collaborations

CDF has measured the lifetimes of the B^0, B^+, B_s and Λ_b hadrons. In this report, we will discuss a recent update of the semi-exclusive B_s lifetime measurement, using a total of 110 pb^{-1} of data. A summary of other CDF results on B hadron lifetimes can be found in [5-8].

The lifetime of the B_s meson is measured at CDF using the semileptonic decay $B_s \to D_s l \nu X$. D_s mesons are reconstructed in a cone around the lepton using the following channels:

(a) $D_s^- \to \phi \pi^-$, $\phi \to K^+ K^-$

(b) $D_s^- \to K^{*0} K^-$, $K^{*0} \to K^+ \pi^-$

(c) $D_s^- \to K_S^0 K^-$, $K_S^0 \to \pi^+ \pi^-$

(d) $D_s^- \to \phi \mu^- \nu$

For the first three decay modes the analysis starts with a single lepton trigger data set, while the semileptonic D_s decay mode is based on a dimuon data sample obtained with a trigger requirement of $M(\mu\mu) < 2.8$ GeV/c^2. A secondary vertex is defined at the intersection of the D_s and lepton trajectories in the plane transverse to the beam axis, and a transverse decay distance, L_{xy}, as the projection of the the vector difference of the secondary and primary vertex positions onto the direction of the $D_s l$ system:

$$L_{xy} = \frac{(\vec{x}_{sec} - \vec{x}_{prim}) \cdot \vec{p}_T(D_s l)}{p_T(D_s l)} \quad (1)$$

To extract the proper decay length, $c\tau$, from L_{xy} we need to correct with the appropriate $\beta\gamma$ factor:

$$c\tau = L_{xy} \cdot \frac{M_{B_s}}{p_T(B_s)} \quad (2)$$

where M_{B_s} is the mass of the B_s meson and $p_t(B_s)$ its transverse momentum. However, it is not possible to calculate the $\beta\gamma$ factor exactly: at least a neutrino is missing. $c\tau$ is therefore calculated as follows:

$$c\tau = L_{xy} \cdot \frac{M_{B_s}}{p_T(D_s l)} \cdot K \quad (3)$$

where K is an average correction factor calculated with a Monte Carlo simulation.

About 600 B_s candidates have been reconstructed in the four D_s decay channels, where the $D_s \to \phi\pi$ mode contributes the largest statistics with 220 ± 21 events (Figure 1 (left)). Figure 1 (right) shows the corresponding decay length distributions. From all four D_s decay modes a B_s lifetime of

$$\tau(B_s) = 1.39 \pm 0.09 \ (stat.) \pm 0.05 \ (syst.) \ \text{ps} \quad (4)$$

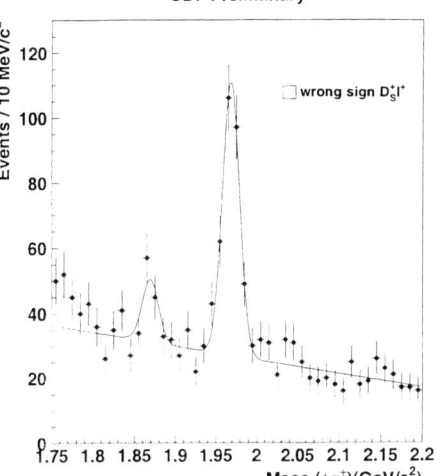 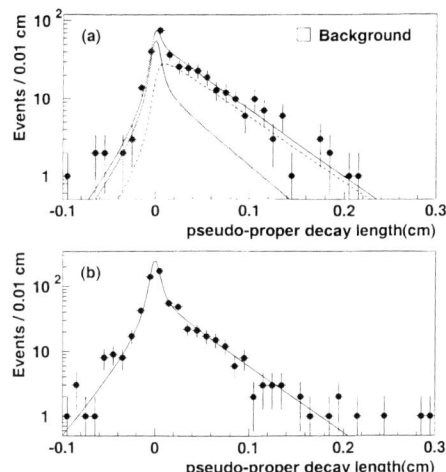

FIGURE 1. (Left) Invariant mass of the $\phi\pi$ system. The points are for right sign $D_s l$ combinations. The shaded histogram shows the wrong sign combination distribution. (Right) Corresponding decay length distributions for the signal (top) and background (bottom) samples.

has been measured. This measurement is still statistically limited. The main systematic errors arise from the background shape and normalization.

Another possible measurement using the same data sample is to look for a difference in the lifetime of the two B_s mass eigenstates. Theoretical estimates predict $\Delta\Gamma/\Gamma$ to be on the order of $10 - 20\%$ [9,10]. In the standard model, $\Delta m/\Delta\Gamma$ is related to the ratio of the Kobayashi-Maskawa matrix elements $|V_{cb}V_{cs}|/|V_{tb}V_{ts}|$ which is quite well known, and depends only on the size of QCD corrections. If these QCD corrections can be precisely calculated, a measurement of $\Delta\Gamma$ would imply a determination of Δm, and thus a way to infer the existence of $B_s - \bar{B}_s$ oscillations. We fitted the proper decay length distribution allowing for two different lifetime components (Γ_H and Γ_L), with $\Gamma_{H,L} = \Gamma \pm \Delta\Gamma/2$. Fixing the mean lifetime to its PDG average value [11]

$$\tau_{mean} = \Gamma^{-1}(1 - \frac{1}{4}\frac{\Delta\Gamma^2}{\Gamma^2})^{-1} = 1.57 \pm 0.08 \text{ ps} \quad (5)$$

a preliminary fit result is

$$\frac{\Delta\Gamma}{\Gamma} = 0.48 \, ^{+0.26}_{-0.48} \quad (6)$$

indicating that the current statistics is not sensitive to a B_s lifetime difference. Based on the fit, a limit on $\Delta\Gamma/\Gamma < 0.81$ (95% CL) can be set. Using the value of $\Delta\Gamma/\Delta m = (5.6 \pm 2.6) \times 10^{-3}$ [10], an upper limit on the B_s mixing frequency Δm_s can be obtained

$$\Delta m_s < 92 \times (5.6 \cdot 10^{-3})/(\Delta\Gamma/\Delta m) \times (1.57 \text{ ps}/\tau_{B_s}) \quad (95\% \text{ CL}) \quad (7)$$

OBSERVATION OF THE B_C MESON

The B_c^+ meson is the lowest-mass bound state of a family of quarkonium states containing a charm quark and a bottom antiquark. It decays weakly yielding a large branching fraction to final states containing a J/ψ [12-15]. Non-relativistic potential models predict its mass in the range 6.2-6.3 GeV/c^2 [16,17]. In these models, the c and the $\bar{b}$ are tightly bound in a very compact system and have a rich spectroscopy of excited states. There are three major contributions to the B_c decay width: $\bar{b} \to \bar{c}W^+$ leading to final states like $J/\psi\pi$ or $J/\psi l\nu$; $c \to sW^+$ leading to final states like $B_s\pi$ or $B_s l\nu$; and $c\bar{b} \to W^+$ annihilation, leading to final states like DK, $\tau\nu$, or multiple pions. The predicted lifetime is in the range 0.4-1.4 ps [12,18-22]. Because of the wide range of predictions, a B_c lifetime measurement is a test of the different assumptions made in the various calculations.

Limits on B_c production have been placed by various searches at LEP [23-25]. A prior CDF search placed a limit on B_c production in the $B_c^+ \to J/\psi\pi^+$ mode [26].

We report here the observation of B_c mesons produced at the Tevatron using 110 pb^{-1} of data collected by CDF. A more detailed description of this work can be found in Ref. [27]. An online dimuon trigger yielded a sample of about 196,000 $J/\psi \to \mu^+\mu^-$ events. We searched for the B_c using the decays $B_c^\pm \to J/\psi l^\pm \nu$. These decays have a very simple topology: a decay point for $J/\psi \to \mu^+\mu^-$ displaced from the primary interaction point and a third track emerging from the same decay point. A measure of the time between production and decay of a B_c candidate is the quantity

$$ct^* = L_{xy} \cdot \frac{M(J/\psi l)}{p_T(J/\psi l)} \quad (8)$$

where L_{xy} in the distance of the B_c candidate decay vertex to the beam center in the transverse plane. We required $ct^* > 60$ μm.

$B^\pm \to J/\psi K^\pm$ events were identified as a peak in the $\mu^+\mu^-K^\pm$ mass distribution; the fitted peak signal contained 290 ± 19 events. This signal provided a valuable rate calibration, as discussed below. However, for the B_c search, the $B^\pm$ signal events were excluded using a ± 50 MeV/c^2 cut around $M(B^+)$. Finally, the third track had to satisfy a number of electron or muon standard identification cuts.

A Monte Carlo calculation of B_c production and decay to $J/\psi l\nu$ showed that, for an assumed mass of 6.27 GeV/c^2, 93% of the final states would have $J/\psi l$ masses with $4.0 < M(J/\psi l) < 6.0$ GeV/c^2. We refer to this as the signal region, but candidates with masses in the range 3.35 - 11 GeV/c^2 were accepted. We found 23 $B_c^\pm \to J/\psi\, e^\pm \nu$ candidates of which 19 were in the signal region, and 14 $B_c^\pm \to J/\psi\, \mu^\pm \nu$ candidates of which 12 were in the signal region.

TABLE 1. B_c Signal and Background Summary

	$3.35 < M(J/\psi l) < 11.0$ GeV/c^2	
	$J/\psi\, e$ Events	$J/\psi\, \mu$ Events
False Electrons	4.2 ± 0.4	
Undetected Conversions	2.1 ± 1.7	
False Muons		11.4 ± 2.4
$B\bar{B}$ bkg.	2.3 ± 0.9	1.44 ± 0.25
Total Background	8.6 ± 2.0	12.8 ± 2.4
Background (fit)	9.2 ± 2.0	10.6 ± 2.3
Signal (fit)	$12.0\, ^{+3.8}_{-3.2}$	$8.4\, ^{+2.7}_{-2.4}$
Signal + Background	21.2 ± 4.3	19.0 ± 3.5
Candidates	23	14

Backgrounds are dominated by fake leptons and by random combinations of real leptons with J/ψ mesons. Table 1 summarizes the results of the background calculation and of a simultaneous fit for the muon and electron channels to the mass spectrum over the region 3.35 - 11 Gev/c^2 [27]. Figure 2 (left) shows the mass spectra for the combined candidate sample, combined backgrounds and fitted B_c contribution. The fitted number of B_c events is $20.4\, ^{+6.2}_{-5.5}$. To test the significance of the result, we generated a number of Monte Carlo trials with the statistical properties of backgrounds, but no B_c contribution. The probability of obtaining a yield of 20.4 or more events is 0.63×10^{-6}, equivalent to a 4.8 sigma effect.

To check the stability of the signal, we generated Monte Carlo signal templates for $5.52 < M(B_c) < 7.52$ GeV/c^2 and repeated the fit. The study showed that the magnitude of the signal is stable over the range of theoretical predictions for $M(B_c)$, and the dependence of the log-likelihood function on mass yielded $M(B_c) = 6.40 \pm 0.39\,(stat.) \pm 0.13\,(syst.)$ GeV/c^2.

We obtained the mean proper decay length $c\tau$ of the B_c meson from the distribution of ct^*, using only events in the mass signal region and after changing the $ct^* > 60$ μm requirement to $ct^* > -100$ μm. This yielded a sample of 42 $J/\psi\, e$ and 29 $J/\psi\, \mu$ events. We determined a functional form for the shapes in ct^* for each of the backgrounds, and added a resolution- and $\beta\gamma$-smeared exponential decay distribution for the B_c contribution. An unbinned likelihood fit to the data (Figure 2 (right)) yielded the result:

$$c\tau = 137\, ^{+53}_{-49}\,(stat.) \pm 9\,(syst.)\ \mu\text{m} \tag{9}$$

$$\tau = 0.46\, ^{+0.18}_{-0.16}\,(stat.) \pm 0.03\,(syst.)\ \text{ps} \tag{10}$$

From the 20.4 B_c events and the 290 $B^\pm \to J/\psi K^\pm$ events, we calculated the B_c production cross section times the branching fraction $\mathcal{B}(B_c^+ \to J/\psi l^+ \nu)$, relative to that for the topologically similar decay $B^+ \to J/\psi K^+$. Several systematic uncertainties cancel in the ratio. Since the detection efficiency for $B_c^+ \to J/\psi l^+ \nu$ depends on $c\tau$ because of the ct^* requirement, we quote a separate systematic error

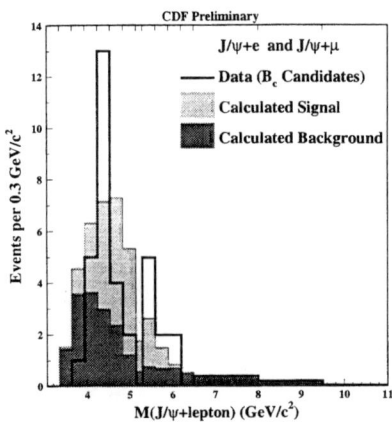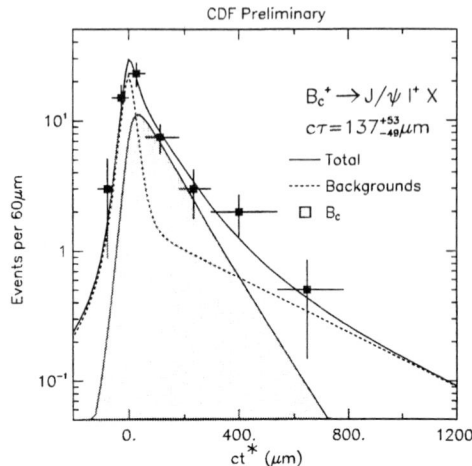

FIGURE 2. (Left) Invariant mass of the $J/\psi l$ system, comparing the data to the signal and background contributions determined in the fit. (Right) Corresponding decay length distributions.

because of the lifetime uncertainty. Finally, we multiply the 20.4 events by a factor 0.85 ± 0.15 to correct for other decay channels such as $B_c \to \psi' l \nu$ [27]. The result is

$$\frac{\sigma(B_c) \cdot \mathcal{B}(B_c \to J/\psi l \nu)}{\sigma(B^+) \cdot \mathcal{B}(B \to J/\psi K)} = 0.132 \,^{+0.041}_{-0.037} \,(stat.) \pm 0.031 \,(syst.) \,^{+0.032}_{-0.020} \,(life.) \quad (11)$$

for mesons with $p_T > 6.0$ GeV/c and $|y| < 1.0$. This result is consistent with limits from previous searches [23-25].

$B^0 - \bar{B}^0$ OSCILLATIONS

$B^0 - \bar{B}^0$ transitions are allowed in the Standard Model via higher order weak interaction diagrams. Since the flavour eigenstates are not exactly the mass eigenstates, a B^0 produced at time $\tau = 0$ has a certain probability to turn (mix) into a $\bar{B}^0$ at a later time τ. Defining $x = \Delta m \, \tau_B$, where Δm is the mass difference and τ_B the average lifetime of the eigenstates of the mass matrix, the mixing probability is given by

$$\mathcal{P}(B^0(0) \to \bar{B}^0(\tau)) = \frac{e^{-\tau/\tau_B}}{2\tau_B} \cdot (1 - \cos(x \frac{\tau}{\tau_B})) \quad (12)$$

The mixing parameters $x_{d,s}$, for the $B^0_{d,s}$ mesons, are related directly to the elements $V_{t(d,s)}$ of the CKM matrix.

In general, a time dependent mixing analysis requires knowledge of the flavour of the B meson at production and decay times. Experimentally, to measure the decay time implies the use of some kind of vertexing algorithm. The flavour at the decay time is determined from the decay products. All the analyses reported here use semileptonic B decays. The B meson decay vertex is measured at the intersection of the lepton and reconstructed charm trajectories. The charm signal "D" can be fully reconstructed or inclusively tagged using a secondary vertex algorithm. In analogy with the semi-exclusive B_s lifetime analysis, the proper decay length is calculated using

$$c\tau = L_{xy} \cdot \frac{M_B}{p_T("D"l)} \cdot K \qquad (13)$$

where K is an average kinematical Monte Carlo correction factor. In all cases, the flavour at decay time is determined using the charge of the lepton.

More challenging is to know the flavour at production time. Several approaches are possible. One possibility is to use the charge correlations of the B meson and other particles produced in the same jet (same-side tagging). These correlations are expected to appear in the fragmentation process or B^{**} decays. The second possibility is to look at the other B meson in the event (opposite-side tagging), that can (for instance) also decay semileptonically, or using a jet charge algorithm.

We present herein three recent measurements of the B_d frequency from CDF.

$B - \bar{B}$ mixing in $D^{(*)}l$ events

In this analysis the charm is reconstructed explicitly using the following channels:

(a) $D^0 \to K^-\pi^+$, where the D^0 is not from a D^{*+}

(b) $D^{*+} \to D^0\pi_s^+$, $D^0 \to K^-\pi^+$

(c) $D^{*+} \to D^0\pi_s^+$, $D^0 \to K^-\pi^+X$

(d) $D^{*+} \to D^0\pi_s^+$, $D^0 \to K^-\pi^+\pi^+\pi^-$

(e) $D^+ \to K^-\pi^+\pi^+$

Tracks with impact parameters significantly displaced from the primary vertex are selected in order to decrease combinatorial backgrounds. The signals are identified as peaks in the invariant mass distributions.

Same-side tagging is used. The momentum of the B meson is approximated by the momentum of its reconstructed portion. Charged tracks within a cone around the reconstructed B meson and consistent with the hypothesis that they originate from the primary vertex of the event are considered. Of the candidate tracks we select as the tag the track with minimum p_T^{rel} relative to the sum of the momenta of the B and that track. The efficiency for finding a tag is about 72%.

Next, the number of right-sign (RS) correlations (i.e. $B^0\pi^+$, $B^+\pi^-$) is compared to the number of wrong sign correlations (WS) (i.e. $B^0\pi^-$, $B^+\pi^+$) as a function of the proper decay time. For the B^0 meson the following asymmetry is expected:

$$A(\tau) = \frac{N_{RS}(\tau) - N_{WS}(\tau)}{N_{RS}(\tau) + N_{WS}(\tau)} = D \cos(\Delta m_d \, \tau) \qquad (14)$$

where D is the so-called dilution of the tagging algorithm. D is related to the mistag fraction w by $D = 1 - 2w$. After correcting for channel cross-talk, the values of D and Δm_d are extracted from a fit to the data. The results are shown in Figure 4 (left):

$$\Delta m_d = 0.471 \,^{+0.078}_{-0.068} \, (stat.) \pm 0.034 \, (syst.) \, \text{ps}^{-1} \qquad (15)$$

and a dilution $D(B^0) = 0.18 \pm 0.03 \pm 0.02$. The systematic error is dominated by the uncertainty in the fraction of D^{**} in semileptonic B decays.

$B - \bar{B}$ mixing in $e - \mu$ events

For this analysis, we trigger on leptons from the semileptonic decay of both B hadrons in an event: $B_1 \to eX$ and $B_2 \to \mu X$. Sequential decays from one B hadron are rejected with the requirement $M(e\mu) > 5$ GeV/c^2. An inclusive secondary vertex is reconstructed in association with one of the leptons. The vertexing algorithm has been tuned for high efficiency near $c\tau = 0$, with the efficiency reaching a plateau of about 40% for $c\tau > 500$ μm. The boost resolution is about 22%. The charge of the lepton associated to the displaced vertex gives the flavour at the decay time. The other lepton provides the flavour tag at the production time.

The challenge of this analysis is to determine the sample composition. It can be estimated from several kinematical quantities, like p_T^{rel} or the invariant mass of the tracks that form the displaced vertex. Here p_T^{rel} is defined as the transverse momentum with respect to the lepton direction of the hardest track in a cone around the lepton. About 86% of the sample is made of $b\bar{b}$ events (10 − 15% of these events contain sequential leptons), around 11% are events with at least a fake lepton, and the rest comes from $c\bar{c}$ events.

The final sample is formed by 6025 events with a secondary vertex around the electron (electron tags) and 5819 muon tags. Approximately 16% of these events contain both an electron tag and a muon tag. Figure 3 (left) shows the dependence on $c\tau$ of the like-sign fraction of events, defined as $N_{LS}(\tau)/(N_{LS}(\tau) + N_{OS}(\tau))$. A fit to the data is performed including components for direct and sequential b decays, $c\bar{c}$, and fake events. The result is:

$$\Delta m_d = 0.450 \pm 0.045 \, (stat.) \pm 0.051 \, (syst.) \, \text{ps}^{-1} \qquad (16)$$

where the dominant systematic error arises from the uncertainties in the sample composition.

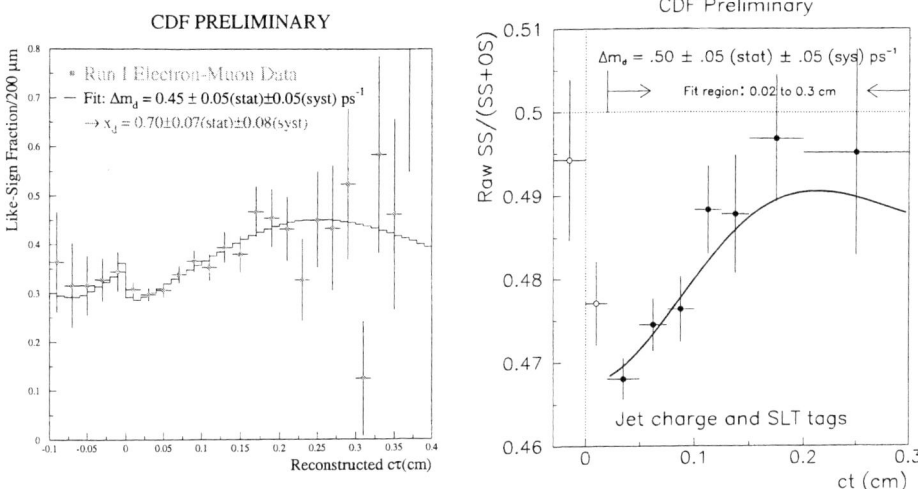

FIGURE 3. Fitted like-sign fraction as a function of the proper decay time for the $e - \mu$ (left), and inclusive lepton (right) mixing analyses.

$B - \bar{B}$ mixing in inclusive lepton events

Starting from events that satisfy an inclusive lepton trigger with $p_T > 8$ GeV/c, opposite-side flavour tagging is implemented using jet charge and soft leptons. The result of this analysis (Figure 3 (right)) is:

$$\Delta m_d = 0.496 \pm 0.052 \ (stat.) \pm 0.048 \ (syst.) \ \text{ps}^{-1} \qquad (17)$$

The effective tagging efficiency of this algorithm has been studied in detail by CDF. Values of $\epsilon D^2 = 1.07 \pm 0.09 \pm 0.10\%$ for lepton tagging, and $0.78 \pm 0.12 \pm 0.09\%$ for the jet charge algorithm are found respectively.

There are still two other CDF opposite-side tagging mixing analyses. The first uses $D^{*+}l^-$ combinations in dilepton events ($\Delta m_d = 0.512 \ ^{+0.095}_{-0.093} \ ^{+0.031}_{-0.038}$ ps^{-1}), and the second, dimuon data ($\Delta m_d = 0.503 \pm 0.064 \pm 0.071$ ps^{-1}). After taking into account the statistical overlap between the samples and common systematic errors, the CDF average result is

$$\Delta m_d = 0.481 \pm 0.028 \ (stat.) \pm 0.027 \ (syst.) \ \text{ps}^{-1} \qquad (18)$$

This number is competitive with the LEP determinations [2].

PROSPECTS FOR RUN II

In 1999, the Tevatron together with the Main Injector is supposed to deliver 2 fb^{-1} in two years. By then, the CDF [28] detector will be upgraded with a new

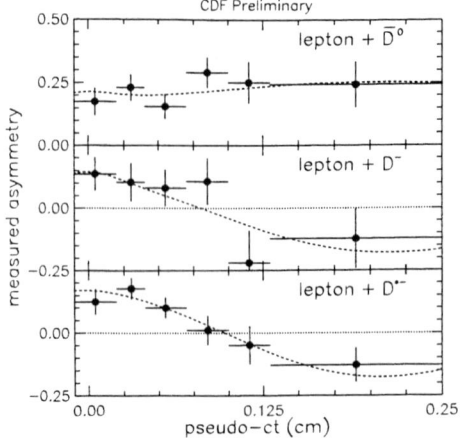
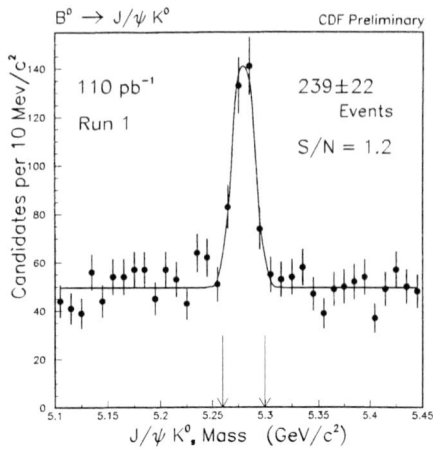

FIGURE 4. (Left) Time dependent asymmetry for B^+ (top) and B^0 mesons in the $D^{(*)}l$ mixing analysis. (Right) $B^0 \to J/\psi K_S^0$ signal collected at CDF during the 1992-96 run.

silicon vertex detector, which doubles the fiducial volume of the current SVX and provides 3-d tracking. A new central tracking system will be in place, designed to handle higher rates and shorter bunch crossing times, while maintaining the excellent momentum resolution and dE/dx capabilities of the Run I detector. With an upgraded trigger and DAQ system CDF plans to operate a fully hadronic trigger for the first time. D0 [29] will enhance considerably its tracking capabilities with an inner silicon vertex detector, surrounded by four superlayers of a scintillating fiber tracker. These detectors will be located inside a 2 Tesla superconducting solenoid.

sin 2β

CP asymmetries in the decay $B^0 \to J/\psi K_S^0$ determine the value of sin 2β. The 1992-96 CDF signal (Figure 4 (right)) is formed by 239±22 events with a signal-to-noise ratio of 1.2. Taking into account the improvements in luminosity, lower trigger thresholds, etc, we expect to collect 15,000 such events. This results in an error of $\Delta \sin 2\beta = 0.09$, assuming $\epsilon D^2 = 6\%$.

sin 2α

CP asymmetries in the decay $B^0 \to \pi^+ \pi^-$ are related to the value of sin 2α. Here, the challenge is to be able to trigger on this decay. CDF plans to require two charged tracks at L1, and use impact parameter information at L2 (20 Hz). We expect about 10,000 events, that will produce an uncertainty of $\Delta \sin 2\alpha = 0.10$ for $\epsilon D^2 = 6\%$.

B_s mixing

Here the reach on Δm_s is limited by the proper time resolution. Monte Carlo studies show that the experiment will be sensitive to values up to 10 ps^{-1}.

CONCLUSIONS

CDF has produced very competitive measurements of the B hadron lifetimes. The B_c meson has been observed with a significance at the 4.8 sigma level. Mixing results are still dominated by statistics. In the frame of these analyses, the feasibility of a number of flavour tagging techniques has been demonstrated for the first time at a hadron collider. For Run II, CDF and D0 will be significantly upgraded. The experiments will focus on the discovery of CP violation in the B sector.

REFERENCES

1. S. Abachi et al. (D0 Collaboration), Fermilab preprint FERMILAB-Conf-96-248-E, paper submitted to the 28th International Conference on High Energy Physics, Warsaw, Poland, July 1996.
2. P. Perret, these proceedings.
3. I.I. Bigi, *Nuovo Cim.* **A 109** (1996) 713.
4. M. Neubert, *Int. J. Mod. Phys.* **A11** (1996) 4173.
5. J.F. de Trocóniz, Fermilab preprint FERMILAB-Conf-97-165-E, Proceedings of the 32nd Rencontres de Moriond, Electroweak Interactions and Unified Theories, Les Arcs, France, March 1997.
6. F. Abe et al. (CDF Collaboration), *Phys. Rev. Lett.* **72** (1994) 3456.
7. F Abe et al. (CDF Collaboration), *Phys. Rev. Lett.* **76** (1996) 4462.
8. F. Abe et al. (CDF Collaboration), *Phys. Rev. Lett.* **77** (1996) 1439.
9. I. Dunietz, *Phys. Rev.* **D52** (1995) 3048.
10. M. Beneke, G. Buchalla, and, I. Dunietz, *Phys. Rev.* **D54** (1996) 4419.
11. R.M. Barnett et al. (Particle Data Group), *Phys. Rev.* **D54** (1996) 1.
12. M. Lusignoli and M. Masetti, *Z. Phys.* **C51** (1991) 549.
13. N. Isgur et al., *Phys. Rev.* **D39** (1989) 799.
14. D. Scora and N. Isgur, *Phys. Rev.* **D52** (1995) 2783.
15. C.H. Chang and Y.Q. Chen, *Phys. Rev.* **D49** (1994) 3399.
16. W. Kwong and J. Rosner, *Phys. Rev.* **D44** (1991) 212.
17. E. Eichten and C. Quigg, *Phys. Rev.* **D49** (1994) 5845.
18. I.I. Bigi, *Phys. Lett.* **371B** (1996) 105.
19. M. Beneke and G. Buchalla, *Phys. Rev.* **D53** (1996) 4991.
20. S.S. Gershtein et al., *Int. J. Mod. Phys.* **A6** (1991) 2309.
21. P. Colangelo et al., *Z. Phys.* **C57** (1993) 43.
22. C. Quigg, Fermilab preprint FERMILAB-Conf-93-267, 1994.
23. P. Abreu et al. (DELPHI Collaboration), *Phys. Lett.* **398B** (1997) 207.
24. K. Ackerstaff et al. (OPAL Collaboration), *Phys. Lett.* **420B** (1998) 157.

25. R. Barate et al. (ALEPH Collaboration). *Phys. Lett.* **402B** (1997) 213.
26. F. Abe et al. (CDF Collaboration), *Phys. Rev. Lett.* **77** (1996) 5176.
27. F. Abe et al. (CDF Collaboration), Fermilab preprint FERMILAB-Pub-98-121-E, 1998.
28. F. Abe et al. (CDF Collaboration), Fermilab preprint FERMILAB-Pub-96-390-E, 1996.
29. S. Abachi et al. (D0 Collaboration) PAC-Report, 1996.

FIELD THEORY

Chiral Symmetry Breaking in an External Field

C. N. Leung

Department of Physics and Astronomy, University of Delaware
Newark, DE 19716

Abstract.
The effects of an external field on the dynamics of chiral symmetry breaking are studied using quenched, ladder QED as our model gauge field theory. It is found that a uniform external magnetic field enables the chiral symmetry to be spontaneously broken at weak gauge couplings, in contrast with the situation when no external field is present. The broken chiral symmetry is restored at high tempeatures as well as at high chemical potentials. The nature of the two chiral phase transitions is different: the transition at high temperatures is a continuous one whereas the phase transition at high chemical potentials is discontinuous.

I would like to report here some recent work [1,2] done in collaboration primarily with D.-S. Lee and Y. J. Ng. The objective is to understand how an external field may affect the dynamics of chiral symmetry breaking in gauge theories. Quantum electrodynamics (in 3 + 1 dimensions), treated in the quenched, ladder (or planar) approximation, will serve as our model gauge theory. At present our formalism is applicable only for a constant (in space as well as in time) external field and I shall discuss results pertinent to the case of a constant magnetic field.

The problem of dynamical chiral symmetry breaking in a uniform magnetic field in quenched, planar QED has received some attention in recent years [1] - [13]. Following the pioneering works [14] on the study of chiral symmetry breaking in quenched, ladder QED, our approach is based on the Schwinger-Dyson (SD) equation for the fermion self-energy. In this approach, one starts with a Lagrangian with a zero bare fermion mass and looks for nontrivial solutions to the SD equation which signal a dynamically generated fermion mass and a possible spontaneous breakdown of the chiral symmetry.

Let us consider QED with a single charged fermion having a zero bare mass and charge g. In the quenched, ladder approximation, the SD equation in the x-representation reads

$$M(x,x') = ig^2 \gamma^\mu G_A(x,x') \gamma^\nu D_{\mu\nu}(x-x'), \qquad (1)$$

where $M(x,x')$ is the fermion mass operator in the x-representation, $M(x,x') = \langle x|\hat{M}|x'\rangle$, $D_{\mu\nu}(x-x')$ is the bare photon propagator, and $G_A(x,x')$ is the fermion propagator in the presence of an external field represented by the vector potential $A_\mu(x)$. We adopt the metric with the signature $g_{\mu\mu} = (-1,1,1,1)$. G_A satisfies the equation,

$$\gamma^\mu \Pi_\mu G_A(x,y) + \int d^4x' M(x,x') G_A(x',y) = \delta^{(4)}(x-y), \qquad (2)$$

where $\Pi_\mu \equiv -i\partial_\mu - gA_\mu(x)$. Schwinger [15] was the first to obtain an exact analytical expression for the fermion Green's function in the presence of a constant electromagnetic field of arbitrary strength. However, we find the alternative representation of $G_A(x,y)$ proposed by Ritus [16] more convenient for our purpose. The essence of this approach is explained below.

Since, in the presence of a constant external field, the fermion asymptotic states are no longer free particle states represented by plane waves, but are described by wavefunctions consistent with the particular external field configuration, namely, eigenfunctions of $(\gamma^\mu \Pi_\mu)^2$:

$$-(\gamma \cdot \Pi)^2 \psi_p(x) = p^2 \psi_p(x). \qquad (3)$$

Instead of the usual momentum space, it is more convenient to work in the representation spanned by these eigenfunctions. Another advantage of using this representation is that, for constant external fields, the mass operator is diagonal [16].

If we work in the chiral representation of the Dirac matrices in which γ_5 and $\Sigma_3 = i\gamma_1\gamma_2$ are both diagonal with eigenvalues $\chi = \pm 1$ and $\sigma = \pm 1$, respectively, the eigenfunctions $\psi_p(x)$ has the general form

$$\psi_p(x) = E_{p\sigma\chi}(x)\omega_{\sigma\chi}, \qquad (4)$$

where $\omega_{\sigma\chi}$ are bispinors which are the simultaneous eigenvectors of Σ_3 and γ_5. The exact functional form of the $E_{p\sigma\chi}(x)$ will depend on the specific external field configuration.

In the case of a constant magnetic field of strength H pointing in the z-direction, the vector potential may be taken to be $A_\mu = (0,0,Hx_1,0)$ and one finds that the eigenfunctions $E_{p\sigma\chi}(x)$ do not depend on χ:

$$E_{p\sigma}(x) = N e^{i(p_0 x^0 + p_2 x^2 + p_3 x^3)} D_n(\rho). \qquad (5)$$

Here N is a normalization factor and $D_n(\rho)$ are the parabolic cylinder functions [17] with argument $\rho \equiv \sqrt{2|gH|}(x_1 - \frac{p_2}{gH})$ and index n which labels the Landau levels:

$$n = n(k,\sigma) \equiv k + \frac{gH\sigma}{2|gH|} - \frac{1}{2}, \quad n = 0, 1, 2, \ldots \qquad (6)$$

The eigenvalue p now stands for the four quantum numbers (p_0, p_2, p_3, k), where k is the discrete quantum number of the quantized squared transverse momentum:

$$-(\gamma \cdot \Pi_\perp)^2 \psi_p(x) \equiv -(\gamma^1 \Pi_1 + \gamma^2 \Pi_2)^2 \psi_p(x)$$
$$= 2|gH|k\psi_p(x). \tag{7}$$

For a given n, the allowed values for k are $k = n, n+1$.

The eigenfunction-matrices $E_p(x)$ defined as

$$E_p(x) \equiv \sum_\sigma E_{p\sigma}(x) \mathrm{diag}(\delta_{\sigma 1}, \delta_{\sigma -1}, \delta_{\sigma 1}, \delta_{\sigma -1})$$
$$\equiv \sum_\sigma E_{p\sigma}(x) \Delta(\sigma) \tag{8}$$

satisfy the orthonormality and completeness relations ($\bar{E}_p \equiv \gamma^0 E_p^\dagger \gamma^0$):

$$\int d^4x \bar{E}_{p'}(x) E_p(x) = (2\pi)^4 \hat{\delta}^{(4)}(p - p')$$
$$\equiv (2\pi)^4 \delta_{kk'} \delta(p_0 - p'_0) \delta(p_2 - p'_2) \delta(p_3 - p'_3), \tag{9}$$

$$\oint d^4p\, E_p(x) \bar{E}_p(y) \equiv \sum_k \int dp_0 dp_2 dp_3\, E_p(x) \bar{E}_p(y) = (2\pi)^4 \delta^{(4)}(x - y), \tag{10}$$

provided that the normalization constant in Eq.(5) is taken to be $N(n) = (4\pi|gH|)^{1/4}/\sqrt{n!}$. They also satisfy the useful relation [16]:

$$\gamma \cdot \Pi\, E_p(x) = E_p(x)\, \gamma \cdot \bar{p}, \tag{11}$$

where $\bar{p}_0 = p_0$, $\bar{p}_1 = 0$, $\bar{p}_2 = -\mathrm{sgn}(gH)\sqrt{2|gH|k}$, $\bar{p}_3 = p_3$. Note that, in terms of the momentum $\bar{p}$, the system is effectively a (2+1)-dimensional one.

These properties of the E_p-functions enable us to introduce the E_p-representation of the fermion Green's function:

$$G_A(p, p') \equiv \int d^4x d^4y\, \bar{E}_p(x) G_A(x, y) E_{p'}(y)$$
$$= (2\pi)^4 \hat{\delta}^{(4)}(p - p') \frac{1}{\gamma \cdot \bar{p} + \tilde{\Sigma}_A(\bar{p})} \tag{12}$$

where $\tilde{\Sigma}_A(\bar{p})$ represents the eigenvalue matrix of the mass operator:

$$\int d^4x' M(x, x') E_p(x') = E_p(x) \tilde{\Sigma}_A(\bar{p}). \tag{13}$$

It is straightforward to verify that the inverse transform,

$$G_A(x, y) = \oint \frac{d^4p}{(2\pi)^4} E_p(x) \frac{1}{\gamma \cdot \bar{p} + \tilde{\Sigma}_A(\bar{p})} \bar{E}_p(y), \tag{14}$$

satisfies Eq.(2). Eqs.(12) and (14) are the generalization of the well-known relations between coordinate space and momentum space Green's functions, with the plane wave eigenfunctions in the Fourier transform replaced here by the E_p-eigenfunctions in order to account for the presence of the external field. Eq.(12) shows explicitly that the fermion propagator is diagonal (in momentum) in the E_p-representation. As stated earlier, the mass operator is also diagonal in this representation:

$$M(p,p') = \int d^4x d^4x' \bar{E}_p(x) M(x,x') E_{p'}(x')$$
$$= (2\pi)^4 \hat{\delta}^{(4)}(p-p') \tilde{\Sigma}_A(\bar{p}). \qquad (15)$$

Transforming to the E_p-representation, the SD equation, Eq.(1), becomes [2]

$$\tilde{\Sigma}_A(\bar{p}) \delta_{kk'} = ig^2 \sum_{k''} \sum_{\{\sigma\}} \int \frac{d^4q}{(2\pi)^4} \frac{e^{i \text{sgn}(gH)(n-n''+\tilde{n}''-n')\varphi}}{\sqrt{n!n'!n''!\tilde{n}''!}} \frac{e^{-\hat{q}_\perp^2}}{q^2}$$
$$\cdot J_{nn''}(\hat{q}_\perp) J_{\tilde{n}''n'}(\hat{q}_\perp) \cdot \Delta \gamma^\mu \Delta'' \frac{1}{\gamma \cdot \bar{p}'' + \tilde{\Sigma}_A(\bar{p}'')} \tilde{\Delta}'' \gamma_\mu \Delta', \qquad (16)$$

where

$$J_{nn''}(\hat{q}_\perp) \equiv \sum_{m=0}^{\min(n,n'')} \frac{n! n''!}{m!(n-m)!(n''-m)!} [i \text{sgn}(gH) \hat{q}_\perp]^{n+n''-2m}, \qquad (17)$$

$$\hat{q}_\perp^2 \equiv \frac{q_1^2 + q_2^2}{2|gH|}, \qquad \varphi \equiv \arctan\left(\frac{q_2}{q_1}\right), \qquad (18)$$

and the momentum $\bar{p}''$ is given by: $\bar{p}_0'' = p_0 - q_0$, $\bar{p}_1'' = 0$, $\bar{p}_2'' = -\text{sgn}(gH)\sqrt{2|gH|k''}$, $\bar{p}_3'' = p_3 - q_3$. Here $n' = n(k',\sigma')$, $n'' = n(k'',\sigma'')$, $\tilde{n}'' = n(k'',\tilde{\sigma}'')$, $\Delta' = \Delta(\sigma')$, $\Delta'' = \Delta(\sigma'')$, $\tilde{\Delta}'' = \Delta(\tilde{\sigma}'')$, and the summation over $\{\sigma\}$ means summing over σ, σ', σ'', and $\tilde{\sigma}''$. Eq.(16) is valid in the Feynman gauge. The issue of gauge dependence has been addressed recently in Ref. [12] which shows that, within the approximations used, the solution [1,2] to this equation which will be discussed below satisfies the Ward-Takahashi identities.

A general analytic solution to Eq.(16) for $\tilde{\Sigma}_A(\bar{p})$ is not yet available. However, one can obtain an approximate infrared solution by the following simplifications. First we observe that, due to the factor $e^{-\hat{q}_\perp^2}$ in the integrand, only the contributions from small values of $\hat{q}_\perp$ are important. We may therefore truncate the $J_{nn''}$ series and keep only the terms with the smallest power of $\hat{q}_\perp$, i.e., $J_{nn''}(\hat{q}_\perp) \to n! \, \delta_{nn''}$; and similarly for the $J_{\tilde{n}''n'}(\hat{q}_\perp)$. This will be referred to as the small $\hat{q}_\perp$ approximation and is valid only for weak couplings ($g^2/4\pi \ll 1$) [2].

Next, we sum over the spin indices and note that the remaining summation over k'' involves at most three terms: for $k > 0$, $k'' = k$, $k \pm 1$. In the limit $k = 0 = \bar{p}_\perp$, we keep only the dominant $k'' = 0$ term. This is known as the lowest Landau level approximation [3]. The SD equation is now simplified to

$$\Sigma_A(\bar{p}_\|) \simeq 2g^2 \int \frac{d^4q}{(2\pi)^4} \frac{e^{-\hat{q}_\perp^2}}{q^2} \frac{\Sigma_A(\bar{p}_\| - q_\|)}{(\bar{p}_\| - q_\|)^2 + \Sigma_A^2(\bar{p}_\| - q_\|)}, \quad (19)$$

where we have made a Wick rotation to Euclidean space: $p_0 \to ip_4$, $q_0 \to iq_4$. Note that the fermion wavefunction renormalization vanishes in the Feynman gauge, hence the self-energy $\hat{\Sigma}_A$ has been replaced by the dynamically generated fermion mass Σ_A in Eq.(19). Except for the exponential factor in the integrand, Eq.(19) has the same form as the corresponding SD equation when the external field is absent. The difference is that only the longitudinal momentum is relevant here. This reduction of dimensions from 4 to 2 has been stressed in Ref. [3].

Finally, we consider the $\bar{p}_\| = 0$ limit and approximate the $\Sigma_A(q_\|)$ in the resultant integrand by $\Sigma_A(0) \equiv m$ to secure the gap equation,

$$\begin{aligned} 1 &\simeq 2g^2 \int \frac{d^4q}{(2\pi)^4} \frac{e^{-\hat{q}_\perp^2}}{q^2} \frac{1}{q_\|^2 + m^2} \\ &\simeq \frac{g^2}{4\pi^2} |gH| \int_0^\infty d\hat{q}_\perp^2 \frac{e^{-\hat{q}_\perp^2} \ln(2|gH|\hat{q}_\perp^2/m^2)}{2|gH|\hat{q}_\perp^2 - m^2}. \end{aligned} \quad (20)$$

The solution to Eq.(20) has the form

$$m \simeq a \sqrt{|gH|}\, e^{-2\pi b/g}, \quad (21)$$

where a and b are positive constants of order 1. The g-dependence of m clearly indicates the nonperturbative nature of this result.

The above solution for the fermion dynamical mass is consistent with that found by Gusynin et al. [3], who studied the Bethe-Salpeter equation for the bound-state Nambu-Goldstone bosons of the spontaneously broken chiral symmetry. The order parameter of this symmetry breakdown is computed to be [2,8]

$$\begin{aligned} \langle \bar{\psi}\psi \rangle &\simeq -\frac{|gH|}{2\pi^2} m \ln\left(\frac{|gH|}{m^2}\right) \\ &\simeq -\frac{2ab}{g\pi} |gH|^{3/2} e^{-2\pi b/g}. \end{aligned} \quad (22)$$

Note that, by allowing the fermion field ψ to carry quantum numbers of internal symmetries, $\langle \bar{\psi}\psi \rangle \to \langle \bar{\psi}_i \psi_j \rangle$, the formalism developed above may be applied to study the dynamical breaking of internal gauge symmetries (in quenched, ladder approximation) in the presence of an external field.

Currently there is no experimental evidence for the magnetic field induced dynamical chiral/gauge symmetry breaking found in the above solution. However, suggestive hints are available in excitonic systems. In their recent experiment with the coupled AlAs/GaAs quantum wells [18], Butov et al., found evidence of exciton condensation (condensation of electron-hole pairs analogous to the fermion-antifermion pairing in the spontaneously broken chiral vacuum) in the form of a huge broad band noise in the photoluminescence intensity when a sufficiently strong magnetic field is applied, while no exciton condensation occurs in the absence of the external magnetic field.

As an application, the method developed in the study of chiral symmetry breaking in an external magnetic field has been employed in the study of the effects of external magnetic fields on high-T_c superconductors [19]. The dynamical chiral symmetry breaking solution found above may also be relevant for the chiral phase transition in heavy-ion collisions or in the electroweak phase transition [3] in the early universe when a large primordial magnetic field is expected to be present [20]. In these cases, the effects of temperature and chemical potential need be incorporated.

It is straightforward to generalize the above formalism to include nonzero temperature ($T \neq 0$) and nonzero chemical potential ($\mu \neq 0$) effects. One finds that the gap equation, Eq.(20), becomes [2]

$$1 \simeq \frac{g^2}{8\pi^2}|gH|\int_{-\infty}^{\infty} dq_3 \int_0^{\infty} d\hat{q}_\perp^2 \frac{e^{-\hat{q}_\perp^2}}{Q_1 Q_2}$$
$$\cdot \left\{ Q_1 \left[\frac{\coth(\frac{Q_2}{2T})}{Q_1^2 - (Q_2 + \mu - i\pi T)^2} + \frac{\coth(\frac{Q_2}{2T})}{Q_1^2 - (Q_2 - \mu + i\pi T)^2} \right] \right. $$
$$\left. + Q_2 \left[\frac{\tanh(\frac{Q_1+\mu}{2T})}{Q_2^2 - (Q_1 + \mu - i\pi T)^2} + \frac{\tanh(\frac{Q_1-\mu}{2T})}{Q_2^2 - (Q_1 - \mu + i\pi T)^2} \right] \right\} \quad (23)$$

where $Q_1^2 \equiv q_3^2 + m_{T\mu}^2$, $Q_2^2 \equiv q_3^2 + 2|gH|\hat{q}_\perp^2$, and $m_{T\mu}$ is the infrared dynamical fermion mass which depends on both the temperature and the chemical potential. We use units in which the Boltzmann constant equals 1. Aside from the quenched, ladder approximation, and the small $\hat{q}_\perp$ and lowest Landau level approximations, Eq.(23) is exact in its dependence on the coupling constant, the magnetic field, the temperature, and the chemical potential.

We have considered separately the $\mu = 0$ and $T = 0$ limits of Eq.(23). We examine both analytically and numerically the behavior of the dynamical mass as T (or μ) is varied. In the ($T \neq 0$, $\mu = 0$) case, we find that m_{T0} decreases monotonically as the temperature is raised and eventually vanishes above a critical temperature, indicating that the chiral symmetry is restored at high temperatures. The critical temperature at which this continuous phase transition takes place is estimated to be [2,6,7]

$$T_c \sim 2\sqrt{2}\, m_{00}, \quad (24)$$

where m_{00} is the ($T = 0$, $\mu = 0$) solution found in Eq.(21). The order parameter for this phase transition exhibits similar behaviors:

$$\langle\bar{\psi}\psi\rangle_{T0} \simeq -\frac{|gH|}{2\pi^2} m_{T0} \ln\left(\frac{|gH|}{m_{T0}^2}\right)$$

$$\sim |gH|^{3/2}\left(1-\frac{T}{T_c}\right)^{1/2} \ln\left(1-\frac{T}{T_c}\right)^{1/2}, \quad \text{as } T \to T_c^-. \tag{25}$$

The last expression reflects the behavior of m_{T0} near T_c:

$$m_{T0} \sim |gH|^{1/2}\left(1-\frac{T}{T_c}\right)^{1/2}, \quad \text{as } T \to T_c^-. \tag{26}$$

In the ($T=0$, $\mu \neq 0$) case, $m_{0\mu}$ also vanishes as μ is increased beyond a critical value, thus restoring the chiral symmetry at high chemical potentials. However, this chiral phase transition is discontinuous [2]:

$$\langle\bar{\psi}\psi\rangle_{0\mu} \simeq -\frac{|gH|}{2\pi^2} m_{0\mu} \ln\left(\frac{|gH|}{m_{0\mu}^2}\right),$$

$$\begin{aligned} m_{0\mu} &> 0, & \mu &< \mu_c, \\ &= 0, & \mu &> \mu_c. \end{aligned} \tag{27}$$

The critical chemical potential is approximately

$$\mu_c \simeq \frac{m_{00}}{\sqrt{1-\frac{2I_2}{I_1}}} \tag{28}$$

where

$$\begin{aligned} I_1 &= \int_{-\infty}^{\infty} dq_3 \int_0^{\infty} d\hat{q}_\perp^2 \frac{e^{-\hat{q}_\perp^2}}{Q^2 Q_2(Q+Q_2)}\left(\frac{1}{Q}+\frac{1}{Q+Q_2}\right), \\ I_2 &= \int_{-\infty}^{\infty} dq_3 \int_0^{\infty} d\hat{q}_\perp^2 \frac{e^{-\hat{q}_\perp^2}}{QQ_2(Q+Q_2)^3}, \end{aligned} \tag{29}$$

and $Q^2 \equiv q_3^2 + m_{00}^2$. The integrals I_1 and I_2 are both positive and finite.

With these results, we can evaluate whether the dynamics of chiral symmetry breaking in a magnetic field discussed here may be relevant for the electroweak phase transition in the early universe. For instance, the electroweak phase transition took place at a temperature of order 100 GeV. ¿From Eq.(24), this requires a magnetic field of order 2×10^{41} gauss if we take $a = b = 1$ and $4\pi/g^2 \simeq 137$. This is much larger than any estimates of the magnetic field strength at the time of the electroweak phase transition [20]. We may therefore conclude that the chiral symmetry breaking solution considered here does not play any role in the electroweak phase transition.

On the other hand, as suggested earlier, the formalism described here can be used to study the dynamical breaking of internal gauge symmetries, abelian as well as nonabelian, in the presence of a magnetic field. Within the quenched, planar approximation, we expect the same generic results as obtained here to be applicable in those situations. We may therefore entertain the possibility that the coupling constant is relatively large, e.g., $4\pi/g^2$ of order 0.1 (As concrete examples, we may consider electroweak symmetry breaking in technicolor models or chiral symmetry breaking in QCD by colored fermions belonging to large representations of SU(3)). In this case the required magnetic field could be of order 10^{28} gauss or less, which makes it an interesting possibility for the study of phase transitions in the early universe.

ACKNOWLEDGEMENTS

This work was supported in part by the U.S. Department of Energy under Grant No. DE-FG02-84ER40163. I would like to thank the faculties at the Physics Departments of the Tokyo Metropolitan University and of the Academia Sinica in Taipei, especially H. Minakata, O. Yasuda, and H.-L. Yu, for their hospitality during my visit to their institutions where this article was written.

REFERENCES

1. C. N. Leung, Y. J. Ng, and A. W. Ackley, Phys. Rev. D **54**, 4181 (1996).
2. D.-S. Lee, C. N. Leung, and Y. J. Ng, Phys. Rev. D **55**, 6504 (1997); Phys. Rev. D **57**, 5224 (1998).
3. V. P. Gusynin, V. A. Miransky, and I. A. Shovkovy, Phys. Rev. D **52**, 4747 (1995); Nucl. Phys. **B462**, 249 (1996).
4. D. K. Hong, Y. Kim, and S.-J. Sin, Phys. Rev. D **54**, 7879 (1996).
5. A. V. Shpagin, preprint in the Los Alamos archive, hep-ph/9611412.
6. V. P. Gusynin and I. A. Shovkovy, Phys. Rev. D **56**, 5251 (1997).
7. D. Ebert and V. Ch. Zhukovsky, Mod. Phys. Lett. A **12**, 2567 (1997).
8. I. A. Shushpanov and A. V. Smilga, Phys. Lett. B **402**, 351 (1997).
9. D. K. Hong, Phys. Rev. D **57**, 3759 (1998).
10. K. Farakos, G. Koutsoumbas, and N. E. Mavromatos, preprint in the Los Alamos archive, hep-lat/9802037 (February, 1998).
11. Y. J. Ng, preprint in the Los Alamos archive, hep-th/9803074 (March, 1998).
12. E. J. Ferrer and V. de la Incera, preprint in the Los Alamos archive, hep-th/9803226 (March, 1998).
13. V. A. Miransky, preprint in the Los Alamos archive, hep-th/9805159 (May, 1998).
14. K. Johnson, M. Baker, and R. Willey, Phys. Rev. Lett. **11**, 518 (1963) and Phys. Rev. **136**, B1111 (1964); T. Maskawa and H. Nakajima, Prog. Theor. Phys. **52**, 1326 (1974); R. Fukuda and T. Kugo, Nucl. Phys. **B117**, 250 (1976).
15. J. Schwinger, Phys. Rev. **82**, 664 (1951).
16. See the contributions of V. I. Ritus in *Issues in Intense-Field Quantum Electrodyanamics*, ed. V. L. Ginzburg (Nova Science, Commack, 1987), and references therein.
17. See, e.g., *Handbook of Mathematical Functions*, eds. M. Abramowitz and I. A. Stegun (Dover, New York, 1964).
18. L. V. Butov *et al.*, Phys. Rev. Lett. **73**, 304 (1994).
19. K. Farakos and N. E. Mavromatos, preprint in the Los Alamos archive, cond-mat/9710288 (October, 1997).
20. See, e.g., M. S. Turner and L. M. Widrow, Phys, Rev. D **37**, 2743 (1988); J. Ambj/orn and P. Olesen, hep-ph/9304220, in the *Proceedings of the 4th Hellenic School on Elementary Particle Physics*, Corfu, September 1992; G. Baym, D. Bodeker, and L. Mclerran, Phys. Rev. D **53**, 662 (1996); A. Brandenburg, K. Enqvist, and P. Olesen, Phys. Rev. D **54**, 1291 (1996); Phys. Lett. B **392**, 395 (1997); G. Sigl, A. Olinto, and K. Jedamzik, Phys. Rev. D **55**, 4582 (1997); J. Ahonen and K. Enqvist, Phys. Rev. D **57**, 664 (1998).

Yukawa Interactions and Dynamical Generation of Mass in an External Magnetic Field[1]

E. J. Ferrer and Vivian de la Incera

Department of Physics
State University of New York at Fredonia
NY 14063, USA

Abstract. In this work we study the dynamical generation of a fermion mass induced by a constant and uniform external magnetic field in an Abelian gauge model with a Yukawa term. We show that the Yukawa coupling not only enhances the dynamical generation of the mass, but it substantially decreases the magnetic field required for the mass to be generated at temperatures comparable to the electroweak critical temperature. These results indicate that if large enough primordial magnetic fields were present during the early universe evolution, the field-induced generation of fermion masses, which in turn corresponds to the generation of fermion bound states, may play an important role in the electroweak phase transition.

In this talk I would like to speak about a recently found phenomenon known as the catalysis of chiral symmetry breaking due to a magnetic (chromomagnetic, hypermagnetic) field and its possible implications for the electroweak phase transition. The essence of the catalysis of chiral symmetry breaking lies in the dimensional reduction in the dynamics of fermion pairing in the presence of a magnetic field [1]. Due to such a dimensional reduction, the magnetic field catalyses the generation of a fermion condensate, and consequently, of a dynamical fermion mass, even in the weakest attractive interaction between fermions.

An important aspect of the magnetic field induced generation of a dynamical mass (MIGDM) is related to its possible cosmological consequences. The existence in the early universe of a magnetic field induced dynamical mass, as well as the fermion condensate associated with it, would require the presence of very large primordial magnetic fields during the early stages of the universe evolution. It is worth to note, however, that at present large primordial magnetic fields in the early universe do not seem an impossible option. As it is known, such large fields may be needed to explain the large-scale galactic magnetic fields $\sim 10^{-6}$G observed in our own, as well as in other galaxies. There are several primordial field generating

---

[1] Talk presented by Vivian de la Incera

mechanisms which typically predict fields as large as $10^{24}G$ during the electroweak phase transitions. Moreover, Ambjørn and Olesen [2] have claimed that seed primordial fields even larger, $\sim 10^{33}G$, would be necessary at the electroweak scale to explain the observed galactic fields.

It has been speculated [1] that the character of electroweak phase transition could be affected by the MIGDM. However, the results of Lee, Leung and Ng [3], and of Gusynin and Shovkovy [4] seem to indicate that the fields required for the MIGDM to be important at the electroweak scale are too large ($\sim 10^{42}G$) to be realistically attainable. It must be pointed out however that their results were found from the study of MIGDM in QED at finite temperature, and it is reasonable to expect that their conclusions may change in the context of the electroweak model when a richer set of interactions enters in scene.

In the present paper, we show that even in a simpler toy model, which retains some of the attributes of the electroweak theory, the effect of new couplings, like Yukawa couplings, can substantially change the order of magnitude of the magnetic field required for the dynamical fermion mass to be nonzero at the electroweak scale.

The model under study will be an Abelian gauge model of massless fermions with Yukawa interaction in the presence of a constant magnetic field. By solving the Schwinger-Dyson equation for the fermion propagator in the ladder approximation, we prove that the Yukawa interaction enhances the dynamical generation of a magnetic-field-induced fermion mass. We study the same model at finite temperature, calculating the critical temperature at which the field-induced fermion mass disappears. For a Yukawa coupling of order of the top coupling, the field strength, required to obtain a critical temperature comparable to the electroweak critical temperature, is decreased in 10 orders of magnitude as compared to the corresponding field strength in QED.

These results indicate that if large enough primordial magnetic (or hypermagnetic) fields were present at the electroweak scale, the field-induced generation of fermion masses, which in turn corresponds to the generation of fermion bound states, may play an important role in the electroweak phase transition. Our main conclusion is that the Yukawa interactions enhance the dynamical generation of fermion bound states and masses in the presence of external magnetic (or hypermagnetic) fields, and therefore, it is worth to study this effect in the unbroken phase of the electroweak system.

Let us consider an Abelian gauge model with a Yukawa interaction described by the Lagrangian

$$L = \frac{1}{4}F^{\mu\nu}F_{\mu\nu} + i\overline{\psi}\gamma^\mu\partial_\mu\psi - e\overline{\psi}\gamma^\mu\psi A_\mu - \frac{1}{2}\xi(\partial_\mu A^\mu)^2 + \frac{1}{2}\partial_\mu\phi\partial^\mu\phi - \frac{\lambda}{4}\phi^4 - \sqrt{2}\lambda_y\phi\overline{\psi}\psi$$

(1)

Note that this Lagrangian has a U(1) gauge symmetry and a fermion number global symmetry, but it does not have chiral symmetry, so the appearance of a dynamical mass in this model cannot be linked to chiral symmetry breaking. This is fine since our ultimate goal is to get insight of a possible magnetic field induced dynamical mass in the electroweak theory, where anyway there is no chiral symmetry to break. We are interested in the study of this theory in the presence of a constant and uniform external magnetic field H. Our aim is to find nonperturbative solutions of the Schwinger-Dyson equation for the fermion propagator to investigate the dependence of the dynamically generated mass on the Yukawa coupling. We must point out however that, even though the appearance of the dynamical mass can be traced to the existence of a fermion-antifermion condensate, there is no Goldstone field produced in this case, because there is no continuous symmetry broken by the fermion condensate in this simple model. This would not be the case in the electroweak model, on which, if a fermion condensate were catalyzed by the magnetic field, it would give rise to a nonzero vev of the scalar field and hence to a Higgs-like spontaneous gauge symmetry breaking.

The Schwinger-Dyson equation derived from this theory takes the form

$$\overline{G}^{-1}(x,y) = G^{-1}(x,y) - ie\int d^4u d^4w \gamma \overline{G}(x,u)\overline{D}(x-w)\overline{\Gamma}_{\psi\psi A}+$$

$$+i\sqrt{2}\lambda_y \int d^4u d^4w \overline{G}(x,u)\overline{S}(x-w)\overline{\Gamma}_{\psi\psi\phi} \qquad (2)$$

Here G refers to fermions, D to gauge bosons and S to scalar bosons. $\overline{\Gamma}_{\psi\psi A}$ and $\overline{\Gamma}_{\psi\psi\phi}$ are three-fields vertex functions. The bar indicates full Green functions.

Assuming that the couplings are small, we can take eq.(2) in the ladder approximation. In this approximation the vertices will be taken bare (we assume no coupling is running), the fermion propagator is taken full, and the gauge and scalar boson propagators are taken in the tree approximation. The change to momentum coordinates can be done with the help of Ritus [6] E_p functions. Then the SD equation becomes

$$(2\pi)^4 \delta_{kk'}\delta(p_0-p_0')\delta(p_2-p_2')\delta(p_3-p_3')\widetilde{\Sigma}_A(\overline{p}) =$$

$$= ie^2 \int d^4x d^4x' \sum_{k''}\int \frac{dp''_0 dp''_2 dp''_3}{(2\pi)^4}\{\overline{E}_p(x)\gamma^\mu E_{p''}(x)\frac{1}{\gamma\cdot\overline{p}''-\widetilde{\Sigma}_A(\overline{p}'')}\times$$

$$\times\overline{E}_{p''}(x')\gamma^\nu E_{p'}(x')D_{\mu\nu}(x-x')\} - i2\lambda_y^2\int d^4x d^4x'\sum_{k''}\int \frac{dp''_0 dp''_2 dp''_3}{(2\pi)^4}\{\overline{E}_p(x)E_{p''}(x)\times$$

$$\times\frac{1}{\gamma\cdot\overline{p}''-\widetilde{\Sigma}_A(\overline{p}'')}\overline{E}_{p''}(x')E_{p'}(x')S(x-x')\} \qquad (3)$$

where $\widetilde{\Sigma}_A(\overline{p})$ is the fermion mass operator. The E_p matrix is defined as

$$E_p(x) = \sum_\sigma [N(n)e^{i(p_0x^0+p_2x^2+p_3x^3)}D_n(\rho)]diag(\delta_{\sigma 1},\delta_{\sigma -1},\delta_{\sigma 1},\delta_{\sigma -1})$$

$$= \sum_\sigma N(n)e^{i(p_0x^0+p_2x^2+p_3x^3)}D_n(\rho)\Delta(\sigma) \qquad (4)$$

with $D_n(\rho)$ being the parabolic cylinder functions [7] with argument $\rho = \sqrt{2|eH|}(x_1 - \frac{p_2}{eH})$ and positive integer index

$$n = n(k, \sigma) \equiv k + \frac{eH\sigma}{2|eH|} - \frac{1}{2}, \quad n = 0, 1, 2, ...; \quad \sigma = -1, 1 \tag{5}$$

Using the properties of the E_p functions, the expression (3) can be reduced to

$$\delta_{kk'}\widetilde{\Sigma}_A(\bar{p}) = ie^2 2|eH| \sum_{k''} \sum_{\{\sigma\}} \int \frac{d^4\hat{q}}{(2\pi)^4} \{ \frac{e^{isgn(eH)(n-n''+\tilde{n}''-n')\varphi}}{\sqrt{n!n''!\tilde{n}''!n'!}} e^{-\hat{q}_\perp^2} J_{nn''}(\hat{q}_\perp) J_{\tilde{n}''n'}(\hat{q}_\perp) \frac{1}{\hat{q}^2} \times$$

$$\times \left(g_{\mu\nu} - (1-\xi)\frac{\hat{q}_\mu\hat{q}_\nu}{\hat{q}^2}\right) \Delta\gamma^\mu\Delta'' \frac{1}{\gamma\cdot\bar{p}'' - \widetilde{\Sigma}_A(\bar{p}'')} \tilde{\Delta}''\gamma^\nu\Delta' -$$

$$i2\lambda_y^2(2|eH|) \sum_{k''} \sum_{\{\sigma\}} \int \frac{d^4\hat{q}}{(2\pi)^4} \{ \frac{e^{isgn(eH)(n-n''+\tilde{n}''-n')\varphi}}{\sqrt{n!n''!\tilde{n}''!n'!}} e^{-\hat{q}_\perp^2} J_{nn''}(\hat{q}_\perp) J_{\tilde{n}''n'}(\hat{q}_\perp) \frac{1}{\hat{q}^2} \times$$

$$\times \Delta\Delta'' \frac{1}{\gamma\cdot\bar{p}'' - \widetilde{\Sigma}_A(\bar{p}'')} \tilde{\Delta}''\Delta' \tag{6}$$

where

$$J_{n_p n_r}(\hat{q}_\perp) \equiv \sum_{m=0}^{\min(n_p,n_r)} \frac{n_p! n_r!}{m!(n_p-m)!(n_r-m)!} [isgn(eH)\hat{q}_\perp]^{n_p+n_r-2m} \tag{7}$$

$$\hat{q}_\mu \equiv \frac{q_\mu\sqrt{2|eH|}}{2eH}, \quad \mu = 0, 1, 2, 3 \tag{8}$$

$$\hat{q}_\perp \equiv \sqrt{\hat{q}_1^2 + \hat{q}_2^2}, \quad \varphi \equiv \arctan(\hat{q}_2/\hat{q}_1) \tag{9}$$

$$\bar{p}'' = (p_0 - q_0, 0, -sgn(eH)\sqrt{2|eH|k''}, p_3 - q_3) \tag{10}$$

To solve equation (6) we must use the structure of the mass operator. Although in the presence of the external field the mass operator's structure is quite rich(see ref. [8]), we can use a more simple structure, which is in agreement with the solution of the Ward identity within the present approximation [8]. Thus, we consider

$$\widetilde{\Sigma}_A(\bar{p}) = Z_\parallel \gamma\cdot\bar{p}_\parallel + Z_\perp \gamma\cdot\bar{p}_\perp + m(\bar{p}) \tag{11}$$

Note the separation between parallel- and perpendicular- to-the-magnetic-field variables.

Using the structure of $\widetilde{\Sigma}_A(\bar{p})$ given in (11), taking into account that the contributions of large $\hat{q}_\perp$ in (6) are suppressed by the factor $e^{-\hat{q}_\perp^2}$, and considering the infrared region $\bar{p}^2 << |eH|$ and in particular the lower Landau level contributions $\bar{p}_\perp = 0$, equation (6) (taken in the Feynman gauge) can be simplified to lead to the following two equations in Euclidean space

$$Z_\| \gamma \cdot \bar{p}_\| = -4\lambda_y^2(|eH|) \int \frac{d^4\hat{q}}{(2\pi)^4} \frac{e^{-\hat{q}_\perp^2}(1+Z_\|)\gamma \cdot (\bar{p}_\| - q_\|)}{\hat{q}^2 \left[(1+Z_\|)^2(\bar{p}_\| - q_\|)^2 + m^2(\bar{p}_\| - q_\|)\right]} \quad (12)$$

$$m(\bar{p}_\| - q_\|) = 4|eH|(e^2 + \lambda_y^2) \int \frac{d^4\hat{q}}{(2\pi)^4} \frac{e^{-\hat{q}_\perp^2} m(\bar{p}_\| - q_\|)}{\hat{q}^2 \left[(1+Z_\|)^2(\bar{p}_\| - q_\|)^2 + m^2(\bar{p}_\| - q_\|)\right]} \quad (13)$$

The first equation has solution $Z_\| = 0$ if $\lambda_y^2 << 16\pi^2$. The second is the gap equation, which, in the infrared limit that we are considering, has solution

$$m \simeq \sqrt{2|eH|} \exp\left[-\sqrt{\frac{\pi}{\alpha + \frac{\lambda_y^2}{4\pi}}}\right] \quad (14)$$

α is the fine structure constant. The consistency of the approximation requires $m << \sqrt{|eH|}$, which is satisfied if $\alpha + \frac{\lambda_y^2}{4\pi} << 1$, so the dynamical mass appears in the weak coupling region of the theory. Note that because of the exponential function in (14), small changes in the exponent can yield substantial changes in the mass. For instance, for $\lambda_y \simeq 0.7$, a value comparable to the top Yukawa coupling, the dynamical mass (14) is five orders of magnitude larger than the mass found in QED [1,3], which is given by $m \simeq \sqrt{2|eH|} \exp\left[-\sqrt{\frac{\pi}{\alpha}}\right]$.

Finite temperature calculations can be done using the well known imaginary time formalism. In that case the gap equation takes the form

$$m(\omega_{n'}, p) = (\alpha + \frac{\lambda_y^2}{4\pi}) \frac{T}{\pi} \sum_{n=-\infty}^{\infty} \int_{-\infty}^{\infty} \frac{dk\, m(\omega_n, k)}{\omega_n^2 + k^2 + m^2(\omega_n, k)} \int_0^\infty \frac{dk \exp(-\frac{x}{2|eH|})}{(\omega_n - \omega_{n'})^2 + (k-p)^2 + x} \quad (15)$$

with $\omega_n = (2n+1)\pi T$. From this equation one can show that the critical temperature at which the dynamical mass m vanishes is

$$T \simeq m(T=0) \simeq \sqrt{2|eH|} \exp\left[-\sqrt{\frac{\pi}{\alpha + \frac{\lambda_y^2}{4\pi}}}\right] \quad (16)$$

Therefore, for $\lambda_y \simeq 0.7$, we can estimate the magnetic field required to have a critical temperature (the temperature at which the dynamical mass vanishes) comparable to the electroweak critical temperature. Such a critical field is $H \approx 10^{32} G$. That is, thanks to the Yukawa interaction the critical field has decreased in 10 orders of magnitude as compared to its corresponding value in QED ($\lambda_y \simeq 0$).

We conclude that the Yukawa interactions enhance the dynamical generation of fermion bound states and masses in the presence of external magnetic (or hypermagnetic) fields, and therefore, it is worth to study this effect in the unbroken phase of the electroweak system.

Acknowledgments

It is a pleasure to thanks D. Caldi, C. N. Leung, Y. J. Ng and I. A. Shovkovy for very useful discussions. Our special thanks to V. A. Miransky for enlightening discussions and for calling our attention to the basic papers of ref.[3].

This work has been supported in part by NSF grant PHY-9722059.

REFERENCES

1. V. P. Gusynin, V. A. Miransky, and I. A. Shovkovy, Phys. Rev D 52, 4747 (1995); Nucl. Phys. B462,249 (1996); Phys.Lett. B 349, 477 (1995); Phys. Rev. Lett., 73, 3499 (1994).
2. J. Ambjørn and P. Olesen, hep-ph/9304220
3. D.-S Lee, C. N. Leung, and Y. J. Ng, Phys. Rev D 55, 6504 (1997), C. N. Leung, Y. J. Ng, and A. W. Ackley, Phys. Rev D 54, 4181 (1996).
4. V. P. Gusynin and I. A. Shovkovy, Phys. Rev D 56, 5251 (1997)
5. V. A. Miransky, *Dynamical Symmetry Breaking in Quantum Field Theories* (World Scientific Publ., Singapore, 1993).
6. V. I. Ritus in Issues in Intense-Field Quantum Electrodynamics, ed. V. L. Ginzburg (Nova Science, Commack, 1987).
7. *Handbook of Mathematical Functions*, eds. M. Abramowitz and I. A. Stegun (Dover, New York, 196).
8. E. J. Ferrer and V. de la Incera, hep-th/9703226

Magnetic Response in Anyon Fluid at High Temperature [1]

E. J. Ferrer and Vivian de la Incera

Department of Physics
State University of New York at Fredonia
NY 14063, USA

Abstract. The magnetic response of the charged anyon fluid at temperatures lower and larger than the fermion enery gap ω_c is investigated in the self-consistent field approximation. We prove that the anyon system with boundaries exhibits a total Meissner effect at temperatures smaller than the fermion energy gap ($T \ll \omega_c$). The London penetration length at $T \sim 200K$ is of the order $\lambda \sim 10^{-5} cm$. At $T \gg \omega_c$ a new phase, characterized by an inhomogeneous magnetic penetration, is found. We conclude that the energy gap, ω_c, defines a scale that separates two phases: a superconducting phase at $T \ll \omega_c$, and a non-superconducting one at $T \gg \omega_c$.

Anyons [1,2] are particles with fractional statistics in (2+1)-dimensions. The anyon description within the Chern-Simons (CS) gauge theory is equivalent to attaching flux tubes to the charged fermions. The Aharonov-Bohm phases resulting from the adiabatic transport of two anyons is the source of the fractional exchange statistics [2].

It has been argued that strongly correlated electron systems in two dimensions can be described by an effective field theory of anyons [3]. Anyons can be also obtained as solitons with fractional spin in electron systems. Excitations with fractional spin in two dimensional systems necessarily obey fractional statistics [4].

As it is known, anyon superconductivity has an origin different from the Nambu-Goldstone-Higgs like mechanism. The genesis of the anyon superconductivity is given by the spontaneously violation of commutativity of translations in the free anyon system [5]. This new mechanism might find wide applications in new physical studies.

Anyon superconductivity at $T = 0$ is a well establish result [6,5]. However, at $T \neq 0$ several authors [7] have advocated that the superconducting phase evaporates at any finite temperature. The reasons is that at $T \neq 0$ there exists a long-range electromagnetic mode inside the infinite bulk [8]. This long range mode is the

---

[1] Talk presented by E.J. Ferrer

consequence of the existence of a pole $\sim \left(\frac{1}{k^2}\right)$ in the fermion polarization operator component Π_{00} at finite temperature.

In previous works [10] we found that, contrary to some authors' belief, the superconducting behavior, manifested through the Meissner effect in the charged anyon fluid at $T = 0$, does not disappear as soon as the system is heated. In Ref. [10] we showed that the presence of boundaries affects the dynamics of the two-dimensional system in such a way that the long-range mode, that accounts for a homogeneous field penetration [8], cannot propagate in the bulk. According to these results, the anyon system with boundaries exhibits a total Meissner effect at temperatures smaller than the fermion energy gap ($T \ll \omega_c$).

It is natural to expect that at temperatures larger than the energy gap this superconducting behavior should not exist. At those temperatures the electron thermal fluctuations should make accessible the free states existing beyond the energy gap. These heuristic arguments were corroborated in Ref. [11]. There, we proved that at $T \gg \omega_c$ the charged anyon fluid does not exhibit a Meissner effect.

We can conclude that the energy gap ω_c defines a scale that separates two phases in the charged anyon fluid: a superconducting phase at $T \ll \omega_c$, and a non-superconducting one at $T \gg \omega_c$.

We must emphasize that the scenario we have found for the anyon superconductivity at finite temperature is in agreement with the rationale of anyon superconductivity given by Wilczek [2]. Wilczek has pointed out that the London arguments, which start from the role of the energy gap as an essential fact in the theory of superconductivity, seem to provide the base for anyon superconductivity. In the charged anyon fluid, there is no a charge-violating local order parameter so familiar in the theories with spontaneously broken symmetry. In this system, instead, it is the background CS magnetic field $\bar{b}$ what determines the energy gap ($\omega_c = \bar{b}/m$) and plays the role of the order parameter in the anyon gas [5].

In what follows we present the results for the linear magnetic response of the charged anyon fluid to an applied constant and uniform magnetic field, at temperatures lower and larger than the energy gap.

The linear response of the medium can be found under the assumption that the quantum fluctuations of the gauge fields about the ground-state are small. In this case the one-loop fermion contribution to the effective action, obtained after integrating out the fermion fields, can be evaluated up to second order in the gauge fields. The effective action in terms of the quantum fluctuation of the gauge fields within the linear approximation [8,9] takes the form

$$\Gamma_{eff}(A_\nu, a_\nu) = \int dx \left(-\frac{1}{4}F_{\mu\nu}^2 - \frac{N}{4\pi}\varepsilon^{\mu\nu\rho}a_\mu\partial_\nu a_\rho\right) + \Gamma^{(2)} \tag{1}$$

$\Gamma^{(2)}$ is the one-loop fermion contribution to the effective action in the linear approximation

$$\Gamma^{(2)} = \int dx dy \left[a_\mu(x) + eA_\mu(x)\right] \Pi^{\mu\nu}(x,y) \left[a_\nu(y) + eA_\nu(y)\right]. \tag{2}$$

In (2) $\Pi_{\mu\nu}$ represents the fermion one-loop polarization operator in the presence of the CS background magnetic field $\bar{b}$.

Taking into account that we will investigate the magnetic response of the charged anyon fluid to a uniform and constant applied magnetic field, we need the $\Pi_{\mu\nu}$ leading behavior for static ($k_0 = 0$) and slowly ($\mathbf{k} \sim \mathbf{0}$) varying configurations. In this limit, and using the frame on which $k^i = (k, 0)$, $i = 1, 2$, the polarization operator takes the form

$$\Pi^{\mu\nu} = \begin{pmatrix} -(\Pi_0 + \Pi_0' k^2) & 0 & -i\Pi_1 k \\ 0 & 0 & 0 \\ i\Pi_1 k & 0 & \Pi_2 k^2 \end{pmatrix} \qquad (3)$$

The leading contributions of the one-loop polarization operator coefficients Π_0, Π_0', Π_1 and Π_2 in the static limit ($k_0 = 0$, $\mathbf{k} \sim \mathbf{0}$) at low temperatures ($T \ll \omega_c$) are

$$\Pi_0 = \frac{2\beta\bar{b}}{\pi} e^{-\beta\bar{b}/2m}, \qquad \Pi_0' = \frac{mN}{2\pi\bar{b}}\Lambda, \qquad \Pi_1 = \frac{N}{2\pi}\Lambda, \qquad \Pi_2 = \frac{N^2}{4\pi m}\Lambda',$$

$$\Lambda = \left[1 - \frac{2\beta\bar{b}}{m} e^{-\beta\bar{b}/2m}\right], \qquad \Lambda' = \left[\Lambda - \frac{2\beta\bar{b}}{mN^2} e^{-\beta\bar{b}/2m}\right] \qquad (4)$$

and at high temperatures ($T \gg \omega_c$) are

$$\Pi_0 = \frac{m}{2\pi}\left[\tanh\frac{\beta\mu}{2} + 1\right], \qquad \Pi_0' = -\frac{\beta}{48\pi}\operatorname{sech}^2\left(\frac{\beta\mu}{2}\right), \qquad \Pi_1 = \frac{\bar{b}}{m}\Pi_0',$$

$$\Pi_2 = \frac{1}{12m^2}\Pi_0 \qquad (5)$$

In these expressions μ is the chemical potential and $m = 2m_e$ (m_e is the electron mass). These results are in agreement with those found in Ref.[8].

The extremum equations obtained from the effective action (1) for the Maxwell and CS fields are

$$\nabla \cdot \mathbf{E} = e J_0, \qquad -\partial_0 \mathbf{E}^k + \varepsilon^{kl}\partial_l \mathbf{B} = e \mathbf{J}^k$$

$$\frac{eN}{2\pi} b = \nabla \cdot \mathbf{E}, \qquad \frac{eN}{2\pi} f_{0k} = \varepsilon_{kl}\partial_0 \mathbf{E}^l + \partial_k \mathbf{B} \qquad (6)$$

$f_{\mu\nu}$ is the CS gauge field strength tensor, defined as $f_{\mu\nu} = \partial_\mu a_\nu - \partial_\nu a_\mu$, and J_{ind}^μ is the current density induced by the many-particle system.

$$J_{ind}^0(x) = \Pi_0 \left[a_0(x) + eA_0(x)\right] + \Pi_0' \partial_x (\mathcal{E} +]\mathcal{E}) + \Pi_1 (b + eB) \qquad (7)$$

$$J_{ind}^1(x) = 0, \qquad J_{ind}^2(x) = \Pi_1 (\mathcal{E} +]\mathcal{E}) + \Pi_2 \partial_x (b + eB) \qquad (8)$$

In the above expressions we used the following notation: $\mathcal{E} = \{i_\infty, E = F_{01}, b = f_{12}$ and $B = F_{12}$. We confine our analysis to gauge field configurations which are static and uniform in the y-direction. Within this restriction we are taking a gauge in which $A_1 = a_1 = 0$.

The magnetic field solution obtained from eqs. (6)-(8) is

$$B(x) = -\gamma_1 \left(C_1 e^{-x\xi_1} - C_2 e^{x\xi_1}\right) - \gamma_2 \left(C_3 e^{-x\xi_2} - C_4 e^{x\xi_2}\right) + C_5 \qquad (9)$$

where $\gamma_1 = (\xi_1^2 \kappa + \eta)/\xi_1$, $\gamma_2 = (\xi_2^2 \kappa + \eta)/\xi_2$, $\kappa = \frac{2\pi}{N\delta}\Pi_2$, $\eta = -\frac{e^2}{\delta}\Pi_1$. As can be seen from the magnetic field solution (9), the real character of the inverse length scales ξ_1 and ξ_2 is crucial for the realization of the Meissner effect.

At temperatures much lower than the energy gap ($T \ll \omega_c$) the inverse length scales are given by the following real functions

$$\xi_1 \simeq \sqrt{\frac{e^2 \Pi_1}{\pi \Pi_2}} = e\sqrt{\frac{m}{\pi}}\left[1 + \frac{\pi n_e}{2m}\beta \exp - \left(\frac{\pi n_e \beta}{2m}\right)\right] \qquad (10)$$

$$\xi_2 \simeq \sqrt{\frac{e^2 \Pi_1}{\pi \Pi_0{}'} + \frac{\Pi_0}{\Pi_0{}'}} = e\sqrt{\frac{n_e}{m}}\left[1 + \frac{\pi^2 n_e}{e^2}\beta \exp - \left(\frac{\pi n_e \beta}{2m}\right)\right] \qquad (11)$$

While at temperatures much larger than the energy gap ($T \gg \omega_c$) the inverse length scales are given by

$$\xi_1 \simeq e\sqrt{\Pi_0} = e\sqrt{m/2\pi}\left(\tanh\frac{\beta\mu}{2} + 1\right)^{\frac{1}{2}} \qquad (12)$$

$$\xi_2 \simeq \frac{1}{\pi}(\Pi_2 \Pi_0{}')^{-1/2} = 24i\sqrt{2m/\beta}\cosh\frac{\beta\mu}{2}\left(\tanh\frac{\beta\mu}{2} + 1\right)^{-\frac{1}{2}} \qquad (13)$$

The imaginary value of the inverse length ξ_2 at $(T \gg \omega_c)$ is due to the fact that at those temperatures, $\Pi_2 > 0$ and $\Pi_0{}' < 0$ (see eq. (5)). An imaginary ξ_2 implies that the term $\gamma_2 \left(C_3 e^{-x\xi_2} - C_4 e^{x\xi_2}\right)$, in the magnetic field solution (9), does not have a damping behavior, but an oscillating one.

On the other hand, the presence of the constant coefficient C_5 in the magnetic field solution (9) means that there exists a magnetic long-range mode. Nevertheless, to completely determine the characteristics of the magnetic response in this case, it is needed to find the values of the $C's$ unknown coefficients which are in agreement with the problem boundary conditions and the minimization of the system free-energy density. Considering that the anyon fluid is confined to a half plane $-\infty < y < \infty$ with boundary at $x = 0$. The boundary conditions for the magnetic field are $B(x = 0) = \bar{B}$ ($\bar{B}$ constant), and $B(x \to \infty)$ finite. Because no external electric field is applied, the boundary conditions for this field are, $E(x = 0) = 0$,

$E(x \to \infty)$ finite. The leading contribution of the stable magnetic configurations which satisfy the problem boundary conditions are given at $T \ll \omega_c$ by [10]

$$B(x) = \overline{B} e^{-\xi_2 x} \tag{14}$$

while at $T \gg \omega_c$ it is [11]

$$B(x) = \overline{B} \cos(|\xi_2| x) \tag{15}$$

From (14) and (11) we have that at temperatures lower than the energy gap ($T \ll \omega_c$) a constant and uniform applied magnetic field cannot penetrate the anyon fluid (i.e. the Meissner effect takes place in that superconducting phase) since it exponentially decays with a London penetration length $\lambda \sim 10^{-5} cm$ at $T \sim 200K$. On the other hand, from (15) and (13) we have that at $T \gg \omega_c$ there is not Meissner effect, but an inhomogeneous magnetic penetration. Hence, we can conclude that the energy gap ω_c defines a scale that separates two phases in the charged anyon fluid: a superconducting phase at $T \ll \omega_c$, and a non-superconducting one at $T \gg \omega_c$.

ACKNOWLEDGMENTS

This work has been supported in part by NSF grant PHY-9722059.

REFERENCES

1. F. Wilczek, Phys. Rev. Lett. **48** (1982) 1144; **49** (1982) 957.
2. F. Wilczek (ed.), *"Fractional Statistics and Anyon Superconductivity,"* (World Scientific, Singapore 1990).
3. P. W. Anderson, in: Physics of low-dimensional systems, Proc. Nobel symp. 73, eds. S. Lundquist and N. R. Nilsson (North Holland, Amsterdam, 1989).
4. J. Frohlich, F. Gabbiani and P.-A. Marchetti, in: Physics, Geometry and Topology , ed. H. C. Lee (Plenum Press, NY, 1990).
5. Y.-H Chen, F. Wilczek, E. Witten and B. I. Halperin, Int. J. Mod. Phys. **B3** (1989) 1001.
6. D. P. Arovas, J. R. Schrieffer, F. Wilczek, and A. Zee, Nucl. Phys. **B251** (1985) 117; A. Goldhaber, R. MacKenzie and F. Wilczek, Mod. Phys. Lett. **A4** (1989) 21; R.B. Laughlin, Phys. Rev. Lett. **60**; X.G. Wen and A. Zee, Phys. Rev. **B41** (1990) 240; G. S. Canright, S. M. Girvin and A. Brass, Phys. Rev. Lett. **63** (1989) 2291, 2295; E. Fradkin, Phys. Rev. Lett. **63** (1989) 322; Y. Hosotani and S. Chakravarty, Phys. Rev. **B42** (1990) 342.
7. J.D. Lykken, J. Sonnenschein and N. Weiss, Phys. Rev. **D42** (1990) 2161; Int. J. Mod. Phys. **A6** (1991) 1335; Y. Leblanc, and J.C. Wallet, Mod. Phys. Lett. **B5** (1991) 211; **B6** (1992) 1623; I.E. Aronov, E.N. Bogachek, I.V. Krive and S.A. Naftulin, JETP Lett. **56** (1992) 283; S.S. Mandal, S. Ramaswamy and V. Ravishankar; Int. J. Mod. Phys. **B8** (1994) 3095.

8. S. Randjbar-Daemi, A. Salam and J. Strathdee, Nucl. Phys. **B340** (1990) 403.
9. Y. Hosotani, Int. J. Mod. Phys. **B7** (1993) 2219; J.E. Hetric, Y. Hosotani and B.-H Lee, Ann. Phys **209** (1991) 151.
10. E.J. Ferrer, R. Hurka and V. de la Incera, Mod. Phys. Lett. **B11** (1997) 1; E.J. Ferrer and V. de la Incera, Int. J. Mod. Phys. **B12** (1998) 63.
11. E.J. Ferrer and V. de la Incera, *Fhase Transition in Anyon Superconductivity at Finite Temperature*, (1998) SUNY-FRE-98-04, hep-th/9805039

FUTURE ACCELERATORS

CMS
Concept and Physics Potential

Claudia-Elisabeth Wulz

*Institut für Hochenergiephysik, Österreichische Akademie der Wissenschaften,
Nikolsdorfergasse 18, A-1050 Vienna, Austria*

Abstract. CMS (Compact Muon Solenoid) will be one of two general purpose detectors at the CERN Large Hadron Collider. Its main feature is a strong solenoidal magnetic field ensuring high momentum resolution for charged particles. The detector consists of an inner tracker with an embedded pixel detector, a crystal electromagnetic calorimeter, a copper-scintillator hadron calorimeter and a dual muon system made up of tracking chambers and special trigger chambers. Forward calorimetry is also foreseen.

The discovery potential of CMS for the Standard Model Higgs, the SUSY Higgses and other supersymmetric particles is presented.

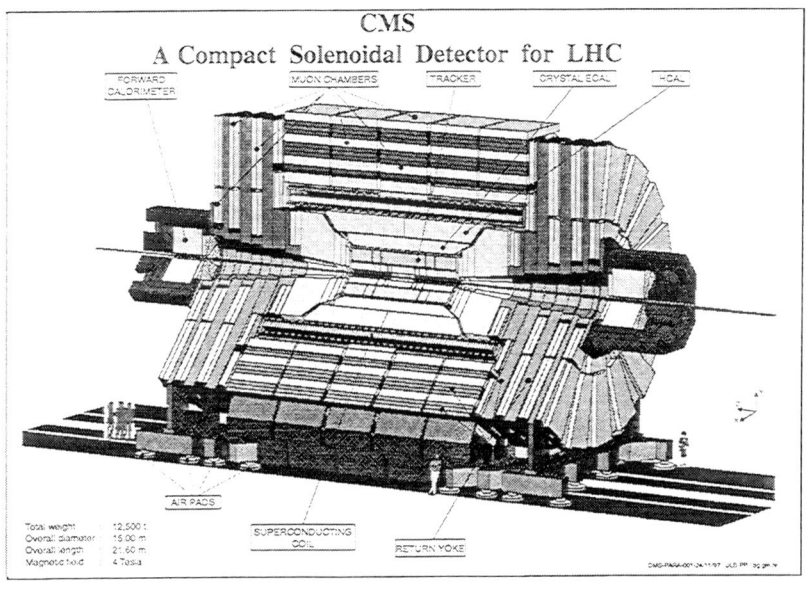

FIGURE 1. General view of CMS.

INTRODUCTION

CMS (Compact Muon Solenoid) is a general purpose experiment designed to explore physics at the planned Large Hadron Collider (LHC) at CERN. It is expected to go into operation in 2005. Proton-proton collisions as well as heavy ion collisions will be available. More than 150 institutions with 1700 physicists and engineers are presently taking part in the collaboration.

The design concept of CMS was first presented at the LHC Workshop at Aachen in 1990 [1]. It is based on a strong solenoidal magnetic field of 4 Tesla generated by a superconducting coil. The inner tracking system, the electromagnetic calorimeter, and the hadron calorimeter with the exception of a tail catcher in the central region, are inside the magnetic field volume. The muon chambers are embedded in the return iron yoke. Forward and Very Forward Calorimetry complete the apparatus in order to detect non-interacting particles.

A perspective view of CMS is shown in Fig. 1.

DETECTOR SETUP

Following the completion of the Technical Design Reports [2–6] of all subdetectors the CMS layout has been essentially finalized in April 1998.

The long superconducting coil is the heart of CMS. It provides a solenoidal field of 4 Tesla parallel to the beam direction. It is essential that the coil be completed before most other detector parts. The design is well advanced and real construction has started. The finished magnet is expected to be tested in 2003.

Inner Tracking

The active volume of the CMS inner tracker is a cylinder with a radius of 115 cm and a length of 270 cm on each side of the interaction point. Three different detectors well suited to the high, medium and low occupancy regions have been chosen to satisfy the stringent resolution and granularity requirements: a silicon pixel detector up to a radius of approximately 20 cm, a silicon strip detector in the region between 20 and 60 cm, and Micro Strip Gas Chambers (MSGC's) from 70 to 120 cm. The setup is shown in Fig. 2. The tracker geometry has been chosen such that typically 13 high resolution measurement planes for high-p_T tracks are available up to $|\eta| \approx 2$, gradually falling off to a minimum of 8 planes at $|\eta| \approx 2.5$. Overall, the silicon and MSGC trackers consist of more than ten thousand independent detector modules instrumented with 12×10^6 channels. The occupancy of each channel will be about 1 to 2 percent at high luminosity.

High-p_T isolated tracks are reconstructed with a transverse momentum resolution of better than $\delta p_T/p_T \approx (15 p_T \oplus 0.5)\%$, with p_T in TeV, in the central region of $|\eta| \leq 1.6$, degrading to $\delta p_T/p_T \approx (60 p_T \oplus 0.5)\%$ as $|\eta|$ approaches 2.5.

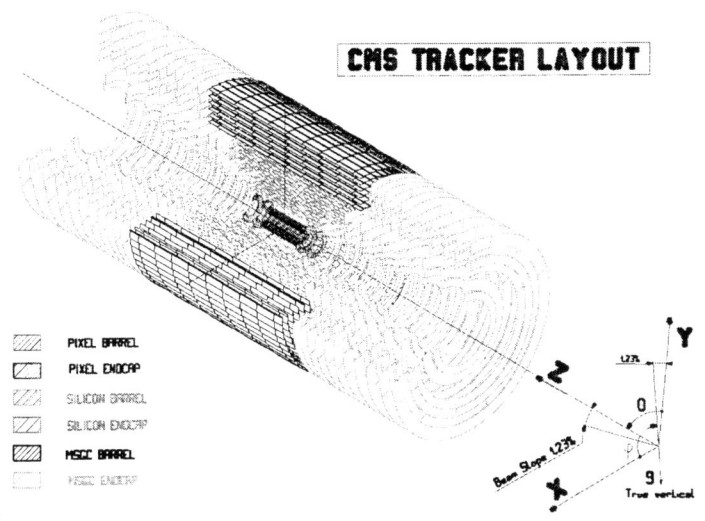

FIGURE 2. The CMS Inner Tracker.

Electromagnetic Calorimeter

CMS has chosen an electromagnetic calorimeter made out of scintillating lead tungstate crystals (PbWO$_4$) because it offers the best prospects of identifying and measuring precisely the energies of photons and electrons in a hostile environment with a magnetic field of 4 Tesla, a time of 25 ns between bunch crossings and radiation doses of 1 to 2 kGy per year at maximum LHC luminosity. The choice was based on the considerations that PbWO$_4$ has a short radiation length of 0.89 cm and Molière radius of 2.19 cm and a short light decay time. The initial drawback of low light yield has been overcome by progress in crystal growth and through the development of large-area silicon avalanche photodiodes. The geometrical coverage extends to $|\eta| = 3$. Precision energy measurement of photons and electrons will be carried out to $|\eta| = 2.63$. A total thickness of about 26 radiation lengths at $|\eta| = 0$ is required to limit the longitudinal shower leakage of high-energy electromagnetic showers to a reasonable level. This corresponds to a crystal length of 23 cm in the barrel region. In the endcap region a $\gamma - \pi_0$ separating preshower detector corresponding to 3 X_0 of lead allows the use of slightly shorter crystals.

For the energy range of about 25 to 500 GeV, typical for photons from the $H \to \gamma\gamma$ decay, the energy resolution can be parametrized as:

$$(\sigma/E)^2 = (a/\sqrt{E})^2 + (\sigma_n/E)^2 + c^2$$

where a is the stochastic term, σ_n the noise, and c the constant term. Fig. 3 shows the different contributions to the energy resolution. Depending on luminosity a

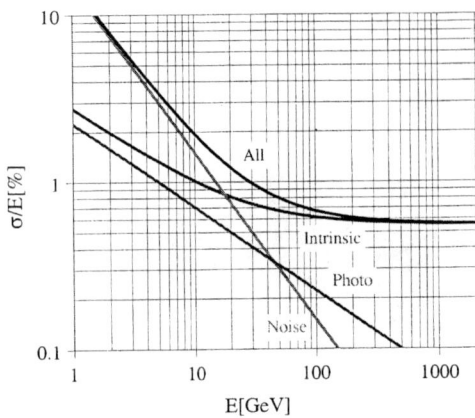

FIGURE 3. Energy resolution of the CMS electromagnetic calorimeter.

mass resolution of 650 to 700 MeV can be obtained for a 100 GeV Standard Model Higgs decaying into two photons.

Hadron Calorimeter

Together with the electromagnetic calorimeter the hadron calorimeter will be essential to measure jets and missing energy, crucial for the discovery of many new particles or phenomena. The targeted energy resolution is:

$$\sigma/E = 65\%/\sqrt{E} \oplus 5\% \text{ (E in GeV)}.$$

In addition to the barrel and endcap hadron calorimeters (HB, HE) extending to $|\eta| \approx 3$ a separate forward calorimeter (HF) covering the region $3 < |\eta| < 5$ is foreseen to maximize hermeticity (Fig. 4). HB and HE are sampling calorimeters consisting of 4 mm thick plastic scintillator tiles read out with wavelength-shifting plastic fibres inserted between copper plates. The barrel hadron calorimeter has only 5.15 nuclear interaction lengths at $\eta = 0$. To ensure adequate sampling depth the first muon absorber layer is instrumented with scintillator tiles to form a tail catcher.

The HF calorimeter which has to withstand high radiation doses uses quartz fibres as the active medium, embedded in a copper absorber matrix. It is not only important for the measurement of missing energy as required for example in Standard Model and SUSY Higgs searches or top quark physics, but also for forward jet detection needed in the search for the heavy Higgs boson in the TeV mass region.

In order to get an idea of the physics performance the process $t \to Wb$ with W decaying into jets was simulated. The obtained dijet mass resolution was 12 GeV with pileup and 8 GeV without.

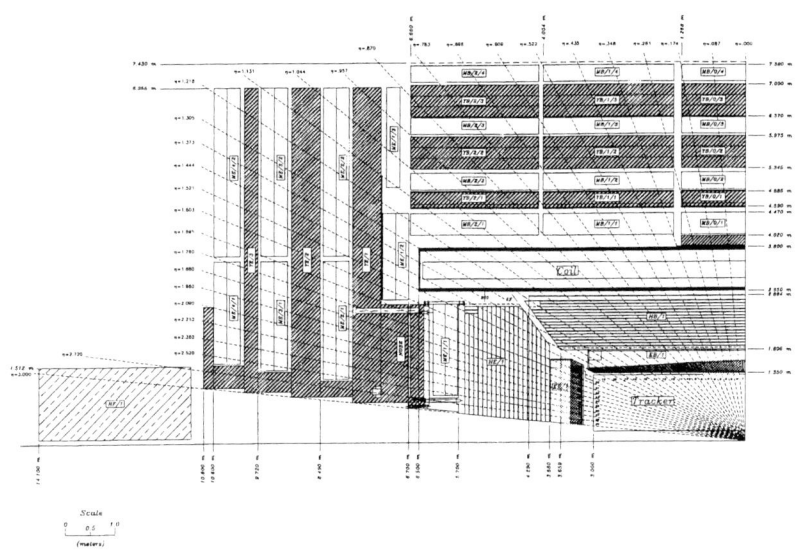

FIGURE 4. rz-projection of CMS.

Muon System

In the barrel region the muon system consists of drift chambers with bunch crossing identification capability (DTBX) to reconstruct muon tracks and of resistive plate chambers (RPC) to detect muon hits for trigger purposes. In the forward region the cathode strip chambers (CSC) perform the tasks of reconstructing the tracks. RPC's are also available.

The chambers are arranged in four stations interleaved with the iron return yoke plates as shown in Fig. 4. They are arranged in concentric cylinders around the beam line in the barrel region, and in disks perpendicular to the beam in the endcaps. The momentum resolution of charged tracks for the muon system alone and combined with the inner tracker at $\eta = 0.1$ is shown in Fig. 5.

Trigger and Data Acquisition

At LHC the collision rate will be 40 MHz. 16 million channels will have to be processed in total. One event is expected to contain 1 Mbyte of data on average. Filtered events will be written to a storage medium with a frequency of 100 Hz. The trigger system has therefore to perform a sizable reduction of data. The first-level trigger, a partly programmable hard-wired system, will run at a frequency of

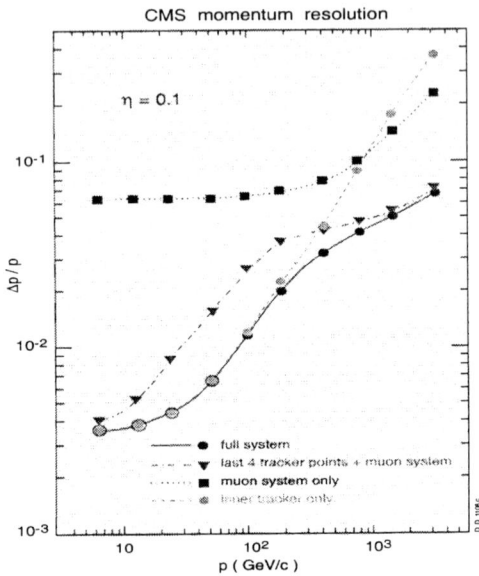

FIGURE 5. Momentum resolution of charged tracks.

100 kHz in pipeline mode. Data at the level one trigger stage will be stored in 500 readout memories with 200 Gbyte storage capacity. The event builder is a large switching network with a total throughput of about 500 Gbit per second. The following event filter consists of a set of high performance commercial processors organized into many farms convenient for on-line and off-line applications. The farms replace the traditional second and higher level triggers. One event will be processed by a single CPU.

PHYSICS PERFORMANCE

CMS has been designed in order to give answers to or shed light on the most important open questions in high energy physics. The understanding of the origin of mass is certainly one of the major problems. Other problems are the verification of Grand Unification Theories, the explanation of dark matter and the matter-antimatter asymmetry in the universe. CMS can also probe if there are really only three generations of quarks and leptons, if elementary particles of today have substructure or if the quark-gluon plasma exists.

We will concentrate here only on Standard Model Higgs [7] and supersymmetry searches [8]. CMS's B-physics capabilities are described in another contribution to these proceedings [9]. Heavy ion physics will not be dealt with either.

Standard Model Higgs searches

In the framework of the Standard Model particles acquire mass through the interaction with the Higgs field. This implies the existence of the Higgs boson. Theory does not predict its mass, but it does predict production rates and decay modes as a function of the Higgs mass. CMS has been optimized to detect the Higgs over the

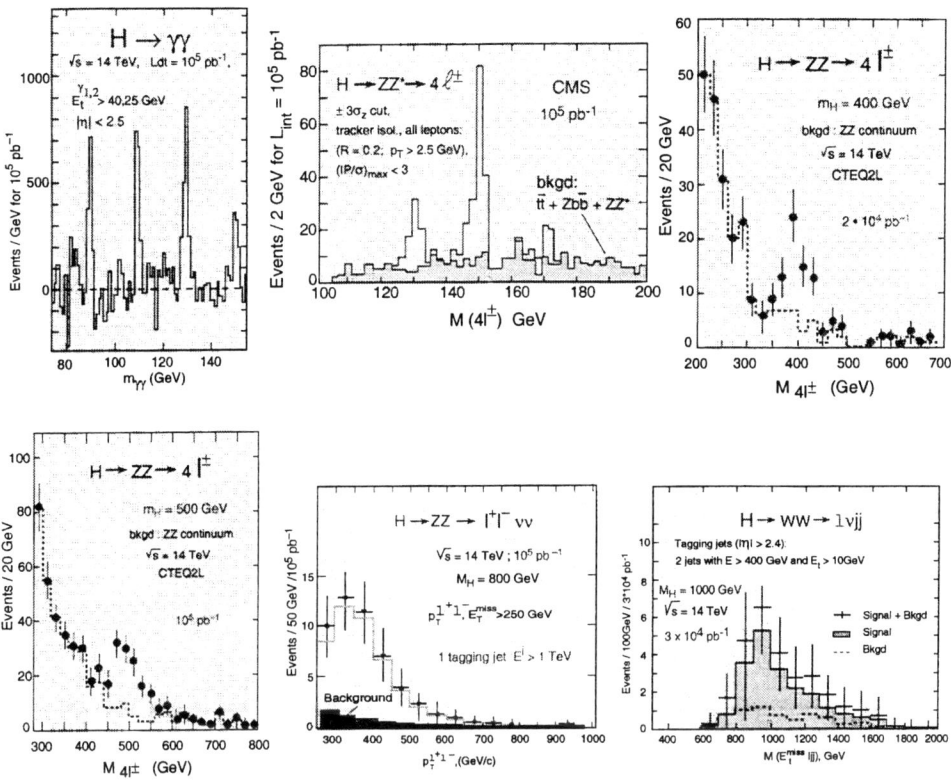

FIGURE 6. Standard Model Higgs in CMS.

entire possible mass range. The most promising channels after taking into account branching ratios and background are (l denotes either electron or muon):

$H \to \gamma\gamma$ for 80 GeV $< m_H <$ 140 GeV
$H \to ZZ^* \to 4l$ for 130 GeV $< m_H <$ 200 GeV
$H \to ZZ \to 4l$ for 200 GeV $< m_H <$ 700 GeV
$H \to ZZ \to 2l + 2\nu$ for 0.5 TeV $< m_H <$ 1 TeV
$H \to WW \to l\nu jj$ for $m_H \approx$ 1 TeV
$H \to ZZ \to lljj$ for $m_H \approx$ 1 TeV

It should be noted that in the lower mass region ($m_H < 130$ GeV) the branching ratio for $H \to b\bar{b}$ is close to one, but due to the large dijet background this channel seems only usable together with an additional lepton signature (e.g. $pp \to WH \to l\nu b\bar{b}$).

Fig. 6 depicts Higgs signals for the different mass ranges.

For the $H \to \gamma\gamma$ channel the diphoton mass resolution is essential. Calorimeter granularity is crucial for photon isolation measurements to suppress the $\pi^0 \to \gamma\gamma$ background. The $\gamma\gamma$ mass resolution at $m_{\gamma\gamma} \approx 100$ GeV is better than 1%, resulting in a signal to background ratio of approximately 1/20.

In the mass range 130 MeV $< m_H < 700$ GeV the most promising channel is the Higgs decay to two Z's, one of them being off-shell for masses smaller than 200 GeV. The detection relies on the excellent performance of the muon chambers, the tracker and the electromagnetic calorimeter. For $m_H < 170$ GeV a mass resolution of about 1 GeV should be achieved.

For the highest Higgs masses, in the range 0.5 to 1 TeV, high luminosity is needed. One also has to exploit decays of Z's and W's into jets and neutrinos. Hadron calorimeter performance is very important. At the very highest masses, above 800 GeV, the signal to background ratio has to be improved by requiring a central jet veto. Thus the $t\bar{t}$ and Z,W to jets backgrounds are reduced considerably. At the highest luminosities one must also take into account pile-up from minimum bias events. Double forward jet tagging will be necessary.

To summarize, CMS can detect a Standard Model Higgs boson in the entire mass range, from the LEP2 limit up to approximately 1 TeV with a significance of at least 5 σ.

Supersymmetry searches

Supersymmetry predicts a number of particles in addition to the Standard Model ones. Fermions have boson super-partners and bosons have fermion super-partners. We use the minimal supergravity-inspired standard model (mSUGRA) with a stable lightest supersymmetric particle (LSP) as a benchmark model [10]. The particle spectrum one expects consists of squarks ($\tilde{q}$), gluinos ($\tilde{g}$), sleptons ($\tilde{l}$), neutralinos ($\tilde{\chi}_i^0$ [i=1,4]) and charginos ($\tilde{\chi}_j^\pm$ [j=1,2]). There is also a Higgs sector with five SUSY Higgses, three neutral (h^0, H^0, A^0) and two charged ones ($H^\pm$).

mSUGRA is determined by only five parameters, the universal scalar (m_0) and gaugino masses ($m_{1/2}$), the SUSY breaking universal trilinear coupling A_0, the ratio of the vacuum expectation values of the Higgs fields $tan\beta$ and the sign of the Higgsino mixing parameter $sign(\mu)$.

Squarks and Gluinos

The total SUSY particle production cross-section is dominated by strongly interacting gluinos and squarks, which through their cascade decays can produce many jets and leptons with missing energy due to escaping LSP's and possibly neutrinos. Due to these escaping particles a complete mass reconstruction of squarks and gluinos is impossible. However, the presence of SUSY can be established by an excess of events of a given topology over known Standard Model backgrounds such as $t\bar{t}$, W + jets, Z + jets, WW, ZZ, ZW, $Zb\bar{b}$ and QCD. In order to establish the limits of the SUSY reach in the $(m_0, m_{1/2})$ parameter space the signal was generated at more than 100 points. $tan\beta = 2$, $A_0 = 0$ and $\mu < 0$ have been assumed. Fig. 7 shows the expected sparticle reach in various channels, for signatures containing leptons in different charge combinations. For $10^5 pb^{-1}$ integrated luminosity the ultimate mass reach for gluinos would be $m_{\tilde{q}} \approx 2.5$ TeV for small m_0 (below 400 GeV) and up to 2 TeV for any reasonable value of m_0 (below 2000 GeV). Squark masses can be probed for values in excess of 2 TeV. The cosmologically interesting region within the relic neutralino dark matter density contour of $\Omega h^2 \leq 1$ can be probed entirely already with an integrated luminosity around $10^3 pb^{-1}$.

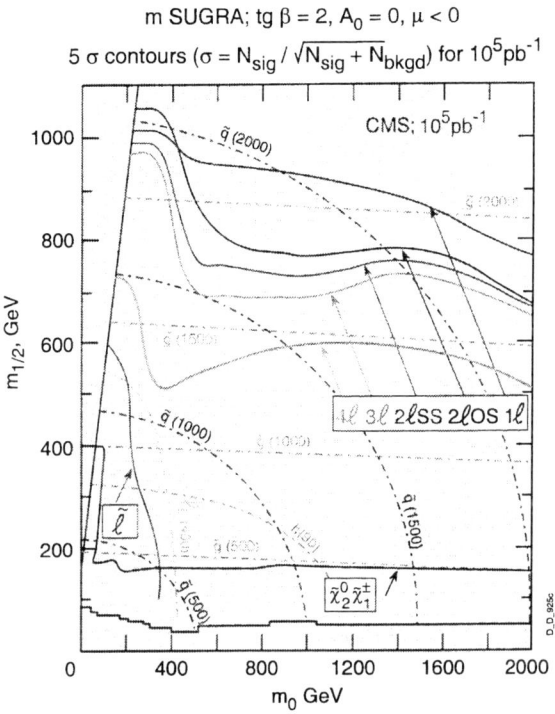

FIGURE 7. Expected sparticle reach in various channels.

Chargino/neutralino pair production

Direct production of $\tilde{\chi}_1^{\pm}\tilde{\chi}_2^0$ with leptonic decays of both sparticles gives three high-p_T isolated leptons accompanied by missing energy. These events have no jet activity except from initial-state QCD radiation. A central jet veto is therefore appropriate. WZ, ZZ and $Zb\bar{b}$ backgrounds can be removed by a Z mass cut. Other backgrounds are $t\bar{t}$, $b\bar{b}$ and SUSY channels ($\tilde{g}$, $\tilde{q}$, $\tilde{l}$, $\tilde{\chi}^0$, $\tilde{\chi}^{\pm}$). From Fig. 8 it can be

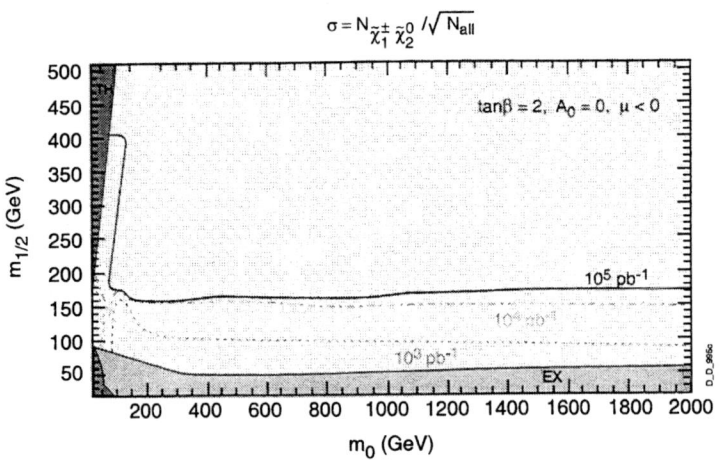

FIGURE 8. Chargino/neutralino pair production.

concluded that $\tilde{\chi}_1^{\pm}\tilde{\chi}_2^0$ direct production can be investigated up to $m_{1/2} \approx 170$ GeV for all m_0 with $10^5 pb^{-1}$ and $m_{1/2} \approx 150$ GeV with $10^4 pb^{-1}$. With $10^5 pb^{-1}$ the discovery region extends up to $m_{1/2} \approx 420$ GeV for $m_0 < 120$ GeV. It is possible to measure the mass of the lightest neutralino for $m_0 \gtrsim 160$ GeV by using the fact that the dilepton mass distribution has a sharp cutoff which is approximately equal to the mass of $\tilde{\chi}_1^0$ for the three-body decay process $\tilde{\chi}_2^0 \to ll\tilde{\chi}_1^0$ [11].

Sleptons

To search for direct slepton production the most appropriate signature is 2 leptons + missing energy + no jets. Backgrounds are expected to come from $\tau\tau$, WW, $t\bar{t}$, $b\bar{b}$ and other SUSY channels. Fig. 9 shows the slepton mapping of the mSUGRA parameter space. With $10^4 pb^{-1}$ luminosity, CMS is sensitive up to $m_{\tilde{l}_L} \approx 160$ GeV. With $10^5 pb^{-1}$ the reach extends up to $m_{\tilde{l}_L} \approx 340$ GeV for all allowed LSP masses (< 200 GeV), and up to $m_{\tilde{l}_L} \approx (340...440)$ GeV if $m_{LSP} \approx (0.45...0.6)\, m_{\tilde{l}_L}$ for a given $m_{\tilde{l}_L}$.

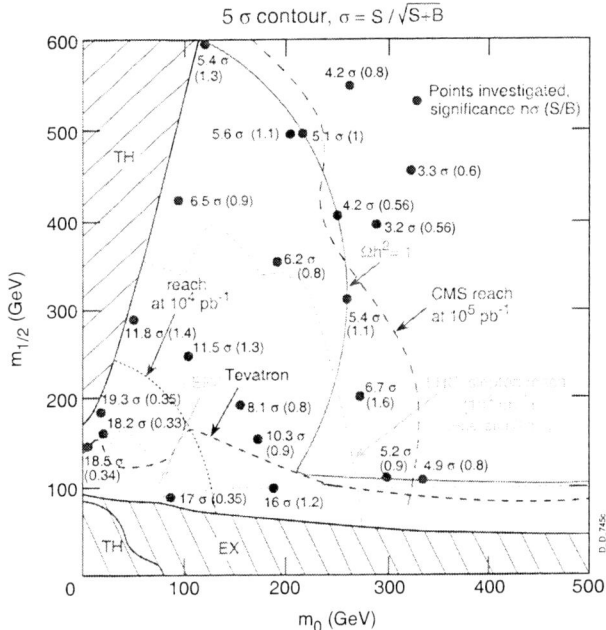

FIGURE 9. Slepton mapping of mSUGRA parameter space.

SUSY Higgs searches

The principal decay modes of the five SUSY Higgses are the following:
$h, H \to \gamma\gamma$
$h \to \gamma\gamma$ in Wh, $t\bar{t}H \to l\gamma\gamma$ events
$H \to ZZ, ZZ^* \to 4l^\pm$
$h, H, A \to \tau\tau \to l^\pm + h^\pm + E_T^{miss}$
$h, H, A \to \tau\tau \to e + \mu$
$h, H, A \to \mu^+\mu^-$
$H^\pm \to \tau\nu$ from $t\bar{t}$

For the first three channels the procedures used in the Standard Model Higgs search can be repeated. A summary plot of the significance contours for SUSY Higgses for $10^5 pb^{-1}$ and assuming no stop mixing is shown in Fig. 10.

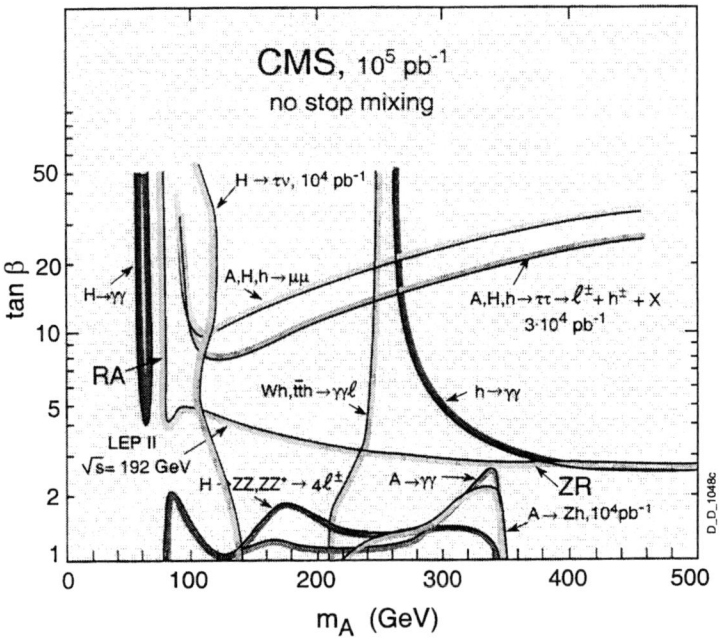

FIGURE 10. 5σ significance contours for SUSY Higgses.

ACKNOWLEDGMENTS

I am grateful to all CMS colleagues who have participated in the detector design and physics simulations.

REFERENCES

1. M. Della Negra et al., Proceedings of the Large Hadron Collider Workshop, Aachen, Vol. III, pp. 547-563 (1990).
2. CMS - The Magnet Project, CERN/LHCC 97-10 (1997).
3. CMS - The Tracker Project, CERN/LHCC 98-6 (1998).
4. CMS - The Electromagnetic Calorimeter, CERN/LHCC 97-33 (1997).
5. CMS - The Hadron Calorimeter, CERN/LHCC 97-31 (1997).
6. CMS - The Muon Project, CERN/LHCC 97-32 (1997).
7. See for example R. Kinnunen, D. Denegri, CMS Note 1997/057 (1997).
8. S. Abdullin et al., CMS Note 1998/006 (1998).
9. F. Charles, B physics with the CMS detector, these proceedings.
10. For a review see e.g. H. P. Nilles, Phys. Rep. 110 (1984) 1.
11. D. Denegri, W. Majerotto, L. Rurua, CMS Note 1997/094 (1997).

THE BTeV PROGRAM at Fermilab

SHELDON STONE

Physics Department, 201 Physics Building, Syracuse Univerisity,
Syracuse, NY 13244-1130, USA

Abstract. A description is given of BTeV, a proposed program at the Fermilab collider sited at the C0 intersection region. The main goals are measurement of mixing, CP violation and rare decays in both the b and charm systems. The detector is a two-arm-forward spectrometer capable of triggering on detached vertices and dileptons, and possessing excellent particle identification, electron, photon and muon detection.

I INTRODUCTION

BTeV is a Fermilab collider program whose main goals are to measure mixing, CP violation and rare decays in the b and c systems. Using the new Main injector, now under construction, the collider will produce on the order of 4×10^{11} b hadrons in 10^7 sec. of running. This compares favorably with e^+e^- colliders operating at the $\Upsilon(4S)$ resonance. These machines, at their design luminosities of $3 \times 10^{33}\mathrm{cm}^{-2}\mathrm{s}^{-1}$ will produce 6×10^7 B mesons in 10^7 seconds [1].

II IMPORTANCE OF HEAVY QUARK DECAYS

The physical point-like states of nature that have both strong and electroweak interactions, the quarks, are mixtures of base states described by the Cabibbo-Kobayashi-Maskawa matrix: [2]

$$\begin{pmatrix} d' \\ s' \\ b' \end{pmatrix} = \begin{pmatrix} V_{ud} & V_{us} & V_{ub} \\ V_{cd} & V_{cs} & V_{cb} \\ V_{td} & V_{ts} & V_{tb} \end{pmatrix} \begin{pmatrix} d \\ s \\ b \end{pmatrix} \qquad (1)$$

The unprimed states are the mass eigenstates, while the primed states denote the weak eigenstates. A similar matrix describing neutrino mixing is possible if the neutrinos are not massless.

There are nine complex CKM elements. These 18 numbers can be reduced to four independent quantities by applying unitarity constraints and using the fact that the phases of the quark wave functions are arbitrary. These four remaining numbers are fundamental constants of nature that need to be determined experimentally, like any other fundamental constant such as α or G. In the Wolfenstein approximation the matrix is written as [3]

$$V_{CKM} = \begin{pmatrix} 1 - \lambda^2/2 & \lambda & A\lambda^3(\rho - i\eta) \\ -\lambda & 1 - \lambda^2/2 & A\lambda^2 \\ A\lambda^3(1 - \rho - i\eta) & -A\lambda^2 & 1 \end{pmatrix}. \quad (2)$$

The constants λ and A have been measured [4].

The phase η allows for CP violation. CP violation thus far has only been seen in the neutral kaon system. If we can find CP violation in the B system we could see if the CKM model works or perhaps go beyond the model. Speculation has it that CP violation is responsible for the baryon-antibaryon asymmetry in our section of the Universe. If so, to understand the mechanism of CP violation is critical in our conjectures of why we exist [5].

Unitarity of the CKM matrix leads to the constraint triangle shown in Fig. 1. The left side can be measured using charmless semileptonic b decays, while the right side can be measured by using the ratio of B_s to B_d mixing. The angles can be found by measuring CP violating asymmetries in hadronic B decays.

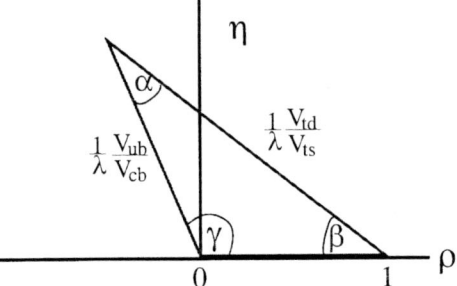

FIGURE 1. The unitarity triangle shown in the $\rho - \eta$ plane. The left side is determined by measurements of $b \to u/b \to c$ and the right side can be determined using mixing measurements in the B_s and B_d systems. The angles can be found by making measurements of CP violating asymmetries in hadronic B decays.

The current status of constraints on ρ and η is shown in Fig. 2. One constraint on ρ and η given by the K_L^o CP violation measurement (ϵ) [6], where the largest error arises from theoretical uncertainty. Other constraints come from current measurements on V_{ub}/V_{cb}, and B_d mixing [4]. The width of both of these bands are dominated by theoretical errors. Note that the errors used are $\pm 1\sigma$. This shows that the data are consistent with the standard model but do not pin down ρ and η.

It is crucial to check if measurements of the sides and angles are consistent, i.e., whether or not they actually form a triangle. The standard model is incomplete. It has many parameters including the four CKM numbers, six quark masses, gauge boson masses and coupling constants. Perhaps measurements of the angles and

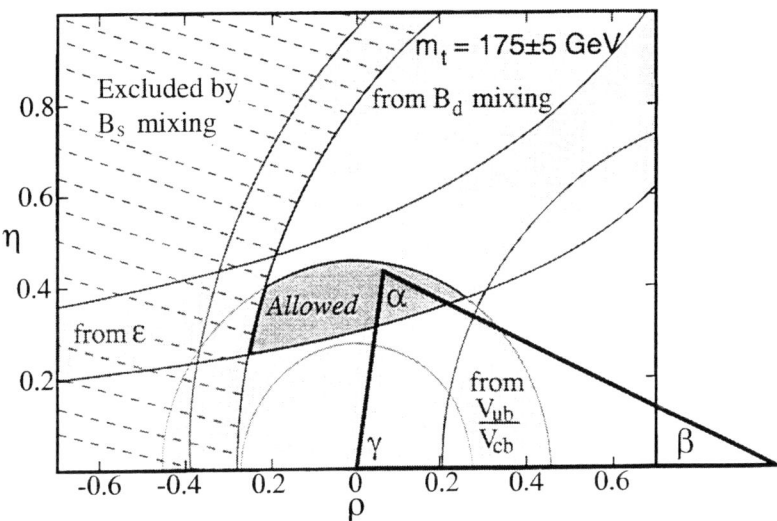

FIGURE 2. The regions in $\rho - \eta$ space (shaded) consistent with measurements of CP violation in K_L^o decay (ϵ), V_{ub}/V_{cb} in semileptonic B decay, B_d^o mixing, and the excluded region from limits on B_s^o mixing. The allowed region is defined by the overlap of the 3 permitted areas, and is where the apex of the CKM triangle sits. The bands represent $\pm 1\sigma$ errors. The error on the B_d mixing band is dominated by the parameter f_B. Here the range is taken as $240 > f_B > 160$ MeV.

sides of the unitarity triangle will bring us beyond the standard model and show us how these paramenters are related, or what is missing.

III THE MAIN PHYSICS GOALS OF BTEV

A Physics Goals For B's

Here we briefly list the main physics goals of BTeV for studies of the b quark.

• Precision measurements of B_s mixing, both the time evolution x_s and the lifetime difference, $\Delta\Gamma$, between the positive CP and negative CP final states.

• Measurement of the "CP violating" angles α and γ. We will use $B^o \to \pi^+\pi^-$ for α [7] and measure γ using several different methods including measuring the time dependent asymmetry in $B_s^o \to D_s^{\pm} K^{\mp}$, or measuring the decay rates $B^+ \to D^o K^+$, and $B^- \to \overline{D}^o K^+$, where the D^o can decay directly or via a doubly Cabibbo suppressed decay mode [8,9].

• Search for rare final states such as $K\mu^+\mu^-$ and $\pi\mu^+\mu^-$ which could result from new high mass particles coupling to b quarks.

• We assume that the CP violating angle β will have already been measured by using $B^o \to \psi K_s$, but we will be able to significantly reduce the error.

B The Main Physics Goals for charm

According to the standard model, charm mixing and CP violating effects should be "small." Thus charm provides an excellent place for non-standard model effects to a appear. Specific goals are listed below.

- Search for mixing in D^o decay, by looking for both the rate of wrong sign decay, r_D and the width difference between positive CP and negative CP eigenstate decays, $\Delta\Gamma$. The current upper limit on r_D is 3.7×10^{-3}, while the standard model expectation is $r_D < 10^{-7}$ [10].
- Search for CP violation in D^o. Here we have the advantage over b decays that there is a large D^{*+} signal which tags the initial flavor of the D^o through the decay $D^{*+} \to \pi^+ D^o$. Similarly D^{*-} decays tag the flavor of initial $\overline{D}^o$. The current experimental upper limits on CP violating asymmetries are on the order of 10%, while the standard model prediction is about 0.1% [11].
- Search for direct CP violation in charm using D^+ and D_s^+ decays.
- Search for rare decays of charm, which if found would signal new physics.

IV CHARACTERISTICS OF HADRONIC B PRODUCTION

It is often customary to characterize heavy quark production in hadron collisions with the two variables p_t and η. The later variable was first invented by those who studied high energy cosmic rays and is assigned the value

$$\eta = -ln\left(\tan\left(\theta/2\right)\right), \qquad (3)$$

where θ is the angle of the particle with respect to the beam direction.

According to QCD based calculations of b quark production, the b's are produced "uniformly" in η and have a truncated transverse momentum, p_t, spectrum, characterized by a mean value approximately equal to the B mass [12]. The distribution in η is shown in Fig. 3.

There is a strong correlation between the B momentum and η. Shown also in Fig. 3 is the $\beta\gamma$ of the B hadron versus η. It can clearly be seen that near η of zero, $\beta\gamma \approx 1/2$, while at larger values of $|\eta|$, $\beta\gamma$ can easily reach values of 6. This is important because the observed decay length varies with $\beta\gamma$ and furthermore the absolute momenta of the decay products are larger allowing for a suppression of the multiple scattering error.

The "flat" η distribution hides an important correlation of $b\bar{b}$ production at hadronic colliders. In Fig. 4 the production angles of the hadron containing the b quark is plotted versus the production angle of the hadron containing the $\bar{b}$ quark according to the Pythia generator. There is a very strong correlation in the forward direction (the direction of the p beam at 0°-0°), where both B and $\overline{B}$ hadrons are going in the same direction. The same strong correlation is present in the $\overline{p}$ direction. This correlation is not present in the central region (near 90°). By

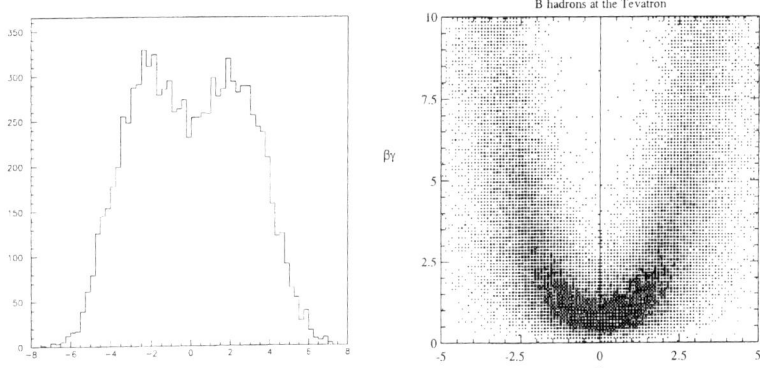

FIGURE 3. The B yield versus η (left). $\beta\gamma$ of the B versus η (right). Both plots are for the Tevatron.

instrumenting a relative small region of angular phase space, a large number of $b\bar{b}$ pairs can be detected. Furthermore the B's populating the two "forward" regions have large values of $\beta\gamma$.

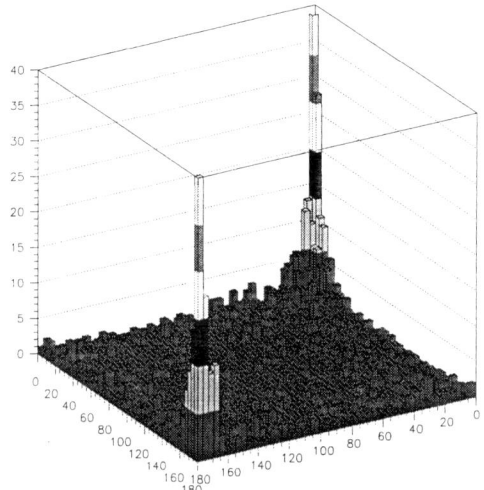

FIGURE 4. The production angle (in degrees) for the hadron containing a b quark plotted versus the production angle for a hadron containing $\bar{b}$ quark.

Charm production is similar to b production but much larger. Current theoretical estimates are that charm is 1-2% of the total $p\bar{p}$ cross-section. Table 1 gives the relevant Tevatron parameters. We expect eventually run at a luminosity of $2 \times 10^{32} \text{cm}^{-2}\text{s}^{-1}$. A machine design that holds the luminosity constant at this value, called "luminosity leveling," has been developed. We plan to adopt this design.

TABLE 1. The Tevatron as a b and c source for C0 in Run II.

Luminosity	$2 \times 10^{32} \text{cm}^{-2}\text{s}^{-1}$
b cross-section	$100\mu b$
# of b's per 10^7 sec	4×10^{11}
b fraction	0.2%
c cross-section	$> 500 \ \mu b$
Bunch spacing	132 ns
Luminous region length	$\sigma_z = 30$ cm
Luminous region length	$\sigma_x \ \sigma_y = \approx 50 \ \mu m$
Interactions/crossing	$< 2 >$

V THE EXPERIMENTAL TECHNIQUE: A FORWARD TWO-ARM SPECTROMETER

A sketch of the apparatus is shown in Fig. 5. The two-arm spectrometer fits in the expanded C0 interaction region, which is being excavated. The magnet that we will use, called SM3, exists at Fermilab. The other important parts of the experiment include the vertex detector, the RICH detectors, the EM calorimeters and the muon system.

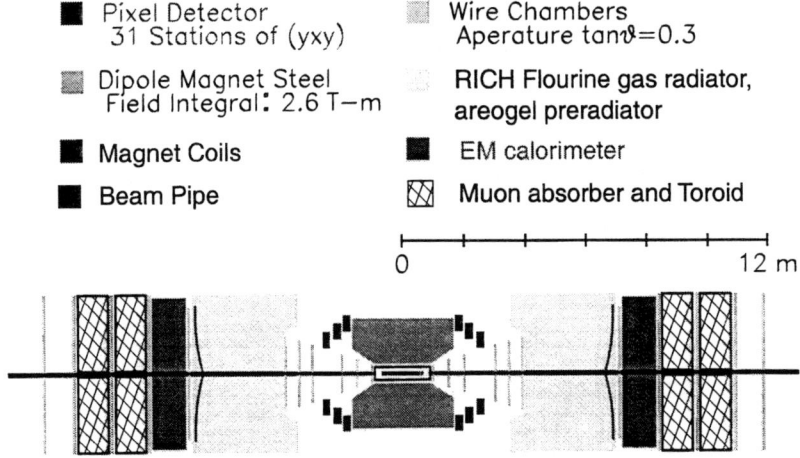

FIGURE 5. Sketch of the BTeV spectrometer.

The angle subtended is approximately ±300 mr in both plan and elevation views. The vertex detector is a multiplane pixel device which sits inside the beam pipe.

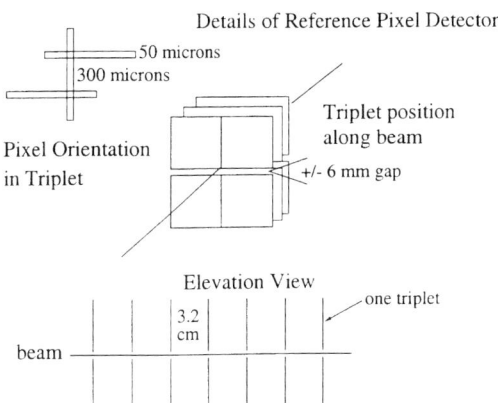

FIGURE 6. Layout of the BTeV pixel detector. There are 31 stations of triplets with the narrow pixel dimension in the bend plane. The most recent version replaces the 12 mm gap between detector planes with a 12 mm × 12 mm square hole centered on the beam.

The baseline design has 31 stations with triplets in each station. The detector is sketched in Fig. 6. Our new baseline detector has a square hole, 12 mm × 12 mm around the beam instead of a 12 mm gap between top and bottom halves. (Some of our simulations have been done with the detector with the gap, called "EOI, and some have been done with the "square hole.) The triggering concept is to pipeline the data and to detect detached b or c vertices in the first trigger level. The vertex detector is put in the magnetic field in order to insure that the tracks considered for vertex based triggers do not have large multiple scattering because they are low momentum.

The RICH detector [13] has a gas radiator, either C_4F_{10} or C_5F_{12} and mirrors which focus the Cherenkov light outside of the fiducial volume of the detector. This system will provide K/π separation in the momentum range between 3 and 70 GeV/c. To resolve protons from kaons below the kaon threshold of 9 GeV/c, an thin aerogel radiator may be placed in front of the gas volume. The same photon detector would be utilized.

The muon detector consists of position measuring chambers placed around and between an iron slab followed by another slab which is used as a magnetized torroid. This system is used both to trigger on final states with dimuons and to identify muons in the final analysis.

VI SIMULATIONS

A Introduction

We have developed several fast simulation packages to verify the basic BTeV concepts and aid in the final design. The trigger simulations, discussed below, are done with full pattern recognition. The input consists only of hits which are smeared by their resolution. To simulate backgrounds in the final physics analysis, we use a fast simulation which simulates track resolutions but not the pattern recognition. This is done because we have to simulate backgrounds in processes with branching ratios in the 10^{-5}-10^{-6} range and we cannot afford the computer time. The key program in our system is MCFast [14]. Charged tracks are generated and traced through different material volumes including detector resolution, multiple scattering and efficiency. This allows us to measure acceptances and resolutions in a fast reliable manner.

B Trigger Simulations

We simulate the trigger using the baseline pixel detector shown in Fig. 6. The triplet stations each provide a three-dimensional space point as well as a track direction mini-vector. This is useful for fast pattern recognition. The trigger simulations are carried out by doing the complete pattern recognition from the hits left in the detector by tracks and converted photons.

Our baseline trigger algorithm works by first determining the main event vertex and then finding how many tracks miss this vertex by $n\sigma$, where σ refers to the impact parameter divided by its error. Furthermore, a requirement is then placed on the track momentum in the bend plane, p_y, as determined on line. The preliminary results of simulating this algorithm are shown in Fig. 7 (right) for a cut $p_y > 0.5$ GeV/c [15]. The choice of the number of tracks and the impact parameter requirement must eventually be fixed, but what is shown here is the efficiency for accepting light quark events (u, d, and s) for various choices on the number of tracks (curves) and the size of their required impact parameter divided by the error in impact parameter. The efficiency for accepting $B^o \to \pi^+\pi^-$ is shown in the left side. Here the efficiency is given after requiring that both tracks are in the spectrometer and accepted for further analysis. For a "typical" $n\sigma$ cut of 3 and track requirement of 2, the $\pi^+\pi^-$ trigger efficiency is about 45%, while The light quark background has an efficiency of about 0.8%. Note, that we do not consider c to be a background in this experiment. For "typical" charm reaction the same trigger gives about a 1% efficiency on charm.

At the BTeV design luminosity of 2×10^{32}cm^{-2}s^{-1} there is an average of two interactions per beam crossing. The interactions are spread out over the long (σ=30 cm) interaction region. The trigger must not fire merely due to the presence of two near by interactions. To insure this we have imposed a requirement that the maximum impact parameter of a track not be larger than 2 mm. The yield for events containing a $B^o \to \pi^+\pi^-$ decay as a function of luminosity is shown in Fig. 8 (left). Here we do not want the trigger rate to increase as a function of luminosity, even though this means that the efficiency on this rare final state increases. Therefore, a linear rise would be ideal. On the right side we show the probability to trigger on light quark background. We would like this to remain

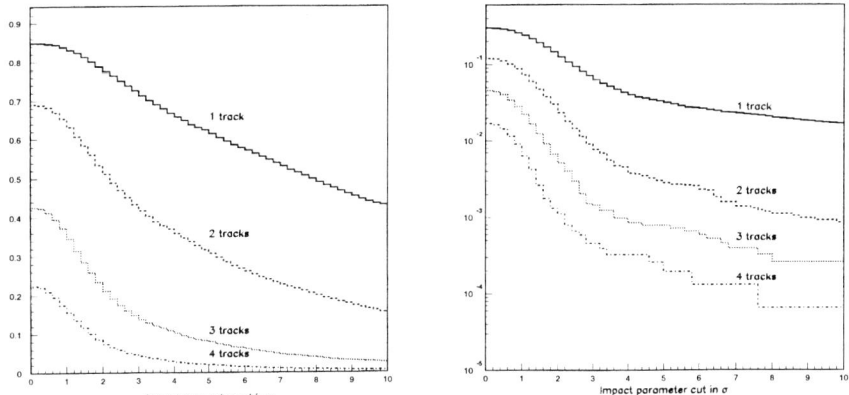

FIGURE 7. (left) Trigger efficiency for $B^o \to \pi^+\pi^-$ for pion tracks in the spectrometer.(right) Trigger efficiency for light quark events. The ordinate gives the choice of cut value on the impact parameter in terms of number of standard deviations (σ) of the track from the primary vertex. The curves show the effect of requiring different numbers of tracks.

constant with increasing luminosity. No increase occurs up to a luminosity of $10^{32}\text{cm}^{-2}\text{s}^{-1}$, after that the probability for this particular trigger condition increases mildly. However, the first level trigger rate is clearly much lower than the 1% we require until we exceed a luminosity of $\approx 3 \times 10^{32}\text{cm}^{-2}\text{s}^{-1}$.

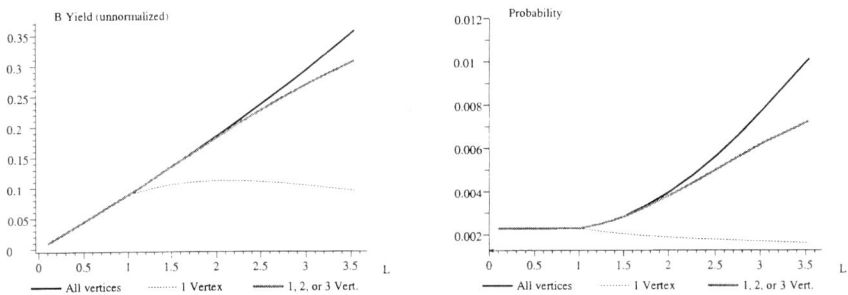

FIGURE 8. Luminosity dependent trigger efficiencies for $B^o \to \pi^+\pi^-$ (left), and light quark events (right). The abscissa is in units of $10^{32}\text{cm}^{-2}\text{s}^{-1}$.

C The CP violating asymmetry in $B^o \to \pi^+\pi^-$

The trigger efficiency for this mode has already been discussed. For the $B^o \to \pi^+\pi^-$ channel BTeV has compared the offline fully reconstructed decay length distributions in their forward geometry with that of detector configured to work in the central region. Fig. 9 shows the B momentum distribution and decay distance error as a function of b momentum (left).

The right side shows the Fig. 9 shows the normalized decay length expressed in terms of L/σ where L is the decay length and σ is the error on L for the $B^o \to \pi^+\pi^-$ decay [16].

The forward detector clearly has a much more favorable L/σ distribution, which is due to the excellent proper time resolution. Being able to keep high efficiency in the trigger and analysis levels and being able to decimate the backgrounds relies mainly on having the excellent L/σ distribution shown for the forward detector.

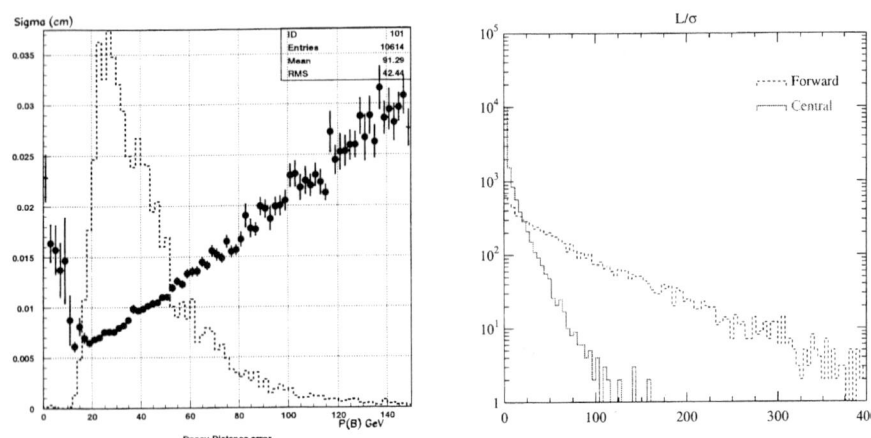

FIGURE 9. (left) The B momentum distribution for events in the detector acceptance and the error on the spatial distance of the $\pi^+\pi^-$ decay vertex from the primary vertex. (right) Comparison of the L/σ distributions for the decay $B^o \to \pi^+\pi^-$ in central and forward detectors produced at a hadron collider with a center of mass energy of 1.8 TeV.

For this analysis L/σ is required to be > 15. Each pion track is required to miss the primary vertex by a distance/error $> 5\sigma$ and that the B^o point back to the primary by a distance/error $< 2\sigma$. Furthermore, each track is required to be identified as a pion and not a kaon in the RICH detector. Without particle identification it is impossible to distinguish $B^o \to \pi^+\pi^-$ from the combination of $B^o \to K^\pm\pi^\mp$, $B_s \to K^+K^-$ and $B_s \to K^\pm\pi^\mp$, as is shown on Fig. 10. Here $\mathcal{B}(B^o \to K^\pm\pi^\mp)$ is taken as 1.5×10^{-5} and $\mathcal{B}(B^o \to \pi^+\pi^-)$ is taken as 0.75×10^{-5}, from recent CLEO measurements [17]. The B_s decay into K^+K^- is assumed to have the same rate as the B^o decay into $K^\pm\pi^\mp$, and the B_s decay into $K^\pm\pi^\mp$ is assumed to have the same rate as the B^o decay into $\pi^+\pi^-$.

Using the good particle identification, BTeV predicts that they can measure the CP violating asymmetry in $\pi^+\pi^-$ to ± 0.013 as detailed in Table 2.

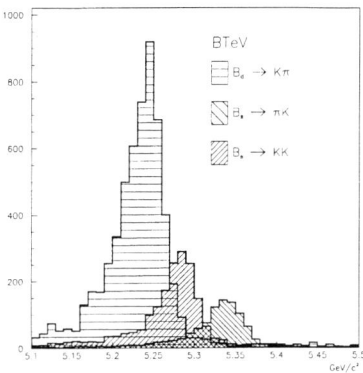 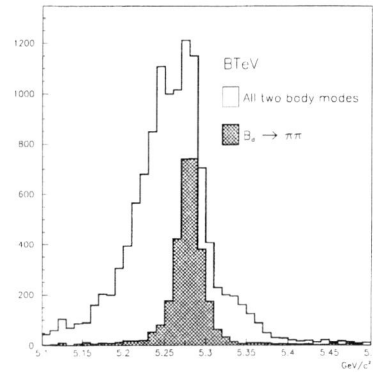

FIGURE 10. Invariant mass distributions of all $B \to h^+h^-$ final states, where h denotes either a pion or kaon, and the mass is computed assuming that both tracks are pions. The plot on the left shows the individual background channels and the one on the right shows the sum of all channels properly normalized (see text) to the $\pi^+\pi^-$ signal.

D Flavor tagging

We have assumed a flavor tagging efficiency of 10%. Actually our studies show that we probably can achieve a higher efficiency. The usual definitions are: N is the number of reconstructed signal events, N_R is the number of right sign flavor tags, N_W is the number of wrong sign flavor tags, ϵ is the efficiency (given by $(N_R + N_W)/N$) and D is the dilution (given by $(N_R - N_W)/(N_R + N_W)$. The quantity of is ϵD^2 which when multiplied by N gives the effective number of events useful for the calculation of an asymmetry error.

We have investigated the feasibility of tagging kaons using a gas Ring Imaging Cherenkov Counter (RICH) in a forward geometry and compared it with what is possible in a central geometry using Time-of-Flight counters with good, 100 ps, resolution. For the forward detector the momentum coverage required is between 3 and 70 GeV/c. The lower momentum value is determined by our desire to tag charged kaons for mixing and CP violation measurements, while the upper limit comes from distinguishing the final states $\pi^+\pi^-$, $K^+\pi^-$ and K^+K^-. The momentum range is much lower in the central detector but does have a long tail out to about 5 GeV/c. Either C_4F_{10} or C_5F_{12} have pion thresholds of about 2.5 GeV/c. The kaon and proton thresholds for the first gas are 9 and 17 GeV/c, respectively.

The BTeV RICH was simulated using the current C0 geometry with MCFast. Fig. 11 shows the number of identified kaons plotted versus their impact parameter

TABLE 2. Numbers entering into the accuracy in measuring the CP violating asymmetry in $B^\circ \to \pi^+\pi^-$.

Quantity	Value
Cross-section	100 μb
Luminosity	2×10^{32}
# of $B^\circ/2\times 10^7$s, $\mathcal{L}$ leveled	2.8×10^{11}
$\mathcal{B}(B^\circ \to \pi^+\pi^-)$	0.75×10^{-5}
Reconstruction efficiency	0.08
Triggering efficiency (after all other cuts)	0.72
# of $\pi^+\pi^-$	128,000
ϵD^2 for flavor tags ($K^\pm$, $\ell^\pm$, same + opposite sign jet tags)	0.1
# of tagged $\pi^+\pi^-$	12,800
Signal/Background	0.9
Error in asymmetry (including background)	± 0.013

divided by the error in the impact parameter for both right sign and wrong sign kaons. A right sign kaon is a kaon which properly tags the flavor of the other B at production. We expect some wrong sign kaons from mixing and charm decays. Many others just come from the primary. A cut on the impact parameter standard deviation plot at 3.5σ gives an overall ϵD^2 of 6%. This number needs to be reduced to 5% because of $b\bar{b}$ mixing [19]. Without the aerogel preradiator to distinguish protons from kaons below threshold we would experience an additional reduction down to about 4%. These numbers are for a perfect RICH system. Putting in a fake rate of several percent, however, does not significantly change this number.

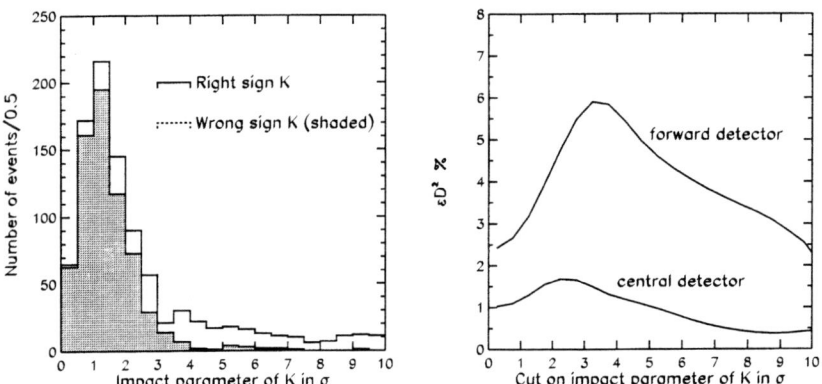

FIGURE 11. (left) L/σ distributions in BTeV for $K^\pm$ impact parameters for right sign (unshaded) and wrong sign (shaded) tags. (right) Overall ϵD^2 values from kaon tagging for a forward detector containing a flourine based RICH versus a central detector with 100 ns time of flight resolution as a function of kaon impact parameter in units of L/σ. (Protons and $b\bar{b}$ mixing have been ignored in both cases.)

TABLE 3. The projected flavor tagging efficiencies for a central detector similar to CDF and BTeV in units of ϵD^2.

	$K^\pm$	$\mu^\pm$	$e^\pm$	SST	Jet Charge	Sum
BTeV	5%	2.0%	1.5%	>2%	5%	>10%
Central	0%	1.0%	0.7%	2%	3%	3%

The simulation of the central detector gives much poorer numbers. In Fig. 11 ϵD^2 for both the forward and central detectors are shown as a function of the kaon impact parameter (protons have been ignored). It is difficult to get ϵD^2 of more than 1.5% in the central detector.

Now let us consider other tags. We have simulated muon and electron flavor tags in our system. Although this technique is very useful at e^+e^- colliders operating at the $\Upsilon(4S)$, it is less useful here because it is difficult to distinguish leptons from the $b \to c \to \ell^+$ decay from the primary leptons from the b quark decay. Our estimates are given in Table 3 along with those from a central detector.

The two other methods considered are "jet charge" and "same side" tagging (sst). We have not yet studied sst, which is using the charge of a track closest in phase space to the reconstructed B. However, CDF has measured ϵD^2 for it to be $(1.5\pm0.4)\%$ and take 2% as their future projection using an improved vertex detector. We have studied jet charge, which involves taking a weighted measure of the charge of the tagging b jet. However, we incorporate information on the detachment of the tracks to help us define the jet. CDF extrapolates to 3% while we expect 5%.

E Measurement of B_s mixing

BTeV has studied the feasability of measuring the B_s mixing parameter $x_s = \Delta m_s/\Gamma_s$. This measurement is key to obtaining the right side of the unitarity triangle shown in Fig. 1. Current limits on B_s mixing from LEP give $x_s > 15$ [20]. Recall that for B_d mesons, $x = 0.73$. The oscillation length for B_s mixing is at least a factor of 10 shorter and may approach a factor of 100!

BTeV has investigated two final states which can be used. The first ψK^{*o}, $\psi \to \mu^+\mu^-$ and $K^{*o} \to K^-\pi^+$ has several advantages. It can be selected using either a dilepton or detached vertex trigger. Backgrounds can be reduced in the analysis by requiring consistency with the ψ and K^{*o} masses. Furthermore, it should have excellent time resolution as there are four tracks coming directly from the B decay vertex. The resolution in proper time is 42 fs. The one disadvantage is that the decay is Cabibbo suppressed, the Cabibbo allowed channel being $\psi\phi$ which is useless for mixing studies. The branching ratio therefore is predicted to have the low value of 8.5×10^{-5}.

The time distributions of the unmixed and mixed decays are shown in Fig. 12, along with a calculation of the likelihood of there being an oscillation as determined by fits to the time distributions. Background and wrong tags are included. The

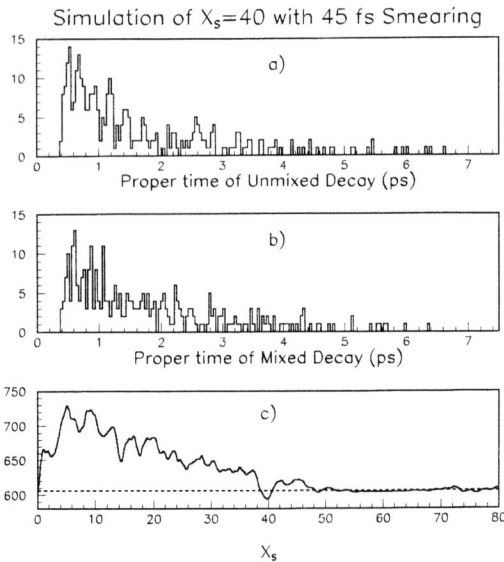

FIGURE 12. The observed decay time distributions for $B_s \to \psi K^{*o}$ generated with $x_s = 40$. Unmixed decays are shown in (a), mixed in (b). Background and mistagging has been included. In (c) the results of a likelihood fit to the time distributions are shown. The dashed line shows a 5σ difference from the best solution.

fitting procedure correctly finds the input value of $x_s = 40$. The danger is that a wrong solution will be found. The dashed line shows the change in likelihood corresponding to 5 standard deviations. If our criteria is that the next best solution be greater than 5σ, then this is the best that can be done with one years worth of data in this mode. Once a clean solution is found, the error on x_s is quite small, being ±0.15 in this case.

BTeV has also investigated the $D_s^+ \pi^-$ decay of the $\overline{B}_s$, with $D_s^+ \to \phi \pi^+$. It turns out that the lifetime resolution is 45 fs, almost the same as for the ψK^{*o} decay mode. Since the predicted branching ratio for this mode is 0.3%, we obtain 19200 events in one year of running, with a signal to background of 3:1. Fig. 13 shows the x_s reach obtainable for a 5σ discrimination between the favorite solution and the next best solution, for both decay modes. The background is assumed to be 20% and the flavor mistag fraction is taken as 25%. The tagging efficiency is taken as 10%. The dashed lines show the number of years of running, where one year is 10^7 seconds. The pixel system with the 12 mm gap is called the EOI detector here.

The other detector configuration that we simulated has the pixels configured around the beam leaving a 12 mm × 12 mm square hole. This detector has better efficiency and time resolution (see section V) and now has become the BTeV

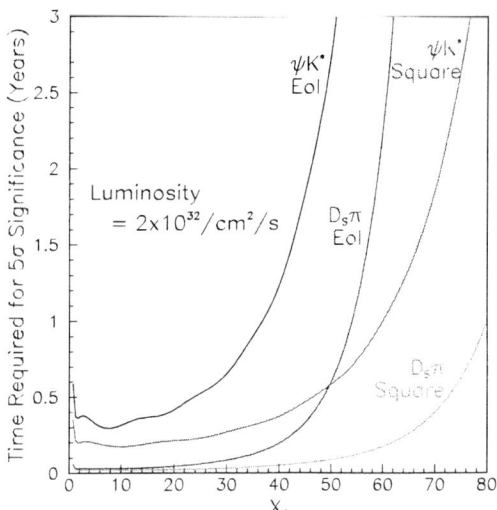

FIGURE 13. The x_s reach for both $D_s^+\pi^-$ and ψK^{*o} decays of the $\overline{B}_s$. The EOI detector is for a 12 mm gap between upper and lower halves of the pixel detector, while the square hole refers to a 12x12 mm^2 hole. The calculation is for non-leveled luminosity.

baseline.

The x_s reach is excellent and extends over the entire predicted Standard Model range.

F Measurement of γ

The angle γ could in principle be measured using a CP eigenstate of B_s decay that was dominated by the $b \to u$ transition. One such decay that has been suggested is $B_s \to \rho^o K_s$. However, there are the same "Penguin pollution" problems as in $B^o \to \pi^+\pi^-$, but they are more difficult to resolve in the vector-pseudoscalar final state. (Note, the pseudoscalar-pseudoscalar final state here is $\pi^o K_s$, which does not have a measurable decay vertex.)

Fortunately, there are other ways of measuring γ. CP eigenstates are not used, which introduces discrete ambiguities. However, combining several methods should remove these.

We have studied three methods of measuring γ. The first method uses the decays $B_s \to D_s^{\pm} K^{\mp}$ where a time-dependent CP violation can result from the interference between the direct decay and the mixing induced decay [21]. Fig. 14 shows the two direct decay processes for $\overline{B}_s^o$.

Consider the following time-dependent rates that can be separately measured

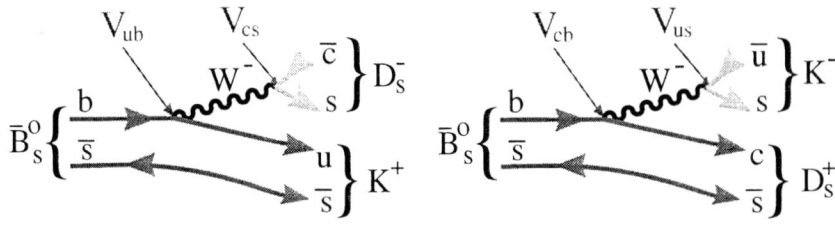

FIGURE 14. Two diagrams for $\bar{B}_s^o \to D_s^\pm K^\mp$.

using flavor tagging of the other b:

$$\Gamma(B_s \to f) = |M|^2 e^{-t}\{\cos^2(xt/2) + \rho^2 \sin^2(xt/2) - \rho\sin(\phi+\delta)\sin(xt)\}$$

$$\Gamma(\bar{B}_s \to \bar{f}) = |M|^2 e^{-t}\{\cos^2(xt/2) + \rho^2 \sin^2(xt/2) + \rho\sin(\phi-\delta)\sin(xt)\}$$

$$\Gamma(B_s \to \bar{f}) = |M|^2 e^{-t}\{\rho^2 \cos^2(xt/2) + \sin^2(xt/2) - \rho\sin(\phi-\delta)\sin(xt)\}$$

$$\Gamma(\bar{B}_s \to f) = |M|^2 e^{-t}\{\rho^2 \cos^2(xt/2) + \sin^2(xt/2) + \rho\sin(\phi+\delta)\sin(xt)\},$$

where $M = \langle f|B \rangle$, $\rho = \frac{\langle f|\bar{B}\rangle}{\langle f|B\rangle}$, ϕ is the weak phase between the 2 amplitudes and δ is the strong phase between the 2 amplitudes. The three parameters ρ, $\sin(\phi+\delta), \sin(\phi-\delta)$ can be extracted from a time-dependent study if $\rho = O(1)$.

In the case of B_s decays where $f = D_s^+ K^-$ and $\bar{f} = D_s^- K^+$, the weak phase is γ. The decay modse $B_s \to D_s^+ K$, $D_s^+ \to \phi\pi^+$, $\phi \to K^+K^-$, or $D_s \to K^{*o}K^+$, were simulated. For the $\phi\pi^+$ mode, the combined geometric acceptance and reconstruction efficiency is 5.2% with S/B=10 [22], and the trigger efficiency is 67%. In the $K^{*o}K^+$ mode the geometric and reconstruction efficiency is 5.9% and the trigger efficiencies and signal to background are same as in the $\phi\pi^+$ mode. Using the branching fractions predicted by Aleksan [21] and assuming a tagging efficiency $\epsilon = 15\%$ we expect 8900 events.

The decay time resolution and the detachment of the decay vertex from the primary production vertex are shown in Fig. 15 for the $D_s^+ \to \phi\pi^+$ decay mode. (The distributions in the $K^{*o}K^+$ mode are similar.)

Using the measured values of S/B and time resolution, a mistag rate of 25%, and $x_s=20$, a mini-Monte Carlo was used to generate the extracted value of γ for an ensemble of experiments each with 8900 signal events, for various sets of input parameters ρ, $\sin(\gamma+\delta), \sin(\gamma-\delta)$. A maximum likelihood fit was then used to extract fitted values of the parameters.

Fig. 16 shows the distributions of the parameters for a signal of 8900 events with input values ρ=0.5, $\sin\gamma = 0.866$ and $\cos\delta = 0.7$. Assuming that $\sin\gamma > 0$ then $\sin\gamma$ can be determined up to a two-fold ambiguity, hence γ up to a four-fold ambiguity.

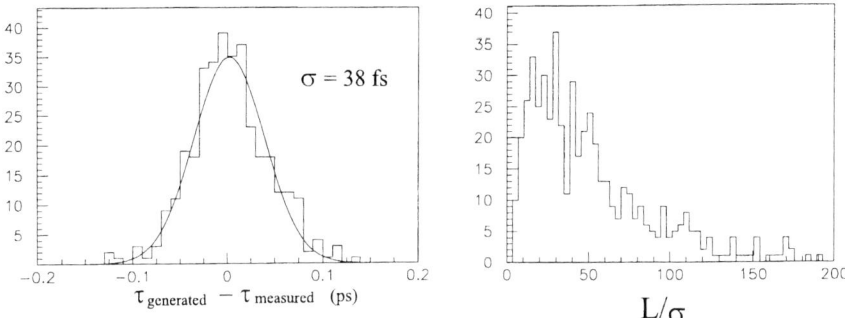

FIGURE 15. (left) The generated minus measured proper time distribution for $\overline{B}_s^0 \to D_s^\pm K^\mp$, $D_s^+ \to \phi\pi^+$. (right) The distribution in L/σ for this mode.

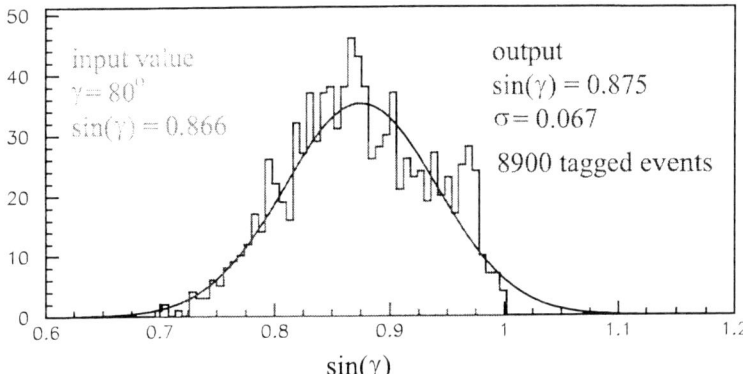

FIGURE 16. Results of determining γ for many different "experiments" with input value of $\gamma=80°$ using the decay time distributions for $\overline{B}_s^0 \to D_s^\pm K^\mp$.

Another method for extracting γ has been proposed by Atwood, Duneitz and Soni [9], who refined a suggestion by Gronau and Wyler [8]. A large CP asymmetry can result from the interference of the decays $B^- \to K^- D^0, D^0 \to f$ and $B^- \to K^- \overline{D}^0, \overline{D}^0 \to f$, where f is a doubly Cabibbo suppressed decay of the D^0 (for example $f = K^+\pi^-, K\pi\pi$, etc.) Since $B^- \to K^- \overline{D}^0$ is color-suppressed and $B^- \to K^- D^0$ is color-allowed, the overall amplitudes for the two decays are expected to be approximately equal in magnitude. The weak phase difference between them is γ. To observe a CP asymmetry there must also be a non-zero strong phase between the two amplitudes. It is necessary to measure the branching ratio $\mathcal{B}(B^- \to K^- f)$ for at least 2 different states f in order to determine γ up to discrete ambiguities. We have examined the decay modes $B^- \to K^-[K^+\pi^-]$ and $B^- \to K^-[K^+3\pi]$. The combined geometric acceptance and reconstruction efficiency was found to be 6.6% for the $K\pi$ mode and 55% for $K3\pi$ with a signal to background of about 1:1. The trigger efficiency is approximately 70% for both modes. The expected number of $B^\pm$ events in 10^7 s is 2400 in the $K\pi$ mode and 4200 in the $K3\pi$ mode. With this number of events we expect to be able to measure γ (up to discrete ambiguities) with a statistical error of about $\pm 8°$ in one year of running at $\mathcal{L} = 2 \times 10^{32} \text{cm}^{-2}\text{s}^{-1}$. The overall sensitivity depends on the actual values of γ and the strong phases.

The next method, described by Gronau and Rosner [23] and Fleischer and Mannel [24], uses $B^0 \to K^+\pi^-$ and $B^+ \to K^0\pi^+$ decays. It is particularly promising as it may complement other methods by excluding some of the region around $\gamma = \pi/2$. We expect to reconstruct 3600 $B^\pm \to K_s\pi^\pm$ with S/B=0.5 and 29000 $B^0/\overline{B}^0 \to K^\pm\pi^\mp$ with S/B=3. Gronau and Rosner estimate a measurement of γ to $10°$ with 2400 events in each channel [25], however there has been much theoretical discussion about the effects of isospin conservation and rescattering which casts doubt on this method [26] [27] [28] [29].

VII DECAY TIME RESOLUTION

In all of these studies we have assumed that we would have 9 μm spatial resolution in each track hit in the pixel plane. We now address the question of whether or not this is reasonable.

The parameters affecting the pixel resolution include the size of the pixel, the use of digital (one bit) versus analog (4 bits) information, the threshold, gain variations, the use of electrons or holes as charge carriers, since the drift velocity for electrons is three times that for holes. For different incident angles of tracks on the pixels the charge sharing is affected by the magnetic field. In BTeV we use a 1.5 T dipole field.

In Fig. 17(a) we show the track angle distribution for two B final states, the two-body state $\pi^+\pi^-$ and the four-body state $\pi^+K^+\pi^-\pi^-$. In both cases the angular distribution peaks at small angles, about 50 mr and then falls slowly towards larger angles. The spatial resolution has been simulated as a function of angle for various pixel sizes. The baseline size is 50 μm $\times$ 300 μm. The results of the simulation for this size are shown in Fig. 17(b). The magnetic field is in the y direction in this case. So the tracks hitting the x layers are bent. The resolution using digital electronics is about 10 μm, while it is about 5 μm using 4-bit analog. Note that the poorer resolution peak near zero degrees in the non-bend plane does exist in the bend plane, but it is shifted toward negative incident angles. Similar results are obtained with holes as charged carriers.

The resolution in proper time is affected by several factors. One is the inherent pixel resolution, as discussed above. Others include the amount of material in the pixel system, and the distance the pixel detector is placed from the beam line. We take the latter as 6 mm. This distance is limited by the maximum amount of radiation damage we are willing to sustain. (The system is retractable during machine injection.) In Fig. 18 we show the proper time resolution achievable on the ψK^{*o} decay of the B_s for several different detector geometries as a function of the spatial resolution. The circles represent a geometry with a 12 mm gap and equivalent silicon thickness of 600, 500, and 600 μm for the three layers. These

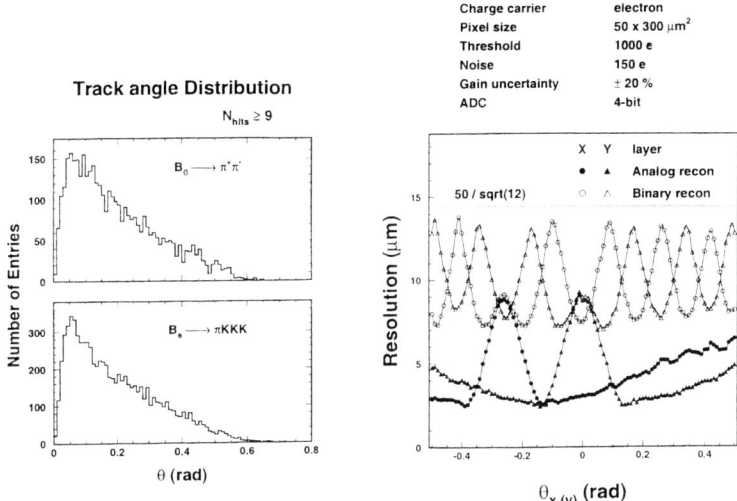

FIGURE 17. The calculated proper time resolutions for the decay mode $B_s \to \psi K^{*o}$, $K^{*o} \to K^- \pi^+$ as a function of spatial resolution in the pixel system for different detector geometries and thicknesses. The two geometries considered are a gap of 12 mm between two halves of the pixel system and square hole of 12mm × 12 mm. The equivalent silicon thicknesses for the three planes are listed.

include the 300 μm of silicon for each layer, an radio-frequency shield of 100 μm of Al and material for electronics and cooling.

The simulations presented here have assumed a 9 μm spatial resolution, even though we believe that 5 μm is possible. The equivalent silicon thickness of a three plane station is taken at 1700 μm. A detector of twice the material thickness and 5 μm resolution would have the same time resolution as the one we have been using. This points out the need to minimize the material, a well known lesson.

VIII COMPARISONS WITH OTHER EXPERIMENTS

A Comparisons with e^+e^- B-factories

Most of what is know about b decays has been learned at e^+e^- machines [30]. Machines operating at the $\Upsilon(4S)$ found the first fully reconstructed B mesons (CLEO), B^o-$\overline{B}^o$ mixing (ARGUS), the first signal for the $b \to u$ transition (CLEO), and Penguin decays (CLEO). Lifetimes of b's were first measured by experiments at PEP, slightly later at PETRA, and extended and improved by LEP [30].

The success of the $\Upsilon(4S)$ machines has led to the construction at KEK and SLAC of two new $\Upsilon(4S)$ machines with luminosity goals in excess of $3 \times 10^{33} \mathrm{cm}^{-2}\mathrm{s}^{-1}$. These machines will asymmetric beam energies so they can measure time dependent

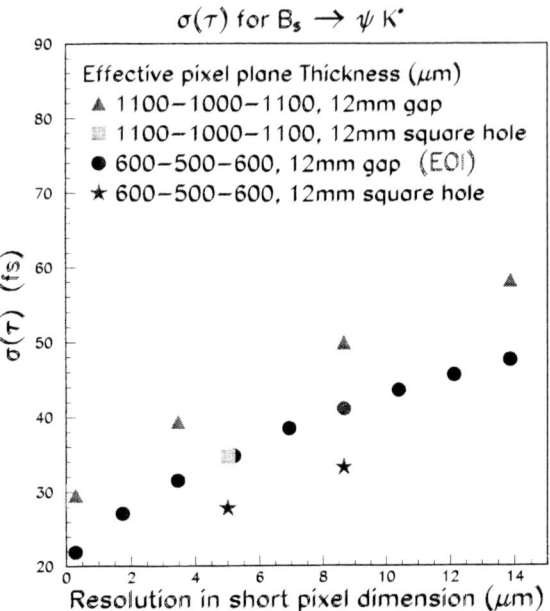

FIGURE 18. The calculated proper time resolutions for the decay mode $B_s \to \psi K^{*o}$, $K^{*o} \to K^-\pi^+$ as a function of spatial resolution in the pixel system for different detector geometries and thicknesses. The two geometries considered are a gap of 12 mm between two halves of the pixel system and square hole of 12mm × 12 mm. The equivalent silicon thicknesses for the three planes are listed.

CP violation. They will join an upgraded CESR machine at Cornell with symmetric beam energies. These machines will investigate only B^o and $B^\pm$ decays, they will not investigate B_s, B_c or Λ_b decays.

Table 4 shows a comparison between BTeV and an asymmetric e^+e^- machine for measuring the CP violating asymmetry in the decay mode $B^o \to \pi^+\pi^-$. It is clear that the large hadronic b production cross-section can overwhelm the much smaller e^+e^- rate.

B Comparisons with Tevatron Central Detectors

Both CDF and D0 have measured the b production cross-section [31] and CDF has contributed to our knowledge of b decay mostly by its measurements of the lifetime of b-flavored hadrons [32], which are competitive with those of LEP [33] and recently through its discovery of the B_c meson [34]. These detectors were designed for physics discoveries at large transverse momentum. It is remarkable that they have been able to accomplish so much in b physics.

However, these dectectors are very far from optimal for b physics. BTeV has been designed with b physics as its primary goal. To have an efficient trigger based on separation of b decays from the primary, BTeV uses the large η region where the

TABLE 4. Number of tagged $B^o \to \pi^+\pi-$ ($\mathcal{B}=0.75 \times 10^{-5}$)

	$\mathcal{L}(\mathrm{cm}^{-2}\mathrm{s}^{-1})$	σ	# $B^o/10^7$s	efficiency	ϵD^2	# tagged
e^+e^-	3×10^{33}	1 nb	3.0×10^7	0.4	0.4	46
BTeV†	2×10^{32}	100μb	1.1×10^{11}	0.06	0.1	6400
BTeV‡	2×10^{32}	100μb	2.1×10^{11}	0.06	0.1	12800

†This is for gap detector, expect increase with square hole
‡Luminosity leveled, use 2×10^7 s/year

b's are boosted. The detached vertex trigger allows collection of interesting purely hadronic final states such as $\pi^+\pi^-$, $D_s^+\pi^-$ and $D_s^+K^-$. It also allows us to collect enough charm to investigate mixing and CP violation.

The use of the foward geometry also allows for excellent charged hadron identification with a gaseous RICH detector. This is crucial for many physics issues such as separating $K\pi$ from $\pi\pi$, $D_s\pi$ from D_sK, kaon flavor tagging etc...

C Comparison with LHC-B

LHC-B is an experiment proposed for the LHC with almost the same physics goals as BTeV [35]. LHC-B has two advantages: the b cross-section is five times higher than at the Tevatron while the total cross-section is only 1.6 times as large, and the mean number of interactions per crossing is three times lower, because the LHC has bunches every 25 ns, while the Tevatron bunches come every 132 ns.

There are, however, many advantages which accrue to BTeV. Let us first consider the machine specific ones. The 132 ns bunch spacing at the Tevatron makes first level detached vertex triggering easier. It is difficult for the vertex detector electronics in LHC-B to settle in 25 ns. The seven times larger energy at the LHC results in a larger track multiplicity per collision which causes trigger and tracking problems and larger range of energy track momenta that need to be analyzed. The interaction region at the LHC is relatively short, σ=5 cm, compared with the 30 cm long region at Fermilab. This somewhat compensates for the larger number of interactions per crossing, since the interactions are well separated.

There are detector specific advantages for BTeV as well. BTeV is a two-arm spectrometer, resulting a factor of two advantage. BTeV has the vertex detector in the magnetic field which allows the rejection of high multiple scattering (low momentum) tracks in the trigger. Furthermore, BTeV is designed around a pixel vertex detector while LHC-B has a silicon strip detector. BTeV can put the detector closer to the beam (6 mm versus 1 cm), and has a much more robust tracking system which can trigger on detached verticies in the first trigger level, while LHC-B triggers on tracks of moderate transverse momentum in their first trigger level.

We feel that we have more than compensated for LHC-B's initial advantages.

IX CONCLUSIONS

Hadron colliders have large b and c cross-sections allowing the opportunity experiments for precision measurents of CP Violation and B_s mixing. In our view this requires high density tracking and triggering information that can be provided by a state of the art pixel system. BTeV has been designed to fit in the new C0 interaction region at the Tevatron and incorporates a pixel vertex detector, downstream tracking, charged paricle identification, lepton identification and photon detection. The vertex detector enables Level I vertex triggering and excellent time resolution on heavy hadron decays [36].

A summary of the physics reach is shown in Table 5. Those simulations that have been upgraded by using the square hole detector are so indicated.

TABLE 5. BTeV Physics Reach

Measurement	Accuracy in 10^7 s $\mathcal{L} = 2 \times 10^{32}$, $\mathcal{L}$ leveled
x_s (square hole)	up to 80 & beyond
$A_{CP}(B^0 \to \pi^+\pi^-)$ (gap)	± 0.013
γ using $D_s K^-$ (square hole)	$\pm 8°$†
γ using $D^0 K^-$ (square hole)	$\pm 8°$‡
$\mathcal{B}(B^- \to K^-\mu^+\mu^-)$ (gap)	4σ at $\mathcal{B}$ of 5.4×10^{-8}
$\sin(2\beta)$ using $B^o \to \psi K_s$ (square hole)	± 0.013

†Assumes ρ=0.7, $\cos\delta$=0.7, $\sin\gamma$=0.5, x_s=20
‡For most values of strong phases and γ

BTeV is an officially recognized R&D project at Fermilab. Development has started on the pixel, trigger, RICH, muon, forward tracking and Electromagnetic calorimeter systems. More information on BTeV can be found on the world-wide-web [36].

REFERENCES

1. The CESR B Physics Working Group, K. Lingel *et al*, "Physics Rationale For a B Factory", Cornell Preprint CLNS 91-1043 (1991); SLAC Preprint SLAC-372 (1991); "Progress Report on Physics and Detector at KEK Asymmetric B Factory," KEK Report 92-3 (1992)
2. N. Cabibbo, *Phys. Rev. Lett.* **10**, 531 (1963); M. Kobayashi and K. Maskawa, *Prog. Theor. Phys.* **49**, 652 (1973).
3. L. Wolfenstein, *Phys. Rev. Lett.* **51**, 1945 (1983).
4. S. Stone, "Prospects For B-Physics In The Next Decade," presented at *NATO Advanced Study Institute on Techniques and Concepts of High Energy Physics*, Virgin Islands, July 1996, to be published in procedings.

5. P. Langacker, "CP Violation and Cosmology," in *CP Violation*, ed. C. Jarlskog, World Scientific, Singapore p 552 (1989).
6. A. J. Buras, "Theoretical Review of B-physics," in *BEAUTY '95* ed. N. Harnew and P. E. Schlein, *Nucl. Instrum. Methods* **A368**, 1 (1995).
7. Measuring the CP violating asymmetry in the $B^o \to \pi^+\pi^-$ channel is insufficient to determine the angle α since there are other diagrams, called penguins, which can contribute to this decay process. There are many suggestions of how to extract α using additional measurements. One such theoretically rigorous suggestions requires the measurment of $\pi^+\pi^o$ and $\pi^o\pi^o$ rates. M. Gronau and D. London, *Phys. Rev. Lett.* **65**, 3381 (1990); N.G. Deshpande, X-G. He, and S. Oh, *Phys. Lett.* B **384**, 283 (1996); A. Buras and R. Fleischer, *Phys. Lett.* B **360**, 138 (1995); M. Gronau and J. L. Rosner, *Phys. Lett.* B **76**, 1200 (1996); A. S. Dighe, M. Gronau, and J. L. Rosner, *Phys. Rev.* D **54**, 3309 (1996); A. S. Dighe and J. L. Rosner, *Phys. Rev.* D **54**, 4677 (1996). R. Fleischer and T. Mannel, *Phys. Lett.* B **397**, 269 (1997); C. S. Kim, D. London and T. Yoshikawa, "Using B_s^o Decays to Determine the CP Angles α and γ, hep-ph/9708356 UdeM-GPP-TH-97-43 (1997).
8. M. Gronau and D. Wyler, *Phys. Lett.* B **265**, 172 (1991).
9. D. Atwood, I. Dunietz and A. Soni, *Phys. Rev. Lett.* **78**, 3257 (1997).
10. Private communication from I. Bigi and G. Burdman.
11. M. Golden and B. Grinstein *Phys. Lett.* B **222**, 501 (1989); F. Buccella *et al*, *Phys. Rev.* D **51**, 3478 (1995).
12. M. Artuso, "Experimental Facilities for b-Quark Physics," in *B Decays* revised 2nd Edition, Ed. S. Stone, World Scientific, Sinagapore (1994).
13. T. Skwarnicki, "The BTeV RICH," presented at *Beauty '97*, UCLA, Los Angeles, CA, to appear in proceedings.
14. P. Avery *et al*, "MCFast: A Fast Simulation Package for Detector Design Studies," Presented at The International Conference on Computing in High Energy Physics, Berlin 1997. To appear in the proceedings.
15. These results are based on the work of R. Isik, W. Selove, and K. Sterner, "Monte Carlo Results for a Seconday-vertex Trigger with On-line Tracking," Univ. of Penn. preprint UPR-234E (1996); D. Husby, W. Selove, K. Sterner, P. Chew, *Nucl. Inst. Meth.* **A383** 193 (1996); R. A. Isik, "Real-Time Pattern-Recognition for HEP," Univ. of Pa. Report, UPR-233E, July 27, 1996.
16. M. Procario, "*B* Physics Prospects beyond the Year 2000," invited talk at 10th Topical Workshop on Proton-Antiproton Physics, Fermilab-CONF-95/166 (1995).
17. R. Godang *et al*(CLEO), *Phys. Rev. Lett.* **27**, 1522 (1997).
18. P. McBride and S. Stone, *Nucl. Instr. and Meth.* A368, 38 (1995).
19. About 20% of the B_d meson mix and 50% of the B_s. However, the final state in most B_s decays contains an s quark from the $b \to c \to s$ decay chain and an $\bar{s}$ spectator quark, negating any effects from mixing. In fact, the B_s does not usefully contribute to kaon tagging in the first place. Therefore we are left with a dilution of 8% from mixing in kaon tagging, taking the B_d fraction as 40%.
20. M. Jimack, "LEP Results on Oscillation and Mixing," presented at *Beauty '97*, UCLA, Los Angeles, CA, to appear in proceedings.
21. R. Aleksan, I. Dunietz, B. Kayser, "Determining the CP-violating phase γ", Z. Phys

C **54**, 653-659 (1992).
22. Private Communication from P. A. Kasper.
23. M. Gronau and J. Rosner, CALT-68-2142, hep-ph/9711246
24. R. Fleischer and T. Mannel, hep-ph/9704423
25. J. Rosner, private communication
26. J.-M. Gerard and J. Weyers, "Isospin amplitudes and CP violation in $(B \to K\pi)$ decays," hep-ph/9711469 (1997).
27. A. Falk, A. Kagan, Y. Nir and A. Petrov, JHU-TIPAC-97018 (December 1997).
28. M. Neubert, "Rescattering Effects, Isospin Relations and Electroweak Penguins in $B \to \pi K$ Decays," hep-ph/9712224 (1997).
29. D. Atwood and A. Soni, "The Possibility of Large Direct CP Violation in $B \to K\pi$-Like Modes Due to Long Distance Rescattering Effects and Implications for the Angle γ," hep-ph/9712287 (1997).
30. See *B Decays, revised 2nd Edition* ed. S. Stone, World Scientific, Singapore, (1994).
31. K. Abe *et al*, (CDF), *Phys. Rev. Lett.* **75**, 1451 (1995); S. Abachi *et al*, (D0), *Phys. Rev. Lett.* **74**, 3548 (1995). See also the UA1 measurement C. Albajar *et al*, *Phys. Lett.* B**186**, 237 (1987); B**213**, 405 (1988); B**256**, 121 (1991).
32. K. Abe *et al*, (CDF), *Phys. Rev. Lett.* **76**, 4462 (1996); ibid **77**, 1945 (1996); K. Abe *et al*, (CDF), *Phys. Rev. D* **57**, 5382 (1998).
33. T. Junk, "A Review of B Hadron Lifetime Measurements from LEP, the Tevatron and SLC," in Proceedings of the 2nd Int. Conf. on *B Physics and CP Violation*, Univ. of Hawaii, (1997), ed. T. E. Browder *et al*, World Scientific, Singapore (1998).
34. K. Abe *et al*, (CDF), "Observation of B_c Mesons in $p - \bar{p}$ Collisions at $\sqrt{s} = 1.8$ TeV," hep-ex/9804014 (1998).
35. "LHC-B Letter of Intent," CERN/LHCC 95-5, LHCC/18 (1995), which can be viewed at http://www.cern.ch/LHC-B/loi/loi_old.html .
36. For more information on BTeV see http://fnsimu1.fnal.gov/btev.html

Muon Colliders: New Prospects for Precision Physics and the High Energy Frontier

Bruce J. King[1]

Brookhaven National Laboratory
email: bking@bnl.gov
web page: http://pubweb.bnl.gov/people/bking/

Abstract. An overview is given of muon collider technology and of the current status of the muon collider research program. The exciting potential of muon colliders for both neutrino physics and collider physics studies is then described and illustrated using self-consistent collider parameter sets at 0.1 TeV to 100 TeV center-of-mass energies.

INTRODUCTION

Muon colliders appear to be emerging as a promising complement and/or alternative to proton and electron colliders for experimental high energy physics (HEP) studies at the high energy frontier. They also provide some interesting possibilities for precision studies in HEP, particularly in neutrino physics.

This paper consists of three main sections. The first section gives a very brief description of muon collider technology then two longer sections give introductions to the neutrino physics potential and collider physics potential of muon colliders, respectively.

The two physics sections use, as examples, the muon collider parameter sets of table 1, at center of mass (CoM) energies from 0.1 TeV to 100 TeV. The parameter set at 0.1 TeV CoM energy, which is intended as an s-channel Higgs factory, was constrained to essentially reproduce one of the parameter sets currently under study [1] by the Muon Collider Collaboration (MCC). In contrast, the other sets represent speculation by the author on how the parameters might evolve with CoM energy. A discussion and assessment of the technical challenges associated with these specific parameter sets is given in [2]. It should be stressed that they are all still rather speculative (additional to the rather immature status of the entire muon collider technology) and have not been studied or discussed in detail within

---

[1] This work was performed under the auspices of the U.S. Department of Energy under contract no. DE-AC02-98CH10886.

TABLE 1. Self-consistent parameter sets for muon colliders at CoM energies ranging from 0.1 TeV to 100 TeV. For completeness, beam parameters and collider ring parameters have been included along with the physics parameters, and the generation and optimization of these parameter sets is described in reference [2]. Except for the first parameter set, which has been studied in some detail by the Muon Collider Collaboration, these parameters represent speculation by the author on how muon colliders might evolve with energy.

center of mass energy, E_{CoM} description	0.1 TeV Higgs factory	1 TeV LHC complement	4 TeV E frontier	10 TeV 2nd gen.	100 TeV ult. E scale
collider physics parameters:					
luminosity, $\mathcal{L}$ [cm^{-2}.s^{-1}]	1.0×10^{31}	1.0×10^{34}	6.2×10^{33}	1.0×10^{36}	4.0×10^{36}
$\int \mathcal{L} dt$ [fb^{-1}/det/year]	0.1	100	62	10 000	40 000
No. of $\mu\mu \to ee$ events/det/year	870	8700	340	8700	350
No. of 100 GeV SM Higgs/det/year	3700	69 000	69 000	1.4×10^7	8.3×10^7
fract. CoM energy spread, σ_E/E [10^{-3}]	0.02	1.6	1.6	1.0	1.0
neutrino physics parameters:					
fract. str. sect. length, f_{ss}	0.15	0.10	0.05	0.04	0.02
neutrino ang. divergence, $\theta_\nu[1/\gamma]$	1	10	10	10	10
high rate det: events/yr/(g.cm^{-2})	8.1×10^6	1.9×10^7	1.5×10^6	1.3×10^8	2.5×10^7
long baseline: events/yr/(kg.km^{-2})	1.8×10^5	4.2×10^5	5.3×10^5	2.9×10^8	5.6×10^9
collider ring parameters:					
circumference, C [km]	0.3	2.0	7.0	15	100
ave. bending B field [T]	3.5	5.2	6.0	7.0	10.5
beam parameters:					
(μ^- or) μ^+/bunch, N_0[10^{12}]	4.0	3.5	3.1	2.4	0.18
(μ^- or) μ^+ bunch rep. rate, f_b [Hz]	15	15	0.67	15	60
6-dim. norm. emittance, ϵ_{6N}[10^{-12}m^3]	170	170	170	50	2
x,y emit. (unnorm.) [$\pi.\mu$m.mrad]	710	12	3.0	0.55	0.0041
x,y normalized emit. [π.mm.mrad]	340	57	57	26	1.9
fract. mom. spread, δ [10^{-3}]	0.03	2.3	2.3	1.4	1.4
relativistic γ factor, E_μ/m_μ	473	4732	18 929	47 322	473 220
ave. current [mA]	20	10	0.46	24	4.2
beam power [MW]	1.0	8.4	1.3	58	170
decay power into magnet liner [kW/m]	1.1	0.58	0.03	1.4	1.3
time to beam dump, $t_D[\gamma\tau_\mu]$	no dump	0.5	0.5	no dump	0.5
effective turns/bunch	519	493	563	1039	985
interaction point parameters:					
spot size, $\sigma_x = \sigma_y$[μm]	270	7.6	1.9	0.78	0.057
bunch length, σ_z [mm]	11	4.7	1.2	1.1	0.79
β^* [mm]	11	4.7	1.2	1.1	0.79
ang. divergence, σ_θ [mrad]	2.6	1.6	1.6	0.71	0.072
beam-beam tune disruption parameter, $\Delta\nu$	0.013	0.066	0.059	0.100	0.100
pinch enhancement factor, H_B	1.000	1.040	1.025	1.108	1.134
beamstrahlung fract. E loss/collision	5×10^{-16}	1.2×10^{-10}	2.3×10^{-8}	2.3×10^{-7}	3.2×10^{-6}
final focus lattice parameters:					
max. poletip field of quads., $B_{4\sigma}$ [T]	6	10	10	15	20
max. full aperture of quad., $A_{\pm 4\sigma}$ [cm]	14	13	30	20	13
β_{max}[km]	0.4	22	450	1100	61 000
final focus demagnification, $\sqrt{\beta_{max}/\beta^*}$	60	2200	19 000	31 000	280 000
synchrotron radiation parameters:					
syn. E loss/turn [MeV]	0.0008	0.01	0.9	17	25 000
syn. rad. power [kW]	0.0002	0.13	0.4	400	110 000
syn. critical E [keV]	0.0006	0.09	1.6	12	1700
neutrino radiation parameters:					
collider reference depth, D[m]	10	125	300	300	300
ave. rad. dose in plane [mSv/yr]	3×10^{-5}	9×10^{-4}	9×10^{-4}	0.66	6.7
str. sect. length for 10x ave. rad., $L_{\times 10}$[m]	1.9	1.3	1.1	1.0	2.4
ν beam distance to surface [km]	11	40	62	62	62
ν beam radius at surface [m]	24	8.4	3.3	1.3	0.13

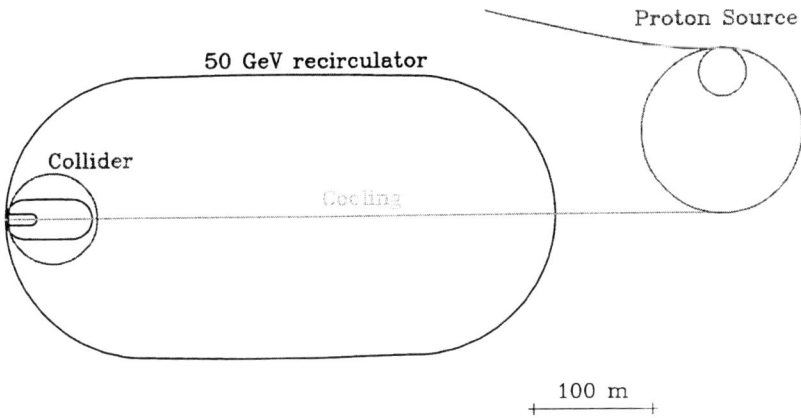

FIGURE 1. Schematic footprint of a 100 GeV muon collider (reproduced from reference [1]).

the MCC. This applies even more strongly to the final parameter set, at 100 TeV, which might represent the ultimate energy scale for muon colliders and which assumes technological extrapolations (in magnets, etc.) that might not come to pass for at least another couple of decades.

AN OVERVIEW OF MUON COLLIDERS

The technology of muon colliders is relatively new. The possibility of muon colliders was introduced by Budker [3], Skrinsky et al. [4] and Neuffer [5] and has been aggressively developed over the past four years in a series of collaboration meetings and workshops [6–9]. A detailed feasibility study for a 4 TeV muon collider [10] was presented at Snowmass96 and, since then, progress has continued on studies for both this collider and others at lower energies. The current status of MCC studies is summarized in [1]. The Muon Collider Collaboration now consists of over 100 physicists and engineers from the U.S.A., Europe and Japan, largely based in the U.S.A. and mainly at three U.S. national laboratories: Brookhaven National Laboratory (BNL), Fermi National Accelerator Laboratory (FNAL) and Lawrence Berkeley National Laboratory (LBNL).

Figure 1 illustrates the basic layout of a $\mu^+\mu^-$ collider using, as an example, the schematic footprint of a 0.1 TeV collider. Initially, large bunches of low energy muons are produced by targeting proton bunches from a high intensity proton source onto a pion production target inside a solenoidal capture and decay channel. The relatively diffuse muon bunches from the decay channel then enter an ionization cooling channel, which shrinks them down to a suitable emittance for fast acceleration and injection, at full energy, into a collider storage ring. (The final acceleration stage is anomalously larger than the collider ring for the low energy collider of figure 1. Depending on design and technology choices, the final acceleration stage may well remain larger than the collider ring for higher energy colliders, but likely by a lesser margin.)

The ionization cooling channel is the most novel and characteristic feature of a muon collider, and also the biggest technical challenge. As a general outline of the cooling process, the muons in each bunch lose both transverse and longitudinal momentum in passing through a material medium and are then reaccelerated in radiofrequency (rf) cavities, restoring the longitudinal momentum but leaving a reduced transverse momentum spread in the bunch. Also, the momentum spread of the bunch can be reduced by using wedges of material in a dispersive section of a magnet lattice to reduce preferentially the momenta of the munos with higher momenta. A large amount of cooling is required – current scenarios give a factor of 10^6 reduction in the invariant 6-dimensional phase space – so the cooling channel will probably be a repetitive structure with perhaps 20 to 30 stages. The MCC is pursuing a vigorous theoretical and experimental program to develop and test the components of the cooling channel.

Because of the short muon lifetime – 2.2 microseconds in the muon rest frame – the muon cooling and acceleration must be done very quickly. Current scenarios envisage about a 50% decay loss in the cooling channel and a 25% loss of the remaining muons during acceleration. Also, the muons survive for only of order 1000 turns in the collider ring (almost independent of the collider energy), so the muon bunches must be frequently replenished. Undesirable consequences of the large bunches of muons decaying to electrons are the resulting large and difficult background in the collider detectors, the radiation heat load on the collider ring and, surprisingly, the potential radiation hazard from the intense neutrino beams. The neutrino radiation hazard becomes an important design constraint for high energy colliders, and is discussed in more detail in a later section.

PROSPECTS FOR NEUTRINO PHYSICS

This section gives an overview of the neutrino physics possibilities at a future muon storage ring, which can be either a muon collider ring or a ring dedicated to neutrino physics that uses muon collider technology to store large muon currents. It summarizes a previous more detailed description of these topics by this author [11] (using a generalized description of neutrino production and event rates that is now applicable to all muon colliders).

The section begins with a characterization of the neutrino beam and predictions for neutrino event rates in both general purpose and long-baseline neutrino detectors, then follows with a description of a specific design for a general baseline detector. Finally, an overview is given of some of the important physics analyses that could be performed at such "muon ring neutrino experiments" (MURINE's).

Neutrino Beam and Experimental Overview

Neutrinos are emitted from the decay of muons in the collider ring:

$$\mu^- \to \nu_\mu + \overline{\nu}_e + e^-,$$
$$\mu^+ \to \overline{\nu}_\mu + \nu_e + e^+. \quad (1)$$

The thin pencil beams of neutrinos for experiments will be produced from long straight sections in either the collider ring or a ring dedicated to neutrino physics. From relativistic kinematics, the forward hemisphere in the muon rest frame will be boosted, in the lab frame, into a narrow cone with a characteristic opening half-angle, θ_ν, given in obvious notation by

$$\theta_\nu \simeq \sin\theta_\nu = 1/\gamma = \frac{m_\mu}{E_\mu} \simeq \frac{10^{-4}}{E_\mu(\text{TeV})}. \quad (2)$$

The final focus regions around collider experiments are important exceptions to equation 2 since the muon beam itself will have an angular divergence in these regions that is large enough to spread out the neutrino beam by at least an order of magnitude in both x and y. It is likely that neutrino experiments at sub-TeV CoM energy muon colliders will use the beams from either dedicated or utility straight sections opposite the collider detector while those at higher energy muon colliders – where neutrino radiation is an important design constraint – will use the more divergent beam emanating from the final focus region. A dedicated storage ring could avoid the problem of neutrino radiation by using a long downward-tilting long straight section.

The dominant interaction of TeV-scale neutrinos is deep inelastic scattering (DIS) off nucleons with the production of several hadrons. This is reinterpreted in the quark-parton model as elastic or quasi-elastic scattering off the quark constituents of the nucleons followed by hadronization of the final state quark. Charged current (CC) DIS scattering, which is mediated by a charged W boson and comprises about 75% of the total cross-section, may be represented as

$$\nu + q \to l^- + q',$$
$$\overline{\nu} + q' \to l^+ + q, \quad (3)$$

where l is an electron/muon for electron/muon neutrinos and the quarks, (q) and (q'), have charges differing by one unit. Neutral current (NC) DIS scattering,

$$\nu + q \to \nu + q, \quad (4)$$

which is interpreted as neutrino-quark elastic scattering with the exchange of a neutral Z boson, makes up the remaining 25% of the cross-section.

For TeV-scale neutrinos, the neutrino cross-section is approximately proportional to the neutrino energy, E_ν, and the charged current (CC) and neutral current (NC) interaction cross sections for neutrinos and antineutrinos have numerical values of [12]:

$$\sigma_{\nu N} \text{ for } \begin{pmatrix} \nu - CC \\ \nu - NC \\ \overline{\nu} - CC \\ \overline{\nu} - NC \end{pmatrix} \simeq \begin{pmatrix} 0.72 \\ 0.23 \\ 0.38 \\ 0.13 \end{pmatrix} \times E_\nu[\text{TeV}] \times 10^{-35} \text{ cm}^2. \quad (5)$$

Using these cross-section values, it is straightforward to derive predictions for the approximate neutrino event rates at a neutrino detector. For a general purpose detector subtending the boosted forward hemisphere of the neutrino beam:

$$\text{Number of } \nu \text{ events/yr} \simeq 1.8 \times 10^7 \times l[\text{g.cm}^{-2}]$$
$$\times f_b[\text{Hz}] \times N_0[10^{12}] \times E_\mu[\text{TeV}] \times f_{ss} \times (1 - e^{-t_D[\gamma\tau_\mu]}), \quad (6)$$

with notation as in table 1, where l is the detector length, 10^7 seconds of running time per year are assumed and the fractional breakdown into interaction types is as in equation 5.

The analagous equation for a long baseline detector in the center of the neutrino beam is:

$$\text{Number of } \nu \text{ events/yr} \simeq 9 \times 10^6 \times \frac{M[\text{kg}]}{(L[\text{km}])^2 \times (\gamma\theta_\nu)^2}$$
$$\times f_b[\text{Hz}] \times N_0[10^{12}] \times E_\mu[\text{TeV}] \times f_{ss} \times (1 - e^{-t_D[\gamma\tau_\mu]}), \quad (7)$$

where M is the detector mass, L the distance from the neutrino source and the factor $(\gamma\theta_\nu)^{-2}$ allows for the possibility that the divergence, θ_ν, of the neutrino beam is larger than $1/\gamma$. Using these equations, table 1 gives numerical predictions of event rates for each of the parameter sets. Clearly, these can be several orders of magnitude higher than at today's neutrino beams, even when using less massive targets.

A General Purpose Neutrino Detector

Figure 2 is an example of the sort of high rate general purpose neutrino detector that would be well matched to the intense neutrino beams at muon colliders. The neutrino target is a 1 meter long stack of CCD tracking planes with a radius of 10 cm chosen to match the beam radius at approximately 200 meters from production for a 250 GeV muon beam. It contains 750 planes of 300 micron thick silicon CCD's, corresponding to a mass per unit area of approximately 50 g.cm^{-2}, about 2.5 radiation lengths and 0.5 interaction lengths. (Note the contrast with the kilotonne-scale calorimetric targets used in today's high rate neutrino experiments.) For this target, it is seen that the parameter sets in table 1 typically correspond to several hundred million neutrino interactions per year, and the rate could be even higher for a dedicated muon storage ring or more massive target.

Besides providing the mass for neutrino interactions, the tracking target allows precise reconstruction of the event topologies from charged tracks, including event-by-event vertex tagging of those events containing charm or beauty hadrons or tau

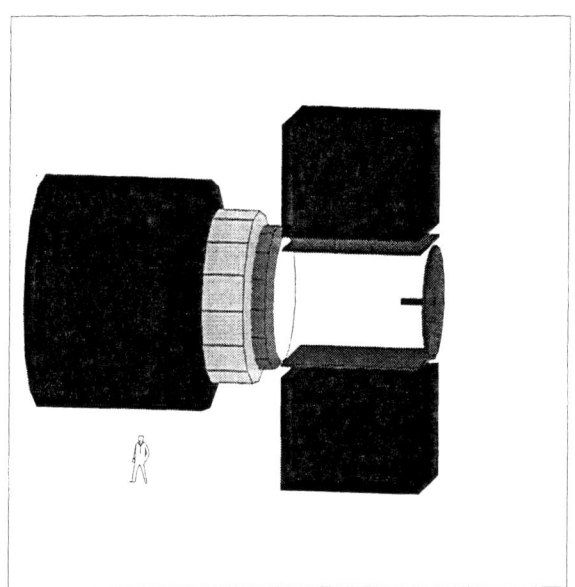

FIGURE 2. Example of a general purpose neutrino detector, reproduced from reference [12]. A human figure in the lower left corner illustrates its size. The neutrino target is the small horizontal cylinder at mid-height on the right hand side of the detector. Its radial extent corresponds roughly to the radial spread of the neutrino pencil beam, which is incident from the right hand side. Further details are given in the text.

leptons. Given the favorable vertexing geometry and the few-micron typical CCD hit resolutions, it is reasonable to expect almost 100 percent efficiency for b tagging, perhaps 70 to 90 percent efficiency for charm tagging and excellent discrimination between b and c decays.

The target in figure 2 is surrounded by a time projection chamber (TPC) tracker in a vertical dipole magnetic field. The characteristic dE/dx signatures from the tracks would identify each charged particle. Further particle ID is provided by the Cherenkov photons that are produced in the TPC gas then reflected by a spherical mirror at the downstream end of the tracker and focused onto a read-out plane at the upstream end of the target. The mirror is backed by electromagnetic and hadronic calorimeters and, lastly, by iron-core toroidal magnets for muon ID.

The relativistically invariant quantities that are routinely extracted in DIS experiments are 1) Feynman x, the fraction of the nucleon momentum carried by the struck quark, 2) the inelasticity, $y = E_{\mathrm{hadronic}}/E_\nu$, which is related to the scattering angle of the neutrino in the neutrino-quark CoM frame, and 3) the momentum-transfer-squared, $Q^2 = 2M_{\mathrm{proton}}E_\nu xy$. As a significant advance, this detector will have the further capability of accurately reconstructing the hadronic 4-vector, resulting in a much better characterization of each interaction, particularly for NC interactions.

Another big improvement over today's neutrino detectors is the vastly improved ability to reconstruct the flavor of the final state quark. Final state c and b quarks can be identified by vertex tagging of the decaying charm or beauty hadrons that contain them, and some statistically based flavor tagging will also be available for u, d or s final state quarks, taking advantage of the "leading particle effect" that is used, for example, in LEP analyses of hadronic Z decays.

Neutrino Physics Opportunities

Neutrino interactions are interesting both in their own right and as probes of the quark content of nucleons, so a MURINE has wide-ranging potential to make advances in many areas of elementary particle physics. This section gives an overview for measurements involving the Cabbibo-Kobayashi-Maskawa (CKM) quark mixing matrix, nucleon structure and QCD, electroweak measurements, neutrino oscillations and, finally, studies of charmed hadrons.

With huge samples of flavor tagged events, a MURINE should be able to make impressive measurements of the absolute squares of several of the elements in the CKM quark mixing matrix. The analyses would be analagous to, but vastly superior to, current neutrino measurements of $|V_{cd}|^2$ that use dimuon events for final state tagging of charm quarks. The current, experimentally determined values for the 9 mixing probabilities are given in table 2 [13], along with their current percentage uncertainties and speculative projections [11] for how 4 of the 9 uncertainties could be reduced by a MURINE at a 500 GeV CoM muon collider. Additionally, if muon colliders eventually reach the 100 TeV energy scale then the associated neutrino beams will even produce final states containing a top quark, almost certainly resulting in uniquely precise determinations of $|V_{td}|^2$ and $|V_{ts}|^2$.

Another major motivation for MURINE's is the potential for greatly improved measurements of nucleon structure functions (SF). Knowledge of these SF's is crucial for precision measurements in neutrino physics, charged lepton scattering experiments and some precision analyses at proton-proton and lepton-proton colliders. Further, they provide important tests of quantum chromodynamics (QCD), and a MURINE might well be the best single experiment of any sort for the examination of perturbative QCD [11].

Neutrino physics has also had an important historical role in measuring the electroweak mixing angle, which is simply related to the mass ratio of the W and Z intermediate vector bosons:

$$\sin^2\theta_W \equiv 1 - \left(\frac{M_W}{M_Z}\right)^2. \tag{8}$$

(To be precise, this is the Sirlin on-shell definition of $\sin^2\theta_W$.)

Now that M_Z has been precisely measured at LEP, measurements of $\sin^2\theta_W$ in neutrino physics can be directly converted to predictions for the W mass. The

TABLE 2. Absolute squares of the elements in the Cabbibo-Kobayashi-Maskawa (CKM) quark mixing matrix. The second row for each quark gives current percentage uncertainties in quark mixing probabilities and speculative projections of the uncertainties after analyses on 10^{10} events from a MURINE at a 500 GeV CoM muon collider. The two uncertainties in brackets have not been measured directly from tree level processes. The uncertainties assume that no unitarity constraints have been used.

	d	s	b
u	0.95 ±0.1%	0.05 ±1.6%	0.00001 ±50% → 1-2%
c	0.05 ±15% → 0.2-0.5%	0.95 ±35% → ~1%	0.002 ±15% → 3-5%
t	0.0001 (±25%)	0.001 (±40%)	1.0 ±30%

comparison of this prediction with direct M_W measurements in collider experiments constitutes a precise prediction of the SM and a sensitive test for exotic physics modifications to the SM [14]. Reference [11] estimates that the predicted uncertainty in M_W from a MURINE analysis might be of order 10 MeV, which improves by an order of magnitude on today's neutrino experiments [14,15] and is approximately equal to the projected best direct measurements from future collider experiments.

There are currently several experimental indications [16] that neutrinos might have non-zero masses and oscillate in flight between the flavor eigenstates. The probability for an oscillation between two of the flavors is given by [17]:

$$\text{Oscillation Probability} = \sin^2\theta \times \sin^2\left(1.27\frac{\Delta m^2[\text{eV}^2].L[\text{km}]}{E_\nu[\text{GeV}]}\right), \tag{9}$$

where the first term gives the mixing strength and the second term gives the distance dependence. Reference [11] obtains the following order-of-magnitude mass limit for an assumed long-baseline detector with reasonable parameters and with full mixing:

$$\Delta m^2|_{\min} \sim O(10^{-4})\,\text{eV}^2, \tag{10}$$

relatively independent of the distance to the detector. Similarly, a mixing probability sensitivity for 10^{10} events in a short-baseline detector is found to be as low as

$$\sin^2\theta|_{\min} \sim O(10^{-7}), \tag{11}$$

for the most favorable value of Δm^2. Both of these estimates apply generically to all 3 possible mixings between 2 flavors: $\nu_e \leftrightarrow \nu_\mu$, $\nu_e \leftrightarrow \nu_\tau$ and $\nu_\mu \leftrightarrow \nu_\tau$. (See also reference [18] for another discussion of neutrino oscillations at a MURINE.)

The Δm^2 estimate is more than an order of magnitude better than any proposed accelerator or reactor experiments for $\nu_\mu \leftrightarrow \nu_\tau$ and $\nu_e \leftrightarrow \nu_\tau$, and is competitive with the best such proposed experiments for $\nu_e \leftrightarrow \nu_\mu$. The estimated value for $\sin^2 \theta|_{min}$ is even more impressive – orders of magnitude better than in any other current or proposed experiment for each of the three possible oscillations. Such an experiment would either convincingly refute or accurately characterize the claimed observations of oscillations by both the LSND and Super-Kamiokande collaborations.

As an interesting final topic, MURINE's should be rather impressive factories for the study of charm – with a clean, well reconstructed sample of several times 10^8 charmed hadrons produced in 10^{10} neutrino interactions. There are several interesting physics motivations for charm studies at a MURINE [19]. As an example, particle-antiparticle mixing has yet to be observed in the charm sector [20], and it is quite plausible [11] that a MURINE would provide the first observation of $D^0 - \overline{D^0}$ mixing.

MUON COLLIDER SCENARIOS

This section explores the collider physics opportunities at muon colliders through reference to the example collider parameter sets of table 1. Complementary discussion on the collider physics aspects of these parameters can be found in [2], where it is opined that each parameter set has some aspects that appear challenging but none of the parameter sets are obviously implausible. Admittedly, table 1 gives a rather incomplete sampling of the possibilities and, for example, discussions of additional physics options with sub-TeV muon colliders may be found in [21].

The section begins with a discussion on muon collider design constraints due to the potential neutrino radiation hazard – a serious problem that is unique to muon colliders – before examining, in turn, the physics potential of each of the parameter sets in table 1.

The Potential Radiation Hazard from Neutrinos

A serious and unexpected problem that has arisen for multi-TeV $\mu^+\mu^-$colliders is the potential radiation hazard posed by neutrinos emitted from muon decays in the collider ring [22,23]. These neutrinos produce a
"radiation disk" in the plane of the ring, and the potential radiation hazard results from the showers of ionizing particles from occasional neutrino interactions in the soil and other objects bathed by this disk. Although the neutrino cross-section is tiny, this is greatly compensated by the enormous number of tightly collimated high energy neutrinos produced at the collider ring.

With some reasonable assumptions, the approximate average numerical value for the annual radiation dose in the plane of the collider ring is easily derived to be [23]:

$$D_{ave}[mSv/yr] \simeq 0.044 \times \frac{f_b[Hz] \times N_0[10^{12}] \times (1 - e^{-t_D[\gamma\tau_\mu]}) \times (E_\mu[TeV])^3}{D[m]}, \quad (12)$$

with notation as in table 1 and assuming an accelerator running time of 10^7 seconds per year. For comparison, the U.S. federal off-site radiation limit is 1 mSv/year, which is of the same order of magnitude as the typical background radiation from natural causes (i.e. 0.4 to 4 mSv/yr [17]).

To explain the form of equation 12, the inverse dependence of the neutrino radiation on the collider depth arises because the radiation levels fall as the inverse square of the distance from the ring while the distance to reach the Earth's surface, assuming a spherical Earth, goes as the square root of the depth. Also, the cubic dependence on the collider energy comes from combining the approximately linear rises with energy of a) the neutrino cross section b) the energy deposited per interaction, and c) the beam intensity due to the decreasing angular divergence of the neutrinos in the vertical plane(equation 2). (There are actually some mitigating factors that come into play at the highest energies and are not included in equation 12 [23].)

This equation is not intended to be accurate at much better than an order of magnitude level and is deliberately conservative, i.e. it may well overestimate the radiation levels. Because of the energy dependence, the radiation levels rapidly become a serious design constraint for colliders at the TeV scale and above.

The radiation intensity may be greatly enhanced downstream from straight sections in the collider ring, with the additional intensity rising in proportion to the length of the straight section. As a benchmark, the length of straight section to produce ten times the planar average dose, L_{x10}, may be shown [23] to be approximately:

$$L_{x10}[meters] \simeq 0.3 \times \frac{C[km]}{E_\mu[TeV]}. \quad (13)$$

This equation shows that the intensity from the straight section picks up another power of the collider energy, which is due to the falling horizontal angular divergence, but this is approximately compensated for by the collider circumference also rising in approximate proportion to the beam energy. As can be seen from table 1, L_{x10} is only of order a meter at all collider energies, so great care must be taken in the design of the collider ring to minimize or eliminate long straight sections.

Because of the cubic rise with energy of the neutrino radiation intensity, muon colliders at CoM energies of beyond a few TeV will probably have to be constructed at isolated sites where the public would not be exposed to the neutrino radiation disk. Such sites clearly exist, perhaps even with useful existing infrastructure. (An extreme example would be close to a nuclear test site, such as in Nevada, U.S.A.) These will presumably be "second generation" machines, arriving after the technology of muon colliders has been established in one or more smaller and less expensive machines built at existing HEP laboratories.

An S-Channel Higgs Factory

Besides exploring the physics at the energy frontier, muon colliders with very narrow CoM energy spreads are particularly suited to both resonance production and threshold studies of elementary particles. The principal example of such a resonant process is the s-channel production of Higgs bosons. The relatively strong coupling strength of muons to the Higgs channel – approximately 40 000 times that for electrons – gives $\mu^+\mu^-$ colliders a unique potential to study this process.

The first parameter set in table 1 is intended for precision studies of a 100 GeV SM-like Higgs boson, hypothesized to have been discovered previously at either LEP, the Tevatron or the LHC. (Of course, the CoM energy of the collider would actually be fixed at the true Higgs mass.) The low CoM energy spread has been chosen to reproduce the predicted width of a SM Higgs at this energy: 2 to 3 MeV. After an initial coarse scan to find the exact energy of the resonance, a fine scan of the resonance would provide uniquely precise measurements of the Higgs mass, width and cross-section.

The technological issues specific to these Higgs factory parameters have been evaluated in some detail over the past year by the MCC [1]. For example, a collider magnet lattice has been designed for the narrow momentum spread beam and the required precise beam calibration was found to be possible by measuring the rate of the muon spin precession.

The physics case for an s-channel Higgs factory has also been studied in some detail [21]. The effectiveness of the collider obviously depends on the existence of a Higgs boson in the appropriate mass range. If the Higgs is too light then, at currently assumed luminosities, the signal will be buried in the backgrounds from the Z resonance. On the other hand, the Higgs width increases with mass, becoming too broad for effective study beyond about 150 GeV. The following approximate scenarios emerge for a SM or SM-like (e.g. supersymmetric) Higgs:

1. $M_H < 105$ GeV: probable discovery at LEP. Backgrounds probably too high for an s-channel Higgs factory.

2. 105 GeV $< M_H < 150$ GeV: fairly likely to be discovered at FNAL but with poor mass resolution. An s-channel Higgs factory will become useful following more precise M_H measurements from the LHC and/or a future lepton collider with a CoM energy of a few hundred GeV.

3. $M_H > 150$ GeV (this is now experimentally disfavored): the resonance would be too broad for an s-channel factory.

4. (for completeness) no Higgs. A Higgs factory is obviously not useful.

As an example of a detailed study for the Higgs mass in a favorable region, reference [21] makes predictions for 0.4 fb^{-1} of on-peak data and a SM-like Higgs with $M_H = 110$ GeV. They predict a resolution of approximately 0.1 MeV on the Higgs mass, 0.5 MeV on the width and branching ratio determinations as accurate as 3 percent (for the H → b$\bar{\text{b}}$ channel). These precise measurements on such an important elementary particle clearly provide strong motivation for this muon collider option, either as a stand-alone "first muon collider" or as a relatively inexpensive add-on to a complex with a higher energy machine.

A 1 TeV Muon Collider to Complement the LHC

The motivation for a 1 TeV muon collider would be roughly the same as that of proposed e^+e^- linear colliders at the 1 TeV energy scale – that is, to perform precision studies on whatever elementary particles are discovered at the LHC hadron collider and to search for new particles that will not be evident in the physical and experimental conditions of the LHC. Thus, a 1 TeV muon collider may be considered as a valuable back-up technology in case electron colliders at this energy either run into unforeseen technical difficulties or are found to be unacceptably expensive. Further, such a muon collider may have a role to play even if a 1 TeV e+e- collider is built, due to potentially different physics processes (e.g. Higgs-type particle production) and also to differences in the beam specifications, as follows.

One TeV electron colliders should be able to achieve higher levels of polarization than their muon collider counterparts, which may have polarization levels in the region of 20% [10]. On the other hand, beamstrahlung at 1 TeV electron colliders will result in roughly a 10% fractional spread in collision energy rather than the parts-per-mil spreads assumed for muon colliders. Thus, electron colliders will be favored for studies where high polarization is important while muon colliders should do better in studies of resonances and in other processes where the CoM energy constraint is important.

The 1 TeV parameter set of table 1 would give about the same luminosity as, for example, the design for the proposed NLC linear electron collider [24] at the same energy, and the physics motivation and capabilities might be relatively similar. Placement of the collider at 125 meters depth (the approximate depth of the existing LEP/LHC tunnel at CERN) reduces the average neutrino radiation in the collider plane to less than one thousandth of the U.S. federal off-site radiation limit. (As already mentioned, attention would still need to be paid to minimizing the length of any low-divergence straight sections in the collider ring.)

A Muon Collider at the Energy Frontier: 4 TeV

Muon colliders appear to have much greater potential than electron colliders to push to lepton collision energies above the LHC mass reach (which might be roughly 1 to 2 TeV, depending on the process). The 4 TeV parameter set was chosen as being at about the highest energy that is practical for a "first generation" muon collider on an existing laboratory site, due to neutrino radiation.

The same comments about neutrino radiation apply as in the 1 TeV design and, in addition, it is necessary to greatly reduce the muon current, accepting the consequent loss in luminosity. The assumed 300 meter depth happens to correspond approximately to appropriate bedrock formations at both the BNL and Fermilab HEP laboratories.

Even the reduced luminosity of this parameter set, $6.2 \times 10^{33} \text{cm}^{-2}.\text{s}^{-1}$, appears sufficient to discover whatever elementary particles lie in the mass range of 1 to 4 TeV (i.e. beyond the reach of the LHC), provided only that the experimental signature for production is not particularly obscure and the production cross-section is not greatly suppressed relative to typical SM couplings (as exemplified by the benchmark process, $\mu\mu \to ee$).

An added motivation for building a "first generation" muon collider at the highest possible energy is that this would provide the best technical foundation for construction of the very high energy, high luminosity muon colliders (10 TeV and above) that are the ultimate goal of muon collider technology.

A Second Generation Muon Collider at 10 TeV

The 4th parameter set of table 1 specifies a "second generation" muon collider at 10 TeV CoM energy, assumed to be constructed at a site where neutrino radiation is not a constraint (see the previous subsection on neutrino radiation). It is seen that the relaxed neutrino radiation constraint might allow an exciting luminosity of $1.0 \times 10^{36} \text{cm}^{-2}.\text{s}^{-1}$ at several times the discovery mass reach of the LHC, making this collider an exciting prospect for the future progress of HEP.

Besides mapping out the spectrum of elementary particles in the energy decade up to 10 TeV, it is further reasonable to assume that anything already discovered at the LHC could be more fully studied in the much cleaner physics environment of such a lepton collider. Particles in the 100 GeV to 1 TeV range should be copiously produced through higher order processes in a 10 TeV muon collider, as evidenced by the production, via the WW-fusion process, of order ten million SM Higgs particles per year (assuming it exists with a mass below 1 TeV).

The technical difficulties specific to muon colliders at this energy scale and above have yet to be assessed in detail. It is comforting that relativistic kinematics makes the acceleration of the muons progressively easier at higher energies due to a rising muon lifetime, shrinking transverse bunch size and reduced sensitivity to disruptive influences such as wake fields. On the other hand, detector backgrounds involving high energy muons will clearly become more challenging, as will the design and layout of the final focus magnets around the ip (see [2] for details). To put this in perspective, these technical challenges will need to be compared with the considerable challenges that are essentially independent of the collider energy – particularly the construction and operation of the muon cooling channel.

The Ultimate Energy Scale for Muon Colliders: 100 TeV

The highest energy parameter set in table 1, at 100 TeV, represents what is likely the ultimate energy scale for muon colliders, with a mass reach for discovering elementary particles that is probably inaccessible even to hadron colliders.

The parameter set assumes technical extrapolations beyond today's limits and presents easily the most difficult design challenge of all the parameter sets, for the following reasons:

- cost reductions will be needed to make a machine of this size affordable

- siting will be more difficult than at 10 TeV, since the neutrino radiation is now well above the U.S. federal limit

- the final focus design is much more difficult even than at 10 TeV, as illustrated by the much larger demagnification factor (see [2] for further discussion)

- the muon bunches, although much smaller than in the other sets, are also much cooler (again, see [2] for further discussion)

- the beam power has risen to 170 MW, with synchrotron radiation rising rapidly to contribute a further 110 MW.

It seems reasonable to assume that the rapid rise to prominence of the synchrotron radiation will effectively prohibit muon colliders at the PeV energy scale, even if the other challenges could be negotiated.

A 100 TeV muon collider is clearly not a near-term prospect. However, the unique opportunity to explore the physics at this energy scale could well turn out to be crucial in unlocking the profound mysteries of the elementary particle spectrum and its role in the universe. With such compelling motivation, it is certainly not ruled out that a muon collider at this energy scale could become achievable after a couple of decades of dedicated research and development.

SUMMARY

An overview has been given of the potential prospects for neutrino physics and collider physics at muon colliders and it has been shown that muon colliders may well come to assume a central role in the future of experimental high energy physics. Their discovery reach for new elementary particles might eventually be in the region of 100 TeV and they could also open up exciting new vistas in neutrino physics and other precision studies.

This provides strong motivation for a continuing and expanding vigorous research and development program in muon collider technology, and such a program will be needed to make muon colliders a reality on an attractive timescale.

REFERENCES

1. The Muon Collider Collaboration, "Status of Muon Collider Research and Development and Future Plans", to be submitted to Phys. Rev. E.
2. B.J. King, *Discussion on Muon Collider Parameters at Center of Mass Energies from 0.1 TeV to 100 TeV*, BNL CAP-223-MUON-98C, submitted to Proc. Sixth European Particle Accelerator Conference (EPAC'98), 22-26 June,1998, Stockholm, Sweden, available at http://pubweb.bnl.gov/people/bking/.
3. G.I. Budker, *Accelerators and Colliding Beams*, talk at 7th Internat. Accelerator Conference, Erevan (1969); talk at the Internat. High Energy Physics Conference, Kiev (1970).
4. E. A. Perevedentsev and A. N. Skrinsky, Proc. 12th Int. Conf. on High Energy Accelerators, F. T. Cole and R. Donaldson, Eds., (1983) 485; A. N. Skrinsky and V.V. Parkhomchuk, Sov. J. of Nucl. Physics **12**, (1981) 3; *Early Concepts for $\mu^+\mu^-$ Colliders and High Energy μ Storage Rings, Physics Potential & Development of $\mu^+\mu^-$ Colliders. 2nd Workshop*, Sausalito, CA, Ed. D. Cline, AIP Press, Woodbury, New York, (1995).
5. D. Neuffer, IEEE Trans. **NS-28**, (1981) 2034.
6. *Proceedings of the Mini-Workshop on $\mu^+\mu^-$ Colliders: Particle Physics and Design*, Napa CA, Nucl Inst. and Meth., **A350** (1994) ; Proceedings of the Muon Collider Workshop, February 22, 1993, Los Alamos National Laboratory Report LA-UR-93-866 (1993) and *Physics Potential & Development of $\mu^+\mu^-$ Colliders 2nd Workshop*, Sausalito, CA, Ed. D. Cline, AIP Press, Woodbury, New York, (1995).
7. *Transparencies at the 2 + 2 TeV $\mu^+\mu^-$ Collider Collaboration Meeting*, Feb 6-8, 1995, BNL, compiled by Juan C. Gallardo; transparencies at the *2 + 2 TeV $\mu^+\mu^-$ Collider Collaboration Meeting*, July 11-13, 1995, FERMILAB, compiled by Robert Noble; Proceedings of the 9th Advanced ICFA Beam Dynamics Workshop, Ed. J. C. Gallardo, AIP Press, Conference Proceedings 372 (1996).
8. D. V. Neuffer and R. B. Palmer, Proc. European Particle Acc. Conf., London (1994); M. Tigner, in Advanced Accelerator Concepts, Port Jefferson, NY 1992, AIP Conf. Proc. **279**, 1 (1993).
9. R. B. Palmer et al., *Monte Carlo Simulations of Muon Production, Physics Potential & Development of $\mu^+\mu^-$ Colliders 2nd Workshop*, Sausalito, CA, Ed. D. Cline, AIP Press, Woodbury, New York, pp. 108 (1995); R. B. Palmer, et al., *Muon Collider Design*, in Proceedings of the Symposium on Physics Potential & Development of $\mu^+\mu^-$ Colliders, Elsevier.
10. $\mu^+\mu^-$ *Collider: A Feasibility Study*, BNL-52503, Fermilab-Conf-96/092, LBNL-38946, July 1996.
11. B.J. King, *Neutrino Physics at a Muon Collider*, Proc. Workshop on Physics at the First Muon Collider and Front End of a Muon Collider, Fermilab, November 6-9, 1997.
12. See, for example, Chris Quigg, *Neutrino Interaction Cross Sections*, FERMILAB-Conf-97/158-T.
13. Values extracted from Andrzej J. Buras, *CKM Matrix: Present and Future*, TUM-HEP-299/97.

14. Janet M. Conrad, Michael H. Shaevitz and Tim Bolton, *Precision Measurements with High Energy Neutrino Beams*, hep-ex/9707015, submitted to Rev. Mod. Phys. (1997)
15. K.S. McFarland et al. (CCFR/NuTeV Collaboration) *A Precision Measurement of Electroweak Parameters in Neutrino-Nucleon Scattering*, FNAL-Pub-97/001-E. B.J. King, Columbia University Ph.D. Thesis, 1994; Nevis Report: Nevis-283, CU-390, Nevis Preprint R-1500 (1994).
16. LSND Collaboration, *Evidence for numu to nue Oscillations from Pion Decay in Flight Neutrinos* (Updated June 12,1997), submitted to PRC, LA-UR-97-1998, UCRHEP-E191; Super-Kamiokande Collaboration, *Measurement of a small atmospheric ν_μ/ν_e ratio* (February 12, 1998), submitted to Phys. Lett. B.
17. R.M. Barnett et al., Physical Review D54, 1 (1996) and 1997 off-year partial update for the 1998 edition available on the PDG WWW pages (URL: http://pdg.lbl.gov/).
18. S. Geer, *The Physics Potential of Neutrino Beams From Muon Storage Rings*, Proc. Workshop on Physics at the First Muon Collider and Front End of a Muon Collider, Fermilab, November 6-9, 1997.
19. I.I Bigi, *Open Questions in Charm Decays Deserving an Answer*, CERN-TH.7370/94, UND-HEP-94-BIG08 (1994). I.I Bigi, *The Expected, The Promised and the Conceivable - on CP Violation in Beauty and Charm Decays.*, UND-HEP-94-BIG11 (1994).
20. Tiehui (Ted) Liu, *The D0-D0bar Mixing Search – Current Status and Future Prospects*, HUTP-94/E021 (1994). Gustavo Burdman, *Charm Mixing and CP Violation in the Standard Model*, FERMILAB-Conf-94/200 (1994).
21. Proc. Workshop on Physics at the First Muon Collider and Front End of a Muon Collider, Fermilab, November 6-9, 1997, Ed. S. Geer and R. Raja.
22. B.J. King, *Assessment of the prospects for muon colliders*, paper submitted in partial fulfillment of requirements for Ph.D., Columbia University, New York (1994).
23. B.J. King, *A Characterization of the Neutrino-Induced Radiation Hazard at TeV-Scale Muon Colliders*, BNL Center for Accelerator Physics internal report 162-MUON-97R, to be submitted for publication.
24. The NLC Design Group, *Zeroth-Order Design Report for the Next Linear Collider* (May, 1996), LBNL-5424, SLAC-474, UCRL-ID-124161, UC-414.

B physics expected performances with the Compact Muon Solenoid detector

François Charles

Groupe de Recherche en Physique des Hautes Energies
Université de Haute Alsace, 61 rue A. Camus 68093 Mulhouse, France

Abstract. We present here the future performances of the Compact Muon Solenoid detector for B physics. We show that CMS will contribute significantly to the CP violation parameter $sin2\beta$ measurement with a precision of $\delta sin2\beta \simeq 0.02$ (1 year of integrated luminosity). The asymetry in the channel $B_s^0 \to J/\psi\phi$ will be tested to the $2-5\%$ level. The mixing parameter x_s of B_s^0 oscillations will be measured up to 40. Finaly the rare B decay should be searched down to the SM expectation and in the case of the semileptonic rare decays will provide enough statistics to performed detailed studies.

I INTRODUCTION

The Compact Muon Solenoid is a multipurpose experiment for the Large Hadron Collider which is expected to collide protons at $E_{cm} = 14\ TeV$ with a design luminosity of $10^{34}\ cm^{-2}s^{-1}$. The large $b\bar{b}$ cross section expected will enable us to address many aspects of B-physics in the first years (low luminosity period) including the precise determination of the CKM matrix elements which is one major goal of High Energy Physics. In the Wolfenstein parametrisation [1] the CKM matrix parameters can be represented in a complex plane by a triangle. It is expected that the measurement of the angle β of the unitarity triangle will be performed in the channel $B_d^0 \to J/\psi K_s^0$ while the B_s^0 oscillation in the channel $B_s^0 \to D_s^\mp \pi_1^\pm \to \phi \pi_2^\mp \pi_1^\pm$

Changing Neutral Current are forbidden at the tree level and expected to occur only through loop diagram via the exchange of W. This is particularly interesting as these decays are sensitive to new physics via the exchange of heavy new particles.

II CMS DETECTOR

A Description

A detailed description of the CMS detector can be found in [2]. In CMS, a superconducting solenoid of 6 m diameter providing a field of 4 T contains the inner tracker, the electromagnetic and hadronic calorimeters and is surrounded by the muon chambers (figure 1). I will concentrate my description on the tracking devices which have been used for this study. The calorimeter description and performances can be found in [3].

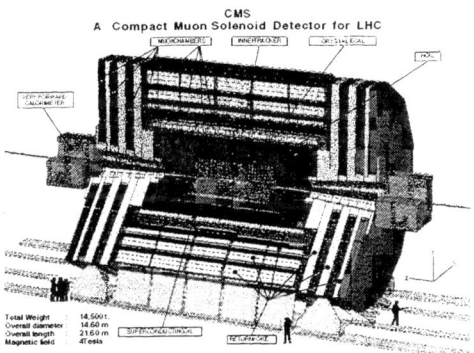

FIGURE 1. The Compact Muon Solenoid Detector

measurement(x_s parameter) will provide us with the length of one side of the triangle. CMS will also look for rare B decays. In the Standard Model, Flavour

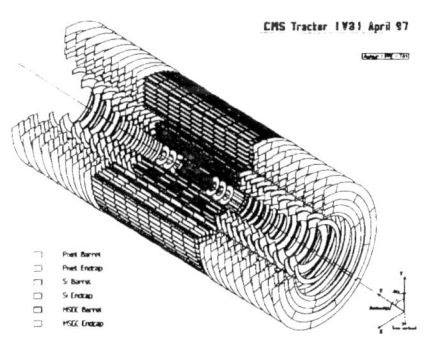

FIGURE 2. CMS Inner tracker

The vertex detector of the inner tracker (figure 2) of CMS is constituted of silicon pixels. In the barrel region, 2 layers located at radii 7.5 cm and 11.5 cm are made of pixels of 125 by 125 μm^2, and provide a space point resolution of 15 μm, while the forward region is equipped with 3 layers. The layers can be placed closer to the beam pipe ($r = 4\ cm$) in the early phase of LHC at low luminosity. The silicon strips (3 layers) that surround the vertex detector will provide a high spatial and time resolution. The outer tracking layers employ MicroStrip Gas Chamber (7 layers) with a (r,ϕ) resolution of 40 μm and extend up to a radius of 1.3 m.

The rapidity coverage extends up to 2.4. The tracking devices will provide on average 12 hits per high p_T track, while the reconstruction efficiency for an isolated track is expected to be about 98%.

B Tracking performances

Figure 3 shows the impact parameter resolution in the central and forward regions for 3 different positions of the first layer. At high p_T a resolution of 20 μm can be achieved everywhere with the present design and could reach 10 μm with a first layer at $r = 3\ cm$.

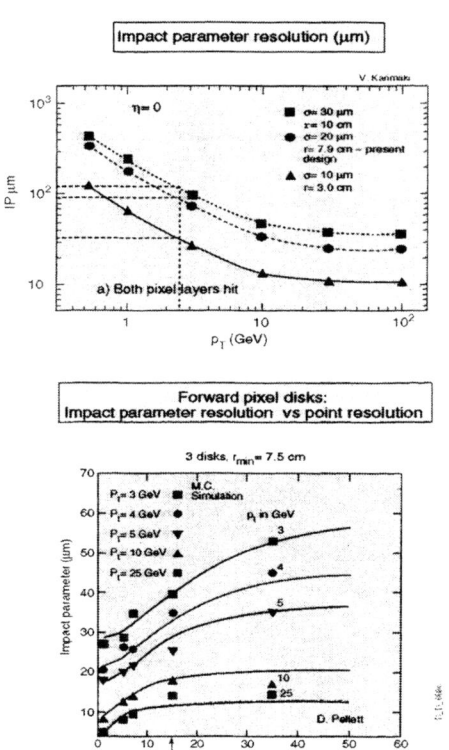

FIGURE 3. Impact parameter resolution

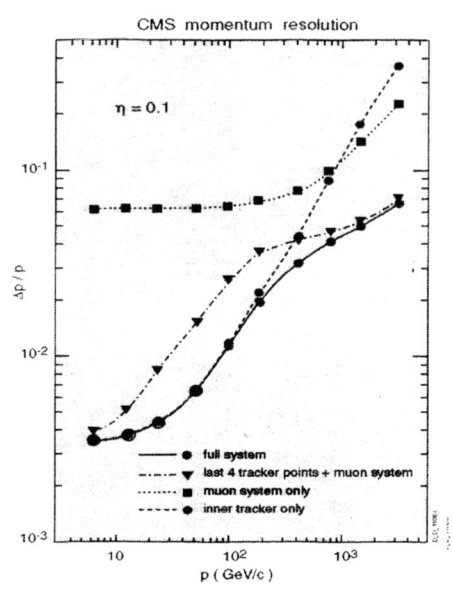

FIGURE 4. Momentum resolution

The expected momentum resolution $\frac{\Delta p}{p}$ (see figure 4) in the central region ($\eta = 0.1$) and for low p_T tracks ($p_T < 15\ GeV/c$) is about 0.5% and degrades to 1% at $p_T = 100 GeV/c$.

522

III RARE DECAYS

A $B_s^0 \to \mu^+\mu^-$

The decay $B_s^0 \to \mu^+\mu^-$ presents several interesting aspects. Among them, its branching ratio is proportionnal to x_s, secondly it is sensitive to new physics. The Standard Model branching ratio for this channel is : $3.5 \pm 1.0 \; 10^{-9}$ [4]. It will be a challenge for the CMS detector to separate the huge background essentially originating from $b\bar{b}$ pair production. For this channel the $b\bar{b}$ pairs generation includes gluon fusion and gluon splitting process. The basic event selection is the following : we apply a two muons trigger selection: $p_t^\mu > 4.3 GeV, |\eta| < 2.4$ we impose a cut on the B_s^0 transverse momentum: $p_t^{\mu\mu} > 12 GeV/c$ and $0.4 < \Delta R_{\mu\mu} < 1.2$ where $\Delta R_{\mu\mu} = \sqrt{\Delta\eta_{\mu\mu}^2 + \Delta\phi_{\mu\mu}^2}$. This last cut is very effective against gluon fusion due to the large opening angle expected in this process. The expected number of signal events after this first selection is $N = 66$ while the background remains at $2.9\;10^7$ events. To suppress furthermore the background one makes use of the capacity of the CMS detector to isolate the dimuon system, precisely reconstruct the secondary vertex and at the end to obtain a good dimuon mass resolution. The isolation selection was performed on the tracking by requiring $I = \Sigma p_t^{ch}/p_t^{\mu\mu}$ for $p_t^{ch} > 1 GeV/c$ to be lower than 0.05 for the charged particles in a cone: $R < 0.5 \times \Delta R_{\mu\mu} + 0.4$ and also on the calorimeter by imposing $E_t < 4 GeV$ from neutral particles in the same ECAL+HCAL cone. The reconstruction of the secondary vertex was done with full pattern recognition, track and secondary vertex reconstruction in the low luminosity pixel configuration (first layer at radius of 4 cm) or in the high luminosity condition (first layer at radius of 7 cm).

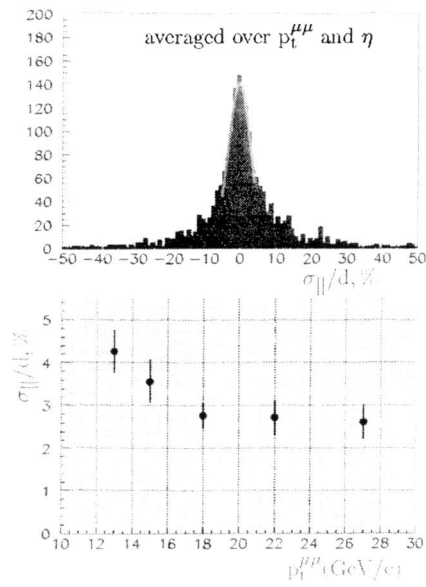

FIGURE 5. Secondary vertex resolution

We can observe on figure 5 the longitudinal resolution divided by the flight distance. The gaussian part of the distribution exhibit a resolution: $\sigma = 3.2\%$, slightly improving with the p_t^B. The selection which was applied makes use of this resolution: the flight distance was required to be : $d > 3 \times \sigma_\parallel$ but also of the geometrical vertex reconstruction: $\delta\alpha < 0.04 \; rad$ where $\delta\alpha$ is the angle between the reconstructed momentum of the B_s^0 meson and the line joining the primary and secondary vertex. The mass resolution can be observed on figure 6. We notice that the resolution improves in the central η region down to $\sigma = 23 MeV$.

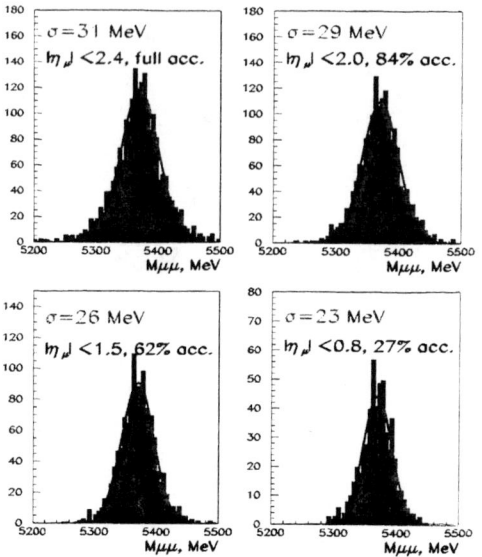

FIGURE 6. Mass resolution in $B_s^0 \to \mu^+\mu^-$ channel

Finaly we have summarized the expected number of events on table 1 where for the luminosity of $10^4 pb^{-1}$ the first pixel layer is at a radius of 4 cm while for the luminosity of $10^5 pb^{-1}$ the position is 7.5 cm. The CMS experiment should be able to observe this rare decay.

Condition	Signal	Bkgd	$\frac{S}{\sqrt{B}}$
$10^4 pb^{-1}$	13	55	1.8
$10^5 pb^{-1}$	81	260	5.0
$3 \times 10^5 pb^{-1}$	243	780	8.7

TABLE 1. Expected samples of reconstructed events

B Semimuonic rare B decays

We have considered here the rare B meson decays due to the $b \to dl^+l^-$ and $b \to sl^+l^-$ transition. Explicitly 3 channels were studied : $B_d^0 \to K^{*0}\mu^+\mu^-$, $B_d^0 \to \phi\mu^+\mu^-$ and $B_d^0 \to \rho\mu^+\mu^-$.

The expected branching ratios are [5]:

$$B(B_s^0 \to \mu^+\mu^-\phi) = 1.5 \times 10^{-6}$$
$$B(B_d^0 \to \mu^+\mu^- K^{*0}) = 1.5 \times 10^{-6}$$

For the $B_d^0 \to \mu^+\mu^-\rho$ channel we have choosen an optimistic expectation according to the value for $B(B_d^0 \to \mu^+\mu^- K^{*0})$. This leads to the following value :

$$B(B_d^0 \to \mu^+\mu^-\rho) = 1.5 \times 10^{-7}$$

Recently several theoretical papers [6,7,8] have appeared dealing with these decays and showing that 3 distributions might be of special interest: the invariant dimuon mass spectrum, the Forward-Backward asymmetry (where the asymmetry is estimated from the angle between the B meson and the positive muon) as a function of the invariant dimuon mass and the normalized longitudinal lepton polarization defined as :

$$\bar{A}_{LP}(\hat{s}) = \frac{\frac{dN(\hat{s},\lambda=+1)}{d\hat{s}} - \frac{dN(\hat{s},\lambda=-1)}{d\hat{s}}}{\frac{dN(\hat{s},\lambda=+1)}{d\hat{s}} + \frac{dN(\hat{s},\lambda=-1)}{d\hat{s}}}$$

where λ is the helicity of the positive lepton.

This last observable requires the determination of the lepton polarization which can only be expected for the lepton τ. Using these observables it is expected that we can fully determine the

3 Wilson coefficients involved[9] in the matrix elements calculation. The ratio of the branching ratios :

$$\frac{BR(B_d^0 \to \rho\mu^+\mu^-)}{BR(B_d^0 \to K^{*0}\mu^+\mu^-)}$$

is also of interest as it is proportional to the CKM matrix elements ratio $\frac{|V_{td}|^2}{|V_{ts}|^2}$ [10]. Explicitly we have:

$$\frac{BR(B_d^0 \to \rho\mu^+\mu^-)}{BR(B_d^0 \to K^{*0}\mu^+\mu^-)} = \frac{F^{B_d^0-\rho}}{F^{B_d^0-K^{*0}}} \frac{|V_{td}|^2}{|V_{ts}|^2}.$$

The form factor $\frac{F^{B_d^0-\rho}}{F^{B_d^0-K^{*0}}}$ can be obtained from QCD sum rules for other rare decays (namely $B_d^0 \to K^{*0}\gamma$ and $B_d^0 \to \rho\gamma$) . The expected error on the ratio of the form factor is expected to be of the order of 10% [11]. In order to discriminate the signal events from the background we have applied several cuts. The following elements are common to all 3 channels. For the hadrons we required $p_t^{had.} > 2$ GeV/c with $|\eta| < 2.4$. The π or K mass was assigned to the hadrons accordingly to the seeked channels. The muons were required to fulfill the 2 muon trigger conditions for which the thresholds are defined as a function of η :

$p_t > 4.5\ GeV/c$	$0.0 < |\eta| < 1.5$
$p_t > 3.6\ GeV/c$	$1.5 < |\eta| < 2.0$
$p_t > 2.6\ GeV/c$	$2.0 < |\eta| < 2.4$

We have excluded the ψ and ψ' resonances by demanding:

$$|m_{\mu^+\mu^-} - m_{J/\psi}| > 84 MeV/c^2$$
$$|m_{\mu^+\mu^-} - m_{\psi'}| > 84 MeV/c^2.$$

A good ($\chi^2 < 5$) reconstructed 4 particles vertex was required and the flight distance of B^0 should be above 1 mm. We also constrained the angle between the vector defined by the geometrical reconstruction (primary vertex to secondary vertex) and the track reconstruction (B^0 momentum reconstructed as the sum of the 4 particles momentum) to be less than 0.1 rad.

The mass constraints for the ρ, ϕ and K^{*0} were decided specifically according to the channel and the expected reflection from the 2 other rare decay channels. The 4 particles invariant masses were required to be within 2 σ of the B_d^0 or B_s^0 mass. On figure 7 we present the invariant $\mu^+\mu^- K\pi$ mass for one year of LHC running at low luminosity ($L = 10^4 pb^{-1}$)

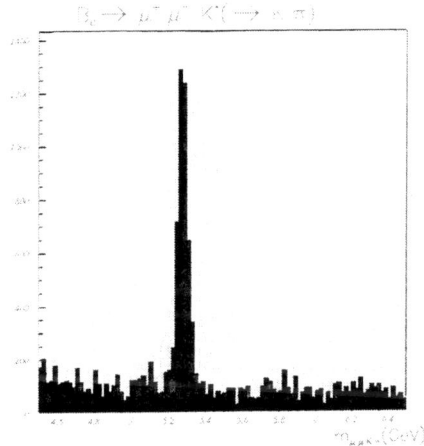

FIGURE 7. $B_d^0 \to K^{*0}\mu^+\mu^-$

We expect about 4200 events with a S/B ratio of 9. The channel $B_s^0 \to \phi\mu^+\mu^-$ turns out to be cleanest among the 3 rare decays investigated here. This is essentialy due to the narrowness of the reconstructed ϕ (see figure 8).

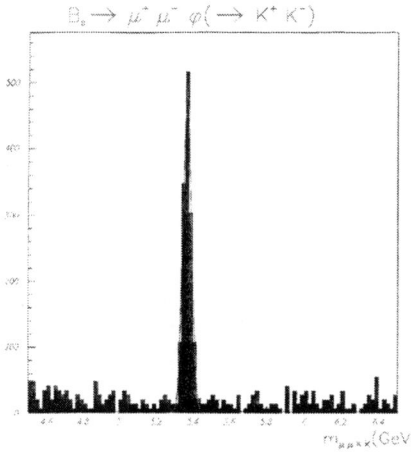

FIGURE 8. $B_s^0 \to \phi \mu^+ \mu^-$

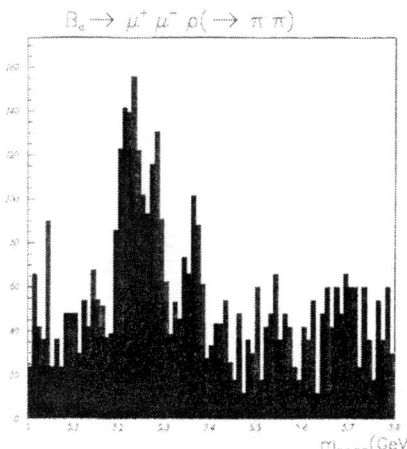

FIGURE 9. $B_d^0 \to \rho \mu^+ \mu^-$

The analysis of $B_d^0 \to \rho \mu^+ \mu^-$ channel is the most difficult and probably the most interesting in the point view of CP violation estimation. The contamination of $B_d^0 \to \rho \mu^+ \mu^-$ channel from $B_d^0 \to K^{*0} \mu^+ \mu^-$ constitutes one third of the background, the remaining originates from $b \to \mu$ transition. We can see on figure 9, the two reflections of the other rare channels while the signal is the central peak. We obtain the following result for the selected number of signal and background events:

channel	Signal Events	Total bkgd
$B_s^0 \to \phi\mu\mu$ (1)	350	340
$B_d^0 \to \rho\mu\mu$ (2)	1200	70
$B_d^0 \to K^{*0}\mu\mu$ (3)	4200	435

TABLE 2. Expected signal and sum of background events for one year of LHC running at low luminosity

channel	$b \to \mu$ $\bar{b} \to \mu$	$B \to \mu\mu$ +X	Int. Cont.
(1)	100	120	120
(2)	10	40	20
(3)	120	260	55

TABLE 3. Expected detailled background events for one year of LHC running at low luminosity

The expected statistical error on the measurement of the ratio $\frac{|V_{td}|^2}{|V_{ts}|^2}$ is of the order of 7%. We also expect that CMS will be able to make precise measurement of two differential distributions: the dimuon invariant mass and the forward backward differential asymmetry

IV CP VIOLATION

A Sensitivity in $B_d^0 \to J/\psi K_s^0$

$B_d^0 \to J/\psi K_s^0$ is considered as the gold plated channel [12] to obtain the angle β. The purpose is to measure the asymetry: $A = \frac{N_+ - N_-(t)}{N_+ + N_-} \propto D \times sin2\beta$

where N_+ and N_- are the number of events coming from respectively B_d^0 or $\bar{B}_d^0$ and D is the dilution factor (see section V). To determine the sign of the b quark , the classical method is to tag the sign of the lepton in the semileptonic decay of the accompanying b in the pair production.

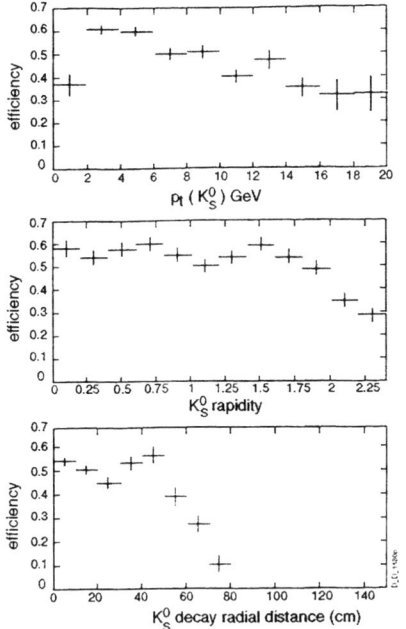

FIGURE 10. K_s^0 reconstruction efficiency

Another technic [13] makes use of the possibility for a B_d^0 to be produced by an orbitally excited meson: $B^{**\pm}$ which decays in $\pi^\pm$ and B_d^0 where the sign of the pion is tagging the sign of the b quark. The selection for this channel is the following: two charged hadrons with $p_t > 0.7 GeV/c$ and $|\eta| < 2.4$ (K_s^0 decay), mass cuts within $\pm 2\sigma$ of the invariant mass of J/ψ and K_s^0 , and a requirement on the transverse decay length of the K_s^0 to be more than 1 cm.

The figure 10 shows the K_s^0 reconstruction efficiency as a function of the transverse momentum, the rapidity and the radial distance of the decay position. We can see that the efficiency remains beyond 50% dropping drastically at larger radius due to the requirement to reconstruct at least 5 hits. On the next figure (11) we can observe the mass resolution for K_s^0 and B_d^0 with the background included. The background to signal ratio is about 13%.

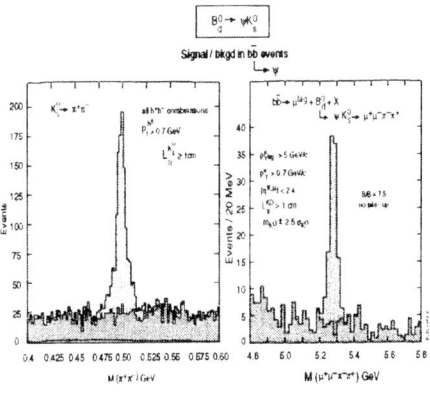

FIGURE 11. Mass resolution

The next figure (12) describes the nature of the wrong tagging as a function of the p_t of the tagged lepton. We can notice that the fraction of wrong tag is about 25% level with the mixing and the b to c transition accounting for 10% each. We can see the invariant reconstructed mass of the B^{**} excited meson (figure 13) where the charged particles associated with the B_d^0 meson to form the B^{**} is required to have $p_t > 1 GeV/c$ and $\Delta R_{B_d^0 \pi} < 0.7$ where $\Delta R = \sqrt{\Delta \eta^2 + \Delta \phi^2}$.

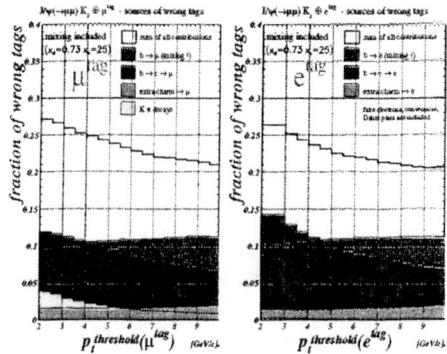

FIGURE 12. Fraction of wrong tag

B^{**} tagging is illustrated by the figure 13.

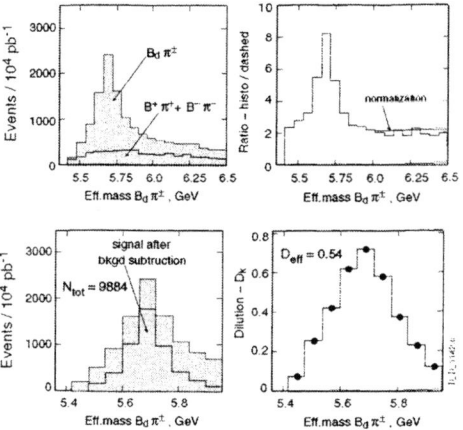

FIGURE 13. B^{**} tagging

The first plot shows the efffective $B_d^0 \pi^\pm$ reconstructed mass with the efffective $B^-\pi^-$ and $B^+\pi^+$ reconstructed mass superimposed. This last one is used to substract statistically the coincidence. As expected the normalization (second plot) is a factor 2 accounting for $\pi^\pm$ sign. The next plot shows the resonnance with the background substracted and the final one shows the resulting dilution factor (average 0.54).

The sensitivity obtain for this channel is $\delta sin2\beta \simeq 0.037$. Combining all methods (lepton and B^{**} tagging) we obtain a final sensitivity of $\delta sin2\beta \simeq 0.02$ see table 4.

channel	$sin2\beta$ sens.
$B_s^0 \to J/\psi(\mu^+\mu^-)K_s \oplus \mu^{tag}$	0.046
$B_s^0 \to J/\psi(\mu^+\mu^-)K_s \oplus e^{tag}$	0.048
$B_s^0 \to J/\psi(e^+e^-)K_s \oplus \mu^{tag}$	0.08
$B_s^0 \to J/\psi(e^+e^-)K_s \oplus e^{tag}$	0.11
$B_s^0 \to J/\psi(\mu^+\mu^-)K_s \, (B^{**})$	0.037
Combined	0.023

TABLE 4. Expected detailled background events for one year of LHC running at low luminosity

B Sensitivity in $B_s^0 \to J/\psi\phi$

CP violation effects in the channel $B_s^0 \to J/\psi\phi$ are expected to be very small. The asymetry is here [14] :

$$A(t) = \frac{N+(t)-N-(t)}{N+(t)+N-(t)}$$
$$= 2|V_{cd}|\frac{|V_{ub}|}{|V_{cb}|}sin\gamma sin(x_s\frac{t}{\tau})$$
$$\propto sin(2\phi_{ts})sin(x_s\frac{t}{\tau})$$

The factor multiplying $sin\gamma$ is about 0.03. We can observe on the figure 14 the 4 particles ($\mu\mu hh$) invariant mass peaking on the B_s^0 mass. The selection includes here single and two muon triggers, a cut on the kaons of $p_t > 1 GeV/c$, a cut on the B meson of $p_t > 10 GeV/c$, a requirement on the secondary vertex to be reconstructed 3σ away from the interaction point, a vertex pointing cut $\delta\alpha < 0.1 \, rad$. Applying this selection, the background is negligible (about 1%).

x_s	10	20	30
$\delta sin(2\phi_{ts})$	0.021	0.028	0.041

TABLE 5. Sensitivity on $sin(2\phi_{ts})$

V B_S^0 OSCILLATIONS

B_s^0 mixing is estimated through the measurement of $x_s = \frac{\Delta m_s}{\overline{\Gamma}}$ (where Δm_s is the mass difference of the eigenstates and $\overline{\Gamma}$ the B_s^0 width). The mixing parameter x_s is proportional to the oscillation frequency of the $B_s^0/\bar{B}_s^0$ system and to the length of the triangle side opposite to the γ angle. In the standard model frame its value is expected to be within 10 to 40 [15], while direct experimental constraints imply $x_s > 10$ [16]. The direct measurement is performed in the channel: $\mu^{tag} \leftarrow b\bar{b} \rightarrow B_s^0 \rightarrow D_s^\mp \pi_1^\pm \rightarrow \phi \pi_2^\mp \pi_1^\pm$ where the charge of the π_1 is tagging the charge of the b quark at the decay while the muon is tagging at the production, see reference [17] for details. We then can measure the asymmetry:

$$A(t) = \frac{dn_{++}}{dt} - \frac{dn_{+-}}{dt} \propto Dcos(x_s\frac{t}{\tau})e^{-\Gamma t}$$

where D is the overall dilution factor:

$$D = D_{tag} * D_{mix} * D_{back} * D_{time} \sim 0.3$$

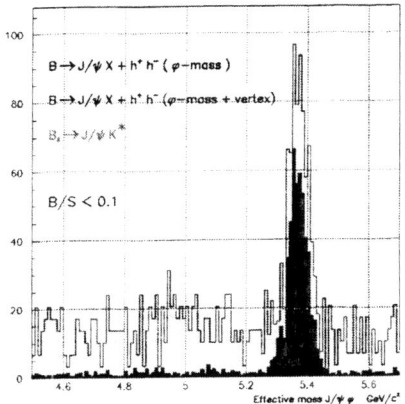

FIGURE 14. $\mu\mu hh$ Invariant mass

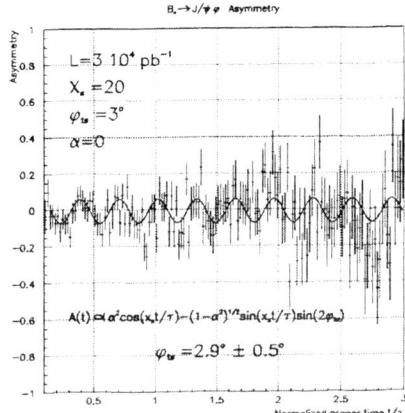

FIGURE 15. Asymetry

The next figure (15) present the expected asymetry for an angle $\phi_{ts} = 3$ degree overestimated (with respect to SM expectations) in order to observe the oscillation. We obtain the following results on the sensitivity on $sin(2\phi_{ts})$:

including respectively the uncertainty coming from the tagging efficiency (cascade decay $b \rightarrow c \rightarrow \mu$, wrong B-hadron determination), the mixing (we assume $x_d = 0.73$), the remaining background and the measurement of the decay time. From $A(t)$ we derive x_s by applying a Fourier transform. The following selection is used for the $B_s^0 \rightarrow D_s^\mp \pi^\pm$ decay: we require 4 charged particles with $p_t^{had.} > 1\ GeV/c$, all in the range $|\eta| < 2.4$. The invariant mass of two particles (respectively 3 and 4) should be reconstructed in the window $\Delta m < \pm 2\sigma_m$ of the ϕ (respectively D_s and B_s^0). We impose a vertex pointing cut, $\Delta\alpha < 0.1\ rad$ where $\Delta\alpha$ is the angle difference between the B_s^0 direction reconstructed from the 4 particle momenta and the one reconstructed from the secondary vertex determination. The B_s^0 decay distance in the transverse plane is required to exceed 240 μm. For the tagging side we impose $p_T(\mu^{tag}) > 7\ GeV/c$. For an integrated luminosity of $10^4\ pb^{-1}$ we obtain 2200 signal events (we assume $B(B_s^0 \to D_s\pi) = 3.5 \times 10^{-3}$ and a second level trigger efficiency of 0.5) and a background of less than 2000 events. The dilution factor obtained is 0.2. The sensitivity is $x_s = 22$. It is expected that using the decay $D_s \to K^*K$ and electron tagging a value for x_s of 38 can be reached.

VI CONCLUSION

CMS should be able to test the rare FCNC decay $B_s^0 \to \mu^+\mu^-$ down to the expected SM level; further rare semileptonic FCNC decays accessible only at a hadron machine, can also be investigated with significant statistics. CMS can contribute significantly to the CP violation parameter $sin2\beta$ expected precision: $\delta sin2\beta \simeq 0.02$ for $L = 10^4 pb^{-1}$. The asymmetry in the $B_s^0 \to J/\psi\phi$ decay could be tested to the $2-5\%$ level, depending on x_s, i.e. to the SM expectations level. The direct measurement of the B_s^0 oscillation parameter x_s should be possible up to $\simeq 40$ provided an efficient second-level trigger algorithm can be developed.

ACKNOWLEDGEMENTS

I would like to thank C. Racca for reading the manuscript.

REFERENCES

1. L.Wolfenstein, Phys. Rev. Lett. 51 (1983) 1945
2. CMS Techn. Des. Rep., CERN/LHCC 97-10,32
3. CMS Techn. Des. Rep., CERN/LHCC 97-33
4. A. Ali, hep-ph 9610333
5. D. Melikhov et al. hep-ph/9711332
6. T.M. Aliev et al. hep-ph/9612480
7. T.M. Aliev et al. hep-ph/9704323
8. L.T. Handoko hep-ph/9707222
9. J.L. Hewett, Phys. Rev. D53 (1996) 4964
10. A. Ali, DESY 95-157 (1995)
11. A. Ali et al., Z.phys. C63 (1994) 437
12. A.Buras, R. Fleischer, Heavy flavours II, World Scient.(1997)
13. ALEPH Coll., 'Resonnant Structure and Flavour-tagging in $B\pi^\pm$ .. ,EPS Conference (1996)
14. I. Dunietz, Phys. Lett. B270 (1991) 75.
15. ALEPH Col., Phys. Lett. B356 (1995) 409
16. A.Starodumov, CMS TN/96-109
17. A. Ali and D. London, Z. Phys. C65 (1995) 341

SEMINARS

Upper bound on the neutrino magnetic moment from collisions induced by Landau damping in supernovae

Alejandro Ayala

Instituto de Ciencias Nucleares
Universidad Nacional Autónoma de México
Aptdo. Postal 70-543. México D.F. 04510.

Abstract.
For neutrinos with a magnetic moment, we show that the Landau damping mechanism acts as an efficient process for the conversion of left-handed neutrinos into right-handed ones in a dense ultrarelativistic plasma such as the core of a supernova. The effect can be used to place an upper bound on the neutrino magnetic moment $\mu < (0.1 - 0.4) \times 10^{-11} \mu_B$.

The non-standard properties of neutrinos have become the subject of an increasing research effort over the last years. Amongst these properties, the neutrino magnetic moment has received attention in connection with various helicity-flipping processes that could have important consequences as efficient cooling mechanisms for supernovae cores [1], [2]. The most recently proposed of such processes is the helicity-flipping emission or absorption of longitudinal Čerenkov radiation either in a degenerate or in a high temperature electromagnetic plasma [3] whose origin is the existence of a space-like branch in the dispersion relation of the longitudinal photon, the so called plasmon. As it is well known, this mode develops a large imaginary part [4] which means, first of all, that the Landau damping mechanism is acting to preclude its propagation. Nevertheless, the authors in ref. [3] have used the properties of this branch to place a stringent upper bound on the neutrino magnetic moment. They argue that the only important fact is that processes like $\nu_L \longrightarrow \nu_R + \gamma$ and $\nu_L + \gamma \longrightarrow \nu_R$ can take place, ignoring any possible consequences that the imaginary part of the dispersion relation could introduce. In this paper we show that the above is not the correct picture for the occurrence of neutrino helicity-flipping processes and that the true scenario is the emission and absorption of both, longitudinal and transverse space-like (Čerenkov) photons by neutrinos through the Landau damping mechanism. These rates can then be used to place an upper bound on the neutrino magnetic moment μ. By considering a supernova core as a degenerate high-temperature plasma, we find that μ is in the range $\mu < (0.1 - 0.4) \times 10^{-11} \mu_B$, for a core temperature T between $T = 30 - 60$ MeV, where μ_B is the Bohr magneton.

Consider a left-handed neutrino traveling with energy E and momentum $\vec{p}$ in a medium at high temperature T. We are first interested in computing the Čerenkov reaction rate Γ which is equal to the sum of the Čerenkov emission Γ_e and absorption Γ_a rates. We know that Γ is given in terms of the neutrino self energy Σ by means of [5]

$$\Gamma = \Gamma_e + \Gamma_a = \frac{1}{2E} \frac{\text{Tr}\,[\text{Im}\,\Sigma \slashed{P}]}{(e^{E/T} + 1)}. \tag{1}$$

The one loop contribution to the neutrino self energy is illustrated in Fig. 1 where the kinematical variables involved are also defined.

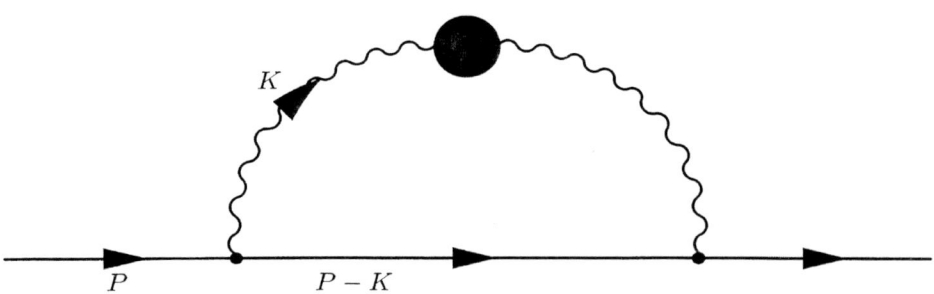

FIGURE 1. Feynman diagram for the self energy Σ of a left-handed neutrino to one-loop. The heavy dot indicates that the photon propagator has been resummed in the HTL approximation

The intermediate right-handed neutrino line should be taken as a free, zero temperature fermion propagator S_F, since the right-handed neutrino does not interact with the particles in the medium. The photon line however is the propagator $^*D^{\mu\nu}$, which is generated through interactions and we first consider its Hard Thermal Loop (HTL) approximation. Since the effective vertices preserve chirality [6], the neutrino-photon vertex is simply the magnetic dipole interaction

$$\Gamma_\mu = \mu \sigma_{\mu\nu} K^\nu. \tag{2}$$

In Euclidean space, the one loop expression for Σ is

$$\Sigma = \frac{1}{2} \mu^2 T \sum_n \int \frac{d^3k}{(2\pi)^3} K_\alpha \sigma^{\alpha\mu} \frac{(\slashed{P} - \slashed{K})}{(P-K)^2} K_\beta \sigma^{\beta\nu} \, ^*D^{\mu\nu}(K), \tag{3}$$

where the factor of 1/2 accounts only for left-handed neutrinos and we have neglected the neutrino mass. Hereafter, capital letters are used to refer to momentum four-vectors in Euclidean space. In a covariant gauge, the HTL approximation to $^*D^{\mu\nu}$ is given explicitly by

$$^*D^{\mu\nu}(K) = \frac{P_L^{\mu\nu}}{\left[K^2 + 2m^2\frac{K^2}{k^2}(1 - (\frac{iw_n}{k})Q_0(\frac{iw_n}{k}))\right]}$$
$$+ \frac{P_T^{\mu\nu}}{\left[K^2 + m^2(\frac{iw_n}{k})\left([1 - (\frac{iw_n}{k})^2]Q_0(\frac{iw_n}{k}) + (\frac{iw_n}{k})\right)\right]}, \quad (4)$$

where we dropped the term proportional to the gauge parameter since it does not contribute to Σ, as can be easily checked. Also here,

$$m^2 = \frac{e^2}{2\pi^2}\left(\tilde{\mu}_e^2 + \frac{\pi^2 T^2}{3}\right) \quad (5)$$

is the photon thermal mass with $\tilde{\mu}_e$ the electron's chemical potential and $P_L^{\mu\nu}$, $P_T^{\mu\nu}$ are the four-dimensional longitudinal and transverse projectors, respectively. Q_0 is a Legendre function of the second kind given explicitly by $Q_0(x) = \frac{1}{2}\ln\left(\frac{x+1}{x-1}\right)$.

In order to compute the imaginary part of Σ, we use the identity [7]

$$\text{Im } T \sum_n g(iw_n)\tilde{g}(i(w - w_n)) = \pi(e^{E/T} + 1)$$
$$\times \int_{-\infty}^{\infty}\frac{dk_0}{2\pi}\int_{-\infty}^{\infty}\frac{dp_0}{2\pi}f(k_0)\tilde{f}(p_0)\delta(E - k_0 - p_0)\rho(k_0)\tilde{\rho}(p_0). \quad (6)$$

In our case, $f(k_0) = (e^{k_0/T} - 1)^{-1}$ and $\tilde{f}(p_0) = \varepsilon(p_0)$ –where $\varepsilon(p_0)$ is the sign function– are the photon and right-handed neutrino distributions, respectively, whereas $\rho(k_0)$ and $\tilde{\rho}(p_0)$ are the corresponding spectral densities, given by

$$\tilde{\rho}(p_0) \equiv 2\text{Im}\left[w^2 + (\vec{p} - \vec{k})^2\right]^{-1}_{iw \to p_0 + i\epsilon}$$
$$= (2\pi)\delta(p_0^2 - (\vec{p} - \vec{k})^2), \quad (7)$$

$$\rho_L(k_0) \equiv -2\text{Im}\left[K^2 + 2m^2\frac{K^2}{k^2}(1 - (\frac{iw_n}{k})Q_0(\frac{iw_n}{k}))\right]^{-1}_{iw_n \to k_0 + i\epsilon} \quad (8)$$

$$\rho_T(k_0) \equiv 2\text{Im}\left[K^2 + m^2(\frac{iw_n}{k})\left([1 - (\frac{iw_n}{k})^2]Q_0(\frac{iw_n}{k}) + (\frac{iw_n}{k})\right)\right]^{-1}_{iw_n \to k_0 + i\epsilon}. \quad (9)$$

The spectral densities $\rho_{L,T}(k_0)$, contain the discontinuities of the photon propagator across the real-k_0 axis. Their support depends on the magnitude of the ratio between k_0 and k. For $|k_0/k| > 1$, $\rho_{L,T}(k_0)$ has support on the points $w_{L,T} = \pm w_{\pm L,T}(k)$, i.e., the time-like quasiparticle poles. For $|k_0/k| < 1$, $\rho_{L,T}(k_0)$ has support on the whole interval $-k < k_0 < k$ but the contribution to the discontinuity comes from the branch cut of Q_0. For emission or absorption of Čerenkov photons, for which the medium's index of refraction $n = k/k_0 > 1$, the kinematically allowed region is the interval $-k < k_0 < k$. In this case, Eqs. (8) and (9) become

$$\rho_L(k_0) = -\frac{2\pi(m^2 k_0/k)(\frac{k^2}{k^2-k_0^2})\theta(k^2-k_0^2)}{\left[k^2 + 2m^2(1 - \frac{k_0}{2k}\ln\left|\frac{k_0+k}{k_0-k}\right|)\right]^2 + \left[\pi m^2 \frac{k_0}{k}\right]^2}, \quad (10)$$

$$\rho_T(k_0) = \frac{2\pi(m^2 k_0/k)(k^2-k_0^2)/(2k^2)\theta(k^2-k_0^2)}{\left[(k^2-k_0^2) + m^2\left((\frac{k_0}{k})^2 + (\frac{k_0}{2k})(\frac{k^2-k_0^2}{k^2})\ln\left|\frac{k_0+k}{k_0-k}\right|\right)\right]^2 + \left[\pi m^2 \frac{k_0(1-k_0^2/k^2)}{2k}\right]^2}. \quad (11)$$

We now need the trace of the numerator in (3), which can be easily shown to yield, after analytic continuation,

$$\frac{1}{2}K_\alpha K_\beta P_T^{\mu\nu}\text{Tr}[\sigma^{\alpha\mu}(P-K)\sigma^{\beta\nu}P] \to \frac{(n^2-1)^2}{n^2}k_0^2\left[(2E+k_0)^2 - k^2\right]$$

$$\frac{1}{2}K_\alpha K_\beta P_L^{\mu\nu}\text{Tr}[\sigma^{\alpha\mu}(P-K)\sigma^{\beta\nu}P] \to \frac{(n^2-1)^2}{n^2}k_0^2(2E+k_0)^2. \quad (12)$$

By means of Eqs. (10) and (12) and after performing the angular integration, we can write the expression for the Čerenkov reaction rate

$$\Gamma = \frac{\mu^2}{32\pi^2 E^2}\int_0^\infty dk\, k \int_{-k}^k dk_0 \theta(2E+k_0-k)$$
$$f(k_0)\frac{(n^2-1)^2}{n^2}k_0^2(2E+k_0)^2\left[\rho_L(k_0,k) + \left(1 - \frac{k^2}{(2E+k_0)^2}\right)\rho_T(k_0,k)\right], \quad (13)$$

where the the theta function results from imposing the consistency condition $|\cos\Theta| \leq 1$, with Θ the angle between the incoming neutrino and the photon. Thus, we see that the contribution to the Čerenkov rate does not come from the spacelike quasiparticle pole but instead it comes from the propagator branch cut which is associated with Landau damping [7] and that both, longitudinal and transverse photons, contribute to this rate. We also note from Eq. (13), that the Čerenkov absorption rate, with the photon in the initial state, corresponds to the interval $0 \leq k_0 < k$, whereas the Čerenkov emission rate, with the photon in the final state, corresponds to the interval $-k < k_0 \leq 0$ as can be checked by means of the identity

$$1 + f(k_0) + f(-k_0) = 0 \quad (14)$$

and the substitution $k_0 \to -k_0$ in this second interval.

A word should be said about the contribution from soft transverse photons. It is well known that the production of these photons is enhanced by the Bose-Einstein statistical factor [9] which could potentially lead to a divergent contribution. Fortunately, in this case, there are enough powers of k coming from the vertex factors to render this contribution finite. The details of the calculation as well as a more thorough discussion can be found elsewhere [6].

To find the rates of energy emitted and absorbed in the Čerenkov processes, we should insert a factor of $(E - k_0)$ and $(E + k_0)$ in the expressions for the Čerenkov emission and absorption rates, respectively

$$\dot{S}_e(E) = \frac{\mu^2}{32\pi^2 E^2} \int_0^\infty dk k \int_0^k dk_0 \theta(2E - k_0 - k)[1 + f(k_0)]\frac{(n^2-1)^2}{n^2}$$
$$k_0^2(2E-k_0)^2(E-k_0)\left[\rho_L(k_0,k) + \left(1 - \frac{k^2}{(2E-k_0)}\right)\rho_T(k_0,k)\right], \quad (15)$$

$$\dot{S}_a(E) = \frac{\mu^2}{32\pi^2 E^2} \int_0^\infty dk k \int_0^k dk_0 \theta(2E + k_0 - k)f(k_0)\frac{(n^2-1)^2}{n^2}$$
$$k_0^2(2E+k_0)^2(E+k_0)\left[\rho_L(k_0,k) + \left(1 - \frac{k^2}{(2E+k_0)}\right)\rho_T(k_0,k)\right]. \quad (16)$$

Finally, the total neutrino luminosity is given by

$$Q = \frac{V}{2\pi^2} \int_0^\infty dE E^2 \tilde{f}_\nu(E)[\dot{S}_e(E) + \dot{S}_a(E)]. \quad (17)$$

where V is the plasma volume, $\tilde{f}_\nu(E) = (\exp(E - \tilde{\mu}_\nu)/T + 1)^{-1}$ is the left-handed neutrino distribution and $\tilde{\mu}_\nu$ its chemical potential. Taking $\tilde{\mu}_\nu = 160$ MeV and T in the range $T = 30 - 60$ MeV, with corresponding electron's chemical potential $\tilde{\mu}_e$ between $\tilde{\mu}_e = 307 - 280$ MeV, inside a supernova core with a volume $V = 4.19 \times 10^{18}$cm^3 and a density $\varrho = 8 \times 10^{14}$gr/cm^3 with a relative proton density number $Y_p = 0.3$ [2] and performing the integrals in Eqs. (15)–(17) numerically, we find

$$Q = \mu^2((0.4 - 1.8) \times 10^{62} \text{ MeV}^4). \quad (18)$$

Assuming that not all of the energy of the core collapse is taken away by the right-handed neutrinos, i.e.

$$Q < 10^{53} \text{ergs/sec}, \quad (19)$$

then, Eq. (18) places an upper bound to the neutrino magnetic moment

$$\mu < (0.1 - 0.4) \times 10^{-11} \mu_B, \quad (20)$$

which is a slight improvement on previously found bounds [2].

The Čerenkov rate is proportional to the photon's thermal mass squared m^2. To see this, let us define the dimensionless variables $x = k_0/k$, $y = k/T$ and $z = E/T$ to extract the dimensions out of, for instance, the contribution to the luminosity form the longitudinal piece of Eq. (16),

$$Q_a^L \sim \mu^2 m^2 T^5 \int_0^\infty dz \tilde{f}_\nu(z) \int_0^\infty dy y^4 \int_0^1 dx \theta(2z + yx - y)$$
$$\frac{[1 + f(yx)]x(1 - x^2)(2z - yx)^2(z - yx)}{\left[y^2 + \frac{2m^2}{T^2}\left(1 - \frac{x}{2}\ln\left(\frac{1+x}{1-x}\right)\right)\right]^2 + \left[\pi \frac{m^2}{T^2} x\right]^2}. \quad (21)$$

Thus, the integral in Eq. (21) is proportional to $m^2 T^5 \sim e^2 T^7$ and not just to T^7. This is also the case for the rest of the terms that make up the total luminosity Q. The result can be understood by recalling that the interaction is governed by the space-like branch cut of the photon propagator. This implies, for example, that once a Čerenkov photon is produced, it quickly decays back into the plasma. The decay channel is provided by hard electrons traveling at the photon's phase velocity [8] and the true amplitude to compute for the decay rate is the sum of collision processes mediated by space-like photons, where the electron-photon vertex provides the extra powers of the electric charge upon squaring. We notice that the longitudinal Čerenkov rate is thus suppressed, with respect to a calculation that ignores the damping mechanism, by an extra power of e^2. This suppression is however partially compensated by the rates of emission and absorption of transverse Čerenkov photons allowed by the same mechanism.

We also enphasize that the Landau damping mechanism is the correct physical picture behind the helicity flip scattering process $\nu_L e \to \nu_R e$. Plasma effects are consistently taken incto account to leading order by means of the resummation method of Braaten and Pisarski and in particular, there is no need of including an ad hoc constant mass prescription for screening purposes.

In conclusion, we have shown that the Landau damping mechanism allows for the emission and absorption of longitudinal and transverse Čerenkov photons in an ultrarelativistic plasma such as supernovae cores. The Čerenkov rates can be used to establish an upper bound on the neutrino magnetic moment which slightly improves the one one obtained from the cooling of SN1987 by helicity flip scattering [2].

Support for this work has been received in part by CONACyT-México under grant No. I27212-E.

REFERENCES

1. S. Nussinov and Y. Raphaeli, Phys. Rev. D **36**, 2278 (1987).
   J.M. Lattimer and J. Cooperstein, Phys. Rev. Lett. **61**, 23 (1988).
   S. Mohanty and M.K. Samal, Phys. Rev. Lett. **77**, 806 (1996).
   G. Raffelt, Phys. Rev. Lett. **79**, C773 (1997).
   J.C. D'Olivo and J.F. Nieves, *Nucleon effects on the photon dispersion relations in matter*, hep-ph/9710305, to be published in Phys. Rev. D.
2. R. Barbieri and R.N. Mohapatra, Phys. Rev. Lett. **61**, 27 (1988)
3. S. Mohanty and S. Sahu, *Neutrino helicity flip by Čerenkov emission and absorption of plasmons in supernova*, hep-ph/9710404.

4. H.A. Weldon, Phys. Rev. D **26**, 1394 (1982).
5. M. Le Bellac, *Thermal Field Theory* Cambridge University Press (1996).
6. A. Ayala, J.C. D'Olivo and M. Torres, *Neutrino chirality flip through photon Landau damping in supernovae* hep-ph/9804230, submitted to Phys. Rev. Lett.
   A. Ayala, J.C. D'Olivo and M. Torres, work in progress.
7. E. Braaten and R.D. Pisarski, Phys. Rev. D **46**, 1829 (1992).
8. E.M. Lifshitz and L.P. Pitaevskii, *Physical Kinetics, Course of Theoretical Physics* Vol. 10, Pergamon Press (1981).
9. J.-P. Blaizot and E. Iancu, Phys. Rev. Lett. **76**, 3080 (1996).

Asymmetry studies in $\Lambda^0/\bar{\Lambda}^0$, Ξ^-/Ξ^+ and Ω^-/Ω^+ production

J.C. Anjos, J. Magnin [1], F.R.A. Simão and J. Solano

Centro Brasileiro de Pesquisas Físicas - CBPF
Rua Dr. Xavier Sigaud 150, CEP 22290-180, Rio de Janeiro, Brazil.

Abstract. We present a study on hyperon/anti-hyperon production asymmetries in the framework of the recombination model. The production asymmetries for $\Lambda^0/\bar{\Lambda}^0$, Ξ^-/Ξ^+ and Ω^-/Ω^+ are studied as a function of x_F. Predictions of the model are compared to preliminary data on hyperon/anti-hyperon production asymmetries in 500 GeV/c $\pi^- p$ interactions from the Fermilab E791 experiment. The model predicts a growing asymmetry with the number of valence quarks shared by the target and the produced hyperons in the $x_F < 0$ region. In the positive x_F region, the model predicts constant asymmetries for $\Lambda^0/\bar{\Lambda}^0$ and Ω^-/Ω^+ production and a growing asymmetry with x_F for Ξ^-/Ξ^+. We found a qualitatively good agreement between the model predictions and data, showing that recombination is a competitive mechanism in the hadronization process.

I INTRODUCTION

In hadron interactions, the leading particle effect manifests as an enhancement in the production rate of particles which share valence quarks with the initial hadrons. As a consequence of the leading particle effect, strong asymmetries are expected in the x_F ($= 2p_L/\sqrt{s}$) inclusive differential cross sections for particles and anti-particles when the content of valence quarks shared by the produced particle and anti-particle with the initial hadrons is different.

This effect has been extensively studied, from both the experimental [1] and theoretical [2,3] points of view in charm hadron production.

In the production of strange baryons, the same type of leading effects are expected. Indeed, there is some evidence of asymmetries in $\Lambda^0/\bar{\Lambda}^0$ production in π^-Cu interactions at 230 GeV/c [4] and in $\Lambda^0/\bar{\Lambda}^0$ and Ξ^-/Ξ^+ production in 250 GeV/c $\pi^- p$ interactions [5]. Some additional evidence for $\Lambda^0/\bar{\Lambda}^0$ asymmetry can be found in Ref. [6], but, in general, hyperon production asymmetries in $\pi^- p$ interactions were not systematically studied until recently, when the Fermilab E791

---

[1] supported by FAPERJ, Fundação de Amparo à Pesquisa do Estado de Rio de Janeiro.

experiment presented preliminary results on hyperon production in $\pi^- p$ interactions at 500 GeV/c [7]. The E791 experiment has measured particle/anti-particle production asymmetries in both the small $x_F < 0$ and $x_F > 0$ regions for the $\Lambda^0/\bar{\Lambda}^0$, Ξ^-/Ξ^+ and Ω^-/Ω^+ hyperons.

The results obtained by the E791 experiment show a large asymmetry in the $x_F < 0$ region for $\Lambda^0/\bar{\Lambda}^0$ production and a lower asymmetry, in the same region, for Ξ^-/Ξ^+ production. In the $x_F > 0$ region, an approximately constant asymmetry with x_F is observed for Ξ^-/Ξ^+ and $\Lambda^0/\bar{\Lambda}^0$ hyperons. The asymmetry measured for Ω^-/Ω^+ production is approximately constant in the whole $-0.12 \leq x_F \leq 0.12$ region.

These results in the $x_F < 0$ region are consistent with the fact that Λ^0 hyperons share a ud diquark whereas the Ξ^-s share a d quark with the target particles (protons and neutrons), so a lower asymmetry is expected for the later since the Λ^0 is a double leading whereas the Ξ^- is a leading particle. In the $x_F > 0$ region, the Ξ^- share a d quark with the initial π^-, being a leading particle, while Ξ^+ shares none. Then a growing asymmetry with x_F is expected in Ξ^-/Ξ^+ production. The Λ^0 and $\bar{\Lambda}^0$ each have one valence quark in common with the beam π^- and, therefore, have equal enhancement i.e., no asymmetry is expected from this effect.

The Ω^-/Ω^+ hyperons are both non-leading in all the x_F regions studied and, consequently, no asymmetry is expected in Ω^-/Ω^+ production in $\pi^- p$ interactions.

Due to the smallness of the strange quark mass and the p_T^2 values involved, hyperon production can not be accounted for in the usual framework of perturbative QCD.

The recombination scheme, initially introduced by Das and Hwa [8], appears to be a possible framework to deal with the non-perturbative QCD aspects involved in hadron and, in particular, in hyperon production. Indeed, this type of model has been used succesfuly to describe charmed particle/anti-particle asymmetries [3,9] in hadroproduction.

In this work, we compare the predictions of a simple version of the recombination model on hyperon/anti-hyperon production asymmetries with the preliminary results of the E791 experiment. The good qualitative agreement found between model predictions and experimental data even at small values of x_F shows that it might be interesting to make efforts to improve the outcome of the model.

II HYPERON PRODUCTION BY RECOMBINATION IN $\pi^- P$ INTERACTIONS

The recombination model was introduced long time ago by Das and Hwa [8] to describe meson production in hadron-hadron collisions. A simple extension of the model was made by Ranft [10] to calculate baryon production. From those first attempts up to now, several modifications have been introduced trying to improve the outcome of the model [11]. Nevertheless, the recombination model remains a simple approach to deal with some non-perturbative aspects of QCD involved in hadron-hadron interactions.

The basic idea behind recombination is that the produced hadrons are formed from the debris of the fragmented beam (in the forward region) or target (in the backward direction) particles in such a way that partons initially in the incoming particles *recombine* into the final hadrons. All that is needed to deal with the problem is to know the distribution of partons in the initial particles, which are measured in Deep Inelastic Scattering experiments, and the so-called recombination function which will take into account all aspects involved in the recombination of partons into a hadron. Of course, the recombination function has a phenomenological origin, since no calculation from first principles is yet possible to obtain it.

For a generic hyperon H, the x_F inclusive distribution in recombination is given by

$$\frac{2E}{\sigma^{rec}\sqrt{s}}\frac{d\sigma^{rec}}{d|x_F|} = \int_0^1 \frac{dx_1\, dx_2\, dx_3}{x_1\, x_2\, x_3} F_3^H(x_1, x_2, x_3)\, R_3(x_1, x_2, x_3, x_F) \,, \tag{1}$$

where $\sqrt{s}$ is the center of mass energy in the $\pi^- p \to H + X$ reaction, E is the energy of the outgoing hyperon and σ^{rec} is a normalization constant. In eq. (1), $F_3^H(x_1, x_2, x_3)$ is the three-quark distribution, which contains the distribution of valence quarks in the final particle inside the beam or target hadrons, $R_3(x_1, x_2, x_3, x_F)$ is the recombination function and x_i; $i = 1, 2, 3$ is the momentum fraction of the ith quark with respect to the initial particle.

Following the approach of Ref. [10], the three quark distribution function is assumed to be of the form

$$F_3^H(x_1, x_2, x_3) = \beta g(x_1, x_2, x_3)(1 - x_1 - x_2 - x_3)^\gamma \,, \tag{2}$$

where

$$g(x_1, x_2, x_3) = F_{q_1}(x_1)\, F_{q_2}(x_2)\, F_{q_3}(x_3) \tag{3}$$

contains the single quark distribution, $F_{q_i} = x_i q_i(x_i)$, of the q_i valence quark of the final hyperon in the initial particle. Note that F_{q_i} includes valence as well as sea quark contributions from the initial hadron. The coefficients β and γ are fixed using the consistency condition

$$F_q(x_i) = \int_0^{1-x_i} dx_j \int_0^{1-x_i-x_j} dx_k\, F_3^H(x_1, x_2, x_3)$$
$$i, j, k = 1, 2, 3 \tag{4}$$

which must be valid for the valence quarks in the initial particle.

For the recombination function we use [13]

$$R_3(x_1, x_2, x_3) = \alpha \frac{(x_1 x_2)^{n_1} x_3^{n_2}}{x_F^{n_1+n_2-1}} \delta(x_1 + x_2 + x_3 - x_F) \tag{5}$$

TABLE 1. $g_H(x_1,x_2,x_3)$ and $g_{\bar{H}}(x_1,x_2,x_3)$ used in the calculation of asymmetries. q^n ($\bar{q}^n$) is the quark (anti-quark) distribution in nucleons, q^π ($\bar{q}^\pi$) is the quark (anti-quark) distribution in the π^-. The individual parton distributions were taken from Ref. [12].

	\multicolumn{2}{c}{$x_F < 0.$}	\multicolumn{2}{c}{$0. < x_F$}	
	$g_H(x_1,x_2,x_3)$	$g_{\bar{H}}(x_1,x_2,x_3)$	$g_H(x_1,x_2,x_3)$	$g_{\bar{H}}(x_1,x_2,x_3)$
$\Lambda^0/\bar{\Lambda}^0$	$u^n(x_1)d^n(x_2)s^n(x_3)$	$\bar{u}^n(x_1)\bar{d}^n(x_2)\bar{s}^n(x_3)$	$u^\pi(x_1)d^\pi(x_2)s^\pi(x_3)$	$\bar{u}^\pi(x_1)\bar{d}^\pi(x_2)\bar{s}^\pi(x_3)$
Ξ^-/Ξ^+	$d^n(x_1)s^n(x_2)s^n(x_3)$	$\bar{d}^n(x_1)\bar{s}^n(x_2)\bar{s}^n(x_3)$	$d^\pi(x_1)s^\pi(x_2)s^\pi(x_3)$	$\bar{d}^\pi(x_1)\bar{s}^\pi(x_2)\bar{s}^\pi(x_3)$
Ω^-/Ω^+	$s^n(x_1)s^n(x_2)s^n(x_3)$	$\bar{s}^n(x_1)\bar{s}^n(x_2)\bar{s}^n(x_3)$	$s^\pi(x_1)s^\pi(x_2)s^\pi(x_3)$	$\bar{s}^\pi(x_1)\bar{s}^\pi(x_2)\bar{s}^\pi(x_3)$

allowing in this way a different weight for the heavier s ($\bar{s}$) quark than for the light u ($\bar{u}$) and d ($\bar{d}$) quarks.

The constant α in eq. (5) is fixed by the condition [14]

$$\frac{1}{\sigma^{rec}} \int_0^1 dx_F \frac{d\sigma^{rec}}{dx_F} = 1 , \qquad (6)$$

then σ^{rec} is the recombination cross section of the hyperon H in $\pi^- p \to H + X$ in the forward ($x_F > 0$) or backward ($x_F < 0$) region. σ^{rec} may be fixed from experimental data.

The asymmetry as a function of x_F is defined by

$$A(x_F) = \frac{d\sigma_H/d|x_F| - d\sigma_{\bar{H}}/d|x_F|}{d\sigma_H/d|x_F| + d\sigma_{\bar{H}}/d|x_F|} \qquad (7)$$

where H is the Hyperon and $\bar{H}$ is the anti-Hyperon.

Replacing eq. (1), with eqs. (2) to (6), into eq. (7) for the hyperons and anti-hyperons we obtain

$$A(x_F) = \frac{\int_0^{|x_F|} dx_1 \int_0^{|x_F|-x_1} dx_2 \left[g_H(x_1,x_2,x_3) - \sigma g_{\bar{H}}(x_1,x_2,x_3)\right]}{\int_0^{|x_F|} dx_1 \int_0^{|x_F|-x_1} dx_2 \left[g_H(x_1,x_2,x_3) + \sigma g_{\bar{H}}(x_1,x_2,x_3)\right]}$$
$$x_3 = |x_F| - x_1 - x_2 \qquad (8)$$

with $\sigma = \sigma^{\bar{H}}/\sigma^H$ the relative normalization between the hyperon and anti-hyperon distributions. The delta function of eq. (5) has been used to do one of the integrals in eq. (8).

As in Ref. [13], we have used $n_1 = 1$, $n_2 = 3/2$ and $\gamma = -0.3$ in our calculations.

In Figs. 1 and 2 we show the predictions of the recombination model compared to the E791 measurements. In order to obtain the theoretical curves shown in the figures, the asymmetry as given by eq. (8) has been calculated for each x_F region independently. In Table 1, the g_H and $g_{\bar{H}}$ distributions used in each region and for each one of the produced hyperons and anti-hyperons are displayed. The relative normalization between the particle and anti-particle x_F distributions, σ, has been chosen to fit the experimental data (See Table 2).

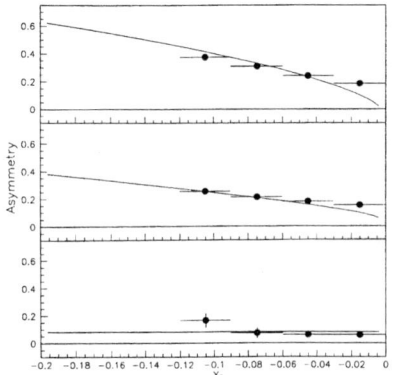

FIGURE 1. $\Lambda^0/\bar\Lambda^0$ (upper), Ξ^-/Ξ^+ (middle) and Ω^-/Ω^+ (lower) asymmetries as a function of x_F in the negative x_F region. Full line is the prediction from recombination model. Black dots are the preliminary E791 results taken from Ref. [7].

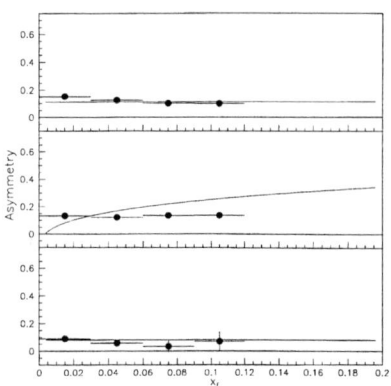

FIGURE 2. $\Lambda^0/\bar\Lambda^0$ (upper), Ξ^-/Ξ^+ (middle) and Ω^-/Ω^+ (lower) asymmetries as a function of x_F in the positive x_F region. Full line is the prediction from recombination model. Black dots are the preliminary E791 results taken from Ref. [7].

For Ω^-/Ω^+ production, since both hyperon and anti-hyperon are non-leading in the $x_F < 0$ as well as in the $x_F > 0$ regions, the recombination model predicts a constant asymmetry arising only from the difference between the global normalization of the particle and anti-particle x_F distributions.

III CONCLUSIONS

As can be seen in Figs. 1 and 2, the predictions of the recombination model agree qualitatively well with the experimental data.

In the negative x_F region, the growth of the asymmetry predicted by the model with the number of valence quarks shared by the leading hyperons and the target particles is in a remarkable agreement with the E791 data. In this region, the

TABLE 2. Relative normalization, σ, between hyperon and anti-hyperon cross sections

	$x_F < 0.$	$0. < x_F$
$\Lambda^0/\bar\Lambda^0$	3.5	0.8
Ξ^-/Ξ^+	1.9	2.2
Ω^-/Ω^+	0.85	0.85

asymmetry as a function of x_F is also qualitatively well described by the model. The agreement between data and model predictions is better as $|x_F|$ rises. This is possibly due to the fact that, at very low values of $|x_F|$, other mechanisms than recombination are competitive in the hadronization.

In the positive x_F region, although the leading effect in Ξ^-/Ξ^+ production is qualitatively accounted for by the model, in general, the agreement between data and recombination model predictions is poorer than in the negative x_F region. Note, however, that the quark distributions in pions are not as well known as in nucleons, being one of the possible causes of the discrepancies between model prediction and data in this region.

The asymmetry predicted for Ω^-/Ω^+ production is constant over all the x_F region under study. This is consistent with the fact that the Ω^- and the Ω^+ are both non-leading particles over the whole x_F region from -1 to 1.

In conclusion, the recombination model, although simple, is able to reproduce qualitatively the behaviour of the experimental data on hyperon/anti-hyperon asymmetries, so it might be possibly of interest to make efforts in order to improve the outcome of the model.

For a meaningful quantitative comparison between data and model predictions, however, the individual inclusive x_F distributions for leading and non-leading particles must be taken into account.

ACKNOWLEDGEMENTS

Two of us, J.C.A. and F.R.A.S. would like to thank the organizing committee and the Centro Latino Americano de Fisica, CLAF, for financial support to attend the SILAFAE II. The authors also would like to acknowledge J.A. Appel, B. Meadows and D.A. Sanders for useful comments.

REFERENCES

1. WA92 Collaboration (M. Adamovich et al.), Nucl. Phys. **B495**, 3 (1997); E791 Collaboration (E.M. Aitala et al.), Phys. Lett. **B371**, 157 (1996); E769 Collaboration (G. A. Alves et al.), Phys. Rev. Lett. **77**, 2388 (1996) and Phys. Rev. Lett. **72**, 812 (1994); WA82 Collaboration (M. Adamovich et al.), Phys. Lett. **B305**, 402 (1993).
2. R. Vogt and S.J. Brodsky, Nucl. Phys. **B478**, 311 (1996)
3. G. Herrera and J. Magnin, Eur. Phys. J. **C2**, 477 (1998), E. Cuautle, G. Herrera and J. Magnin, ibid. 473.
4. S. Barlag et al., Phys. Lett. **B325**, 531 (1994).
5. D. Bogert et al., Phys. Rev. **D16**, 2098 (1977).
6. S. Mikocki et al., Phys. Rev. **D34**, 42 (1986), R.T. Edwards et al., Phys. Rev. **D18**, 76, (1978) and N.N. Biswas, Nucl. Phys. **B167**, 41 (1980).

7. E791 Collaboration and J. Solano, J. Magnin and F.R.A. Simão, Fermilab-Conf-97/368-E and CBPF-NF-No 072/97, to be published in the RANP97 proceedings (hep-ex/9710033).
8. K.P. Das and R.C. Hwa, Phys. Lett. **B68**, 459 (1977).
9. R.C. Hwa, Phys. Rev. **D 51**, 85 (1995).
10. J. Ranft, Phys. Rev. **D18**, 1491 (1978).
11. E. Takasugi and X. Tata, Phys. Rev. **D23**, 2573 (1981), R.C. Hwa, Phys. Rev. **D22**, 759 (1980), R.C. Hwa, Phys. Rev. **D22**, 1593 (1980).
12. R.D. Field and R.P. Feynman, Phys. Rev. **D 15**, 2590 (1977).
13. G. Herrera, J. Magnin, Luis M. Montaño and F.R.A. Simão, Phys. Lett. **B382**, 201 (1996).
14. J.C. Anjos, G. Herrera, J. Magnin and F.R.A. Simão, Phys. Rev. **D56**, 394 (1997).

Λ^0 Polarization in Exclusive pp Reactions at 27.5 GeV/c

J. Félix[1], C. Avilez[1‡], D.C. Christian[3], M.D. Church[4b], M. Forbush[5g], E.E. Gottschalk[4d], G. Gutierrez[3], E.P. Hartouni[2a], S.D. Holmes[3], F.R. Huson[5], D.A. Jensen[2b], B.C. Knapp[4], M.N. Kreisler[2a], G. Moreno[1], J. Uribe[2c], B.J. Stern[4d], M.H.L.S. Wang[2a], A. Wehmann[3], L.R. Wiencke[4f], J.T. White[5]

[1] *Universidad de Guanajuato, León, Guanajuato, México,* [2] *University of Massachusetts, Amherst, Massachusetts, USA,* [3] *Fermilab, Batavia, Illinois, USA,* [4] *Columbia University, Nevis Labs, New York, USA* [5] *Department of Physics, Texas A&M University, College Station, Texas 77843,USA.*

Abstract

In Λ^0's created from the specific reactions: $pp \to p\Lambda^0 K^+\pi^+\pi^-$, $pp \to p\Lambda^0 K^+\pi^+\pi^-\pi^+\pi^-$, $pp \to p\Lambda^0 K^+\pi^+\pi^-\pi^+\pi^-\pi^+\pi^-$, and $pp \to p\Lambda^0 K^+\pi^+\pi^-\pi^+\pi^-\pi^+\pi^-\pi^+\pi^-$, using 27.5 GeV/c proton beam incident on a liquid hydrogen target, we measured Λ^0 polarization as a function of x_F and P_T. We found that Λ^0 polarization is independent of any particular final state and of any specific mechanism responsible of producing Λ^0.

Introduction

The discovery that Λ^0 hyperons are polarized when produced in high energy pp collisions [1] has posed an interesting puzzle to theories of particle production. That discovery for Λ^0's, and subsequent observations that other hyperons were polarized as well [2,3], has called into question the generally accepted assumption that spin plays no role in high energy multi-particle production.

Despite many works on this phenomenon, both experimental and theoretical, an understanding of the source of the polarization remains elusive. Models proposed to date do not fit all of the data well and tend not to have predictive power [4].

In this paper, we report a study of Λ^0 polarization, performed in a high statistics exclusive sample, of the particular reactions:

$$pp \to p\Lambda^0 K^+\pi^+\pi^-, \tag{1}$$

$$pp \to p\Lambda^0 K^+\pi^+\pi^-\pi^+\pi^-, \tag{2}$$

$$pp \to p\Lambda^0 K^+\pi^+\pi^-\pi^+\pi^-\pi^+\pi^-, \tag{3}$$

$$pp \to p\Lambda^0 K^+\pi^+\pi^-\pi^+\pi^-\pi^+\pi^-\pi^+\pi^-. \tag{4}$$

We investigated the kinematic dependence of the polarization in each reaction, as function of x_F and P_T.

Λ^0 Polarization

The data for this study come from the experiment BNL E766. Details of the experiment and analysis procedures can be found elsewhere [6–9,12,13]. The numbers of exclusive events selected for these measurements are 5421, 51195, 48195, 14582 for the Reaction (1), (2), (3), and (4), respectively. This study of Λ^0 polarization explores its dependence on the specific final states (1), (2), (3) and (4), and on the kinematic variables: P_T, the transverse momentum of Λ^0 with respect to the incident proton beam, and x_F, defined by $x_F = \frac{P_Z}{P_{Zmax}}$, where P_Z is the longitudinal Λ^0 momentum with respect to the beam proton momentum, in the event's center of mass, and P_{Zmax} is the maximum value of P_Z in this frame.

In the method we have used to determine Λ^0 polarization, $\mathcal{P}$, the decay angular distribution is assumed to be described by the expression:

$$\mathrm{dN}/\mathrm{d}\Omega = N_0(1 + \alpha \mathcal{P} cos\theta), \tag{5}$$

where $\mathrm{dN}/\mathrm{d}\Omega$ is the angular distribution of the proton from the Λ^0 decay in the Λ^0 rest frame, N_0 is a normalization constant, α is the asymmetry parameter (0.642±0.013) [11] and $\mathcal{P}$ is the polarization. θ is the angle between the direction of the proton from the decay of the Λ^0 and the normal to the production plane, $\hat{n} \equiv \frac{\vec{P}_{beam} \times \vec{P}_\Lambda}{|\vec{P}_{beam} \times \vec{P}_\Lambda|}$ where $\vec{P}_{beam}$ and $\vec{P}_\Lambda$ are the momentum vectors of the Λ^0 and the incident beam proton respectively.

Results

Λ^0 polarization has been observed to be odd in x_F in Reaction (2) [6]. We have observed that to be the case as well in Reactions (1), (3), and (4). We determined $\mathcal{P}$ as a function of x_F and P_T, in each final state; the obtained results agree, inside statistics, each other. In order to improve the statistical power of this measurement, we have combined the whole data from $x_F > 0$ and $x_F < 0$ by multiplying $cos\theta$ by the sign of x_F, without caring about the multiplicity. In what follows, we present our discussion in terms of $|x_F|$, for whole sample, and without correction by acceptance for we have found that the acceptance is flat in the variable we used to determine Λ^0 polarization ($cos\theta$).

To determine $\mathcal{P}$, the data are separated into x_F and P_T bins and histogrammed in $cos\theta$. These histograms are then fit to Eq. (5) with N_0 and $\mathcal{P}$ as free parameters. This method is described in detail in Ref. [12].

The polarization results, as a function of x_F are shown in Figure 1; the straight line fits are shown. The polarization results, as a function of P_T are shown in Figure 2; the straight line fits are shown also. In both cases the χ^2/dof of the straight line fits are around 1, showing that the straight line hypothesis are acceptable.

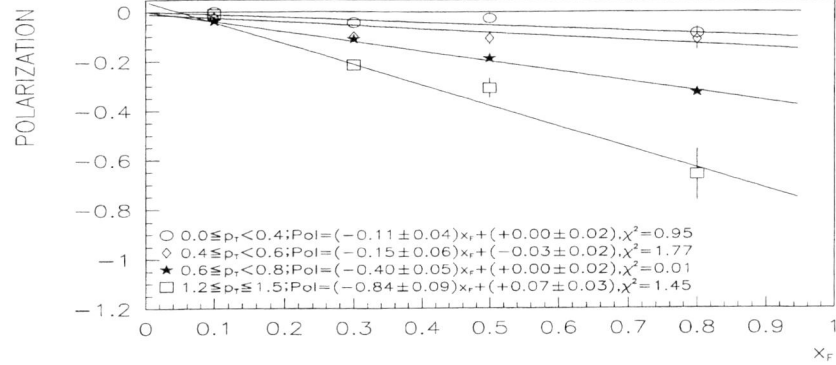

FIGURE 1. Λ^0 polarization as a function of x_F, for the entire sample.

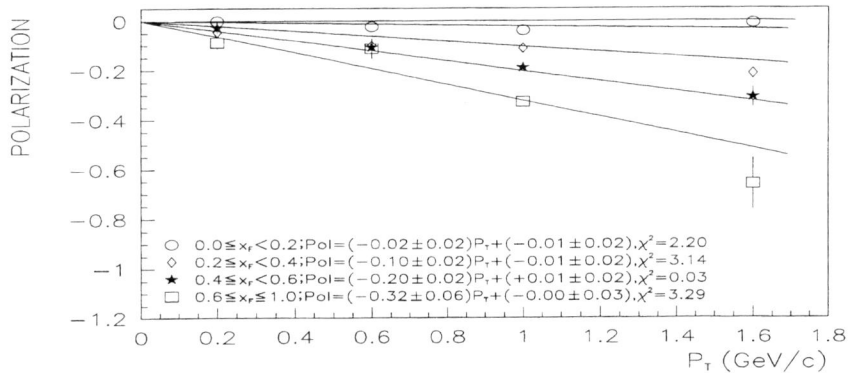

FIGURE 2. Λ^0 polarization as a function P_T, for the entire sample.

These results agree with the previous ones, inside statistics, as a function of both X_F and P_T and in each multiplicity. We conclude that the polarization is independent of the final state; and since the production mechanism that generates Reactions (1) and (2) is different from the one that produces Reaction (3) and (4) [13], we conclude that the mechanism that produces Λ^0 polarization is independent of the Λ^0 production mechanism. Or we can state it in other words, Λ^0 polarization is independent of the number of pions in the final state.

From the above fits, it seems that the distributions of the polarization, both as a function of x_F and P_T, can be described by $\mathcal{P} = (-0.443 \pm 0.037)x_F P_T$. Also, inside statistical errors, it seems that the polarization distributions can be described by $\mathcal{P} = (-0.485 \pm 0.041)x_F P_T^2$; the two distributions, inside statistical errors, fit the $\mathcal{P}$ distribution well. The last expression is the first order approximation of that empirical one used in Ref. [14] to fit Λ^0 polarization distributions from inclusive pp reactions; i.e, both exclusive and inclusive Λ^0 polarization distributions, from pp reactions, are fitted by the same expression. We conclude that Λ^0 polarizing mechanism is independent of that particular mechanism that produces Λ^0.

Conclusions

The agreement of Λ^0 polarization between the four specific final states indicates that Λ^0 polarization is independent of a specific final state; in other words, Λ^0 polarization is independent of the number of pions in the final state. Since the mechanism producing the final state particles changes significantly from Reaction (1) through Reaction (4) [13], our results suggest that the Λ^0 polarizing mechanism is independent of the specific Λ^0 associated production mechanism.

The Λ^0 polarization can be described as either $\mathcal{P} = (-0.443 \pm 0.037)x_F P_T$ or $\mathcal{P} = (-0.485 \pm 0.041)x_F P_T^2$. The agreement of these parameterizations between specific final states and inclusive final states [14] strongly suggests that the mechanism which produces the polarization is independent of a particular final state.

Acknowledgments

This work was supported in part by National Science Foundation Grants No. PHY90-14879 and No. PHY89-21320, by the Department of Energy Contracts No. DE-AC02-76 CHO3000, No. DE-AS05-87ER40356 and No. W-7405-ENG-48, and by CoNaCyT of México under Grant 458100-5-4009PE.

REFERENCES

[‡] Deceased.
[a] Present address: Lawrence Livermore National Laboratory. Livermore CA 94550.
[b] Present address: Fermilab, Batavia, IL 60510.
[c] Present address: University of Texas, M.D. Anderson Cancer Center, Houston, TX 77030.
[d] Present address: University of Illinois, Urbana, Illinois.
[e] Present address: AT&T Research Laboratories, Murray Hill, NJ 07974.
[f] Present address: University of Utah, Salt Lake City, UT 84112.
[g] Present address: University of California, Davis, CA 95616.

1. A. Lesnik et al., Phys. Rev. Lett. **35**, 770 (1975); G. Bunce et al., Phys. Rev. Lett. **36**, 1113 (1976); K. Heller et al., Phys. Lett. **68B**, 480 (1977).
2. G. Bunce et al., Phys. Lett. **86B**, 386 (1979); J. Duryea et al., Phys. Rev. Lett. **67**, 1193 (1991); R. Rameika et al., Phys. Rev. **D 33**, 3172 (1986); C. Wilkinson et al., Phys. Rev. Lett. **58**, 855 (1987); B. Lundberg et al., *Phys. Rev. D* **40**, 39 (1989).
3. G. Bunce et al., *Phys. Rev. Lett.* **36**, 1113 (1976); F. Lomanno et al., *Phys. Rev. Lett.* **43**, 1905 (1979); S. Erhan et al., *Phys. Lett.* **82B**, 301 (1979); F. Abe et al., Phys. Rev. Lett. **50**, 1102 (1983); K. Raychaudhuri et al., Phys. Lett. **90B**, 319 (1980); K. Heller et al., Phys. Rev. Lett. **41**, 607 (1978); F. Abe et al., *J. of the Phys. S. of Japan.* Vol. **52**, 12 (1983) 4107-4117; P. Aahlin et al., *Lettere al Nuovo Cimento* Vol. **21**, No. 7, (1978); A. M. Smith et al., Phys. Lett. *B* **185**, 209 (1987); V. Blobel et al., *Nuclear Physics B* **122**, 429 (1977); K. Heller et al., Phys. Lett. **68B**, 480 (1977).
4. T. A. DeGrand et al., Phys. Rev. **D 24**, 2419 (1981); B. Andersson et al., Phys. Lett. **85B**, 417 (1979); J. Szweed et al., Phys. Lett. **105B**, 403 (1981); K. J. M. Moriarty et al., *Lett. Nuovo Cimento*, **17** 366 (1976); S. M. Troshin and N. E. Tyurin *Sov. J. Nucl. Phys.* **38**(4), Oct. 1983; J. Soffer and N.E. Törnqvist *Phys. Rev. Lett.* **68**, 907 (1992); Y. Hama and T. Kodama *Phys. Rev. D* 48, 3116 (1993); R. Barni et al. *Phys. Lett. B* 296 (1992) 251-255; W. G. D. Dharmaratna and G. R. Goldstein *Phys. Rev. D* **53** 1073 (1996); W. G. D. Dharmaratna and G. R. Goldstein *Phys. Rev. D* **41** 1731 (1990); S. M. Troshin and N. E. Tyurin *Phys. Rev. D* **55** 1265 (1997); L. Zuo-Tang and C. Boros *Phys. Rev. Lett.* **79** 3608 (1997).
5. T. Henkes et al., Phys. Lett. **B 283**, (1992) 155.
6. J. Félix et al., Phys. Rev. Lett. **76**, 22 (1996).
7. J. Uribe et al., Phys. Rev. **D 49**, 4373 (1994), and References 10 and 12 therein.
8. E. P. Hartouni et al., Phys. Rev. Lett. **72**, 1322 (1994).
9. D. C. Christian et al., Nucl. Instr. and Meth. **A345**, 62 (1994).

10. B. C. Knapp and W. Sippach, IEEE Trans. on Nucl. Sci. **NS-27**, 578 (1980); E. P. Hartouni *et al.*, *ibid.* **NS-36**, 1480 (1989); B. C. Knapp, Nucl. Instrum. Methods A **289**, 561 (1990).
11. Particle Data Group, Phys. Rev. **D 50**, 1 (1994).
12. J. Félix, Ph.D. thesis, Universidad de Guanajuato, México, 1994.
13. E. E. Gottschalk,*et al.*, Phys. Rev. **D 53**, 4756 (1996).
14. L. G. Pondrom, Phys. Rep. **122**, 57 (1985).

A model for baryon structure and its application to magnetic moments and semileptonic decays

V. Gupta[a], R. Huerta[a], and G. Sánchez-Colón[a,b]

[a]*Departamento de Física Aplicada*
Centro de Investigación y de Estudios Avanzados del IPN. Unidad Mérida
A.P. 73, Cordemex. Mérida, Yucatán 97310, MEXICO
[b]*Department of Physics, University of California*
Riverside, CA 92521-0413, U.S.A.

Abstract. The spin 1/2 baryons are pictured as a composite system made out of a "core" of three valence quarks (as in the simple quark model) surrounded by a "sea" (of gluon and $q\bar{q}$ pairs) which is specified by its total quantum numbers. We assume the sea is a $SU(3)$ flavor octet with spin 0 or 1 but no color. This model, considered earlier, is used to obtain simultaneous fits for magnetic moments and G_A/G_V for semileptonic decays. These fits give predictions for nucleon spin distributions in reasonable agreement with experiment.

I INTRODUCTION

The simple quark model (SQM) though qualitatively successful, fails to account for low energy properties of baryons quantitatively. Experimentaly [1], it is found that quarks cannot even account for the proton spin and thus it is necessary to go beyond SQM. Since quarks interact through strong color forces mediated by gluons, a physical hadron, in reality, consists of valence quarks surrounded by a "sea" of gluons and quark-antiquark ($q\bar{q}$) pairs. The effect of the sea contribution to hadron structure has been considered by several authors [2–4].

In this paper, we study the static properties of the spin 1/2 baryons (p, n, Λ, ...) following Refs. [3] and [4] where the general sea is specified by its total flavor, spin and color quantum numbers. The baryons are pictured as a composite system made out of a baryon "core" of the three valence quarks (as in SQM) and a flavor octet sea with spin 0 and 1 but no color. The purpose of this paper is to use this wavefunction to obtain a simultaneous fit to magnetic moments and semileptonic decays.

II BARYON WAVE FUNCTIONS WITH SEA

The physical baryon octet states, denoted by $B(1/2 \uparrow)$ are obtained by combining the "core" wavefunction $\tilde{B}(8, 1/2)$ (the usual SQM spin 1/2 baryon octet wave function) with the sea wavefunction with specific properties given below. We assume the sea is a color singlet but has flavor and spin properties which when combined with those of the core baryons $\tilde{B}$ give the desired properties of the physical baryon B. Since both the physical and core baryon have $J^P = \frac{1}{2}^+$, this implies that the sea has even parity and spin 0 or 1. The spin 0 and 1 wavefunctions for the sea are denoted by H_0 and H_1, respectively. We also refer to a spin 0 (1) sea as a scalar (vector) sea. For $SU(3)$ flavor we assume the sea has a $SU(3)$ singlet component and an octet component described by wavefunctions $S(\mathbf{1})$ and $S(\mathbf{8})$, respectively. The color singlet sea in our model is thus described by the wavefunctions $S(\mathbf{1})H_0$, $S(\mathbf{1})H_1$, $S(\mathbf{8})H_0$, and $S(\mathbf{8})H_1$.

The total flavor-spin wavefunction of a spin up ($\uparrow$) physical baryon which consists of 3 valence quarks and a sea component (as discussed above) can be written schematically as

$$B(1/2 \uparrow) = \tilde{B}(\mathbf{8}, 1/2 \uparrow)H_0 S(\mathbf{1}) + b_0 \left[\tilde{B}(\mathbf{8}, 1/2) \otimes H_1\right]^\uparrow S(\mathbf{1})$$
$$+ \sum_N a(N) \left[\tilde{B}(\mathbf{8}, 1/2 \uparrow)H_0 \otimes S(\mathbf{8})\right]_N \quad (1)$$
$$+ \sum_N b(N) \left\{\left[\tilde{B}(\mathbf{8}, 1/2) \otimes H_1\right]^\uparrow \otimes S(\mathbf{8})\right\}_N.$$

The first term is the usual q^3-wavefunction of the SQM (with a trivial sea) and the second term (coefficient b_0) comes from spin-1 (vector) sea which combines with the spin 1/2 core baryon $\tilde{B}$ to a spin $1/2\uparrow$ state. So that, $\left[\tilde{B}(\mathbf{8}, 1/2) \otimes H_1\right]^\uparrow = \sqrt{\frac{2}{3}}\tilde{B}(\mathbf{8}, 1/2 \downarrow)H_{1,1} - \sqrt{\frac{1}{3}}\tilde{B}(\mathbf{8}, 1/2 \uparrow)H_{1,0}$. In both these terms the sea is a flavor singlet. The third (fourth) term in Eq. (1) contains a scalar (vector) sea which transforms as a flavor octet. The various $SU(3)$ flavor representations obtained from $\tilde{B}(\mathbf{8}) \otimes S(\mathbf{8})$ are labelled by $N = \mathbf{1}, \mathbf{8_F}, \mathbf{8_D}, \mathbf{10}, \mathbf{\bar{10}}, \mathbf{27}$. As it stands, Eq. (1) represents a spin $1/2\uparrow$ baryon which is not *a pure flavor octet* but has an admixture of other $SU(3)$ representations weighted by the unspecified constants $a(N)$ and $b(N)$. It will be a flavor octet if $a(N) = b(N) = 0$ for $N = \mathbf{1}, \mathbf{10}, \mathbf{\bar{10}}, \mathbf{27}$. The color wavefunctions have not been indicated as the three valence quarks in the core $\tilde{B}$ and the sea (by assumption) are in a color singlet state. The sea isospin multiplets contained in the $SU(3)$ flavor octet $S(\mathbf{8})$ are denoted as $(S_{\pi^+}, S_{\pi^0}, S_{\pi^-})$, (S_{K^+}, S_{K^0}), $(S_{\bar{K}^0}, S_{K^-})$, and S_η. The familiar pseudoscalar mesons are used here as subscripts to label the isospin and hypercharge quantum numbers of the sea states. Details of the wavefunction have been given earlier [3,4]. However, for completeness the explicit physical baryon states in terms of the core and sea states are given in Table 1. The normalization of a given baryon state (not indicated in

TABLE 1. Contribution to the physical baryon state $B(Y, I, I_3)$ formed out of $\tilde{B}(Y, I, I_3)$ and flavor octet states $S(Y, I, I_3)$ (see 3rd term in Eq. (1)). The core baryon states $\tilde{B}$ denoted by $\tilde{p}$, $\tilde{n}$, etc. are the normal 3 valence quark states of SQM. The sea octet states are denoted by $S_{\pi^+} = S(0, 1, 1)$, etc. Further, $(\tilde{N}S_\pi)_{I,I_3}$, $(\tilde{\Sigma}S_{\bar{K}})_{I,I_3}$, $(\tilde{\Sigma}S_\pi)_{I,I_3}$, ..., stand for total I, I_3 *normalized* combinations of $\tilde{N}$ and S_π, etc. Only the contribution from the 3rd term is given. 4th term has exactly the same flavor symmetries with coefficients $(\bar{\beta}_i, \beta_i, \gamma_i, \delta_i) \to (\bar{\beta}'_i, \beta'_i, \gamma'_i, \delta'_i)$. See Ref. [3] for explicit expressions for the coefficients $\bar{\beta}_i$, β_i, ..., and $\bar{\beta}'_i$, β'_i, ...

$B(Y, I, I_3)$	$\tilde{B}(Y, I, I_3)$ and $S(Y, I, I_3)$
p	$\bar{\beta}_1 \tilde{p} S_\eta + \bar{\beta}_2 \tilde{\Lambda} S_{K^+} + \bar{\beta}_3 (\tilde{N}S_\pi)_{1/2,1/2} + \bar{\beta}_4 (\tilde{\Sigma}S_K)_{1/2,1/2}$
n	$\bar{\beta}_1 \tilde{n} S_\eta + \bar{\beta}_2 \tilde{\Lambda} S_{K^0} + \bar{\beta}_3 (\tilde{N}S_\pi)_{1/2,-1/2} + \bar{\beta}_4 (\tilde{\Sigma}S_K)_{1/2,-1/2}$
Ξ^0	$\beta_1 \tilde{\Xi}^0 S_\eta + \beta_2 \tilde{\Lambda} S_{\bar{K}^0} + \beta_3 (\tilde{\Xi}S_\pi)_{1/2,1/2} + \beta_4 (\tilde{\Sigma}S_{\bar{K}})_{1/2,1/2}$
Ξ^-	$\beta_1 \tilde{\Xi}^- S_\eta + \beta_2 \tilde{\Lambda} S_{\bar{K}^-} + \beta_3 (\tilde{\Xi}S_\pi)_{1/2,-1/2} + \beta_4 (\tilde{\Sigma}S_{\bar{K}})_{1/2,-1/2}$
Σ^+	$\gamma_1 \tilde{p} S_{\bar{K}^0} + \gamma_2 \tilde{\Xi}^0 S_{K^+} + \gamma_3 \tilde{\Lambda} S_{\pi^+} + \gamma_4 \tilde{\Sigma}^+ S_\eta + \gamma_5 (\tilde{\Sigma}S_\pi)_{1,1}$
Σ^-	$\gamma_1 \tilde{n} S_{K^-} + \gamma_2 \tilde{\Xi}^- S_{K^0} + \gamma_3 \tilde{\Lambda} S_{\pi^-} + \gamma_4 \tilde{\Sigma}^- S_\eta + \gamma_5 (\tilde{\Sigma}S_\pi)_{1,-1}$
Σ^0	$\gamma_1 (\tilde{N}S_{\bar{K}})_{1,0} + \gamma_2 (\tilde{\Xi}S_K)_{1,0} + \gamma_3 \tilde{\Lambda} S_{\pi^0} + \gamma_4 \tilde{\Sigma}^0 S_\eta + \gamma_5 (\tilde{\Sigma}S_\pi)_{1,0}$
Λ	$\delta_1 (\tilde{N}S_{\bar{K}})_{0,0} + \delta_2 (\tilde{\Xi}S_K)_{0,0} + \delta_3 \tilde{\Lambda} S_\eta + \delta_4 (\tilde{\Sigma}S_\pi)_{0,0}$

Eq. (1)) depends on the parameters which enter in the wavefunction and is different for different isospin multiplets (see Ref. [3]).

For applications, we need the quantities $(\Delta q)^B$, $q = u, d, s$; for each spin-up baryon B. These are defined as $(\Delta q)^B = n^B(q \uparrow) - n^B(q \downarrow) + n^B(\bar{q} \uparrow) - n^B(\bar{q} \downarrow)$, where $n^B(q \uparrow)$ $(n^B(q \downarrow))$ are the number of spin-up (spin-down) quarks of flavor q in the spin-up baryon B. Also, $n^B(\bar{q} \uparrow)$ and $n^B(\bar{q} \downarrow)$ have a similar meaning for antiquarks. However, these are zero as there are no explicit antiquarks in the wavefunctions given by Eq. (1). The expressions for $(\Delta q)^B$ in terms of $\bar{\beta}_i$, $\bar{\beta}'_i$, β_i, etc. are given in Table 3 of Ref. [3].

III MAGNETIC MOMENTS (MM) AND SEMILEPTONIC DECAYS (SLD)

For any operator $\hat{O}$ which depends only on quarks, the matrix elements are easily obtained using the orthogonality of the sea components. Clearly $\langle B \uparrow | \hat{O} | B' \uparrow \rangle$ will be a linear combination of the matrix elements $\langle \tilde{B} \uparrow | \hat{O} | \tilde{B}' \uparrow \rangle$ (known from SQM) with coefficients which depend on the coefficients in the wavefunction, Eq. (1).

We assume the baryon magnetic moment operator, $\hat{\mu}$, to act solely on the valence quarks in $\tilde{B}$, so that $\hat{\mu} \equiv \sum_q \mu_q \sigma_z^q$ where $\mu_q = e_q/2m_q$ and e_q and m_q are quark charge and mass for $q = u, d, s$.

It is possible to show that the MM of the spin 1/2 baryons, μ_B ($B = p, n, \Lambda, \ldots$), and the transition magnetic moment, $\mu_{\Sigma^0 \Lambda}$, can be written as

$$\mu_B = \sum_{q=u,d,s} (\Delta q)^B \mu_q \quad \text{and} \quad \mu_{\Sigma^0 \Lambda} = \sum_{q=u,d} (\Delta q)^{\Sigma^0 \Lambda} \mu_q, \qquad (2)$$

where the $(\Delta q)^B$ are defined above. Expressions for $(\Delta q)^B$ in terms of the parameters b_0, β_i and β_i' are given in Ref. [3]. From Eqs. (2) we see that the MM depend on the quark masses (or quark MM) and on the parameters b_0, $a(N)$, $b(N)$ which determine the sea.

For SLD, the detailed expressions for $G_{V,A}(B \to B') = \langle B' | J_{V,A} | B \rangle$ of the charge changing hadronic vector (J_V) and axial vector (J_A) currents using our wavefunction (Eq. (1)) are given in Ref. [4]. Here we briefly summarize how they were calculated.

The $\Delta S = 0$ and $\Delta S = 1$ vector currents are the total isospin raising ($I_+ = I_+^{(q)} + I_+^{(s)}$) and V-spin lowering ($V_- = V_-^{(q)} + V_-^{(s)}$) operators [4]. The operators $I_+^{(q)}$ and $V_-^{(q)}$ act on the quarks in the core baryons and $I_+^{(s)}$ and $V_-^{(s)}$ act on the sea states in the wavefunction. However, the axial vector current has a quark part $J_A^{(q)}$ and a sea part $J_A^{(s)}$ which may, in general, have different relative strengths, so that

$$J_A(\Delta S = 0, 1) = J_A^{(q)}(\Delta S = 0, 1) + A_{0,1} J_A^{(s)}(\Delta S = 0, 1) \qquad (3)$$

where the constants A_0 and A_1 specify the strength of $J_A^{(s)}$ relative to $J_A^{(q)}$ for $\Delta S = 0$ and $\Delta S = 1$ transitions respectively. In SQM, $J_A^{(q)}(\Delta S = 0) = \sum_q I_+^{(q)} \sigma_z^q$ and $J_A^{(q)}(\Delta S = 1) = \sum_q V_-^{(q)} \sigma_z^q$ so that, in analogy, we took $J_A^{(s)}(\Delta S = 0) = 2 I_+^{(s)} S_z^{(s)}$ and $J_A^{(s)}(\Delta S = 1) = 2 V_-^{(s)} S_z^{(s)}$ where $S_z^{(s)}$ is the spin operator acting only on the sea states in the wavefunction. For $\Delta S = 0$ transitions, the quark part was sufficient so that $A_0 = 0$ for all the fits. For $\Delta S = 1$ transitions, a direct sea contribution through $J_A^{(s)}$ is needed when the theoretical error on the MM is very small.

IV COMBINED FITS AND RESULTS

The excellent fits for the MM given in Ref. [3] can straightaway be used to predict the G_A/G_V for the 4 SLD's $n \to p$, $\Lambda \to p$, $\Sigma^- \to n$, and $\Xi^- \to \Lambda$ for which data are available. The numerical predictions are poor especially for $n \to p$ and $\Sigma^- \to n$. The minimum χ^2-fits to MM alone do not give acceptable fits to the SLD data. This situation changes profoundly when a combined fit to *both* the 8 magnetic moment data and 4 SLD G_A/G_V data is made with 1 or 2 more parameters to describe the sea, but reducing one of the μ_q's as a parameter, for example, by having $m_u = m_d$.

IV A. Fits with experimental errors for all data. A combined fit to the SLD and MM data with 5 parameters to describe the sea and 2 parameters μ_u and μ_s (we put $m_u = m_d$) give an excellent fit (given in Table 2) with $A_0 = 0$ and $A_1 = -1$, which determine the strength of the direct sea contribution to $G_A(\Delta S = 0)$ and $G_A(\Delta S = 1)$ respectively. Treating A_0 and A_1 as free parameters does not affect the χ^2 as their values come to be $A_0 = -0.016$ and $A_1 = -1.01$. Thus our combined fit gives $\chi^2_{MM} = 0.7$, $\chi^2_{SLD} = 0.3$, with total $\chi^2/DOF = 1/5$. If one considers A_0 and A_1 as parameters then $\chi^2/DOF = 1/3$. The values of the sea parameters, μ_u, and μ_s are given at the end of Table 2.

For a comparison of this 7 parameter fit with the earlier 6 parameter fits of Ref. [3] see Ref. [4].

In summary, at the expense of an extra parameter overall one obtains a better fit to MM data than before as well as fit the known SLD data *using experimental errors throughout.*

IV B. Fits with theoretical errors of $0.1\mu_N$ for MM. We consider such fits because all the fits in the literature (unlike our fits above) add an arbitrary theoretical error. The motivation for adding this error is that all MM are treated "democratically". Otherwise, the extremely accurately measured μ_p and μ_n act as inputs to a minimum χ^2-fit. We add a theoretical error of $0.1\mu_N$ in quadratures to the experimental errors for all the MM data. This is a popular choice [6,7]. An error of $0.1\mu_N$ is fairly large (compared to the actual experimental errors) and facilitates a good fit with a few parameters only. This is true in our model also! A 3 parameter fit with inputs $m_u = m_d = 0.6m_s$ and $A_0 = A_1 = 0$ is given in column 4 of Table 2. In this fit, the scalar and vector seas are described by one parameter each, namely $a(\mathbf{10})$ and $b(\mathbf{8_D})$ respectively.

How does our fit compare to other fits with $0.1\mu_N$ theoretical error? We give a comparison with the most recent fits [7] refered to as CS below. Unlike us the model of CS does not fit $\mu_{\Sigma^0\Lambda}$ as it is not clear how to include it in their picture of 3-quark correlation within a baryon. For MM alone our fit give $\chi^2_{MM}/DOF = 3.8/5$ compared to $4.4/4$ for Models AII and AIII of CS, their best fits. An important difference in their and our model is reflected in the phenomenological values of $(\Delta q)^B$. In particular, CS obtain (their Model AIII) $(\Delta u)^p = 0.783$, $(\Delta d)^p = -0.477$, and $(\Delta s)^p = -0.147$. This is to be contrasted with the fact that our fits yield $(\Delta u)^p = 0.964$, $(\Delta d)^p = -0.296$, and $(\Delta s)^p = 0.008$. Physically, our fits require a very tiny strange-quark content in the nucleon compared to their and other similar fits [6,7]. Another physical difference is that in our case the valence quarks carry 67% of the proton spin compared to about 16% in Model AIII of CS.

IV C. Spin Distributions. The spin distribution, I_{1B}, for baryon B is defined as $I_{1B} \equiv \int_0^1 g_{1B}(x)dx$, where the spin structure function g_{1B} occurs in polarized electron-baryon scattering.

In SQM, I_{1B} is given by the expectation value $I_{1B} \equiv \langle B|\hat{I}_1^{(q)}|B\rangle$ where the quark operator $\hat{I}_1^{(q)} = (1/2)\sum_q e_q^2 \sigma_Z^q$. This gives

TABLE 2. Combined fits to the SLD and MM data. All the MM values are given in nuclear magnetons, μ_N. a) Fit with experimental errors for all data, see Sec. IV A (column 3). b) Fit with theoretical errors of $0.1\mu_N$ added in quadratures for MM, see Sec. IV B (column 4).

	Data [5]	a) Experimental errors	b) Theoretical errors
$\mu(p)$	$2.79284739 \pm 6 \times 10^{-8}$	2.79284739	2.79239
$\mu(n)$	$-1.9130428 \pm 5 \times 10^{-7}$	-1.9130428	-1.96330
$\mu(\Lambda)$	-0.613 ± 0.004	-0.613	-0.608
$\mu(\Sigma^+)$	2.458 ± 0.010	2.458	2.538
$\mu(\Sigma^0)$	——	0.6396	0.7186
$\mu(\Sigma^-)$	-1.160 ± 0.025	-1.179	-1.101
$\mu(\Xi^0)$	-1.250 ± 0.014	-1.251	-1.151
$\mu(\Xi^-)$	-0.6507 ± 0.0025	-0.6506	-0.5331
$|\mu(\Sigma\Lambda)|$	1.61 ± 0.08	1.59	1.55
$G_A/G_V(n \to p)$	1.2601 ± 0.0025	1.2599	1.2598
$G_A/G_V(\Lambda \to p)$	0.718 ± 0.015	0.719	0.739
$G_A/G_V(\Sigma^- \to n)$	-0.340 ± 0.017	-0.338	-0.304
$G_A/G_V(\Xi^- \to \Lambda)$	0.25 ± 0.05	0.22	0.22
χ^2/DOF	——	1.02/5	10.70/9
Inputs	——	$m_d = m_u$ $A_0 = 0$ $A_1 = -1$	$m_d = m_u$ $m_s = (5/3)m_u$ $A_0 = A_1 = 0$
Fitted parameters	——	$\mu_u = 2.4900$ $\mu_s = -0.7785$ $a(\mathbf{8_F}) = -0.1465$ $a(\mathbf{10}) = 0.5130$ $b_0 = 0.3060$ $b(\mathbf{8_F}) = -0.3296$ $b(\mathbf{\overline{10}}) = 0.2442$	$\mu_u = 2.5166$ $a(\mathbf{10}) = 0.5280$ $b(\mathbf{8_D}) = 0.5658$

$$I_{1B}^{(q)} = \frac{1}{18}\left[4(\Delta u)^B + (\Delta d)^B + (\Delta s)^B\right]. \tag{4}$$

In our model in addition to the quarks there can be a direct sea contribution $I_{1B} \equiv \langle B|\hat{I}_{1B}^{(s)}|B\rangle$ where by analogy we take $\hat{I}_{1B}^{(s)} = e_s^2 S_Z^{(s)}$. Thus only the charged states in the vector sea will contribute to $I_{1B}^{(s)}$. For the nucleons, one obtains

$$I_{1p}^{(s)} = \frac{1}{3N_1^2}\left(\bar{\beta}_2'^2 + \frac{2}{3}\bar{\beta}_3'^2 + \frac{1}{3}\bar{\beta}_4'^2\right), \quad I_{1n}^{(s)} = \frac{1}{3N_1^2}\left(\frac{2}{3}\bar{\beta}_3'^2 + \frac{2}{3}\bar{\beta}_4'^2\right). \tag{5}$$

Putting the two contributions together we have $I_{1B} = I_{1B}^{(q)} + B_1 I_{1B}^{(s)}$, where B_1 determines the strength of the direct sea contribution to the valence quark contribution. Since the value of B_1 is not known á priori, so phenomenologically it may be treated as a parameter.

Experiment [1,8] gives $I_{1p} = 0.126 \pm 0.018$ and $I_{1n} = -0.08 \pm 0.06$ which are very different from the SQM predictions $I_{1p} = 5/18 = 0.2778$ and $I_{1n} = 0$. One must note that the EMC experiment gives I_{1p} for $\langle Q^2 \rangle = 10.7$ (GeV/c)^2 and this could be very different for the very low Q^2 (≈ 0) result predicted by SQM or other theoretical models. This could mean that a model which gives values for I_{1B} differing by 2–3 standard deviations from experiment may be quite acceptable.

Using the fit to MM and SLD data with experimental errors given in Table 2 we can predict I_{1B}. One obtains $I_{1p}^{(q)} = 0.205$, $I_{1n}^{(q)} = -0.005$, while, $I_{1p}^{(s)} = 0.044$ and $I_{1n}^{(s)} = 0.057$, where we have used $(\Delta u)^p = 0.989$, $(\Delta d)^p = -0.271$, and $(\Delta s)^p = 0.009$. If one keeps only the quark part, that is $B_1 = 0$, then our I_{1p} is much lower than the SQM value but still 4σ higher than experiment. This may be due to large $\langle Q^2 \rangle$ in the experiment. Another possibility is to invoke the direct sea contribution. For example, with $B_1 = -1$ one obtains $I_{1p} = 0.161$ and $I_{1n} = -0.062$ in good agreement with experiment.

V SUMMARY

In summary, we have shown that our model of the sea component in spin-1/2 baryons can fit their MM, weak decay constants G_A/G_V for both $\Delta S = 0$ and 1 SLD as well as nuclear spin distributions *using experimental errors*. To accomplish this, one has to invoke a direct sea contribution for $\Delta S = 1$ decays and nucleon spin distribution. The sea was found to be both scalar (spin 0) and vector (spin 1). Two physical features of our fits are that about 70% of the proton spin resides with the valence quarks and they give a tiny strange-quark content to the nucleon.

Acknowledgments. This work was partially supported by CONACyT (México).

REFERENCES

1. J. Ashman *et al.*, Nucl. Phys. **B328**, 1 (1989).

2. J. F. Donoghue and E. Golowich, Phys. Rev. **D15**, 3421 (1977). E. Golowich, E. Haqq and G. Karl, Phys. Rev. **D28**, 160 (1983). He Hanxin, Zhang Xizhen and Zhuo Yizhang, Chinese Phys. **4**, 359 (1984). J. Franklin, Phys. Rev. **D30**, 1542 (1984). F. E. Close and Z. Li, Phys. Rev. **D42**, 2194 (1994). F. E. Close, Rep. Prog. Phys. **51**, 833 (1988). Z. Li, Phys. Rev. **D44**, 2841 (1991). Z. Li and G. Karl, Phys. Rev. **D49**, 2620 (1994). V. Gupta and X. Song, Phys. Rev. **D49**, 2211 (1994). R. L. Jaffe and H. J. Lipkin, Phys. Lett. **B226**, 458 (1991).
3. V. Gupta, R. Huerta and G. Sánchez-Colón, Int. J. of Mod. Phys. **A12**, 1861 (1997).
4. V. Gupta, R. Huerta and G. Sánchez-Colón, preprint hep-ph/9803317; to be published in Int. J. of Mod. Phys. **A** (1998).
5. Particle Data Group, Phys. Rev. **D54**, 619 (1996).
6. G. Karl, Phys. Rev. **D45**, 247 (1992).
7. M. Casu and L. M. Sehgal, Phys. Rev. **D55**, 2644 (1997).
8. SMC, B. Adeva *et. al.*, Phys. Lett. **B302**, 533 (1993).

SPECIAL TALKS

The National Astronomy and Ionosphere Center's (NAIC) Arecibo Observatory in Puerto Rico

Daniel R. Altschuler

NAIC Arecibo Observatory
PO Box 995, Arecibo PR, 00614

It is a pleasure to welcome you to the Arecibo Observatory, site of the largest radio telescope on Earth! I would like to provide you with some historical facts about the observatory and in particular tell you about the recent upgrade to the telescope and our site which allows me to talk about the "new Arecibo"

The Arecibo Observatory is part of the National Astronomy and Ionosphere Center (NAIC), a national research center operated by Cornell University under a cooperative agreement with the National Science Foundation (NSF).

The Observatory operates on a continuous basis, 24 hours a day every day, providing observing time, electronics, computer, travel and logistic support to scientists from all over the world. All results of research are published in the scientific literature which is publicly available.

As the site of the world's largest single-dish radio telescope, the Observatory is recognized as one of the most important national centers for research in radio astronomy, planetary radar and terrestrial aeronomy. Use of the Arecibo Observatory is available on an equal, competitive basis to all scientists from throughout the world. Observing time is granted on the basis of the most promising research as ascertained by a panel of independent referees who review the proposals sent to the Observatory by interested scientists. Every year about 200 scientists visit the Observatory facilities to pursue their research project, and numerous students perform observations that lead to their master and doctoral dissertations.

The Observatory had its origins in an idea of Professor William E. Gordon, then of Cornell University, who was interested in the study of the Ionosphere. Gordon's research during the fifties led him to the idea of radar back scatter studies of the Ionosphere. Gordon's persistence culminated in the construction of the Arecibo Observatory which began in the Summer of 1960. Three years later the Arecibo Ionospheric Observatory (AIO) was in operation under the direction of Gordon. The formal opening ceremony took place on November 1, 1963. Soon after its inauguration the Arecibo radar made its first surprising discovery: the rotation

FIGURE 1. A view of the construction site in 1961

rate of Mercury was not 88 days as previously thought but 59 days.

From the beginning there were certain requirements for the site. It had to be near the equator, since there, a radar capable of studying the ionosphere could also be used to study nearby planets which pass overhead. The Arecibo site offered the advantage of being located in Karst terrain, with large limestone sinkholes which provided a natural geometry for the construction of the 305 meter reflector.

The Observatory also maintains a Ionospheric Interactions Facility which consists of thirty-two log-periodic antennas and transmitters capable of concentrating energy in the ionosphere. The waves from the facility energize electrons in the ionosphere and produce a number of interactions that are studied using the 305 meter telescope, in effect providing a unique capability as a "laboratory" for studies of plasma physics.

In addition an Optical Laboratory with a variety of instrumentation used for the passive study of terrestrial airglow is located at the Observatory. A lidar (Light Detection And Ranging) together with a Fabry-Perot interferometer is primarily used to measure neutral winds and temperatures of the middle atmosphere This capability complements that of the incoherent scatter radar, and gives Arecibo a unique capability in the world in terms of aeronomic research.

On October 1, 1969 the National Science Foundation took over the facility from the Department of Defense and the Observatory was made a national research center. On September 1971 the AIO became the National Astronomy and Ionosphere Center (NAIC).

In 1974 a new high precision surface for the reflector (the current one) was installed together with a high frequency planetary radar transmitter. The surface

FIGURE 2. The replacement of the original wire mesh surface of the reflector by aluminium panels.

is made of 38,772 perforated aluminum panels, each measuring about 3 feet by 6 feet, supported by a network of steel cables strung across the underlying karst sinkhole. It is a spherical (not parabolic) reflector.

This allowed observations at higher frequencies, in particular at a wavelength of 21 cm. The fact that the primary reflector is spherical provides the means to point the telescope without moving the reflector. Waves arriving from a particular direction will be focused along a radial line, and are collected by a "line feed". Changing the direction of this line feed allows for pointing to different directions in the sky. However, ohmic losses inherent to line feeds limit the sensitivity of the telescope, they produce relatively high side lobes and operate only at one specific frequency over a narrow range in frequencies.

Over the next twenty years the study of the "21 cm line" of neutral hydrogen, originating from the hyperfine transition became one of the most important areas of Arecibo research. It was also in 1974 that a professor and his graduate student from the University of Massachusetts arrived at Arecibo to pursue a search for Pulsars. The results of the study of one of those new pulsars discovered, the binary pulsar PSR1257+12 as it is known, led to the eventual confirmation of the existence of gravitational waves as predicted by Einstein's theory of gravitation. The 1993 Nobel prize in physics was awarded to Joseph Taylor and Russell Hulse for their work at Arecibo. We are very proud of that achievement.

The second and major upgrade to the telescope was completed in 1997 and provides a geometrical optics correction for the spherical aberration of the primary reflector. Two subreflectors in the gregorian dome are used to bring radiation

FIGURE 3. On May 16 1996 the new gregorian dome was lifted to the platform.

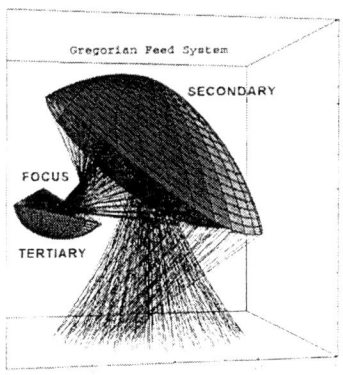

FIGURE 4. Two subreflectors correct for the spherical aberration of the primary as shown.

to a focus where a set of receivers can be positioned. This eliminates the above mentioned limitations of the line feeds. A ground screen around the perimeter of the reflector was also installed to shield the feeds from ground radiation. The gregorian dome with its subreflectors and new electronics greatly increases the capability of the telescope. A new more powerful radar transmitter was also installed.

Those who see the Arecibo radio telescope for the first time are astounded by the enormity of the reflecting surface, or radio mirror. The huge "dish" is 305 m (1000 feet) in diameter, 167 feet deep, and covers an area of about twenty acres.

Suspended 450 feet above the reflector is the 900 ton platform. Similar in design to a bridge, it hangs in midair on eighteen cables, which are strung from three reinforced concrete towers. One is 365 feet high, and the other two are 265 feet high. All three tops are at the same elevation. The combined volume of reinforced concrete in all three towers is 9,100 cubic yards. Each tower is back-guyed to ground anchors with seven 3.25 inch diameter steel bridge cables. Another system of three pairs of cables runs from each corner of the platform to large concrete blocks under the reflector. They are attached to giant jacks which allow adjustment of the height of each corner with millimeter precision.

Just below the triangular frame of the upper platform is a circular track on which the azimuth arm turns. The azimuth arm is a bow shaped structure 328 feet long. The curved part of the arm is another track, on which a carriage house on one side and the gregorian dome (not named for a pope or for music but in honor of James Gregory, one of the foremost mathematicians of the seventeenth century being the first professor of mathematics at the University of Edinburgh) on the other side can be positioned anywhere up to twenty degrees from the vertical. Inside the gregorian dome two subreflectors (secondary and tertiary) focus radiation to a point in space where a set of horn antennae can be positioned to gather it. Hanging below the carriage house are various linear antennas each tuned to a narrow band

FIGURE 5. A panoramic view of the upgraded telescope.

of frequencies. The antennas point downward and are designed specially for the Arecibo spherical reflector. By aiming a feed antenna at a certain point on the reflector, radio emissions originating from a very small area of the sky in line with the feed antenna will be focused on the feed antenna.

Attached to the antennas are very sensitive and highly complex radio receivers. These devices are cooled to 16 degrees above absolute zero. At such low temperatures the electron noise in the receivers is very small, and only the incoming radio signals, which are very weak, are amplified. The Arecibo system operates at frequencies from 50 megahertz (6 m wavelength) up to 10,000 megahertz (3 cm wavelength).

A total of 26 electric motors control the platform. These motors drive the azimuth and the gregorian dome and carriage house to any position with millimeter precision. The tertiary reflector can be moved to improve focusing, receivers are moved into focus on a rotating floor inside the gregorian and the dynamical tie downs activate as needed to maintain platform position. The 1 MW planetary radar transmitter located in a special room inside the dome, directs radar waves to objects in our solar system. Analyzing the echoes provides information about surface properties and object dynamics.

This giant telescope has scrutinized our atmosphere from a few kilometers to a few thousand kilometers where it smoothly connects with interplanetary space. With its radar vision it studies the properties of planets, comets and asteroids. In our Galaxy it detects the faint pulses emitted hundreds of times per second from pulsars. And from the farthest reaches of the Universe quasars and galaxies emit radio waves which arrive at earth 100 million years later as signals so weak that they can only be detected by a giant eye like this one.

The giant size of the reflector is what makes the Arecibo Observatory so special to scientists. It is the largest curved focusing antenna on the planet, which means it is the world's most sensitive radio telescope. Other radio telescopes may require several hours observing a given radio source to collect enough energy for analysis whereas at Arecibo this may require just a few minutes of observation.

About 140 persons are employed by the Observatory providing everything from food to software in support of the operation. A scientific staff of about 16 divide their time between scientific research and assistance to visiting scientists. Engineers, computer experts, and technicians design and build new instrumentation and keep it in operation. A large maintenance staff keeps the telescope and associated instrumentation as well as the site in optimal condition. A staff of telescope operators support observing twenty -four hour per day.

Finally let me also point out that this auditorium where we are sitting today is part of a new facility inaugurated a year ago, constructed for the benefit of those who visit us. After several years of fund-raising, planning and construction the new "Angel Ramos Foundation" visitor center is the only facility of its kind serving the general public and the public and private schools of Puerto Rico, and we believe sets an example for other similar initiatives at other research centers throughout the Nation. After one year of operation we have received over 120,000 visitors,

FIGURE 6. The new *Angel Ramos Foundation* visitor center.

about one third of them of school age. Pride in the Observatory and an effective fund raising campaign, caused local Puerto Rican organizations to contribute the funds necessary for the construction of the Center, the National Science Foundation providing the funds for the exhibits. In this complex society of changing priorities we must do all we can to help people understand the value of what we scientists do, to convey the importance and excitement of the sciences, and encourage young people to consider a career in the sciences. We owe it to ourselves and to our society.

Radio Astronomy Highlights at Arecibo

C. J. Salter

Arecibo Observatory
P.O. Box 995, Arecibo
Puerto Rico, PR 00613

I admit to having been a little taken aback when asked to address a workshop on "Cosmology & Elementary Particles". However, it did occur to me that many in the audience would not be experimentalists, and if I were to be speaking about the giant Arecibo telescope, it might be a good idea to put "nuts and bolts radio astronomy" into context first. Well, like most Englishmen, I am a compulsive tea drinker, so I thought it might be of interest to consider how long it would take to boil the water for a single cup of the "holy liquid", were it possible to harness for this purpose all the energy collected by our telescope from the standard calibrator radio source, quasar 3C286, using our 1.4-GHz receiver. Well, I do not recommend rushing off to try this experiment, as the answer is 10^{10} yr, or about the projected age of the Universe. Now this may not be much of a cosmological revelation, but I trust it does illustrate how tiny are the amounts of energy with which we are attempting to work, even in the case of a "strong" radio source. (In passing, it also illustrates that a watched pot never boils!)

Dr. Altschuler has already described the history of our observatory, and told you a great deal about the telescope upgrade which it has undergone recently. The upgrade itself will help us greatly in collecting more efficiently those meager photons with which the heavens provide us. Today, I plan to say a little of what our telescope has achieved over the years past, and I will also try to hint why we expect the future to be just as fruitful.

I PULSARS

The news of the discovery of pulsating radio sources – pulsars – burst on our community almost exactly 30 years ago. Since then, it has been demonstrated over and over again that this is a subfield where nothing rivals sheer antenna size. Arecibo's first pulsar coup was to demonstrate that a "sporadic radio source" at the center of the Crab Nebula supernova remnant was in fact the then fastest repeating pulsar, emitting 30 radio pulses per second. Here was direct evidence for the contention that pulsars are the rapidly rotating super-dense collapsed stellar cores left behind when a massive star ends its life in a supernova explosion. The year 1968 was also when Arecibo discovered its own first pulsars, launching a research endeavor that has productively filled large amounts of telescope time right up to the present day, and presumably far on into the future.

In 1975, an Arecibo search for new pulsars at low galactic latitudes not only discovered 40 new objects, but revealed one of these to be the first known "binary pulsar", with two neutron stars in orbit about each other. With its highly elliptical orbits, its strong gravitational fields, and despite its remoteness, this system provides a remarkable physics laboratory. Over the years, Joe Taylor and his team from Princeton have shown that the secular changes of the orbits of the pair are completely consistent with Einstein's predictions for the emission of gravitational waves from this system. This is the only evidence to date for the existence of gravitational waves, and it was a source of pride to us here when both Joe and his collaborator, Russell Hulse, received the 1993 Nobel Prize for Physics for their discovery.

Arecibo again entered the history books in 1982 when the telescope was used to reveal a pulsar which blinked 642 times every second, representing a rotation speed at its stellar surface of 0.1c. This is predicted to be near the fastest a neutron star can spin before disintegrating. This particular pulsar is still the most rapid repeater known, and has become the prototype for the growing class of recycled pulsars, spun up by accreting material from a companion.

One of Arecibo's most exciting discoveries came in 1990 with the detection of the first star outside the Solar System to be circled by planets. That this star is a millisecond pulsar was certainly a big surprise to most of us. Ironically, the pulsar in question was discovered while the telescope was temporarily crippled by a broken drive system, and observations only possible for those random positions which the Earth's rotation drifted through the observing beam! This "planets pulsar" is certainly orbited by two, more controversially three, planets, and the presence of a further planetary body may be becoming apparent through recent observations.

The search for new pulsars continues unabated at Arecibo. Following Alex Wolszczan's 1990 discovery of the planets pulsar, and a further body which is potentially the best laboratory yet for research in relativistic physics, there was a sudden interest in further drift-scan surveys. These have been pursued right through our recent upgrade period, with five teams attempting to cover the entire Arecibo-accessible sky between them, spending just 40 sec searching each independent position. By now, these are some 70% complete, and a milestone was passed last weekend when the hundredth Arecibo drift-scan survey pulsar was confirmed. Further, many more candidates await confirmation with the upgraded instrument. During the life of these upgrade surveys, the observing equipment has been completely upgraded too, and now for the first time Arecibo pulsar searches are sensitive to the presence of any sub-millisecond period pulsars that might exist. If they do, their presence will tell much about the equation of state of neutron star material.

When a new pulsar is discovered, the first thing that is usually done is to time it. Apart from a spin-down rate, this timing contains a host of additional information. For example, should the pulsar be in a binary system, timing reveals the detailed orbital parameters, and even allows improved positional determinations, plus accurate measures of proper motions for nearby pulsars. One exciting property measured through Arecibo timing for a few binary pulsars whose orbital planes are seen closely edge-on is the existence of a general-relativistic "Shapiro Delay" caused when the pulsar signals traverse the gravitational well of the stellar companion. This has the useful property of permitting an estimate of companion mass to be made.

Millisecond pulsars are exquisitely accurate clocks, and this has interested a number of our users in employing them to attempt detection of the long-wavelength component of any gravitational-wave background that exists. This is the gravitational analog of the Cosmic Microwave Background. Its detection is now a possibility because of the improved accuracy of timing equipment, the availability of timing at higher frequencies reducing the effects of propagation through the interstellar medium (ISM), and the ever increasing number of known "millisecond pulsar clocks".

II SPECTRAL LINES

As with pulsar research, Arecibo has made a huge impact in the field of radio spectroscopy. Here, the most studied spectral line has been the 1420-MHz hyperfine line of neutral hydrogen (HI). The huge Arecibo collecting area has made the dish unique for detecting this emission from the ISMs of distant spiral and irregular galaxies. In the case of the optically Low Surface Brightness (LSB) galaxies, a constituent of the Universe only recently recognized as being fundamental to our understanding of its make up, Arecibo often represents the sole practical hope of measuring redshifts. Indeed, using our new, recently commissioned spectrometer, a number of observing teams have already dived into the determination of redshifts for catalogs of LSB galaxies that they have compiled optically. One important development has been the first radio detection of red LSB galaxies.

During our upgrade, "blind radio searches" were made in drift-scan mode to provide a census of the present HI content of galaxies, independent of the biases that come from prior optical selection. These have found that there has to have been considerable evolution in the HI content of the Universe since the epoch near a redshift of 3, as there are insufficient counterparts for the high column density end of the Lyman-α quasar absorption-line systems to provide the high cross section for absorption found at high z. Two further upgrade "blind" line searches looked for HI detections at redshifts of $z = 3.5$ and 5, hoping to detect emission form massive proto-cluster clouds near the era of galaxy formation. We await their results with excitement.

Arecibo is a unique redshift engine, and redshift measurements allow the local structure of the Universe to be mapped out via spiral galaxies. The complicated filaments, walls and shells, plus enclosed voids, that are found, reveal a situation with direct cosmological implications. Further, the Tully-Fisher relation allows redshift-independent distance estimates for galaxies from their observed HI spectra, permitting estimates of peculiar velocities relative to a smoothly expanding Universe to be obtained. These peculiar velocities can be used to draw conclusions concerning the large-scale distribution of mass in the Universe. In this way, Greg Bothun and collaborators have recently showed the "Great Wall", the largest coherent feature in the galaxy distribution, to indeed be a thin 2-D structure in real, as well as redshift, space. They could detect no shear across the angular extent of the Wall, constraining the amplitude of any large-scale density fluctuations.

Apart from HI studies, molecular species in galaxies have been studied from Arecibo. Fifteen years ago, Willem Baan detected strong maser emission from the OH radical in the galaxy, Arp 220, with over a million times the luminosity of the OH masers found in the star-forming regions of the Milky Way. Over 50 such OH megamasers are now known, with the masers being pumped by far infra-red radiation, and amplifying the continuum emission from the active nuclei of the galaxies. Later, Baan again used Arecibo to add a new inmate to the "megamaser zoo", detecting somewhat weaker formaldehyde (H_2CO) maser emission from a number of galaxies.

Of course, the ISM of our own Galaxy contains "molecular soup" on its menu, though until now Arecibo has been mostly restricted to studying emission from the OH radical. A notable effort here has been Murray Lewis's systematic search for OH maser emission from color-selected IRAS sources. This has added a remarkable 400 new OH/IR stars to the annals. However, the post-upgrade opening up of the Arecibo accessible spectrum through to 10 GHz brings many other molecular transitions within our grasp. Examples of these are the ubiquitous CH, H_2CO, Methanol and HC_3N. This will allow Arecibo to join in the exciting investigations of star-forming regions where protostars are in the process of collapsing towards the moment when they begin to shine as true stars, and blow away the cocoons of dust and gas that obscure their births from our optical gaze.

III VERY LONG BASELINE INTERFEROMETRY (VLBI)

Finally, I will mention the technique of Very Long Baseling Interferometry (VLBI) which is used to combine the signals received at telescopes separated by intercontinental distances or, within the past year, including antennas situated on spacecraft in high Earth orbit. ¿From these observations, it is possible to to construct images of radio galaxies and quasars with milliarcsecond resolution, or finer. Arecibo's VLBI history goes back over 30 years. However, our recording equipment became outdated a decade ago, and no VLBI observations were made at Arecibo between 1993 and 1997. in practise, that 1993 observation was particularly interesting, imaging the nucleus of the nearby quasar, 3C273, using a record 23 telescopes. The image created by this "Global Array" was compared with an earlier one, and shows apparent "superluminal motion" out to 200 pc from the core. This optical illusion of superluminal motion is caused by highly relativistic bulk motion in a nuclear jet aligned close to the line of sight.

Arecibo's VLBI future is now bright. We recently obtained a modern VLBI recorder on loan from ISTS, Toronto, Canada, that allows us to participate in Space VLBI with the Japanese orbiting antenna, HALCA. We obtained our first interferometric fringes with HALCA in July last year, and have recently become a regular participant in experiments with this space antenna. We will soon be making our huge area even more available to the VLBI user community through collaboration with the US Very Long Baseline Array (VLBA), consisting of 10 dedicated VLBI antennas distributed from the Virgin Islands to Hawaii. Our special contribution is that the existing VLBA antennas are just 25 m in diameter, while we offer a 305-m diameter, making many interesting investigations possible that would not be so without our presence.

IV A THREAT TO THE FUTURE OF RADIO ASTRONOMY

After the optimism I have exhibited above, however, let me close with a word of warning. The prospects for the future are indeed bright, but there is the dark cloud of radio frequency interference (RFI) casting its long shadow over Arecibo. At the Observatory, we continually monitor the complete radio spectrum, and are pained to note how quickly interference-free parts of the spectrum are vanishing. Even within our own internationally protected radio-astronomy bands, there is need for constant vigilance. Puerto Rico contains the second highest density of radio transmitters within the wider territory of the USA, and new licenses to transmit are being granted daily. Sadly, the danger is one that radio astronomers encounter continually as they attempt to wrest the Universe's secrets from the sea of terrestrially generated disturbances. It would be a tragedy if the national treasure which is the Arecibo telescope ever had to admit final defeat before an ever-greater welter of man-made RFI.

LIST OF PARTICIPANTS

Ali, Tariq <t.ali@ic.ac.uk>
  Imperial College, England

Avila, Manuel A. <manuel@servm.fc.uaem.mx>
  Universidad de Morelos, Mexico

Ayala, Alejandro <ayala@xochitl.nuclecu.unam.mx>
  Universidad Nacional Autónoma de Mexico, Mexico

Baker, Richard G. <baker@lns62.lns.cornell.edu>
  Cornell University, USA

Balbinot, Roberto <balbinot@bologna.infn.it>
  Universita di Bologna, Italy

Blum, Lesser <lblum@rrpac.upr.clu.edu>
  Universidad de Puerto Rico - Río Piedras, Puerto Rico

Braibant, Sylvie <sylvie.braibant@cern.ch>
  CERN, Switzerland

Brook, Nicholas H. <n.brook@physics.gla.ac.uk>
  University of Glasgow, United Kingdom

Caldwell, David O. <caldwell@slac.stanford.edu>
  University of California - Santa Barbara, USA

Carrillo, Salvador <salvador@fnal.gov>
  CINVESTAV, Mexico

Casey, Dylan P. <casey@pa.msu.edu>
  Michigan State University, USA

Casini, Horacio G. <casini@cab.cnea.edu.ar>
  CONICET, Argentina

Castilla Valdez, Heriberto <castilla@fis.cinvestav.mx>
  CINVESTAV, Mexico

Chang, Lay Nam <laynam@vt.edu>
  Virginia Tech., USA

Charles, Francois <charles@sbghp3.in2p3.fr>
  GRPHE-Mulhouse, France

Dos Anjos, Joao C. <janjos@lafex.cbpf.br>
  CBPF - Rio de Janeiro, Brazil

D'Olivo, Juan Carlos <dolivo@nuclecu.unam.mx>
Universidad Nacional Autónoma de Mexico, Mexico

Eckerlin, Guenter <eckerlin@mail.desy.de>
DESY, Hamburg, Germany

Edgecock, Thomas R. <rob.edgecock@rl.ac.uk>
Rutherford Appleton Laboratory, England

Esteban, Ernesto P. <e_esteban@cuhac.u.pr.clu.edu>
Universidad de Puerto Rico - Humacao, Puerto Rico

Faccini, Riccardo <riccardo.faccini@cern.ch>
INFN Rome, Italy and University of California - San Diego, USA

Felcini, Marta <marta.felcini@cern.ch>
ETH - Zurich, Switzerland

Felix, Julian <felix@ifug1.ugto.mx>
Universidad de Guanajuato, Mexico

Ferrer, Efrain J. <ferrer@oak.ait.fredonia.edu>
SUNY at Fredonia, USA

Formaggio, Joseph A. <josephf@cuphy3.phys.columbia.edu>
Columbia University, USA

Frampton, Paul H. <frampton@physics.unc.edu>
University of North Carolina, USA

Fuster, Juan <fuster@evalo1.ific.uv.es>
Universitat de Valencia, Spain

Glashow Sheldon L. <glashow@physics.harvard.edu>
Harvard University, USA

Golob, Bostjan <bostjan.golob@ijs.si>
J. Stefan Institute, Ljubljana, Slovenia

Gupta, Virendra <virendra@gema.cieamer.conacyt.mx>
CINVESTAV - Merida, Mexico

Halprin, Arthur <halprin@udel.edu>
University of Delaware, USA

Hedin, David <hedin@niu.edu>
Northern Illinois University, USA

Huerta, Rodrigo <rhuerta@gema.cieamer.conacyt.mx>
CINVESTAV - Merida, Mexico

Incera, Vivian F. <incera@oak.ait.fredonia.edu>
    SUNY at Fredonia, USA

Jungmann, Klaus P. <jungmann@physi.uni-heidelberg.de>
    University of Heidelberg, Germany

Kayser, Boris <bkayser@nsf.gov>
    US National Science Foundation, USA

Kim, Doris Y. <doris.kim@cern.ch>
    INFN Roma I, Italy

King, Bruce J. <bking@sun2.bnl.gov>
    Brookhaven National Laboratory, USA

Krauss, Lawrence M. <krauss@theory1.phys.cwru.edu>
    Case Western Reserve University, USA

Lane, Charles E. <lane@duphy4.physics.drexel.edu>
    Drexel University, USA

Leung, Chung Ngoc <leung@physics.udel.edu>
    University of Delaware, USA

Lopez, Angel M. <angel@fnal.gov>
    Universidad de Puerto Rico - Mayagüez, Puerto Rico

Magnin, Javier E. <jmagnin@lafex.cbpf.br>
    Centro Brasileiro de Pesquisas Fisicas, Brazil

Masiero, Antonio <masiero@sissa.it>
    SISSA - Trieste, Italy

Mirles, Alejandro <mirles@charma.upr.clu.edu>
    Universidad de Puerto Rico - Mayagüez, Puerto Rico

Mishra, Sanjib R. <mishra@physics.harvard.edu>
    Harvard University, USA

Nang, Freedy <nang@fnal.gov>
    University of Arizona, USA

Nieves, Jose F. <nieves@ltp.upr.clu.edu>
    Universidad de Puerto Rico - Río Piedras, Puerto Rico

Nomerotski, Andrei <nomerot@fnal.gov>
    University of Florida, USA

Pakvasa, Sandip <pakvasa@uhheph.phys.hawaii.edu>
    University of Hawaii, USA

Perret, Pascal P. <perret@in2p3.fr>
Universite Blaise Pascal, France

Petcov, Serguey T. <Petcov@susy.sissa.it>
SISSA/INFN, Italy and INRNE, BAS, Bulgaria

Ponce de Leon, Jaime A. <jponce@upracd.upr.clu.edu>
Universidad de Puerto Rico - Río Piedras, Puerto Rico

Ponce, William A. <wponce@fisica.udea.edu.co>
Universidad de Antioquia, Colombia

Sadoulet, Bernard <sadoulet@physics.berkeley.edu>
University of California - Berkeley, USA

Sanchez-Colon, Gabriel <gsanchez@galaxy.ucr.edu>
University of California - Riverside, USA

Schwickerath, Ulrich <Ulrich.Schwickerath@cern.ch>
CERN, Switzerland

Selsby, Ronald <rselsby@rrpac.upr.clu.edu>
Universidad de Puerto Rico - Río Piedras, Puerto Rico

Shaevitz, Michael H. <shaevitz@nevis.nevis.columbia.edu>
Columbia University, USA

Shafi, Qaisar <shafi@bartol.udel.edu>
Bartol Research Institute, University of Delaware, USA

Sikivie, Pierre <sikivie@phys.ufl.edu>
University of Florida, USA

Silverman, Dennis <djsilver@uci.edu>
University of California - Irvine, USA

Simao, Fernando R. A. <simao@cat.cbpf.br>
CBPF, Brazil

Solano, Carlos J. <javier@lafex.cbpf.br>
Centro Brasileiro de Pesquisas Fisicas, Brazil

Sotnikova, Natalia A. <nata@nata.npi.msi.su, natalia@fnal.gov>
Moscow State University, Rusia

Spallucci, Euro <spallucci@trieste.infn.it>
University of Trieste, and I.N.F.N., Italy

Stone, James <stone@superk.bu.edu>
Boston University, USA

Stone, Sheldon L. <stone@suhep.phy.syr.edu>
Syracuse University, USA

Suarez, Yefer M. <mauricio@charma.upr.clu.edu>
Universidad de Puerto Rico - Mayagüez, Puerto Rico

Taylor, Joe <joe@puppsr12.Princeton.edu>
Princeton University, USA

Tessarotto, Massimo M. <max@cmfd.univ.trieste.it>
University of Trieste, Italy

Tharrats, Jesus <jtharrats@rrpac.upr.clu.edu>
Universidad de Puerto Rico - Río Piedras, Puerto Rico

Thurman-Keup, Randy M. <keup@fnal.gov>
Argonne National Laboratory

Tikhonin, Feodor F. <tikhonin@mx.ihep.su>
Institute for High Energy Physics, Moscow, Rusia

Torruella, Alfredo J. <atorruell@rrpac.upr.clu.edu>
Universidad de Puerto Rico - Río Piedras, Puerto Rico

Troconiz, Jorge <troconiz@hepdc1.ft.uam.es>
Universidad Autónoma de Madrid, Spain

Ubriaco, Marcelo R. <ubriaco@ltp.upr.clu.edu>
Universidad de Puerto Rico - Río Piedras, Puerto Rico

Vazquez, Fabiola <fabiola@fnal.gov>
CINVESTAV, Mexico

Vazquez-Bello, Jose Luis <vbello@tunku.uady.mx>
Universidad Autónoma de Yucatán, Mexico

Verzocchi, Marco <marco.verzocchi@cern.ch>
Albert-Ludwigs Universitaet Freiburg, Germany

Weiler, Tom J. <weilertj@macpost.vanderbilt.edu>
Vanderbilt University, USA

Wulz, Claudia-Elisabeth <claudia.wulz@cern.ch>
Insitute for High Energy Physics, Vienna, Austria

Zavrtanik, Danilo <danilo.zavrtanik@ses-ng.si>
J. Stefan Institute, Ljubljana, Slovenia

Tropical Workshop on Particle Physics and Cosmology

April 1-7, 1998
San Juan, Puerto Rico

SCHEDULE

Tuesday March 31 - Arrival

Wednesday April 1

8:00	8:45	Registration
8:45	9:00	Welcome Chancellor Jorge L. Sánchez Dean Gladys Escalona de Motta

Moderator: Halprin

9:00	10:00	J. Stone - Atmospheric Neutrinos and Proton Decay in Super-Kamiokande
10:00	10:30	Coffee Break
10:30	11:30	A. Masiero - Highlights on Dark Matter and Particle Physics
11:30	4:15	Lunch (and informal discussions)

Moderator: Nieves

4:15	5:15	S. Pakvasa - Neutrino Oscillations
5:15	5:30	Break
5:30	6:00	D. Caldwell - Neutrino Mass and Dark Matter
6:00	6:30	D. Silverman - Predictions of New Physics Models for CP Violating B Decay Asymmetries

Thursday April 2

Moderator: Petcov

9:00	10:00	B. Kayser - CP Violation and Beauty
10:00	10:30	Coffee Break
10:30	11:30	C. Lane - Some Rsults from Neutrino Oscillation Experiments
11;30	12:30	J. Stone - Solar Neutrino Results from Super-Kamiokande
12:30	4:15	Lunch (and informal discussions)

Moderator: Leung

4:15	5:15	P. Sikivie - Searches for Axion Dark Matter
5:15	5:30	Break
5:30	6:00	P. Frampton - Spontaneous CP violation
6:00	6:30	L. N. Chang - Gravity couplings and CPT in the Standard Model

Friday April 3

Moderator: Stone

9:00	10:00	S. Mishra - Review of Neutrino Oscillation Results at Accelerators
10:00	10:30	Coffee Break
10:30	11:30	K. Jungmann - Leptonic Flavor Violation Experiments
11:30	12:30	J. Taylor - Neutron Stars: Probes and Properties
12:30	4:15	Lunch (and informal discussions)
4:15	6:30	Reception at Chancellor's House (UPR)

Saturday April 4

10:00		Departure for Visit to Arecibo
12:00		Arrival in Arecibo
12:00	1:00	Lunch

Moderator: Pantoja

1:00	1:30	Daniel Altschuler - The Arecibo Observatory
1:30	2:00	Chris Salter - Radioastronomy Highlights of the Arecibo Observatory
2:00	2:30	Break
2:30	3:30	S. L. Glashow - Cosmic Ray Tests of Lorentz Invariance
3:30	5:00	Scientific and Recreational Activities
5:00	7:00	Dinner in Arecibo
7:00		Return from Arecibo
9:00		Arrival in San Juan

Sunday April 5

Free day. Some excursions will be organized and/or suggested.

Monday April 6

Moderator: Shafi

9:00	10:00	L. M. Krauss - Dark Matter, the Cosmological Constant and the Age of the Universe
10:00	10:30	Coffee Break
10:30	11:00	T. Weiler - Physics Possibilities with the Highest Energy Neutrino Cosmic Rays
11:00	11:30	M. Shaevitz - Neutrino Oscillation Searches in the NuTeV and BooNE Experiments at Fermilab
11:30	4:15	Lunch (and informal discussions)

Moderator: Masiero

4:15	5:15	S. Petcov - Matter-Enhanced Neutrino Transitions and the Solar Neutrino Problem: Some Recent Results
5:15	5:30	Break
5:30	6:00	M. Felcini - Search for Dark Matter at LEP 2
6:00	6:30	T. Ali - The search for WIMPS by the UK Dark Matter Experiment
7:30		Banquet

Tuesday April 7

Moderator: Halprin

9:00	10:00	B. Sadoulet - Deciphering the nature of Dark Matter
10:00	10:30	Coffee Break
10:30	11:30	Discussion
11:30	11:45	Closing remarks
12:00		Farewell Cocktail

Second Latin American Symposium on High Energy Physics

April 8-11, 1998
San Juan, Puerto Rico

SCHEDULE

Tuesday April 7 - Arrival

Wednesday April 8

8:00	8:45	Registration
8:45	9:00	Welcome
		Dean Gladys Escalona

Moderator: T. Weiler

9:00	9:45	Masiero - Seeking SUSY - a theoretical outlook
9:45	10:30	Nomerotski - Supersymmetry searches at the Tevatron
10:30	11:00	Coffee Break
11:00	11:45	Schwickerath - Higgs searches at LEP
11:45	12:30	Faccini - Electroweak tests and New Physics at LEP
12:30	1:00	Eckerlin - Searches for New Particles and New Interactions at HERA
1:00	4:00	Lunch

Moderator: J. Dos Anjos

4:00	4:45	Shafi - The MSSM and Beyond
4:45	5:30	Braibant - SUSY Searches at LEP
5:30	5:45	Break
5:45	6:30	Hedin - Tevatron Searches for Leptoquarks
6:30	6:50	Felix - Exclusive Λ^0 polarization produced in pp collisions at 27.5 GeV

Thursday April 9

Moderator: J. F. Nieves

9:00	9:45	Halprin - Neutrino Oscillations with Non-universal Gravitational Interactions
9:45	10:30	Petcov - The Solar and Atmospheric Neutrino Problems: Current Status and Neutrino Physics Interpretations
10:30	11:00	Coffee Break
11:00	11:45	Weiler - Box-parameters for Neutrino Oscillations
11:45	12:30	Zavrtanik - Extremely High Energy Cosmic Rays and the Auger Observatory
12:30	1:15	Leung - Chiral Symmetry Breaking in an External Field
1:15	1:35	Ayala - Upper Bound on the Neutrino Magnetic Moment from Landau Damping in supernovae
1:35	4:00	Lunch

Moderator: J. C. D'Olivo

4:00	4:40	Nang - QCD at the Tevatron
4:40	5:20	Fuster - High Precision Tests of QCD at LEP
5:20	5:40	Incera - Yukawa Interaction and Chiral Symmetry Breaking in an External Magnetic Field
5:40	6:00	Break
6:00	6:40	Casey - QCD at the Tevatron: W, Z and Photon Production
6:40	7:00	Gupta - A Model for Baryon Structure and its Application to Magnetic Moments and Semileptonic Decays
7:00	7:20	Ferrer - Magnetic Response in Anyon Fluid at High Temperature
8:00		Dinner

Friday April 10

Moderator: J. Fuster

9:00	9:30	Troconiz - B physics at the Tevatron
9:30	10:00	Kim - Tau physics from LEP
10:00	10:30	Baker - Recent Tau Results from CLEO
10:30	11:00	Coffee Break
11:00	11:30	Perret - B Physics at LEP
11:30	12:00	Sotnikova - Tevatron Results on the Top Quark
12:00	12:30	Brook - QCD at HERA
12:30	1:00	Dos Anjos - Hyperon Production Asymmetries in 500 GeV Pion Nucleon Interactions
1:00	1:20	Ponce - Non-SUSY and SUSY one-step unification
1:20	4:00	Lunch

Moderator: A. Lopez

4:00	4:45	Wulz - CMS - Concept and Physics Potential
4:45	5:30	King - Muon Colliders: New Prospects for Precision Physics and the High Energy Frontier
5:30	5:45	Break
5:45	6:30	S. Stone - The BTeV program
6:30	7:00	Charles - B hysics with the CMS detector
7:00	7:20	Tikhonin - Some Features of Muon Collider Physics Potential

Saturday April 11

Moderator: R. Faccini

9:00	9:45	Castilla - Gauge couplings at the Tevatron
9:45	10:30	Verzocchi - W boson couplings measurements at LEP
10:30	11:00	Coffee Break
11:00	11:45	Edgecock - W mass from LEP
11:45	12:30	Keup - The mass of the W boson
12:30	1:00	Golob - W Decays at Delphi
1:00		Farewell Cocktail

AUTHOR INDEX

A

Altschuler, D. R., 563
Anjos, J. C., 540
Avilez, C., 547
Ayala, A., 533

B

Baker, R. G., 380
Braibant, S., 211
Brook, N. H., 357

C

Caldwell, D. O., 82
Casey, D. P., 317
Chang, L. N., 130
Charles, F., 520
Christian, D. C., 547
Church, M. D., 547

D

de la Incera, V., 452, 458
de Troconiz, J. F., 429

E

Eckerlin, G., 253

F

Faccini, R., 271
Félix, J., 547
Ferrer, E. J., 452, 458
Forbush, M., 547
Frampton, P. H., 160
Fuster Verdú, J., 329

G

Glashow, S. L., 119
Golob, B., 306
Gottschalk, E. E., 547
Gupta, V., 553
Gutierrez, G., 547

H

Halprin, A., 40
Hartouni, E. P., 547
Hedin, D., 223
Holmes, S. D., 547
Huerta, R., 553
Huson, F. R., 547

J

Jensen, D. A., 547
Jungmann, K. P., 148

K

Kim, D. Y., 368
King, B. J., 503
Knapp, B. C., 547
Krauss, L. M., 59
Kreisler, M. N., 547

L

Leung, C. N., 443

M

Magnin, J., 540
Masiero, A., 179
Moreno, G., 547

N

Nang, F., 345
Nomerotski, A., 228

P

Pakvasa, S., 17
Perret, P., 395

S

Salter, C. J., 571
Sánchez-Colón, G., 553
Schwickerath, U., 237
Shaevitz, M. H., 28
Shafi, Q., 196
Sikivie, P., 70
Silverman, D., 165
Silvestrini, L., 179
Simão, F. R. A., 540
Solano, J., 540
Soo, C., 130
Sotnikova, N., 416
Stern, B. J., 547
Stone, J., 3
Stone, S., 479

T

Thurman-Keup, R., 282

U

Uribe, J., 547

V

Verzocchi, M., 294

W

Wagner, D. J., 46
Wang, M. H. L. S., 547
Wehmann, A., 547
Weiler, T. J., 46, 105
White, J. T., 547
Wiencke, L. R., 547
Wulz, C.-E., 467

Z

Zavrtanik, D., 95, 138